5th International Scientific Conference on Information, Control, and Communication Technologies (ICCT-2021)

Journal of Physics: Conference Series Volume 2091

Astrakhan, Russia
4-7 October 2021

Editors:

E. Barabanova
V. Vishnevsky
K. Vytovtov

ISBN: 978-1-7138-5062-5
ISSN: 1742-6588

Printed from e-media with permission by:

Curran Associates, Inc.
57 Morehouse Lane
Red Hook, NY 12571

Some format issues inherent in the e-media version may also appear in this print version.

Institute of Physics
Dirac House, Temple Back
Bristol BS1 6BE UK

Phone: 44 1 17 929 7481
Fax: 44 1 17 920 0979

techtracking@iop.org

Additional copies of this publication are available from:

Curran Associates, Inc.
57 Morehouse Lane
Red Hook, NY 12571 USA
Phone: 845-758-0400
Fax: 845-758-2633
Email: curran@proceedings.com
Web: www.proceedings.com

TABLE OF CONTENTS

SESSION 2. BIOMEDICINE INFORMATION SYSTEMS AND BIOMEDICAL INFORMATICS (ICCT, ARMIMP)

SESSION 3. HARDWARE AND SOFTWARE FOR INFORMATION AND COMMUNICATION SYSTEMS

SESSION 4. RELIABILITY, DIAGNOSTICS AND NON-DESTRUCTIVE TESTING (ICCT, ARMIMP)

SESSION 5. CONTROL AND AUTOMATION SYSTEMS

SESSION 6. DIGITAL ECOSYSTEMS, PRODUCTION AND LOGISTICS MANAGEMENT

SESSION 7. METHODS OF MATHEMATIC MODELING OF PHYSICAL PROCESSES AND MATERIALS FOR INFOCOMMUNICATION SYSTEMS

Preface

These are the proceedings of the 5th International Scientific Conference on Information, Control, and Communication Technologies (ICCT-2021) which had been held October 4-7, 2021 in Astrakhan, Russian Federation. The main technical areas and applications covered by the proceedings are optoelectronics, acousto-optic, microwave technology, antenna systems, measuring technology, metamaterials, nanostructures, nanofilms, photonic crystals, biology and medicine, biophotonics, bioengineering, neural networks in communication technologies; 5G/6G systems; optical and millimeter-wave communication systems; defectoscopy and non-destructive testing of materials and devices for radio- and microelectronics; problems of reliability; automated control systems; diagnostics, etc. The Proceedings can be interesting for specialists in the communications and control systems.

The conference had been held in a mixed format (online and offline) due to COVID-19. The organizing committee has chaired the meeting from Astrakhan (Astrakhan State Technical University). The conference included plenary talks (35 min for presentation and 5 min for questions for each speaker) and session talks (10 min for presentation, 5 min for questions). Fifty one speaker participated in the conference in person and ninety five speakers participated online via Zoom.

ORGANISING COMMITTEE

Co-Chairs

I. Kvyatkovskaya, Prof., Astrakhan, Russia
V. Vishnevsky, Prof., Moscow, Russia

Vice-Chair

K. Vytovtov, Prof., Moscow, Russia

Members

E. Barabanova, Prof., Moscow, Russia
M. Bulatov, Prof., Moscow, Russia
D. Churikov, Dr., Moscow, Russia
D. Kovalenko, Prof., Gomel, Belarus
N. Maltseva, Dr., Astrakhan, Russia
A. Machihin, Dr., Moscow, Russia
O. Pischin, Dr., Astrakhan, Russia
V. Pozhar, Prof., Moscow, Russia

I. Semchenko, Prof., Gomel, Belarus
M. Sysoeva, Dr., Saratov, Russia
S. Khakhomov, Prof., Gomel, Belarus
M. Scherbakov, Prof., Volgograd, Russia

PROGRAM COMMITTEE
Chairman
D. Novikov, A.m., Moscow, Russia
Co-chairmans
E. Barabanova, Prof., Moscow, Russia
V. Vishnevsky, Prof., Moscow, Russia
K. Vytovtov, Prof., Moscow, Russia
A. Machihin, Dr., Moscow, Russia
Members
S. Abramov, A.m., Moscow, Russia
N. Bahtadze, Prof., Moscow, Russia
P. Belov, Prof., St. Petersburg, Russia
A. Bogolyubov, Prof., Moscow, Russia
A. Bugaev, Acad., Moscow, Russia
M. Bulatov, Prof., Moscow, Russia
D. Churikov, Dr., Moscow, Russia
A. Dvorkovich, A.m., Moscow, Russia
M. Kafesaki, Prof., Crete, Greece
I. Kvyatkovskaya, Prof., Astrakhan, Russia
S. Khakhomov, Prof., Gomel, Belarus
V. Kravchenko, Prof., Moscow, Russia
B. Kutuza, Prof., Moscow, Russia
N. Kuznetsov, Acad., Moscow, Russia
S. Maksimenko, Prof., Minsk, Belarus
M. Pagano, Prof., Pisa, Italy
V. Ponomarev, Prof., Mexico, Mexico
G. Samsonov, Prof., Moscow, Russia
E. Stoikova, Prof., Sofia, Bulgaria
K. Talypov, Prof., Beshkek, Kyrgystan
O. Tretiakov, Prof., Sydney, Australia
S. Tretyakov, Prof., Aalto, Finland
S. Zouhdi, Prof., Paris, France

Editorial Committee
E. Barabanova
V. Vishnevsky
K. Vytovtov

Invited Speakers
Alexander Dvorkovich, Moscow, Russia
Michele Pagano, Piza, Italy
Elena Stoykova, Sofia, Bulgaria
Pavel Belov, St. Petersburg, Russia
Sergei Maksimenko, Minsk, Belarus
Rakesh Kumar, Namibia, Africa
Sergei Bozhevolnyi, Odense, Denmark
Andrey Sukhorukov, Sydney, Australia

Alexander Zagoskin, Loughborough, United Kingdom
Oleg Tretiakov, Sydney, Australia

Sponsors
V. A. Trapeznikov Institute of Control Sciences of RAS
Astrakhan State Technical University
Scientific and Technological Center of Unique Instrumentation of RAS
Kotelnikov Institute of Radioengineering and Electronics of RAS

V.A. Trapeznikov Institute of Control Sciences of RAS

Astrakhan State Technical University

V.A. Kotelnikov Institute
of Radio Engineering
and Electronics of RAS

STC UI RAS
Scientific and Technological Centre of
Unique Instrumentation of RAS

The contact details
Vladimir Vishnevsky, V. A. Trapeznikov Institute of Control Sciences of Russian Academy of Sciences, 65 Profsoyuznaya street, Moscow 117997, Russia, vishn@ipu.ru

Peer Review Declaration Form

All papers published in this volume of Journal of Physics: Conference Series have been peer reviewed through processes administered by the Editors. Reviews were conducted by expert referees to the professional and scientific standards expected of a proceedings journal published by IOP Publishing.

• **Type of peer review:** Single-blind

• **Describe criteria used by Reviewers when accepting/declining papers. Was there the opportunity to resubmit articles after revisions?**

Reviewers used following criteria when accepting/declining papers: scientific merit (notably scientific rigour, accuracy and correctness); clarity of expression; communication of ideas; readability and discussion of concepts; sufficient discussion of the context of the work, and suitable referencing; originality; motivation; repetition, length of paper; presentation criteria (title, abstract, diagrams, figures, tables and captions, text and mathematics; conclusion)

• **Conference submission management system:** https://uconfy.com/

• **Number of submissions received:** 146

• **Number of submissions sent for review:** 146

• **Number of submissions accepted:** 78

• **Acceptance Rate** (Number of Submissions Accepted / Number of Submissions Received X 100): 78/146=0.53

• **Average number of reviews per paper:** 2

• **Total number of reviewers involved:** 10

• **Any additional info on review process:** Plagiarism check system was used

• **Contact person for queries (Full name, affiliation, institutional email address)**

Vladimir Vishnevsky, V. A. Trapeznikov Institute of Control Sciences of Russian Academy of Sciences, 65 Profsoyuznaya street, Moscow 117997, Russia, vishn@ipu.ru

On the proper choice of datasets and traffic features for real-time anomaly detection

C Callegari, S Giordano and M Pagano

Dept. of Information Engineering, University of Pisa
Via Caruso 16, Pisa 56122, Italy

E-mail: {c.callegari, s.giordano, m.pagano}@iet.unipi.it

Abstract. Thanks to its ability to face unknown attacks, Anomaly-based Intrusion Detection is a key research topic in network security and different statistical methods, fed by *suitable traffic features*, have been proposed in the literature. The choice of a proper dataset is a critical element not only for performance comparison, but also for the correct identification of the normal traffic behaviour. In this paper we address the general problem of selecting traffic features from recent real traffic traces (MAWI dataset) and verify how the real-time constraint impacts on the general performance. Although a state-of-the-art IDS (Intrusion Detection System) based on deep neural networks is considered, our conclusions can be extended to *any* anomaly detection algorithm and advocate for a fair comparison of IDSs using representative datasets and traffic features that can be extracted on-line (and do not depend on the entire dataset).

1. Introduction

In network anomaly detection one of the key issues is the availability of suitable public datasets that are representative of current traffic mix (including up-to-date applications and state-of-the-art attacks) and for which accurate ground truth is available. Indeed, these data are essential not only for the repeatability of experiments, the validation of new approaches and their comparison with old ones, but also for the construction of the underlying statistical model (training phase) that will permit to discriminate the presence of anomalies from normal network behaviour.

This issue has been addressed already at the end of the '90s and the DARPA 1998/99 datasets, created at the MIT Lincoln Lab within an emulated network environment [1], are still the most popular datasets for intrusion detection. The DARPA 1998 and DARPA 1999 datasets contain seven and five weeks of network traffic in packet-based (pcap) format, including various kinds of attacks like DoS (Denial of Service), R2L (Remote to Local), U2R (User to Root) and probing.

For scalability reason quite often IDSs (Intrusion Detection Systems) do not work at packet level, but at connection level (a connection is defined as a sequence of packets starting and ending at some well-defined times, between which data flow to and from two pairs of IP address and port number, under some well-defined protocol). Hence, it is not surprising that by processing the data captured in DARPA'98 traces a new dataset, known as KDDcup'99 [2], has been created and is still widely used for IDS performance comparisons despite two fundamental weaknesses: the extracted features refer to traffic in an emulated network and, above all, data are more than 20 years old, so *normal* traffic as well as anomalies are not representative of what can be found nowadays in the Internet. In other words, the estimated performance, although statistically exact as data have been obtained in a controlled environment and the ground truth is available, may significantly differ from the ones obtained when the IDS is deployed in a working network.

One of the most interesting alternatives to these datasets is represented by packet traces from the MAWI (Measurement and Analysis on the WIDE Internet) archive (sample-points F), publicly available at [3]. Each trace contains the traffic captured for 15 minutes in a specific day, since 2001 until nowadays, on a trans-Pacific link between Japan and the USA; hence, unlike DARPA and KDD datasets, the network is not emulated, and the traffic is representative of the current mixtures of network services and applications. Moreover, unlike other datasets (such as CIDDS-001 [4], UNSW-NB 15 [5] and the Kyoto honeypot [6]) the traffic is captured over a real backbone network and offers a good variety of both normal and attack flows. Finally, although an exact description of the attacks is not available, in the framework of the successive project MAWILab [7], every traffic flow is classified by means of labels, which indicate the probability (according to well-known anomaly detection algorithms, based respectively on the Hough transform, the Gamma distribution, the Kullback-Leibler divergence and the Principal Component Analysis [8]) that an anomaly is present. Since these labels are available together with the traces, they can be used as a common reference for testing a new IDS.

It is important to point out that the MAWI archive contains packet traces, i.e., the data that a traffic analyser can collect in real time and pass to the IDS as raw input data. If the IDS works at connection level, a pre-processing phase is needed to extract the relevant traffic features. It is quite natural to start from the ones collected in KDD since the latter are widely used in the literature; however, as it will be highlighted in the following section, not all of them can be extracted in real-time since they depend on all the collected data (or at least on a significant portion of them).

In this work we address the problem related to *real-time* features extraction and check how this limitation impacts on IDS performance. To this aim, we consider the system recently described in [9], which makes use of three different kinds of deep neural networks, since approaches based on artificial intelligence are getting quite popular not only in intrusion detection, but also in a wide range of applications, ranging from image recognition to chess and go programs. However, we expect that our conclusions are more general, and can be applied to any feature-based anomaly detection system.

The rest of the paper is organized as follows. In Section 2 we recall the KDD traffic features and then deeply detail the corresponding features we extracted from MAWI traces, highlighting that only a subset of them can be calculated in real-time. Then, in Section 3 we briefly describe the considered IDS and compare the results depending on the selected set of traffic features. Finally, in the Conclusion we summarize the main findings of the paper.

2. Traffic features

The use of traffic features instead of raw packet data is an important issue for the scalability of IDSs working in high-speed networks. Real-time data processing requires the "causality" of such features and, possibly, fast update methods (as highlighted in [9], the latter issue may be addressed using hash tables) with a stateless approach. Although a complete agreement has not been reached, we assume as starting point the KDD features, since they are considered the "de-facto" standard, and then describe in more detail the features we extracted from MAWI traces, focusing on the ones that can be calculated in real-time.

2.1. KDD cup'99

In KDD each single connection is represented by a vector of 41 features and a label that states if the connection is normal or the category of the attack. The considered features can be classified into four groups (see [10] and [11] for more details):

- *Basic features:* all the attributes that can be extracted from a TCP/IP connection. Most of these features can lead to an implicit delay in detection.
- *Content related features*: features related to the data portion, e.g., number of failed login attempts.
- *Time related traffic features*: features computed with respect to a window interval of two seconds. These features are divided into two groups:

  - *"Same host" features*: statistics related to protocol behaviour, service, etc..., for the connections in the past two seconds that share the same destination host as the current connection.
  - *"Same service" features*: features of the connections in the past two seconds that have the same service as the current connection.
- *Host-based traffic features*: features computed on a window interval of 100 connections. As the previous group, this category is divided into two classes:
  - *"Same host" features*: statistics related to protocol behaviour, service, etc..., for the connections in window interval of 100 connections that share the same destination host as the current connection.
  - *"Same service" features*: features of the connections in window interval of 100 connections that have the same service as the current connection.

It is worth noting that content related features are necessary to identify R2L and U2R attacks, which do not have any specific traffic patterns, unlike DoS attacks, which involve many connections to some host in a very short interval. R2L and U2R attacks are embedded in the data portions of the packets, and normally involve only a single connection; thus, in order to detect these kinds of attacks, some features capable of looking for suspicious behaviour in the data portion are needed.

Time related and host-based features are basically the same but are computed on different window intervals. The reason is related to different intrusion patterns of DoS and some probing attacks. As mentioned above, DoS attacks involves many connections to the same host in a very short period of time, producing intrusion patters in the time window of two second, and so they can be easily identified using time related features. However, there are several slow probing attacks that scan the hosts (or ports) using a much larger time interval. As a result, these attacks do not produce intrusion patterns in a time window of two seconds, and they can be identified only using host-based features that consider longer observation intervals. Hence it is reasonable to expect that real-time traffic analysis can worsen the IDS performance.

2.2. MAWI traces: KDD-like features

The first original contribution of our work is the extraction of KDD-like features from the MAWI dataset to have the possibility of testing IDs with more realistic and recent traffic data.

In more detail, 41 features have been extracted per each bidirectional flow. The first 21 features, called *basic features* (see Table 1), are related to the content of packet headers and are based on the KDD basic features, while the others 20, named *time-related features* (see Table 3), are aggregated values, computed over a time-bin of two seconds, as KDD time related traffic features.

Note that two groups of KDD features are not considered. Indeed, it is not possible to extract content related features since payload data are not available (and, moreover, it would be not compliant with privacy rules) and we also excluded host-based features because a window interval of 100 connections with the same source or destination IP address would cause too much traffic processing (and make too unfair the comparison with real-time features).

Finally, the STATE feature is created combining the flags in the original unidirectional flows (see Table 2), where the state ``OTH2" has been added wrt KDD.

2.3. MAWI traces: Real time feature

Among the 41 features previously discussed, only 17 can be extracted in *quasi* real time (see Table 4): the first 13 features are related to the packet headers, while the other four are aggregated values, computed over a time-bin of two seconds. Note that all the features, apart from DUR (for its calculation it is enough to consider the timestamps of the first and last packets of the connection), are additive and can be easily updated for each arriving packet that matches the feature definition.

Table 1. KDD-like basic features

Name	Description
SPORT	Source port
DPORT	Destination port
PROT	IP Protocol
SPKTS	Number of source packets
DPKTS	Number of destination packets
SBYTES	Number of data bytes sent by source to destination
DBYTES	Number of data bytes sent by destination to source
FLAGS	Bitwise OR of TCP flags on all packets sent by source to destination
FLAGD	Bitwise OR of TCP flags on all packets sent by destination to source
DURS	Duration of source flow in milliseconds
DURD	Duration of destination flow in milliseconds
ITYPE	ICMP type value for ICMP or ICMPv6 flows and -1 for non-ICMP flows
ICODE	ICMP code value for ICMP or ICMPv6 flows and -1 for non-ICMP flows
INITFS	TCP flags on first packet in the source flow
INITFD	TCP flags on first packet in the destination flow
SESSFS	Bitwise OR of TCP flags on the second through final packets in the source flow
SESSFD	Bitwise OR of TCP flags on the second through final packets in the destination flow
URG	If URG flag is set
PUSH	If PUSH flag is set
ECN	If ECN is used
STATE	State of TCP connection (see Table 2)

Table 2. Values of the STATE feature

Value	Description
SF	Normal SYN/FIN completion
REJ	Connection rejected, initial SYN elicited an RST in reply
S0	State 0: initial SYN seen but no reply
S1	State 1: connection established (SYN's exchanged), nothing further seen
S2	State 2: connection established, initiator has closed its side
S3	State 3: connection established, responder has closed its side
RSTO	Connection reset by the originator
RSTR	Connection reset by the responder
OTH	Other, a state not contemplated here
OTH2	Non-TCP flows or TCP flows without any flag

Journal of Physics: Conference Series 2091 (2021) 012001 doi:10.1088/1742-6596/2091/1/012001

Table 3. KDD-like time-related features

Name	Description
COUNT	Number of connections to the same destination host as in the current connection
SRV_COUNT	Number of connections to the same destination port number as in the current connection
SERROR	Percentage of connections that are in state S0, S1, S2, or S3 among the connections aggregated in COUNT
SRV_SERROR	Percentage of connections that are in state S0, S1, S2, or S3 among the connections aggregated in SRV_COUNT
RERROR	Percentage of connections that are in state REJ among the connections aggregated in COUNT
SRV_RERROR	Percentage of connections that are in state REJ among the connections aggregated in SRV_COUNT
SAME_SRV	Percentage of connections that are to the same service among the connections aggregated in COUNT
DIFF_SRV	Percentage of connections that are to different services among the connections aggregated in COUNT
SRV_SAME	Percentage of connections that are to same destination host as the current connection among all the connections aggregated in SRV_COUNT
SRV_DIFF	Percentage of connections that are to different destination hosts among all the connections aggregated in SRV_COUNT
SCOUNT	Number of connections to the same source host as the current connection
S_SRV_COUNT	Number of connections to the same source port number as in the current connection
S_SERROR	Percentage of connections that are in state S0, S1, S2, or S3 among the connections aggregated in SCOUNT
S_SRV_SERROR	Percentage of connections that are in state S0, S1, S2, or S3 among the connections aggregated in S_SRV_COUNT
S_RERROR	Percentage of connections that are in state REJ among the connections aggregated in SCOUNT
S_SRV_RERROR	Percentage of connections that are in state REJ among the connections aggregated in S_SRV_COUNT
S_SAME_SRV	Percentage of connections that are to the same service among the connections aggregated in SCOUNT
S_DIFF_SRV	Percentage of connections that are to different services among the connections aggregated in SCOUNT
S_SRV_SAME	Percentage of connections that are to same destination host as the current connection among all the connections aggregated in S_SRV_COUNT
S_SRV_DIFF	Percentage of connections that are to different destination hosts among all the connections aggregated in S_SRV_COUNT

Table 4. Traffic features

Name	Description
SPORT	Source port of the connection
DPORT	Destination port of the connection
PROT	IP Protocol
NPKTS	Number of packets
NBYTES	Number of bytes
NSYN	Number of packets with the flag SYN set
NACK	Number of packets with the flag ACK set
NRST	Number of packets with the flag RST set
NFIN	Number of packets with the flag FIN set
NPUSH	Number of packets with the flag PUSH set
NURG	Number of packets with the flag URG set
NECN	Number of packets with the flag ECN set
DUR	Duration of the connection
SAMEAP1	Number of connections to same SOURCEADDR and SPORT of the current flow
SAMEAP2	Number of connections to same DESTADDR and DPORT of the current flow
SAMEA1	Number of connections to the same SOURCEADDR of the current flow
SAMEA2	Number of connections to the same DESTADDR of the current flow

3. Experimental results

The performance comparison has been carried out considering three different Deep Neural Networks (DNNs):

- Multi-Layer Perceptron (MLP)
- Recurrent Neural Network (RNN)
- Convolutional Neural Network (CNN)

For each DNN all its parameters have been chosen after a tuning phase (see [9] for the implementation details), with a number of input neurons equal to the amount of considered features (41 for the KDD-like case and 17 for the real-time features) and two output nodes (binary classification). Moreover, in order to eliminate the effect of the features range, the min-max normalization has been applied:

$$x_i^* = \frac{x_i - x_{min}}{x_{max} - x_{min}}$$

where x_i denotes the actual value of a feature, x_{max} and x_{min} represent respectively the maximum and the minimum values of the feature in the time-bin, and x_i^* is the value after normalization.
In all cases, the analysed traffic is split into time-bins of two seconds, according to the duration of the windows in *time related traffic features* in KDD. Note that the time-bins can be easily chosen longer (for more accurate detection performance) or shorter (for being closer to real time processing).
The system performance has been evaluated in terms of

- True Positives (*TP*), attacks correctly detected
- True Negatives (*TN*), normal traffic correctly classified
- False Positives (*FP*), normal traffic classified as an attack
- False Negatives (*FN*), undetected attacks.

In more detail, we have considered three classical performance indexes:

- Detection Rate

$$DR = \frac{TP}{TP + FN}$$

- False Alarm Rate

$$FAR = \frac{FP}{FP + TN}$$

- Accuracy

$$ACC = \frac{TP + TN}{TP + TN + FP + FN}$$

As already mentioned, as input data for tuning, training and testing we considered the MAWI dataset. In more detail, we selected five consecutive days of traffic, collected in July 2018, each of them roughly composed of 30 million traffic flows (so, in average, each two second time-bin contains roughly 65000 flows). The dataset is then split into training set, composed of the first 400 seconds of traffic from the first two days and test set, composed of the remaining traffic.

Moreover, the tuning phase has been run on a reduced dataset, composed of only two days of traffic: the first 400 seconds of both days were used for training and the remaining 500 seconds for testing. Note that both training datasets are balanced, so the number of normal samples is equal to the number of attack samples.

At first, we considered the 41 KDD-like features and determined the optimal DNNs architectural parameters over the reduced dataset according to the methodology described in [9]. The corresponding performance (in terms of DR, ACC and FAR) obtained by the three DNNs are shown in Table 5, while the results for the entire dataset (five days of traffic) are reported in Table 6.

Table 5. Results on the reduced dataset using KDD-like features

DNN	DR	ACC	FAR
MLP	0.944	0.956	0.042
RNN	0.975	0.944	0.058
CNN	0.978	0.965	0.047

Table 6. Results on five days of traffic using KDD-like features

DNN	DR	ACC	FAR
MLP	0.934	0.878	0.126
RNN	0.943	0.846	0.161
CNN	0.952	0.885	0.119

Although CNN behaves slightly better, all the selected DNNs perform in a similar way on each dataset, indicating that such method is quite robust. Also, small changes in the hyperparameters of the networks do not significantly affect the final performance of the system, as can be seen by comparing the two tables (the architectural parameters are not changed, just the weights are adapted to the new training set).

Finally, Table 7 highlights the performance of the three DNNs using the 17 real-time features with the entire dataset (after a architectural parameter optimization over the reduced dataset). Unlike the KDD-like features, in this case CNNs, with 89.4% of detection rate, perform much better than the other DNNs and, in general, performance is significantly worse.

Table 7. Results on five days of traffic using real-time features

DNN	DR	ACC	FAR
MLP	0.603	0.889	0.087
RNN	0.646	0.893	0.086
CNN	0.894	0.871	0.130

4. Conclusions
The choice of the traffic features is an important issue in the design of new algorithms for anomaly detection. In this paper we highlighted that the requisite of *quasi real-time* detection (in our case after the next two second time-bin ends) limits the available features and this impacts on the overall system performance. As a side effect of our highly parametric IDS architecture, we noticed that not only the overall performance worsens, but also the relative performance among different classes of DNNs strongly changes, making critical the testing of the proposed algorithm on up-to-date traffic datasets and features that can really be extracted without introducing significant delays.

Acknowledgments
This work has been partially supported by the Italian Ministry of Education and Research (MIUR) in the framework of the CrossLab project (Departments of Excellence – LAB Cloud Computing, Big data & Cybersecurity).

References
[1] "DARPA Intrusion Detection Evaluation Datasets," https://www.ll.mit.edu/r-d/datasets/
[2] "KDD Cup 1999 Data," http://kdd.ics.uci.edu/databases/kddcup99/
[3] "MAWI Working Group Traffic Archive," http://mawi.wide.ad.jp/mawi/
[4] Ring M, Wunderlich S, Gruedl D, Landes D and Hotho A 2017 Flow-based benchmark data sets for intrusion detection *Proceedings of the 16th European Conference on Cyber Warfare and Security (ECCWS)* ACPI pp 361-9
[5] Moustafa N and Slay J 2015 UNSW-NB15: A comprehensive data set for network intrusion detection systems *Proceedings of the 2015th Conference on Military Communications and Information Systems (MilCIS)* IEEE pp 1-6
[6] "Traffic Data from Kyoto University's Honeypots," http://www.takakura.com/Kyoto_data/
[7] "MAWILab," http://www.fukuda-lab.org/mawilab/
[8] Fontugne R, Borgnat P, Abry P and Fukuda K 2010 MAWILab: combining diverse anomaly detectors for automated anomaly labeling and performance benchmarking *CoNEXT: Proceedings of the 6th international conference* ACM pp 1-12
[9] Callegari C, Bucchianeri E, Giordano S and Pagano M 2019 Real time attack detection with deep learning 16th Annual IEEE International Conference on Sensing, Communication, and Networking (SECON) IEEE pp 1–5
[10] Tavallaee M, Bagheri E, Lu W and Ghorbani A A 2009 A detailed analysis of the KDD cup 99 data set *Computational Intelligence for Security and Defense Applications* IEEE pp 1–6
[11] Lee W and Stolfo S J 2000 A framework for constructing features and models for intrusion detection systems *ACM transactions on Information and system security (TiSSEC)* **3 (4)** pp 227–61

Normalization in dynamic speckle analysis for non-destructive monitoring of speed of processes

Elena Stoykova, Dimana Nazarova, Lian Nedelchev, Mikhail Levchenko, Nataliya Berberova-Buhova and Branimir Ivanov

Institute of Optical Materials and Technologies - Bulgarian Academy of Sciences, Acad. Georgi Bonchev street, Bl.109, Sofia 1113, Bulgaria

E-mail: elena.stoykova@gmail.com

Abstract. The paper is dedicated to analysis of normalized intensity-based pointwise algorithms for processing dynamic speckle images with spatially varying speckle statistics in non-destructive visualization of regions of faster or slower changes across an object. Both existing and newly proposed algorithms are analyzed. Extraction of speed of changes is done by acquiring correlated in time speckle images formed on the object surface under laser illumination. The studied algorithms have been applied to simulated low and high contrast speckle data. Their performance has been compared to processing of binary patterns as another approach for dealing with varying speckle statistics in the acquired images. The efficiency of the algorithms have been checked on the experimental data, including data in a compressed format. We have proven that the algorithms with normalization at successive instants by a sum of two intensities or a single intensity outperform as a whole the algorithms which apply the time-averaged estimates of the mean value and the variance of speckle intensity.

1. Introduction

Speed of a process, which leads to micro-changes of topography or a refractive index of an object, is encoded in temporal variation of speckle patterns formed on the object surface under laser illumination [1,2]. Dynamic laser speckle, called also biospeckle in some applications, has been used for non-destructive testing of industrial or biological objects. Monitoring of blood flow perfusion in human tissues [3-5], seeds viability [6,7], plants growing and leaves contamination with chemical agents [8,9], and penetration of cosmetic ingredients in human skin [10], analysis of ear biometrics [11] and bacterial response [12,13], study of animal reproduction [14], food assessment [15-17], drying of paints, coatings and polymer thin films [18,19] have been reported.

Usage of a two-dimensional (2D) optical sensor enables pointwise application of Dynamic Speckle Analysis (DSA), which yields speed distribution in space as a 2D map of a statistical parameter. The intensity-based 2D DSA relies on a comparatively large number of correlated in time speckle images to produce a single 2D map of activity for visualizing regions of faster or slower intensity fluctuations across the objects. This DSA implementation has serious advantages as simplicity of the acquisition set-up, high spatial resolution and spatial analysis applicable to 3D objects. The main challenge of the 2D DSA is the speckle nature of the raw data. The small size of speckle grains in the acquired images results in strong spatial intensity fluctuations within the recorded patterns. The relevant information is buried in signal-dependent noise, and the choice of the processing algorithm is crucial. The algorithm is expected to be fast and robust at non-uniform illumination and to provide a weakly fluctuating estimate for constant speed of changes. Another substantial issue in the 2D DSA is storage, transfer and processing of huge amount of data for monitoring a process when a sequence of activity maps is built in time. This makes necessary compression of the speckle data and raises the question of the efficiency of the algorithms in processing compressed data.

We focus in this work on normalization approaches for processing data with spatially varying speckle statistics. A typical example of such statistics is acquisition of speckle images under non-uniform illumination. Both existing and newly proposed algorithms are analyzed. We compared performance of the analysed algorithms by processing simulated and experimental data.

2. Normalized processing

2.1. Description of the raw data

For the intensity-based dynamic speckle measurement, a CMOS or a CCD camera records speckle images of size $N_x \times N_y$ pixels at a pixel interval Δ and a time interval Δt between the frames. This time interval should ensure recording of correlated patterns. Exposure time is small to avoid speckle averaging during the recording. The dynamic range of the camera is adjusted to the dynamic range of the recorded data. The raw data usually are 8-bit encoded intensity images. An object on a vibration-insulated table is illuminated by the linearly polarized laser light. The illuminating beam is spatially filtered, expanded and collimated. Nevertheless, the intensity may be different at its centre and periphery. This results in speckle statistics which varies not only in time but also in space. The non-normalized processing encounters difficulties in correct detection of activity, and some normalization is required.

A 2D activity map is built from N speckle images. The input data for a pointwise processing algorithm are sequences $I_{ik,1}, I_{ik,2}...I_{ik,N}; i = 1...N_x, k = 1...N_y$ where $I_{ik,n}$ is the intensity at pixel $(i\Delta, k\Delta)$ and instant $n\Delta t$. The activity estimate is obtained by averaging within the time interval, $T = N\Delta t$. Temporal correlation function, $R_{ik}(\tau)$, where $\tau = m\Delta t$ is the time lag between the compared intensities and $m \geq 0$ is an integer, characterizes intensity fluctuations at pixel $(i\Delta, k\Delta)$. The width of $R_{ik}(\tau)$ gives the temporal correlation radius $\tau_c(i\Delta, k\Delta)$. For non-uniform illumination, the variance of intensity fluctuations, σ_{ik}^2, in $R_{ik}(\tau) = \sigma_{ik}^2 \rho_{ik}(\tau)$ varies from point to point; $\rho_{ik}(\tau)$ is the normalized correlation function respectively.

To check the performance of different normalized algorithms, we generated speckle patterns for a synthetic object consisting of four rectangular regions Z1, Z2, Z3 and Z4 of the same size with temporal correlation radii of intensity fluctuations equal to 10 Δt, 20 Δt, 40 Δt and 80 Δt respectively (figure 1a). Simulation was based on producing correlated wrapped phase distributions in the plane of the object that were propagated to the sensor plane through the objective lens. A 2D array of random delta-correlated phase values uniformly distributed from 0 to 2π was used as an initial phase, and normal distribution was accepted for the phase change in time due to normal movement of the scattering centres with respect to the surface. Mutual independence was accepted for the amplitudes and phases of the scattered light at each scattering centre and between the centres. We used $\rho_{ik}(\tau) = \exp\{-\tau / \tau_c(i\Delta, k\Delta)\}$ for description of the modelled processes. The standard deviation of the phase change between successive images was $\sigma_\varphi = N(0,1)\left[\frac{\Delta t}{\tau_c(i\Delta, k\Delta)}\right]^{1/2}$. Speckle integration by the camera pixels was also taken into account. Simulation procedure is described in more details in [20]. We generated speckle patterns of size 256×256 pixels for both low and high contrast speckle at wavelength 532 nm for uniform and non-uniform illumination. A Gaussian intensity distribution $I_0(i\Delta, k\Delta) = I_0 \exp\{-\alpha(i\Delta, k\Delta)\}$ where $I_0 = const$ and $\alpha(i\Delta, k\Delta) = \frac{(i-128)\Delta^2 + (k-128)\Delta^2}{\Omega^2}$ with $\Omega = 210\Delta$ is used for simulation. The histograms of intensity fluctuations across the object and the speckle patterns for the simulated four cases are given in figure 1(b) and figure 1(c). At uniform illumination, the speckle statistics is the same in the areas Z1, Z2, Z3 and Z4 and is described by symmetric or asymmetric intensity distributions depending on the contrast. Speckle has the same statistics in the areas of higher or lower activity, and averaging in time is required to visualize these areas. At non-uniform illumination, the histograms of intensity in the acquired patterns become asymmetric both for the low and the high contrast due to darker areas in the acquired images.

2.2. Normalized processing of a test object

A standard approach to deal with non-uniform illumination is to normalize using the estimates of the variance or standard deviation of temporal intensity fluctuations at each point. Besides the normalized temporal correlation function [21]:

$$\hat{C}(i,k,m) \equiv \hat{C}(i\Delta,k\Delta,m\Delta t) = \frac{1}{N\hat{\upsilon}_{ik}} \sum_{n=1}^{N-m} \left(I_{ik,n+m} - \hat{I}_{ik}\right)\left(I_{ik,n} - \hat{I}_{ik}\right) \tag{1}$$

where $\hat{\upsilon}_{ik} = N^{-1}\sum_{n=1}^{N}\left(I_{ik,n} - \hat{I}_{ik}\right)^2$ and $\hat{I}_{ik} = N^{-1}\sum_{n=1}^{N} I_{ik,n}$ are the estimates of the variance and the average intensity at point $(i\Delta, k\Delta)$, effective evaluation of activity is provided by the normalized estimates of the structure function, $\hat{S}_1$, and modified structure function, $\hat{S}_2$, as follows [20,21]:

$$\hat{S}_1(i,k,m) = \frac{1}{N\hat{\upsilon}_{ik}} \sum_{n=1}^{N-m} \left(I_{ik,n+m} - I_{ik,n}\right)^2 \quad , \quad \hat{S}_2(i,k,m) = \frac{1}{N\hat{\upsilon}_{ik}^{1/2}} \sum_{n=1}^{N-m} \left|I_{ik,n+m} - I_{ik,n}\right| \tag{2}$$

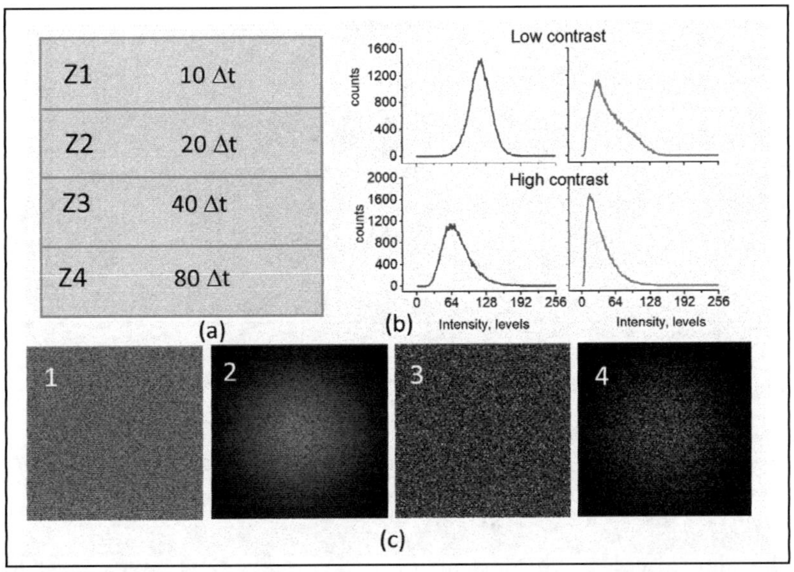

Figure 1. Synthetic object with 4 activity regions Z1, Z2, Z3, Z4 with temporal correlation radii of intensity fluctuations equal to 10Δt, 20Δt, 40Δt and 80Δt (a), speckle intensity histograms (b) for patterns (c) obtained for the object at low contrast and uniform (1) and non-uniform (2) illumination and high contrast and uniform (3) and non-uniform (4) illumination; wavelength 532nm, image size 256×256 pixels.

The need to evaluate the average value and the variance from the sequences of intensity values at all pixels, however, complicates these algorithms. Computation of the variance, $\hat{\upsilon}_{ik}$, by averaging in time is reliable at comparatively high ratios between the acquisition time

$T = N\Delta t$ and the correlation radius, τ_c. In general, the 2D distribution $\hat{\upsilon}_{ik}, i = 1..N_x, k = 1..N_y$ may strongly differ from $\sigma_{ik}, i = 1..N_x, k = 1..N_y$. Instead of using $\hat{\upsilon}_{ik}$, one may take the average intensity value or its square for normalization. Efficiency of such an approach was studied in [22].

More flexible algorithms can be built by time-averaging of a fraction whose nominator and denominator are composed from one or two intensity values acquired at different instants. We checked efficiency of several estimates $\hat{S}_l(i,k,m) = (N-m)^{-1}\sum_{n=1}^{N-m} f^l_{ik,n}(m)$ with $f^3_{ik,n}(m) = \frac{I_{ik,n}}{I_{ik,n+m}}$,

$$f^4_{ik,n}(m) = \left|\frac{I_{ik,n}}{I_{ik,n+m}} - 1\right|,\ f^5_{ik,n}(m) = \left|\frac{I_{ik,n} - I_{ik,n+2m}}{I_{ik,n+m}}\right|,\ f^6_{ik,n}(m) = \frac{|I_{ik,n} - I_{ik,n+m}|}{I_{ik,n} + I_{ik,n+m}},\ f^7_{ik,n}(m) = \left|\frac{I_{ik,n}}{I_{ik,n+m}} - \frac{I_{ik,n+m}}{I_{ik,n}}\right|.$$

The estimate $\hat{S}_6$ was proposed earlier [22]; $\hat{S}_6(m=1)$ coincides with Fuji's algorithm [3]. All estimates $\hat{S}_{1-7}$ increase at higher activity and are equal to 0 at zero activity with exception of $\hat{S}_3$ which is equal to 1 in the latter case.

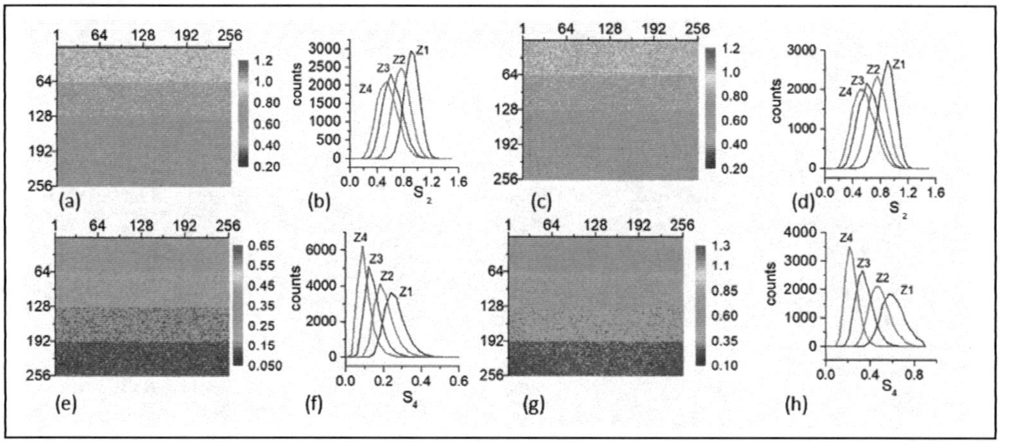

Figure 2. Activity maps for a synthetic object with 4 activity regions Z1, Z2, Z3, Z4 (a,c,e,g) and the histograms of the estimates in these regions (b,d,f,h) for estimates $\hat{S}_2$ (a,b,c,d) and $\hat{S}_4$ (e,f,g,h); (a,b,e,f) - low contrast speckle, (c,d,g,h) - high contrast speckle; $N = 256$, wavelength 532 nm, $m = 10$.

We applied the algorithms $\hat{S}_{1-7}$ to a sequence of 256 speckle images with low and high contrast under non-uniform illumination at 532 nm. The exemplary images taken from the processed sequences are given as images 2 and 4 in figure 1(c). The time lag was taken equal to 10Δt. For illustration, we presented in figure 2 the activity maps obtained for $\hat{S}_2$ and $\hat{S}_4$. The four regions of different activity are clearly seen for both estimates. The contrast of the map corresponding to $\hat{S}_4$ seems better, especially for the regions Z3 and Z4 of lower activity. This is understandable in view that accuracy of determination of the variance estimate, $\hat{\upsilon}_{ik}$, in these regions is rather low at $T\tau_c^{-1}$ equal to 6.4 and 3.2 respectively. The large number of entries in each region (256×64) makes possible evaluation of the probability density function of the estimate at a given activity. The histograms of $\hat{S}_2$ and $\hat{S}_4$ for the four activity regions are also shown in figure 2. As is seen, due to strong fluctuations of the estimates, the histograms are rather wide and overlap. The greater the overlap, the lower the sensitivity of the algorithm. For $\hat{S}_{3-7}$, the higher the activity, the wider the histogram. The result is opposite for $\hat{S}_{1-2}$ due to usage of the variance estimate determined at decreasing accuracy when activity goes to a lower level.

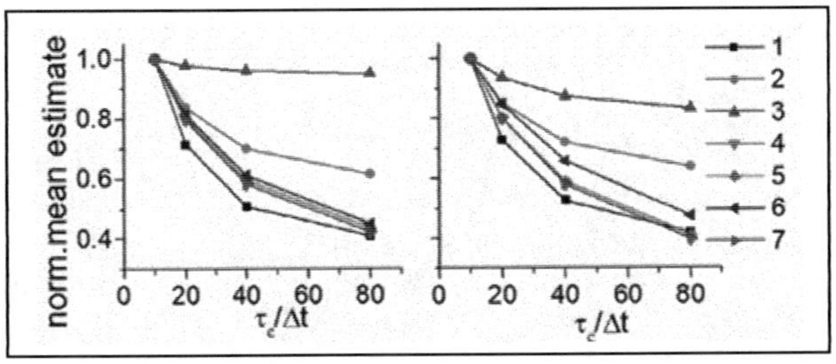

Figure 3. Mean values of the estimates $\hat{S}_{1-7}$ normalized to the mean values in Z1, as a function of the temporal correlation radius; left – low contrast, right – high contrast; $N = 256$, wavelength 532 nm, $m = 10$.

The mean values of the estimates $\hat{S}_{1-7}$ normalized to the mean value for the highest activity in Z1, are shown in figure 3. The mean values as well as the locations of the maximal frequency counts in the histograms depend on activity. As expected, the smallest decrease compared to the value in Z1 is observed for $\hat{S}_3$. This is the fastest algorithm, and despite the low contrast of the map it provided, it can be used for monitoring of objects with large variation of activity across their surface. The largest decrease corresponds to $\hat{S}_1$. The fall of the mean values, when activity is decreasing, is practically the same for the algorithms, $\hat{S}_{4-6}$. They show rather close behavior to $\hat{S}_1$.

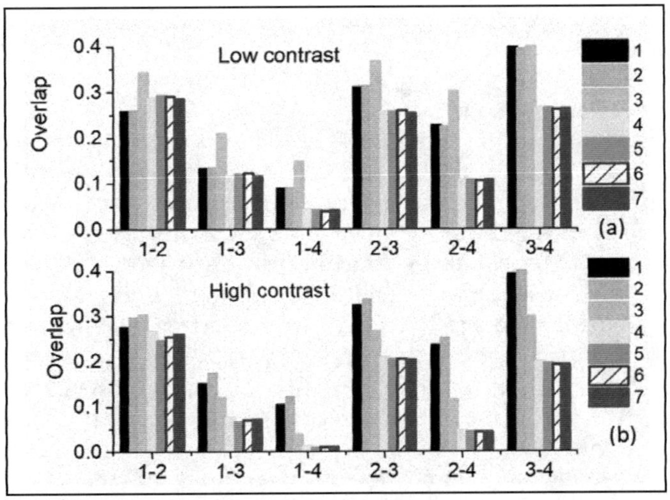

Figure 4. Overlap of histograms of the estimates for activity regions Z1, Z2, Z3, Z4; (1-2) gives the overlap between the histograms in Z1 and Z2, (1-3) – between Z1 and Z3 etc.; $N = 256$, wavelength 532 nm, $m = 10$.

More important parameter for quantitative characterization of sensitivity of the algorithms is the area of overlap of two histograms as a fraction of the total number of entries in a single histogram [20]. Both compared histograms have the same number of entries, and, in our case, it is equal to 256×64. The smaller the overlap, the higher the sensitivity of the algorithms. The overlap as a fraction of 256×64 entries for all algorithms $\hat{S}_{1-7}$ in the cases of low and high contrast speckle is shown in figure 4. The most interesting are the groups of results denoted as 1-2, 2-3 and 3-4 that show comparison of estimates in adjacent spatial regions for the considered test object. The results in the group 1-2 correspond to two spatial areas with close values of the correlation radii – 10 Δt and 20 Δt. For the low contrast case, the results for the algorithms $\hat{S}_{1-2}$ are better than $\hat{S}_{4-7}$ in the group 1-2 , but the result is opposite for the high contrast case. As expected, performance of the algorithm $\hat{S}_3$ is the worst in all cases. Comparison of the estimates for evaluation of activity at τ_c equal to 20Δt and 40Δt (group 2-3) or 40Δt and 80Δt (group 3-4) proves that the algorithms $\hat{S}_{4-7}$ slightly outperform the algorithms $\hat{S}_{1-2}$. They substantially outperform $\hat{S}_{1-2}$ for detection of a high activity area with respect to the low activity area (groups of results 1-3, 1-4 and 2-4). Taking in view the additional advantage of faster computation, these algorithms can be a preferable choice for processing speckle data with spatially varying statistics.

Effective solution for visualization of activity for speckle data with spatially varying statistics is processing of binary patterns [23]. The mean intensity value, $\hat{I}_{ik}$, at each point $(i\Delta, k\Delta)$, can be used as a threshold for binarization. Sequences are formed as $\vartheta_{ik,n} = \pm 1, n = 1,2...N$ depending on $I_{ik,n}$ value being less or larger than $\hat{I}_{ik}$. A polar correlation function is calculated as:

$$\hat{P}(i,k,m) = (N-m)^{-1} \sum_{n=1}^{N-m} \vartheta_{ik,n} \vartheta_{ik,n+m} \qquad (3)$$

Contrary to the estimates, $\hat{S}_{1-7}$, the polar correlation function gets a lower value at higher activity. For the test object in figure 1, we obtained the following values for the overlap between the histograms corresponding to the four different activities: 0.2894 (0.3067) for Z1 and Z2, 0.1757 (0.2013) and 0.1248 (0.1484) for Z1 compared to Z3 and Z4 respectively, 0.3446 (0.3592) for Z2 and Z3, 0.2703 (0.2915) for Z2 and Z4 and 0.4140 (0.4175) for Z3 and Z4. The values in the brackets correspond to the high contrast case. As is seen, this estimator exhibits similar behaviour as the normalized structure and modified structure functions.

3. Processing of experimental data

For experimental verification of the algorithms, we chose a specially designed circular metal object composed from flat annular regions with alternating depths (0 and -2 mm) with respect to the upper surface. There are two flat regions at zero depth and two hollow regions - a central circular section and an annular region. Different activity on the object surface was created by evaporation of polyester paint, which covered the object to form a flat layer. In this way, the hollow regions contained larger quantity of paint compared to the flat surface of the other two annular regions. The object was positioned on a vibration-insulated table and illuminated with a linearly polarized light from a He-Ne laser at 632.8 nm. The drying of the paint produced a dynamic speckle. The speckle images were recorded as 8-bit encoded bitmap images by a color CMOS camera X06c-s (manufactured by Baumer) at interval Δt = 250 ms between the frames. The size of the captured images was 780 × 582 pixels. A single image occupied 1.29 MB, the exposure time was 100 μs.

An exemplary speckle image in bitmap format is shown in figure 5(a). As is seen, the intensity distribution is not uniform within the image. The regions of different activity are in practice indistinguishable. We compared normalized processing provided by $\hat{S}_1$, $\hat{S}_2$ and $\hat{S}_4$. The maps of the estimates after processing 256 patterns for a time lag equal to 10Δt are presented in figure 5 (b-d). Despite the comparatively long sequence of images, the estimate $\hat{\upsilon}_{ik}$ introduces additional fluctuations in $\hat{S}_1$ and $\hat{S}_2$ and in their maps respectively. This decreases the sensitivity of both estimators. Usage of non-normalized structure and modified structure function provides smoother maps at the expense of erroneous determination of activity due to the non-uniformity of illumination. The algorithm $\hat{S}_4$ gives an activity map with enhanced contrast. Evaluation of the processing time on MatLab shows that building a map of size 600×582 pixels for $\hat{S}_1$ and $\hat{S}_2$ requires about 3 times more time than computing of the map of $\hat{S}_4$.

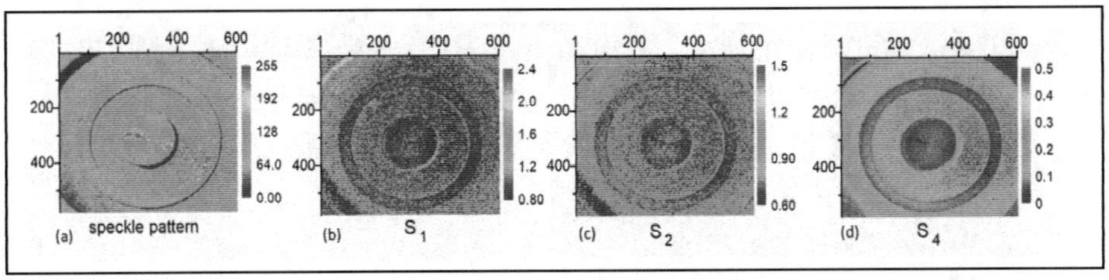

Figure 5. Speckle pattern acquired for a circular metal object with annular regions of different depths in a bitmap format (a) and activity maps for $\hat{S}_1$ (b), $\hat{S}_2$ (c) and $\hat{S}_4$ (d); $N = 256$, wavelength 632.8 nm, $m = 10$.

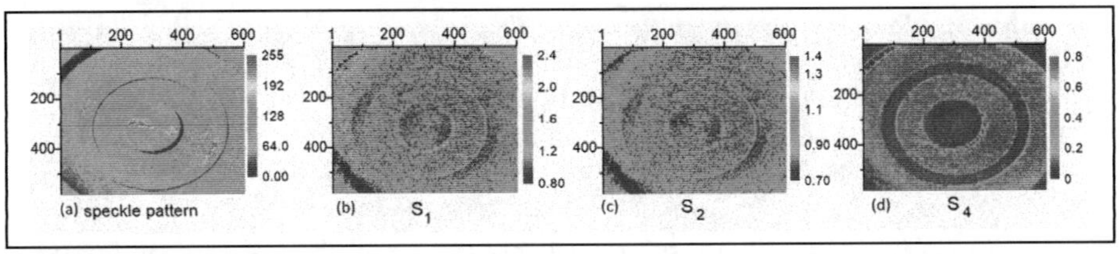

Figure 6. Speckle pattern acquired for a circular metal object with annular regions of different depths in a JPEG format (a) and activity maps for $\hat{S}_1$ (b), $\hat{S}_2$ (c) and $\hat{S}_4$ (d); $N = 256$, wavelength 632.8 nm, $m = 10$.

Storage of big data arrays for monitoring of processes entails usage of data compression. For the intensity-based DSA implementation, we have recently proposed binarization or coarse quantization of the acquired images [23,24]. Another solution is to store the data in the JPEG format. A speckle image decompressed from a JPEG image stored at quality parameter equal to 70 is shown in figure 6(a). Actually, this is the same image as that in figure 5(a). The size of a single image in the JPEG format is 60.4 KB. The results of processing 256 decompressed JPEG images by using $\hat{S}_1$, $\hat{S}_2$ and $\hat{S}_4$ are presented in figure 6(c-d). The contrast of the maps for the normalized structure and modified structure functions is not very good. However, the result provided by $\hat{S}_4$ is acceptable because the regions of different activity are clearly visualized. The only drawback is an artefact expressed as a regular grid of slightly higher values of the estimate. This grid related to the JPEG compression scheme is obscured in the maps in figure 6(b) and figure 6(c) due to different normalization. Since the presence of the grid does not interfere with characterization of activity, one may conclude that JPEG compression is applicable for the DSA data.

In summary, we have studied normalization in intensity-based DSA for processing raw speckle data with spatially varying statistics, as e.g. under non-uniform illumination. We analysed the existing correlation-based algorithms in which normalization is done by using the estimate of the variance at each point. We proposed and studied modifications of algorithms which use for normalization at a given instant a sum of two intensities or a single intensity. We checked efficiency of the algorithms by applying them to low and high contrast simulated data and also compared them for processing of binary patterns as another approach for dealing with varying speckle statistics in the acquired images. The efficiency of the algorithms have been checked for processing experimental data, including data in a compressed format. We have proven that the pointwise algorithms in which normalization is done at each instant by using one or two intensity values are more efficient than the algorithms applying the time-averaged estimates of the mean value and the variance of speckle intensity.

Acknowledgements.
The reported study was funded by the National Science Fund of Bulgaria (contract КП-06-Russia/7). E. Stoykova thanks European Regional Development Fund within the Operational Programme "Science and Education for Smart Growth 2014–2020" under the Project CoE "National center of Mechatronics and Clean Technologies" BG05M2OP001-1.001-0008. M. Levchenko thanks 2020 Plenoptic Imaging H2020 Marie Sklodowska-Curie Innovative Training Network, project No 956770.

References

[1] Goodman J 2007 *Speckle Phenomena in Optics: Theory and Applications* (Roberts and Company Publishers)
[2] Rabal H J and Braga R A Jr 2009 *Dynamic Laser Speckle and Applications* (CRC Press).
[3] Fujii H, Nohira K, Yamamoto Y, Ikawa H and Ohura T 1987 *Appl. Opt.* **26** 5321–5
[4] Serov A and Lasser T 2005 *Opt. Express* **13** 6416–28
[5] Rajan V, Varghese B, van Leeuwen T and Steenbergen W 2006 *Opt. Lett.* **31** 468–70
[6] Braga R, Fabbro I, Borem F, Rabelo G, Arizaga R, Raba H andTrivi M 2003 *Biosystems Eng.* **86**, 287-94
[7] Braga R, Rabelo G, Granato L, Santos E, Machado J, Arizaga R, Rabal H and Trivi M 2005 *Biosystems Eng.* **91** 465-9
[8] Braga R A, Dupuy L, Pasqual M and Cardoso R R 2009 *Eur.Biophys. J.* **38** 679–86
[9]. Ivanov B, Stoykova E, Berberova N, Nikova T, Krumov E and Malinowski N 2013 *Bulg. Chem. Commun.* **45** 149 – 53
[10] Stoykova E, Blagoeva B, Nazarova D, Nedelchev L, Nikova T, Berberova N, Kim , and Kang H 2018 *Opt. Quant.Electron.* **50** 191-201
[11] Chatterjee A, Singh P, Bhatia V and Prakash S 2019 *Opt. Las.Technol.*, **112** 368-78
[12]. Murialdo S E, Sendra G H, Passoni L I, Arizaga R, Gonzalez J F, Rabal H and Trivi M 2009 *J. Biomed. Opt.* **14** 064015
[13]. Mandracchia B, Palpacuer J, Nazzaro F, Bianco V, Rega R, Ferraro P 2018 *IEEE J. Selected Topics in Quant. Electron.* **25** 1-6
[14]. Macedo R, Barreto Filho J, Braga R Jr, Rabelo G 2009 *Reprod. Fertil. Dev.* **22** 170–1
[15]. Ansari M Z, Minz P D and Nirala A K 2012 *Proc. IEEE Conf.on Recent Advances in Information Technology* (IEEE) 873–6
[16]. Mulone C, Budini N, Vincitorio F, Freyre C, López Díaz A and Ramil Rega O 2013 *Proc. SPIE* **8785**, 87851X
[17]. Lyubenova T, Stoykova E, Nacheva E, Ivanov B, Panchev I and Sainov V 2013 *Proc. SPIE* **8770**, 87700S
[18] Harizanova R, Gaydarov V, Zamfirova G, Stoykova E, Nazarova D, Blagoeva B and Nedelchev L 2019 *Thin Solid Films* **687** 137441
[19]. Martí-López L, Cabrera H, Martínez-Celorio R A and González-Peña R 2010 *Opt. Commun.* **283** 4972-7
[20] Stoykova E, Nazarova D, Berberova N and Gotchev A 2015 Opt. Express 23 25128-42
[21] Stoykova E, Ivanov B, Nikova T 2014 *Opt. Lett.* **39** 115-8
[22] Stoykova E 2018 *Proc. SPIE* **10834**, 108341O
[23] Stoykova E, Kang H, Kim Y, Nazarova D, Nedelchev L, Ivanov B, Berberova N and Mateev G 2018 *Opt. Las. Eng.* **111** 50-57
[24] Stoykova E, Nazarova D, Nedelchev L, Ivanov B, Blagoeva B, Oh K and Park J 2020 *Appl. Opt.* **59** 2810-19

Performance Analysis of a Cloud Computing System using Queuing Model with Correlated Task Reneging

Rakesh Kumar[1,2], Bhavneet Singh Soodan[1], Godlove Suila Kuaban[3], Piotr Czekalski[4] and Sapana Sharma[5]

[1] School of Mathematics, Shri Mata Vaishno Devi University, Katra, Jammu and Kashmir, 182320, India
[2] Department of Mathematics and Statistics, Namibia University of Science and Technology, Private Bag 13388 Windhoek, Namibia, 13 Jackson Kaujeua Street,Windhoek, Namibia
[3] Institute of Theoretical and Applied Informatics Polish Academy of Sciences, Baltycka 5,44-100 Gliwice, Poland
[4] Department of Computer Graphics, Vision and Digital Systems, Faculty of Automatic Control, Electronics and Computer Science, Silesian University of Technology
[5] Department of Mathematics, Cluster University of Jammu, Jammu and Kashmir

E-mail: rakesh.kumar@smvdu.ac.in

Abstract. Queuing theory has been extensively used in the modelling and performance analysis of cloud computing systems. The phenomenon of the task (or request) reneging, that is, the dropping of requests from the request queue often occur in cloud computing systems, and it is important to consider it when developing performance evaluations models for cloud computing infrastructures. Majority of studies in the performance evaluation of cloud computing data centres with the use of queuing theory do not consider the fact that the tasks could be removed from queue without being serviced. The removal of tasks from the queue could be due to the user impatience, execution deadline expiration, security reasons, or as an active queue management strategy. The reneging could be correlated in nature, that is, if a request is dropped (or reneged) at any time epoch, and then there is a probability that a request may or may not be dropped at the next time epoch. This kind of dropping (or reneging) of requests is referred to as correlated request reneging. In this paper we have modelled a cloud computing infrastructure with correlated request reneging using queuing theory. An M/M/1/N queuing model with correlated reneging has been used to study the performance analysis of the load balancing server of a cloud computing system. The steady-state as well as the transient performance analyses have been carried out. Important measures of performance like average queue size, average delay, probability of task blocking, and the probability of no waiting in the queue are studied. Finally, some comparisons are performed which describe the effect of correlated task reneging over simple exponential reneging.

1. Introduction

Cloud computing is a distributed [5], dynamic, cost-effective, and scalable computing paradigm that enables on-demand remote access of computing resources such as software, storage, and infrastructure over the internet. The service-oriented architecture of cloud computing can be broadly classified into three categories such as Software-as-a-Service (SaaS), Infrastructure-as-a-Service (IaaS), and Platform-as-a-Service (PaaS) [10]. The SaaS service model provides software applications directly to the end-user as a service. The IaaS service model provides computing infrastructure such as physical machines, virtual machines, virtual storage, and network to the user on-demand. The PaaS service model provides a runtime development environment and tools to developers as a service.

The most popular configuration for cloud computing data centres that has been proposed and discussed by research scholars and practitioners consists of a master server or load balancer, several computing servers, and high-speed transmission systems [19]. The load balancer receives all requests and then schedules them to the appropriate computing servers, which performs the required computation and then send the result of the computation to the transmission system, which transmits it to the users, or the result of the computation is sent to a storage device. Since requests or tasks are queued in buffers when they arrive, and the servers are busy, cloud computing networks have been abstracted into a queuing network, and then used queuing theory to analyze its performance [8]. The performance indicators considered includes response time (delay due to waiting), task blocking probability (due to buffer overflow), probability of immediate service, and the mean number of tasks in the system [13].

Analytical modelling techniques offer less expensive methods for the design and evaluation of cloud computing systems as there is no need for carrying out expensive test-bed experiments [4] or discrete event simulations which are time-consuming and expensive. Analytical models can be used to have high-level insights into the behavior of the systems within short time periods. But, the results obtained from analytical models are approximations of the relative trends of the performance indicators and may deviate from exact values [1]. The prediction and the estimation of the cost benefit of a strategy and the corresponding acceptable quality of service (QoS) may not be feasible by simulation or field measurements [2]; therefore there is a need of analytical methods for analyzing cloud computing systems.

The majority of studies in the performance evaluation of cloud computing data centre with the use of queuing theory do not consider the fact that tasks could be removed from queue without being serviced. The removal of a task from the queue could result from the impatience of the users. The removal of a task from the queue could be due to the expiration of its execution deadline, due to security reasons, or as an active queue management strategy (to avoid saturation or overflow of buffers). The dropping of requests before they are serviced is called the request or task reneging [11]. Request or task reneging in the context of cloud computing has been studied in [12, 7, 8, 3], but all the studies are based on steady-state queuing models, which necessitates the study of the transient behavior of cloud computing queuing systems with task reneging. Also, when a task is removed from the queue, other tasks in the queue that depend on it will equally be removed. Therefore, the removal or reneging of the task from the queue is correlated, and hence, a queuing model with task reneging can be applied to model correlated removal or reneging of tasks from queuing in a cloud computing data centre. The authors have discussed an analytical approach to model queuing systems with correlated reneging in [6, 9].

In this paper, we model a cloud computing data centre into a queuing network model with task reneging. The rest of the paper is organized as follows. In section 2, we present stochastic queuing model. Mathematical model is developed in section 3. Section 4 deals with steady-state solution. In section 5 the transient analysis is carried out. The conclusion is presented in section

2. Stochastic queuing model

We consider a finite capacity single server Markovian queuing model with correlated reneging. The model under investigation is based on the following assumptions:

- The customers arrive at a service facility one by one in accordance with Poisson process with parameter λ.
- There is a single queue and a single server. The service-times are independently, identically and exponentially distributed with parameter μ.
- The capacity of the system is finite (say, K), where $K=N+1$ and N is the queue capacity.

- After joining the queue and waiting for some time, a customer may get impatient and leave the queue (renege) without getting service. The reneging of customers can take place only at the transition marks $t_0, t_1, t_2, \ldots$, where $\vartheta_r = t_r - t_{r-1}$, $r = 1,2,3\ldots$, are random variables with $P[\vartheta_r \le x] = 1 - \exp(-\xi x)$; $\xi \ge 0$, $r = 1,2,3,\ldots$ i.e. the distribution of inter-transition marks is negative exponential with parameter ξ.
- The reneging at two consecutive transition marks is governed by the following transition probability matrix:

$$\text{from } t_{r-1}\ \begin{array}{c}0\\1\end{array}\overset{\text{to } t_r}{\overset{\begin{array}{cc}0 & \quad 1\end{array}}{\left\|\begin{array}{cc}p_{00} & p_{01}\\ p_{10} & p_{11}\end{array}\right\|}}, \text{ where } p_{00} + p_{01} = 1 \text{ and } p_{10} + p_{11} = 1$$

0 refers to no reneging and 1 refers to the occurrence of reneging. Thus, the reneging at two consecutive transition marks is correlated.

3. Mathematical model

Let us define the following probabilities as:

$Q_{0,r}(t)$ = Probability that at time t the queue length is zero, the server is idle, and r is an indicator whether a customer has reneged or not in previous transition mark ($r = 0$ refers to no reneging and $r = 1$ refers to the occurrence of reneging at previous transition mark).

$P_{0,r}(t)$ = Probability that at time t the queue length is zero, the server is not idle, and r is an indicator whether a customer has reneged or not in previous transition mark ($r = 0$ refers to no reneging and $r = 1$ refers to the occurrence of reneging at previous transition mark).

$P_{n,r}(t)$ = Probability that at time t the queue length is n ($1 \le n \le N$), the server is not idle, and r is an indicator whether a customer has reneged or not in previous transition mark ($r = 0$ refers to no reneging and $r = 1$ refers to the occurrence of reneging at previous transition mark).

The differential equations of the model are:

$$\frac{d}{dt}Q_{0,0}(t) = -\lambda Q_{0,0}(t) + \mu P_{0,0}(t) \tag{1}$$

$$\frac{d}{dt}Q_{0,1}(t) = -\lambda Q_{0,1}(t) + \mu P_{0,1}(t) \tag{2}$$

$$\frac{d}{dt}P_{0,0}(t) = -(\lambda + \mu)P_{0,0}(t) + \mu P_{1,0}(t) + \lambda Q_{0,0}(t) \tag{3}$$

$$\frac{d}{dt}P_{0,1}(t) = -(\lambda + \mu)P_{0,1}(t) + \mu P_{1,1}(t) + \lambda Q_{0,1}(t) + \xi[p_{11}P_{1,1}(t) + p_{01}P_{1,0}(t)] \tag{4}$$

$$\begin{aligned}\frac{d}{dt}P_{n,0}(t) &= -(\lambda + \mu + n\xi)P_{n,0}(t) + \mu P_{n+1,0}(t) + \lambda P_{n-1,0}(t)\\ &\quad + n\xi\left[p_{00}P_{n,0}(t) + p_{10}P_{n,1}(t)\right] \quad ,1 \le n < N\end{aligned} \tag{5}$$

$$\begin{aligned}\frac{d}{dt}P_{n,1}(t) &= -(\lambda + \mu + n\xi)P_{n,1}(t) + \mu P_{n+1,1}(t) + \lambda P_{n-1,1}(t)\\ &\quad + (n+1)\xi\left[p_{01}P_{n+1,0}(t) + p_{11}P_{n+1,1}(t)\right] \quad ,1 \le n < N\end{aligned} \tag{6}$$

$$\frac{d}{dt}P_{N,0}(t) = -(\mu + N\xi)P_{N,0}(t) + \lambda P_{N-1,0}(t) + N\xi\left[p_{00}P_{N,0}(t) + p_{10}P_{N,1}(t)\right] \tag{7}$$

$$\frac{d}{dt}P_{N,1}(t) = -(\mu + N\xi)P_{N,1}(t) + \lambda P_{N-1,1}(t) \tag{8}$$

The initial condition is $P_{0,0}(t) = 1$.

4. Steady-state solution

We obtain the steady-state solution of the model from the equations (1) to (8) by using the matrix-decomposition method as:

$$\boldsymbol{P}_0 = \boldsymbol{\psi}_1 P_{0,1}, \boldsymbol{\psi}_1 = \frac{(\lambda + \boldsymbol{A}_{34}\boldsymbol{A}^{-1}{}_{44}\boldsymbol{A}_{43})}{\boldsymbol{A}_{23} - \boldsymbol{A}_{24}\boldsymbol{A}^{-1}{}_{44}\boldsymbol{A}_{43}}, P_{0,0} = \frac{\boldsymbol{\psi}_1 \boldsymbol{A}_{21}}{\lambda} P_{0,1}, \boldsymbol{P}_1 = -(\boldsymbol{A}_{34} + \boldsymbol{\psi}_1 \boldsymbol{A}_{24})\boldsymbol{A}^{-1}{}_{44} P_{0,1},$$

$$Q_{0,0} = \frac{\mu}{\lambda^2} \boldsymbol{\psi}_1 \boldsymbol{A}_{21} P_{0,1}, Q_{0,1} = \frac{\mu}{\lambda} P_{0,1}, P_{0,1} = \frac{1}{\frac{\mu}{\lambda^2} \boldsymbol{\psi}_1 \boldsymbol{A}_{21} + \frac{\mu}{\lambda} + \frac{\boldsymbol{\psi}_1 \boldsymbol{A}_{21}}{\lambda} + \boldsymbol{\psi}_1 \boldsymbol{e} + \mathbf{1} - (\boldsymbol{A}_{34} + \boldsymbol{\psi}_1 \boldsymbol{A}_{24})\boldsymbol{A}^{-1}{}_{44}\boldsymbol{e}}$$

5. Transient Analysis

Now, we perform the transient analysis of the model. Due to the complex nature of the model equations, it is quite difficult to obtain the transient solution analytically. Therefore, we use the Runge-Kutta method of fourth-order to obtain the transient solution. The "ode45" function of MATLAB software is used to compute the transient numerical results.

Key performance indicators

We study the following key performance indicators:

- Expected queue size $L_q(t) = \sum_{n=0}^{N}(n)[P_{n,0}(t) + P_{n,1}(t)]$
- Expected delay in the queue $W_q(t) = \frac{L_q(t)}{\mu[1-Q_{0,0}(t)-Q_{0,1}(t)-P_{0,0}(t)-P_{0,1}(t)]}$

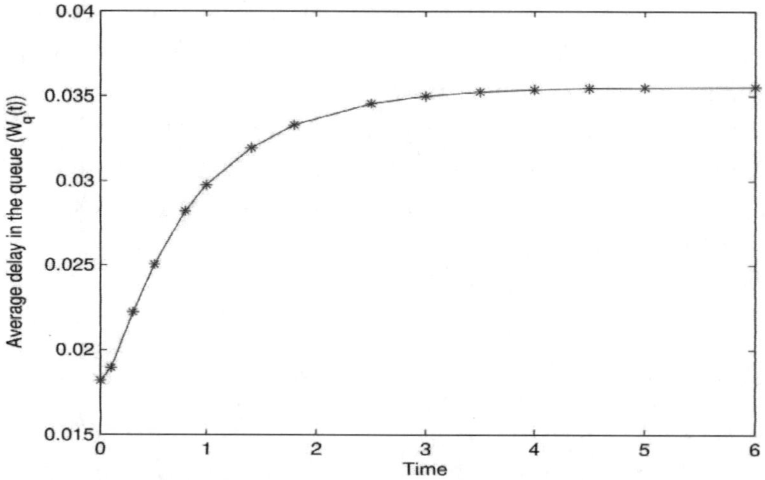

Figure 1: Average delay in the queue vs. Time.

The variation in average delay in the queue with time is presented in figure 1. One can observe that the average delay in the queue also increases as time progresses. But, after some time there is no change in average delay with time, that is, the system has achieved steady-state. In the figures 2 and 3, we compare three queuing models: M/M/1/N queuing model with correlated reneging, M/M/1/N queuing model with reneging and M/M/1/N queuing model. The above mentioned three models are compared with respect to the variation in average delay in the queue as shown in figure 2 with average arrival rate. It can be seen from the figure that the average delay in the queue of the M/M/1/N queuing model without reneging is higher than that of the M/M/1/N queuing model with correlated reneging. Further, the average delay in the queue in the case of M/M/1/N queuing model with reneging is lowest to the other two models. Thus, if the request dropping (reneging) in cloud computing systems is considered as correlated, then the average delay in the queue will be higher as compared to exponential reneging of requests.

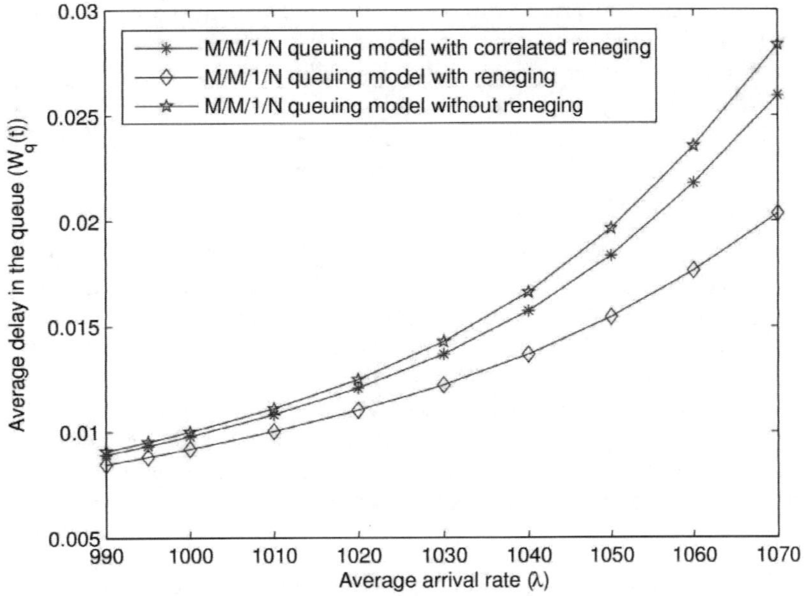

Figure 2: Average delay in the queue vs. Average arrival rate.

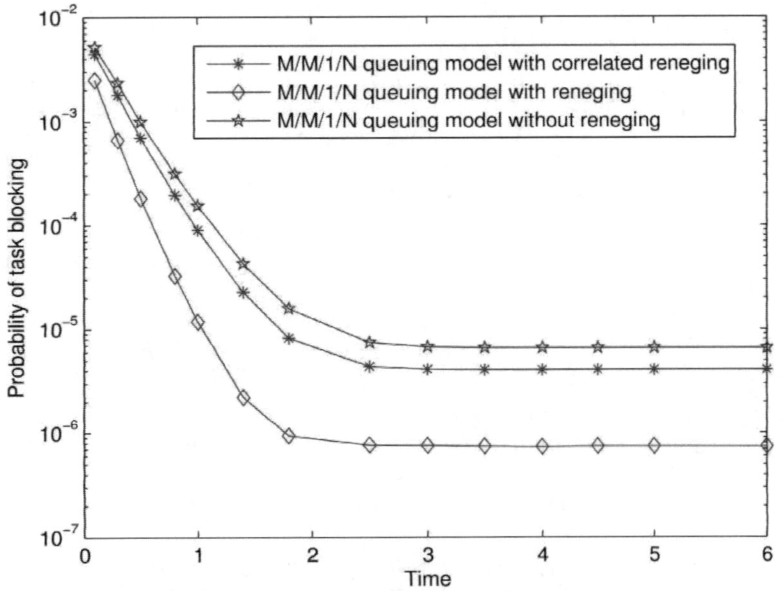

Figure 3: Probability of task blocking vs. Time

The variation in average delay in the queue with respect to the arrival rate as shown in figure 2 indicates that the performance indicators increases slowly with arrival rate until a certain value beyond which a small increase in the arrival rate or traffic load will result in the sharp or rapid increase in the average delay in the queue. Therefore, in designing, sizing and provisioning the load balancing server, it must be ensured that the average delay in queue be maintained within the slow variation regime to ensure QoS guarantee. The second comparison is carried out with respect to variation in the probability of task (request) blocking (P_{N,0}\left(t\right)+P_{N,1}\left(t\right)) with time as shown in figure 3. Initially, the task blocking probability in all the three models is high; it decreases slowly and after some time becomes stationary. The higher value of the task blocking probability at the initial stage is because of the initial condition we have taken\ P_{80,0}\left(0\right)=1, that is, there are 80 requests in the queue initially. Further, task blocking probability always remains higher in the case of M/M/1/N queuing model without reneging as compare to the M/M/1/N queuing model with correlated reneging. The task blocking probability is lowest in the case of M/M/1/N queuing model with reneging. This type of variation is due to the average queue sizes possessed by three queuing models.

6. Conclusion

In this paper, we have studied a single server, finite capacity Markovian queuing model with correlated reneging. We then demonstrate how they can be used in the design and performance evaluation of cloud computing data centres with correlated reneging of requests. The steady-state and the transient-state behaviours of the model are studied. We also investigated the influence of the design parameters such as the buffer sizes, the mean arrival rate, the mean processing speed, the average reneging rate on the key performance indicators (KPIs) such as response time (delay due to waiting), task blocking probability (due to buffer overflow), probability of immediate service, and the mean number of tasks in the system. Finally, some comparisons are performed which describe the effect of correlated request reneging over simple exponential reneging.

References

[1] Paya A and Marinescu D 2017 Energy-Aware Load Balancing and Application Scaling for the Cloud Ecosystem *IEEE Transactions on Cloud Computing* **1(5)** pp 15-27

[2] Bruneo D 2014 A Stochastic Model to Investigate Data Center Performance and QoS in IaaS Cloud Computing Systems *IEEE Transactions on Cloud Computing* **25** pp 560-69

[3] Chiang Y et al 2016 Performance and Cost-Effectiveness Analyses for Cloud Services Based on Rejected *IEEE Transaction on Services Computing* **9** pp 446-55

[4] Duan D, Yu S and Zhang A 2017 Cloud service performance evaluation: status, challenges, and opportunities – a survey from the system modeling perspective *Digital Communications and Networks* **3** pp 101-11

[5] Zanoon N, Kar J and Mishra M 2016 Mitigating Threats and Security Metrics in Cloud Computing *Internation Journal of Information Processing* **12** pp 226-33

[14] ElKafhali S and Salah K 2018 Modelling and Analysis of Performance and Consumption in Cloud Data Centers *Arabian Journal of Science and Enginering* **43** pp 7789-802

[6] Rodriguez V and Guillemin F 2016 Performance analysis of resource pooling for network function virtualization *17th International Telecommunications Network Strategy and Planning Symposium (Networks)* pp 158-63

[7] Chiang Y and Ouyang Y 2014 Profit Optimization in SLA-Aware Cloud Services with a Finite Capacity Queuing Model Mathematical Problems in Engineering *Hindawi Publishing Corporation*

[8] Nan X, He Y and Guan L 2014 Queueing model based resource optimization for multimedia cloud *J. Vis. Commun. Image R* **25** pp 928-42

[9] Kuaban G et al. 2020 A Multi-Server Queuing Model with Balking and Correlated Reneging With Application in Health Care Management *IEEE Access* **8** pp 169623-39

[10] Gupta S and Arora S 2018 Queueing system in cloud services management: a survey *International Journal of Pure and Applied Mathematics* **119** pp 12741-53

[11] Al-Seedyet R al 2009 Transient solution of the M/M/c queue with balking and reneging: a survey *Computers and Mathematics with Applications* **57** pp 1280-85

[12] Homsiet S al 2017 Workload Consolidation for Cloud Data Centers with Guaranteed QoS Using Request Reneging *IEEE Transactions on Parallel and Distributed Systems* **28** pp 2103-16

[13] Yang B et al 2009 Performance Evaluation of Cloud Service Considering Fault Recovery *IEEE International Conference on Cloud Computing* pp 571-76

Journal of Physics: Conference Series **2091** (2021) 012004 doi:10.1088/1742-6596/2091/1/012004

Experimental Research of Static Characteristics of Microfluidic Trigger

A V Balabanov[1], A M Kasimov[2] and A I Popov[3]

V. A. Trapeznikov Institute of Control Sciences of Russian Academy of Sciences 65 Profsoyuznaya street, Moscow 117997, Russia

E-mail: fca07@mail.ru[1], as.kasimov@yandex.ru[2], daten2012@yandex.ru[3]

Abstract. Experimental research of the main static characteristics of the microfluidic trigger of the 100-micrometer feature size is carried out. In order to research the trigger, the experimental (physical) model has been especially made. As the main result of the research, the transfer and flow-rate characteristics of the trigger are obtained in relation to the two connection schemes: 1) the relay scheme with the two-input control, 2) the comparator scheme with the control at a first input and with a reference setpoint at a second input.

1. Introduction

There are a number of the researches devoted different aspects of creating the microfluidics. These are the schematic development, the investigation of microfluidics features, the modelling, and the manufacturing. In this context, what should be noted firstly are [1-10] and others. High resistance of the microfluidics in relation to many destabilizing factors (primarily, electromagnetic and corpuscular emissions) is the main prerequisite to create the reserve microfluidic control systems and carry out the new researches in order to improve the microfluidic element base [e.g. 1-4]. It is of special importance regarding the critical objects because their failures may result in catastrophic consequences. In order to develop the highly reliable reserve fluidic control systems it is required that the full and true performance data of the fluidic elements (especially ones that are fundamental in relation to the digital circuits) be available. Contributing the satisfaction of the above mentioned requirement, this article represents results of the experimental researches of the transfer and flow-rate characteristics of the microfluidic trigger of the 100-micrometer power channel (figure 1).

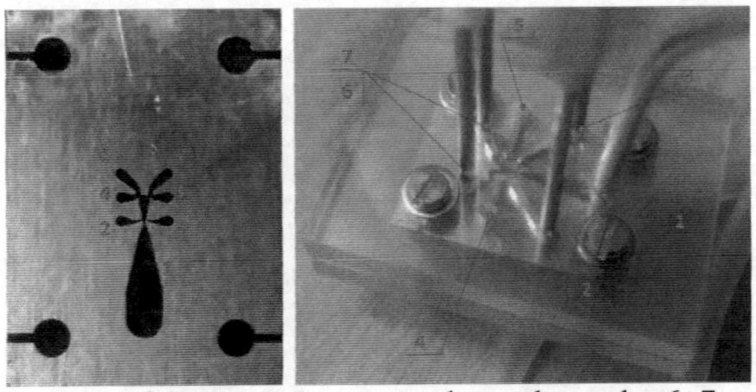

1 – supply; 2, 3 – inputs; 4, 5 – atmosphere channels; 6, 7 – outputs

Figure 1. Research trigger of 100-micrometer feature size.

The trigger represented in figure 1 is an element of the discrete microfluidics. Marked in the figure are the following trigger features: 1 is the channel intended to supply the gas into the trigger, 2, 3 are the input channels to control the input supply flow, 4, 5 are output channels throw out excess volume of the gas, and 6, 7 are the output channels providing the information signals. It is the trigger that is the object of research of this article.

The discrete microfluidics is one of the branches of the micro-jet technology, which had been founded at the V. A. Trapeznikov Institute of Control Sciences and was initially applied to develop miniature reserve control systems of critical objects such as nuclear power plants, chemical plants, aircrafts, etc. This is substantiated by the resistance of the discrete microfluidics to impaction of the destabilizing factors resulting in failures of the electronics. At the present time, this field also includes microanalytical, micromechanical, biotechnical, and other systems. The main current lines of developing the discrete microfluidics are improvement of the weight-size characteristics, and transition to the integrated technology. All the above mentioned in this section causes the necessity for improving the scientific and experimental base to develop the discrete microfluidics more effectively.

2. Test bench and measuring technique

Basing on the experimental model, the test bench to investigate the transfer and flow-rate static characteristics of the microfluidic trigger has been created. The schematic diagram of the test bench is represented in figure 2. The power of the scheme is provided by the pneumatic supply, using the filter F1. The investigation has been performed for both small and large signals. Therefore, the pressure supplies are of 300 and 3000 mm H_2O. The regulators R1-R3 are intended to set levels of the pressure supply as well as the input signal of the trigger A1.

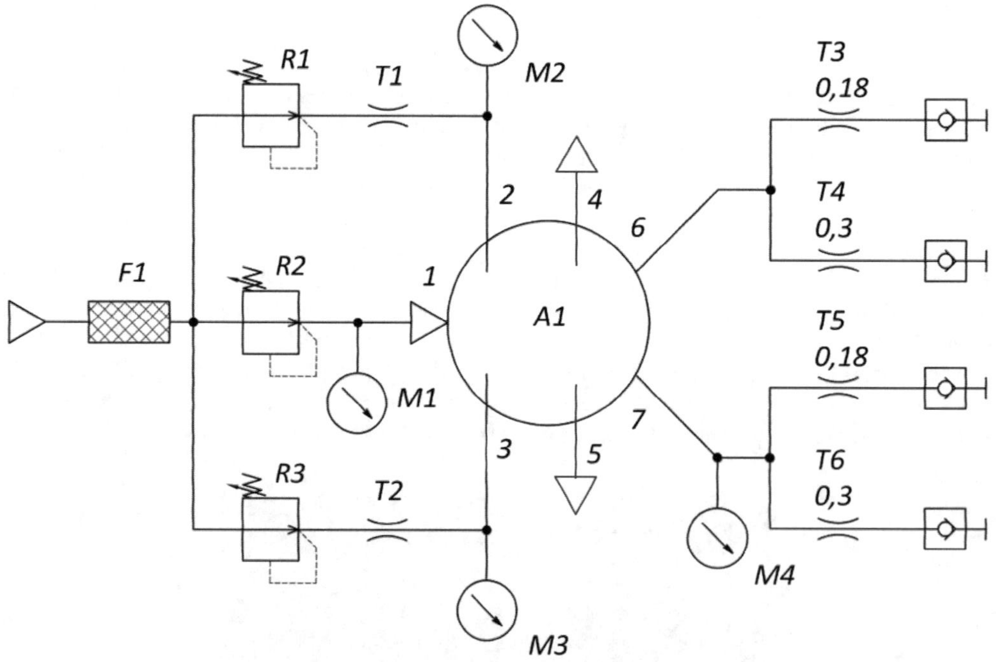

Figure 2. Schematic diagram of test bench.

The pressure monitoring is carried out by means of the devices M1, M2, and M4. These are manometers, when operating at the pressure supplies of 1500 and 3000 mm H_2O. When operating at the pressure supply of 300 mm H_2O, the piezometers are used.

The throttles T1, T2 are necessary in order to provide smooth adjustment of the input pressures of the trigger A1. Each of the outputs is loaded with the throttles T3, T4 and T5, T6 connected in parallel pairwise. The capillaries of T3 and T5 are 0.18 mm in diameter. The capillary diameters of T4, T6 are of 0.3 mm. Levels of the output signals are monitored at the output loaded with the throttles T5, T6 by means of the device M4. The other output of the trigger is loaded with the throttles T3, T4 to provide the symmetry of the loads for the obtain data to be correct. During the experiments, the output stoppers of T3-T6 are opened/closed in the combinations that provide the load range of 0, 0.18, 0.3, and (0.18||0.3) mm.

To determine the transfer characteristics at the relay mode, to begin with, the pressure supply is set equal to 300 mm H_2O. Then, using the input 3, the flow is switched to the output 6 – the initial state. Further, the manometric pressures of the inputs are set to zero. The above mentioned procedures having been performed, the pressure at the input 2 is smoothly increased by means of the regulator R1 until the trigger switches in the forward direction (the output 7), and, again, smoothly decreased to zero. Then, the pressure at the input 3 is smoothly increased until the trigger switches in the reverse direction. The switching procedures being performed, the values of the output pressure are monitored and recorded. The measuring for the pressure supply equal to 3000 mm H_2O is the same one. At the comparator mode, one performs the similar procedures. However, the input 3 (setpoint) is constant (nonzero) value: 50 mm H_2O in the case of P_{supply} = 300 mm H_2O, and 500 mm H_2O in the case of P_{supply} = 3000 mm H_2O.

To determine the flow-rate characteristics, the input pressure is set equal to the direct switching pressure of the trigger. Next, by means of M4, the levels of pressure are measured at the output 7 loaded with throttles T5, T6, and T5||T6, as well as closed with the stoppers. On the base of the four measuring points obtained experimentally, the flow-rate characteristic is approximated by means of the following function:

$$Q = kS\sqrt{\frac{2\Delta P}{\rho}} \qquad (1)$$

where Q is the flow-rate, m^3/s; k is the discharge coefficient; S is the cross-sectional area of the throttle capillary, m^2; ΔP is the differential pressure of the throttle, Pa; ρ is the density of the operating gas, kg/m^3.

3. Results of Experimental Researches

The diagrams represented in figure 3 correspond to the transfer characteristics of the trigger for the relay and comparator modes, with the pressure supply equal to 300 and 3000 mm H2O. These diagrams reflect the function Pout = f(Pin): the values of the pressure at the control inputs 2, 3 are arranged along the axis of abscissas, the values of the pressure at the measuring output of the trigger are arranged along the axis of ordinates. The direct transfer characteristics are imaged with the solid lines, which correspond to the switching of the measuring output of the trigger from '0' to '1'. The dashed lines reflect the reverse transfer characteristics corresponding to the switching of the output 7 from '1' to '0' due to oppositely regulating the control signal.

Basing on the experimental data obtained, the following conclusions may be made. At the relay mode, the pressure supply being reduced, the hysteresis loop moves to/into the area of the negative half-axis of the input pressures, which points to that the stability of the microfluidic trigger increases. However, the stability of the trigger decreases with increasing the pressure supply. It is obvious from the diagrams for P_{supply} = 3000 mm H_2O, the hysteresis loop is only in the positive area. Therefore, the trigger cannot operate stably. Increasing the load at the range of 0, 0.18, 0.3, and (0.18||0.3) mm advantageously influences the stability of the trigger, but, however, decreases the output amplitude.

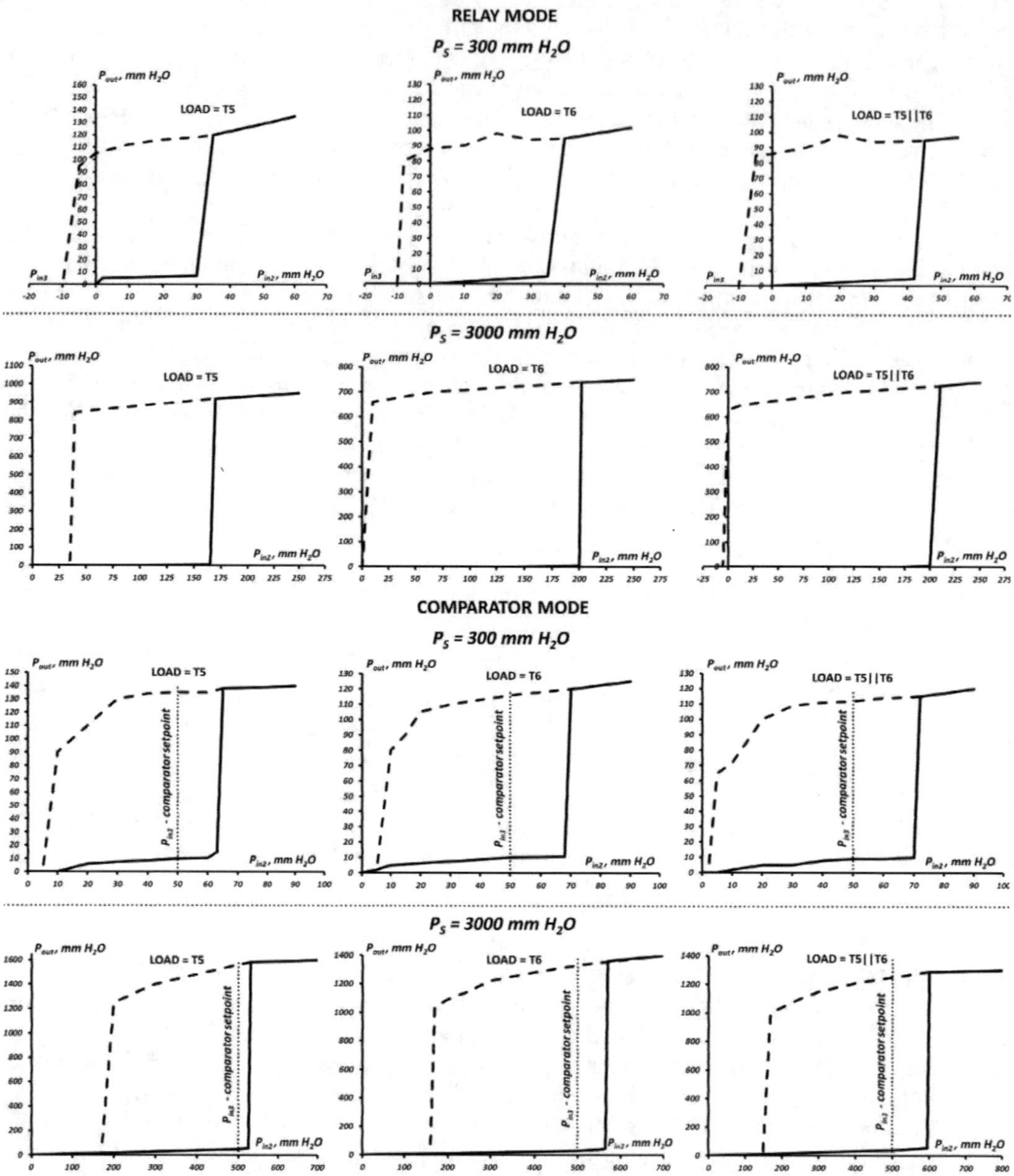

Figure 3. Transfer characteristics of trigger.

At the comparator mode, the stability of the trigger also becomes higher as the load increases at the above range. However, decreasing the output load provides improved precision of the comparator, i.e. the difference between the setpoint at the reference input and the switching pressure at the control input reduces. The comparator mode provides a higher level of the input signal as compared to the relay mode, with the pressure supplies for these modes equal. At the relay mode, however, the steepness of the reverse characteristic is more than at the comparator mode. Both at the relay mode and at the comparator mode, increasing the load gives rise to a wider hysteresis loop of the trigger, with the output pressure reduced. In addition, the increasing results in declined threshold sensitivity of the comparator.

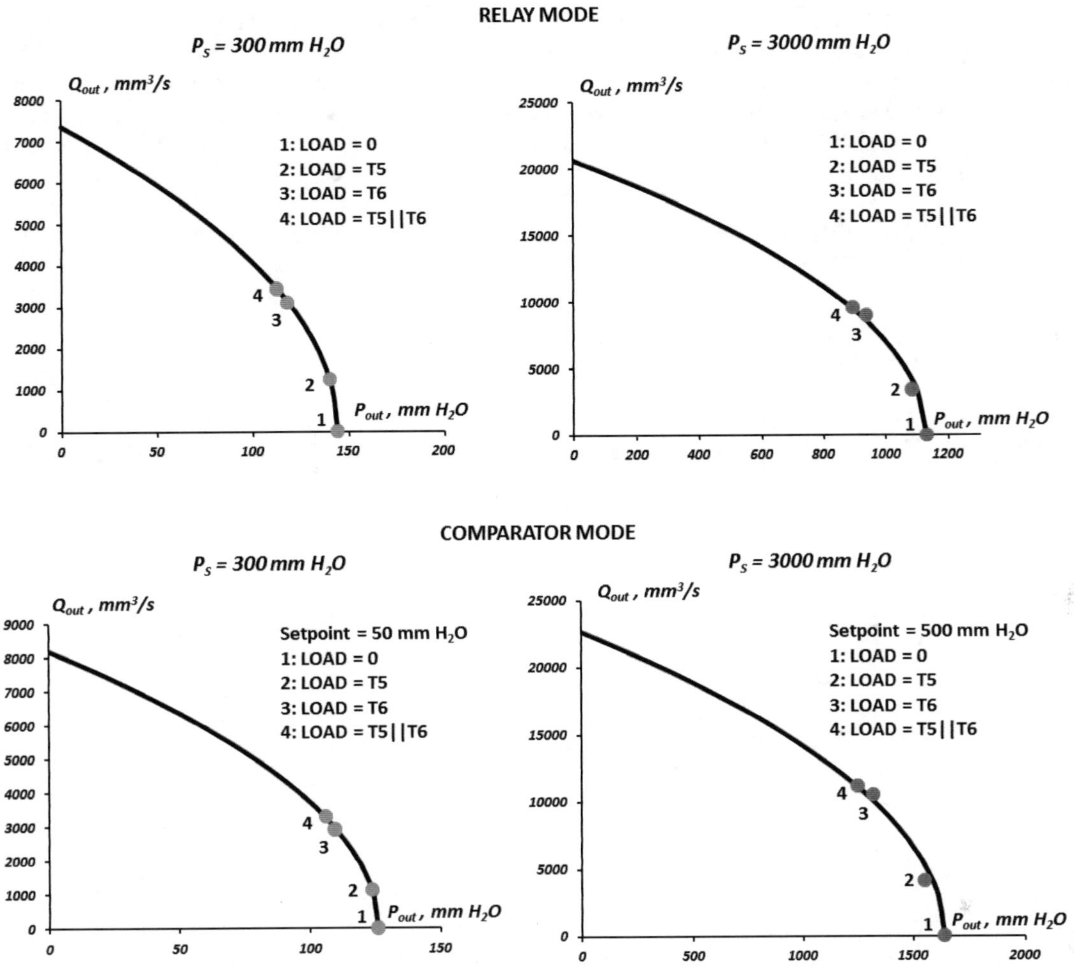

Figure 4. Flow-rate characteristics of trigger.

The diagrams represented in figure 4 correspond to the flow-rate characteristics of the trigger for the relay and comparator modes, with the pressure supply equal to 300 and 3000 mm H_2O. These diagrams reflect the functional dependence of the output flow-rate Q_{out} on the output pressure P_{out}. The four experimental points corresponding to the four loads of the trigger are depicted on the $Q_{out} = f(P_{out})$ curve. Using the measuring data obtained at the four points, the *k*-coefficients for (1) and, thus, the approximating functions for the flow-rate characteristics have been determined.

Basing on the results of the experimental researches of the flow-rate characteristic of the microfluidic trigger, the conclusion is made that the steepness of the characteristic decreases with increasing the pressure supply. Thus, changing the load of the trigger operating at the lower pressure supply results in greater variation of the output flow-rate as compared to the variation when the trigger operating at the higher pressure supply. However, the efficiency factor increases with reducing the pressure supply.

Another result significant to using the trigger at the comparator mode is given in figure 5 in the form of the diagram. This diagram reflects the functional dependence of the relative hysteresis on the relative amplitude of the reference signal, with the output flow-rate equal to zero. The indices *ctrl* and *ref* are identifiers of the pressures at the control and reference inputs correspondingly. The pressures of the direct and reverse switchings are marked by means of the *dir* and *rev* indices. The range of the pressure supply includes 0.1, 0.2, 0.3, 0.4, and 0.5 kgF/sm^2. The diagram shows that the best operational modes from the viewpoint of the precision of controlling the pressure setpoint at the reference input are those of the pressure supply of 0.1 and 0.2 kgF/cm^2 as these modes provide the placement of the curves, mostly, within the colored area. The modes of the pressure supply of 0.3 and 0.4 kgF/cm^2 provide the intersection of the colored area just at the small intervals. The trigger at the comparator mode being supplied by means of the 0.5 kgF/cm^2 pressure, the setpoint is unreachable, for the direct switching occurs before the pressure at the control input becomes equal to the pressure at the reference input. If the curve had located under the colored area, the reverse switching would have occurred before the setpoint.

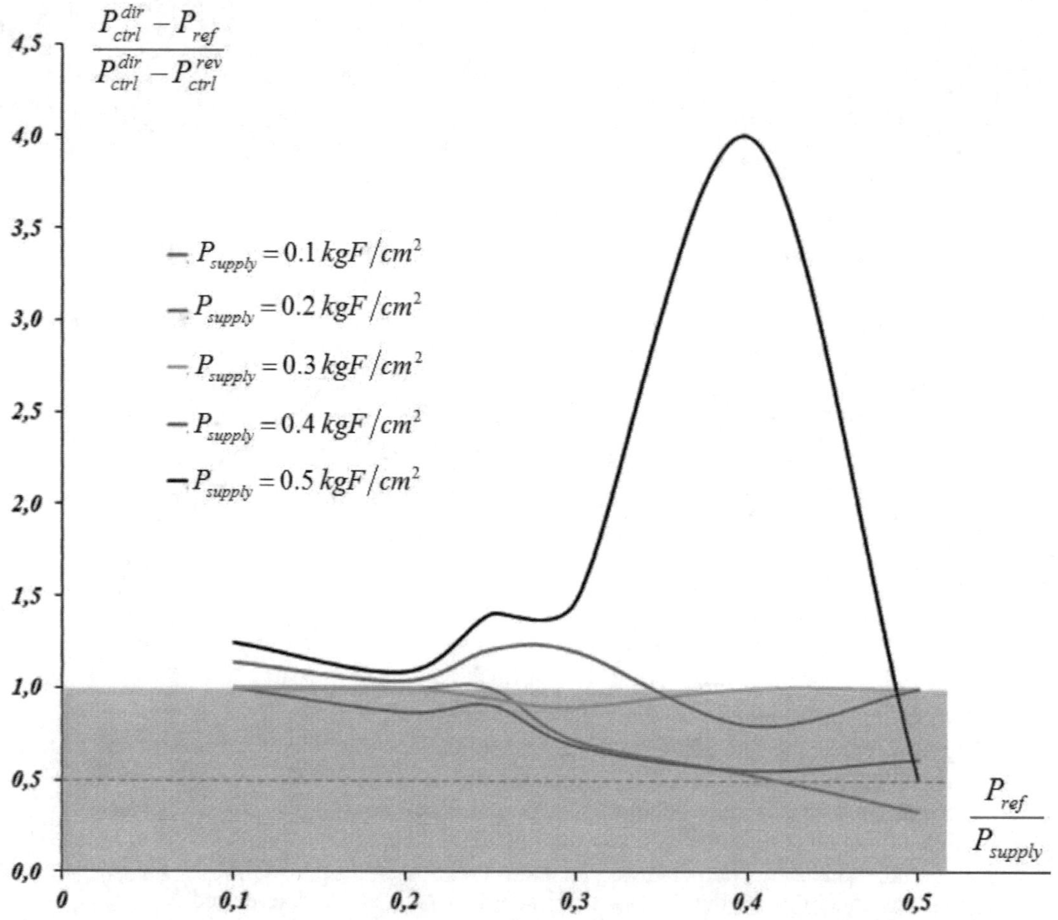

Figure 5. Relative-hysteresis characteristics of trigger.

4. Conclusion

Experimental research of the microfluidic trigger of the 100-micrometer feature size is carried out in this paper. The main results of the experimental researches are the transfer and flow-rate characteristics of the microfluidic trigger, obtained at the two modes of operation, as well as the relative-hysteresis characteristic. The analysis of these characteristics is performed. Features of operating the trigger are revealed in relation to both the relay mode and the comparator mode. The above reported results provide a possibility of improving the quality of digital microfluidic units developed for a wide range of applications in the control, the medicine, the bioengineering, the integrated technologies, and other fields.

5. References

[1] Channon R B, Menger R F, Wang W, Carrão D B, Vallabhuneni S, Kota A K and Henry C S 2021 Design and application of a self-pumping microfluidic staggered herringbone mixer *Microfluidics and Nanofluidics* vol 25(4) article number 31

[2] Huang C, Kuo Y, Chen Y, Huang P and Lee G 2021 A miniaturized, DNA-FET biosensor-based microfluidic system for quantification of two breast cancer biomarkers *Microfluidics and Nanofluidics* vol 25(4) article number 33

[3] Liu X, Dong Z, Zhao Q and Li G 2020 Optimization of micromilled channels for microfluidic applications using gas-blowing-assisted PDMS coating *Microfluidics and Nanofluidics* vol 24(1) article number 11

[4] Şahin M A, Çetin B and Özer M B 2020 Investigation of effect of design and operating parameters on acoustophoretic particle separation via 3D device-level simulations *Microfluidics and Nanofluidics* vol 24(1) article number 8

[5] Lai R L and Huang N 2019 Dimensional analysis and parametric studies of the microwell for particle trapping *Microfluidics and Nanofluidics* vol 23(11) article number 121

[6] Hamilton E S, Ganjalizadeh V, Wright J G, Pitt W G, Schmidt H and Hawkins A R 2019 3D hydrodynamic focusing in microscale channels formed with two photoresist layers *Microfluidics and Nanofluidics* vol 23(11) article number 122

[7] Liu B, Li M, Tian B, Yang X and Yang J 2018 A positive pressure-driven PDMS pump for fluid handling in microfluidic chips *Microfluidics and Nanofluidics* vol 22(9) article number 94

[8] Wu J, Cui Y, Xuan S and Gong X 2018 3D-printed microfluidic manipulation device integrated with magnetic array *Microfluidics and Nanofluidics* vol 22(9) article number 103

[9] Shum C, Rosengarten G and Zhu Y 2017 Enhancing wicking microflows in metallic foams *Microfluidics and Nanofluidics* vol 21(12) article number 177

[10] Yin Z and Zou H 2017 Multilayer patterning technique for micro- and nanofluidic chip fabrication *Microfluidics and Nanofluidics* vol 21(12) article number 174

[11] Balabanov A V and Kasimov A M 2019 Development and Investigation Tests of Microfluidic Generator *Sensors & Systems* vol 7-8 pp 34–40

[12] Balabanov A V and Kasimov A M 2019 Analysis and Synthesis of Fluidic Devices on Basis of Multidimensional Networks *Proceedings of XIII All-Russian Meeting on Control Sciences* (Moscow: V. A. Trapeznikov Institute of Control Sciences) pp 2730–34

[13] Kasimov A M, Balabanov A V and Dolgov I V 2018 Method of Calculating Design Parameters of Communication and Throttle Channels of Microfluidic System *Sensors & Systems* vol 5 pp 39–44

[14] Balabanov A V and Kasimov A M 2018 Method of designing microfluidic operational units of robust reserve control systems of critical objects *Problems of Control of Complex System Safety: Proceedings of XXVI International Conference* (Moscow: V. A. Trapeznikov Institute of Control Sciences) pp 284–8

Acknowledgments

This paper has originated from the 5th International Scientific Conference on Information, Control, and Communication Technologies (ICCT 2021, Astrakhan). The authors very appreciate the efforts spent by the collective of the “Control of Network Systems” laboratory (ICS RAS) headed by V. M. Vishnevsky in order to organize the ICCT 2021 conference at the high level and thank the organizers for the participation to be comfort, useful, and of great interest.

Linear time-delay circuit and stabilizing controller

Yuhao Cong[1,2], Yong Zhang[1] and Guang-Da Hu[1]

[1]College of Sciences, Shanghai University, Shanghai 200444, China.
[2]Shanghai Customs College, Shanghai 201204, China.

E-mail: zhangyong314@shu.edu.cn

Abstract. This paper is concerned with a linear time-delay circuit and its feedback control. We use electronic components such as resistors and capacitors to realize a linear time-delay system. The time-delays are generated by operational amplifiers and single-chip microcomputers. Based on the actual data measured by the oscilloscope, the parameters of the system are estimated using the least square method. Then a comparison study between the waveform image measured by the oscilloscope and the numerical simulation obtained by MATLAB verifies the effectiveness of the parameters estimations of the circuit system. Furthermore, the circuit system is unstable with a large time-delay, a feedback controller is designed to stabilize the circuit system using the optimization method in the literature. Finally, the experimental results in the linear time-delay circuit show the effectiveness of the optimization method.

1. Introduction

Time-delay systems are widely used in engineering, such as automatic control, chemical reactions, biological engineering and so on. Generally speaking, every system has different degrees of delay, and the delay usually leads to poor system stability or even instability. Therefore, it is of great theoretical and practical significance to study the stability and the feedback control of time-delay systems [3].

Linear systems with multiple delays are described as

$$\dot{x}(t) = A_0 x(t) + \sum_{j=1}^{m} A_j x(t - \tau_j) + Bu(t), \tag{1}$$

where $x(t) \in \mathcal{R}^n$ is the state vector, $u(t) \in \mathcal{R}^p$ is the control vector, $A_0, A_j \in \mathcal{R}^{n\times n}$,$B \in \mathcal{R}^{n\times p}$ are constant matrices, $\tau_j > 0$ for $j = 1,2, \dots m$. In [1], a sufficient and necessary condition for the stability of delay systems is studied by the argument principle. In [2,5], a numerical optimization algorithm is presented to seek feedback gain matrices to stabilize system (1) base on an integral performance of the state transition matrix of the system. In this work, we use electronic components such as resistors, capacitors, operational amplifiers and microcontrollers to simulate linear time-delay systems and design controller to stabilize unstable circuit systems.

2. Time-delay Circuit Design

A $R - C$ circuit is shown in Figure 1 [6]. In the circuit, R_1, R_2, Rd_1, Rd_2 are resistances, C_1, C_2 are the capacitances, the voltages of R_1 and R_2 are denoted by $x_1(t)$ and $x_2(t)$ respectively, which are state vectors of the circuit. The voltage signals after τ seconds delay are denoted by $x_1(t - \tau)$ and $x_2(t - \tau)$ respectively.

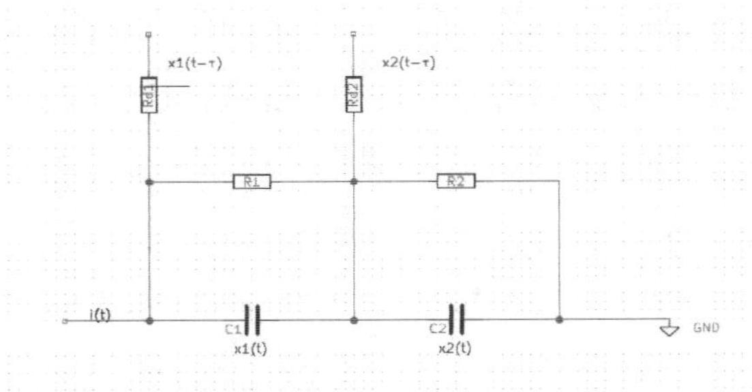

Figure 1. Time-delay circuit model

Kirchhoff's law equations for left nodes of R_1 and R_2 are as follows

$$\begin{cases} i(t)-\dfrac{x_1(t-\tau)}{Rd_1}-\dfrac{x_1(t)}{R_1}-C_1\dfrac{dx_1(t)}{dt}=0, \\ \dfrac{x_1(t)}{R_1}+C_1\dfrac{dx_1(t)}{dt}-\dfrac{x_2(t)}{R_2}-C_2\dfrac{dx_2(t)}{dt}-\dfrac{x_2(t-\tau)}{Rd_2}=0. \end{cases} \tag{2}$$

The signal source used in the experiment is a voltage source, so we designed a circuit to convert voltage into current $i(t)=\frac{1}{1000}v(t)$ and $v(t)$ is control signal. The state differential equations are

$$\begin{pmatrix}\dot{x}_1(t)\\ \dot{x}_2(t)\end{pmatrix}=A_0\begin{pmatrix}x_1(t)\\ x_2(t)\end{pmatrix}+A_1\begin{pmatrix}x_1(t-\tau)\\ x_2(t-\tau)\end{pmatrix}+Bu(t), \tag{3}$$

where

$$A_0=\begin{pmatrix}-\dfrac{1}{C_1R_1} & 0\\ 0 & -\dfrac{1}{C_2R_2}\end{pmatrix}, A_1=\begin{pmatrix}-\dfrac{1}{C_1Rd_1} & 0\\ -\dfrac{1}{C_1Rd_1} & -\dfrac{1}{C_1Rd_2}\end{pmatrix}, B=\begin{pmatrix}\dfrac{1}{1000C_1}\\ \dfrac{1}{1000C_2}\end{pmatrix}, u(t)=v(t).$$

The delayed signal can be generated by the memory function of STM32 single-chip microcosmputer, the delayed waveform is shown in Figure 2.

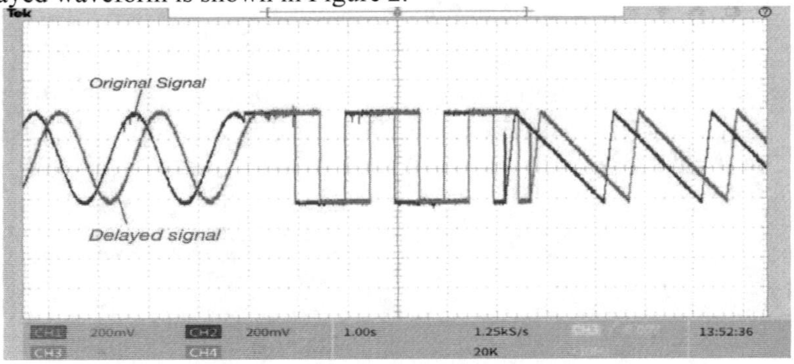

Figure 2. Time-delay signal ($\tau=0.45s$)

Based on the characteristics of high input impedance and low output impedance of the operational amplifier, the circuit is designed to extract the voltage signal from the original system, which guarantees the state of the original system to the greatest extent. The complete circuit diagram is shown in Figure 3. R_1, $Rd_1$and Rd_2 are 510Ω, R_2 is 1000 Ω ,C1 and C2 are 800 μF. Due to the complexity of integrated circuit and the error of resistance and capacitance, it is necessary to make further parameters identification of the circuit system.

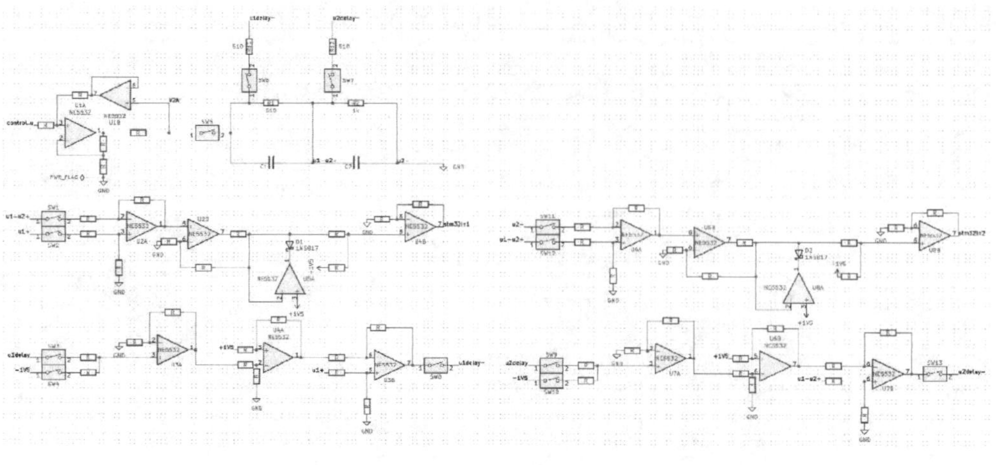

Figure 3. Circuit diagram of the system

3. Parameters Identification

Linear time-delay system (3) with input noise is described by

$$\begin{cases} \dot{x}(t) = A_0 x(t) + \sum_{j=1}^{m} A_j x(t-\tau_j) + B\bar{u}(t), \\ \bar{u}(t) = u(t) + \xi(t), \end{cases} \tag{4}$$

where $u(t)$ is original input, $\xi(t) \in \mathcal{R}^p$ is the white Gaussian noise of the system. The partition $P := \{t_k = kh: k = 0,1,\dots,N\}$ in the time interval $[0,T]$ is defined, where h is the step size and $N = \frac{T}{h}$. Assume that $u(t)$= $u(t_k)$ in for $t \in [t_k, t_{k+1}]$, applying the Euler-Maruyama method to discretize (4), we obtain

$$x(t_{k+1}) - x(t_k) = hA_0 x(t_k) + \sum_{j=1}^{m} hA_j x(t_{k-d_j}) + hBu(t_k) + \sqrt{h}B\xi(t_k), \tag{5}$$

where $u(t_k)$ and $x(t_k)$ are the original input and measured state vector of the system (4) at t_k, and $d_j = \tau_j/h$ is time-delay after discretization. Let $\theta^T = [A_0, A_1, \dots, A_m, B]$ be the parameter matrix, $x(t_k)$ is denoted by x_k, $u(t_k)$ is denoted by u_k, $\sqrt{h}B\xi(t_k)$ is denoted by v_k, $x(t_{k+1}) - x(t_k) = x_{k+1} - x_k$ is denoted by y_k, we have

$$\begin{aligned} y_k &= hA_0 x_k + \sum_{j=1}^{m} hA_j x_{k-d_j} + hBu_k + v_k \\ &= [A_0, A_1, \dots, A_m, B] \begin{bmatrix} hx_k \\ hx_{k-d_1} \\ \vdots \\ hx_{k-d_m} \\ hu_k \end{bmatrix} + v_k \\ &= \theta^T \varphi(k,d) + v_{k,} \end{aligned} \tag{6}$$

where $y_k \in \mathcal{R}^n, \theta^T \in \mathcal{R}^{n\times(2n+r)}, \varphi(k,d) = \left[hx_k^T, hx_{k-d_1}^T, \dots, hx_{k-d_m}^T, hu_k^T\right]^T \in \mathcal{R}^{2n+r}$.

The offline identification algorithm requires to minimize the following criterion function

$$J_k(\theta, d) = \frac{1}{2}\sum_{i=0}^{l-1} e_{k-i}^T e_{k-i}, \tag{7}$$

Where l is the length of the data and $d_j \leq k - l$, e_k is the error at k, that is, $e_k = y_k - \theta^T \varphi(k, d)$. The parameter matrix θ can be estimated by the least squares method [4]

$$\theta_k(d) = [\phi(k, d)\phi^T(k, d)]^{-1}\phi(k, d)Y^T(k), \tag{8}$$

where $\phi(k, d) = [\varphi(k, d), \varphi(k-1, d), \ldots, \varphi(k-m+1, d)]$, $Y^T(k) = (y_k, \ldots, y_{k-m+1})$.

Applying the method mentioned above, the parameters identification results are as follows

$$A_0 = \begin{pmatrix} -2.2626 & 0 \\ 0 & -1.0932 \end{pmatrix}, A_1 = \begin{pmatrix} -2.2761 & 0 \\ -2.1713 & -2.1627 \end{pmatrix}, B = \begin{pmatrix} 1.1494 \\ 1.0965 \end{pmatrix}.$$

The circuit system (3) can be written as

$$\begin{pmatrix} \dot{x}_1(t) \\ \dot{x}_2(t) \end{pmatrix} = \begin{pmatrix} -2.2626 & 0 \\ 0 & -1.0932 \end{pmatrix} \begin{pmatrix} x_1(t) \\ x_2(t) \end{pmatrix} + \begin{pmatrix} -2.2761 & 0 \\ -2.1713 & -2.1627 \end{pmatrix} \begin{pmatrix} x_1(t-\tau) \\ x_2(t-\tau) \end{pmatrix} + \begin{pmatrix} 1.1494 \\ 1.0965 \end{pmatrix} u(t), \tag{9}$$

Let $\tau = 0.45s$, $v(t) = 0.5\sin(2\pi t)$, the comparison between waveform image measured by oscilloscope in Figure 4 and the numerical simulation obtained by MATLAB in Figure 5 verifies the effectiveness of the parameters estimations of the circuit system.

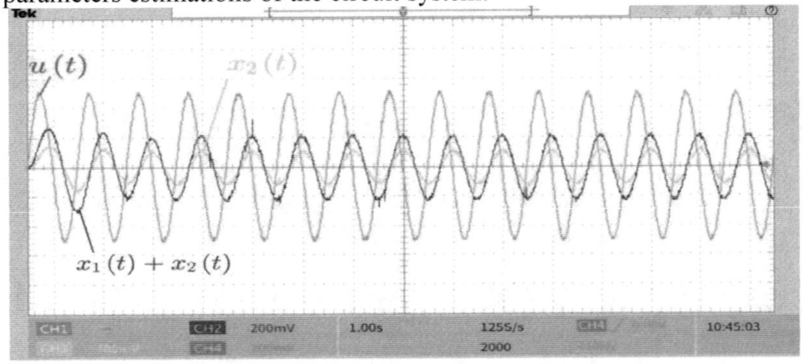

Figure 4. Waveform collected by oscilloscope ($\tau = 0.45s$, $v(t) = 0.5\sin(2\pi t)$)

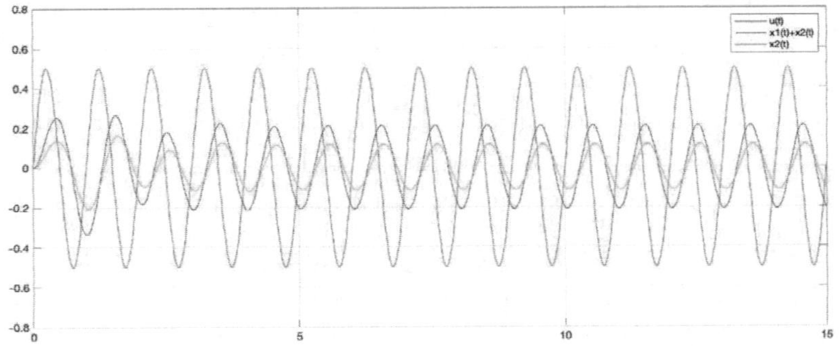

Figure 5. Numerical simulation ($\tau = 0.45s$, $v(t) = 0.5\sin(2\pi t)$)

4. Controller Design

When time-delay increases to 1.2s, the circuit system becomes unstable. In the design of the circuit in section 2, a limiter is designed to protect the circuit [7], thus the voltage will not increase indefinitely. However, it can be seen from the waveform in Figure 6 that the circuit system has become unstable. The main work of this section is to seek the feedback controller's parameters.

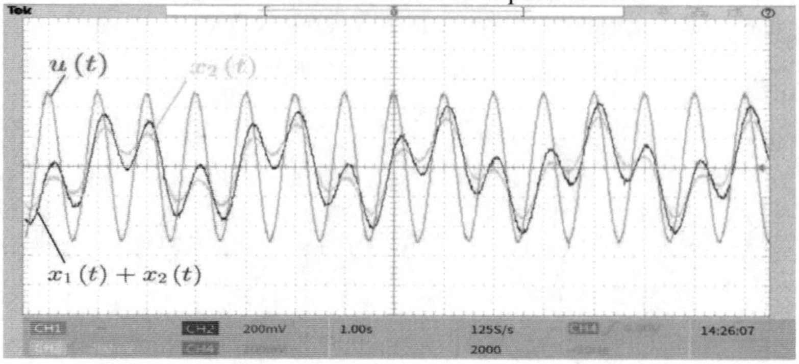

Figure 6. Unstable waveform ($\tau = 1.2s$, $v(t) = 0.5\sin(2\pi t)$)

Let feedback controller structure of (1) be

$$u(t) = K_0 x(t) + \sum_{j=1}^{m} K_j x(t - \tau_j), \quad (10)$$

where $K_0, K_j \in \mathcal{R}^{p\times n}$ are feedback gain matrices. From (3) and (10), the closed-loop system can be described by

$$\dot{x}(t) = (A_0 + BK_0)x(t) + \sum_{j=1}^{m}(A_j + BK_j)x(t - \tau_j). \quad (11)$$

Let $\boldsymbol{K} = [K_0, K_1, \dots K_m]$, the state transition matrix of the closed-loop system (11) is denoted by $F[\boldsymbol{K}, t]$, which is the solution of the matrix delay differential equation

$$\dot{F}[\boldsymbol{K}, t] = (A_0 + BK_0)F[\boldsymbol{K}, t] + \sum_{j=1}^{m}(A_j + BK_j)\, F[\boldsymbol{K}, t - \tau_j] \text{ for } t > 0, \quad (12)$$

under condition

$$F[\boldsymbol{K}, t] = 0 \text{ for } t \le 0 \text{ and } F[\boldsymbol{K}, 0] = I.$$

Lemma 1([2,5]) The closed-loop system (11) is asymptotically stable if and only if there exist feedback gain matrices K_0, and K_j for $j = 1,2, \dots, m$, i.e., K such that

$$\lim_{T\to\infty} \int_0^T \|F[\boldsymbol{K}, t]\|^2 \, d\sigma + \|\boldsymbol{K}\|^2 \quad (13)$$

exists and is finite.

Applying **Lemma 1** to system (9), let $u(t) = K_0 x(t) + K_1 x(t - \tau_j) + v(t)$, T=100 and solve the optimization problem, we can obtain the gain matrices

$$K_0 = [-0.1438 - 1.9887], K_1 = [1.8264\ 1.0752].$$

The closed-loop system of (11) with controller (10) is asymptotically stable.

Let $\tau = 1.2s$, $v(t) = 0.5\sin(2\pi t)$, the waveform of closed-loop system collected by oscilloscope in Figure 7 verifies the effectiveness of controller. The design of controller is in Figure 8 and Figure 9 is the experimental platform of this paper.

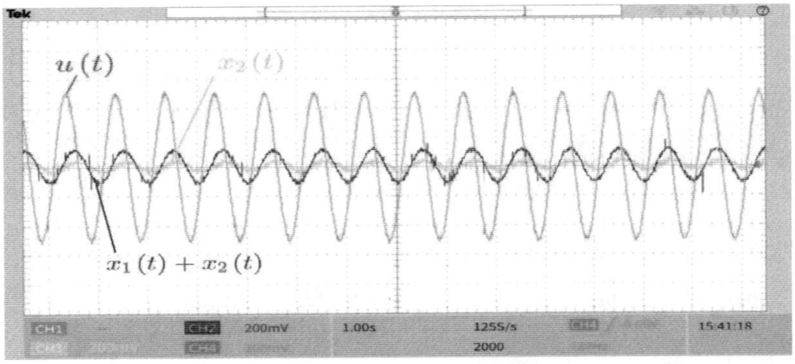

Figure 7. Waveform of closed-loop system ($\tau = 1.2s$, $v(t) = 0.5\sin(2\pi t)$)

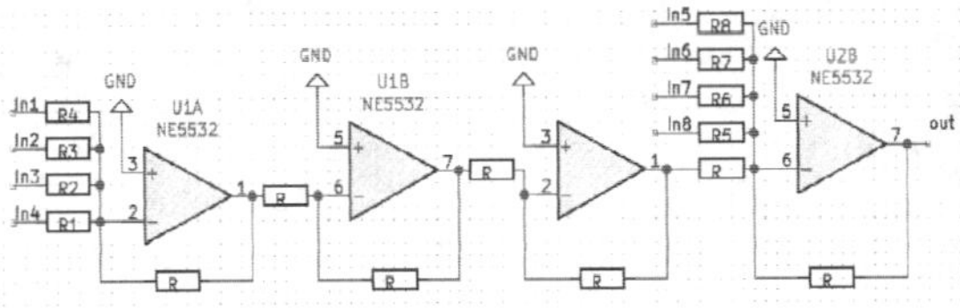

Figure 8. Circuit diagram of the controller

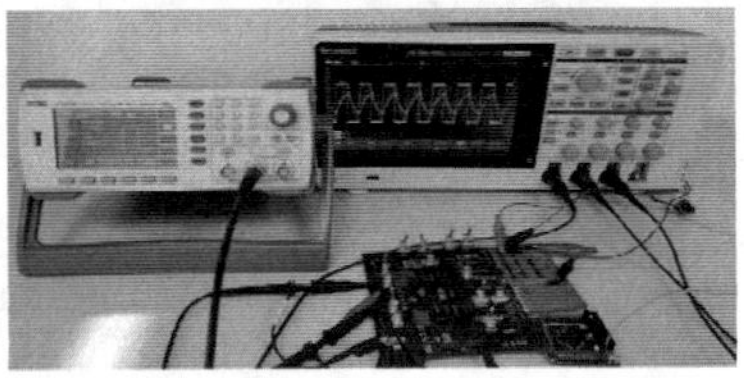

Figure 9. Experimental platform

5. Conclusion

Using the electronic components such as resistors, capacitors, operational amplifiers and STM32H743 single-chip microcomputer, the linear time-delay circuit is designed. Then, the parameters of the system are identified based on the least squares method. And Figure 4 and Figure 5 verify the effectiveness of the parameters estimation of the circuit system. Furthermore, the feedback gain matrix parameters are obtained by **Lemma 1** and the controller circuit is designed to stabilize the unstable circuit system, Figure 8 verifies the effectiveness of the controller.

Acknowledgments

The authors were supported by National Natural Science Foundation of China (No. 11971303 and No. 11871330).

References

[1] Hu G D and Liu M 2007 Stability criteria of linear neutral systems with multiple delays *IEEE Trans. Automat. Contr* **52(4)** 720-4

[2] Hu G D and Hu R 2020 Numerical optimization for feedback stabilization of linear systems with distributed delays *J. Comput. Appl. Math* **371** 112706

[3] Kim A V and Ivanov A V 2015 *Systems with Delays: Analysis, Control, and Computations* (Massachusetts: Scrivener Publishing) pp 29-33

[4] Ljung L 1999 *System Identification, Theory for the user* (Beijing: Tsinghua University Press) pp 203-11

[5] Hu G D and Hu R 2020 An Optimization Method for Feedback Stabilization of Linear Delay Systems *Mathematics and its Applications: International Conference in honor of the 90th birthday of Sergei K* Godunov pp185-91

[6] Wu X and Zhang X 2020 Parameters identification of second order RC equivalent circuit model for lithium batteries *Chin. J. Nanjing Uni.* **56(5)** 754-61

[7] Jia X and Guo Z 1990 High precision active limiter *Chin. J. Semiconductors* **15(1)** 56-7

Analysis on non-stationary gyroscopes and their application in measurement technology

A I Andreev and A P Perekrestov

Astrakhan State Technical University, Tatishcheva 16, Astrakhan, 414056, Russia

E-mail: rap1238@mail.ru

Abstract. In this work, the main subjects of research are: 1) analysis of the process of rotation of a Chinese top on a flat surface, 2) analysis of the main dependencies and an explanation of the rise of the center of gravity of the top during rotation due to nonlinear friction and the resulting force of the overturning top on the upper axis. Also, an analytical study of the motion of the top along the plane and the laws of rotation was carried out, the rotation of the top was analyzed from the point of view of asymmetry during rotation. The top-top spinning top is an interesting case of raising the center of gravity during rotation. The simplest model of a Chinese top can be a dynamic symmetric inhomogeneous ball, the center of mass of which lies on the axis of dynamic symmetry, but does not coincide with its geometric center. The precession of the Chinese top is based on dry friction. The occurrence of a gyroscopic deflection moment is based on the frictional force. Considering the simplest top-type top, one can distinguish the main forces acting on this body, as well as its characteristics that determine the precession. Based on this fact, it is possible to propose the use of this body of revolution as a working medium for instruments such as tribometers and gravimeters. In the first case, this makes it possible to increase the accuracy of the device, in the second, to create a new scheme for measuring gravitational fields. The relevance of this work is the analysis of the rotation of the Chinese top and the possibility of its use in measuring instruments.

1.Introduction

Consider a Chinese spinning top, aka Thomson's spinning top, in particular its procession from coups to the "leg" of the spinning top. The top-top spinning top is an interesting case of raising the center of gravity during rotation. The simplest model of a Chinese top can be a dynamic symmetric inhomogeneous ball, the center of mass of which lies on the axis of dynamic symmetry, but does not coincide with its geometric center [1]. This issue was considered by V.F. Zhuravlev and D.M. Klimov [2] from the point of view of introducing dry friction at the point of contact. This model fully explains the procession of the Chinese spinning top. However, it should be understood that when the Chinese top rotates, the ratio of its moments of inertia relative to various axes does not remain constant, which allows us to make adjustments to the existing models for describing the Chinese top. The forces causing the top to deviate from the vertical position depend on friction and, as a consequence, on the roughness of the surface on which the Spinning top precession occurs, which makes it possible to use this effect in measuring technology, primarily in the design of tribometers. By measuring the forces with which the top tends to move away from the vertical position, you can accurately identify the friction force acting on the base and find the coefficient of friction [3].

Also, the rate of change of the top's position depends on its moment of inertia. So, if we assume that the coefficient of friction and the speed of rotation of the top are known, we can say what mass the top has in a given precession. This simple conclusion makes it possible to use non-stationary gyroscopes in devices for measuring the gravitational field. This class of instruments can be useful both for measuring congestion on aircraft and space vehicles, and for vehicles intended for the study of distant space objects. To explain the operation of such devices, one must initially turn to the theory of gyroscopes, namely, to the precession of the tip-top (Chinese top).

The scientific novelty lies in the analysis of the physics of the tip-top and its application in technology. At the moment, the Chinese spinning top, which has received its final theory in recent years, is very little used in practice. However, the process of its variable precession allows us to speak about its prospects in industry, in particular in measuring instruments.

2.The movement of the center of mass during rotation and deviation of the axis of symmetry

In the general case, the deviation of the Chinese top from symmetry and the rise of its center of gravity is explained by the appearance of a moment of force from the friction force, which in this case is the force that deflects the body from a state of stable equilibrium. Since the center of mass is located above the fulcrum, the moment of force arising from friction tends to raise the center of gravity (figure 1):

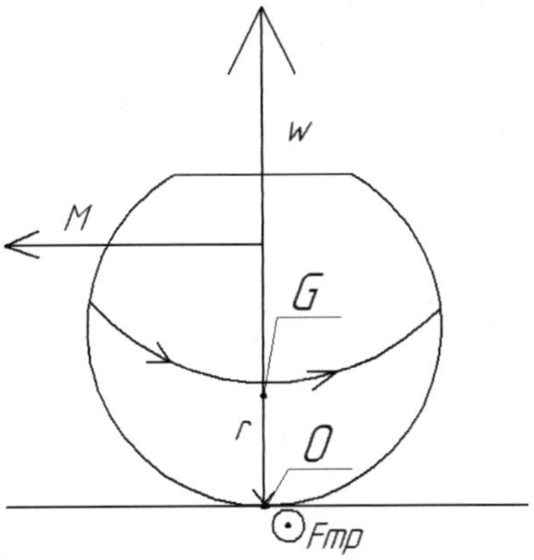

Figure 1 - distribution of the main forces during the rotation of a top type top.

In this figure, point O is the point of contact between the surface and the point of application of the friction force, M is the moment of the friction force, point G is the center of mass of the top, vector r is the radius vector from the point of the center of mass to the point of contact, w is the angular velocity. When the body rotates around its own axis of the figure, the angular momentum is co-directed with it, while the friction force is directed in the direction opposite to the motion, then the gyroscope equation has its classical form of a vector product [4]. Taking a top-type top for a symmetrical gyroscope at the beginning of our movement, we will take the following:

$$\vec{M} = \left[\vec{r} \times \vec{F}_{fr}\right], \quad (1)$$

where F_{fr} - friction force, r - radius vector drawn from the center of mass of the top to the point of the pop;

Let us describe the movement of the center of gravity of the Tip-Top during its procession. We denote by point A - the center of gravity, by point B - the top of the whirligig of the top and the center of application of the external force, by point B - the point of contact of the top with the surface during rotation and the cent of application of the friction force. Draw axes through all points. This procession is shown in figure 1:

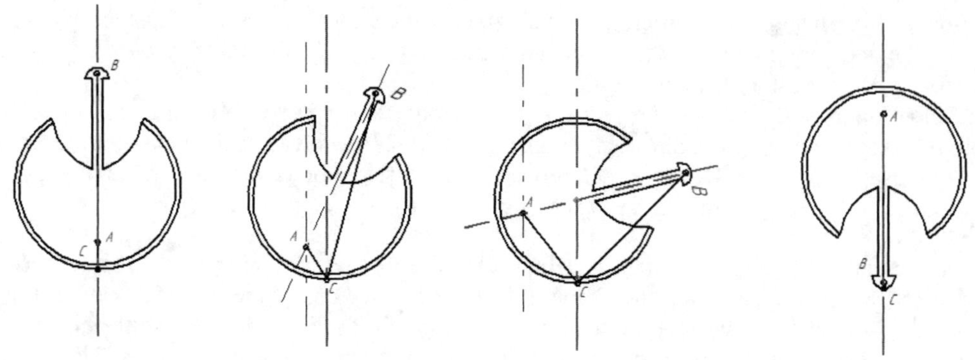

Figure 2 - Deviation of the center of mass during the procession of the top.

As can be seen from Figure 2, the sum of the segments AC and BC remains constant, which is a characteristic distinctive feature of the ellipse.

The moment of frictional force tends to raise the center of gravity and can produce an action only if it has sufficient kinetic energy, i.e. according to Newton's second law for a rotating body.

$$\vec{J} = \frac{d\omega}{dt} = \vec{M} \qquad (2)$$

Then, having entered the distance h from the center of tangency to the center of gravity for a top rotating stably, we obtain a ratio for two axes.

$$\frac{h}{r} = \frac{\vec{J}_z}{\vec{J}_x} \qquad (3)$$

Next, consider the motion of the top's center of mass in the plane from above. It is obvious that the center of gravity moves in a spiral, rising relative to the plane (Figure 3):

Then the initial moment of inertia is taken as:

$$J_a = \sum_{i=1}^{n} m_i r_i^2 \qquad (4)$$

As a model of the process, we will take the rotation of a symmetrical top, increasing the angle of deviation of the axis of the procession (figure 3) [5]:

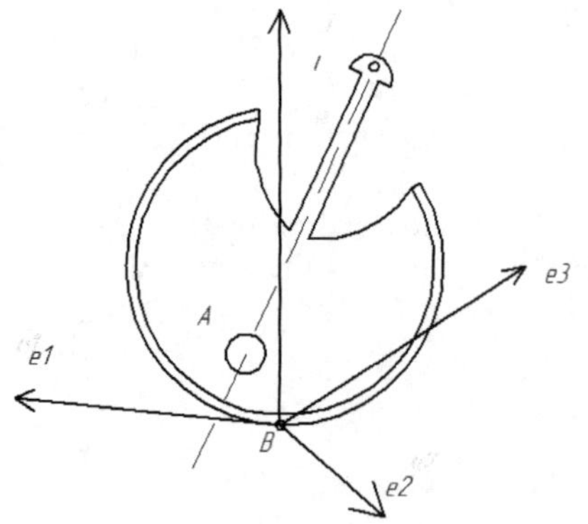

Figure 3 - Application of the model of an asymmetric top.

By B-$e_1e_2e_3$ we denote a basis rigidly connected to the top at a given moment in time. Referring to article, by analogy, let us take the center of gravity for the body being carried at a given moment. If the top rotated in a stable position, then its angular momentum could be considered as:

$$K_o = J\omega + K_o^{rel}, \quad (5)$$

where $J\omega$ - top's inertia tensor in the B-$e_1e_2e_3$ basis. K_o^{rel} is the angular momentum of the center of gravity A relative to the B-$e_1e_2e_3$ basis.

However, the top-type top constantly changes the position of the center of gravity during rotation, which means that equation 5 must be rewritten for each position of the center of gravity during rotation, and have the form:

$$K_o = \frac{d\omega}{d\tau}\frac{dJ}{d\tau} + K_o^{rel}, \quad (6)$$

where $d\tau$ - change in time;

It should be noted that in addition to changing the angular velocity, it is necessary to take into account the change in the axial moment of inertia dJ (change in the moment of action).

The kinetic energy of the top can then be defined as:

$$T = \frac{d(\omega^T)d\omega}{2d\tau}\frac{dJ}{d\tau} + \frac{d(\omega^T)}{d\tau} * K_o^{rel} + T^{rel}, \quad (7)$$

where T^{rel} is the kinetic energy of the top's center of mass;

We apply Köning's theorem to determine the total kinetic energy of the system, during the procession of the center of mass of the top, we obtain the total energy:

$$E = T + U = T = \frac{d(\omega^T)Jd\omega}{2d\tau}\frac{dJ}{d\tau} + \frac{d(\omega^T)}{d\tau} * K_o^{rel} + T^{rel} + mgi_e^T\frac{da}{d\tau}, \quad (8)$$

where m - mass of the top, i - unit vector of the vertical, e - unit vector directed along the vector of the center of mass of the top $a = ae$.

For a symmetric top, it follows from the theorem on the change in the angular momentum that the equations of motion of the system under consideration have a first integral describing the conservation of the projection of the angular momentum onto the vertical. For a tip-top top, the angular momentum changes, and then the formula for a symmetric top can be rewritten:

$$i^T K_o = i^T\left(\frac{d\omega}{d\tau}\frac{dJ}{d\tau} + K_o^{rel}\right) = \frac{dK}{d\tau} \quad (9)$$

From the source [6] , we will show the diagram of the positions of the tip-top top (figure 4):

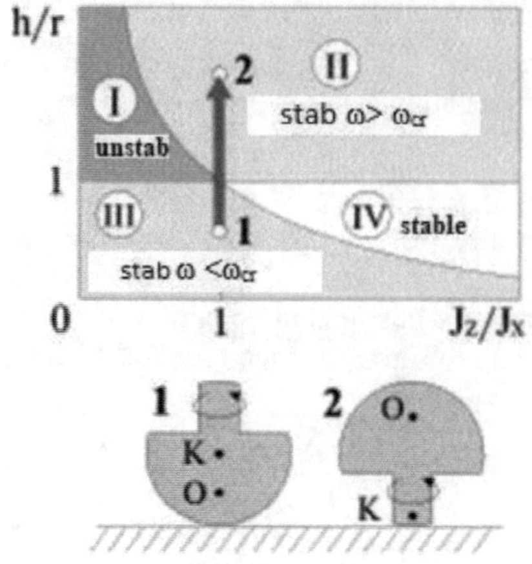

Figure 4 - stable positions of a tip-top top.

In figure 4 J_x is the moment of inertia about the axis of symmetry, J_z is the moment of inertia about the main central axis perpendicular to the axis of symmetry. Region I corresponds to unstable rotation of the top at all angular velocities, region II - to stable rotation at sufficiently high angular velocities $\omega > \omega_{cr}$. Region III corresponds to stable rotation at low angular velocities $\omega < \omega_{cr}$, region IV - to stable rotation at arbitrary ω. The critical angular velocity ω depends on the moments of inertia J_x, J_z of the distances r, h and the body weight $P = mg$ [7].

$$\omega^2 = \frac{(h-r)*P}{J_x\left(\frac{r}{h}\right)*\left(\frac{J_z}{J_x}-\frac{r}{h}\right)} \quad (10)$$

A Chinese top, untwisted to $\omega > \omega_{cr}$ and placed on a plane vertically, will be in the region of instability, and the top will turn over on its leg. The angular velocity ω included in the above formulas is above the critical one, for changing the position of the center of mass of a tip-top [8],[9],[10].

3. Changing the axial moment of inertia and angular momentum of the tip-top top

It should be said that a change in the position of the center of mass of a top-top top during its rotation entails a change in the moment of inertia of the top (the moment of inertia relative to the main axis of rotation is considered).

$$J = \int_0^T mR^2 dT \quad , \qquad (11)$$

where *T* is the exact moment when the top turns over on the stem.

It is also necessary to take into account that R is also changing, since the classical Chinese top is not a perfect ball, but a truncated ball. Then:

$$R = \frac{dr}{d\tau} \quad , \qquad (12)$$

where *r* is the distance to the points of a rigid body at a given time.

After the top is turned over on the leg, its moment of inertia does not change until the moment when the top stops and falls. Then the angular momentum of such a top can be written, using formulas 10 and 11, we obtain the angular momentum of the top.

$$d\vec{L} = \omega * J = \omega \int_0^T mR^2 dT \qquad (13)$$

From this formula, it becomes clear that a change in the moment of inertia of the top, due to the moment of force from friction (as mentioned above), changes its angular momentum, until the top turns over on the leg.

The given regularities are of practical value and can be used in measuring equipment.

4.Practical value.

The use of a Chinese spinning top in technology brings with it a number of advantages and allows you to improve existing and create new devices and methods. So, for example, the measurement of the friction force by the force arising during rotation suggests the creation of a tribometer based on a body similar to a Chinese top (figure 5).

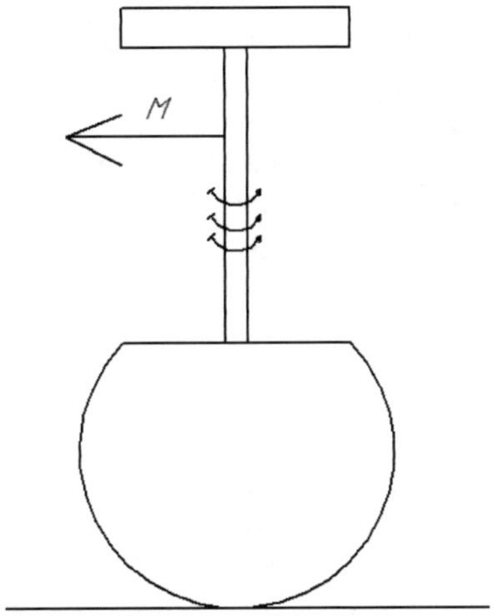

Figure 5 - Circuit of the friction mechanism, based on the Chinese top.

As can be seen from formula (1), the cause of the deflection of the top is the friction force, which makes it possible to create devices for its detection. This can be done as follows. The body is untwisted from the electric motor to the required revolutions and is placed on the object under study, as a result of which, by measuring the moment of force, it is possible to establish the friction force, and through it the desired coefficient of friction. This method of measurement makes it possible to carry out direct studies of surface friction and to increase the accuracy of tribometric measurements in general.

In addition, it is possible to create potentiometers (figure 6).

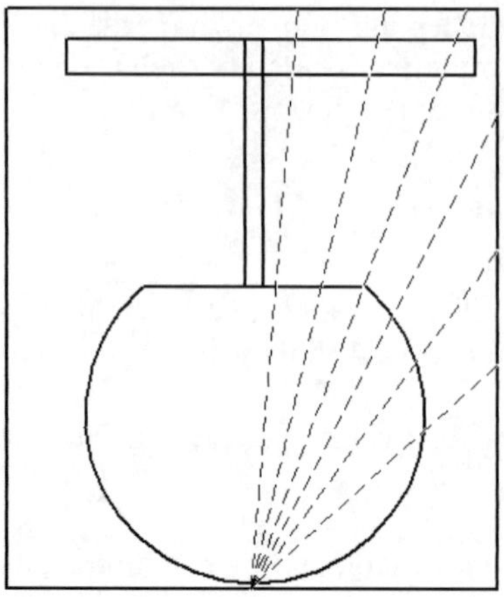

Figure 6 - Potentiometer circuit.

Formulas (8-10) show that the rate of change in the angle of precession of a non-stationary gyroscope also depends on the acceleration of gravity. This means that by measuring the rate of change in the angle of deflection of the top from the initial position during rotation, it is possible to accurately establish the acceleration of gravity, and as a consequence, the mass of the gravitating source or the overload acting on the body, which allows us to speak of the prospects of using tip-top tops in gravimeters.

This can be done in the following way. The axis of the top is fixed and can only move in one plane. By the deflection of the top of spinning top during rotation, it is possible to calculate the reaction of the support and the acceleration of gravity, thereby determining the gravitational constant. This device can be used as a gravitational field meter, and a g-force and weightlessness meter in aviation and space technology.

5. Conclusion

The use of asymmetric non-stationary gyroscopes in measuring instruments will improve their accuracy and pave the way for further research both in the field of instrumentation and in the field of solid mechanics. The use of a tip-top grinder in measuring technology allows direct measurement of the friction force and the coefficient of friction and roughness for various materials with increased accuracy. In addition, based on the physical model of the precession of this top, it becomes possible to measure the force of gravity, which means that it leads to the possibility of creating a device for continuous measurement of changing gravitational fields, which is important both for aviation technology in terms of measuring congestions and for space technology. Further studies of the tip-top top will be able to clarify important aspects of its physics and prospects for application in technology.

References

[1]. N. I. Amel'kin On the Motions of a Rigid Body Containing Two-Degree-of-Freedom Control Moment Gyros with Dissipation in Gimbal Axes, *Izvestiya Rossiiskoi Akademii Nauk Mechanics of Solids*, **4**, pp 19–30

[2]. Zhuravlev, V G The Model of Dry Friction in the Problem of the Rolling of Rigid Bodies, *Journal of Applied Mathematics and Mechanics* 0021-8928 **62(5)** pp 705–710

[3]. Yu. L. Bolotin, A. V. Tur, and V. V. Yanovskii 2002 Nonlinear friction as a mechanism of directed motion, *Technical Physics* 47, 803

[4]. Ambrósio J.A C 2003 Impact of rigid and flexible multibody systems: deformation description and contact model. *Virtual Nonlinear Multibody Systems* 103 pp 57–81

[5]. Karapetyan, A.V 2008 Global Qualitative Analysis of Tippe Top Dynamics, *Izvestiya Rossiiskoi Akademii Nauk Mechanics of Solids* 2008 **3** pp. 33–41

[6]. A.A. Zobova 2011 Nonstationary movements of a two-spherical Chinese top *Lobachevsky State University of Nizhni Novgorod*, 2011 **4 (2)** pp 143-144
[7]. D Featonby Spinning tops revisited // 2019 *Physics Education* 54 015022
[8]. Rabinowicz, E.: The nature of the static and kinetic coefficients of friction. *Journal of Applied Mathematics and Mechanics*. **22** pp 1373–1379 (1951)
[9]. Hsieh C, Pan Y C 2000 Dynamic behavior and modelling of the pre-sliding static friction. *Wear* 242 pp 1–17
[10]. A Fréne, J., Cicone, T.: 2001 Friction in Lubricated Contacts. Handbook of Material Behavior Models pp. 760–767. *Academic publishing from Cambridge University Press*

Parameter identification of permanent magnet synchronous AC motor based on spectral method

Yuhao Cong[1,2], Qingjiao Ji[1,3] and Guang-Da Hu[1]

[1] College of Sciences, Shanghai University, Shanghai 200444, China.
[2] Shanghai Customs College, Shanghai 201204, China.

E-mail: jiqingjiao@shu.edu.cn

Abstract. This paper establishes a mathematical model with random terms for a permanent magnet synchronous AC motor, which is a nonlinear system using d - axis current, q - axis current and the rotor electric angular speed as state variables. The numerical solution obtained by the Euler-Maruyama method is used as the measurement data. Aiming at the parameter identification of the system, a step-by-step identification method based on the spectral method discretize the system first and then using the least squares is proposed. This method is used to identify multiple parameters of the motor in the same model. In step-by-step identification, firstly by fixing the motor speed, the system is transformed into a linear system, which is used to estimate the resistance, inductance and flux linkage. After that, the speed is not fixed, for the electrical parameters are known, we can identify damping and inertia by using mechanical equations. Finally, the experimental results show that the relative errors of the parameters identified by the proposed method are smaller, which shows the effectiveness of this method for multi-parameter identification.

1. Introduction

Profiting from its advantages of high reliability, high power density, excellent control performance and low maintenance cost, the surface-mounted permanent magnet synchronous motor (SPMSM) has been widely used in applications such as robotics industries, petrochemical industries and electric vehicles. In order to achieve high performance control of motors, more and more advanced control algorithms have been studied. However, these control algorithms all have high requirements for the accuracy and stability of the electrical and mechanical parameters of the motor [1]-[2]. Therefore, parameter identification is quite necessary.

For this reason, researchers have studied the identification of motor parameters. The methods of identification of motor parameters mainly include recursive least square method [3] and extended Kalman filter method [4]. In this paper, based on the spectral method [5], a step-by-step identification method is proposed to identify the electrical and mechanical parameters of the motor. Firstly, the system is transformed into a linear system by fixing motor speed. We construct the least square model of electrical parameters after using the Chebyshev orthogonal polynomials to approximate the solution of the system. The least square method can be used to identify the electrical parameters. When the speed is fixed, it is difficult to identify the mechanical parameters of the motor. After that, the motor speed is no longer fixed. For the electrical parameters are known, we construct the least square model of mechanical parameters after approximating the solution of the mechanical equation of the motor by the Chebyshev orthogonal polynomials. The mechanical parameters can be identified by using the least square method.

2. Mathematical model of SPMSM and Euler-Maruyama approximate solution of model

In this section, the surface-mounted permanent magnet synchronous motor can be partitioned into two subsystems, namely, the electrical system and the mechanical system. In order to be more consistent with the practical problem, we consider introducing the random noise term into the system, and use the Euler-Maruyama method to obtain the numerical solution of the system [6].

2.1. Mathematical model of SPMSM

The SPMSM is a nonlinear multivarilable system whose mathematical model of the three-phase stationary coordinate system is relatively complicated. In order to simplify the analysis, coordinate transformation is usually used to transform its model into a two-phase rotating coordinate system. The motor model with the random noise term is shown in equation (1)

$$\begin{cases} \dfrac{di_d}{dt} = -\dfrac{R}{L} i_d + \omega_e i_q + \dfrac{1}{L}(u_d + \xi_1(t)), \\ \dfrac{di_q}{dt} = -\omega_e i_d - \dfrac{R}{L} i_q - \dfrac{\psi_f}{L}\omega_e + \dfrac{1}{L}(u_q + \xi_2(t)), \\ \dfrac{d\omega_e}{dt} = -\dfrac{B}{J}\omega_e - \dfrac{P}{J}T_L + \dfrac{3P^2\psi_f}{2J}(i_q + \xi_3(t)), \end{cases} \quad t \in [0, \tilde{T}], \tag{1}$$

where i_d and i_q are the stator d - and q -axes currents, u_d and u_q are the stator d - and q -axes voltages used as control inputs, ω_e is the electric angular speed, R is the stator resistance, L is the stator inductances, ψ_f is the flux linkage, J is the rotor inertia, B is the damping coefficient , T_L is the load torque. $\xi_i(t), i = 1,2,3$ are all Gaussian white noise with zero mean.

2.2. Euler-Maruyama approximate solution of model

The parameters of the system are based on the experimental data. In the sequel, we use the Euler-Maruyama method to obtain the state data of the motor.

Applying the Euler-Maruyama method to equation (1), we obtain

$$\begin{cases} i_{d,n+1} = i_{d,n} - h\dfrac{R}{L} i_{d,n} + h\omega_{e,n} i_{q,n} + \dfrac{h}{L} u_{d,n} + \dfrac{\sqrt{h}}{L}\xi_{1,n}, \\ i_{q,n+1} = i_{q,n} - h\omega_{e,n} i_{d,n} - h\dfrac{R}{L} i_{q,n} - h\dfrac{\psi_f}{L}\omega_{e,n} + \dfrac{h}{L} u_{q,n} + \dfrac{\sqrt{h}}{L}\xi_{2,n}, \\ \omega_{e,n+1} = \omega_{e,n} + h\dfrac{3P^2\psi_f}{2J} i_{q,n} - h\dfrac{B}{J}\omega_{e,n} - h\dfrac{P}{J}T_L + \sqrt{h}\dfrac{3P^2\psi_f}{2J}\xi_{3,n}, \end{cases} \tag{2}$$

where h is any step size. The partition $H := \{t_n = nh : n = 0,1,\cdots,M\}$ of the time interval $[0,\tilde{T}]$ is defined, where $M = \dfrac{\tilde{T}}{h}$ is a positive integer. The numerical approximation to $i_d(t_n)$, $i_q(t_n)$ and $\omega_e(t_n)$ are denoted $i_{d,n}$, $i_{q,n}$ and $\omega_{e,n}$, respectively. Furthermore, $u_{d,n} = u_d(t_n)$ and $u_{q,n} = u_q(t_n)$. $\{\xi_{i,n}\}, i = 1,2,3$ are independent "pseudo-random" numbers generated from the $N(0,1)$ distribution using matlab, where $N(0,1)$ denotes a normally distributed random variable with zero value and unit variance.

The numerical solutions $\{i_{d,n}\}_{n=0}^{M}$, $\{i_{q,n}\}_{n=0}^{M}$ and $\{\omega_e\}_{n=0}^{M}$ are used as the real state data of the system, combined with the input data $u_{d,n}$ and $u_{q,n}$ to identify the electrical and mechanical parameters of the motor.

3. Identification of electrical and mechanical parameters

For SPMSM, parameter identification is divided into two parts: electrical parameters identification and mechanical parameters identification. Electrical parameters include R , L and ψ_f ; mechanical parameters include J , B . In this section, we design the least squares model of parameters by applying the spectral method to the system.

3.1. Identification of electrical parameters

Now the speed of the motor is fixed, that is, ω_e is a fixed value. In the steady state, the motor equation can be simplified to the following linear equation, which is called the electrical equation of the motor.

$$\dot{I}(t) = AI(t) + BU(t) + D + B\xi(t), \qquad t \in [0, \tilde{T}] , \tag{3}$$

where $I(t) = \begin{pmatrix} i_d & i_q \end{pmatrix}^T$, $U(t) = \begin{pmatrix} u_d & u_q \end{pmatrix}^T$, $\xi(t) = \begin{pmatrix} \xi_1(t) & \xi_2(t) \end{pmatrix}^T$. Let $a = \frac{R}{L}$, $b = \frac{1}{L}$, $c = \frac{\psi_f}{L}$, then the matrices $A = \begin{pmatrix} -a & \omega_e \\ -\omega_e & -a \end{pmatrix}$, $B = \begin{pmatrix} b & 0 \\ 0 & b \end{pmatrix}$, $D = \begin{pmatrix} 0 \\ -\omega_e c \end{pmatrix}$.

3.1.1. Spectral method approximation. First, we convert the solution interval. Let $t = \frac{\tilde{T}}{2}(1+s)$, $t \in [0, \tilde{T}]$, then $s = \frac{2}{\tilde{T}}t - 1$, $s \in [-1,1]$. We obtain $I(t) := x(s)$, $U(t) := v(s)$ and $\xi(t) := \zeta(s)$. The approximation $Y(s)$ is sought in the form of the truncated serie

$$x(s) \approx Y(s) = \tilde{T}(1+s)\sum_{k=0}^{N} \boldsymbol{a_k} P_k(s) , \tag{4}$$

where $\{P_k(s)\}_{k=0}^N$ are the Chebyshev orthogonal polynomials. $\{\boldsymbol{a_k} = (a_{1k} \quad a_{2k})^T : k = 0,1,\cdots,N\}$ are the expansion coefficients vector. Then equation (3) is equivalent to

$$\dot{Y}(s) = \bar{A}Y(s) + \bar{B}v(s) + \bar{D} + \bar{\zeta}(s), \qquad s \in [-1,1] , \tag{5}$$

where $\bar{A} = \frac{\tilde{T}}{2}A$, $\bar{B} = \frac{\tilde{T}}{2}B$, $\bar{D} = \frac{\tilde{T}}{2}D$, $\bar{\zeta}(s) = \bar{B}\zeta(s)$.

We take the Chebyshev-Gauss type integration points $\{s_k\}_{k=0}^N$ as the collocation points. Let $\alpha = \begin{pmatrix} \boldsymbol{a}_1^T & \boldsymbol{a}_2^T & \cdots & \boldsymbol{a}_N^T \end{pmatrix}^T$. E_2 denotes the identity matrix; $\otimes$ denotes the Kronecker product. Let $\lambda(B) = (b \quad 0 \quad 0 \quad b)^T$, this symbol is introduced in [7]. We first obtain the value of the projection coefficient α according to the known data, and from equation (4) we obtain

$$\begin{pmatrix} Y(s_0)^T & Y(s_1)^T & \cdots & Y(s_N)^T \end{pmatrix}^T = \begin{pmatrix} K_{00}^T & K_{01}^T & \cdots & K_{0N}^T \end{pmatrix}^T \alpha = K\alpha , \tag{6}$$

where $K_{0i} = \tilde{T} \cdot E_2 \otimes ((1+s_i)P_k(s_i))_{k=0}^N$, $i = 0,1,\cdots,N$. Let $\hat{m}$ represents the estimated value of m . According to the state data $\{I_n = (i_{d,n} \quad i_{q,n})^T\}_{n=0}^M$, the approximate values $\{\hat{Y}(s_k)\}_{k=0}^N$ are obtained by using cubic spline interpolation. Substituting them into equation (6), the approximate value $\hat{\alpha}$ of the projection coefficient can be obtained.

Substituting the collocation points into equation (5) and then converting its terms into matrix form, we have

$$C\hat{\alpha} = K_0 h_1 + K_0 h_2 a + K_1 h_3 b + E h_4 c + \bar{\zeta} , \tag{7}$$

where $\bar{\zeta} = \begin{pmatrix} \bar{\zeta}(s_0)^T & \bar{\zeta}(s_1)^T & \cdots & \bar{\zeta}(s_N)^T \end{pmatrix}^T$, $K_0 = \left(\left((E_2 \otimes (K_{0i}\hat{\alpha})^T)^T \right)_{i=0}^N \right)^T$, $K_1 = \left(\left((E_2 \otimes v(s_i)^T)^T \right)_{i=0}^N \right)^T$.

$$\left(\hat{Y}(s_0)^T \quad \hat{Y}(s_1)^T \quad \cdots \quad \hat{Y}(s_N)^T\right)^T = (C_1^T \quad C_2^T \quad \cdots \quad C_N^T)^T \hat{\alpha} = C\hat{\alpha},\ C_i = \tilde{T}\cdot E_2 \otimes (P_k(s_i) + (1+s_i)\dot{P}_k(s_i))_{k=0}^N\ ;$$

$$\left((\bar{B}v(s_0))^T \quad (\bar{B}v(s_1))^T \quad \cdots \quad (\bar{B}v(s_N))^T\right)^T = K_1\lambda(\bar{B}) = K_1 h_3 b\ ;\ \left(\bar{D}^T \quad \bar{D}^T \quad \cdots \quad \bar{D}^T\right)^T = E\lambda(\bar{D}) = Eh_4 c\ ,$$

$$E = \left(E_2 \quad E_2 \quad \cdots \quad E_2\right)^T;\ \left((\bar{A}\hat{Y}(s_0))^T \quad (\bar{A}\hat{Y}(s_1))^T \quad \cdots \quad (\bar{A}\hat{Y}(s_N))^T\right)^T = K_0\lambda(\bar{A}) = K_0(h_1 + h_2 a)\ .$$

3.1.2. Least squares method. The equation (7) can be written to the linear regression matrix

$$Z = S\theta + \bar{\zeta}\ , \tag{8}$$

where $Z = C\hat{\alpha} - K_0 h_1$, $S = \left(K_0 h_2 \quad K_1 h_3 \quad Eh_4\right)$, $\theta = (a \quad b \quad c)^T$.

According to the least squares principle [8], a cost function is defined as follows

$$J(\theta) = \frac{1}{2}\|Z - S\theta\|_2^2. \tag{9}$$

Further minimizing the cost function $J(\theta)$, we obtain $J(\hat{\theta}) = \min_{\theta} J(\theta)$. Thus the least-squares estimate of θ is $\hat{\theta} = (S^T S)^{-1} S^T Z$.

3.2. Identification of mechanical parameters

Supposed that the electrical parameters of SPMSM are obtained. Now that the speed is not fixed, we focus on the third state equation of equation (1)

$$\frac{d\omega_e}{dt} = -\frac{B}{J}\omega_e - \frac{P}{J}T_L + \frac{3P^2\psi_f}{2J}(i_q + \xi_3(t)), \qquad t \in [0, \tilde{T}]. \tag{10}$$

This is the mechanical equation of the motor, which contains the mechanical parameters of the motor. The following approximates the mechanical equation by spectral method.

3.2.1. Spectral method approximation. Converting the solution interval and the solution space of equation (10), we can obtain $\omega_e(t) := y(s)$, $i_q(t) := v_q(s)$, $\xi_3(t) := \zeta_1(s)$. The spectral approximation $W(s)$ is sought in the form of the truncated serie

$$y(s) \approx W(s) = \tilde{T}(1+s)\sum_{k=0}^{N} b_k P_k(s) + \omega_e(0)\ , \tag{11}$$

where $\{b_k\}_{k=0}^N$ are the expansion coefficients, $\omega_e(0)$ is the electric angular speed at $t = 0$. Then equation (10) is equivalent to

$$\dot{W}(s) = -\frac{\tilde{T}}{2}\bar{a}W(s) + \frac{\tilde{T}}{2}\frac{3P^2\hat{\psi}_f}{2}\bar{b}v_q(s) - \frac{\tilde{T}}{2}PT_L\bar{b} + \bar{\zeta}_1(s), \qquad s \in [-1,1], \tag{12}$$

where $\bar{a} = \frac{B}{J}$, $\bar{b} = \frac{1}{J}$, $\bar{\zeta}_1(s) = \frac{\tilde{T}}{2}\frac{3P^2\hat{\psi}_f}{2}\bar{b}\zeta_1(s)$. In addition, $\hat{\psi}_f$ is obtained in the previous subsection.

Let $\beta = \left(b_1 \quad b_2 \quad \cdots \quad b_N\right)^T$ and $d_0 = \omega_e(0)e$, where e is an all-one $n \times 1$ column vector. The equation (12) uses the same method as the previous subsection, we can get

$$F\hat{\beta} = -\frac{\tilde{T}}{2}\bar{a}(G\hat{\beta} + d_0) + \frac{\tilde{T}}{2}\frac{3P^2\hat{\psi}_f}{2}\bar{b}\hat{v}_q - \frac{\tilde{T}}{2}PT_L\bar{b}\cdot e + \bar{\zeta}_1, \tag{13}$$

where $\bar{\zeta}_1 = \left(\bar{\zeta}_1(s_0) \quad \bar{\zeta}_1(s_1) \quad \cdots \quad \bar{\zeta}_1(s_N)\right)^T$. $\left(\hat{W}(s_0) \quad \hat{W}(s_1) \quad \cdots \quad \hat{W}(s_N)\right)^T = G\hat{\beta} + d_0, G = \left(\left(G_{0i}^{\ T}\right)_{i=0}^N\right)^T$;

$\left(\hat{\hat{W}}(s_0) \ \ \hat{\hat{W}}(s_1) \ \ \cdots \ \ \hat{\hat{W}}(s_N)\right)^T = F\hat{\beta}$, $F = \left(\left(F_i^T\right)_{i=0}^N\right)^T$. The matrices $G_{0i} = \tilde{T}\cdot((1+s_i)P_k(s_i))_{k=0}^N$ and $F_i = \tilde{T}\cdot(P_k(s_i)+(1+s_i)\dot{P}_k(s_i))_{k=0}^N$, $i = 0,1,\cdots,N$. And $\left(\hat{v}_q(s_0) \ \ \hat{v}_q(s_1) \ \ \cdots \ \ \hat{v}_q(s_N)\right)^T = \hat{v}_q$, where $\{\hat{v}_q(s_k)\}_{k=0}^N$ are obtained by using cubic spline interpolation on $\{i_{q,n}\}_{n=0}^M$.

3.2.2. Least squares method. The equation (13) can be written to the linear regression matrix

$$Z_1 = S_1\theta_1 + \bar{\zeta}_1, \tag{14}$$

where $Z_1 = F\hat{\beta}$, $S_1 = \left[-\frac{\tilde{T}}{2}(G\hat{\beta}+d_0) \ \ \frac{\tilde{T}}{2}(\frac{3P^2\hat{\psi}_f}{2}\hat{v}_q - PT_L\cdot e)\right]$, $\theta_1 = \left[\bar{a} \ \ \bar{b}\right]^T$.

According to the least squares principle [8], we minimize the following criterion function $J_1(\theta_1) = \frac{1}{2}\|Z_1 - S_1\theta_1\|_2^2$. And hence the least-squares estimate of θ_1 is $\hat{\theta}_1 = (S_1^TS_1)^{-1}S_1^TZ_1$.

4. Experimential results and analysis

In order to verify the effectiveness and accuracy of the proposed identification method, the method is implemented on the SPMSM with the specifications in Table 1.

Table 1. Electrical and mechanical specifications of experimental motor [9]-[10].

Symbol	Parameter	Value
R	stator resistance	$0.331\,\Omega$
L	stator inductance	$2.1\,mH$
ψ_f	flux linkage	$0.3537\,Wb$
J	rotor inertia	$0.6329\,kg\cdot ms^2$
B	damping coefficient	$0.3035\,S^{-1}$
T_L	load torque	$75.0\,N\cdot m$

After Clark transformation and Park transformation, the three-phase AC voltages are equivalent to the two-phase DC voltages as the inputs of the system. To show the superiority of the proposed method, it is compared with the Milne method [11]. The criteria for evaluation is the relative error.

Table 2. Results of electrical parameters identification method.

Items \ Methods	Reference values	Spectral method		Milne method	
		Estimated values	Relative errors	Estimated values	Relative errors
Stator inductance (L/mH)	0.00210000	0.00210001	0.00038646%	0.00210008	0.0040243%
Stator resistance (R/Ω)	0.33100000	0.33099991	0.000026973%	0.33105018	0.01516%
Flux linkage (ψ_f/Wb)	0.35370000	0.35369865	0.00038207%	0.35368549	0.0040999%

In the experiment, the motor runs at a constant speed of $n = 1500r/\min$ and the running time is $0.3s$. When $N = 80$, the spectral method is used to obtain the identification results of electrical parameters. After $0.3s$, the speed of the motor is not fixed, and the running time is $0.3s$. When $N = 35$, the spectral method is used to identify mechanical parameters. At the same time, the Milne method also is used to estimate the electrical and mechanical parameters of the motor. Table 2 and Table 3 respectively show the identification results of the motor parameters of the two methods.

Table 3. Results of mechanical parameters identification method.

Items \ Methods	Reference values	Spectral method		Milne method	
		Estimated values	Relative errors	Estimated values	Relative errors
Rotor inertia ($J / kg \cdot ms^2$)	0.63290000	0.63290070	0.00011127%	0.63237186	0.083447%
Damping coefficient (B / S^{-1})	0.30350000	0.30353793	0.012498%	0.30273477	0.25214%

According to the experimental results, comparing our method with Milne method, we can be concluded that the spectral method is more accurate to estimate the parameters.

5. Conclusion

In this paper, the SPMSM multi-parameter identification problem is investgated based on the spectral method. Firstly, the least squares model of electrical parameters is designed. The electrical parameters are identified. After that, for the electrical parameters are known, the least squares model of the mechanical parameters is constructed for the mechanical equation. We can identify the mechanical parameters. The experimental results show that the relative errors of the parameters identified by the spectral method are smaller than other methods, which verifies the good reliability of the method.

Acknowledgments

The authors were supported by National Natural Science Foundation of China (No. 11971303 and No. 11871330).

References

[1] Bae B-H, Sul S-K, Kwon J-H and Byeon J-S 2003 Implementation of sensorless vector control for super-high-speed pmsm of turbo-compressor *IEEE Trans. Ind. Appl.* **39(3)** 811–8

[2] Kim H, Degner M W, Guerrero J M, Briz F and Lorenz R D 2010 Discrete-Time Current Regulator Design for AC Machine Drives *IEEE Trans. Ind. Appl.* **46(4)** 1425–35

[3] Alonge F, D'Ippolito F and Raimondi F M 2001 Least squares and genetic algorithms for parameter identification of induction motors *Control. Eng. Pract.* **9(6)** 647-57

[4] Douiri M R and Cherkaoui M 2014 Induction motor drive's parameters identification using extended kalman filter algorithms *WSEAS Transactions on Circuits and Systems* **13** 29–36

[5] Gottlieb D and Orszag S A 1977 *Numerical analysis of spectral methods: Theory and application* (Philadelphia: Society for Industrial and Applied Mathematics) p 9

[6] Higham D J 2001 An algorithmic introduction to numerical simulations of stochastic differential equations *SIAM Review* **43(3)** 525-46

[7] Horng I-R and Chou J-H 1985 Parameter estimation and optimal control of time-delay systems via chebyshev series *International Journal of Control* **41(5)** 1221-34

[8] Ljung L 1999 *System Identification, Theory for the user* (Beijing: Tsinghua University Press) pp 203-11

[9] Wang Q L , Zhang C W and Zhang X 2008 Variable-structure MRAS Speed Identification for Permanent Magnet Synchronous Motor *Proceedings of the CSEE* **28(9)** 71-5

[10] Yang Z and Wang L 2014 Online multi-parameter identification for surface-mounted permanent magnet synchronous motors *Transactions of China Electrotechnical Society* **29(3)** 111-8

[11] Butcher J C 1976 *Numerical methods for ordinary differential equations* (Chichester: Wiley) pp111-2

Experimental demonstration of ghost imaging with speckled light in long-wavelength infrared range

I Sh Khasanov

Scientific and Technological Centre of Unique Instrumentation of the Russian Academy of Sciences, Butlerova, 15, Moscow, 117342, Russia

E-mail: khasanov@ntcup.ru

Abstract. Infrared imaging with a single-pixel camera can achieve radiometric accuracy not currently available with multi-pixel thermal cameras. One of the actively investigated single-pixel camera technology is ghost imaging. We propose an optical design to experimentally demonstrate classical ghost imaging in the long-wave infrared frequency range with speckled light without using a spatial light modulator. We obtained an example low-quality ghost image and then used the cross-correlation function to identify and explain the main difficulties in ghost imaging with a thermal camera.

1. Introduction

The long-wavelength infrared range (LWIR) is a spectral range used in thermography from around 8 to 14 μm. The advantage of LWIR measurements is the high integral energy efficiency of blackbody radiation in the temperature range of 300 to 400 K (peaking at 350 K) as compared to other spectral ranges (such as mid-wave or short-wave infrared) [1]. That's why most thermal imaging cameras that measure at temperatures close to room temperature operate in this range. LWIR infrared cameras are used for remote temperature monitoring and their radiometric accuracy is very important, especially during pandemics (such as COVID-19) to identify people with fever during mass non-contact monitoring [2,3].

Despite the widespread use of commercially available thermal imaging cameras, infrared imaging currently has a number of challenges. Some of them are caused by imperfections in the optics of the infrared imaging lens. Thus, many common materials used to make IR optics are only conventionally transparent and have a low transmittance. This leads to greater energy losses compared to visible optics. Furthermore, infrared cameras are limited in optics in the sense that the size of the scattering spot (or diameter of the Airy disk) is many times larger than the camera pixel size. Consequently, the camera spatial resolution is primarily determined by the fabrication of the lens optical system, not by the number of pixels. There is also an optical reflection effect that is unknown from imaging in the visible spectral range, the so-called Narcissus effect. This effect occurs when the camera detects the reflection of its own heat from the camera lens, a window, or an object and which appears as a dark (or bright) spot in the image. The narcissus effect becomes pronounced if the reflecting object is in the focus of the camera.

An even more significant limitation of IR focal plane array (FPA) receivers is the non-uniformity of the response due to the inefficient photodetection process. The FPA consists of individual detector elements, causing spatial or fixed structural noise to their non-uniform properties. These difficulties are caused by the complexity of manufacturing matrix pixels, requiring the electrical integration of two different classes of materials. So the response curve can be significantly nonlinear in areas with high or weak signals. Also, temperature drift caused by the heating of neighboring pixels can change the pixel response curve. Non-uniformity correction (NUC) algorithms are used to match the response curves of individual pixels to each other. However, the NUC cannot properly handle low-contrast images or images with strong light inhomogeneity. If NUCs are performed during a long shooting session, they lead to jerks, as all values (relative to the average) can change abruptly during recalibration.

The problem of non-uniformity can be circumvented by using single-pixel scanning systems. This leads to radiometric accuracy due to the fact that radiation from all image pixels is detected by only one detector element. But existing scanning systems are based on mechanical elements, which leads to a number of drawbacks, such as low scanning speed, low reliability, inability to obtain an instantaneous field of view.

The disadvantages of mechanical scanning can be avoided by applying the method of single-pixel cameras developed in the last decade, known as ghost imaging (GI) [4].

2. Theory

The image of the object $O(x,y)$ in GI is restored by calculating 2nd order correlation between the spatial distribution of the reference beam light intensity P (patterns from speckles) (controlled in second optical path $P(x',y') = P(x,y)$ by a camera) and the total intensity S of the reflected light from the object received by a single-pixel detector [5]:

$$O(x,y) \propto \langle P(x',y')_i \cdot S_i \rangle - \langle P(x',y') \rangle \langle S \rangle, \tag{1}$$

where $\langle ... \rangle = \frac{1}{N} \sum_{i=1}^{N} ...$ — is the averaging operator, N is the number of independent patterns.

Thus, GI assumes the use of spatially modulated light. Depending on the type of light source GI can be classified as quantum, classical or computational. Classical GI uses either true thermal light sources – a non-coherent light, or pseudo-thermal – an ensemble of random interference patterns, so-called "speckled light" [6]. One of the advantages of the GI is the ability to acquire images without the use of lenses. The absence of lenses is important, for example, because the spatial resolution of LWIR cameras is limited by the optics. Moreover, the absence of lenses allows you to create optical imaging systems operating in a wide frequency range without chromatic aberrations, which is important for hyperspectral imaging. Also, we have shown that GI can increase the lateral resolution of surface plasmon resonance wide-field microscopy by an order of magnitude [7], [8].

To analyze the factors affecting the spatial resolution we can represent the estimated image $O_o(x,y)$ of the investigated surface as the sum of two parts [9]:

$$O_o(x,y) = \iint R(x',y')\, C_s(x-x',y-y')\, dx'dy' + \langle P(x,y) \rangle^2 \iint R(x',y') dx'dy', \tag{2}$$

where the correlation function C_s is :

$$C_s(x-x',y-y') = \langle P(x,y)P(x',y') \rangle - \langle P(x,y) \rangle \langle P(x',y') \rangle. \tag{3}$$

The first term in sum (2) is responsible for lateral resolution, and the second term in sum (2) affects contrast of ghost image. The noise in the estimated image can be reduced by the number of measurements (with collection of independent speckle patterns). So with an increase in the number N of measurements signal-to-noise (SNR) improves as $\text{SNR(N)} \propto \sqrt{N}\text{SNR(1)}$.

3. Experiment

In the LWIR only computational GI was previously demonstrated [10], and classical LWIR GI using speckle light has not been realized yet. To demonstrate LWIR GI, we propose a lensless setup with equal optical paths (Figure 1) with one camera observing simultaneously both arms of the scheme (Figure 1). The camera global shutter guarantees synchronization of both signals $P(x,y)$ and S. To simulate a single-pixel receiver signal S, we integrate the signal over the square camera area (128x128 pixels) in which the object image is located. Also, this setup allows determining the cross-correlation function C_s between speckle patterns $P(x,y)$ and $P(x',y')$. As it is known that the resolving power of GI is determined by the form of this function and is fundamentally limited by the half-width of the autocorrelation function, that is, the average speckle size. To create a pseudo-thermal source for demonstration purposes we use radiation passing through a transparent rough object (ground glass in visible range, polypropylene fabric in LWIR), we change the speckles mechanically with the help of a motorized translator.

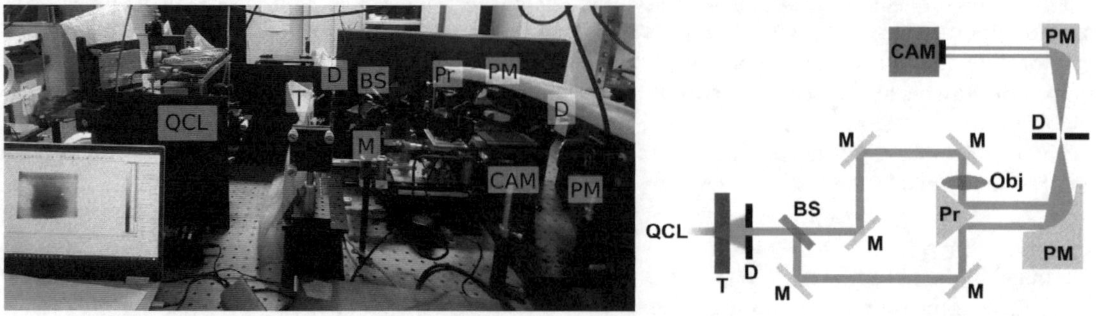

Figure 1. Photo and optical layout of the setup: QCL – IR source Alpes Lasers λ=8μm; BS – beamsplitting high resistivity float zone silicon wedge plate (wedge angle 3°, thickness at the thin part is 1mm); T – a diffuser with a rough surface (polypropene with thickness 1 mm) mounted on a motorized translator Standa 8MT175-200; D – diaphragm (D = 1 mm); M – first surface mirrors; Pr is a first surface mirror prism for converging parallel beams; Obj – object; PM – parabolic mirrors forming a telescopic system (Thorlabs MPD399-M03 with reflected focal length RFL=228.6 mm and Thorlabs MPD239-M03 with RFL=76.2 mm); CAM – LWIR camera COX CG640 (uncooled FPA microbolometer, pixel size 17μm, resolution 640x480, NETD < 50mK @f1.0, 30Hz, 300K).

To test the operability of the optical scheme, the setup was initially tested to obtain a ghost image in the visible range using a red He-Ne laser. We recorded 3000 measurements of various speckles and images of the object in the speckled light. The resulting ghost image (Figure 3d) made it possible to distinguish the contour of the object, even for a small number of measurements compared to the size of the imaging area 160x80 = 12800 pixels.

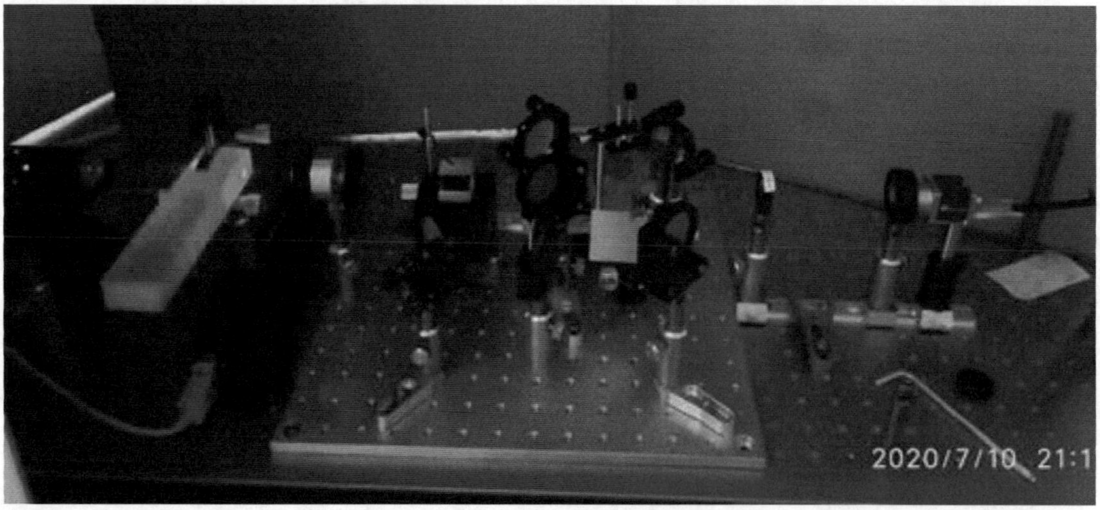

Figure 2. Photo of the setup of preliminary experiment in the visible range. Laser source red He-Ne 632.8 nm, non-polarizing glass 70/30 beamsplitter cube, CMOS camera Mako U-130B (1280x1024, pixel size 4.8 μm).

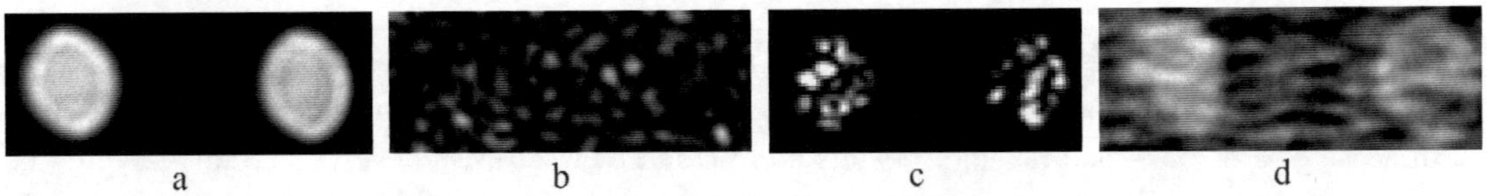

Figure 3. Results of preliminary experiment in the visible range: (a) image of object, (b) example of speckle in reference beam, (c) example of image of the object in speckled light, (d) reconstructed ghost image.

An important condition for the successful obtaining of a ghost image in our demonstration is the similarity of the speckle patterns in the absence of an object. In this case, if the optical system is well aligned, we observe similar speckle patterns on the camera. On the initial alignment, we achieve identical laser beam pictures (Figure 4a), i.e. images without a diffuser T. After adding a light diffuser (Figure 4b) a minimal adjustment of the optical system may be necessary because the speckles are extremely sensitive to the slightest distortion in the course of the optical beam. The measure of speckles similarity is the cross-correlation function between speckle patterns, which in the case of an ideal optical system should coincide with the autocorrelation function.

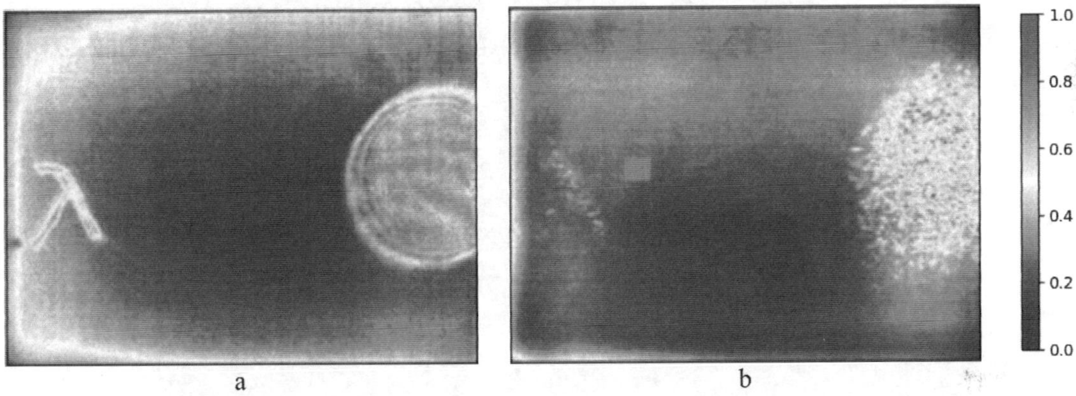

Figure 4. Example measurement of both object and reference beams: (a) in laser light, (b) in speckled light. Presented color bar is used in all figures below.

In the LWIR GI demonstration experiment, we performed 13250 measurements of speckle patterns and their corresponding object images. The figure 5 shows a comparison of the object image in the laser beam and the object image in speckle light, as well as the object image after averaging over all measurements. Due to the fact that the camera pixels experience a temperature drift during series of long-term imaging (20 series about 25 seconds each, 30 frame per second), and the speckle signal itself is small compared to the noise (the signal-to-noise ratio is about 2). We see that the image of the object after averaging over all speckles (thus, the image in incoherent light) turned out to be low-contrast and noisy. However, after the filtering procedure, we managed to achieve a significant reduction in the background noise. But even in this case, the image remained extremely irregularly illuminated.

To restore the ghost image, we applied a filtering procedure, which consisted of the sequential application of a threshold filter and a spatial filter of the high-frequency signal. The latter filter is a standard procedure for removing the noise component from images. For this purpose, a two-dimensional Fourier image was found. Then, to remove the high spatial frequencies, the Fourier image was cropped, leaving only 20% of the central region, where the main low-frequency information of the speckle structure is concentrated. The threshold filter operation consisted in equating to zero the pixels of the image whose value did not exceed the selected threshold. This operation was motivated by the fact that for speckles the intensity distribution should be a Rayleigh distribution [11]. However, in the presence of noise, this distribution becomes close to a skewed Gaussian distribution. By trimming the low-energy tail, choosing the threshold value equal to the mean value, and then subtracting this mean value from the non-zero values, we leave only the high-energy tail, bringing the distribution close to the Rayleigh distribution. To choose the filtering parameters more accurately we can use the autocorrelation function. In the case of a noisy experimental speckle pattern (Figure 6a, 6b), we observe that this function is a triangle. The triangular form of the autocorrelation function means that the spatial correlation length is small, that is, the observed picture is close to the random noise picture. However, after applying the filtering procedure described above, we see that the auto-correlation picture starts to have a Gaussian profile, as it should be in theory for the speckle autocorrelation function. By selecting the parameters so that the maximum of the component of the triangular function is approximately 0.5, which is approximately observed in numerical speckle simulations, we obtain an autocorrelation function similar to the speckle autocorrelation function. Applying the filtering procedure to all images of the object in speckled light and then averaging over all measurements, we are convinced that the filtered image of the object in speckle light is well contoured and the background noise practically disappears.

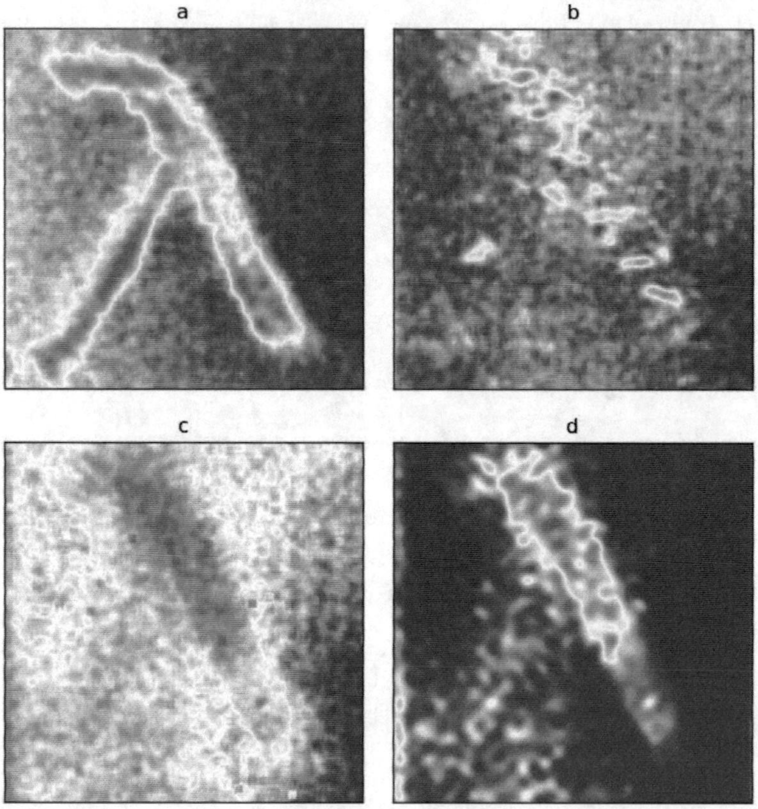

Figure 5. Images of object: (a) in laser light, in speckle light in camera: (b) one shot, (c) averaged, (d) averaged and filtered

We reconstruct the ghost image according to formula (1). On the ghost image obtained from the original measurements, we see (Figure7b) that the object image is not reconstructed and there is a pattern of inhomogeneous background noise. However, by applying the filtering procedure to the original measurements, we see that in the ghost image the outline of the object appears (Figure7c). The quality of the ghost image is comparable with that of the unfiltered averaged image of the object in the light of the speckles (Figure5c). However, this image is significantly worse in quality (sharpness and contrast) than the theoretically expected image (Figure7a). This is due to the peculiarities of IR imaging with a

Journal of Physics: Conference Series **2091** (2021) 012008 doi:10.1088/1742-6596/2091/1/012008

thermal camera. However, we can expect that increasing the number of measurements will produce a clearer image since the signal-to-noise ratio grows as the square root of the number of measurements. The image quality can also be improved by eliminating the non-uniformity in the thermal imaging camera. In particular, this non-uniformity is caused by temperature drift and the observed Narcissus effect (Figure 4).

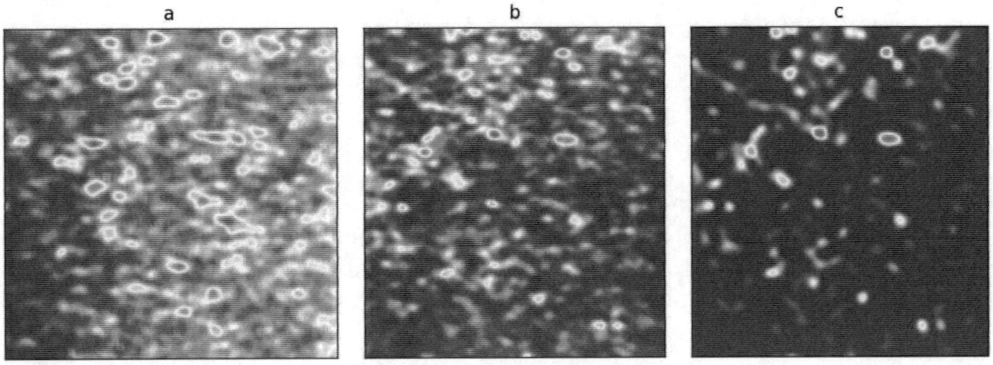

Figure 6. Examples of speckle pattern: (a) in object beam, (b) in reference beam, (c) speckle pattern b after filtration procedure

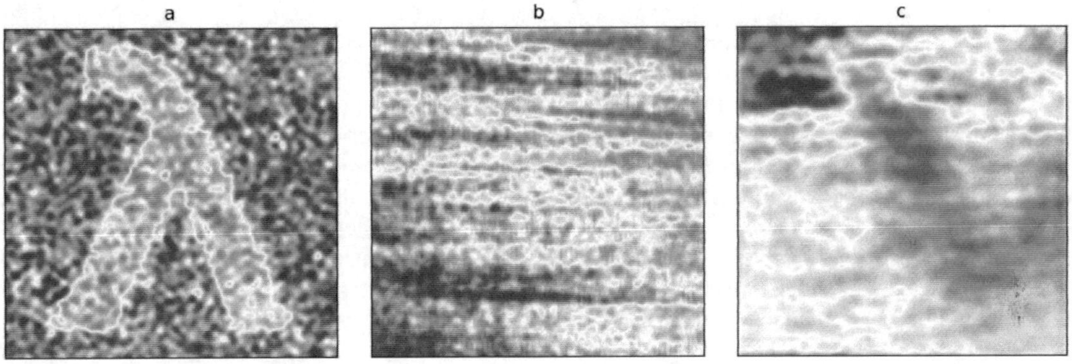

Figure 7. Ghost images: (a) model, (b) experiment, (c) experiment with filtering. Number of measurements in experiment N=13250. Number of pixels 128x128 = 16384.

The fact that the image was worse than the theoretical one could be expected from the calculation of the cross-correlation function for the speckles in the object (Figure 6a) and reference beams (Figure 6b). This function in the experiment turned out to be asymmetric and deformed (Figure 8b), significantly different from the autocorrelation function of the speckle pattern in the reference beam. This is caused both by noise and uneven illumination, as well as by possible aberrations in the optical telescopic system. Therefore, for successful image acquisition in classical GI in the first place the cross-correlation function between the object and reference beams must be controlled. In this case, its calculation can be performed rather quickly using several speckle patterns, and the calculation will not take much time due to the connection of the cross-correlation function with the Fourier transform.

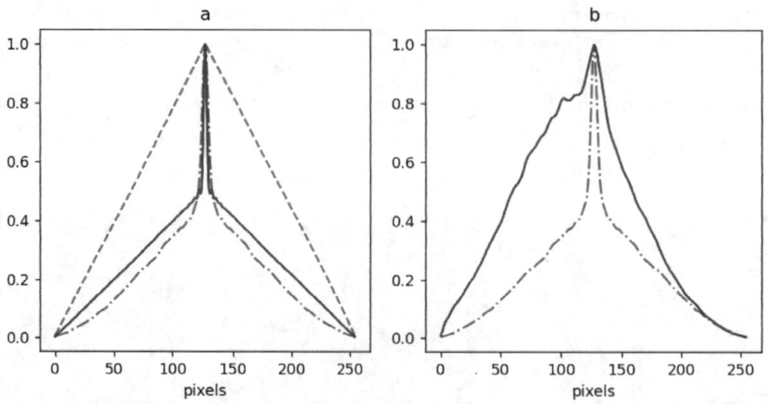

Figure 8. (a) Comparison of normalized speckle autocorrelation functions: (dashed) experiment without filtering, (dot-dashed) experiment with filtering, and (solid) model; (b) Comparison of normalized autocorrelation (dot-dashed) and cross-correlation (solid) functions in experiment with filtering.

4. Discussion

In the LWIR GI demonstration we obtained a low-quality image. In any case, our experiment revealed the main difficulties that must be taken into account when shooting with an IR camera. First of all, it is the non-uniformity of the signal response and temperature drift. In the case of the GI implementation, we do not need the complex mirror system shown in Figure 1, which allows us to exclude the influence of the aberrations of the mirror telescope system. To exclude the influence of the multi-pixel camera, we can pre-scan the speckle patterns from a mechanically non-deformable diffuser (e.g., a rough metal surface). In this case, to avoid temperature drift effects, we can periodically perform NUC during the scan and use an additional single-pixel detector to control the average signal level. After pre-recording we can control the speckle patterns knowing only the position of the light diffuser placed on the mechanical stage. Since the motorized stage has a position reproducibility as good as wavelength, this is sufficient to recover similar speckle patterns [12]. Since the controlled spatial shift does not have to exceed the wavelength to achieve similarity of speckle patterns, having a mechanical displacement with an accuracy not worse than the wavelength, we can reproduce the previously recorded speckles quite accurately. Thus, the mechanical control of speckles is fairly easy to implement in the IR range. By placing an object in place of the camera used to scan the speckles, and positioning behind the object a high-speed single-pixel receiver to record the integral signal, we can obtain enough measurements to reconstruct the ghost image. In addition, in the infrared range, we can use surfaces with controlled relief as a light diffuser. And since the roughness must be of the order of a wavelength to obtain pronounced speckles, it is easily achievable with modern technologies such as 3D printing. Then using numerical methods based on the calculation of the Rayleigh-Sommerfeld integral, one can easily calculate the light field at any point, knowing the illumination parameters of the controlled surface. In this way, it is possible in principle not to use multi-pixel IR cameras to reconstruct ghost images and bypass the difficulties described above.

5. Conclusions

We have developed a simple optical lensless setup to demonstrate and study ghost imaging over a wide frequency range from visible to THz band. For the first time we demonstrated classical GI with speckle light at LWIR. We obtained a ghost image of lower quality than predicted by theory due to non-ideal conditions of the experiment: due to the peculiarities of the thermal camera operation at low signal levels and non-uniform illumination. We have also verified that image filtering, i.e., post-processing, in the case of classical GI allows us to obtain a signal in case of noisy background, if such post-processing leads to the recovery of the form of the auto-correlation function for speckles.

Acknowledgments

The reported experiment was supported by grant of the President of the Russian Federation for state support of young Russian scientists - candidates of sciences (research project MK-3998.2021.1.2). The development of experimental setup was funded by the Russian Foundation for Basic Research (project 20-52-54004). The QCL source was provided by Shared research center SSTRC based on the Novosibirsk FEL complex at BINP SB RAS. The presented optical design was prepared at the Center for Collective Use of the Scientific and Technological Center of Unique Instrumentation of the Russian Academy of Sciences. The author expresses his gratitude to V.V. Gerasimov for his help in setting up the equipment.

References

[1] Vollmer M and Möllmann K-P 2011 *Infrared Thermal Imaging: Fundamentals, Research and Applications* (Wiley-VCH)
[2] Khanam F-T-Z, Chahl L A, Chahl J S, Al-Naji A, Perera A G, Wang D, Lee Y H, Ogunwa T T, Teague S, Nguyen T X B, McIntyre T D, Pegoli S P, Tao Y, McGuire J L, Huynh J and Chahl J 2021 Noncontact Sensing of Contagion *Journal of Imaging* **7** 28
[3] Tay M R, Low Y L, Zhao X, Cook A R and Lee V J 2015 Comparison of Infrared Thermal Detection Systems for mass fever screening in a tropical healthcare setting *Public Health* **129** 1471–8
[4] Moreau P-A, Toninelli E, Gregory T and Padgett M J 2018 Ghost Imaging Using Optical Correlations *Laser & Photonics Reviews* **12** 1700143
[5] Gatti A, Brambilla E, Bache M and Lugiato L A 2004 Correlated imaging, quantum and classical *Phys. Rev. A* **70** 013802
[6] Turiaci G J, Varga J J M and Iemmi C 2013 Ghost imaging with white light speckle patterns *Optik* **124** 6212–5
[7] Khasanov I S and Zykova L A 2020 Can the ghost imaging increase the lateral resolution of surface plasmon resonance microscopy? *J. Phys.: Conf. Ser.* **1636** 012039
[8] Khasanov I Sh, Knyazev B A, Nikitin A K, Gerasimov V V, Zykova L A and Trang T T 2020 Enhancing resolution of terahertz surface plasmon resonance microscopy by classical ghost imaging using free electron laser radiation *AIP Conference Proceedings* **2299** 030008
[9] Khasanov I and Zykova L 2020 Terahertz ghost imaging and surface plasmon resonance microscopy: analysis of factors affecting the image quality *Fourth International Conference on Terahertz and Microwave Radiation: Generation, Detection, and Applications* Fourth International Conference on Terahertz and Microwave Radiation: Generation, Detection, and Applications ed O A Romanovskii and Y V Kistenev (Tomsk, Russian Federation: SPIE) p 54
[10] Liu H-C and Zhang S 2017 Computational ghost imaging of hot objects in long-wave infrared range *Appl. Phys. Lett.* **111** 031110
[11] Goodman J W 2007 *Speckle Phenomena in Optics: Theory and Applications* (Roberts and Company Publishers)
[12] Françon M 1979 *Laser Speckle and Applications in Optics* (Academic Press)

Journal of Physics: Conference Series **2091** (2021) 012009 doi:10.1088/1742-6596/2091/1/012009

Contribution of wave aberrations represented by Zernike polynomials to the cross-correlation function between distorted and actual speckle patterns

T V Blagova[1,2], I Sh Khasanov[1]

[1] Scientific and Technological Centre of Unique Instrumentation of the Russian Academy of Sciences, Butlerova, 15, Moscow 117342, Russia
[2] Bauman Moscow State Technical University, 2-nd Baumanskaya, 5, Moscow 105005, Russia

E-mail: khasanov@ntcup.ru

Abstract. Speckles are sensitive to the slightest inhomogeneities of the medium, which is used in optical research methods such as speckle interferometry. However, the stochastic nature of propagation of speckle fields complicates their accurate detection and processing. For example, aberrations in the optical system result in the decorrelation of the image of speckles with the actual speckles that are observed in free space. The report will consider the main types of wave aberrations of optical system and their influence on the correlation properties of speckle patterns. The research results can be used to optimize optical systems in which speckles play a significant role, for example, in classical ghost imaging.

1. Introduction

Speckles - random interference patterns - began to be studied with the first appearance of the highly coherent sources [1]. And to date, from the point of view of statistical optics, the theory of speckles has been sufficiently developed, including the statistics of speckles of the second order (associated with the interference of intensities, not phases) and higher orders [2]. With the development of the process of spontaneous parametric down-conversion, the questions devoted to higher-order interference were considered again from the standpoint of quantum mechanics [3]. It has been shown that speckles can be used as a model for a thermal source. That is, using a set of speckle patterns, it is possible to simulate the radiation of a thermal source, replacing the ensemble of states in time with an ensemble of states in terms of their number. Such a pseudo-thermal source can be used to obtain images in classical ghost imaging [4]. At the same time, for the formation of ghost images, in contrast to images obtained by averaging speckle patterns, lenses are not required. Secondly they are similar to images in incoherent light, i.e. diffraction effects are small (see Fig. 1). These are especially important in the long-wavelength range, for which the size of the camera pixels is comparable to the Airy disk, for example, for mid-IR and THz radiation [5]. In addition, in ghost imaging the object is observed by a detector without spatial resolution. Thus, the ghost imaging (GI) is not susceptible to aberrations in the optical path between the object and the single-pixel detector. This property was used to image objects through light-scattering media, e.g. for imaging through fog. However, it must be emphasized that aberrations present in the optical system on the optical path to the object still affect the obtained image. This paper is devoted to evaluating the influence of such aberrations.

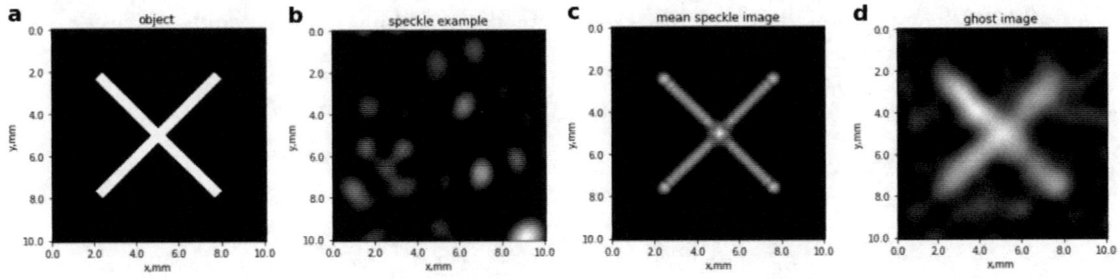

Figure 1. The image (c) of an object (a), obtained by averaging 1000 speckle patterns (an example of a speckle pattern (b) for a laser at a wavelength of $\lambda = 130$ μm), and the corresponding ghost image (d). All images are calculated using the Diffractio software package (https://pypi.org/project/diffractio/).

2. Cross-correlation function between distorted and actual speckle patterns

In GI, the image of the object is restored by calculating the second-order correlation between the intensity pattern $P(x, y)$ in the reference beam and the integral intensity $S = \int P(x', y')dx'dy'$ in the object beam. The resolution of the method is determined by the half-width of the cross-correlation function [6,7]:

$$C_s(x - x', y - y') = \langle P(x, y)P(x', y')\rangle - \langle P(x, y)\rangle\langle P(x', y')\rangle, \quad (1)$$

where $P(x, y)$ and $P(x', y')$ are images of actual and distorted speckle patterns. For similar images the cross-correlation function coincides with the autocorrelation function. But any differences between the speckle patterns in the reference and object beams lead to deterioration of the reconstructed image. These distortions are most often caused by aberrations in the optical system. The most developed approach to wave aberrations is its expansion in a series in terms of Zernike polynomials. The orthogonality of the Zernike polynomials gives them great advantages in the analysis of aberrations in comparison with the power basis, as in the case of Seidel's theory. The absolute values in Zernike polynomials decrease with an increase in the degree of the polynomials, that is, the Zernike series, as a rule, always converges. And most importantly certain types of aberrations, represented by the expansion in terms of Zernike polynomials, affect the image quality completely independently of each other.

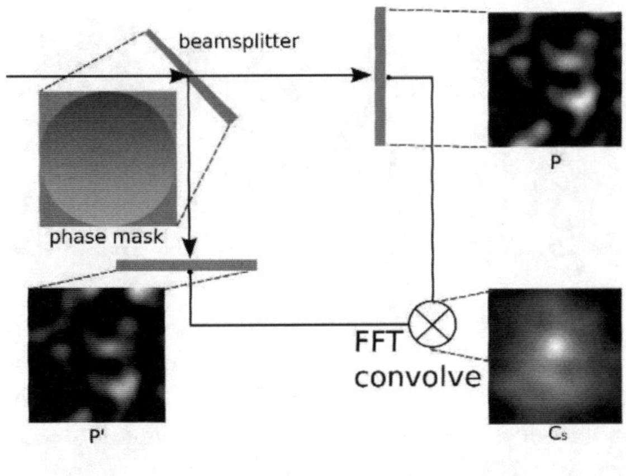

Figure 2. Optical scheme to simulate wave aberrations represented by Zernike polynomials.

In order to investigate how aberrations, affect the quality of the ghost images, let us consider the optical scheme shown in Figure 2. Such aberrations can occur for example after passing through a beam splitter of non-uniform thickness, then the passing beam will have a phase shift, which we can describe using the Zernike aberration terms.

The Zernike aberration term are defined as

$$Z_n^m(\rho,\varphi) = z_n^m R_n^m(\rho)\cos(m\varphi), \tag{2}$$

where Zernike coefficient z_n^m describe deviation from mean phase.

The radial polynomials R_n^m are defined as

$$R_n^m(\rho) = \sum_{k=0}^{\frac{n-m}{2}} \frac{(-1)^k \rho^{-2k+n}(-k+n)!}{k!(\frac{n+m}{2}-k)!(\frac{n-m}{2}-k)!}\rho^{n-2k}. \tag{3}$$

Then the wavefront is described as a sum of Zernike aberration terms by means of aberration function

$$W(\rho,\varphi) = \sum_{n=0}^{\infty}\sum_{m=0}^{\infty} Z_n^m. \tag{4}$$

Table 1. Zernike coefficient and corresponding phase mask in presented optical scheme.

	1	2	3
Piston Z_0^0			
Tip Z_1^1			
Defocus Z_2^0			

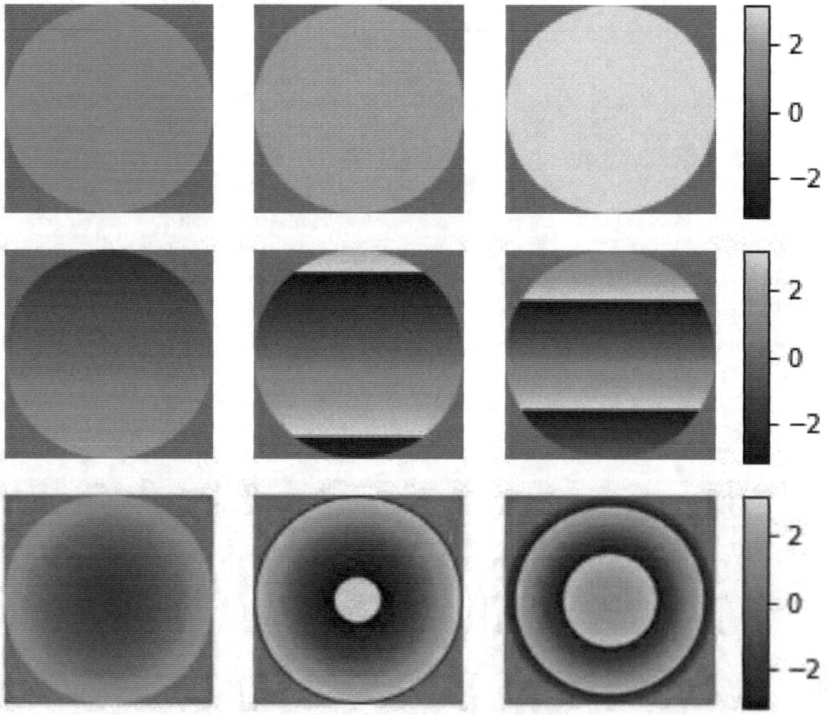

At Table 1 we present several illustration of phase mask corresponding to simplest Zernike polynomials for different Zernike coefficient. It can be seen that Piston is just constant phase. Which gives the differences in optical path between P and P. Z_1^1 gives the tip which leads to to the deflection of the rays to the side from the specified trajectory. Z_2^0 leads to the defocusing effect.

From Table 2 we can see what effects the corresponding wave aberrations lead to. So piston adds blurring and visually enlarges the object (i.e. visually gets closer). That is, piston causes the focal plane of the optical system to shift from the image plane. Tip results in a linear shift of the image from focal plane. And the defocus distorts the image symmetrically relative to the center.

Table 2. Ghost images distorted by phase mask in one of optical path of the optical scheme.

Z_n^m	1	2	3
Z_0^0	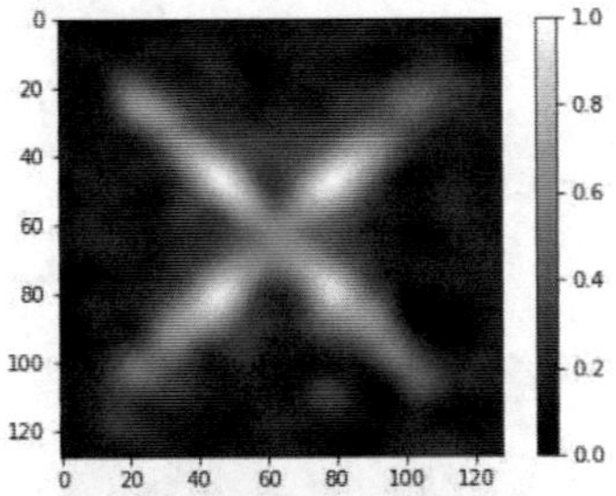	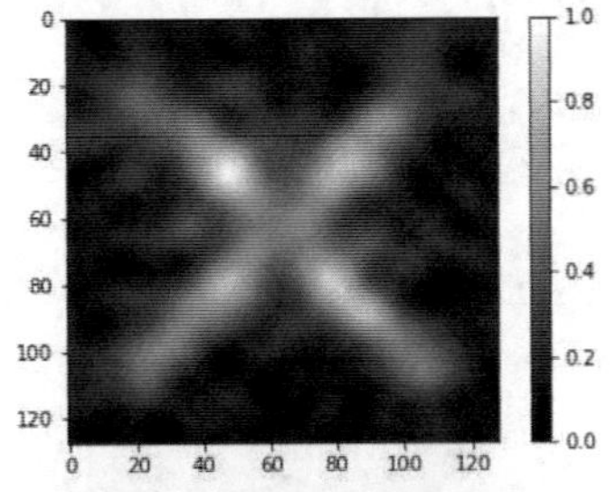	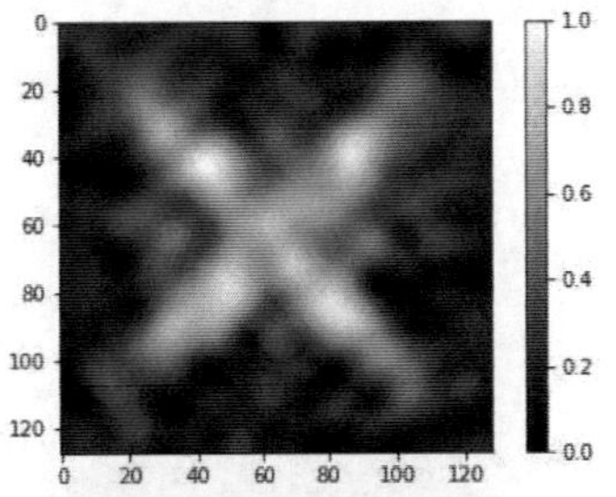
Z_1^1	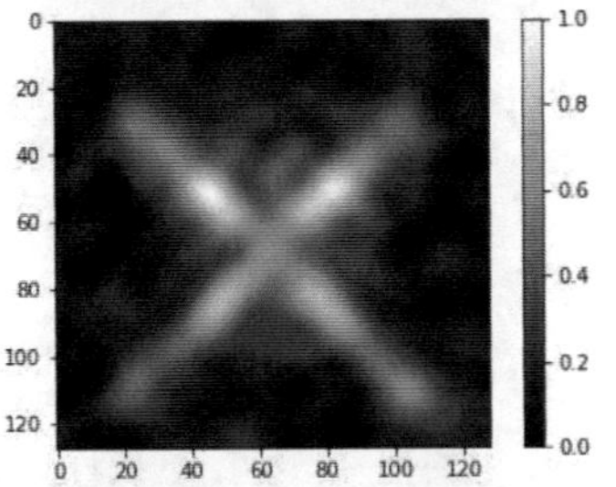	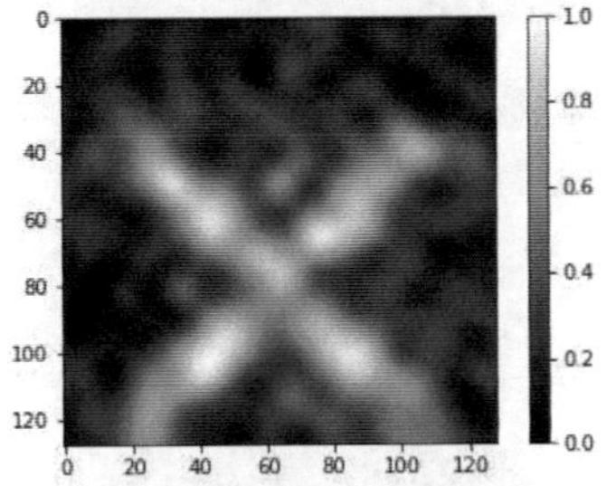	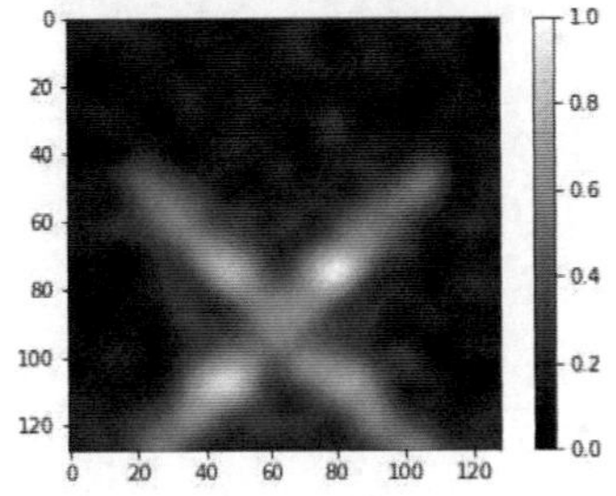
Z_2^0	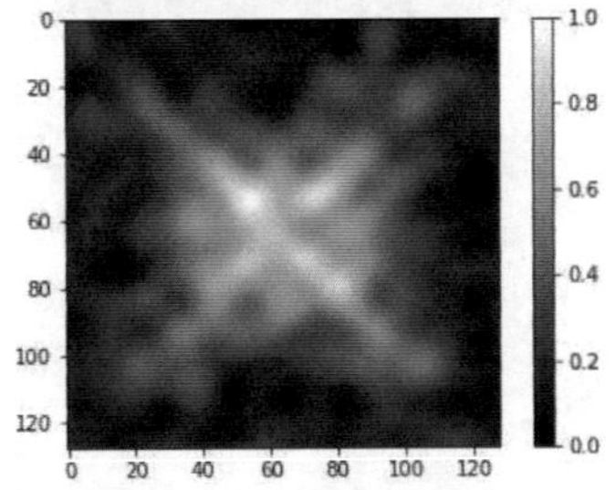	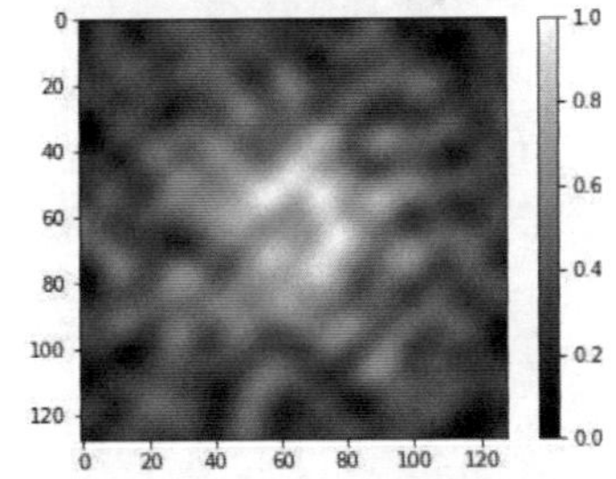	

However, estimating aberrations from ghost images is a difficult task, because when aberrations are added, the noise can increase significantly and the picture is difficult to estimate. A good tool in such a case is the cross-correlation function.

The calculation of the cross-correlation function is the basis of optical methods for measuring positions in 2D and 3D space, known as digital image correlation. Since their introduction in the early 1970s, these methods have spread widely in various fields of science and engineering. These methods allow to measure local displacements and build deformation maps. And therefore can be applied to the estimation of wavefront aberrations. The cross-correlation function can be mathematically represented as a convolution of Fourier images, which is much faster than the formula based on element-by-element multiplication and allows to measure cross-correlation in real time. This principle is the basis, for example, for computer mice that can measure local displacements of speckle pictures obtained by illuminating a surface with a quasi-monochromatic LED.

We have numerically calculated the propagation of speckle patterns through the pupil of an optical system with a given aberration represented by Zernike polynomials using numerical approximation of the Rayleigh-Sommerfeld diffraction integral by a Fast Fourier Transform. Figure 3 (a-c) shows that different types of aberrations give distinguishable distortions of the shape of the cross-correlation function. For example, the polynomial Z_0^0 (piston) at Figure 3a is responsible for thickening the half-width of the C_s function and smoothing its side maxima. In the case of aberration Z_1^1 (horizontal x-tilt) leads to a shift in the position of the maximum C_s (Figure 3b). Aberration Z_2^0 (defocus) leads to an increase in side maxima (Figure 3c). Higher order aberrations also have a discernible effect. At Figure 3d it can be seen that in the presence of both types of aberrations Z_1^1 and Z_2^0 leads to distortions of the cross-correlation function, which can be approximately described as the sum of two cross-correlation functions separately for the Z_1^1 component and the Z_2^0 component.

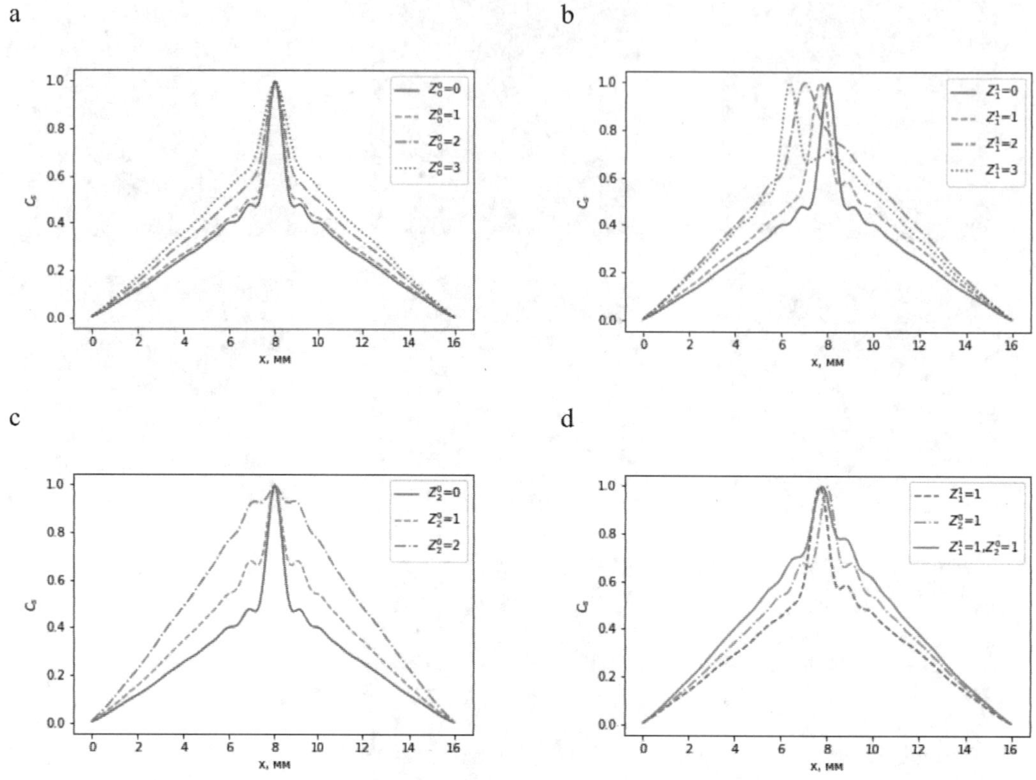

Figure 3. Cross-correlation function of speckle patterns (averaging over 1000 measurements) for various types of aberrations represented by Zernike polynomials (Z_0^0, Z_1^1, Z_2^0).

The maximum shift (as in the case of tip) allows you to determine the speckle pattern shift. But other types of aberrations also introduce specific distortions into the cross-correlation function. For example Z_2^0 adds side maxima to the main one. By magnitude of additional maxima, you can judge e.g. the degree of spherical deformation of the surface after transmission of the plane wavefront through the phase mask. The piston does not distort the cross-correlation function, reducing only the signal-to-noise ratio, which leads to an increase of the trigonal component at the edges caused by diffraction on the aperture and to a broadening of the autocorrelation function of speckle. Knowing the aperture parameters, we can for example find the position where the signal-to-noise ratio will be maximal which will correspond to the equality of the optical paths after the beam splitter. And according to the size of the speckle, we can find the offset in depth. From the Fig 3d we can see that the combination of aberrations does not erase the specific features in the cross-correlation profile, but modifies them, which allows us to distinguish (measure) several types of aberrations simultaneously with a single operation. This can be used both to create simple controllers (e.g. 3D optical mice) and to tune optical systems, such as phantom vision systems consisting of independent optical paths which cannot be matched by conventional interference methods (based on phase comparison, i.e. first order interference).

3. Conclusions

The report considered the main types of wave aberrations introduced by the optical system, presented in the form of Zernike polynomials. We have determined that different types of aberrations introduce distinguishable changes in the shape of the cross-correlation function of the distorted and actual speckle patterns. This means that in the case of second-order interference, the orthogonal Zernike polynomials describe independent contributions for wave aberrations. We have found that aberrations corresponding to high-degree polynomials distort the cross-correlation function of speckle patterns more strongly, i.e. have a significant impact on the quality of the image, which can be obtained, for example, by the ghost imaging.

Acknowledgments
The reported study was supported by grant of the President of the Russian Federation for state support of young Russian scientists - candidates of sciences (research project MK-3998.2021.1.2).

References

[1] Oliver B M 1963 Sparkling spots and random diffraction *Proceedings of the IEEE* **51** 220–1
[2] Goodman J W 2007 *Speckle Phenomena in Optics: Theory and Applications* (Roberts and Company Publishers)
[3] Scarcelli G, Berardi V and Shih Y 2006 Can Two-Photon Correlation of Chaotic Light Be Considered as Correlation of Intensity Fluctuations? *Phys. Rev. Lett.* **96** 063602
[4] Gatti A, Bache M, Magatti D, Brambilla E, Ferri F and Lugiato L A 2006 Coherent imaging with pseudo-thermal incoherent light *Journal of Modern Optics* **53** 739–60
[5] Khasanov I Sh, Knyazev B A, Nikitin A K, Gerasimov V V, Zykova L A and Trang T T 2020 Enhancing resolution of terahertz surface plasmon resonance microscopy by classical ghost imaging using free electron laser radiation *AIP Conference Proceedings* **2299** 030008
[6] Khasanov I and Zykova L 2020 Terahertz ghost imaging and surface plasmon resonance microscopy: analysis of factors affecting the image quality *Fourth International Conference on Terahertz and Microwave Radiation: Generation, Detection, and Applications* Fourth International Conference on Terahertz and Microwave Radiation: Generation, Detection, and Applications ed O A Romanovskii and Y V Kistenev (Tomsk, Russian Federation: SPIE) p 54
[7] Basano L and Ottonello P 2007 Ghost imaging: Open secrets and puzzles for undergraduates *American Journal of Physics* **75** 343–51

Wide-angle acousto-optic devices based on isotropic light scattering in biaxial crystals

M I Kupreychik[1], V I Balakshy[1,2] and V E Pozhar[2]

[1] Faculty of Physics, M. V. Lomonosov Moscow State University
Leninskie Gory, MSU, Moscow 119991, Russia
[2] Scientific and Technological Center of Unique Instrumentation, Russian Academy of Sciences 15 Butlerova street, Moscow 117342, Russia

E-mail: maxim10111992@gmail.com

Abstract. Using the example of a rhombic crystal of alpha-iodic acid, a detailed study of new variants of acousto-optic isotropic light scattering in oblique sections of biaxial crystals, whose refractive surface contains a concave portion, is carried out. For these cases, the frequency dependences of the Bragg angles and the ranges of acousto-optic interaction are calculated. The possibility of implementing a unique, simultaneously broadband and wide-angle variant of light scattering is established. It is shown that the two-dimensional transfer function of such acousto-optic geometry has a complex *X*-shaped structure that has no analogues among uniaxial crystals. The possibility of applying this diffraction regime in light beam spatial filtering devices is studied.

1. Introduction

Acousto-optic (AO) devices are used in various fields of science and technology, favorably differing from analogues, which are built on the basis of other physical principles, in high speed of operation, low energy consumption, reliability and simplicity of design. AO filters, modulators and deflectors allow us quickly and with high accuracy to control frequencies and angular spectra of light waves, as well as change the direction of their propagation [1–3].

The vast majority of modern AO devices are made on the basis of a paratellurite (TeO_2) crystal. A wide range of the material transparency (from 0.3 to 5 μm) combined with its extremely high AO efficiency (for the variant of anisotropic diffraction), allow us to solve most problems of controlling optical radiation in both the visible and IR ranges. Basic AO devices, such as modulators, deflectors and filters, require a special geometry of AO interaction. For example, for AO deflectors, the broadband interaction region with weak frequency selectivity is optimal, whereas in AO video-filters, a wide-angle geometry with low angular selectivity is used [1–3]. In the case of anisotropic diffraction in uniaxial crystals, these requirements are contradictory and cannot be satisfied simultaneously, except of using special AO cells with phased-array piezoelectric transducers [4]. This means that different AO interaction geometries must be used when creating deflectors and filters. It follows from this that the problem of AO controlling light beams having a large divergence has not yet found a satisfactory solution in acousto-optics.

We suppose that a possible solution of this problem concludes in the application of peculiar geometries of AO light scattering in biaxial crystals. A special structure of the optical and acousto-optical anisotropy of these crystals allows for a greater variety of variants of AO diffraction, which permits creating AO devices of a fundamentally new type [5–7]. Thus, in [6], an unusual variant of a broadband AO deflector operating at a quasi-plane area of the optical refraction surface was experimentally investigated, while in a recent work [7], characteristics of simultaneously broadband and wide-angle light scattering regime in the interaction plane containing the optical axes of a non-gyrotropic biaxial crystal were theoretically studied. It was also shown that similar variants of the AO interaction can be realized in oblique cuts of biaxial crystals.

The purpose of this work is a comprehensive study of new variants of isotropic light scattering in oblique sections of a biaxial crystal, including a concave area of the normal optical surface. In doing so, the special attention is paid to the above-mentioned low-selective regime of diffraction. For definiteness, the calculations are performed for a biaxial crystal of alpha-iodic acid (α-HIO_3), although similar results are also true for other biaxial crystals.

2. Computation results

Alpha-iodic acid is a synthetic crystal of rhombic syngony (axial class 222), favorably characterized by good nonlinear optical and AO properties [5,8–11]. One of the transparency windows of the material lies in the visible and near-IR ranges from 0,4 to 1,3 μm. Low acoustic and optical losses along with a noticeable birefringence and the possibility of obtaining large crystals with high optical quality make it a promising material for creating AO filters with ultra-high spectral resolution.

During our calculations, the crystallographic axes *X*, *Y* and *Z*, as well as the values of elastic modules and photoelastic coefficients were selected in accordance with [8]. In this case, for the light wavelength $\lambda = 0{,}633$ μm, the main refractive indices are equal to: $n_x = 1{,}9865$, $n_y = 1{,}9604$ and $n_z = 1{,}8378$ [9]; the angle between the optical axes, lying in the *XZ* plane, is $2\psi = 47^\circ$. The crystal HIO_3 is characterized by a high level of optical activity with a pronounced anisotropy of the gyration surface. For the above wavelength, the components of the gyration tensor are equal to $g_X = -1 \cdot 10^{-4}$, $g_Y = 2{,}6 \cdot 10^{-4}$ and $g_Z = 3{,}3 \cdot 10^{-4}$; rotational capacity along the optical axis is $\rho = 37$ deg/mm.

2.1. Frequency dependencies of Bragg angles

When studying characteristics of new non-trivial regimes of isotropic AO diffraction in the HIO_3 crystal, we will limit ourselves to the scattering of a slow optical mode (α-wave), having the refractive index n_α, by a pure longitudinal acoustic mode propagating along the crystallographic *Y* axis with a phase velocity of $V = 3{,}04$ km/s. This choice is due to both the relative simplicity of the practical implementation of such interaction geometry, and the higher AO efficiency in comparison with similar diffraction regimes in oblique sections of the crystal. In this case, it is convenient to introduce the AO interaction plane by means of the polar angle ϑ, which sets the direction along which the plane under consideration intersects with the main plane *XZ*. All further calculations are carried out for the light wavelength $\lambda = 0{,}633$ μm.

During the calculations, it has been found that non-trivial variants of isotropic AO diffraction are observed in the planes corresponding to the interval of the angle change from $\vartheta_1 = 21{,}9°$ to $\vartheta_2 = 25{,}2°$. The sections of the refraction surface by planes at these boundaries contain an extended flat areas that are optimal for creating a special broadband AO deflector with a close to П-shaped frequency characteristic [6]. It should be noted that due to considerable birefringence of the iodic acid crystal, the low-selective regimes of isotropic α-wave scattering, studied in this paper, are realized at a noticeable angular distance from the optical axes. Therefore, the influence of the optical activity of the material on diffraction characteristics is insignificant.

The frequency dependences of the Bragg angles $\theta_B(f)$ for the process $\alpha \to \alpha$ (isotropic diffraction) in various planes of AO interaction are shown in figure 1. The solid curves on the graphs illustrate scattering of the incident slow optical mode α into the –1st order, while the dashed curves correspond to the diffraction process into the +1st order. Due to symmetry of the refractive surface, the dashed curves are mirror reflections of the solid curves relative to the abscissa axis and vice versa. The calculations are fulfilled for four values of the polar angle ϑ: $\vartheta = \vartheta_2$, $\vartheta = 24{,}9°$, $\vartheta = 24{,}3°$ and $\vartheta = \psi$.

Figure 1(a) shows that the Bragg curves for isotropic diffraction on a flat part of the refraction surface have no singularities. However, even a slight decrease in the angle ϑ changes the situation dramatically. In addition to the usual almost linear curves $\theta_B(f)$, additional oval-like regions of synchronism appear, which correspond to non-trivial variants of AO isotropic interaction (figure 1(b)). At the points of intersection of these branches (denoted by symbol WD), a low-selective light scattering regime is implemented, which can be used for creating a wide-angle deflector described lower. The vector diagram of such an unusual geometry of AO diffraction is presented with a greatly enlarged scale in figure 2(a) by the solid lines. Here $\boldsymbol{k}_i$ and $\boldsymbol{k}_d$ are the wave vectors of the incident and diffracted optical waves respectively, K_0 is the acoustic wave vector. The points D in figure 1 indicate the areas which are optimal for creating another unusual deflector with light scattering from the convex to concave parts of the normal optical surface. The vector diagram of AO interaction for this case is shown by the dashed lines in figure 2(a). The geometry of triple isotropic scattering, which is realized at the angles of light incidence $\theta_i = \pm\theta^{(3)}$ and acoustic frequencies $f^{(3)}$, also deserves a special mention. Such a diffraction regime is interesting because the AO filter, created on its basis, has to have an instrument function with noticeably suppressed side lobes. This opens up new opportunities for creating AO devices with improved characteristics.

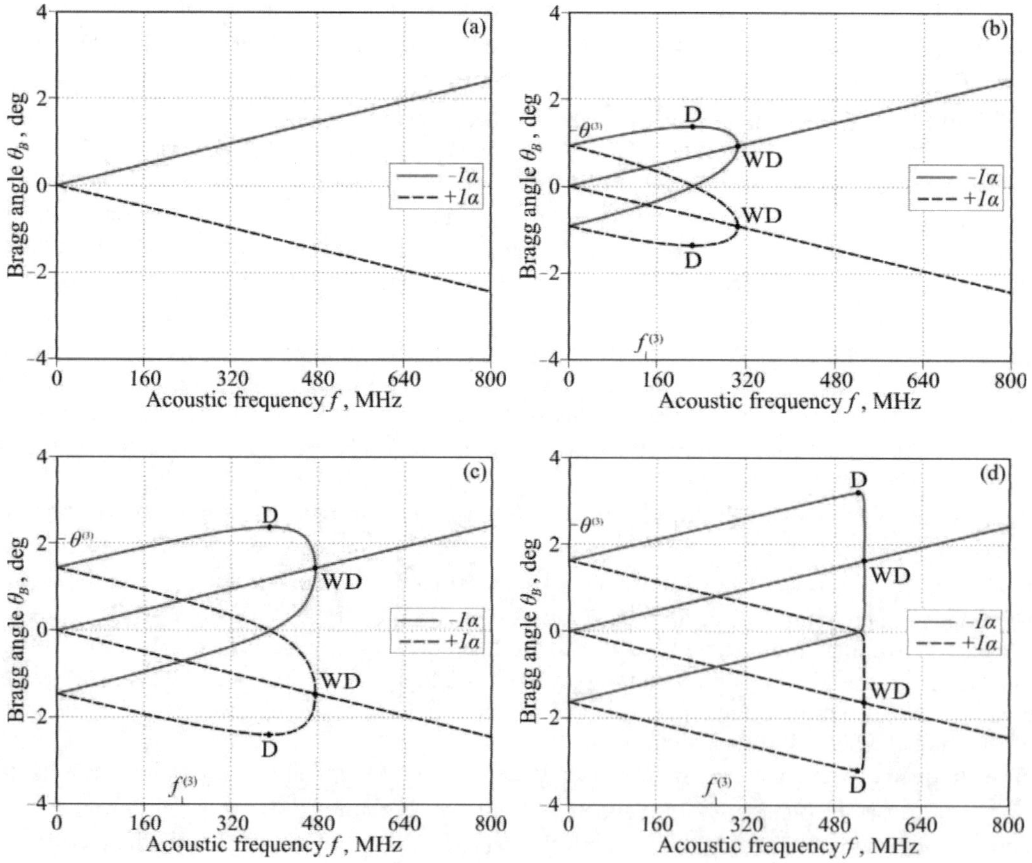

Figure 1. Frequency dependences of Bragg angles for different interaction planes:

$\vartheta = \vartheta_2$ (**a**), $\vartheta = 24{,}9°$ (**b**), $\vartheta = 24{,}3°$ (**c**), $\vartheta = \psi$ (**d**).

Decreasing the angle ϑ shifts the points WD and D to a higher frequency range (figure 1(c)). Finally, when the light wave diffracts in the plane passing through the optical axis, additional regions of the Bragg curves take a shape resembling a parallelogram with rounded edges (figure 1(d)). In this case, the angular selectivity of the diffraction near the WD point turns out to be extremely low, while the frequency selectivity, on the contrary, significantly increases. The subsequent decrease in the angle ϑ leads to deformation of the Bragg curves in the reverse direction. The central frequency of the wide-angle AO deflector $f_0 = K_0 V / 2\pi$ decreases, turning to zero at $\vartheta = \vartheta_1$, which corresponds to the operating point strictly in the center of a flat part of the normal optical surface. It should be emphasized that the shape of the curves $\theta_B(f)$ in figure 1 is caused by the complicated view of the optical wave surface in biaxial crystals near the biradial direction and, as a result, has no analogues among similar dependencies in uniaxial crystals.

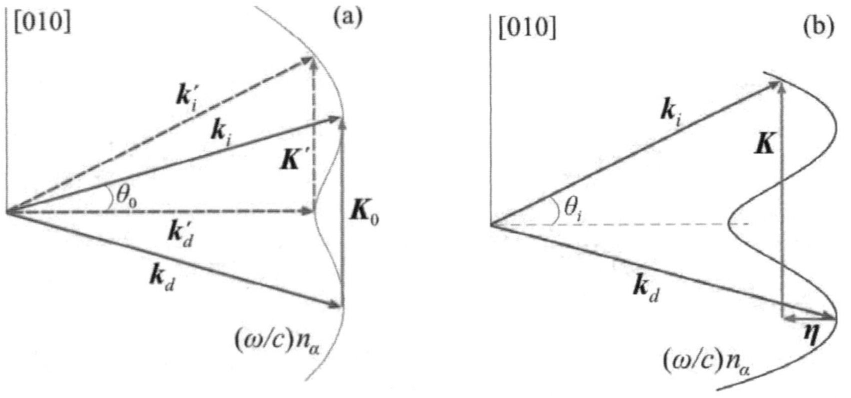

Figure 2. Wave vector diagrams for low-selective regime of AO interaction: (**a**) variant of AO phase matching; (**b**) variant of light scattering with phase mismatch η.

2.2. Wideband deflectors based on iodic acid crystal

Let us consider in more details characteristics of the low-selective light scattering regime near the WD points in various cuts of the HIO_3 crystal from the point of view of its practical use. Figure 3(a) demonstrates spatial distributions of the AO figure of merit M (solid curve) and the operating frequency f_0 (dashed curve) for this diffraction variant. It is seen that the dependencies $M(\vartheta)$ and $f_0(\vartheta)$ differ greatly from each other in their view. So, the maximum AO efficiency equal to $M = 51 \cdot 10^{-18}$ c^3/g is observed when light is scattered on the flat part of the refractive surface at $\vartheta = 25{,}2°$, whereas in the interaction plane with $\vartheta = 23{,}6°$, which corresponds to the maximum operating frequency $f_0 = 537$ MHz, the figure of merit is close to zero. Such a behavior of the value M is typical for rhombic (as well as monoclinic) biaxial crystals with close values of the photoelastic coefficients p_{12}, p_{22} and p_{32}.

Characteristics of AO deflectors functioning in the low-selective diffraction regime are shown in figure 3(b). Here, the solid curve shows the frequency range of the device Δf as a function of the polar angle ϑ, while the dashed curve illustrates the spatial distribution of the central frequency f_c. In these calculations, the length of the piezoelectric transducer was assumed to be equal to $l = 2$ cm, and the angle of incidence θ_i was chosen in such a way that the phase mismatch η in figure 2(b) was

$\eta = 0{,}8\pi/l$. In this case, the diffraction frequency band turns out to be as wide as possible, and the frequency characteristic of the AO deflector has an M-shaped form with a failure of the diffraction efficiency in the center by -3 dB [1]. It should be noted that the central frequencies f_c of such AO deflectors differs noticeably from the operating frequencies f_0 in figure 3(a), which is due to the low angular selectivity of the light scattering. During the calculations, it was found that this frequency shift is the value of about 35% of the frequency band, which has to be taken into account at developing AO devices.

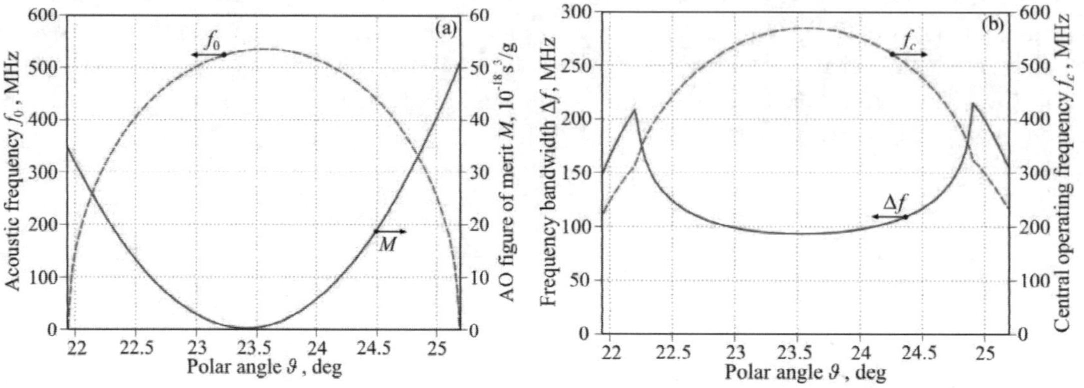

Figure 3. Wideband AO deflectors based on HIO_3 crystal: (**a**) spatial distributions of AO figure of merit M and the operating frequency f_0; (**b**) spatial distributions of frequency bandwidth and central operating frequency.

The graph shows that the value f_c reaches its maximum value of 569 MHz in the cut with $\vartheta = 23{,}5°$ and monotonically decreases as it approaches the boundaries of the angle range ϑ. The deflector has the narrowest frequency band in the plane with $\vartheta = 23{,}5°$, which contains the optical axis. The fractures on the curves $\Delta f(\vartheta)$ and $f_c(\vartheta)$ at $\vartheta = 22{,}2°$ and $\vartheta = 24{,}9°$ are associated with the limitation of the frequency band in the regime of isotropic light scattering by a value of 2/3 f_c (in order to avoid strong optical energy transfer to the higher diffraction orders). As a result of this restriction, the deflector has to have the maximum frequency range $\Delta f = 216$ MHz in the cut with $\vartheta = 24{,}9°$ (in the case of $l = 2$ cm). Such a deflector is capable to provide scanning a collimated laser beam in the angular range of 2,6° (in air) with a speed of $\tau = 6{,}7$ μs at an optical aperture of 2 cm. This corresponds to the number of spatially resolvable positions $N = 1430$. For comparison, the full scanning angle of paratellurite AO deflectors with the same cell parameters does not exceed 1,2°, and their operation speed is almost 5 times lower.

Low-selective regimes with cut angles $\vartheta = 24{,}9°$ and $\vartheta = 24{,}5°$ were studied in detail. The calculations have shown that on the basis of the first cut, it is possible to create special deflectors for deflecting strongly diverging light beams, as well as ultra-wideband devices for scanning collimated laser beams. The second cut, due to the high value of the operating frequency $f_0 = 438$ MHz, is best suited for creating wide-angle video filters. Similar AO filters based on biaxial crystals with high birefringence, transparent in the IR region, can be used to solve the problem of effective image filtering in the midle IR range. The large permissible angular aperture of such devices can provide, with using a two-crystal scheme for compensation of transverse chromatic aberrations [12], a high spatial resolution. Besides, the wide spectral bandwidth of the AO cell will give a noticeable increase in the signal-to-noise ratio.

The frequency characteristics of the diffraction efficiency $\varsigma(f)$ for the crystal cut with $\vartheta = 24{,}5°$ are presented in figure 4(a). The calculations are carried out for the length of the piezoelectric transducer $l = 2$ cm and three different angles of incidence θ_i: 0,97°, 1,33° and 1,67°. Note that the observed asymmetry of the curves $\varsigma(f)$ relative to the central frequency is mainly due to the anisotropy of the figure of merit *M*, which occurs due to a noticeable change in the polarization of the diffracted light wave within the frequency band of the AO cell. The maximum frequency range of diffraction is reached at $\theta_i = 1{,}67°$ and is equal to 160 MHz, which corresponds to the full scanning angle in the air of 2°. The graph shows that in the frequency band $\Delta f_1 = 93$ MHz, the angular range of the AO cell is $\Delta\theta_1 = 0{,}34°$, while in the frequency band $\Delta f_2 = 52$ MHz, the latter expands to $\Delta\theta_2 = 0{,}7°$. For comparison, the maximum permissible angular aperture of paratellurite AO deflectors (in the air) with the same cell parameters is only 0,1°.

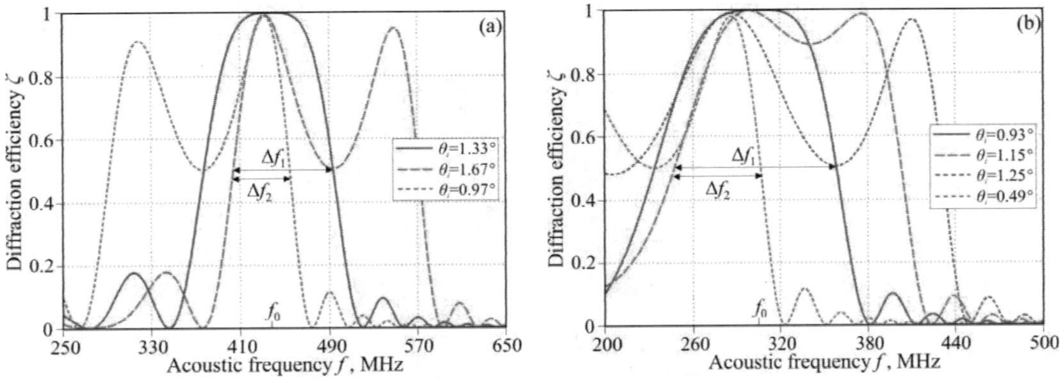

Figure 4. Frequency characteristics of low-selective AO interaction in the angular range near the operating point: **(a)** $\vartheta = 24{,}5°$, **(b)** $\vartheta = 24{,}9°$.

Similar dependences $\varsigma(f)$ in the interaction plane with $\vartheta = 24{,}9°$, calculated for $l = 2$ cm and the four angles of incidence of light: 0,49°, 0,93°, 1,15° and 1,25°, are shown in figure 4(b). In this case, as previously noted, the maximum frequency range is 216 MHz (at $\theta_i = 1{,}25°$). A variant of a deflector operating at the incidence angle $\theta_i = 1{,}15°$ is also of interest. Such a device has to have a high uniformity of the frequency characteristic (the depth of the dip in $\varsigma(f)$ in the center of the frequency range is 0,5 dB), while providing scanning of a collimated light beam in a wide angular range of 1,5°. It is easy to see that the wide-angle deflectors in the crystal cut under consideration also exceed noticeably in their characteristics similar AO devices in the crystal cut with $\vartheta = 24{,}5°$. Thus, the angular aperture of the deflector (inside the crystal) having a frequency range of $\Delta f_1 = 112$ MHz is $\Delta\theta_1 = 0{,}32°$, whereas when the frequency band is narrowed to $\Delta f_1 = 59$ MHz, the latter increases to $\Delta\theta_2 = 0{,}76°$.

2.3. 2D transfer functions of AO cell

Analyzing characteristics of deflectors intended for deflecting collimated laser beams, we can limit our consideration by the problem of AO interaction in a two-dimensional approximation. The basis for this approach is that the main characteristics of such AO devices are determined by the peculiarities of light scattering in the interaction plane. However, in a situation when the AO deflector is used for scanning strongly divergent optical beams, it is necessary to take into account the three-dimensional nature of AO scattering. In these cases, the spatial structure of the diffracted light beam may differ significantly from the structure of the incident beam, since the AO cell, due to the angular selectivity of the interaction, has a different effect on different angular components of the incident optical radiation [13]. This change in the angular spectrum of the light beam is described by a two-dimensional transfer function (2DTF), the form of which depends on the geometry of the diffraction pattern, the crystal cut and the ultrasound frequency [14].

Our analysis has shown that for the regime of low-selective light scattering, the 2DTF has a complex X-shaped form in the working order of diffraction, which has no analogues in uniaxial crystals. Figure 5(a) demonstrates the 2DTF of the α-HIO_3 cell for the case $\vartheta = 24{,}5°$, $l = 1$ cm and $f = f_0$. The module of the 2DTF is shown in color gradations: from zero (dark blue) to one (dark red). The center of the picture corresponds to zero spatial frequency (the axis of the incident beam). It is clear from the graph that the tangents to the branches of the curved cross in the center of the 2DTF coincide with the axes of the abscissae and ordinates, which is due to the equal curvature of the refraction surface for the axial components of the interacting light beams both in the interaction plane and in the orthogonal direction.

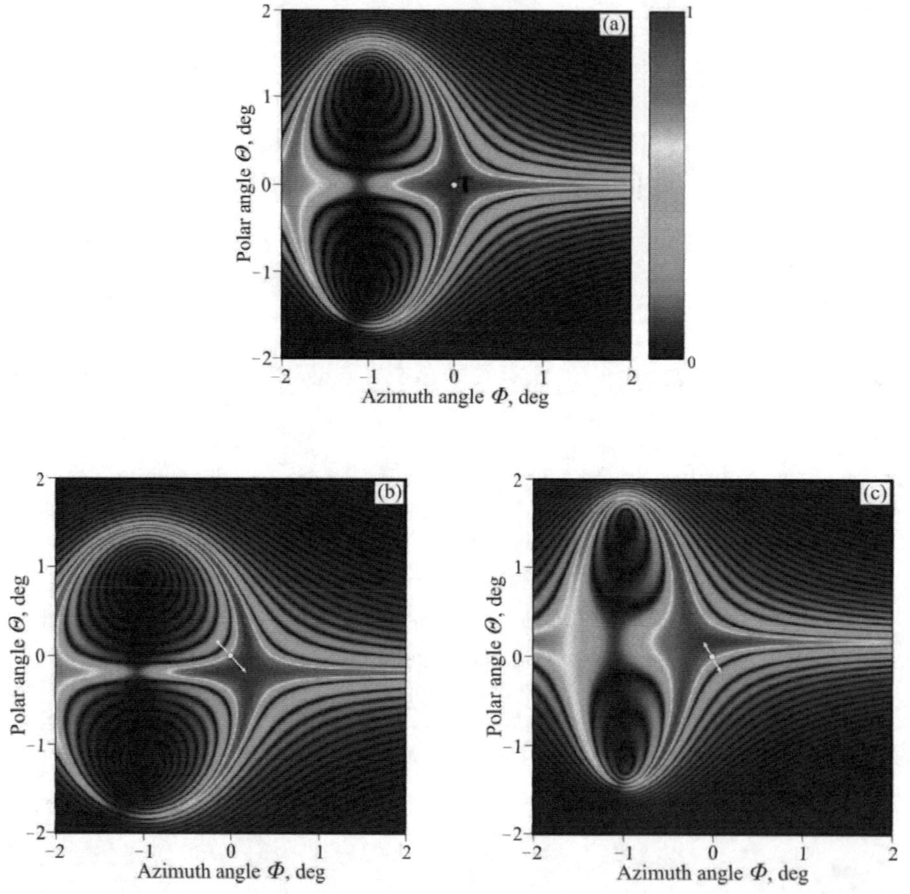

Figure 5. 2D transfer functions of AO cell at various ultrasound frequencies near the frequency f_0 at $\vartheta = 24{,}5°$: **(a)** f = 438 MHz, **(b)** f = 376 MHz, **(c)** f = 494 MHz.

With a decrease in the ultrasound frequency, the cross-shaped region shifts to the right and down, which corresponds to the implementation of a low-selective variant of AO scattering in the plane with a higher angle ϑ (figure 5(b)), whereas with an increase in the frequency (relative to f_0) a similar displacement occurs in the opposite direction (figure 5(c)). It is clear that such a nature of the 2DTF transformation makes it possible to implement a wide-angle AO deflector capable to deflect not only slit beams, but also beams carrying a 2D image.

The pictures in figure 5 also open up new opportunities for AO spatial filtration of images [15]. For example, by choosing a working point in the center of the 2DTF (as shown by the symbol T in figure 5(a)), it is possible to perform integration operations (at $f = f_0$) and directional differentiation of the image (at f = 376 MHz or f = 494 MHz) in one AO cell. The direction of differentiation in each of these cases is illustrated by the arrows in figure 5. As follows from the graphs, these directions differ significantly from each other (by a value of about 15°), which may be useful at solving problems of phase object visualization by the AO method [15].

3. Summary

In this paper we have presented results of a theoretical study of new variants of isotropic AO light scattering in oblique sections of a biaxial iodic acid crystal, which contain a concave area of the refraction surface. It is shown that a unique, simultaneously wide-angle and broadband regime of Bragg diffraction can be implemented in these interaction planes. Based on it, a special wide-angle AO deflector is proposed that can scan highly divergent optical beams. It is proved that for a number of characteristics, such a device will be significantly surpass existing analogues, which allows us to count on its possible practical applications. Characteristics of these deflectors in different cuts of the iodic acid crystal are studied. Optimization of parameters of the AO cells by the values of the frequency range and the AO figure of merit is carried out. The prospects of using the new diffraction variants are shown both at solving problems of AO image filtering and visualizing phase objects by the AO method.

Acknowledgements

The authors express their gratitude to the Russian Science Foundation, through a grant of which (Project № 19-19-00606) the present study was carried out.

References

[1] Balakshy V I, Parygin V N, Chirkov L E 1985 Physical Principles of Acousto-Optics (*Moscow: Radio & Svyaz*)

[2] Das P K, DeCusatis C M. 1991 Acousto-Optic Signal Processing: Fundamentals and Applications (*New York: Artech House*)

[3] Xu J, Stroud R 1992 Acousto-Optic Devices (*New York: Wiley*)

[4] Balakshy V I, Kupreychik M I, Mantsevich S N, Molchanov V Ya 2021 Acousto-optic cells with phased-array transducers and their application in systems of optical information processing *Materials* **14** 451

[5] Kupreychik M I, Balakshy V I 2018 Peculiarities of acousto-optic interaction in biaxial crystal of alpha-iodic acid *Appl. Opt.* **57** pp 5549–55

[6] Milkov M G, Voloshinov V B, Isaenko L I, Vedenyapin V N 2018 An experimental study of ultra-wide-band and ultra-wide-aperture non-collinear acousto-optic diffraction in an optically biaxial potassium arsenate titanyl crystal *Moscow Univ. Phys.* **73** pp 83–8

[7] Kupreychik M I, Balakshy V I, Mantsevich S N 2020 Extraordinary variants of anisotropic acousto-optic interaction in optically biaxial crystals *Proc. Int. Conf 2020 Wave Electronics and its Application in Information and Telecommunication Systems (WECONF) (IEEE Xplore Digital Library)* pp 1–7

[8] Pinnow D A, Dixon R W 1968 Alpha-iodic acid: a solution-grown crystal with a high figure of merit for acousto-optic device applications *Appl. Phys. Lett.* **13** pp 156–8

[9] Naito H, Inaba H Measurement of the refractive indices of α-iodic acid, HIO_3, crystal 1972 *Opt. Electron.* **4** pp 335–7

[10] Balakshy V I, Kupreychik M I 2016 Anisotropic light diffraction in a biaxial crystal of alpha-iodic acid *Phys. Wave Phen.* **24** pp 58–63
[11] Kupreychik M I, Balakshy V I 2019 Investigation of close to collinear anisotropic acousto-optic interaction in a biaxial crystal of alpha-iodic acid *Proc. SPIE* **11210** 112100O
[12] Machikhin A, Pozhar V, Batshev V 2015 Double-AOTF-based aberration-free spectral imaging endoscopic system for biomedical applications *J. Inn. Opt. Health Sci.* **8** 1541009
[13] Balakshy V I, Voloshinov V B, Babkina T M, Kostyuk D E 2005 Optical image processing by means of acousto-optic spatial filtration *J. Mod. Opt.* **52** pp 1–20
[14] Balakshy V I, Kostyuk D E 2006 Spatial structure of acousto-optic phase matching in uniaxial crystals *Opt. Spectrosc.* **101** pp 283–9
[15] Balakshy V I 2018 Acousto-optic visualization of optical wavefronts *Appl. Opt.* **57** pp 56–63

Dependence of the energy efficiency of the acousto-optic modulator of terahertz radiation on the dimensions of the ultrasound transducer

P A Nikitin* and A K Nikitin

Scientific and Technological Centre of Unique Instrumentation RAS, 15 Butlerova str., Moscow 117342, Russia

*Corresponding author's e-mail: nikitin.pavel.a@gmail.com

Abstract. The paper presents results of experimental investigation of an acousto-optic modulator of terahertz radiation based on liquefied sulfur hexafluoride. It was found that the intensity of diffracted radiation at a fixed ultrasound power depends on the dimensions of the ultrasound transducer. The optimal size of the ultrasound transducer in the direction orthogonal to the plane of the acousto-optic interaction was determined, at which the maximum energy efficiency of the acousto-optic modulator was achieved.

1. Introduction

The acousto-optic (AO) effect consists in the interaction of electromagnetic radiation with a phase structure induced in a medium by means of an ultrasonic wave. Therefore, AO devices are compact and allow real-time control of the radiation beam. Most of these devices have been designed to operate in the ultraviolet, visible and infrared spectral range [1]. In the Bragg mode, there are only two diffraction orders: zero, corresponding to the transmitted radiation, and the first or minus the first, into which information from the sound beam is transmitted. The use of the AO effect in the terahertz (THz) region of the spectrum is hindered by low intensity I_1 of the diffracted radiation, which is inversely proportional to the square of the radiation wavelength λ. For convenience, the dimensionless diffraction efficiency ξ is used, which is equal to the ratio of the intensity I_1 of the beam in the first diffraction order to the intensity I_0 of the beam in the zero diffraction order [2]:

$$\xi = \frac{I_1}{I_0}. \quad (1)$$

In the general case, the efficiency of AO diffraction depends in a complex way on the ultrasound power P_a and the radiation wavelength. However, with weak AO interaction (when $\xi \ll 1$), the diffraction efficiency is directly proportional to the ultrasound power density and inversely proportional to the square of the radiation wavelength. In the quasi-orthogonal geometry of the AO interaction, when the angle of deflection of the diffracted radiation is small, the AO diffraction efficiency can be calculated as follows [3]:

$$\xi = \frac{\pi^2}{2\lambda^2} M_2 \frac{L}{d} P_a \exp(-\alpha_s l), \quad (2)$$

where M_2 is the coefficient of the AO figure of merit of the medium, P_a is the acoustical power, L and d are dimensions of the transducer in the plane of the AO interaction and orthogonal to this plane, respectively, α_S is the sound power attenuation coefficient, l is s the distance from the sound transducer at which the THz beam travels. Relation (2) differs from the known relation only by an exponential factor, which takes into account the attenuation of the ultrasonic beam on the way from the ultrasound radiator to the region of AO interaction.

AO figure of merit M_2 of the medium is the universal coefficient characterizing the AO properties of the medium and does not depend on the dimensions of the ultrasound transducer, the geometry of the AO interaction, the ultrasound power and the radiation wavelength. For single crystals it can be calculated as $M_2=p_{\mathrm{eff}}^2 n^6/\rho V^3$, where p_{eff} is the effective photo-elastic constant, n is the refractive index, ρ is the density, V is the sound velocity. For liquids, this relation is more complex and it can be derived directly from the Lorentz-Lorenz equation for the refractive index:

$$M_2 = \left[\frac{(n^2-1)(n^2+2)}{6n}\right]^2 \frac{4}{\rho V^3}. \quad (3)$$

From (2) it follows that the diffraction efficiency ξ is proportional to the power of the sound beam P_a. Therefore, by changing the amplitude of the electrical signal supplied to the sound transducer, it is possible to modulate the intensity of the radiation beam. Note, that relation (2) was derived for the AO diffraction on the sound beam with plane wavefront. However, the real sound beam structure is complex. Since the AO interaction process is cumulative, the efficiency ξ of AO diffraction on the real sound beam is less than one predicted by (2).

As is known, the best characteristics of AO modulators and filters can be achieved by using birefringent crystals as a medium for AO interaction. In addition, a significant reduction in the cost of production can be achieved by using piezoelectric crystals with electrodes deposited on them. However, in the THz range, birefringent and piezoelectric AO crystals such as paratellurite (TeO_2) and lithium niobate ($LiNbO_3$) are opaque [4]. High-resistivity silicon (Si) and germanium (Ge) can be distinguished among AO crystals transparent to THz radiation. They are both optically isotropic cristals. The first of them is characterized by a low value of the coefficient $M_2 \approx 10\cdot10^{-15}$ s^3/kg, and with the use of the second it was possible to achieve a diffraction efficiency of only about 0.01% per 1 W of electric power [5]. Non-polar liquids such as cyclohexane C_6H_{12} can also be used as a medium for AO interaction in the THz range. However, the diffraction efficiency still remains at the level of 0.01%/W [6]. This fact is related to the significant attenuation of ultrasound in liquids.

A breakthrough in the study of the AO effect in the THz range was made when a liquefied inert gas, sulfur hexafluoride or elegase (SF_6), was used as the interaction medium [7]. The very idea of using a liquefied inert gas arose from the fact that this class of substances is characterized by a low speed of sound, and the AO figure of merit M_2 is inversely proportional to the cube of the speed of sound V [8]. Elegase liquefies at room temperature at a pressure of about 20 bar and has a sound speed of about $V \approx 300$ m/s. Due to this, the AO figure of merit of SF_6 M_2=15000·10-15 s^3/kg is more than an order of magnitude higher than the AO figure of merit of TeO_2 M_2=1200·10-15 s^3/kg, which was previously considered as the best AO medium.

Investigations in [7] were carried out at temperatures of 13 and 14°C, and were later repeated in [3] at a temperature of 25°C. The size of the transducer in both studies was about $d \approx 12$ mm and was close to the thickness of 6 mm, corresponding to a resonant frequency of 300 kHz. In this case, a complex type of deformation occurs in the transducer, which is far from piston vibrations [8]. As a consequence, the wavefront of the sound beam is not flat, which affects the efficiency of AO diffraction. Therefore, the purpose of this work is to experimentally determine the optimal size d of an ultrasound transducer, at which the efficiency of AO diffraction per 1 W of electrical power will be maximum.

2. Results and discussion

AO interaction has a resonant character, and to achieve the highest efficiency of AO diffraction, it is necessary to use monochromatic coherent radiation. In addition, for reliable registration of diffracted radiation, according to (1), a significant ultrasound power P_a is required. In our experiments, P_a was about 1 W in order to avoid the cavitation effect near the ultrasound transducer. Therefore, it was necessary to use high-intensity radiation, and the most suitable source was the Novosibirsk free-electron laser (FEL). The radiation wavelength λ=130 μm was chosen to correspond to the optical transparency window in the atmosphere.

Our experiments were carried out at a temperature +22°C and pressure 25 bar. These conditions are the most easily implemented and do not require the use of a complex cooling system. The schematic diagram of experimental setup is shown in Figure 1. The THz beam was generated by FEL 1. Since an AO modulator based on a liquid medium is insensitive to radiation polarization, the intensity of the THz beam was set by wire polarizer 2. The radiation beam was incident on the center of the input optical window of the AO cell 3 at the distance l=5 cm from the sound transducer. A detailed diagram of the AO cell is given in [3].

The AO cell for the implementation of the AO interaction was a round steel tube with the diameter 15 cm, the length 10 cm, and the wall thickness 3 mm. The TPX 6-mm-thick plastic windows were placed inside the cuvette on both flanges. The flanges had the thickness 3 cm, while the openings in the flanges were 100×10 mm^2. It follows from formula (2) that the diffraction efficiency is proportional to the length L of the ultrasound transducer. Therefore, the ultrasound transducers were made of monolithic polycrystalline lead zirconate titanate (PZT) piezoelectric ceramics having the dimensions 80×d×6 mm^3 (the length of the ultrasound transducer was L=80 mm, whereas its width d was from 6 to 14 mm with step 2 mm). The electrodes were set on the opposite faces 80×8 mm^2. The ultrasound transducer was placed inside the AO cell, along the axis of the cylinder, right near the input optical window. Due to this, the AO interaction took place in the horizontal plane.

For observation of the diffracted radiation, the THz beam falls on the AO cell at the Bragg angle, and a signal from generator of electrical signals 4 was applied to the ultrasound transducer via amplifier 5. At a distance of about 30 cm after the AO cell, lens 6 was located, focusing the radiation into the receiver, Golay cell 7. This receiver requires modulation of the radiation at a frequency of 10 Hz. Therefore, the signal from the generator of electrical signals was modulated in amplitude with the specified frequency. The signal from the receiver was isolated from the background noise using lock-in amplifier 8.

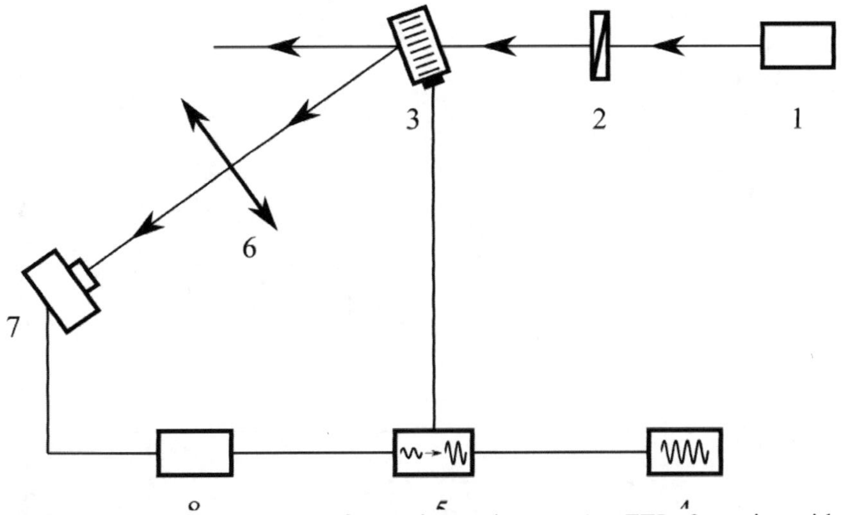

Figure 1. Schematic diagram of experimental setup: 1 – FEL; 2 – wire-grid polarizer; 3 – AO cell; 4 – signal generator; 5 – electrical amplifier; 6 – lens; 7 – Golay cell; 8 – lock-in amplifier.

The acoustic power P_a is usually considered to be approximately equal to the supplied electric power P_{el}, which is proportional to the square of the voltage U. However, it is not correct to compare the efficiency of AO devices at the same amplitude of the electric signal. This is due to the fact that ultrasound transducers with different sizes and shapes have different electrical impedances Z. The simplest interpretation is that the ultrasound transducer can be thought of as a capacitor. Thus, it is necessary to compare the efficiency of AO diffraction per 1 W of electric power, i.e. normalized diffraction efficiency:

$$\xi_{\text{norm}} = \frac{\xi}{P_{el}}. \tag{4}$$

The electric power was determined as:

$$P_{\text{el}} = \frac{U^2}{2|Z|}\cos\phi, \tag{5}$$

where $\cos\phi = \text{Re}(Z)/|Z|$ is the so-called power factor, equal to the ratio of the instantaneous real power used by an electrical load to the apparent power running through the circuit. Therefore, the relation for the electric power reads as:

$$P_{\text{el}} = \frac{1}{2}U^2\frac{\text{Re}(Z)}{|Z|^2}. \tag{6}$$

The complex impedance Z of the sound transducer and amplitude U of the electrical signal was measured at the resonant frequency. The dependence of the diffraction efficiency ξ_{norm} per 1 W of the electrical power on the width d of the sound transducer is shown in Figure 2.

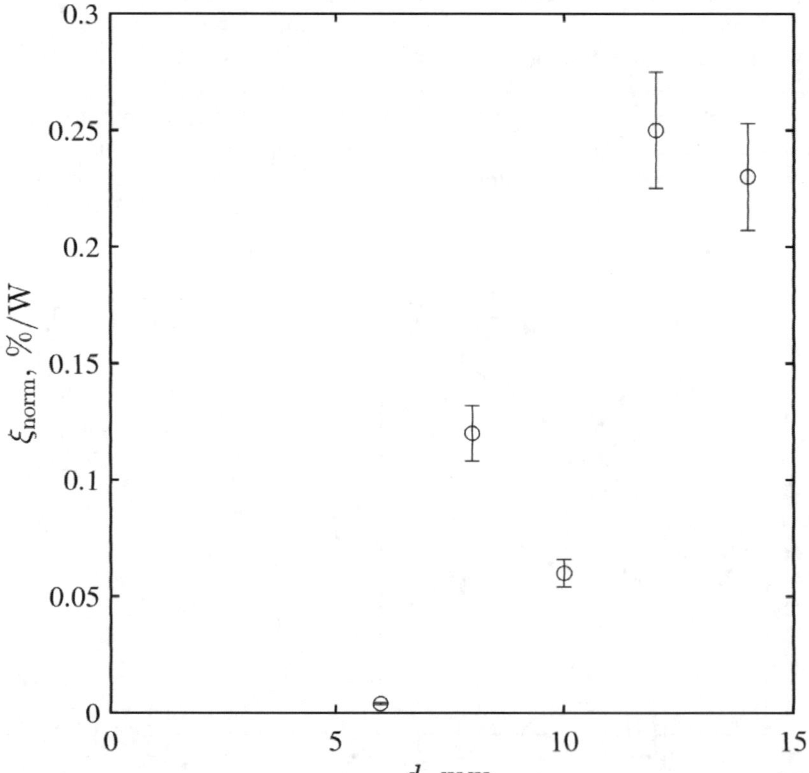

Figure 2. Dependence of the normalized efficiency of AO diffraction on the sound transducer width.

It was found that, the normalized diffraction efficiency increases with an increase in the width of the ultrasound transducer *d*. Obviously, this increase cannot be unlimited, since with an increase in the width of the ultrasound transducer, the acoustic power density decreases. Moreover, it follows from equation (2) that the dependence of the AO diffraction efficiency on the width of the ultrasound transducer is inversely proportional, i.e. with an increase in the width of the transducer, the diffraction efficiency should decrease. In addition, when the width of the ultrasound transducer is greater than the diameter of the radiation beam, the AO interaction occurs only in a small region. As a result, some of the acoustic power is wasted.

In our experiment, the width of the THz radiation beam was 10 mm and was bounded by the size of the optical window of the AO cell. It was found that at d=6 mm, the efficiency of the dependence of the diffraction efficiency is approximately an order of magnitude smaller than at d=10 mm. However, according to relation (2), the diffraction efficiency, on the contrary, should be approximately 2 times higher (since $\xi \propto 1/d$). Note, that when d decreases from 10 mm to 6 mm, the sound beam interacts with only half of the THz radiation beam. Therefore, the intensity of the diffracted radiation should be approximately the same at d=6 mm and at d=10 mm. We attribute such a significant difference between theory and experiment to the strong phase inhomogeneity of the sound beam. This inhomogeneity strongly depends on the width d of the ultrasound transducer, since when d is comparable to the thickness of the transducer, the oscillations of its surface differ from the piston type. The increase in the efficiency of AO diffraction at d>10 mm can be explained by the more uniform phase structure of the sound beam. Calculations show that the efficiency of AO diffraction should be 1.5 times higher. We believe that the discrepancy between theory and experiment with the fact that the structure of the sound beam, even at d=15 mm, remains complex. In addition, not all electrical power is converted into acoustic power due to the significant difference between the acoustic impedances of a piezoelectric ultrasound transducer and liquefied gas.

The experimental results contradict the model (2) of AO interaction, which is usually used for calculations. Hence it follows that there are a number of unaccounted for factors that have led to an anomalous increase in the efficiency of AO diffraction with an increase in the width of the ultrasound transducer. The model used assumes that a plane electromagnetic wave interacts with a sound column with a plane wavefront. As you can see, the model uses the plane-wave approximation. This is usually justified, since laser radiation in the visible range has a low divergence. On the other hand, THz radiation has a wavelength of 100 times greater than that of the visible range. Therefore, due to the phenomenon of diffraction in free space, the THz radiation beam has a 100 times greater divergence. Therefore, to reduce this effect, a wider radiation beam is needed. However, in this case, it is necessary to use a wider ultrasound transducer, which, according to relation (2), will lead to a decrease in the efficiency of AO diffraction. Therefore, we used THz radiation from a Novosibirsk FEL with a wavelength of λ = 130 μm, and the radiation beam width was about D = 1 cm. The advantage of this radiation source is that radiation is monochromatic and its divergence is slightly less than one degree (λ/D). Thus, the plane-wave approximation is valid for the radiation beam. The situation is quite different for an ultrasound beam. The speed of sound in liquefied SF_6 gas is about V = 300 m/s and at a frequency of F = 300 kHz, the ultrasound wavelength is Λ = 1 mm. Therefore, when the width of the transducer is d = 10 mm, the divergence of the ultrasound beam is 5° (Λ/d), and when d = 5 mm, it is about 10°, which is several times larger than the angular bandwidth of the AO interaction. As can be seen, the plane-wave approximation is violated for an ultrasound beam. In addition, it should be noted that the thickness of the ultrasound transducer was about 6 mm (necessary for effective generation at a frequency of 300 kHz). Therefore, in our experiment, the width of the transducer was comparable to its thickness. In this case, the oscillations of the surface of the transducer have a complex shape and very strongly depend on the ultrasound frequency F. Thus, we believe that it is the complex structure of the sound beam that led to an anomalous increase in the AO diffraction efficiency with increasing width of the ultrasound transducer.

3. Conclusion

The operation of the AO modulator of THz radiation based on liquefied sulfur hexafluoride has been investigated. The dependence of the efficiency of AO diffraction on the width of the ultrasound transducer was measured. It was expected that the diffraction efficiency would be approximately constant as the transducer width was reduced from 10 mm to 6 mm. However, a strong decrease in the diffraction efficiency was observed. At an intermediate value of the width d=8 mm, the diffraction efficiency was approximately 2 times higher than at d=10 mm. We believe that the nonmonotonic dependence of the AO diffraction efficiency on the width of the transducer is due to the complex type of acoustic modes of the ultrasound transducer. With the transducer width of 12 and 14 mm, the diffraction efficiency was approximately 5 times higher than with a width of 10 mm, and only 1.5 times less than predicted by theory. In our opinion, this indicates that the wavefront of the acoustic beam can already be considered approximately flat. A more accurate assessment of the phase structure can only be obtained by direct measurement of the amplitude of the sound field at the surface of the ultrasound transducer. It is shown that the highest efficiency of AO diffraction per 1 W of electric power can be achieved using a ultrasound transducer with a width of 12 mm. The results of the work can be used in the design of energy-efficient AO devices operating in the THz range.

The experimentally obtained anomalous dependence of the AO diffraction efficiency on the width of the ultrasound transducer indicates a complex structure of the sound beam. This effect is due to the fact that the width of the ultrasound transducer was comparable to its thickness. The weakening of this effect due to the use of an ultrasound transducer with a significantly smaller thickness is not possible due to an increase in the resonance frequency and, as a consequence, a quadratic increase in the attenuation of ultrasound. Therefore, in the future, we plan to conduct a more complete study of the AO diffraction of THz radiation in liquefied SF6 using transducers with different widths: 1) measure the dependence of the complex electrical impedance of an ultrasound transducer on the ratio of the thickness and width of the transducer; 2) determine the resonant frequency of the transducer at which the effective generation of the ultrasonic beam occurs; 3) measure the acoustic field near the surface of the ultrasound transducer at the resonant frequency and in its vicinity in the effective AO interaction band. The results of these measurements should form the basis of a new model, which will make it possible to determine the optimal ratio of the thickness and width of the ultrasound transducer, at which the maximum efficiency of AO diffraction is achieved.

Acknowledgments

This research was financially supported by a grant from the Russian Science Foundation № 20-72-00184.

References

[1] Voloshinov V, Gupta N 2004 *Applied Optics* **43**(19) 3901-09
[2] Balakshy V, Kupreychik M, Mantsevich S, Molchanov V 2021 *Materials* **14**(2) 451
[3] Nikitin P A, Knyazev B A, Voloshinov V B, Scheglov M A 2020 *IEEE Transactions on Terahertz Science and Technology* **10**(1) 44-50
[4] Unferdorben M, Buzady A, Hebling J, Kiss K, Hajdara I, Kovacs L, Peter A, Palfalvi L 2016 *Journal of Infrared, Millimeter, and Terahertz Waves* **37** 703-09
[5] Voloshinov V B, Nikitin P A, Gerasimov V V, Knyazev B A, Choporova Yu Yu 2013 *Quantum Electronics* **43**(12) 1139-42
[6] Vogel T, Dodel G 1985 *Infrared Physics* **25**(1/2) 315-18
[7] Durr W 1986 *International Journal of Infrared and Millimeter Waves* **7**(10) 1537-58
[8] Holland R 1968 *The Journal of the Acoustical Society of America* **43**(5) 988-97

Analysis of chromatic aberrations influence on operation of the tunable AOTF-based source

A S Beliaeva, G E Romanova and A N Chertov

ITMO. ITMO University, 49, Kronverksky pr., St. Petersburg, 197101, Russia

E-mail: belyaevalina@inbox.ru, romanova_g_e@mail.ru, a.n.chertov@gmail.com

Abstract. Developing a source with the possibility of tuning in wavelengths based on an acousto-optical tunable filter (AOTF), an essential factor is the ability to provide accurate color coordinates of the simulated color (in the case of developing a color standard for colorimetry) or the required wavelength (in the case of an application for spectral studies). As shown earlier, the choice of the principal layout – confocal or parallel beam path - primarily determines the dimensions and efficiency of using the luminous flux. However, these schemes also need to be analyzed for color or wavelength fidelity, considering other components used in the scheme. The analysis performed allows to identify the optimal scheme to ensure the required color reproduction accuracy and establish the requirements for correcting chromatic aberrations of the components.

1. Introduction

Sources with the ability to control the number of spectral lines, their position and width can be used in science and industry, for example, spectral analysis (including food quality control, pharmacy, etc.), microscopy, psychophysiological research, and the calibration of spectral, colorimetric instruments and cameras for machine vision [1,2].

The main element in the system is AOTF, which determines its basic functional properties. In the literature, one can find the results of analysis of the operation of such an element from the point of view of the diffraction efficiency. However, the literature mainly discusses its work as part of imaging systems.

In addition to the value of the provided flux, one of the main characteristics of a tunable source used in spectral measurements and instrument calibration is color mixing quality – uniformity of the color field, color constancy across the field.

Since AOTF can provide the extraction of up to 4 spectral lines, the positions of light beams of different wavelengths in the system may differ. Consequently, the colored edges of the illuminated field can be observed caused by the misalignment of the positions of the beams. The distinguishable boundaries of different wavelengths can affect the analysis and calibration results. The colored edge of the working beam can influence the color assessment when revealing color preferences [3]. In addition, chromatic aberrations of optical elements and systems responsible for the conjugation of AOTF, the light source and the receiving part also influence on these effects.

As shown earlier, the traditionally used schemes - with confocal and parallel beam paths - provide different coupling efficiency of the source and AOTF, which determines the attainable brightness. This paper analyzed the chromatic aberrations of two variants of schemes - with confocal and parallel beam paths.

In a scheme with a confocal beam path (Figure 1 a), AOTF provides the minimum aberration contribution. Therefore, such a beam path is preferable to use in imaging systems [4-6]. The parallel beam path is more compact (Figure 1 b) and has a higher luminous flux utilization rate than the confocal one [7]. However, AOTF adds more significant aberrations [4].

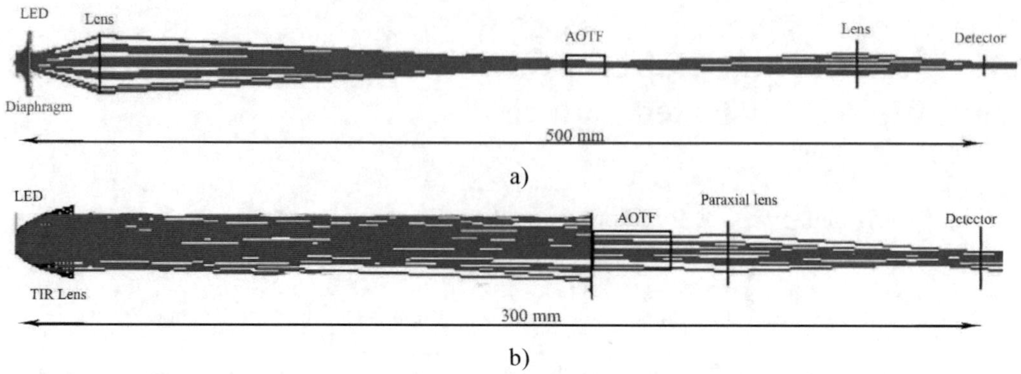

Figure 1. System layout with a) confocal beam path, b) parallel beam path.

To assess the contribution of chromatic aberrations, the schemes were considered with different optical components: in one case, the first component was a single lens which introduces a chromatic aberration, in the other case, an objective with corrected chromatic was used.

2. Theoretical evaluation

Below we provide theoretical expressions for analysis the scheme with parallels beam path as the most prospective case for the development of the tunable source: such type of the systems is the most compact and efficient [7]. In such a system, the aberration contribution of AOTF is more significant than in the confocal beam path [4].

To determine chromatic aberrations, we can evaluate the contribution of chromatic aberrations of individual components of the scheme:

- the lens responsible for the direction of the beam of rays from the source to AOTF, if a thin lens is used as this element, introduces chromatism, which can be estimated by the formula 1 [8]

$$\Delta s'_{chrom} = -\frac{f'}{\nu_1}. \quad (1)$$

Where: f' - is the back focal length of the lens,
ν_1 - is Abbe number of the first component.
The chromatic difference in angular units (in radians) after the lens can be estimated as:

$$\Delta\psi'_{chrom} = -\frac{f'}{\nu_1}\frac{D}{2f'^2_1} = -\frac{D}{2\nu_1 f'_1}. \quad (2)$$

Where: D – AOTF size

When an illumination component is a lens with total internal reflection (TIR), chromatism can be assessed by performing ray tracing at various wavelengths.

- The angle of beam deflection ψ in AOTF in working orders at different wavelengths can be estimated using the formulas [9-10]. Moreover, for the same wavelength, the deflection angle will depend on the incidence angle on the cell (and on the state of polarization).

- Transverse chromatic aberration (in mm) of the system located behind the cell and responsible for matching the beam at the output of the cell and the receiving device, for example, a fiber bundle, when operating in a parallel beam path, can also be estimated by the formula:

$$\Delta y'_{chrom2} = -\frac{D}{2\nu}. \quad (3)$$

For the case under consideration, with chosen elements parameters, we can obtain following numerical results for chromatism values:

- If the condenser with focal length f'=13 mm made from BK7 is used as an illumination lens, the chromatic difference in angular units after the lens is 0.378°.
- The cut angle γ of AOTF is 7 degrees, the calculated angle of incidence of the beam relative to the crystal axis θ, the deflection angle of the beam ψ will depend on the diffraction order and wavelength. Taking into account the chromatic difference from the first component, the rays will fall on the input face of the cell at angles that differ from the nominal (zero) value by the value $\Delta\theta$, so the angle of deflection of the rays will have a value different from the initial one $\psi+\Delta\psi$ (Table 1).

Table 1. Angles of deflection of the diffraction orders.

λ (nm)	450		625	
Order	1	-1	1	-1
ψ (deg.)	2.04	-1.96	1.92	-1.84
ψ+Δψ (deg.)	1.59	-2.38	1.52	-2.26

- The third component – the lens with a focal length of 50 mm made of BK7 material, will have chromatic aberration $\Delta y'_{chrom2}$ = 0.09 mm.

The simultaneous influence of the chromatism of elements can lead to a change in the efficiency of radiation diffraction in a crystal. In addition, it will be complicated to spatially align the diffraction orders at different wavelengths.

3. Scheme with confocal beam

An experimental setup with a confocal beam path was assembled to compare chromatic aberrations in the schemes (figure 2).

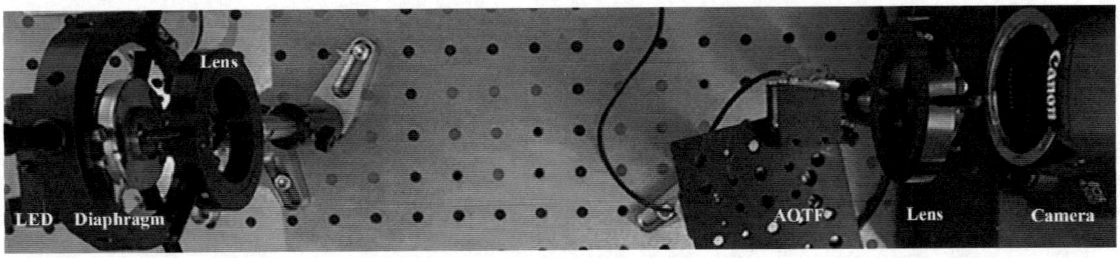

Figure 2. Confocal path scheme setup with a single lens in illumination part.

Figure 3 shows the pattern observed on the detector when working with unpolarized radiation at wavelengths 450 and 625.

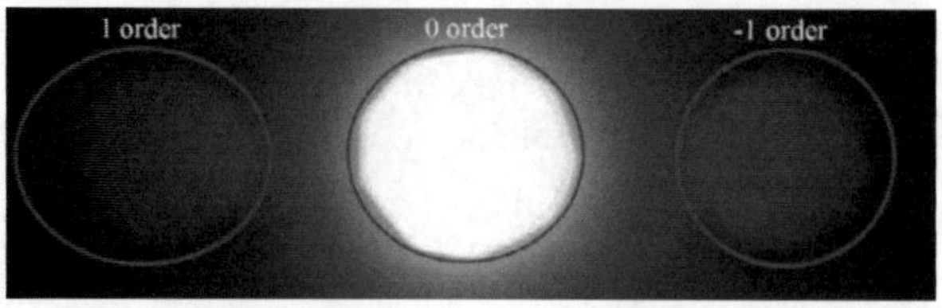

Figure 3. Image of diffraction orders at wavelengths of 450 and 625 nm in a confocal scheme with a single lens.

Figure 4 shows the relative position of the spots in the considered orders for several wavelengths in the visible range.

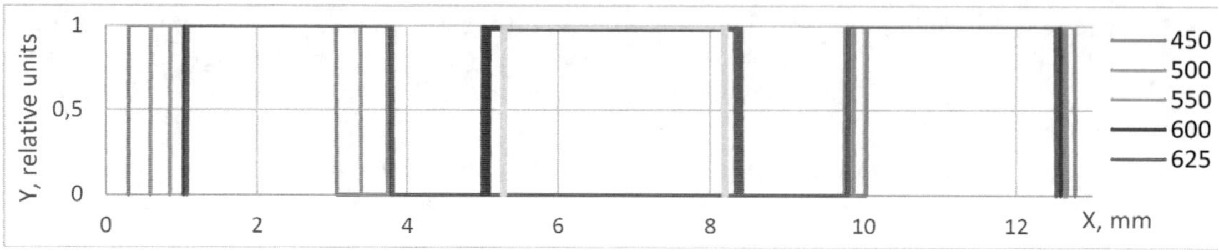

Figure 4. A diagram of the arrangement of orders on the detector at 450-625 nm in a confocal scheme with a single lens.

For comparison, images were also recorded in the same scheme when an objective with corrected axial color was used as a system installed in front of AOTF. Figure 5 shows the positions of the diffraction orders obtained on the detector in this case.

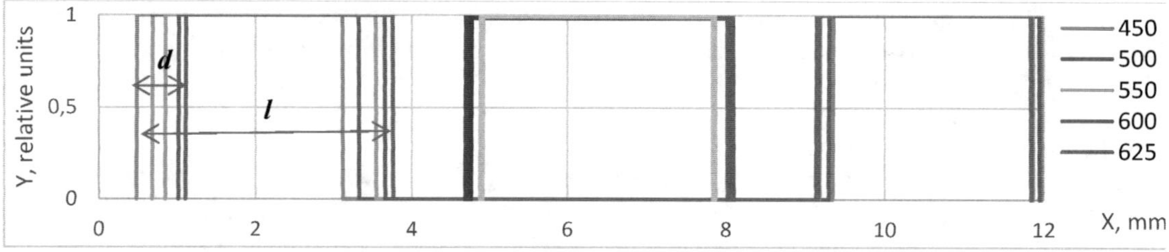

Figure 5. The diagram of the arrangement of orders on the detector at 450-625 nm in a confocal scheme with an objective.

To compare confocal schemes with a single lens and an objective, the relative order shift was calculated as d / l, where d is the shift width, l is the spot size (figure 5). In the system with a single lens, the relative shift for order 1 is 0.28 and relative shift for order – 1 is 0.086. In the system with an objective, relative shift 1st order is 0.23 and in -1st order is 0.068. It can be concluded that these optical elements contribute insignificant chromatic aberration compared to AOTF.

4. Scheme with parallel beam

For the parallel beam scheme, simulation was also performed before the experimental verification stage. Figure 6 shows its results, where it is seen that in the diffraction of unpolarized radiation, one of the observed diffraction orders will have a significant chromatic shift compared to the other orders. In the confocal scheme, such a shift can also be observed but its value is less.

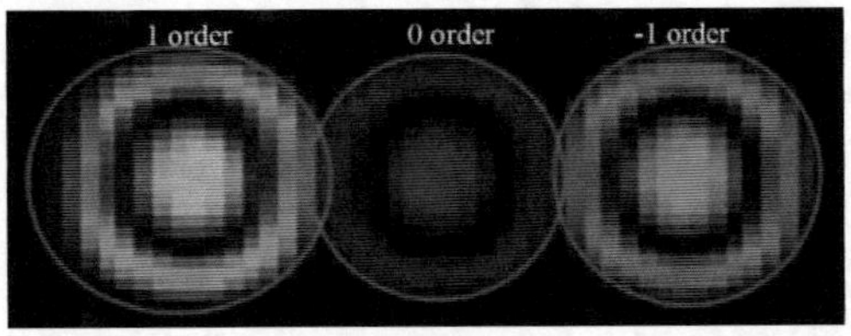

Figure 6. Results of modeling acousto-optic diffraction in a parallel beam path: the pattern on the detector when using a lens based on TIR in the illumination part.

An experimental setup with a parallel beam path is shown in figure 7.

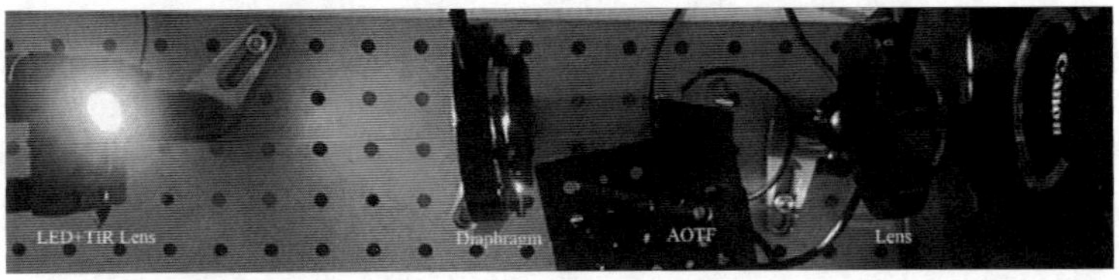

Figure 7. Parallel beam path setup with TIR lens.

In this case, on the detector (figure 8), in addition to chromatic effects, non-uniformity is visible due to the features of the TIR lens.

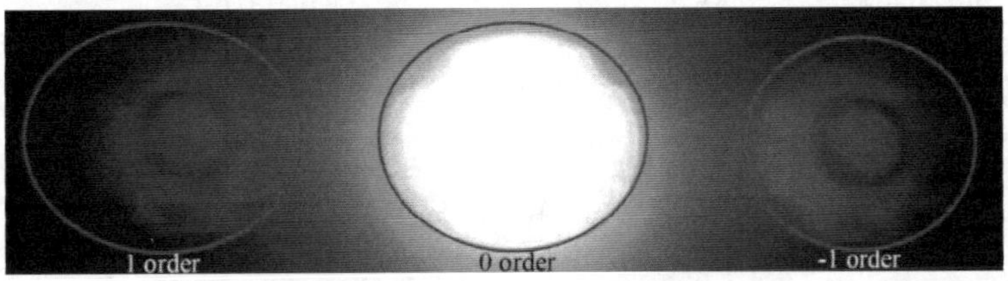

Figure 8. Image of diffraction orders on a detector at 450 and 625 nm wavelengths in the parallel scheme with TIR lens.

Figure 9 shows the relative position of the spots of the diffraction orders for system with objective (continuous line) and TIR lens (dotted line). For comparison of parallel path scheme with lens and objective, the relative order shift was also calculated as d/l. The relative shift 1st order in a system with a lens is 0.26 and -1st order is 0.058. Relative shift 1st order in a system with an objective is 0.18 and -1st order is 0.12. The system with an objective where chromatic aberration of position is corrected has introduced minimum additional shift

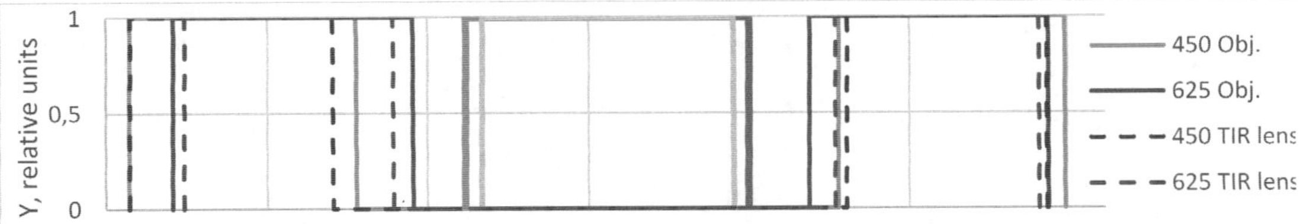

Figure 9. The diagram of the arrangement of orders on the detector at 450 and 625 nm in a parallel beam path for two cases.

As can be seen from the figures, one of the diffraction orders always has higher chromatic shift between the wavelengths both in the scheme with parallel path and confocal.

5. Conclusion

The paper discusses the influence of chromatic aberrations on the operation of a tunable source system with AOTF.

The study has considered two variants: the system with corrected axial color in the illuminating part and a single lens. Calculations and experiments have shown that the confocal scheme and the parallel beam path provide almost the same relative chromatic shift of working orders at different wavelengths. In addition, experiments have shown that the use of a system with corrected axial color in a parallel beam path provides a reducing of the chromatic shift.

Acknowledgments

The work was funded by the Russian Science Foundation (grant No. 21-16-00124, https://rscf.ru/en/project/21-16-00124/).

This work was performed using the equipment of the Shared Research Facilities of the Scientific and Technological Centre of Unique Instrumentation of the Russian Academy of Sciences (STC UI RAS) [http://ckp.ntcup.ru].

References

[1] Berns, R. S. 2005 Color-accurate image archives using spectral imaging *Scientific Examination of Art: Modern Techniques in Conservation and Analysis* **12** 105-119

[2] Zhu Z. et al. 2016 Color calibration forcolorized vision system with digital sensor and LED array illuminator *Act. Passive Electron. Compon.* **2016** 7467165

[3] Endler J. A. and Lainy B. D. 2006 Ornament colour selection, visual contrast and the shape of colour preference functions in great bowerbirds *Chlamydera nuchalis Animal Behaviour* **72(6)** 1405-1416

[4] Machikhin A, Batshev V and Pozhar V 2017 Aberration analysis of AOTF-based spectral imaging systems *J. Opt. Soc. Am. A* **34(7)** 1109-13

[5] Zhao H. et al. 2017 Chromatic aberrations correction for imaging spectrometer based on acousto-optic tunable filter with two transducers *Opt. Express* **25(20)** 23809-23825

[6] Suhre, D. R., Louis J. D. and Neelam G. 2004 Telecentric confocal optics for aberration correction of acousto-optic tunable filters *Applied optics* **43(6)** 1255-1260

[7] Romanova G, Batshev V. and Beliaeva A. 2021 Design of an optical illumination system for a tunable source with acousto-optical filtering *J. of Opt. Technol.* **88(2)** 66-71.

[8] Churilovsky, V.N. 1968 Theory of chromatism and third-order aberrations Mashinostroenie, Moscow (In Russian) 139

[9] Yariv A and Yeh P 1984. Optical waves in crystals. John Wiley Sons, New York 589

[10] Machikhin, A., Pozhar V. 2010 Image aberrations in an acousto-optic tunable filter *Zurnal Technicheskoy phisiki* (Journal of Applied Physics) (In Russian) **80(10)** 101-107

2D scanning system of the acousto-optical deflector with high diffraction efficiency

Y V Pichugina[1,2], S V Garnov[2] and Y N Bulkin[1]

[1]The Russian Federal Nuclear Center – All-Russian Scientific Research Institute of Experimental Physics (FSUE "RFNC-VNIIEF")
37 Mira Ave, Sarov 607188, Russia
[2]Prokhorov General Physics Institute of the Russian Academy of Sciences
38 Vavilov street, Moscow 119991, Russia

E-mail: pichuginaa@yandex.ru

Abstract. The paper considers advantages and special features of an acousto-optic method for the development of a functional device for spatial and temporal control of laser light and reports results of investigations to create a two-directional acousto-optic laser scanner with diffraction efficiency as high as 80 percent or better. An acousto-optic deflector on a paratellurite crystal operating under the Bragg diffraction condition with a lithium niobate piezoelectric acoustic transducer was developed and tested experimentally. The results of this work can serve as a basis for the development of prototype functional acousto-optic devices for spatial and temporal control of laser light and their application to create laser television images.

1. Introduction

At present, development of fast spatial laser beam control devices is a relevant issue of data transmission in modern telecommunication networks [1].

Until recently, the use of electromechanical devices was the most common way of laser beam deflection. The drawbacks of the electromechanical devices are associated with their system inertness, which limits their speed (4 μs for a mirror up to 10 mm in diameter). Electro-optic devices are fast, with a response time of the order of 10 ns, but their angular scanning range is limited to units of angular minutes, while their power consumption is very high. Non-mechanical inertia-free scanning based on the acousto-optic effect offer a more promising way. Acousto-optic deflectors (AODs) cover a range of scanning angles as wide as several degrees, their response time ranges from units to dozens of microseconds, so they are slower than electro-optical devices, but faster than mechanical systems [2,3].

A number of applications involving high-energy lasers require a maximum increase in the optical transmission coefficient (OTC) in the entire angular range of operation. At present, a number of Russian and foreign companies offer a variety of AOD models, but their diffraction efficiency in the angular range of scanning between 1.5° and 3° often does not exceed 35 to 70 percent [4,5]. The highest optical transmission coefficient in the AADTSX-250 two-directional acousto-optic device made by QBICLASER is 70 percent at a scanning angle of 2.7°×2.7°. In the EOPC DTSXY-400 deflector having a scanning angle of 2.8°×2.8°, diffraction efficiency does not exceed 50 percent at a wavelength of 532 nm. In a two-directional system with scanning angles above 3°, the wave vector of radiation leaves the diffraction plane in the second cell, which results in a sharp drop in diffraction efficiency on edges.

Considering this, the objective of this study was to develop, build and test a prototype two-directional acousto-optic system on a paratellurite crystal with a high OTC (at least 80 percent).

2. Acousto-optic interactions in the TeO_2 crystal

To build a two-directional AOD, its structure was calculated to ensure its operation under the acousto-optic diffraction condition [6].

Of interest is the dependence of diffraction efficiency (f) on sound frequency at the exit from the region of acousto-optic interaction, which is determined by the length of the sound wave along the path of light propagation. Efficiency is understood to be the ratio of Bragg order intensity to incident light intensity:

$$\eta(f, L) = \frac{I_1}{I_0}. \quad (1)$$

The solution to the system above is given by

$$I_1 = I_0 \frac{k}{(\Delta K(f) \cdot L)^2 + k} \sin^2 \frac{\sqrt{(\Delta K(f) \cdot L)^2 + k}}{2}, \quad (2)$$

where $k = \frac{2P_{\text{ак}} M_2 \pi^2 L}{\lambda_0^2 \cdot \cos^2\theta \cdot H}$, L is the length of acousto-optic interaction along the Z axis, H is the light beam diameter, λ_0 is the laser light wavelength, and M_2 is the acousto-optic quality factor. The acoustic power P_{ac} is an adjustable parameter controlled by the high-frequency electric power fed to the piezoelectric transducer. Figure 1 shows calculated plots of η(f) for a one-directional deflector for λ_0=0.532 μm with a piezoelectric transducer of L=1.5 mm.

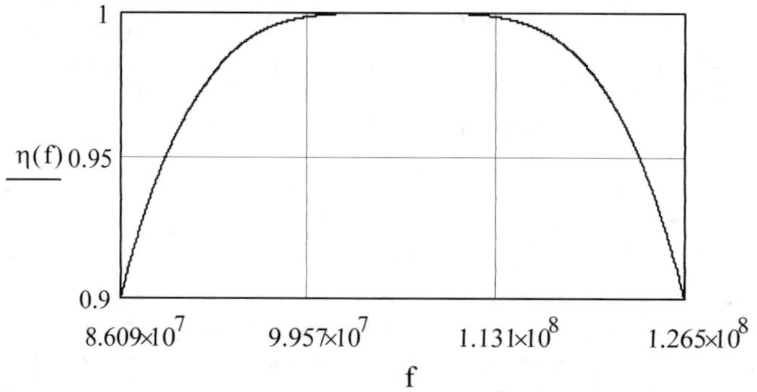

Figure 1. Calculated plot of diffraction efficiency η(f) as a function of sound wave frequency [Hz] of a one-directional deflector

For this model, the calculated scanning angles Δβ(f) for η(f) ≥ 0.9 in the traditional acousto-optic interaction geometry are equal to 1.85°.

To expand the angular range of scanning, in anisotropic deflectors one uses the interaction geometry shown in Figure 2 [7]. Once the deflector is aligned, the angle of incident light θ_0 relative to the Z axis decreases, and the triangle formed by the wave vectors of incident light $\mathbf{k}_0$, sound $\mathbf{q}$(f) and diffracted light $\mathbf{k}_d$(f) gets closed at two frequencies: $\mathbf{q}_1(f_1)$ and $\mathbf{q}_2(f_2)$. The angle of incidence θ_0 is adjusted by precision alignment of the deflector.

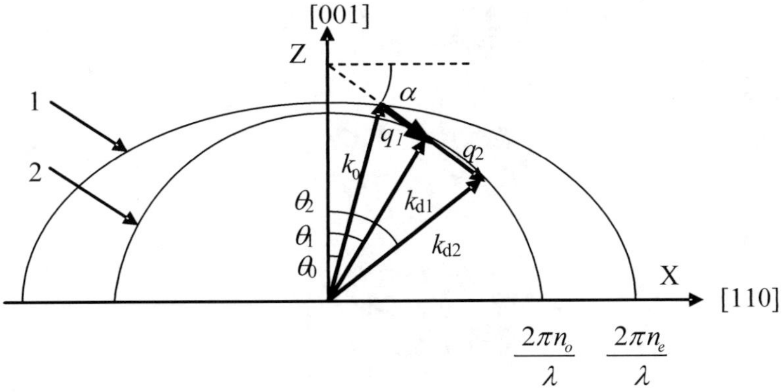

Figure 2. Vector diagram of the one-directional acousto-optic deflector in the expanded angular range of scanning

Once the angle θ_0 is adjusted, the function of $\eta(f)$ is varied: the synchronism condition is satisfied at two frequencies, f_1 and f_2 (Figure 3). As a result, the angular range of scanning increases to 2.62°.

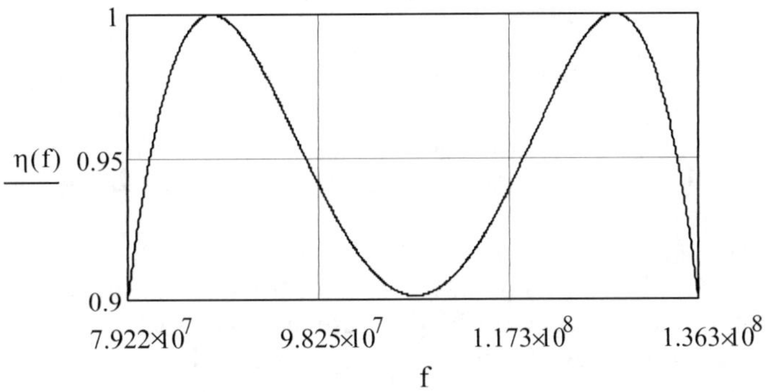

Figure 3. Diffraction efficiency $\eta(f)$ as a function of the sound wave frequency with the adjusted angle θ_0

The resulting functions of diffraction efficiencies $\eta(f)$ are hold true for the one-directional anisotropic deflector, in which diffraction occurs in the $(1\,\bar{1}\,0)$ crystallographic plane [8]. The two-directional deflector consists of two orthogonal acousto-optic cells. In the first cell directed along the path of light propagation, diffraction occurs in this plane. However, as a result of scanning, the beam generally leaves its $(1\,\bar{1}\,0)$ crystallographic plane once it enters the second cell. Therefore, diffraction efficiency in the second cell is a function of not only the sound wave frequency, but also of the scanning angle $\theta(f_Y)$ in the first cell:

$$\eta_{YX}(f_Y, f_X) = \eta_Y(f_Y) \times \eta_X(f_X, \theta(f_Y)). \quad (3)$$

To expand the angular range of scanning and improve diffraction efficiency, we developed a modified two-directional deflector design based on anisotropic diffraction in a single crystal of paratellurite.

3. Results of experiment

The composition and the flow diagram of the acousto-optic system (AOS) are shown in Figure 4. The AOS is controlled by a special program from a PC via a USB port.

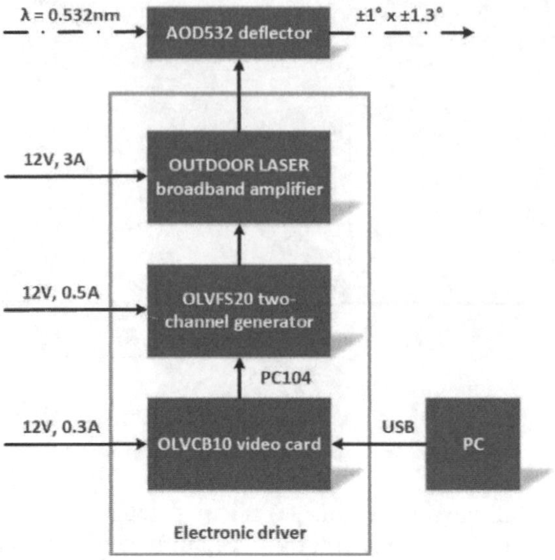

Figure 4. Flow diagram of the AOS

The plots of diffraction efficiency for two-directional scanning in the angular ranges of 2°×2° and 2°×2.6° are shown in Figures 5 and 6.

For the scanning angle of 2°×2°, the frequency range must be 45 MHz for diffraction efficiency to be not less than 95 percent without considering optical losses. To increase the angle of scanning along the X axis to 2.6°, the range of frequencies in cell 2 must be equal to 42 MHz.

The program measures the level of residual optical power in the [00] order in each angular position and adjusts the value of the output power of the driver in the Y coordinate to provide the highest possible attenuation of this order. This creates the driver's response function, at which the highest level of diffraction efficiency $\eta_Y(f_Y)$ is achieved in the entire range of angular scanning.

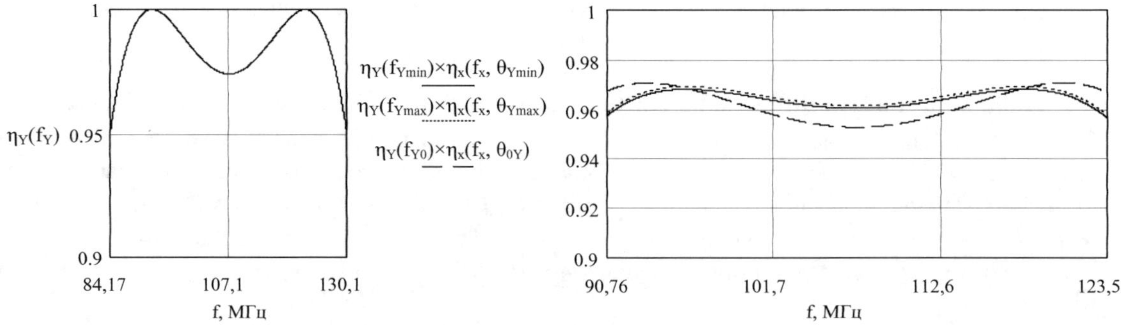

Figure 5. Plots of diffraction efficiency in the two-directional anisotropic deflector with the angular range of scanning of 2°×2°

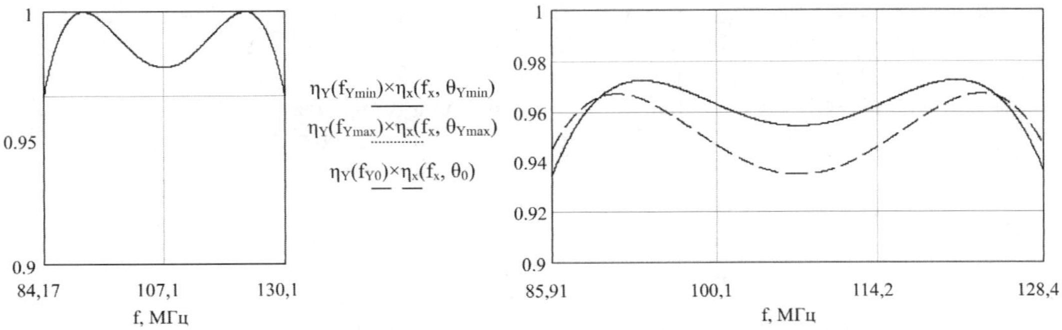

Figure 6. Plots of diffraction efficiency in the two-directional anisotropic deflector with the angular range of scanning of 2°×2.6°

The plot of $\eta_Y(f_Y)$ as a function of the sound wave frequency f_Y for the program-adjusted response function of the driver is shown in Figure 7.

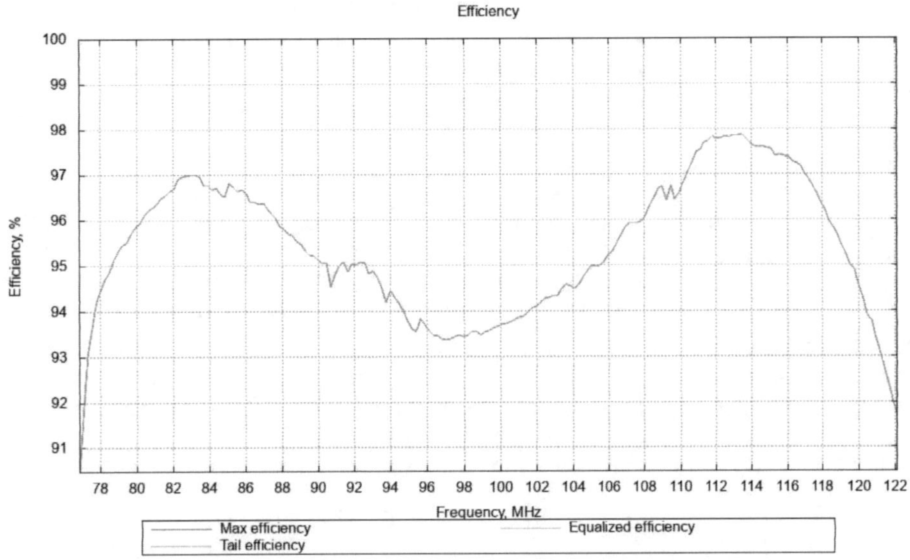

Figure 7. Sound wave frequency as a function of diffraction efficiency $\eta_Y(f_Y)$ in cell 2

Stable diffraction efficiency of the AOD with a scanning angle of 2°×2.6° in one coordinate reached 90 percent or higher (considering optical losses).

The two-directional scanning system consists of two identical cells from TeO_2 crystals providing an optical transmission coefficient of 80 percent or higher in both coordinates with up to 512×512 resolvable spots at a given central frequency of the transducer:

- (AOC 1), vertical scanning in the Y coordinate;
- (AOC 2), vertical scanning in the X coordinate.

Each cell operates under the anisotropic Bragg diffraction condition. The angle of light incidence in the crystal is $\theta_0=5.1°$, and the range of acousto-optic interaction is L=1.5 mm. A slow acoustic S-wave in the crystal propagates in the (001) plane at an angle of $\alpha=7.2°$ to the [110] direction. The X-scanning cell is tilted to 45° relative to the transmitted light. This is explained by the choice for AOC 2 of a special-geometry crystal to increase the angular range of scanning and diffraction efficiency.

The shape of the paratellurite crystal in AOC 1 corresponds to conventional acousto-optic interaction. This is because the incident and diffracted beams are nearly aligned. The X offset in the cell is ~ 8.5 mm, which is admissible, and in the model, an increased angle of scanning is achieved. The incident light is polarized vertically along the Y axis. After the light is diffracted in AOC 1, its polarization turns to 90^0 to the plane of diffraction in AOC 2. At the exit of the deflector, the light recovers its polarization along the Y axis.

4. Conclusion

In the tests of the system, a two-directional scanning pattern of 512×512 spots was produced. The measured optical transmission coefficient at scanning angles of 2°×2° and 2°×2.6° in the central frequency range reached 84 percent and 85 percent, respectively, which validates the predictions and is 1.6 times that of equivalent devices.

References

[1] Machikhin A.S., Pichugina Y.V., Kozlov A.B., Batshev V.I., Pozhar V.E. and Nosov P.A. Acousto-Optical Deflector for Non-Mechanical Manipulating Using Optical Tweezers // *4th International Conference on Metamaterials and Nanophotonics, METANANO 2019, Journal of Physics: Conference Series*, 2020, 1461, 1, 012087

[2] Lei X. Shangian Z., Xiaoli Z. Electro-optical Analog-to-Digital Converter based on LinbO3 Mach-Zehnder Modulators // *The 9th International Conference on Optical Communication and Networks* – Nanjing. — 2010. —pp. 343-346.

[3] Warner A.W., White D.L., Bonner W.A. Acousto-optic light deflectors using optical activity in paratellurite // *Appl. Phys. Letters*. 1972. Vol. 43. No. 11

[4] Peled I., Kaminsky R., Kolter Z. Acousto-optics bandwidth broadening in a Bragg cell based on arbitrary synthesized signal methods // *J. Applied Optics* –vol.54#6. — 2015 — P. 5065-5073.

[5] Gulyaev Yu.V., Kazaryan M.A., Mokrushin Yu.M., Shakin O.V. Acousto-optic laser systems for creating television images. — *M.: FIZMATLIT*. — 2016. — 240 p.

[6] Magdich L.N., Molchanov V.Ya. Acousto-optic devices and their applications, Moscow "Sovetskoe radio 1978.

[7] Guk A.S., Gulyaev Yu. V., Evstigneev V.L., Kazaryan M.A., Mokrushin Yu.M., Talalaev M.A., Shakin O.V. Temperature effects in acousto-optic deflectors on paratellurite. // *Russian Academy of Sciences*. – 2017 ISBN 978–5–906906–41–0.

[8] Binh L.N. Photonic Signal Processing: Techniques and Applications – *CRC Press*. — 2007. — P. 376.

Journal of Physics: Conference Series **2091** (2021) 012014 doi:10.1088/1742-6596/2091/1/012014

Diffuse reflectance photometric system for noninvasive blood glucose control

K V Pozhar, M O Mikhailov, E A Polyakova and E L Litinskaia

National Research University of Electronic Technology, 1 Shokin square, Zelenograd, Moscow 124498, Russia

E-mail: kir-p@rambler.ru

Abstract. The paper considers the method of diffuse reflectance spectroscopy for noninvasive glucose level measuring in biological tissues. It is proposed a portable implementation of the system, based on a laser diode with a wavelength of 1600 nm and a power of 30 mW, operating in a pulsed mode, as well as a set of NIR-photodiodes located around the laser. The measuring system has been tested on model solutions with diffuse reflectance from polystyrene and pigskin. In the course of measurements on fixed system, an error of ~17 mg / dl is observed, which can be reduced by improving miniature laser sources.

1. Introduction

The problem of noninvasive blood glucose measurement has been one of the most urgent in the field of diabetic biomedical engineering for several decades [1]. Patients with type 1 diabetes require continuous monitoring of glucose levels in order to make timely decisions on adjusting the therapy regimen. Noninvasive monitors are the most promising feedback sensors for automated and automatic blood glucose control systems.

None of known noninvasive meters is widely used in practice, which leaves relevant the task of researching of methods and developing of devices for noninvasive blood glucose measurement. Spectroscopic methods are the most studied and currently provide the highest accuracy [2]. At the same time, only methods based on the registration of reflected radiation can potentially achieve the necessary ergonomic characteristics. One of the most promising methods is diffuse scattering spectroscopy, based on the analysis of the dynamics of the intensity of radiation reflected back in the layers of tissue (Fig.1).

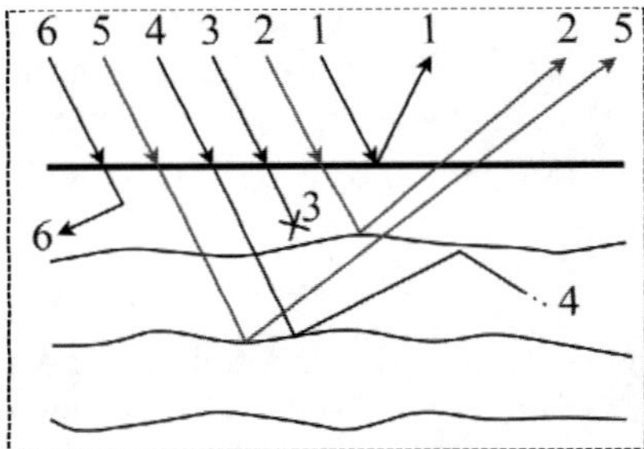

Figure 1. Scheme of radiation propagation in biological tissue. The incident radiation undergoes reflection from the surface (1), diffuse scattering at the layer boundaries (2, 5), absorption (3), and tissue scattering (4, 6).

Numerous studies in this area show that the method is accurate; however, in a number of works as [3], bulky systems with optical elements and fibers are implemented. In the most recent works as [4], a relatively high accuracy of the method was shown for a portable implementation, providing an error of the order of 20 mg / dL under laboratory conditions.

The aim of this study is to investigate the possibility of a portable implementation of a measuring system based on diffuse reflectance spectroscopy.

2. Materials and methods

The work proposes a measuring system based on diffuse reflection photometry. A miniature NIR-laser diode with a power of 30 mW, a wavelength of 1600 nm and a spectrum width of 5 nm is used as a radiation source. The high penetrating ability of radiation at a given wavelength with a high absorption rate of glucose makes it possible to take into account the contribution of deeper tissue layers with a higher glucose content. One of the important advantages is also the predominance of absorption over scattering.

A set of miniature NIR-photodiodes located coplanar with the working body of the laser is used as radiation detectors. The power supply is realized autonomously, and data exchange is carried out wirelessly, which excludes the influence of electrical noise on measurements.

The scheme of the meter is shown in Figure 2. The meter is pressed close to the biological tissue by the optical window, somewhat aligning the layers and causing occlusion. The tissue is irradiated with a laser; part of the radiation is diffusely reflected and recorded by the detector, while the radiation intensity will decrease with an increase of the tissue glucose concentration.

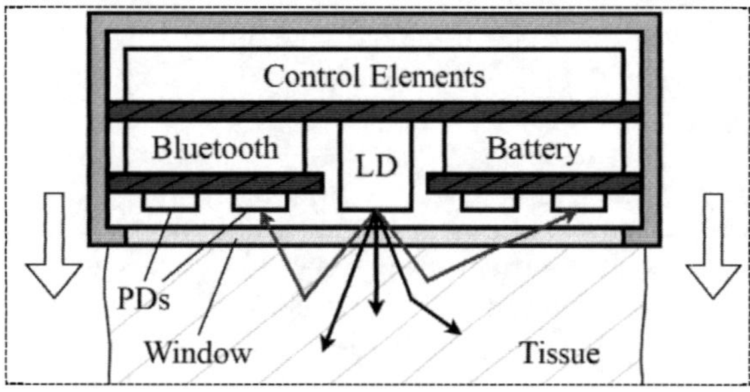

Figure 2. The scheme of a portable implementation of a glucose meter. The device is pressed against biological tissue; the reflected radiation is registered. LD is a laser diode, PD is a photodiode.

The glucose concentration is calculated numerically using the integral equation for the recorded flux F on the photodetector, which is considered in more detail in [5]:

$$F = \sum_{n=1}^{N} \frac{\mathrm{H}_n}{\pi} \iint_{\ddot{\mathrm{y}}2} I_0(x,y) \exp\left(-\left(\sum_{k=1}^{n} \mu_k\left(c_g\right) \frac{z_k}{z_n} \right) l_{n\downarrow} \right) \iint_D \frac{1}{l_{n\uparrow}^{4}} \exp\left(-\left(\sum_{k=1}^{n} \mu_k\left(c_g\right) \frac{h_k}{z_n} \right) l_{n\uparrow} \right) dxdydx_n dy_n$$

where N is the number of fabric layers taken into account, H_n is the coefficient determined by the albedo of n layers of tissue, $I_0(x, y)$ is the distribution of radiation intensity along the profile of the initial beam, c_g is the tissue glucose concentration, μ_k is the extinction coefficient of the k-th tissue layer, h_k is the k-th tissue layer thickness, z_n is the depth of the n-th layer interface, $l_{n\downarrow}$ is the optical path of the beam into the tissue, $l_{n\uparrow}$ is theoptical path of the beam from the point of reflection to the detector, D is the contour of a rectangular photosensitive area.

3. Experiments

Experimental study of the accuracy of the glucose concentration measuring were carried out under several conditions. The meter was installed close to the measured object in a fixed optical system. Various glucose concentrations ranging from 0 to 2 g / dL were formed in the cuvette as an aqueous solution. In another experiment, the model solution was pumped with a membrane pump through a quartz tube. Polystyrene plates and pigskin were used as a diffusely reflecting surface.

During the experiment, the laser operated in a pulsed mode with a period of 60 s and a porosity of 6, the readings of the photodiodes were recorded at a frequency of 40 Hz. The resulting data sets were twice processed with a moving average filter. The data for 5 pulses were averaged and the glucose concentration in the medium was calculated. The system was pre-calibrated.

The experiment showed that vibrations do not affect the measurement accuracy and they can be completely neutralized by digital filters. The dynamics of all photodiodes readings was the same, which makes it possible to use the changes in it to estimate the system displacements. The error of the glucose concentration measuring in model media was 16.7 mg / dl.

4. Discussion

Despite the promise of spectroscopic methods for measuring glucose in biological tissues, their accuracy is still not sufficient to create a completely noninvasive glucometer. This is due to the large number of factors affecting measurements.

The identification of the mathematical model by direct measurement of the object parameters is impossible. As a result, it is necessary to train the system by long-term comparison of the meter readings with the data of a precision invasive glucometer. During ordinary usage, the system requires periodic calibration to operate more accurately, just like invasive glucose monitors.

Despite the fact that the structure of the skin layers can be considered unchanged, the hydration of the tissues can change during the day. And since water is the main attenuator of radiation, the additional control is advised, for example, by using an impedance meter, which in turn will not lead to a significant increase in the size of the system.

5. Conclusion

Diffuse scattering spectroscopy is one of the most promising methods for noninvasive blood glucose measurement. The system presented in this work shows the possibility of a portable ergonomic implementation of a glucose monitor in biological media by measuring diffusely reflected radiation. At the same time, one of the key problems in the implementation of the system is the measurement accuracy, which is not sufficient for using the system in its current form on patients due to the poor optical characteristics of miniature lasers. To solve this problem, it is necessary to develop highly stable laser diodes without the use of bulky optical circuits.

Acknowledgments

The reported study is supported by the Ministry of Science and Higher Education of the Russian Federation, project number 075-15-2020-440.

References

[1] Fabris C and Kovatchev B P, 2020 *Artificial Organs.* **44** 671-679

[2] Lin T, Gal A, Mayzel Y, Horman K, Bahartan K, 2017 *Curr Trends Biomedical Eng & Biosc.* **6.** 555696

[3] Maruo K, Tsurugi M, Tamura M, Ozaki Y, 2003 *Applied Spectroscopy* **57** 1236-1244

[4] Yadav J, Rani A, Singh V, Murari B M, 2018 *Proceedings of Networking Communication and Data Knowledge Conference* 197-216

[5] Pozhar K V, Mikhailov M O, Polyakova E A, Litinskaia E L, Abolemova Yu E, 2020 *Meditsinksaya technika* № **2** 5-7

Methodology for identifying gaseous constituents of the atmosphere

A E Krasnov[1], D Yu Ryzhkova[2] and V A Vagin[2]

[1]Russian State Social University, 4 Wilhelma Pica street, Moscow 129226, Russia
[2]Scientific and Technological Center for Unique Instrumentation of Russian Academy of Sciences, 15 Butlerova street, Moscow 117342, Russia

E-mail: darijez@mail.ru

Abstract. Methodology for the analysis of concentrations of gaseous composite of atmosphere by the corresponding infrared (IR) spectrum, measured with the help of trajectory spectroradiometers (TSR) is observed. The developed algorithm for mathematical processing of the measurement results is briefly described, including the detection and estimation of the concentrations of the sought gases using the notch filtration of their spectral components, which makes it possible to significantly reduce the concentration identification error. The spectra of various substances in the mid-IR range are considered, and the results of approbation of the technique based on the TSR model with an external high-temperature radiation source on a 1 m path are presented.
Keywords: atmospheric monitoring, path mode, spectroradiometer, infrared spectrum, notch filtering, detection, concentration calculation.

1. Introduction

Atmospheric air is a vital component of the environment. Outdoor air pollution is a major environmental health problem affecting everyone in low-, middle-, and high-income countries. To maintain a clean atmosphere, effective control of air pollution is necessary. Typical chemical pollutions include emissions of gases and aerosols into the atmosphere such as carbon monoxide, sulfur dioxide, hydrocarbons, aldehydes, ammonia, etc.

Spectral and chromatographic methods are currently used to analyze polluted air. The detection and measurement of gas concentrations using the characteristic optical absorption of the gas species is important for both understanding and monitoring a variety of phenomena from industrial processes to environmental change [1]. The major advantage of infrared (IR) spectroscopy over other techniques is that all compounds show IR absorption and can thus be analyzed. Besides, IR is a non-destructive, noncontact measurement technique, offering the possibilities to remote detection. From this point of view, IR spectroscopy is the most versatile of all analytical techniques, since a spectrum of almost any sample can be recorded without elaborate preparations within fractions of a second [2].

The measurement principle of Fourier transform infrared (FTIR) spectroscopy is to use the absorption phenomena of specific frequency spectrum of infrared by some specific gases in the air due to molecular vibration. Because every gas has its own specific absorption spectrum, it would be possible to figure out the gas by spectrum analysis. The strength of specific absorption spectrum is proportional to the concentration in the air so that it is also possible to do quantification by this characteristic. The wide spectral range coverage of FTIR (1.5-13.5 μm) allows the measurement of a large range of chemical species simultaneously. On the basis of FTIR spectrometers, including spectroradiometers, a number of laboratory and field automated chemical analysis systems have been created.

The purpose of this work is to outline the fundamentals of the methodology for estimating the concentrations of the gas components of a mixture from its IR spectrum using the method of notch filtration of the spectral components of the sought gases.

2. Laboratory spectral analysis

The spectra of vapors of chemically hazardous and toxic substances can be obtained on a standard laboratory FTIR equipped with gas cells, or in a multi-probe Fourier spectrometer [3]. In the latter case, the measuring head of one of the probes is an optical system for measuring transmission spectra of gas mixtures - an "open" gas cell, which is introduced directly into the objects under study, for example, at key points of the technological process (figure 1).

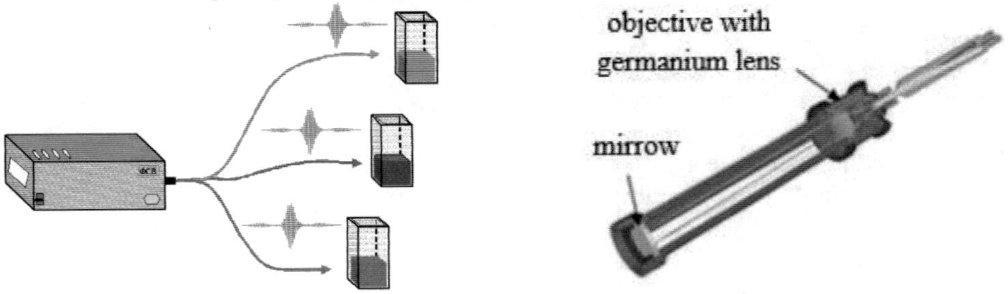

(a) (b)

Figure 1. Multi-probe FTIR spectrometer (a) and measuring gas head of the fiber-optic probe (b).

3. Route chemical monitoring method

Route chemical monitoring is a gas analysis method using FTIR spectroradiometers on fixed monitoring routes using an external source of IR radiation, which does not require sampling [4].

The transmittance spectra of the atmosphere are also measured by means of TSR, which are implemented in two ways (figure 2). The illuminator can be a heated blackbody or a corner reflector system. In the latter case, the radiation source is located in the spectrometer [5].

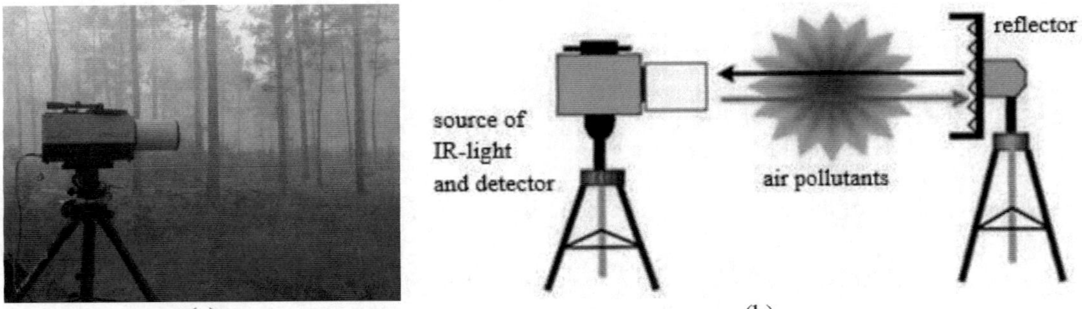

(a) (b)

Figure 2. Bruker OPAG-22 spectrometer system in the field during the fire at Fort Jackson near Columbia, South Carolina (SC), US [6] and TSR principle scheme.

In the present work, we consider the model of a TSR with an external high-temperature radiation source in a generalized form, which allows one to neglect the intrinsic radiation of the device and the radiation of the atmosphere. With this approach, calculations do not require additional mathematical processing, as in the case of passive detection [4].

4. Identification of concentrations of detected gas constituents

The ultimate goal of multicomponent gas analysis is identification of all chemical components present and determination of their concentrations.

Actually, identification of atmospheric gaseous pollutants from measured IR spectra is performed by using the independent component analysis (ICA). ICA is a computational method, which can restore the original source signal from a set of observed mixed signal; it aims at finding a separation matrix that can transform the observed signal into a linear combination of mutually independent components, which are close to the source signal [7].

For quantitative evaluation, the preferred intensity scale is the absorbance scale, in which the band intensities are, in principle, directly proportional to the amount of absorbing material in the optical path, as expressed by the Lambert–Beer law:

$$A(v)=\alpha(v)cl\ , \tag{1}$$

where α is the molar absorption coefficient of the analyte at the given wavenumber, c is its concentration, l is the absorbing path-length.

The absorbance of multicomponent spectrum is the sum of absorbance of each component, and the change of concentration of the component will only cause the change of absorbance amplitude but not the position of absorption peaks and the spectral profile. From the above analysis, the absorption intensity of atmospheric spectra conforms to the linear model, and the ICA algorithm can be used to separate the spectra of mixed gases.

This method is a powerful signal processing tool for laboratory studies of atmospheric impurities. However, it does not take into account environmental conditions, signal interference, etc. Using ICA under actual conditions increases the probability of error in gas identification.

In the present work, the cross section is used instead of the absorption coefficient. Cross section is a measure of the probability that a specific process will take place when some kind of radiant excitation intersects a localized phenomenon. In a gas of finite-sized particles there are collisions among particles that depend on their cross-sectional size. The average distance that a particle travels between collisions depends on the density of gas particles. These quantities are related by

$$\sigma=\frac{1}{n\lambda}, \tag{2}$$

where σ is the cross section of a two-particle collision, λ is the mean free path between collisions, n is the number density of the target particles.

The measured signal of the trace spectroradiometer is described by the expression:

$$I_\Sigma(v)=I_0(v)\exp\left[-\sum_{k=1}^{K}\sigma_k(v)x_kL\right]+H(v), \tag{3}$$

where $I_0(v)$ is the spectral distribution of the emitter intensity (in frequency ν, cm^{-1}), $\sigma_k(v)$ is the known absorption cross section of the kth gas ($k = 1, 2, \ldots, K$), x_k is the desired concentration; L is the length of the track; $H(v)$ - interference signal.

Equation (3) is reduced to the form:

$$\sum_{k=1}^{K}\sigma_k(v)x_kL+\ln\left[\frac{I_\Sigma(v)}{I_0(v)}\right]=\frac{H(v)}{I_\Sigma(v)}=H^*(v). \tag{4}$$

Having squared the right and left sides of equation (4) and multiplied by the notch filter $F_m(v)$, which takes the values "1" or "0", we get:

$$F_m(v)\left\{\sum_{k=1}^{K}\sigma_k(v)x_k L+\ln\left[\frac{I_\Sigma(v)}{I_0(v)}\right]\right\}^2=F_m(v)\left[H^*(v)\right]^2 \quad (5)$$

where $m = 1, 2, ..., K$.

We introduce integral estimates of the error energies $E_m=\int_{v_{\min}}^{v_{\max}}F_m(v)\left[H^*(v)\right]^2 dv$ and transform equation (5). It is possible to estimate the concentration x_m by calculating $\partial E_m/\partial x_k=0$ for $m = 1, 2, ..., K$. As a result, we obtain a system of equations for the unknowns x_k:

$$\sum_{k=1}^{K}\left[\int_{v_{\min}}^{v_{\max}}F_m(v)\sigma_m(v)\sigma_k(v)dv\right]x_k=-\frac{1}{L}\int_{v_{\min}}^{v_{\max}}F_m(v)\ln\left[\frac{I_\Sigma(v)}{I_0(v)}\right]\sigma_m(v), m=1,2,\ldots,K \quad (6)$$

From equation (6) it follows that for the correct conditionality of the matrix inverted when solving the system of equations, it is necessary to select notch filters $F_m(v)$ so that the off-diagonal matrix elements $\int_{v_{\min}}^{v_{\max}}F_m(v)\sigma_m(v)\sigma_k(v)dv$ were minimal for $m\neq k$. For this, we have developed a special procedure for the synthesis of notch filters, which separates the characteristic absorption lines of each gas: $F_m(v)=1$, if the optical density in the separated bands of the m-th gas exceeds this value for all other gases.

As an example, figure 3 shows the optical densities and the corresponding notch filters for various gases.

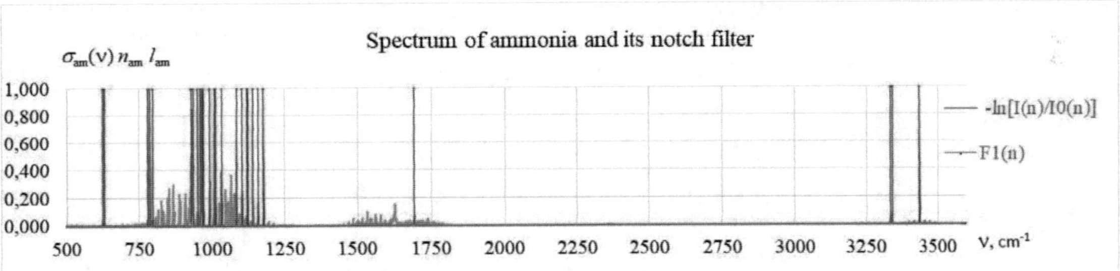

(a)

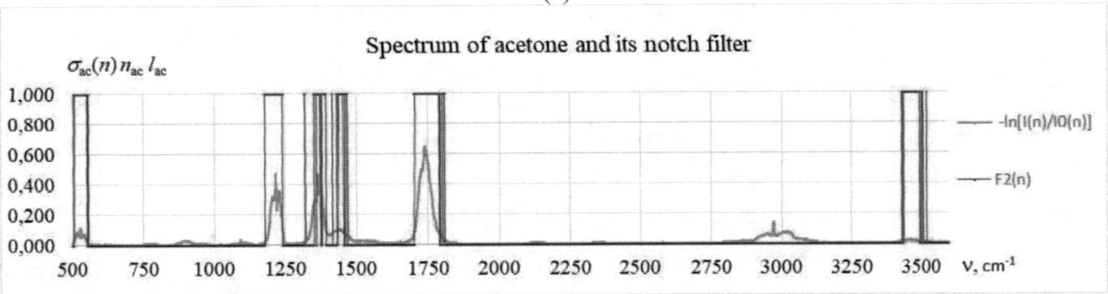

(b)

Figure 3. Optical densities of gases: ammonia (a) and acetone (b).

For the adaptive selection of the reference gases included in equation (4), a procedure for their preliminary detection is proposed, based on the solution of this system for the maximum number of reference gases. Thus, in an uncontrolled experiment, 14 reference gases were used to analyze the spectrum of the prepared test mixture (ammonia, acetone, ethanol), shown in figure 4. The spectral absorption lines of water vapor were preliminarily cut out from the initial spectrum (figure 5).

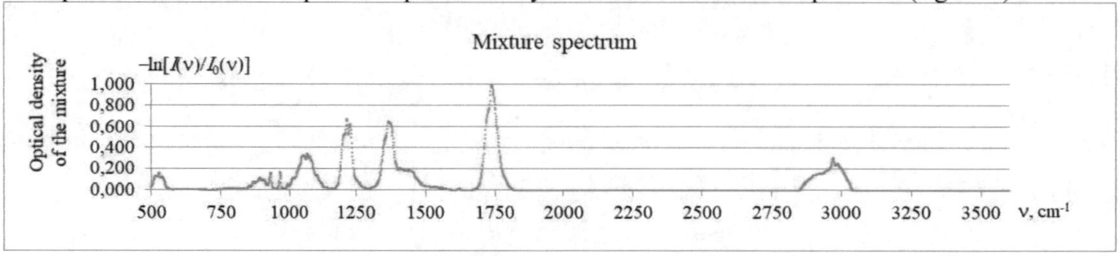

Figure 4. Optical density of the test gas mixture.

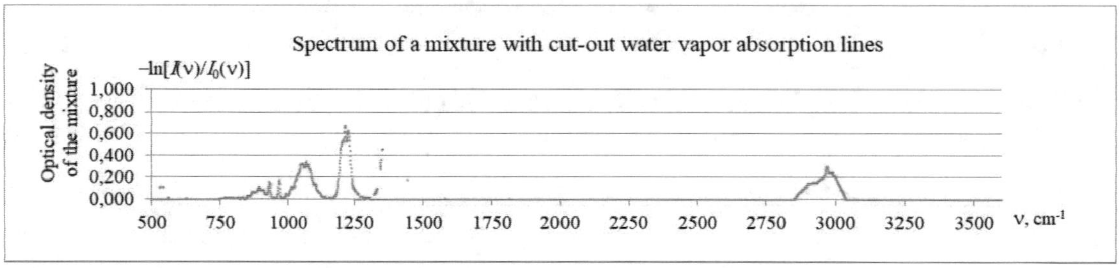

Figure 5. Optical density of the test gas mixture with cut-out spectral absorption lines of water vapor.

Based on this spectrum, the probabilities of detecting the components of the reference gases at the threshold level of 0.05% were calculated with a relative interference ($\Delta I/I$) of the TSR receiver not exceeding 1% (table 1). Further, system (4) was solved on the basis of the absorption cross sections of the detected gases with an accuracy of 0.005 g/m^3 (table 2). As can be seen, along with the test gases introduced into the atmosphere, traces of methane, three fluorides of nitrogen, carbon dioxide and ethylene oxide in low concentrations were found.

Table 1. Detected gases (threshold 0.05%).

	Gases	Probabilities (%)
1	Ammonia	2,95
2	Acetone	2,54
3	Nitrogen dioxide	0,00
4	Sulphur dioxide	0,00
5	Methane	0,14
6	Nitrogen monoxide	0,00
7	Propylene oxide	0,00
8	Hydrogen sulfide	0,00
9	Toluene	0,00
10	Three nitrogen fluoride	0,16
11	Carbon monoxide	0,00
12	Carbon dioxide	0,04
13	Ethanol	93,81
14	Ethylene oxide	0,35

Table 2. Detected gas concentrations.

	Gases	Concentration, (g/m^3)
1	Ammonia	0.382 ± 0,005
2	Acetone	0.328 ± 0,005
5	Methane	0.018 ± 0,005
10	Three nitrogen fluoride	0.020 ± 0,005
12	Carbon dioxide	0.006 ± 0,005
13	Ethanol	12.131 ± 0,005
14	Ethylene oxide	0.046 ± 0,005

Using a numerical experiment, the dependence (figure 6) of the number of false alarms at gas detection on the relative interference $\Delta I / I$ of the TSR receiver was obtained. False alarms are defined as the detection of concentrations of non-existent substances due to interference in the measured signal of the spectrometer. The number of false alarms at $\Delta I / I < 5$ % does not exceed 1 in 100 tests.

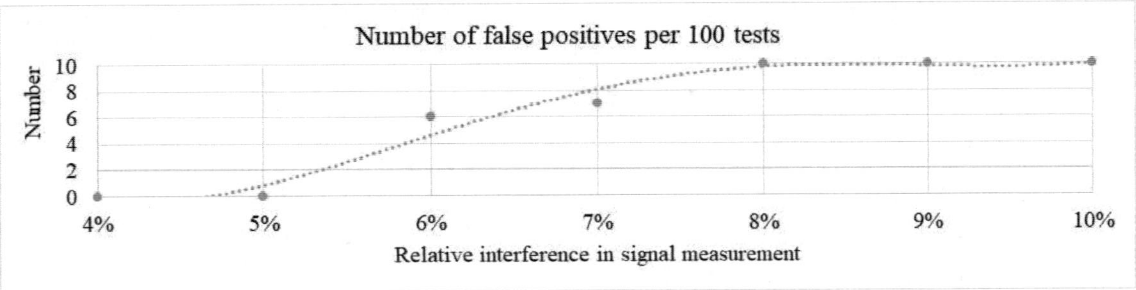

Figure 6. Dependence of the number of false alarms (per 100 tests) on the relative interference of the TSR receiver (when gas is detected for a route with a length of 1 m).

5. Conclusions

A technique for analyzing the concentration of gaseous components of the atmosphere by their IR spectra, obtained on TSR, is considered. An algorithm for mathematical processing of the measurement results, including the detection and calculation of the concentrations of the desired gases using notch filtration, is presented. The results of approbation of the technique based on the TSR model with a high-temperature radiation source on a 1-m-long path for various substances in the mid-IR range are presented. Notch filtering improves the accuracy of concentration estimation. It has been experimentally shown that the number of false alarms with a relative signal measurement interference with a TSR receiver less than 5% does not exceed 1 in 100 tests on a 1 m track.

Acknowledgments

This work was performed using the equipment of the Centre for Collective Use of the Scientific and Technological Centre of Unique Instrumentation of the Russian Academy of Sciences (STC UI RAS) [http://ckp.ntcup.ru].

References

[1] Hodgkinson J, Tatam R P 2013 Optical gas sensing: a review *J. Meas. Sci. Technol.* **24** p 012004
[2] Bacsik Z, Mink J, Keresztury G 2004 FTIR spectroscopy of the atmosphere. I. Principles and methods *J. Applied Spectroscopy Reviews* **39** 3 pp 295-363
[3] Balashov A A, Vaguine V A, Golyak I S, Morozov A N, Khorokhorin A I 2017 Multichannel dynamic IR Fourier-spectrometer *J. Zhurnal Prikladnoii Spektroskopii* **84** 4 pp 643-647
[4] Morozov A N, Fufurin I L 2013 Metodika analiza spektrov, poluchaemyh na infrakrasnom

Fur'e-spektrometre s primeneniem aktivnoj podsvetki (In Russian) *J. Inzhenernyj zhurnal: nauka i innovacii* **8** pp 1-10

[5] Vaguine V A, Khorokhorin A I 2018 Spectral data acquisition and handling system with control function for multy-probe FTIR spectrometer *J. Physical Bases of Instrumentation* **7** 3(29) pp 8–15

[6] Akagi S K, Burling I R, Mendoza A, Johnson T J, Cameron M, Griffith D W, Paton-Walsh C, Weise D R, Reardon J, Yokelson R J 2014 Field measurements of trace gases emitted by prescribed fires in southeastern US pine forests using an open-path FTIR system *J. Atmospheric Chemistry and Physics* **14** 1 pp 199-215

[7] Ju W, Lu C, Liu C, Jiang W, Zhang Y, Hong F 2020 Rapid identification of atmospheric gaseous pollutants using Fourier-transform infrared spectroscopy combined with independent component analysis *Journal of Spectroscopy* **2020** pp 1-14

Optimization of operating modes of the weak absorption spectrometer

P V Korolenko[1,2] and O M Vokhnik[1]

[1]Faculty of Physics M. V. Lomonosov Moscow State University
1-2 Leninskie Gory, Moscow 119991, Russia
[2]P. N. Lebedev Physical Institute of the Russian Academy of Sciences
53 Leninskiy Prospekt, Moscow, 119991, Russia

E-mail: vokhnik@rambler.ru

Abstract. The possibilities of improving the characteristics of a weak absorption spectrometer consisting of a frequency-tunable laser and an external analytical resonator with the test substance are analyzed. The influence of the scanning speed of the laser frequency on the choice of the spectrometer operating modes that provide the required resolution and the required sensitivity of spectral measurements is considered. Particular attention is paid to assessing the role of the interaction of modes in an analytical cavity on the structure of the recorded spectra. It was found that at a high rate of change in the laser frequency and superposition of fields of longitudinal modes, an improvement in the resolving power is combined with a certain decrease in the sensitivity and accuracy of recording weak spectral lines. It is shown that the optimization of the spectrometer operation modes requires correct accounting of the Q-factor of the analytical resonator and the line width of the probe radiation.

1. Introduction

In ecology, biomedicine, as well as in many other fields of science and industries, there is a problem of registering the presence of substances with a low concentration in various media. For its solution, a weak absorption spectrometer using a frequency-tunable laser and an external analytical cavity with the substance under investigation has proven itself well [1-3]. Recording the time variation of the laser beam intensity after passing through the resonator makes it possible to record the absorption spectrum of a substance without using a spectral device. Despite numerous publications on this topic, the problem of resonator excitation by a rapidly frequency-tunable laser beam remains practically unexplored. The aim of this work is to find the regularities that affect the formation of the intracavity light field under conditions when the time interval between the excited modes is comparable with the decay time of radiation in the cavity. Particular attention is paid to assessing the effect of intermode optical coupling arising during fast frequency scanning on the sensitivity of spectral measurements. The approach used is based on the method of summing the amplitudes of partial beams at the resonator output resulting from multiple reflections of the input radiation from the resonator mirrors [4].

2. Calculation model

The scheme used in this work for calculating the characteristics of an analytical resonator was oriented towards identifying the most important effects associated with a change in the scanning speed of the laser frequency. The analysis was carried out by numerical modeling in the most general form using dimensionless quantities in the plane wave approximation. At the first stage of the calculations, it was assumed that the linewidth of the laser radiation is negligibly small in comparison with the width of the absorption lines and the bandwidth of the resonator. The phase incursion of rays per passage Θ is determined by the expression $\Theta = 2\pi L\,(f_0 + vt)$, where the frequency $f_0 = 1/\lambda_0$, L is the cavity length, λ_0 is the starting laser wavelength, $v = df/dt$ is the rate of frequency change, t is the time scanning the laser frequency over the working area. It was assumed that $2\pi Lf_0 = 2\pi N$, where N is a large integer. Let us introduce the quantity $\delta = 2\pi Lv$, which denotes the rate of phase increment of light beams per pass and has the dimension [rad/time]. Considering the phase increment rate proportional to time (usually the frequency changes according to a law close to linear), we can write that $\delta = xk$, where k is an integer index characterizing the degree of sampling of the horizontal axis and determining the number of a significant point on the time scale, x is the proportionality coefficient, with dimension [rad/time]. The value of this coefficient affects the rate of change in frequency. The procedure for summing the partial beams in order to determine the total amplitude S of the output radiation will be carried out according to the formula of a decreasing geometric progression. With its help, the time dependence of the amplitude on time for a symmetric resonator can be represented in the form

$$S(k) = \sum_{s=0}^{Q} \frac{a(1-R^2)e^{i[\delta(k+s\cdot m)d+\Phi]}p(k+s\cdot m,\alpha)}{1-R^2e^{2i[\delta(k+s\cdot m)d+\Phi]}p^2(k+s\cdot m)} e^{-\frac{s\cdot m}{\tau}}, \qquad (1)$$

where R is the amplitude reflection coefficient of the resonator mirrors; a is the amplitude of the incident wave; $k = 0, 1, 2 \dots K$; $p(k)$ is the transmission coefficient of radiation by the intracavity medium, Φ is the phase of the input beam (losses in the mirrors are considered negligible), d is the sampling interval of the time scale, τ - the decay time of radiation in the cavity, the parameter m defines the interval between modes. Note that the range of variation of significant points k can characterize both the time T of one scan of the laser frequency change and the range F of frequency tuning during this time. In the first case, the scale division along the k axis is $l_t = T/K$, in the second - $l_f = F/K$.

The analysis of changes in the spatial-temporal structure of the output radiation depending on the scanning speed of the laser frequency, carried out using formula (1), shows that an increase in the speed leads to the emergence of some specific effects that require additional study and consideration. From the point of view of practical significance, these should primarily include

1) reduction of time and frequency intervals between resonance peaks with their simultaneous narrowing, which helps to increase the resolution of the system;

2) the complication of the structure of the output radiation, which occurs when the degree of sampling of the signal deviates from the optimal one and is a source of additional noise and errors of the measuring system;

3) enhancement of intermode coupling, leading to superposition of fields of neighboring modes and, as a consequence, to distortion of the shape of the recorded absorption lines and a decrease in the sensitivity of the spectral system.

Calculations show that the last two effects begin to manifest themselves noticeably when the value of δ increases by more than an order of magnitude during frequency scanning in the operating spectral range. Let us consider sequentially these factors that affect the characteristics of the resonator.

3. Influence of scan rate on the structure of excited modes

Figures 1*a*, *b* gives a general idea of the effect of frequency rate change on the distribution of excited modes. The dependence plots of the module $S(k)$ are plotted for the parameters $a = 1$, $R = 0.99$, $K = 10^5$, $p(k) = 1$.

It is seen that an increase in the frequency scan rate leads to a decrease in the intermode interval. The value of the initial phase Φ, which causes time shifts in the sequence of resonances, is also important. When the parameter x begins to exceed 10^{-2}, with an unchanged signal sampling procedure, a significant spread in the peak values of resonances is observed. Figure 1*b* corresponds to this case

(Q= 0). The complex nature of the change in the amplitudes of the resonance peaks in the figure is due to the changed distribution of the phases of the interfering partial beams and, as a consequence, to a different condition for the formation of temporal interference maxima

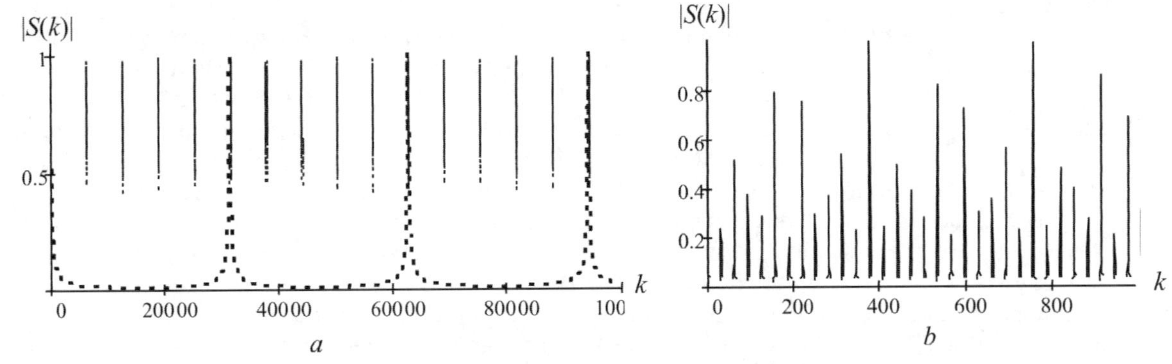

Figure 1. Influence of the frequency scanning rate on the structure of the output radiation (*a*) $x = 10^{-4}$, $\Phi = 0$ (dotted lines); $x = 5\ 10^{-4}$, $\Phi = 0$ (continuous lines); (*b*) $x = 10^{-1}$, $\Phi = 0$.

This can be illustrated by a graphical illustration of the time variation of the total amplitude of two arbitrarily selected interfering partial beams. We will assume that the first beam emerging from the resonator, the amplitude of which is equal to

$$a_0(k) = a\left(1 - R^2\right)e^{i\delta(k)d}, \tag{2}$$

interferes with the beam that has made n double passes through the resonator. Its amplitude is

$$a_n(k) = a\left(1 - R^2\right)e^{i\delta(k)(2n+1)d}R^{2n}, \tag{3}$$

With such two-beam interference, the change in the total amplitude modulus $A(k) = |a_0(k) + a_n(k)|$ at a low scanning speed ($x = 10^{-3}$) is shown in Figure 2*a*.

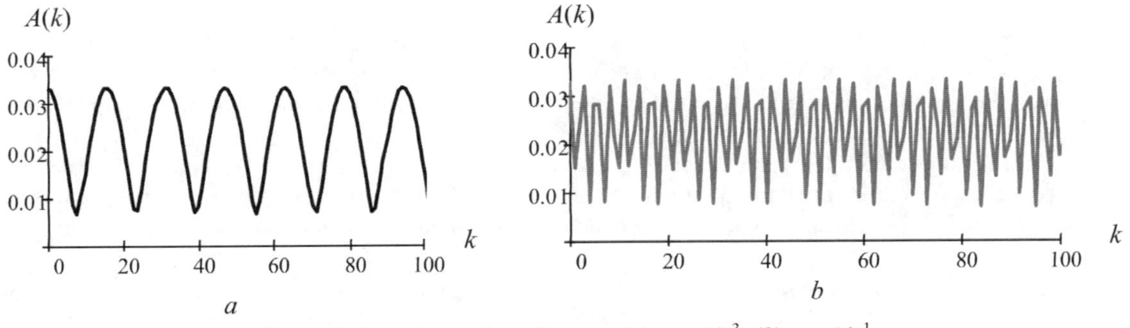

Figure 2. Two-beam interference (*a*) $x = 10^{-3}$, (*b*) $x = 10^{-1}$.

It can be seen that the dependence *A*(*k*) is periodic and its maxima corresponding to the in-phase superposition of the beams have the same magnitude. The indicated dependence will have a different form if the rate of frequency change increases (Figure 2*b*). The maximums of the total amplitude will no longer be fixed, and the interference of the rays will become more complex. The result of summing the amplitudes of not two, but a significantly larger number of beams will lead to a significant deviation from the distribution shown in Figure 1*a*, bringing it closer to the structure in Figure 1*b*. Quasiperiodic distribution of resonances in Figure 1*b* and the presence of fragments similar in shape in it indicates the possibility of assessing the effect of the scanning speed on the complexity of the analyzed distributions by determining the corresponding Hurst parameters. This can be done using the procedure often used for fractal-like structures based on the analysis of the behavior of the structure function C_n [5].

From the slope of the graph of the function C_n, one can find the Hurst parameter H. For the distribution $S(k)$ in Figure 1*b*, $H = 0.235$. Such a signal is antipersistent and has a high degree of randomness. If the value of x becomes less than the value $x = 15\cdot 10^{-3}$, H exceeds 0.5, and the signal becomes persistent and more ordered. The value of the Hurst parameter is also influenced by the Q-factor of the resonator. Thus, a decrease in the reflectance of mirrors from 99% to 95% increased the value of H from 0.235 to 0.423. This is due to a decrease in the spread of the resonance maxima.

The calculation showed that the complication of the output structure of the radiation does not occur if the values of x are multiples of π. In this case, in the absence of an absorbing medium, as the scanning speed increases, the amplitudes of the resonance peaks remain equal to 1.

The phase structure of the output radiation is characterized by sharp changes in the phase of the light oscillations (comparable to the value of π). in the area of resonance peaks. This can be seen from Figure 3, where the amplitude-phase structure of the output beam is shown on an enlarged scale.

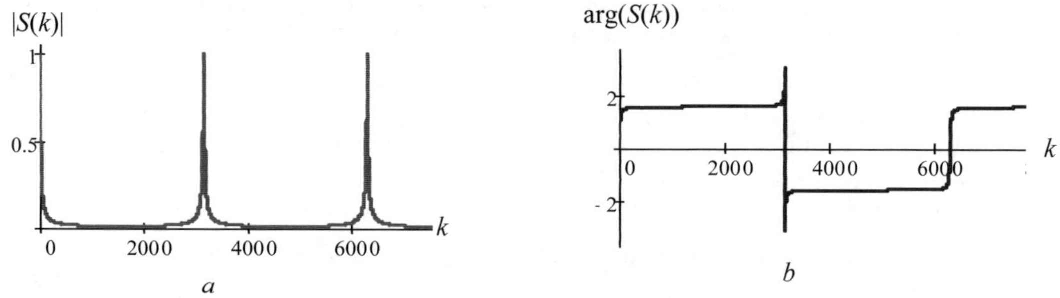

Figure 3. Field structure at the cavity output (enlarged scale) at $x = 10^{-3}$, (*a*) field amplitude, (*b*) field phase.

The presence of rapid changes in the phase of light oscillations near the resonance points on the scale π can play a significant role in the processes of intermode interaction.

4. Intermode interaction

Let us now consider the effect of optical coupling of longitudinal cavity modes on the result of spectral measurements. Such a connection appears when the time interval between the excited modes becomes comparable with the decay time of radiation in the cavity. Due to intermode coupling, the field of each resonance is partially superimposed on the field of the next one.

Taking into account the relaxation of the excitation of longitudinal modes introduces certain changes in the structure of the output radiation. Let us consider this effect for the case when there is an absorbing medium in the analytical resonator. We set its amplitude transmission coefficient using the following relation: $p(k) = \exp\left[-p'(k)\alpha - i\varphi_d(k)\right]$, $p'(k) = \exp\left[-\frac{(k-k_0)^2}{\Delta^2}\right]$ - the form factor of the absorption line, α is the absorption coefficient at the center of the line, k_0 - determines the instant when the laser frequency passes through the center of the line, Δ - characterizes its width, $\varphi_d(k)$ is the additional phase incursion due to a change in the refractive index in the region of the absorption line. Figure 4 shows the sequences of resonances corresponding to the values $Q = 1$, $\Phi = 0$, $k_0 = 5\cdot 10^4$, $\Delta = K/20$, $K = 10^5$, $\alpha = 0.015$. The performed estimates have shown that the quantity $\varphi_d(k)$ does not play a significant role for the indicated parameters. As such, it will be excluded during subsequent consideration.

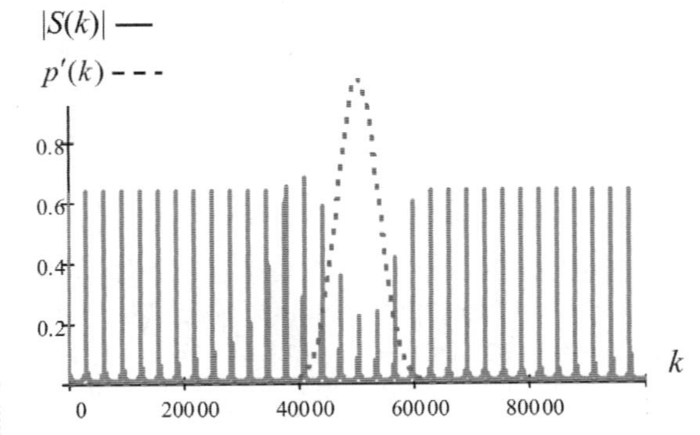

Figure 4. Structure of the output radiation in the presence of intermode interaction: amplitude structure of resonance peaks (continuous lines) and undistorted absorption line (dotted contour).

Analysis of the graphs shows that the amplitude of the resonances has decreased by about 35%. This is a consequence of the phase mismatch of the superimposed resonances. Thus, this fact suggests that intermode interaction can lead to an undesirable decrease in the sensitivity of the method. In addition, the intensity distribution of longitudinal modes in the region of the absorption line turns out to be asymmetric. This distorts the shape of the absorption line, which is recorded at the output of the resonator along the envelope of the resonance peaks.

The effect of intermode coupling manifests itself differently, if we take into account the relationship of each mode not with one, but with two previous modes (Q = 2). In this case, the asymmetry of the dip in the distribution of resonances is preserved, albeit less pronounced. At the same time, the amplitude of resonances in the region of the absorption line is approximately doubled. These changes are due to the difference in the distribution of the oscillation phase of the superimposed resonances. The effect of intermode coupling manifests itself differently, if we take into account the relationship of each mode not with one, but with two previous modes (Q = 2). In this case, the asymmetry of the dip in the distribution of resonances is preserved, albeit less pronounced. At the same time, the amplitude of resonances in the region of the absorption line is approximately doubled. These changes are due to the difference in the distribution of the oscillation phase of the superimposed resonances.

5. Influence of the finite width of the laser radiation line

All the above results were obtained under the assumption that the laser emission line is negligible in comparison with the resonance width and absorption line width. Let's remove this assumption. However, we will continue to assume that the width of the laser radiation spectrum is significantly inferior to the dispersion region of the cavity (the distance between its longitudinal modes). Let us use the result of [6], where it was shown that for a finite radiation line, the beam propagating in the cavity will experience random phase fluctuations. Thus, the value $\delta(k)d$ included in the above formulas should be replaced by the value $\delta'(k) = \delta(k)d + \varphi_k$. The random variable φ_k was set according to the normal law: φ_k = rnorm(K+1,0,0.1). In the presence of phase fluctuations at the cavity output, a random phase will be added to the phase incursion of partial beams per passage. This will result in the destruction of the shape of the resonance peaks and their effective broadening. This broadening can be used to judge the linewidth of the laser radiation.

The indicated process is illustrated in Figure 5, corresponding to one realization of the distribution fluctuations ($\sigma = 0.1$) and the value $x = 10^{-4}$. It shows wide and narrow resonances recorded in the presence and absence of phase fluctuations. Averaging over a number of realizations gives a distribution that is much larger than the width of the resonances under monochromatic illumination.

With a further increase in the frequency scanning rate, the frequency distribution in the region of the absorption line acquires a quasi-continuous character. In this case, asymmetry in the absorption line profile is not observed.

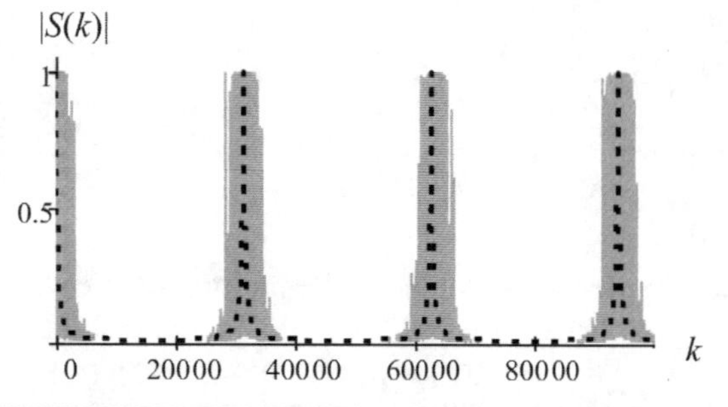

Figure 5. Influence of the laser line width on the structure of the output radiation: grey color - one realization of frequency scanning, dotted line - resonances under monochromatic illumination.

6. Conclusion

The performed analysis of a diode spectrometer operation with an external analytical cavity shows that its operating modes are largely determined by the scanning rate of the master laser frequency. The mode of operation at a high scanning speed is characterized by a number of features caused by the nature of the interference of partial beams at the exit of the resonator and the interaction of modes excited in the resonator. These effects have not received the necessary coverage in the literature. Fast frequency scanning improves the spectral measurement process by increasing the mode density in the operating spectral range. In practice, an additional increase in the mode density can be achieved by repeated scanning of the laser frequency by regularly or randomly changing the initial phase of the probe beam. However, with a significant increase in the frequency scanning speed, when the speed in the working spectral region increases by more than an order of magnitude, some negative phenomena begin to appear. These include the violation of the regularity of the resonant types in the output radiation, associated with the phase shift of the interfering rays. The method used, based on the determination of the Hurst parameter, made it possible to obtain quantitative data on the gradual degradation of the mode structure with an increase in the scanning speed and the transition of the recorded signal to the antipersistence region. Another negative effect is caused by a decrease (by 30–40%) in the measurement sensitivity and distortion of the absorption line contour. The existing experimental data [7] confirm the facts of contour distortion.

7. References

[1] Nadezhdinskii A I, Ponurovskii Y Y 2019 Diode laser spectrometer for high-precision measurements *Quantum Electronics* **49** 7 613 DOI: 10.1070/QEL16776

[2] Korolenko P V, Nikolaev I V, Ochkin V N, Tskhai S N 2014 Recording of absorption spectra by a three-beam integral technique with a tunable laser and external cavity *Quantum Electronics* **44** 4 p 353 DOI: 10.1070/QE2014v044n04ABEH015332

[3] Korolenko P V, Lagunov V V, Nikolaev I V, Ochkin V N, Tskhai S N, Yatskevich A N 2016 Optimisation of external cavity parameters of a weak absorption laser spectrometer *Quantum Electronics* **46** 3 p 255 DOI: 10.1070/QEL15892

[4] Born M, Wolf E 1999 Principles of Optics (*Cambridge University. Cambridge, UK)*

[5] Crownover R M 1995 Introduction to fractals and chaos *(Jones & Bartlett Pub)*

[6] Kasyutich V L, Sigrist M W 2015 Optimisation of laser linewidth and cavity alignment in off-axis cavity-enhanced absorption spectroscopy *Infrared Physics & Technology* **71** pp 179 – 186 DOI: 10.1016/j.infrared.2015.03.011

[7] Lagunov V V, Nikolaev I V, Ochkin V N 2021 High-resolution and high-sensitivity absorption spectroscopy in external optical resonators with fast frequency tuning *Spectrochimica Acta Part A: Molecular and Biomolecular Spectroscopy* **246** 119060 DOI: 10.1016/j.saa.2020.119060

Acknowledgments
The work was carried out with the assistance of the Russian Foundation for Basic Research (project No. 19-02-00540).
The authors thank Professor V. N. Ochkin for a number of valuable tips and comments.

Development of a selective assembly card of the «Koadut» lens

Yu.Yu. Kachurin[1], O.A. Kananykhin[1], I.A. Levin[2]

[1]Bauman Moscow State Technical University, 2-aya Baumanskaya street 5, Moscow, Russia, 105005
[2]JSC "Krasnogorsky Zavod" (KMZ "ZENIT") , Krasnogorsk, Moscow region, Russia

E-mail: kachurin@bmstu.ru

Abstract. The work is devoted to the development of a selective assembly card of the «Koadut» lens, depending on the actual thickness of each of the seven lenses included in the optical scheme of the objective. Criteria for sorting options are substantiated, a macro for processing a file with the results of measurements of actual lens thicknesses, which forms an assembly card, is created. The macro was tested on a test file with a normal distribution of deviations from the nominal value of a batch of one hundred lenses.

Introduction

Modern the assignment of tolerances for optical parts is associated with ensuring the required quality of the designed product. [1,2]. The design requires a balance between production capability and level of quality. In some cases, the resulting tolerance may not be feasible in practice[3,4]. On the other hand, it is possible to increase the tolerance field due to the subsequent selective assembly [5]. This work is devoted to the development of a selective assembly card for the "Koadut" lens, depending on the actual thickness of each of the seven lenses included in the optical layout of the objective. Due to selective assembly, it was supposed to increase the tolerance field for the thickness of each lens from ± 0.003mm to ± 0.005mm. The evaluation criterion was the back focal distance, which determined the position of the CCD matrix from the back face of the frame. It should have changed from the nominal value by no more than a third of the specified tolerance range. Also, the criterion for choosing an option for a complete set was the equality to zero of the sum of deviations from the nominal size for the last four lenses. This eliminated the need for shims or face trimming during assembly.

Also, the initial data for compiling the assembly card were the results of measurements of the thickness of the manufactured lenses, as a result of which the lenses were divided into groups according to deviations from the nominal size with a step of 0.001 from -0.005 to +0.005 for each lens.

1. Compilation of configuration options

The lens layout is shown in Figure 1

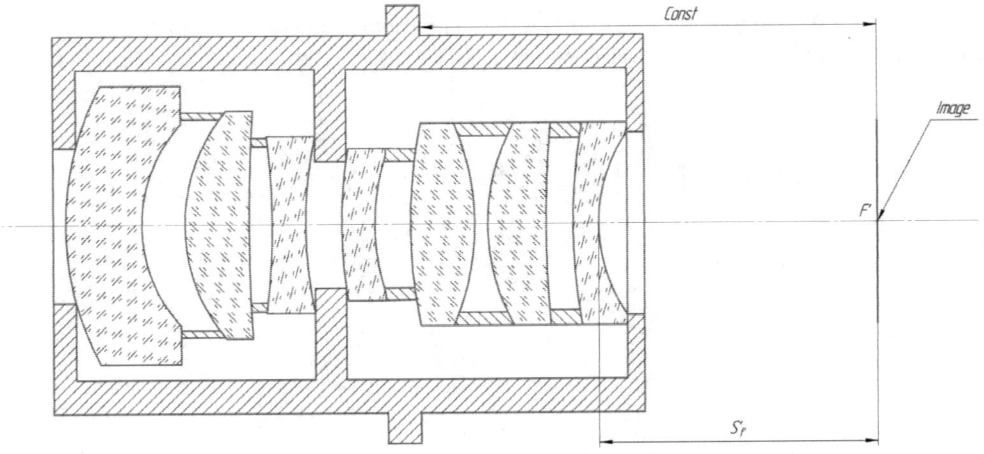

Figure 1. "Koadut" lens layout

To obtain the configuration options, a macro [6, 7, 10] was written, as a result of which 14 files were generated. In these files, for one of the lenses, the thickness was at the maximum deviation value of +0.005 or -0.005, for the rest, the deviation should have been smaller.

Also, 14 files were generated, in which for one of the lenses the deviation was +0.004 or -0.004. The rest had a smaller deviation from the nominal.

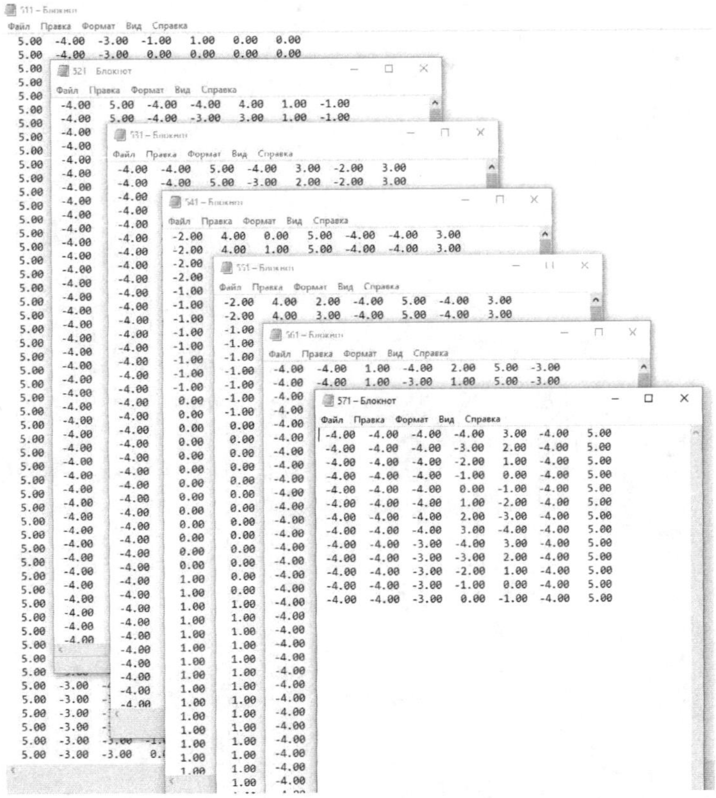

Figure 2. Obtained files with configuration options.

2. Options sorting

Next, the configuration options were sorted so that actual deviations from the nominal value were not repeated on a certain line in all files for each lens. An example of sorted files is shown in Figure 3.

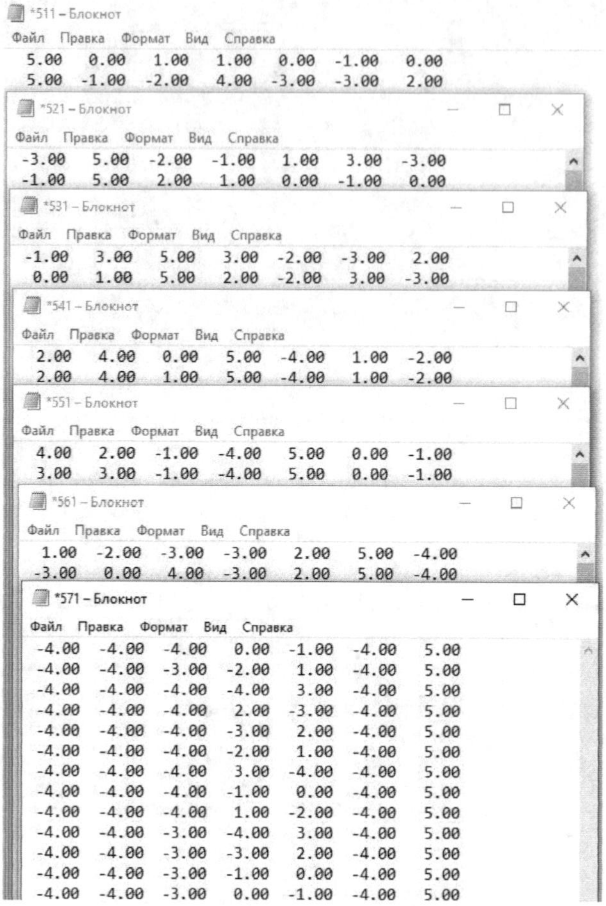

Figure 3. Example of sorted files

3. File processing

At the final stage, a macro was written for processing the file with the results of measurements of the actual lens thicknesses, which forms an assembly card.

The created macro was tested on a test file with a normal distribution of deviations from the nominal value for one hundred lenses. The number of lenses sorted by deviations in their thickness in Table 1 [8, 9] and Figure 4.

Table 1. Number of lenses sorted by deviations in their thickness

Lens 1	Lens 2	Lens 3	Lens 4	Lens 5	Lens 6	Nominal deviation
25	25	25	25	25	25	+0.005
29	29	29	29	29	29	+0.004
33	33	33	33	33	33	+0.003
37	37	37	37	37	37	+0.002
40	40	40	40	40	40	+0.001
40	40	40	40	40	40	0
40	40	40	40	40	40	-0.001
37	37	37	37	37	37	-0.002
33	33	33	33	33	33	-0.003
29	29	29	29	29	29	-0.004
25	25	25	25	25	25	-0.005

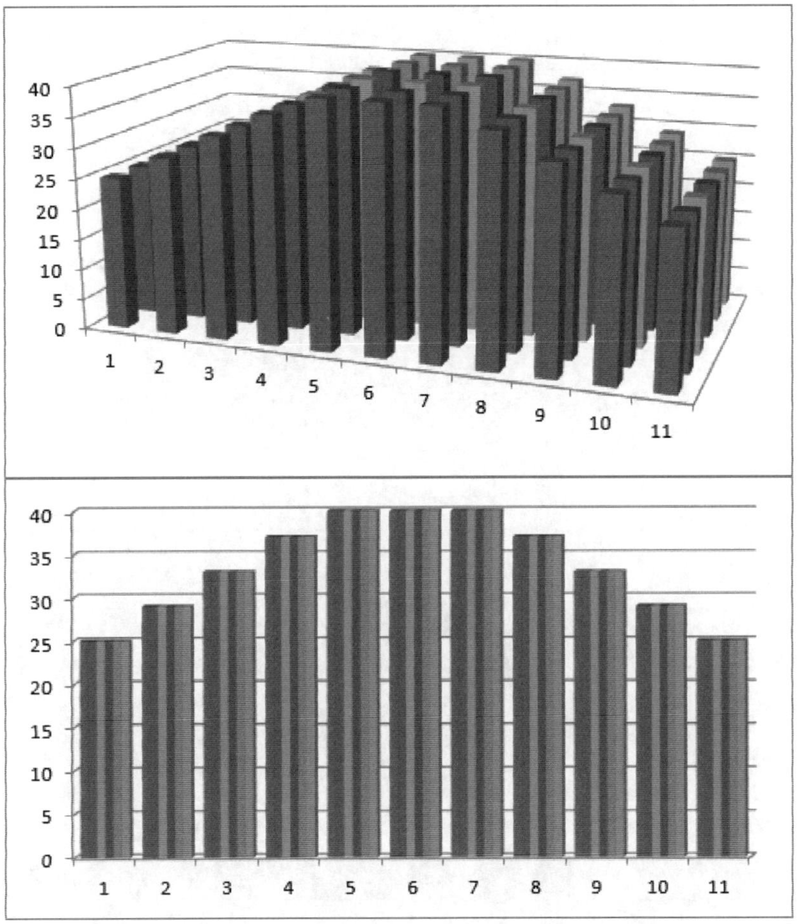

Figure 4. Number of lenses sorted by deviations in their thickness

As a result, an assembly card was obtained in which lenses with an extended thickness tolerance were completely selected from a set of lenses, as shown in Table 2 and Figure 5.

Table 2. The result of the macro

Lens 1	Lens 2	Lens 3	Lens 4	Lens 5	Lens 6	Nominal deviation
0	0	0	0	0	0	+0.005
0	0	0	0	0	0	+0.004
17	18	18	16	15	18	+0.003
22	22	23	21	21	20	+0.002
25	26	24	27	26	26	+0.001
26	26	24	26	27	24	0
23	23	23	21	24	23	-0.001
22	21	22	24	22	22	-0.002
17	16	18	17	17	19	-0.003
0	0	0	0	0	0	-0.004
0	0	0	0	0	0	-0.005

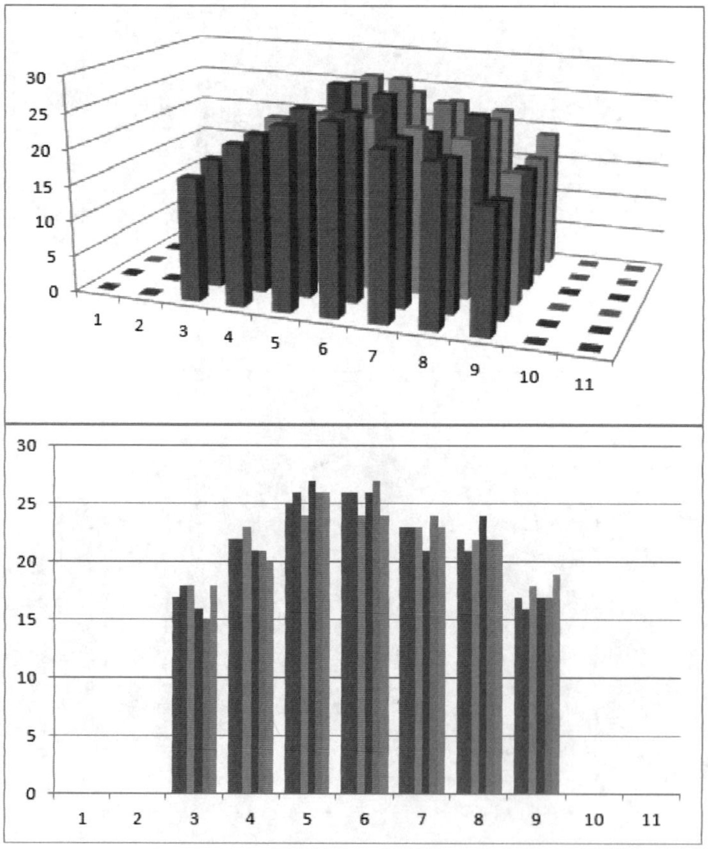

Figure 5. The result of the macro

4. Conclution

Thus, we have developed a database that allows us to form objective lens "coadout" assembly card with the use of lenses with extended thickness deviation while maintaining the specified quality criteria

References

[1] V. Ya. Katkovnik, A. I. Savchenko 1991 *Principles of Selective Assembly Theory* (Leningrad: Politekhnika).

[2] M.D. Mal'tzev 1974 *Tolerances calculation for optical components* (Mascow: Mashinostroenie) p.168

[3] S. M. Latyev, A. P. Smirnov, A. G. Tabachkov, D. N. Frolov, and R. V. Shukhat 2011 Project of automated line for microobjective assembly *Izvestiya vysshikh uchebnykh zavedenii. Priborostroenie* **54** №11 pp 7–13

[4] S. M. Latyev, A. P. Smirnov, D. N. Frolov, A. G. Tabachkov, and R. Tezka, 2010 Providing target performance indices when automating the assembly of microscope objectives *J. Opt. Technol.* **77** №1, pp 38–41

[5] Levin I.A., Kachurin Yu.Yu. 2021 Automation of the batching process of optical elements in the selective assembly of camera lenses *Journal of Optical Technology*, **88** №4, pp. 178–181

[6] ZEMAX 13 2014 Optical Design program / User's Manual (Zemax LLC) p. 879

[7] Goryacheva V.A., Mazaev V.L. 2019 General data of optical system components ZEMAX program *Tula State University Herald. Technical Sciences* **7** pp 194-199

[8] M. I. Kaufman, B. B. Light, R. M. Malone, and M. K. Gregory, 2014 Statistical distributions from lens manufacturing data *Proc. SPIE* **9195**, 919507

[9] M. I. Kaufman, B. B. Light, R. M. Malone, M. K. Gregory, and D. K. Frayer 2015 Techniques for analyzing lens manufacturing data with optical design applications *Proc. SPIE* **9573**, 95730M

[10] Web Site [Electronic resource]. — Access mode: https://osphotonics.wordpress.com/2018/02/10/application-of-zemax-programming-language-paperback-version-available-for-purchase/.

Composite thin films based on multilayer carbon nanotubes and calcium phosphate with electrical conductive properties for bone tissue engineering

S Z Zhovnir[1], M S Savelyev[1,2], A V Kuksin[1,2], U E Kurilova[1], A Yu Gerasimenko[1,2], A Yu Fedotov[3], P V Lobzhanidze[3] and V S Komlev[3]

[1] National Research University of Electronic Technology MIET, Bld. 1, Shokin Square, Zelenograd, Moscow 124498, Russia

[2] I.M. Sechenov First Moscow State Medical University, 8 Trubetskay str., Moscow 119991, Russia

[3] A.A. Baikov Institute of Metallurgy and Materials Science, Russian Academy of Sciences, 49 Leninsky pr., Moscow 119334, Russia

E-mail: zhovnir.sveta@mail.ru

Abstract. This work is aimed at developing of electrically conductive composite thin films based on calcium phosphates (CP) and multi-walled carbon nanotubes (MWCNT) for bone tissue engineering. A method based on the sputtering of an dispersion of CPs and MWCNTs on a substrate with subsequent exposure to a laser with different power values to form connected networks of nanotubes has been proposed. It is shown that the electrical conductivity of CP/MWCNT films reaches high values for dispersions with a MWCNT concentration of 1 mg/ml when exposed to laser radiation with a power of 2.1–3.0 W. It was found that ablation areas of CPs and MWCNTs as a result of exposure to a high-power laser are found in areas of the film with increased resistance.

1. Introduction

Tissue engineering is a traditional method of repairing bone tissue defects [1]. The tissue engineering approach is based on the filling of bone defect with a scaffold material with osteogenic factors. One of the tasks of tissue engineering is the creation of effective scaffolds with electrically conductive properties. Electrical stimulation is successfully used in medical practice as a tool to accelerate the healing process of tissue defects. Electrical stimulation can affect cells: their proliferation [2], migration [3], differentiation and attachment to the surface of the material [4]. Scaffolds must meet a number of requirements e.g.: biocompatibility, certain surface properties, and, if possible, must have electrically conductive properties [5].

The paper presents the results of work on the development of biocompatible electrically conductive materials based on multi-walled carbon nanotubes (MWCNT) and calcium phosphate (CP). Carbon nanotubes have electrical properties and the ability to provide electrical conductivity to various non-conductive materials in the composition [6]. One of the methods for creating connected networks is laser irradiation [7,8]. CP-based ceramics is a synthetic analogue of native bone tissue [9]. Thus, the development of a composite material from nanotubes and CPs will make it possible to achieve acceptable biocompatibility and electrical conductivity of bone tissue engineering constructs.

2. Materials and methods

2.1. Production of CP/MWCNT dispersions

In this work, two dispersions were created, which differed from each other in the concentration of MWCNTs. Thus, dispersions D1 and D2 included 0.1 and 1 mg/ml of nanotubes, respectively. In this case, both dispersions contained 50 mg/ml of CP. They were made by mixing CP and MWCNTs in ethanol. To prepare the dispersions, we used MWCNT Taunit-MD (NanoTechCentre LLC, Russia). The nanotubes had an outer diameter of 8–30 nm, an inner diameter of 5–15 nm, and a length of ≥20 μm. The CP included in the dispersion was synthesized by the liquid-phase method from aqueous-alcoholic solutions of calcium acetate and ammonium hydrogen phosphate while maintaining the pH value of 9 ± 0.1 with 10 % sodium hydroxide solution. The resulting suspension was washed with water and filtered until a thick white paste was obtained. The Ca/P ratio was 1.33 and the particle size reached 0.2 μm. The paste contains adsorbed water 20 times more than the mass of CP. This gives reason to believe that CP is present in the paste in the form of clusters or amorphous aggregates, the transition of which to the crystalline state occurs upon removal of water [10]. This paste was used to prepare dispersions. MWCNTs in an amount of 0.005 g (D1) and 0.05 g (D2) were mixed with both 2.5 g of CP and with ethyl alcohol. The resulting mixtures were exposed to a magnetic stirrer for 40 min, an ultrasonic bath for 1 h at a temperature of 30 °C, and an ultrasonic homogenizer for 35 min until a homogeneous dispersion was formed.

2.2. Formation of CP/MWCNT films on substrates

The production of films from dispersions was carried out using a device that allows sputtering thin layers of a dispersion onto a substrate. The installation consisted of modules for spraying movement, and a heating stage (figure 1). Spraying was carried out with the following parameters: the temperature of the heating table was set at 80 °C, the speed of the spraying module was 10 mm/s, the spray pressure was 3 kg/cm^2. All parameters and trajectory of the spraying module were set using the Laser Studio computer program.

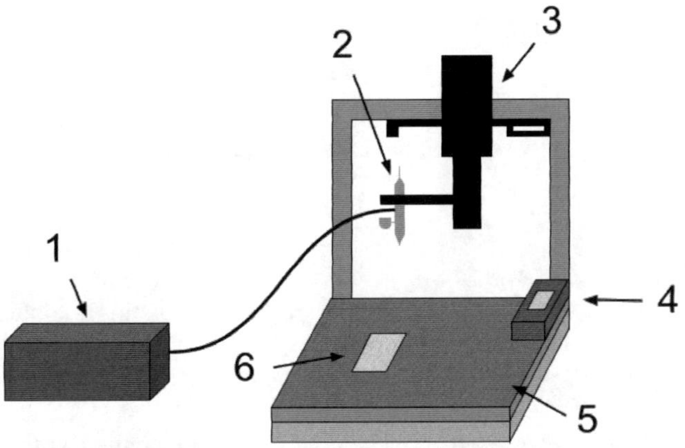

Figure 1. Installation diagram for dispersion spraying: 1 – compressor; 2 – spraying module; 3 – movement module; 4 – heating table control module; 5 – heating table; 6 – substrate.

The homogeneous dispersion was poured into the sputtering module prior to film formation. Then, air under pressure was supplied by a compressor to the spraying module, where the air was mixed with the dispersion and sputtered onto a glass substrate. The substrate was placed on a heating table, due to this, the dispersion instantly evaporated without the formation of drops and formed a homogeneous film. The number of sprayed layers was set by a computer program and was 30, 50, 80, 100, and 150.

2.3. Irradiation of CP/MWCNT films with a laser

Irradiation of the CP/MWCNT films was carried out using a setup based on a Yb laser (figure 2). The exposure parameters were set in the LDesigner 5.0 computer program. Square areas 4×4 mm in size on dispersion films were exposed to a power in the range 2.1–4.5 W.

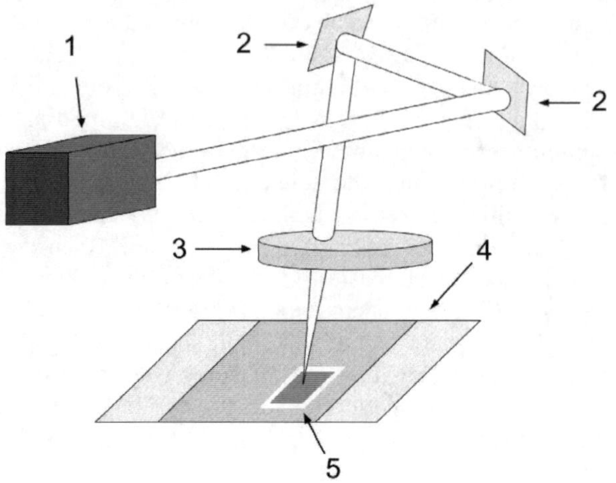

Figure 2. Laser irradiation of of the film areas: 1 – fiber Yb laser; 2 – beam positioning system; 3 – lens; 4 – a substrate with a film; 5 – the area of exposure to a laser with a power in the range 2.1–4.5 W.

2.4. Calculation of the electrical conductivity of CP/MWCNT films

The resistance of the areas of the CP/MWCNT films was measured before and after laser irradiation. Based on the results obtained, the specific electrical conductance of the film areas was calculated by the formula:

$$\sigma = \frac{L}{R \cdot S}, \quad (1)$$

R – measured film resistance, S – cross-sectional area of the measured film area, L – area length.

3. Results

3.1. CP/MWCNT films after laser irradiation

Experimental films of CP/MWCNT were applied on glass slides using a sputtering device: 80 layers (R = 5 MOm), 150 layers (R = 13 MOm), 50 layers (R = 7 MOm) using dispersion D1 and 100 layers (R = 0.3 kOm), 80 layers (R = 0.5 kOm) and 30 layers (R = 1.3 kOm) using D2 dispersion, respectively.

Based on the results of the measured values of the films resistance with a different number of layers, the optimal number of dispersion layers was selected – 80. For the further laser irradiation, samples D1 and D2 with 80 layers were selected. The areas of the CP/MWCNT films after laser irradiation are shown in figure 3.

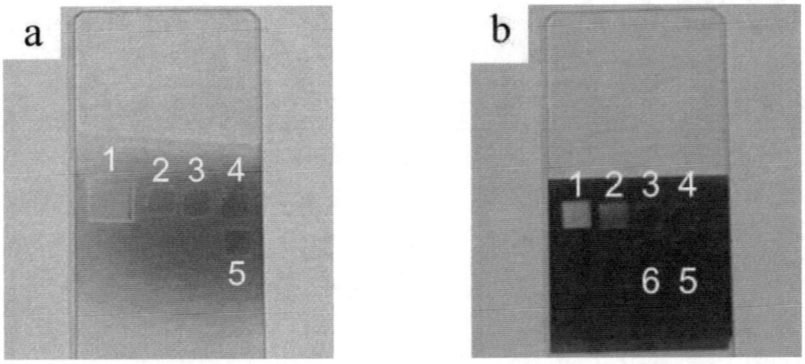

Figure 3. Appearance of films D1 (a) and D2 (b) after laser irradiation (1, 2, 3, 4, 5 – areas with dimensions 4×4 mm, exposed to laser action with different power values).

The dependences of the obtained resistance values on the power of laser irradiation of films D1 and D2 are presented in Table 1. The parameters of laser radiation were: wavelength – 1064 nm, pulse duration – 100 ns, beam diameter at focus – 35 μm, laser beam travel speed – 500 mm/s.

Table 1. Resistance values before and after laser irradiation to film areas D1 and D2.

Film	Area	P, W	R_{before}, Om	R_{after}, Om
	1	3.6	4.9×10^6	5.9×10^6
	2	3.3	5.1×10^6	3×10^6
D1	**3**	3	6×10^6	3.2×10^6
	4	2.7	4.9×10^6	1.5×10^6
	5	2.4	6.1×10^6	2.5×10^6
	1	4.5	0.8×10^3	200×10^3
	2	3.3	0.6×10^3	4×10^3
D2	**3**	3	0.5×10^3	0.45×10^3
	4	2.7	0.55×10^3	0.4×10^3
	5	2.4	0.27×10^3	0.25×10^3
	6	2.1	0.34×10^3	0.28×10^3

As it can be seen from the table, in areas 2, 3, 4, and 5 of film D1 and areas 3, 4, 5 and 6 of film D2, a decrease in the resistance of the film after laser irradiation was obtained, which indicates the formation of bonds between the tubes under the influence of laser radiation. The surface resistance of the films was reduced due to the correctly selected laser power, which was 3.3, 3, 2.7 and 2.4 W for the D1 film and 3, 2.7, 2.4 and 2.1 W for the D2 film.

Using formula (1), the values of the specific electrical conductivity of the film spaces before and after laser irradiation were calculated. The values of specific electrical conductivity for films D1 and D2 are shown in Table 2.

Table 2. Values of specific electrical conductivity before and after laser irradiation on the areas of films D1 and D2.

Film	Area	σ_{before}, S·m^{-1}	σ_{after}, S·m^{-1}
D1	**1**	6.38×10^{-3}	5.65×10^{-3}
	2	6.13×10^{-3}	10.75×10^{-3}
	3	5.21×10^{-3}	9.77×10^{-3}
	4	6.38×10^{-3}	20.83×10^{-3}
	5	5.12×10^{-3}	12.50×10^{-3}
D2	**1**	31.25	0.13
	2	41.67	6.41
	3	50.00	56.26
	4	45.46	62.50
	5	92.59	100.00
	6	73.53	89.29

The highest experimental value of specific electrical conductivity was 100 S·m^{-1}. Laser irradiation with a power from 2.1 to 3.0 W increased the electrical conductivity of the resulting films. The electrical conductivity of the films was directly dependent on the concentration of MWCNTs, while the relative increase in electrical conductivity was large (up to 326 %) for a film with a lower concentration of nanotubes. The electrical conductivity of the human cortical bone is 0.92×10^{-3} S·m^{-1} [11], thus, the electrical conductivity of all obtained layers exceeds the electrical conductivity of the native human bone tissue.

3.2. Surface microstructure of CP/MWCNT films

The microstructure of films with areas irradiated with laser radiation was obtained using optical microscopy and is shown in figure 4. Figure 4a shows the boundary of area 1 after laser exposure to film D1. In this area, the resistance was 10^6 Om, which is a high increase in the resistance value recorded for the D1 film. Due to the high power of the laser (3.6 W), area 1 was ablated on the D1 film. As a result, the resistance of the study area has increased. In the irradiated area 4 of the D1 film, laser irradiation with a power of 2.7 W led to the formation of networks of nanotubes and percolation units, due to which the resistance decreased by 3.4×10^6 Om (figure 4b).

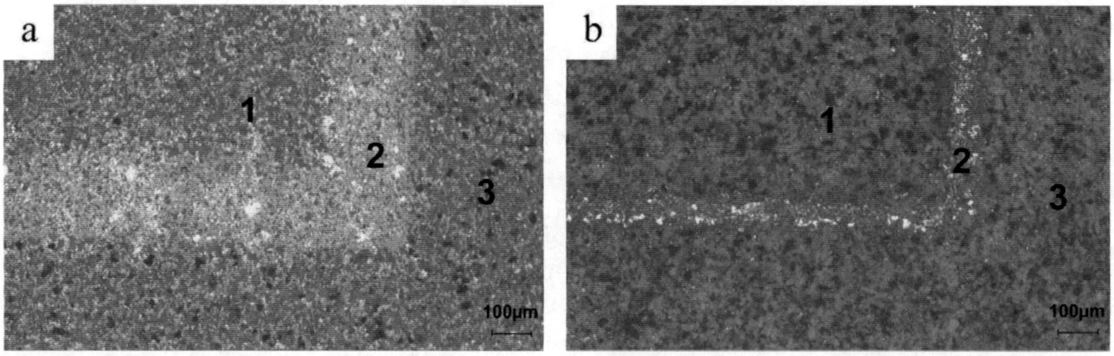

Figure 4. Images of the microstructure of the boundary of area 1 (a) and area 4 (b) on the film D1: 1 – area after laser irradiation; 2 – border between areas; 3 – area before laser irradiation.

Area 1 on film D2 under the influence of a 4.5 W laser became thinner, which indicates an excess laser power (figure 5a). Thus, the number of nanotubes in the composition of the film decreased, due to which the resistance of the film increased. In area 4 of the D2 film, the resistance decreased by

0.15×10^3 Om due to the structuring of MWCNTs with the formation of a connected network of nanotubes (figure 5b).

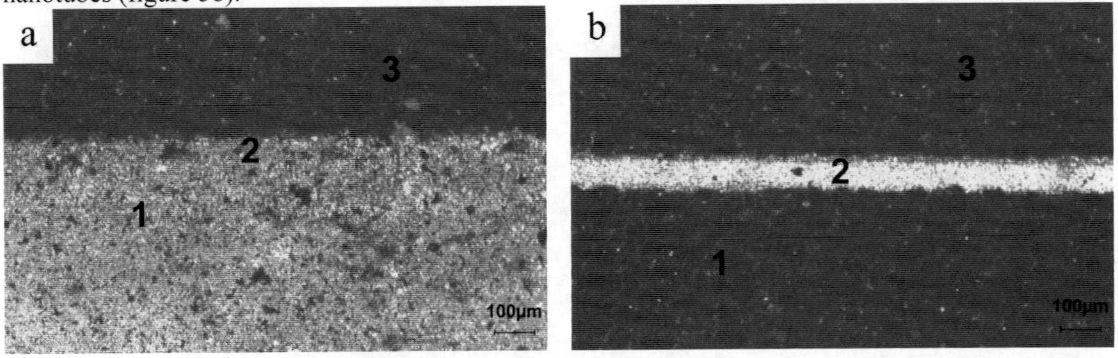

Figure 5. Images of the microstructure of the boundaries of area 1 (a) and area 4 (b) on the film D2: 1 – area after laser irradiation; 2 – border between areas; 3 – area before laser irradiation.

4. Conclusion

This paper presents the results of work on the development of electrically conductive biocompatible material from dispersions of multi-walled carbon nanotubes and calcium phosphate for applications in bone tissue engineering. The films were obtained by sputtering dispersions with nanotube concentrations of 0.1 mg/ml and 1 mg/ml, respectively, onto glass substrates.

Under the action of a laser with a power of 2.7 W, the specific electrical conductivity of a dispersion film with a nanotube concentration of 0.1 mg/ml increased 3.3 times compared with the initial one. The specific electrical conductivity of a film from a dispersion with a nanotube concentration of 1 mg/ml increased 1.4 times as compared to the initial one. Exposure to a higher power led to a decrease in electrical conductivity for both types of films. For dispersions with nanotube concentrations of 0.1 and 1 mg/ml, the maximum specific electrical conductivity was 20.83×10^{-3} and 100 $S{\cdot}m^{-1}$, respectively. The increase in specific electrical conductivity is due to the bonding of carbon nanotubes to each other under the action of laser radiation. As a result, electrically conductive networks are formed in the CP matrix. Excessive laser power leads to the destruction of nanotubes and ablation of a part of the layer of the CP/MWCNT biocompatible material.

The obtained materials can be used as scaffolds for the cultivation of osteogenic cells with electrical stimulation.

Acknowledgements
The work was supported by the Ministry of Science and Higher Education of the Russian Federation no. 075-03-2020-216 and 007-00129-18-00.

References

[1] Leppik L et al. 2018 *Sci. Rep.* **8** 6307
[2] Ercan B and Webster T J 2008 *Int. J. Nanomed.* **3(4)** 477–85
[3] Yuan X, Arkonac D E, Chao P G and Vunjak-Novakovic G 2014 *Sci. Rep.* **4** 3674
[4] Mobini S, Leppik L, Thottakkattumana Parameswaran V and Barker J H 2017 *PeerJ.* **5** e2821
[5] Chen C, Bai X, Ding Y and In-Seop L 2019 *Biomater. Res.* **23(25)**
[6] Kim H D, Park J, Lee E S, Kim E, Lee S S, Yang J K, Lee Y S and Hwang N S 2018 *Biomater. Res.* **22(1)**
[7] Gerasimenko A Y, Glukhova O E, Savostyanov G V, Podgaetsky V M 2017 *J. Biomed. Opt.* **22(6)** 065003
[8] Gerasimenko A Yu, Kurilova U E, Savelyev M S, Murashko D T and Glukhova O E 2021 *Comp. Struct.* **260** 113517

[9] Cairns M L, Meenan B J, Burke G A and Boyd A R 2010 *Colloids and Surfaces B: Biointerfaces* **78(2)** 283–290
[10] Dorvee J R and Veis A 2013 *J. Struct. Biol.* **183(2)** 278–303
[11] Balmer T W, Vesztergom S, Broekmann P, Stahel A, Buchler P 2018 *Sci. Rep.* **8** 8601

Control system for glucose level regulation in peritoneal dialysis

N M Zhilo, E L Litinskaia, and N A Bazaev

National Research University of Electronic Technology,
Bld. 1, Shokin Square, Zelenograd, Moscow 124498, Russia.

E-mail: nikitazhilo@gmail.com

Abstract. Kidney failure leads to the serious health issues associated with abnormal water-salt balance. In this case, peritoneal dialysis therapy is often prescribed: 1-2 liters of dialysis solution is administered in peritoneal cavity for 3-4 hours. During this time, due to diffusion and osmosis, toxins and excess water are transferred from blood to solution. One of the method's downsides is the transition of glucose (osmotic agent) into the bloodstream, which leads to a gradual decrease in the fluid removal rate. To mitigate this problem, one must use the system, which will measure current glucose concentration and inject glucose into solution to compensate absorption. The paper proposes such a control system for automatic regulation of the glucose concentration in peritoneal dialysate solution. Its structure, elements, their functions and characteristics are discussed. Proposed system is capable to work autonomously or can be incorporated into wearable "artificial kidney" device.

1. Introduction

Regulation of body fluid volume is one of the most important metabolism functions. Failure of electrolyte balance regulation (hypohydration and hyperhydration) caused by various conditions (kidney disease, diabetes, burns etc.) are dangerous for the patient and require medical intervention [1], [2]. Disorders of renal excretory function in chronic kidney failure are a frequent cause of lifespan shortening.

To replace the excretory function, different kinds of renal replacement therapy are prescribed. The only adequate and most efficient of them is kidney transplantation, but shortage of donor organs makes it necessary to use methods to support patient's health while he/she waits for transplant. Another one of these methods is hemodialysis; in this case, special hemodialysis machine pumps patient's blood through a dialyzer (device with porous hollow fibres), in which toxins and excess fluids are transported into isotonic solution, and treated blood is returned back to the patient. This kind of procedure requires patient to sit near the hemodialysis machine for 3-4 hours for 2-4 times a week, which is very time-consuming and somewhat unsafe because of the direct contact blood with the blood tubing.

As an alternative to hemosialysis, peritoneal dialysis (PD) method can be prescribed; during procedure, 1-2 liters of solution for peritoneal dialysis (SPD) is administered into the abdominal cavity; due to osmosis, excess fluid passes into solution through the capillaries of the peritoneum (ultrafiltration occurs). Hyperosmolarity relative to blood is provided by an osmotic agent (usually glucose) [3]. This procedure can be performed at the patient's home, and he/she is free to do any kind of activity during the procedure. This method has a number of disadvantages, one of which is a short SPD action duration: after a short time (usually 3 to 5 hours), glucose passes into the blood, and ultrafiltration stops. This leads to the fact that the SPD must be drained and replaced with fresh solution every few hours [4], which negatively affects the safety of the procedure and the patient's quality of life [5].

To solve this problem, one can use a system for maintaining the concentration of glucose in the SPD, which will measure current concentration, calculate and inject the right amount of osmotic agent into solution. This kind of system can be created in the form of an independent apparatus or as part of a wearable "artificial kidney" (WAK) device [6], [7]. In second case, usage of such system is very beneficial, because WAK itself regenerates the SPD (by eliminating toxins and removing excess fluid), and glucose maintaining system extends the ultrafiltration time. This means that time and intensity of artificial blood purification and fluid removal can match those in natural kidneys, i.e. 24 hours a day with low intensity, which increases procedure biocompatibility and safety (lower frequency of SPD exchanges lowers the risk of peritoneal inflammation and infection).

This work describes the possible composition of such system, its functions and system connections. System prototype, its development and first experimental *in vitro* trials are also discussed.

2. System overview

The structure of the proposed control system is shown in figure 1: the reference unit RU sets the target glucose level g(t), the regulator generates a control signals $u_{1,2}(t)$ for the actuators (glucose dispenser A1 and perfusion pump A2) based on signal z(t) from the feedback sensor FS (glucose concentration meter).

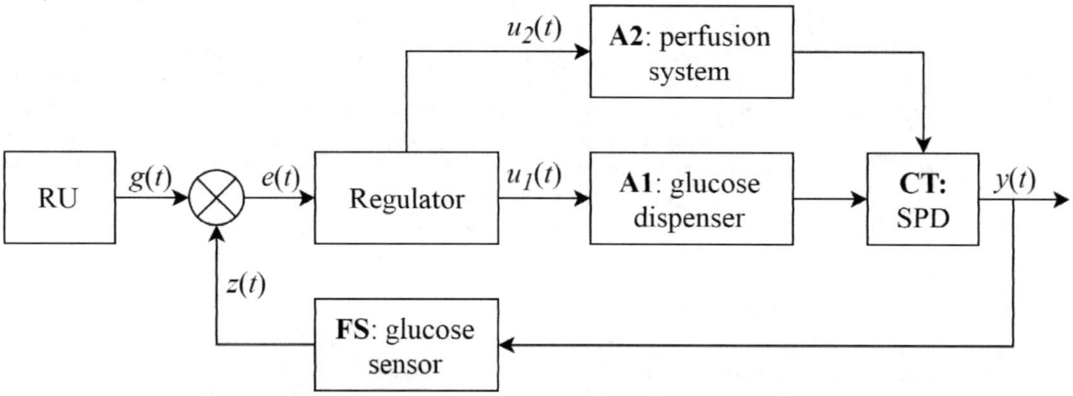

Figure 1. Glucose control system diagram: RU is reference unit; A1 is actuator 1 (glucose dispenser); A2 is actuator 2 (perfusion system); CT is control target (solution for peritoneal dialysis), FS is feedback sensor, g(t) is target glucose concentration, z(t) is feedback signal, e(t) is deviation, $u_{1,2}(t)$ is control signal, y(t) is controlled parameter (glucose concentration)

2.1. Feedback sensor (glucose sensor)

It is obvious that the feedback sensor must show sufficient measurement accuracy to avoid overshoot – high glucose concentrations have a toxic effect on the peritoneum and can cause hyperglycemia in the patient [8], and low glucose concentration lowers the ultrafiltration rate and defeats the system's purpose. In addition, the sensor must provide a high degree of isolation of the solution from the external environment in order to avoid contamination and infection of the solution. These requirements are met by sensors based on infrared photometry – they allow non-contact measurements, with an accuracy higher than that of other methods (acousto- and thermometric) [9-11]. Studies show that absorption spectrum water glucose solution with proteins (which is similar to a spent solution for peritoneal dialysis) allows evaluating glucose concentration by measuring the absorption of infrared radiation [12]. Absorption spectrum of such solution is shown in figure 2.

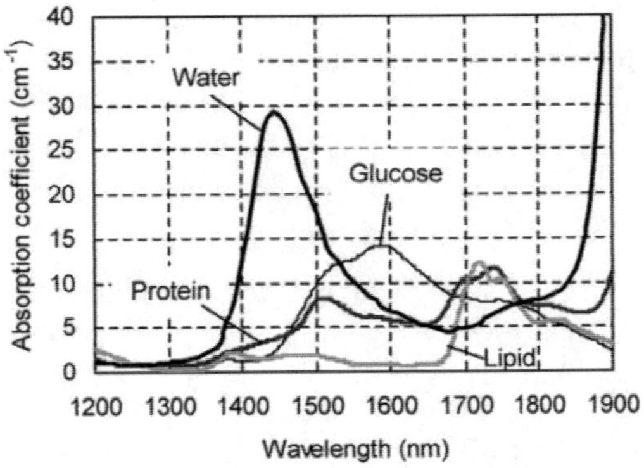

Figure 2. Absorption spectrum of glucose, water, fat and albumin [12]

Experimental studies of IR-photometry system have been carried out. Studied system is shown in figure 3. It consisted of:

- laser diode (optical power 30 mW, center wavelength 1600 nm, spectrum width 5 nm, manufacturer "Dilaz", Russia),
- photodiode DFD1000TO-K (spectral sensitivity 0.9 A/W, sensor diameter 1 mm, dark current 60 nA, manufacturer "Dilaz", Russia),
- analog-to-digital converter AD7689 (16 bit resolution, integral nonlinearity ±0.4 LSB typical, ±1.5 LSB maximum, 250k samples/s, manufacturer "Analog Devices", USA),
- digital-to-analog converter AD5624 (12 bit resolution, relative accuracy ±12 LSB maximum, manufacturer "Analog Devices", USA),
- microcontroller STM32F103RGT6 (32-bit RISC core, 72 MHz frequency, manufacturer "STMicroelectronics", Switzerland),
- USB to UART converter CP2102 ("Silicon Labs", USA) for PC interfacing,
- 3D-printed enclosure holding together electronic components, transparent quartz tube and flexible silicone tubings, connected to pump and flask with model solution.

Experiments were conducted on such fluids as:

- double distilled water,
- double distilled glucose solution (glucose concentration 40-220 mmol/l),
- solution for peritoneal dialysis (Fresenius Balance 4.25% glucose, watered down to glucose concentration 40 mmol/l),
- model solution of spent peritoneal dialysate (peritoneal dialysis solution Fresenius Balance 4.25% glucose, watered down to 40 mmol/L, with addition of urea (14 mmol/l), creatinine (346 μmol/l) and uric acid (100 μmol/l)) to simulate fluid sourced from patients with chronic kidney disease.

To calculate glucose concentration using measurements of light absorption in liquid media, Beer–Lambert–Bouguer law was used, as shown in equation (1):

$$I(t) = I_0 \exp\{-[r + \varepsilon_g c_g(t)]l\} \quad (1)$$

where I_0 is the initial radiation intensity, $I(t)$ is the intensity of the transmitted radiation, l is optical path length, r is attenuation (absorption) coefficient, ε_g is molar glucose attenuation coefficient, $c_g(t)$ is glucose concentration.

Experiments showed that studied system makes it possible to measure the glucose concentration in the spent peritoneal dialysate in the range of 42-220 mmol/L with a relative error of about 15% (compared to lab spectrophotometer StatFax 4500 (intrinsic error 1 %, manufacturer – "Awareness Technology", USA)).

2.2. Actuating device (glucose dispenser)

The executive device must provide a controlled automated injection of glucose into the control object (SPD). The simplest device is a syringe pump filled with glucose concentrate. The disadvantage of this approach is the need for frequent refilling of the system: the solubility of glucose in water under normal conditions is relatively low (47 g / 100 g [13]), and in the case of a wearable system, the use of large syringes (> 20 ml) negatively affects the size and weight of the actuator. Heating of glucose concentrate may be used to reach higher solubility (74 g/100 g at 60 °C), but power consumption must be taken into consideration, as well as heating of circulating dialysate. As an alternative to a syringe pump, systems for administering dry powder or compressed glucose tablets directly into solution may be used, but these systems are not widely available and require additional research and development of mechanics and electronics.

2.3. Perfusion system

The purpose of the perfusion system is to circulate solution from the abdominal cavity through the glucose sensor and to interconnect the glucose dispenser. It can consist of a miniature pump (preferably peristaltic pump, as it allows isolating liquid medium from the pump itself) and a set of tubes (medical grade silicone tubes) that hydraulically connect the elements of the system, as shown in figure 4. The pump must be able to adjust the flow rate, because high circulation rate causes patient discomfort and damage to the peritoneal catheter [14], and low flow rate can cause precipitation of dialysate components and subsequent tube blockage. Additionally, fluid pressure sensors can be added to the system, it will allow detecting tubing occlusion or rupture. If a regulation system is used as part of a wearable "artificial kidney" device, it can be included in the device's hydraulic system, and perfusion parameters can be controlled from the outside.

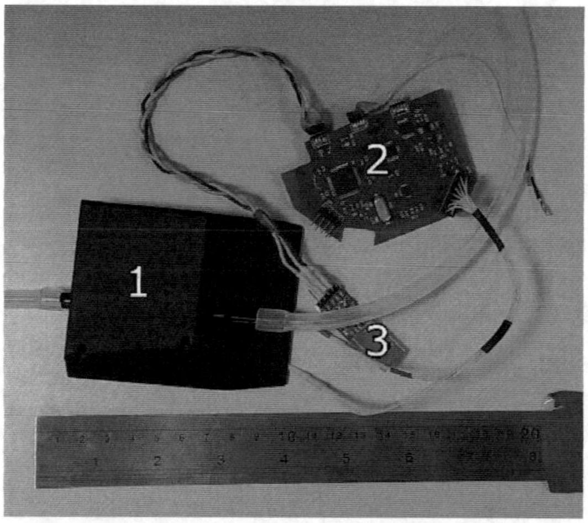

Figure 3. Photo of an experimental glucose sensor: 1 is optical system housing, 2 is electronic control system, 3 is USB interface board

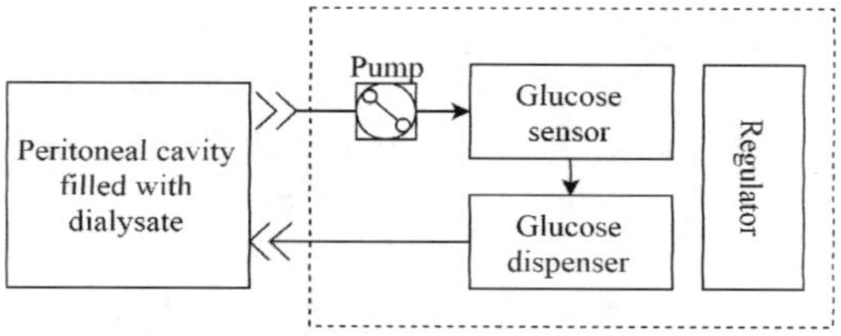

Figure 4. Glucose regulation system structure and fluid flow direction

2.4. Regulator (microprocessor device)

The regulator must provide direct control of the actuators and the feedback sensor, and receive commands from the reference unit. The current level of development of microelectronics allows the use of relatively inexpensive microcontrollers containing a high-performance processor, memory and peripheral devices (digital-to-analog and analog-to-digital converters, data transmission interfaces, etc.) as the central node of the regulating device. Low power consumption and small footprint minimizes device and battery size, which is critical for wearable and portable devices. To interface with external equipment, regulator must use wired communication (USB cable, for example) or wireless interface (Bluetooth, for example). In case of wireless communication, special security measures must be taken, because hacking of such a device can potentially lead to its malfunctioning, failure and/or damage to patient's health.

2.5. Reference unit (PC / smartphone)

The reference unit is necessary to set the target concentration of glucose in the solution, taking into account the physiological parameters of the patient, the type of solution for peritoneal dialysate, and the treatment plan. In the simplest version, this can be a constant level throughout the entire procedure. In the advanced version, it can be a free-form profile (for example, gradually increasing towards the middle of the procedure, and decreasing towards the end). The reference unit can be a PC of the physician, a patient's smartphone or a wearable "artificial kidney" device, in which the system is operated. As stated in previous section, reference unit can communicate with the regulator via wire and/or wireless protocols.

2.6. Miscellaneous system parts

Other parts of the system that were not shown in diagram, but worth mentioning are:

- Eye safety subsystem. Because IR-photometry glucose measurement method requires usage of relatively powerful laser, user eye safety must be implemented. In the investigated system several precautions were implemented, such as: non-transparent, non-reflective enclosure for optical system to contain laser radiation inside chassis; magnetic field sensor in the enclosure base and magnet in lid to detect opened cover and turn off the laser. For improved safety, additional software measures can be used, for example, if signal from optical sensor is above certain threshold (which means that lid was opened or laser light intensity has raised unexpectedly), then laser must be shut off immediately. Other safety measures will need to be implemented according to IEC 60825 standard for safety of laser products [15].
- Battery power subsystem / power management. Because the system is aimed to be used as standalone wearable device or as a part of wearable artificial kidney, energy efficiency must be as high as possible to ensure maximum battery runtime. To achieve that, several techniques can be utilized: 1. minimization of laser duty cycle. 2. Lowering the signal to noise ratio to acceptable minimum level (by adaptive regulation of laser power output). 3. Usage of low-power modes for electronics, especially for microcontroller, analog-to-digital converters and voltage regulators. 4. Emphasis on modern semiconductor technologies with low quiescent current and high power efficiency (for example, GaN DC-DC converters [16]).
- Temperature control and compensation subsystem. During experiments on the system described, large amount of effort was put in research repeatability and reproducibility of measurements because of considerable effect of ambient temperature and laser / photodiode temperatures on acquired data. Experiments were planned in a special manner to ensure that daily ambient temperature fluctuations were minimized; laser had to be warmed up to a certain equilibration point where its temperature would stay relatively constant, and output power variation will be minimal. This problem can be solved in different ways, from using thermoelectric cooled laser (very power consuming, but helps keeping constant temperature [17]) to algorithmical or electronic compensation [18] (by measuring relation between laser optic output and its temperature, and amending calculations accordingly) and digital noise filtering [19].

3. Results and conclusion

The problem of continuous removal of excess body fluid (ultrafiltration) in patients with renal excretory function failure (mostly caused by chronic kidney disease) is one of the long-standing challenges for renal replacement therapy methods. Current procedures such as hemodialysis are uncomfortable for the patient, and require a lot of time and resources. Peritoneal dialysis is a promising type of therapy that can increase patient's quality of life, but it is constricted by rapid reduction of ultrafiltration rate caused by decrease of glucose concentration in peritoneal dialysis fluid. To solve this problem, we propose a wearable system that will prolongate the procedure by measuring glucose concentration on-line and replenishing glucose according to mathematical model of fluid transport through patient's peritoneal membrane.

The proposed control system consists of a reference unit, glucose dispenser, perfusion system, feedback sensor (IR-photometry glucose concentration meter) and regulator. Reference unit is used to set target glucose level and / or profile; perfusion system is used to pump dialysate from patient's peritoneal cavity through feedback sensor (which measures glucose level optically) and glucose dispenser, which adds glucose according to regulator, where peritoneal fluid transport is modeled. Secondary system modules such as eye safety subsystem, battery power subsystem and temperature control / compensation subsystem are also described and discussed.

System prototype was tested on various model solutions, including model solution of spent peritoneal dialysis fluid. Study showed that it allows measuring glucose concentration in the range of 42-220 mmol/L with a relative error of about 15% compared to lab spectrophotometer.

This system may allow extending of the procedure duration and regulating the process of excess fluid removal during peritoneal dialysis. This in turn increases the efficiency and safety of renal replacement therapy (by reducing the number of dialysate exchanges; frequent exchanges cause irritation and possible infection), improves patient's quality of life (less often dialysate exchanges means that patient can spend more time on other activities), lowers the cost of therapy (by reducing the amount of expendable materials needed). The proposed system is capable to work autonomously or can be incorporated into wearable "artificial kidney" device, prolonging its time between exchanges of expendable materials (up to 24 hours and more), increasing its biocompatibility and efficiency.

4. References

[1] Rushing J 2009 Assessing for dehydration in adults *Nursing* **39** p 14
[2] Clark W F, Sontrop J M, Huang S-H and Bankir L 2016 Hydration and Chronic Kidney Disease Progression: A Critical Review of the Evidence *Am. J. Nephrol.* **43** pp 281–92
[3] Szeto C C and Johnson D W 2017 Low GDP Solution and Glucose-Sparing Strategies for Peritoneal Dialysis *Semin. Nephrol.* **37** pp 30–42
[4] Krediet R T 2018 Ultrafiltration Failure Is a Reflection of Peritoneal Alterations in Patients Treated With Peritoneal Dialysis *Front. Physiol.* **9** p 9
[5] Genovesi S, Nava E, Bartolucci C, Severi S, Vincenti A, Contaldo G, Bigatti G, Ciurlino D and Bertoli S V 2019 Acute effect of a peritoneal dialysis exchange on electrolyte concentration and QT interval in uraemic patients *Clin. Exp. Nephrol.* **23** pp 1315–22
[6] Himmelfarb J and Ratner B 2020 Wearable artificial kidney: problems, progress and prospects *Nat. Rev. Nephrol.* **16** pp 558–9
[7] Castro A C, Neri M, Nayak Karopadi A, Lorenzin A, Marchionna N and Ronco C 2019 Wearable artificial kidney and wearable ultrafiltration device vascular access—future d irections *Clin. Kidney J.* **12** pp 300–7
[8] Sun Y, Roumelioti M-E, Ganta K, Glew R H, Gibb J, Vigil D, Do C, Servilla K S, Wagner B, Owen J, Rohrscheib M, Dorin R I, Murata G H and Tzamaloukas A H 2020 Dialysis-associated hyperglycemia: manifestations and treatment *Int. Urol. Nephrol.* **52** pp 505–17
[9] Liu S, Zhang R, Zheng Z and Zheng Y 2018 Electromagnetic–Acoustic Sensing for Biomedical Applications *Sensors* **18** p 3203

[10] Namita T, Sato M, Kondo K, Yamakawa M and Shiina T 2017 Evaluation of blood glucose concentration measurement using photoacoustic spectroscopy in near-infrared region *Photons Plus Ultrasound: Imaging and Sensing* **10064** p 100645A
[11] Jain P, Maddila R and Joshi A M 2019 A precise non-invasive blood glucose measurement systemusing NIR spectroscopy and Huber's regression model *Opt. Quantum Electron.* **51** pp 51
[12] Maruo K., Oota T., Tsurugi M., Nakagawa T., Arimoto H., Tamura M., Ozaki Y. and Yamada Y 2006 New methodology to obtain a calibration model for noninvasive near-infrared blood glucose monitoring *Applied spectroscopy* **60** pp 441-9.
[13] Alves L, Almeida e Silva J and Giulietti M 2007 Solubility of d-Glucose in Water and Ethanol/Water Mixtures *J. Chem. Eng. Data* **52** pp 2166–70
[14] Crabtree J Shrestha B, Chow K-M, Figueiredo A E, Povlsen J V., Wilkie M, Abdel-Aal A, Cullis B and Goh B-L 2019 Creating and Maintaining Optimal Peritoneal Dialysis Access in the Adult Patient: 2019 Update *Perit. Dial. Int. J. Int. Soc. Perit. Dial.* **39** pp 414–36
[15] Safety of laser products (all parts). International Standard IEC 60825:2014. Geneva, Switzerland, IEC
[16] Spaziani L and Lucas L. 2018 Silicon, GaN and SiC: There's room for all: An application space overview of device considerations *2018 IEEE 30th International Symposium on Power Semiconductor Devices and ICs (ISPSD)*. pp 8-11
[17] Golliher E. 2002 Microscale technology electronics cooling overview *AIP Conference Proceedings* **608** pp 250-257
[18] Kumar D, Nauman R, Mohanty R, McGlynn S and Findley G 1983 Laser-power fluctuations: An analog signal-correction technique *Review of Scientific Instruments* **54** pp 463-8.
[19] Riris H, Carlisle C, Warren R and Cooper D 1994 Signal-to-noise ratio enhancement in frequency-modulation spectrometers by digital signal processing *Optics letters* **19** pp 144-6.

Acknowledgments
The reported study was funded by RFBR according to the research project No. 20-37-90049\20.

Problems and methods of a closed-loop blood glucose control system construction

E L Litinskaia, K V Pozhar and N M Zhilo

Institute of biomedical systems, National Research University of Electronic Technology, 1 Shokin Square, Zelenograd, Moscow 124498, Russia

E-mail: zheka101993@yandex.ru

Abstract. The automation of the insulin therapy in the closed-loop blood glucose control system is accompanied by such problems as delays of the feedback and the controlling action effect on blood glucose dynamics, difficulties in low and high glucose levels advance detection and timely compensation, a human factor, complexity of a mathematical model of control object building, evaluation of blood glucose control efficiency. In the paper, these automation problems and methods for their compensation are considered. The closed-loop system based on the model predictive control, adaptive adjustment of the controlling action and decision-making support was developed. Experimental testing with UVA/Padova Type 1 Diabetes Mellitus Simulator showed that the application of suggested methods in the closed-loop allows increasing the efficiency of the blood glucose maintaining in the target range.

1. Introduction

Type 1 diabetes mellitus is the endocrine chronic disease associated with high blood glucose (BG) caused by violation of insulin production in the pancreas. To control BG and maintain it in the target normal range, all over the world personalized closed-loop systems are being developed with different automation levels, control methods and strategies [1]. The main goal of researches and developments in the field is to build an automatic control system that will allow minimizing a patient participation in BG regulation.

Closed-loop control systems consist of an insulin pump as an effector, a continuous glucose monitor (CGM) as a feedback sensor, and a regulator that provides regulation of controlling action (subcutaneous insulin infusion) according to the feedback and built-in control algorithms [2]. Existing and developing systems show promising results in BG maintaining during clinical trials [3–5], however, along with the high cost of devices, the efficiency and safety of most systems today is not sufficient for their widespread use by patients with type 1 diabetes.

A number of difficulties related to the system structure, choosing the control method, physiological features of hormone regulation, mathematical modelling of the system and the efficiency of BG control evaluation accompanies development of an automated closed-loop system. This work is devoted to considering these problems and compensating methods including commonly applied as well as proposed ones.

2. Problems & Methods

2.1. Delays in the closed loop

Accuracy of the feedback sensor is calculated from its error and the feedback delay. Nowadays, invasive CGMs remain the glucose meters with the lowest error (12-25%) and the most effective feedback sensors [6]. However, the monitors measure glucose concentration in the subcutaneous fat and therefore the data is received with 15 min delay in comparison with actual BG. The controlling action also affects BG dynamics with a delay of more than 1.5 hours, depending on the type of insulin. These two problems lead to the lagging in the closed-loop, which decreases the automated control efficiency and causes health risks.

The lagging can be compensated by control methods based on advance regulation of the controlling action. For continuous low-rate insulin infusion (basal delivery), adaptive control algorithms adjusting the rate of infusion by feedback data are being developed [7, 8]. These algorithms are based on usage of optimized thresholds inside the target normal range for advance acceleration, braking or stopping of the controlling action, depending on deviation of measured values from the thresholds.

The proposed method of adaptive basal delivery (figure 1) is based on the use of a nested supernormal range, within which insulin is injected at the standard rate prescribed by the clinician. If the measured BG is in zones 1 or 2, the rate of insulin delivery is being proportionally increased until a certain maximum value reached at the G_{crit} level, and is being decreasing within zone 4. Zones 5 and 6 correspond to stopping of the insulin delivery in order to limit an overshoot.

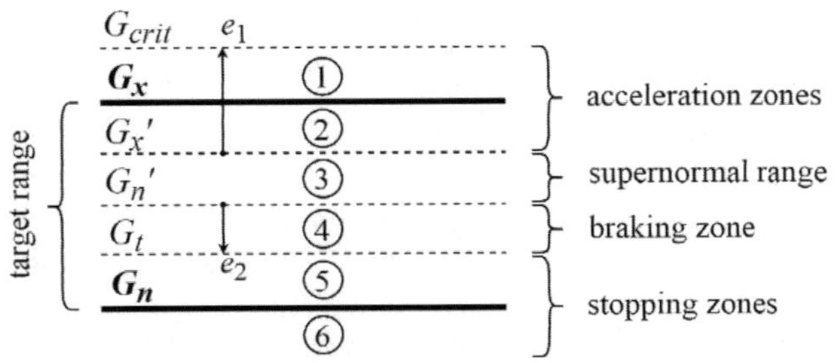

Figure 1. Thresholds and zones of adaptive basal delivery.

Food intake causes a sharp BG increase and requires an impulse controlling action to compensate it – a high-rate insulin infusion (bolus). Taking into account the delay in the insulin effect, to increase the efficiency of BG normalization, prediction algorithms are used to calculate the optimal bolus form and dose according to meal data and to inject insulin in advance. Modern closed-loop systems commonly use such control methods as PID-regulation, model predictive control (MPC), fuzzy logic or their combinations.

PID algorithms [9] respond to measured glucose levels and do not perform direct prediction with taking into account a number of physiological parameters but analysis of the future BG dynamic by its differential component. Therefore, even with well-optimized constants, PID-regulation is usually accompanied by significant overshoot and has a fairly long regulation time. When impulse controlling action, this can lead to a significant sharp decrease of BG and a risk to the patient's health.

A more physiological control method is MPC [10], which predicts the BG dynamics based on the mathematical modeling of the control object. The method takes into account not only current measured BG levels but also meal data and physiological parameters of a particular patient when formatting a controlling action. The disadvantage of the method is the complexity of constructing a sufficiently reliable mathematical model of the object. However, the method implies the possibility of the model parameters optimization during the system operation to improve the prediction accuracy.

Fuzzy logic methods [11] are based on modeling of physician's decisions on real-time adjustments of insulin doses based on CGM data. Compared to MPC and PID-regulation, fuzzy logic control uses more information from experts and relies less on mathematical modeling of the physiological system.

2.2. High and low BG values

Insulin therapy has a unidirectional effect on BG dynamics, allowing only reducing its value. At the same time, control and prevention of excessive overshoots leading to low BG is an important task of the closed-loop control. They usually occur after a meal and are caused by bolus overdosing.

One of the approaches to compensate overshoots is to use a dual-hormone closed-loop system based on additional infusion of glucagon, which increases BG. However, this method adds "complexity, possible risks and expense" [12] to the system and is not widely spread. The second approach to achieve bidirectional effect on BG dynamic is a decision-making support on additional carbohydrates intake [13]. The method also uses prediction to detect the risk of the low BG and to calculate the optimal amount of carbohydrates to intake.

In addition, various factors, such as variability in insulin sensitivity, stress, physical activity, can cause both an unforeseen BG decrease, and its unforeseen increase. High BG can also be observed as a result of insufficient after-meal insulin infusion. In these cases, when only basal delivery is insufficient for BG increase compensation, model predictive control allows calculation of a corrective bolus of a small dose.

2.3. Human factor

Despite the almost automatic BG control, most closed-loop systems require patient to input data about meals, additional insulin injections, current physical state and activities, which are needed to predict the BG dynamics and to calculate recommended controlling action values or corrective carbohydrates amount. Moreover, possible failures of control algorithms and as a result the risk of insufficient or excessive insulin infusion also limits the ability to build the fully automatic system and requires patient confirmation or correction of some controlling actions. Patient's wrong decisions, incorrect or missing data can lead to risks to the patient's health. That is why decision support on control and compensation of the system functioning and operation violations is required.

The proposed concept of such decision support algorithm is based on a comparative analysis of measured and predicted BG [14]. Measured values that deviate from the prognosis by more than the certain value are collected in a buffer. Then, by analyzing the trends in the buffers of predicted and measured values, the most likely events are assumed (figure 2) and advance recommendations to compensate BG dynamics are offered to the patient.

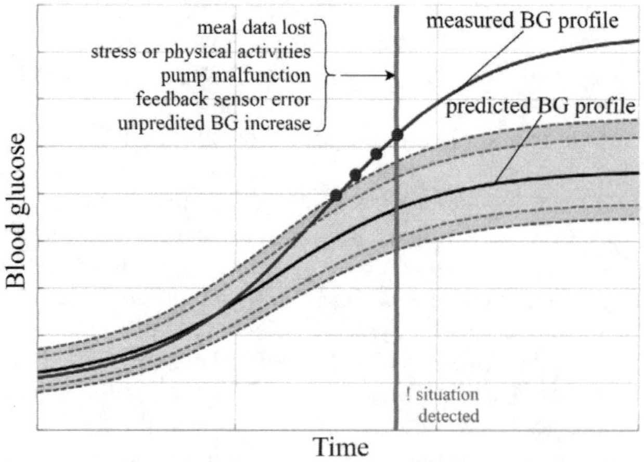

Figure 2. Events detection: one type of trends correlation can correspond to several possible events.

As a result of the trend analysis, such events are detected as the absence or falsity of data about meal or manual insulin injection, the absence of the meal fact, an insulin pump malfunction, a glucose sensor error, as well as stress or physical activity.

2.4. Mathematical modeling of the control object

An important step in a control system synthesis is the construction of mathematical models of system elements, including the control object. These models are used both to optimize system parameters and to form the basis of model predictive control. The mathematical model of the object should describe the controlled parameter dynamic in dependence on the controlling action, external disturbances and other effects arising in the system. In the considering system, the control object is the patient glucose regulation system. The BG dynamics is determined by a variety of physiological processes, the mathematical description of which, due to their multifactorial nature, is a non-trivial task.

Mathematical models aimed at a rigorous description of these processes [15] are usually high-order ODE systems. Some models quite accurately describe particular processes, for example, the delay in the absorption of insulin into the plasma [16], leading to a delay in the control action mentioned in 2.1, and food intake [17]. The work uses a mathematical model based on the integration of these models modified for the case of type 1 diabetes mellitus.

Based on the model, both short-term BG dynamics prediction and the calculation of the optimal values of the recommended corrective boluses and carbohydrates are carried out. Calculations are performed by the Dormand-Prince method in Matlab, while the edges of the food intake and insulin infusion pulses are approximated with steep logistic functions.

2.5. Evaluation of BG control efficiency

When evaluating the efficiency of closed-loop BG control systems in clinical trials, generally accepted criteria are the percentage of measured BG values within the target range (70–180 mg / dL), the glycated hemoglobin level, which correlates with the average BG for 3 months observation period and the number of low and high BG events. However, there is no method to quantify the efficiency of closed-loop BG control at the stage of a system development and its parameters optimization, taking into account not only clinical criteria, but also technical ones related to control efficiency and quality.

An integrated method is proposed for efficiency evaluation of BG control based on criteria of overshoot P ($g_{\min}$), maximum BG M ($g_{\max}$), hyperglycemia integral index J_g (S), regulation time $J_t(\tau)$ and safety of the advance bolus infusion B ($\tilde{g}_{\min}$), illustrated by the figure 3 and calculated as follows:

$$P = \frac{g_{\min} - G_n}{G_{fasting} - G_n}, \quad P \in [0;1] \tag{1}$$

$$M = \frac{G_{crit} - g_{\max}}{G_{crit} - G_{Hb}}, \quad M \in [0;1] \tag{2}$$

$$J_g = \frac{1}{\frac{1}{J_h} \int_{T_{G_{Hb}}} G(t) - G_{Hb} \, dt}, \quad J_g \in (0;1] \tag{3}$$

$$J_t = 1 - \frac{\tau}{\tau_{crit}}, \quad J_t \in [0;1] \tag{4}$$

$$B = P\Big|_{t_m = t_m + 15} = \frac{\tilde{g}_{\min} - G_n}{G_{fasting} - G_n}, \quad B \in [0;1] \tag{5}$$

where $g_{\min}$ is a minimum BG, $G_{fasting}$ is a target BG, G_n is the lower limit of the target range, $g_{\max}$ is a maximum BG, G_{crit} is the upper threshold value, G_{Hb} is the average BG value for a normal glycated hemoglobin index, J_h is the integral value under similar conditions for a healthy person, τ_{crit} is the maximum regulation time, $\tilde{g}_{\min}$ is a minimum BG for B criterion, t_m is the meal time.

The criteria combination using to calculate the overall efficiency function depends on the purpose of the method usage. For example, the criteria J_g and J_t are correlated and it is sufficient to use only one of them. At the same time, the hyperglycemia integral index is more complex in calculation and requires processing a large amount of data and is applicable rather at the stage of optimization of the control system parameters, while the regulation time can be used when testing closed-loop systems during preclinical trials. Criterion B is related to modeling the possibility of meal delay and the predicted risk to the patient's health in this case. As such, it cannot be evaluated when analyzing the results of research trials, but it can be used to determine the optimal and safe bolus form.

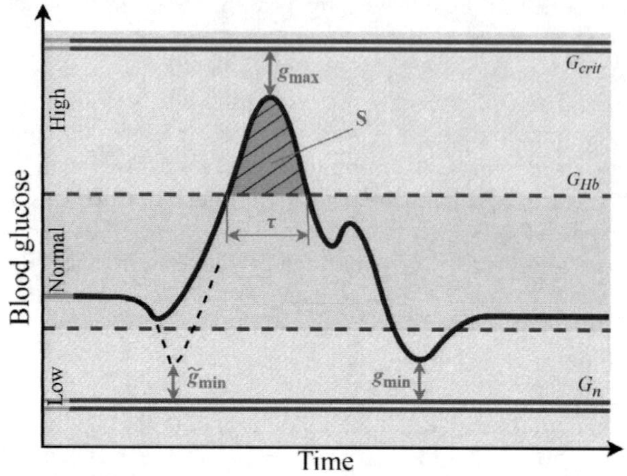

Figure 3. Illustration of efficiency criteria.

3. Results

Taking into account the considered problems and on the basis of proposed methods, the closed-loop BG control system was developed [18]. The system comprises of an automatic adaptive adjustment of basal delivery according to CGM data, and decision support algorithms based on model predictive control for calculation of optimal after-meal boluses, corrective boluses and carbohydrates as well as detecting and compensation of the system functioning and operation violations (figure 4).

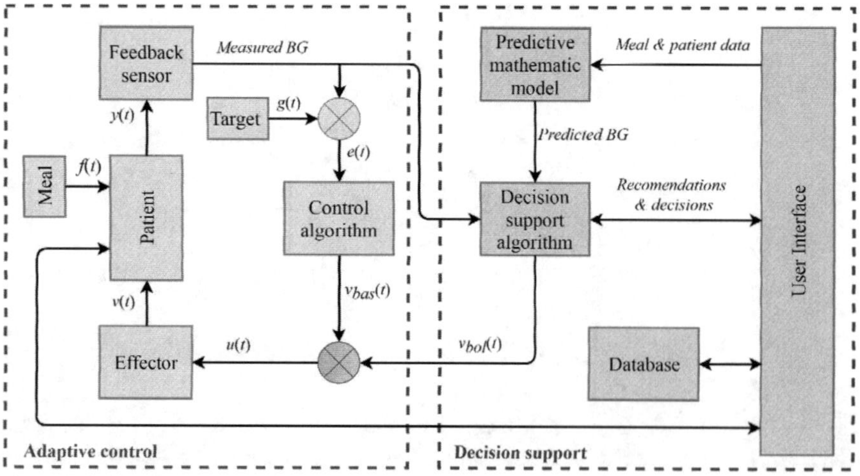

Figure 4. The structure of the closed-loop insulin therapy system with decision support.

In silico testing of the system was performed via UVA/Padova Type 1 Diabetes Mellitus Simulator. The proposed system was compared with a simulated open-loop one that provides instant bolus injection and constant basal delivery. A comparative analysis of the BG control efficiency was carried out according to the method proposed in 2.5. The overall efficiency function was compiled on the basis of overshoot, maximum BG and regulation time criteria. For each BG fluctuation, the efficiency increase for the proposed closed-loop system compared to the open-loop one had been calculated.

The control algorithm showed promising results in BG maintaining in adults and adolescents [19]. The 91 % of measures were in the target range during 72-hours simulation with regular meals from 20 to 75 g (figure 5). The adaptive adjustment of the controlling action allowed decreasing the overshoot during nights in comparison with insulin infusion with constant rate. The average increase in BG control efficiency was 27.3% with a decrease in the maximum value of 10.6%, a decrease in the regulation time of 22.5% and an insignificant increase in overshoot of 4.3%.

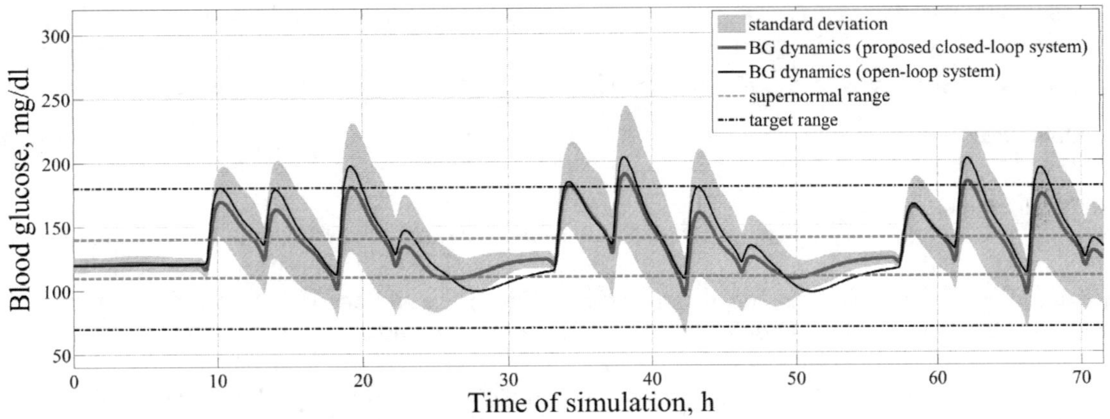

Figure 5. Results of 72-hours simulation with regular meals from 20 to 75 g.

The ability of the system to detect the risk of low and high BG levels and the quality of proposed optimized corrective actions was tested. The system allows advance risks detection and recommendations making, while the simulations shows better BG dynamics when a patient makes a decision to inject a recommended corrective bolus or to intake the calculated corrective amount of carbohydrates than without the decision (figure 6).

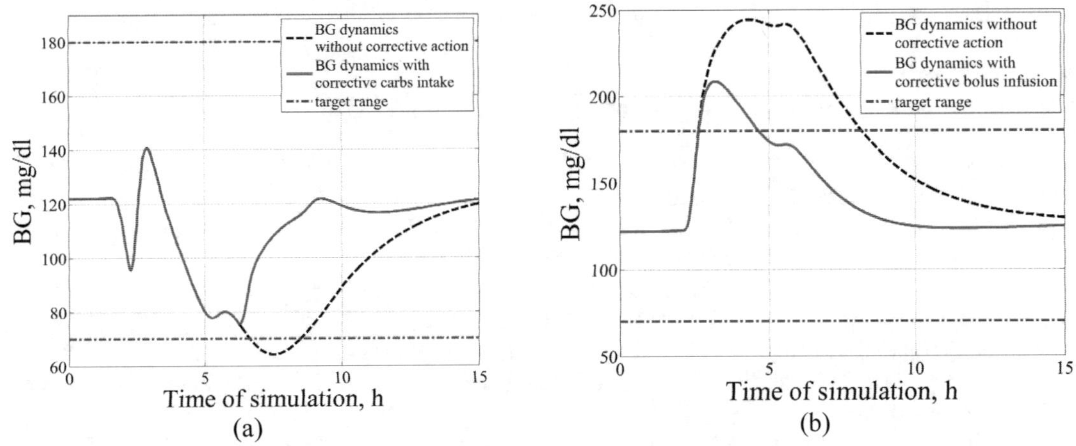

Figure 6. BG dynamics with and without corrective action in cases of low (a) and high (b) BG.

The developed decision support algorithm detects the violations with high sensitivity (96–100 %, CI 95 %, $p < 0.001$) and high specificity (91–100 %, CI 95 %, $p < 0.001$), that allows preventing BG deviations from the normal range caused by the human factor or system functioning violations. The results of evaluating the ability of the algorithm to detect the events of absent meal data and additional insulin injection data, absent meal fact and pump malfunction for a given threshold time T_{thr} are presented in the table 1.

Table 1. Results of developed decision support algorithm testing.

Event	T_{thr}, min	Time of detection, min	Sensitivity, %	Specificity, %
Meal data absence	90	40.0 ± 6.7	100,0 (CI 95 %, 95.8 – 100.0)	91,5 (CI 95 %, 90.1 – 91.5)
Additional insulin data absence	80	42.4 ± 3.0	100,0 (CI 95 %, 96.9 – 100.0)	99.7 (CI 95 %, 98.6 – 99.7)
Meal fact absence	120	74.7 ± 19.5	96,0 (CI 95 %, 90.8 – 98.6)	95.0 (CI 95 %, 93.3 – 95.9)
Pump malfunction	90	71.4 ± 16.5	97,0 (CI 95 %, 93.6 – 97.0)	100.0 (CI 95 %, 98.9 – 100.0)

4. Discussion & Conclusion

The construction of an automated closed-loop BG control system is accompanied by a number of difficulties. This paper considers such problems as delays in a closed-loop associated with the peculiarities of insulin and glucose absorption; problems of control and timely compensation of high and low BG; the human factor; problems of mathematical modeling of the control object and evaluation of BG control efficiency

Control and decision-making support methods, which allow taking into account or compensation for these problems are considered and proposed. On the basis of the proposed methods a closed-loop BG control system was developed and tested via the UVA/Padova Type 1 Diabetes Mellitus Simulator.

Methods of adaptive basal delivery by feedback sensor data based on nested threshold values and methods of BG dynamics prediction make it possible to compensate for the lagging in the closed-loop by advance controlling action. Moreover, a comparative analysis of the predicted and measured profiles, as well as control of their approach to the target range boundaries, provide timely decision-making support on BG normalization and compensation of system functioning and operation violations. Despite the fact that these methods allow minimizing the patient participation in the BG control, they still require additional data input and making corrective and managerial decisions.

Acknowledgments

The reported study was funded by RFBR, project number 19-37-90028.

References

[1] Trevitt S, Simpson S and Wood A, 2016 *J. Diabetes Sci Technol* **10** 714-723
[2] Litinskaia E L and Pozhar K V, 2019 *Proceedings of the 2019 III International Conference on Control in Technical Systems* 152-155
[3] Bergenstal R et al., 2016 *JAMA* **316** 1407-1408
[4] Brown S A et al., 2019 *N. Eng. J. Med.* **381** 1701-1717
[5] Buckingham B A et al., 2018 *Diabetes Technology & Therapeutics* **20** 257-262
[6] Kovatchev B P, Patek S D, Ortiz E A and Breton M D, 2015 *Diabetes Technology & Therapeutics.* **17** 177-86
[7] Forlenza G P et al., 2018 *Diabetes Care* **41** 2155-2161
[8] Litinskaia E L, Pozhar K V and Polyakova E A, 2021 *Biomedical Engineering* **54** 366-370

[9] Wolpert H, Kavanagh M, Atakov-Castillo A and Steil G M, 2016 *Diabetic Med.* **33** 235-242
[10] Hovorka R et al., 2004 *Physiol. Meas.* **25** 905–920
[11] Miller S, Nimri R, Atlas E, Grunberg E A and Phillip M, 2011 *Diabetes Technology & Therapeutics* **13** 983-990
[12] Fabris C and Kovatchev B P, 2020 *Artificial Organs* **44** 671-679
[13] Pérez-Gandía C et al., 2018 *Journal of diabetes science and technology* **12** 243-250
[14] Litinskaia E L and Kozlov M Yu, 2020 *Perspektivy nauki* **12** 54-58
[15] Bazaev N A, Pozhar K V and Rudenko P A, 2014 *Meditsinskaia tekhnika* **6** 8-11
[16] Kraegen E W and. Chisholm D J, 1984 *Diabetologia* **26** 208-213
[17] Dalla Man C, Rizza R.A, and Cobelli C, 2007 *IEEE Trans. Biomed. Eng.* **54**
[18] Litinskaia E L, 2021 *Izvestiya vuzov. Elektronika* **26** 162-171
[19] Litinskaia E L, Pozhar K C and Bazaev N A, 2021 *Proceedings of the 2021 IEEE Conference of Russian Young Researchers in Electrical and Electronic Engineering* 2842-2846

Computational fluid dynamics simulation of hemolysis at different levels of circulatory support in the left ventricular assist device Sputnik

A Romanova[1], D Telyshev[1,2]

[1]Institute of Biomedical Systems, National Research University of Electronic Technology 1 Shokin Square, Zelenograd, Moscow 124498, Russia
[2]Institute for Bionic Technologies and Engineering, Sechenov University 8 Trubetskaya street, Moscow 119991, Russia

E-mail: romanova@bms.zone

Abstract. Designing a ventricular assist device is a complex technological process, and testing a finished product requires a significant investment of money and time. Simulation allows research to conduct research early in the development of a device, thereby reducing time and material costs. In this work, the calculation of hemolysis in the left ventricular assist device Sputnik (Sputnik LVAD) is carried out. Three different levels of circulatory support were chosen. For the first level, the following parameters were selected: blood flow rate of 2 L/min. at a rotor speed of 8000 rpm; for the second - 4 L/min, 8500 rpm; for the third - 6 L/min, 9000 rpm. The distribution of scalar shear stresses and the index of hemolysis were obtained from the pathlines of the particles. When comparing three operating points of LVAD Sputnik, hemolysis indices were obtained using the Lagrangian model. The mean hemolysis indeces were 0.0284%, 0.0210%, 0.0155% for LVAD Sputnik operating at a fixed rate of 2, 4, 6 L/min at a pressure of 100 mm Hg, respectively. The calculation results show that the capacity of 6 L/min is better than 2 and 4 L/min.

1. Introduction

The development of wearable implantable circulatory support devices is an urgent task of the advancement of medicine. This devices enhance the survival rate of patients with acute heart failure (HI), with one of the most dangerous and common diseases in the world. For patients with end-stage heart failure, therapeutic methods cannot provide recovery, and heart transplantation is an effective treatment. Unfortunately, the number of patients requiring transplantation significantly exceeds the number of available donor organs [1]. Since then, the advent of wearable artificial devices that support or partially replace human circulation has become important in the development of medicine, allowing a patient suffering from end-stage heart failure to lead a normal life outside the clinic.

One of the mechanical circulatory support devices is the left ventricular assist device Sputnik (LVAD Sputnik) [2, 3]. It is an axial rotary blood pump (RBP) (figure 1). LVAD replaces the transport function of the left ventricle.

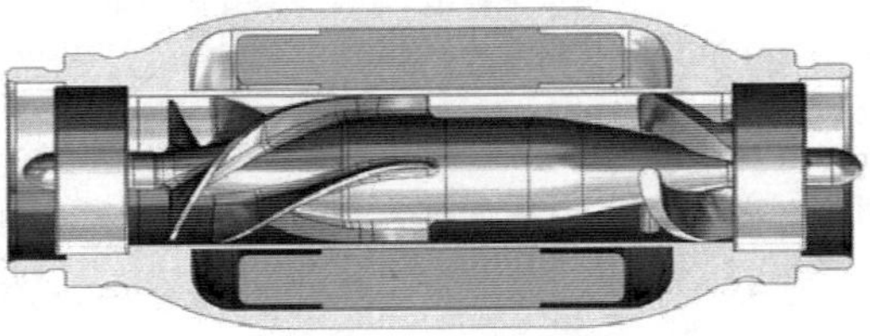

Figure 1. Design of blood pump Sputnik

However, the clinical benefits and cost-effectiveness of LVAD therapy remain limited [4, 5]. In order to expand the use of rotary blood pumps, increase the life expectancy and quality of life of patients, it is required to reduce the side effects arising from these pumps and the cost of therapy.

For the development of LVAD, modern computational fluid dynamics (CFD) methods are used. They help to study the blood flow field inside the device, as well as to predict and evaluate possible blood damage [6–9]. General blood damage alters the clotting process and can lead to life-threatening conditions, such as thrombosis. Therefore, damage to the formed blood elements that occur inside the rotary pump is the main problem. Determining the level of various types of damage inside the device is a difficult task. Various types of damage include hemolysis, platelet activation, and degradation of von Willebrand factor (vWF), which attaches platelets to the damaged vessel. The factor that has a strong influence on the formed elements of the blood is the shear stress in the blood flow field. On the one hand, when the shear stress is low, the blood stagnates, platelet deposition and thrombosis occur. On the other hand, high shear stress can lead to hemolysis (damage of erythrocytes) and platelet activation [8, 10-12].

Excessive exposure to red blood cells (erythrocytes) leads to a reduction in life expectancy or complete damage to red blood cells, which leads to morphological and biochemical changes in the blood.The rupture of red blood cells is accompanied by the release of hemoglobin in the blood plasma. This can lead to a decrease in the content of oxygen and carbon dioxide, and can also cause saturation of the kidneys due to the toxicity of free hemoglobin [13]. Studies of deformation and damage to red blood cells have shown that the most important parameters for assessing blood damage in LVAD are the shear stress and the time at which its effect occurs [14-16].

In this work, a study of the effect of circulatory support different levels on hemocompatibility in the LVAD Sputnik has been carried out. Three different levels of circulatory support were chosen. For the first level, the following parameters were selected: blood flow rate is 2 L/min, at a rotor speed of 8000 rpm; for the second and third - the flow rate is 4 and 6 L/min, at a rotor speed of 8500 and 9000 rpm, correspondingly.

2. Methods

Numerical simulation of the fluid flow in the pumps was carried out using the Fluent ANSYS computational fluid dynamics software. Three-dimensional models were prepared in the SolidWorks environment to build the meshes.

The shear stress transport (SST) k-ω turbulence model was chosen to solve the three-dimensional Reynolds-averaged Navier-Stokes equations of the flow, because this model operates as the k-ω model in zones spaced from the walls and as k-ε model in the near-wall layers, thus, describing the flow behavior more precisely [10, 11]. The trajectory of particles simulating red blood cells was calculated by the discrete phase model (DPM). It allowed to determine flow impact on behavior of RBCs based on the Lagrangian approach.

The mesh was created on the basis of tetrahedrons and prisms. The wall areas between the housing and the impeller blades were formed from 10 prismatic layers. Where the gradient of the flow parameters change was the largest, the cell density was increased for the convergence of the calculation. During the formation of the final version of the mesh, several iterations of the mesh convergence studies were carried out. The convergence criterion was 10-4 for the velocity, turbulent kinetic energy, residual of continuity and specific dissipation rate in this numerical simulation, which is sufficiently accurate for a reasonable computational time.

The hemolysis index determines the number of red blood cells damaged by increased shear stress [17]. In the Lagrangian model, the hemolysis index is integrated along the trajectories of particles [18] or streamlines [7-8, 19]. After solving the constitutive equations of the flow in the computational domain, the trajectories from the inlet to the outlet of the pump are calculated. Then the lines are sampled, and the damage to the particle p in the time step s is apparently calculated as follows:

$$d_{p,s} = 1.800 \cdot 10^{-6} \tau_s^{1.991} \Delta t_s^{0.765}, \tag{1}$$

where $\Delta t_s = t_s - t_{s-1}$ is time step, $s = 1, \ldots, S$ and S is the index corresponding to the last moment in time when the particle leaves the pump; $p = 1, 2, \ldots, N$, where N = 400 is the overall number of particles in the simulation.

Particle damage will intensify through the pump as follows:

$$D_{p,s} = D_{p,s-1} + (1 - D_{p,s-1}) d_{p,s}. \tag{2}$$

$D_p = D_{p,S}$ - maximum damage to the particle *p* after passing through the pump. Since we assume that there is no damage at the initial time, $D_{p,0} = 0$. Thus, the pump hemolysis index E is calculated by the formula 3 as the average damage of all particles [8]:

$$E = \sum_{p=1}^{N} \frac{D_p}{N}. \tag{3}$$

In addition, the assessment of the effect of high shear stress on blood can be realized based on the volumetric shear stress distribution. The influence degree of shear stresses can be characterized by threshold values of 9 Pa, 50 Pa and 150 Pa [12]. The von Willebrand factor deteriorates at 9 Pa, platelets are activated at 50 Pa, and erythrocytes are destroyed at 150 Pa. Calculation of scalar shear stress τ was carried out according to the equations (4) based on viscous stresses by the methods described in literature [6, 15]. Viscous stresses are determined by the values of velocity gradient in the computational domain. The indices *i*, *j*, *k* correspond to directions of *x*, *y*, *z* axes in a fixed Cartesian coordinate system.

$$\tau = \left[\frac{1}{3}(\tau_{ii}^2 + \tau_{jj}^2 + \tau_{kk}^2) - \frac{1}{3}(\tau_{ii}\tau_{jj} + \tau_{jj}\tau_{kk} + \tau_{kk}\tau_{ii}) + (\tau_{ij}^2 + \tau_{jk}^2 + \tau_{ki}^2)\right]^{1/2},$$

$$\tau_{ii} = -2\mu\frac{\partial V_x}{\partial x}, \tau_{jj} = -2\mu\frac{\partial V_y}{\partial y}, \tau_{kk} = -2\mu\frac{\partial V_z}{\partial z}, \tag{4}$$

$$\tau_{ij} = -\mu\left(\frac{\partial V_x}{\partial y} + \frac{\partial V_y}{\partial x}\right), \tau_{jk} = -\mu\left(\frac{\partial V_y}{\partial z} + \frac{\partial V_z}{\partial y}\right), \tau_{ki} = -\mu\left(\frac{\partial V_z}{\partial x} + \frac{\partial V_x}{\partial z}\right),$$

where μ is dynamic fluid viscosity; V_x, V_y, and V_z are components of the flow velocity vector.

Fluid with constant viscosity of 3.5 mPa·s and density of 1055 kg/m^3 was used as it commonly utilized in CFD simulations of RBP operation [12, 20-21].

For the simulations, 400 spherical particles with diameter of 7 μm and density of 1106 kg/m^3 were used as basic RBCs. Mesh nodes at the inlet section of computational domain were the source of particles. The total mass flow of particles was 0.0392 kg/s with initial velocity of 1 mm/s along the normal to the inlet section. The value of time step along the particle trajectories had mean of 0.2 ms.

3. Results

This study focuses on the mechanical destruction of erythrocytes (hemolysis) associated with the exposure to shear stress and the time during which this exposure occurs. Wall shear stress were obtained (figure 2). It can be seen from the figures that the high shear stress is between the rotor blades and the housing.

In the course of the work, the volumetric distribution of scalar shear stresses (SSS) was obtained (figure 3). The following threshold values of the scalar shear stress were selected to evaluate the effect of the rotary blood pump on the formed elements: von Willebrand factor degradation (9 Pa), platelet activation (50 Pa) and hemolysis (150 Pa).
In this study, hemolysis indicators were obtained using the Lagrange method for three LVAD Sputnik working points.The average hemolysis index was 0.0284%, 0.0210%, 0.0155% for LVAD Sputnik operating at a fixed rate of 2, 4, 6 L/min at a pressure of 100 mm Hg, respectively.
The calculation results show that the capacity of 6 L/min is better than that of 2 and 4 L/min. Shear stress above 150 Pa increases as the flow rate increases at operating points 2, 4, 6 L/min. On the contrary, the exposure time decreases.

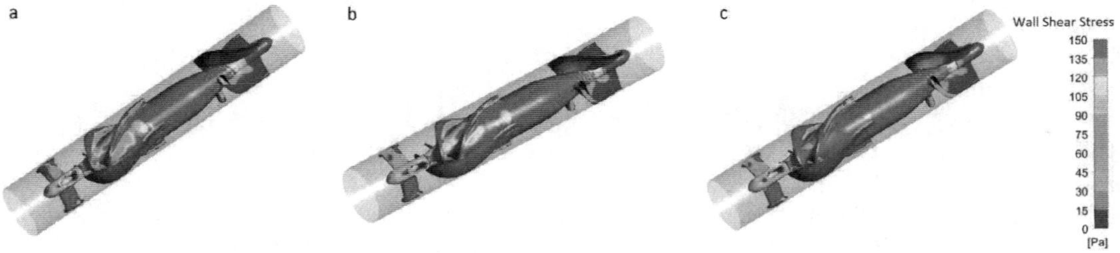

Figure 2. Wall shear stresses for LVAD Sputnik operating at a fixed rate of 2 (a), 4 (b), 6 (c) L/min

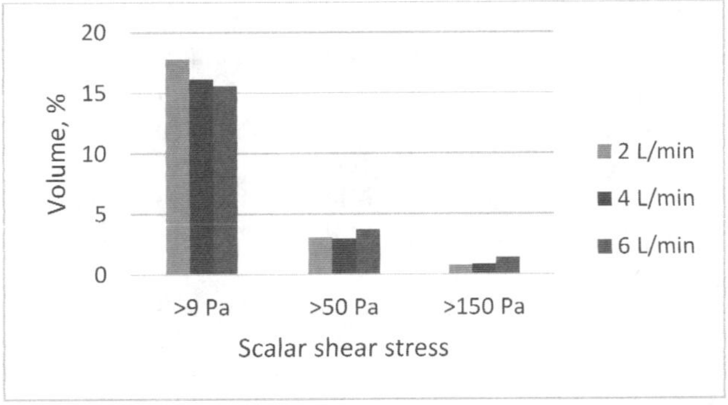

Figure 3. Volumes of blood flow regions subject to specific shear stress thresholds

4. Discussion

In this study, the fields of blood flow and hemocompatibility were considered. The effect on red blood cells was studied in terms of hemolysis indicators commonly used in CFD. Hemolysis is not the only type of blood damage that occurs on the path of RBP. The hemolysis indices are evaluated by the exposure time and the distribution of stresses relative to the established thresholds for various types of blood damage. These thresholds corresponding to vWF degradation (9 Pa), platelet activation (50 Pa) and hemolysis (150 Pa) are taken from the literature [12, 22]. Other threshold values for different types of blood damage are given in the sources [11, 20]. It should be noted that the choice of threshold levels may affect the results. Despite the fact that the threshold values only classify the shear stress and do not take into account the actual damaging effect, this method remains easy, simple and visual for the characteristics of devices. In addition, this method is combined with other indicators of blood damage.
Non-Newtonian blood characteristics were not considered. The viscosity becomes independent of the shear stress at shear rates above 100 s-1 [23], which suggests that blood can be considered Newtonian fluid. In most blood volumes for both Sputnik pumps, the shear rates are higher than 100 s-1.

Therefore, we used the Newtonian blood flow model (constant viscosity of whole blood), which greatly simplified the implementation. Many models have been created that describe the viscous properties of blood in different ways when it behaves like a non-Newtonian liquid [24]. At the moment, there is no single model that would satisfy all researchers, so the assumption that blood has a constant viscosity is generally accepted in practice [25].

5. Conclusions

The study of the effect of circulatory support different levels on hemocompatibility in the LVAD Sputnik has been carried out. The mean hemolysis indices were 0.0284%, 0.0210%, 0.0155% for LVAD Sputnik operating at a fixed rate of 2, 4, 6 L/min at a pressure of 100 mm Hg, respectively. The calculation results show that the capacity of 6 L/min is better than 2 and 4 L/min. Thus, it is impossible to predict the behaviour of hemolysis based on only one parameter, shear stress or exposure time.

Acknowledgments
This study was funded by RFBR and DFG (research project no. 19-51-12005).

References

[1] Genovese E A 2009 Incidence and patterns of adverse event onset during the first 60 days after ventricular assist device implantation *The Annals of thoracic surgery* **88**(4) 1162–70
[2] Selishchev S and Telyshev D 2016 Optimisation of the Sputnik-VAD design *IJAO* **39** pp 407–414
[3] Pugovkin A, Markov A, Selishchev S, Korn L, Walter M, Leonhardt S and Telyshev D 2019 Advances in hemodynamic analysis in cardiovascular diseases investigation of energetic characteristics of adult and pediatric sputnik left ventricular assist devices during mock circulation support *Cardiology Research and Practice* 1–15.
[4] Nunes A J 2016.A systematic review of the cost-effectiveness of long-term mechanical circulatory support *Value in Health* **19** 494–504
[5] Schmier J K, Patel J D, Leonhard M J and Midha P A 2019 A systematic review of cost-effectiveness analyses of left ventricular assist devices: issues and challenges *Appl Health Econ Health Policy* **17** 35–46
[6] Bludszuweit C 1995 Model for a general mechanical blood damage prediction *Artif Organs* **19** 583–589
[7] Song X., Throckmorton A L, Wood H G, Antaki J F and Olsen D B 2003 Computational fluid dynamics prediction of blood damage in a centrifugal pump *Artif Organs* **27** 938–941
[8] Yano T 2003 An estimation method of hemolysis within an axial flow blood pump by computational fluid dynamics analysis *Artif. Organs* **27** 920–925
[9] Hentschel B 2008 Interactive blood damage analysis for ventricular assist devices *IEEE Transactions on Visualization and Computer Graphics* **14** 1515–22
[10] Fraser K H, Taskin M E, Griffith B P and Wu Z J 2011 The use of computational fluid dynamics in the development of ventricular assist devices *Med Eng Phys* **33** 263–280
[11] Su B, Chua L P, Lim T M and Zhou T 2010 Evaluation of the impeller shroud performance of an axial flow ventricular assist device using computational fluid dynamics *Artif Organs* **34** 745 –59
[12] Fraser K H, Zhang T, Taskin M E, Griffith B P and Wu Z J 2012 A quantitative comparison of mechanical blood damage parameters in rotary ventricular assist devices: shear stress, exposure time and hemolysis index *J Biomech Eng* **134** 081002
[13] Franz D D 2019 Long-term changes in kidney function after left ventricular assist device implant: an analysis of the STS Intermacs database. *J. Heart Lung Transplant* **38** S89–S90
[14] Sohrabi S and Liu Y 2017A cellular model of shear-induced hemolysis *Artif Organs* **41** E80–E91

[15] Goubergrits L 2016 Turbulence in blood damage modeling *IJAO* **39** 160–5
[16] Arora D, Behr M and Pasquali M 2004 A tensor-based measure for estimating blood damage *Artif Organs* **28** 1002–15
[17] Garon A and Farinas M 2004 Fast Three-Dimensional Numerical Hemolysis Approximation *Artif Organs* **28** 1016–25
[18] Heuser G and Opitz R 1980 A Couette viscometer for short time shearing of blood *Biorheology* **17** 17–24
[19] Arora D, Behr M and Pasquali M 2006 Hemolysis Estimation In A Centrifugal Blood Pump Using A Tensor-based Measure *Artif Organs* **30** 539–47
[20] Yu J and Zhang X 2016 Hydrodynamic and hemolysis analysis on distance and clearance between impeller and diffuser of axial blood pump *J Mech Med Biol* **16** 1650014
[21] Chan W K, Wong Y W, Ong W, Koh S Y and Chong V 2005 Numerical investigation of the effects of the clearance gap between the inducer and impeller of an axial blood pump *Artif Organs* **29** 250-8
[22] Wiegmann L 2017 Blood pump design variations and their influence on hydraulic performance and indicators of hemocompatibility *Ann Biomed Eng* **46** 417–28
[23] Thamsen B 2015 Numerical analysis of blood damage potential of the HeartMate II and HeartWare HVAD rotary blood pumps *Artif Organs* **39** 651-9
[24] Bird R B, Armstrong R C and Hassager O 1987 Dynamics of Polymeric Liquids *Fluid Mechanics* 1
[25] Papathanasopoulou P 2003 MRI measurement of time-resolved wall shear stress vectors in a carotid bifurcation model, and comparison with CFD predictions *J Magn Reson Imaging* **17** 153–62

Method for dynamic measurement of thermal diffusivity during laser soldering of biological tissues

D I Ryabkin[1], V V Molodykh[2], A Yu. Gerasimenko[3]

[1]National Research University of Electronic Technology, Shokina Square, 1, Zelenograd,Moscow, Russia
[1]I.M. Sechenov First Moscow State Medical University, Trubetskaya 8, Moscow, Russia
[2]National Research University of Electronic Technology, Shokina Square, 1, Zelenograd, Moscow, Russia
[3]National Research University of Electronic Technology, Shokina Square, 1, Zelenograd, Moscow, Russia
[3]I.M. Sechenov First Moscow State Medical University, Trubetskaya 8, Moscow, Russia

E-mail: ryabkin@bms.zone

Abstract. In this paper, we propose a method for dynamic measurement of the thermal diffusivity coefficient during laser soldering of biological tissues. The method is based on modelling the function of temperature dependence on time during cooling of biological tissue after exposure to laser radiation. The simulation is carried out by solving the heat equation for a homogeneous biological tissue and the absence of external heat sources. The desired value of the thermal diffusivity coefficient was determined by optimizing the residual functional of the temperature functions from time obtained experimentally and by solving the thermal diffusivity equation. Experiments were carried out to measure the thermal diffusivity coefficient by the proposed method for myocardial and skin tissues at maximum heating temperatures of 40, 50, 60 °C. The measured values of the thermal diffusivity coefficient for the myocardium are in the range from 2.3 to 2.7 $m^2/s*10^{-6}$, and for the skin from 1.5 to 1.7 $m^2/s*10^{-6}$.

1. Introduction

Laser soldering of biological tissues is an alternative method of joining dissected biological tissues. The using of a laser makes it possible to form a small in size and sealed for liquids weld. This is important for the reconstruction of small-sized vessels. The method of biological tissues restoring by laser soldering has a number of disadvantages. One of the most significant is the temperature necrosis of the connected biological tissues. Various methods are used to reduce the area of temperature necrosis. For example, exogenous chromophores are used to increase the absorption capacity of solders and reduce the overall intensity of laser radiation [1-4]. It is also possible to use the temperature feedback connection of the laser radiation power. To increase the tensile strength of the welds, it is possible to use laser solders based on albumin [5, 6], indocyanine green [7-10] and chromophores based on nanoparticles [11, 12]. In most cases, the temperature feedback is based on a proportional integral differential (PID) controller [13, 14]. The PID controller has three P, I, D, these are the coefficients of the corresponding components. The hyperparameters are unique for each type of fabric and solder. Because of this, there is a need for each use of laser soldering in selection of hyperparameters. The process of selecting P, I, D hyperparameters implies a significant number of experiments on the biological tissue being connected, which is often unacceptable for both technical and ethical reasons.

Hyperparameters P, I, D can be selected by modeling the laser soldering process. But for modeling, it is necessary to know the coefficient of thermal diffusivity. This work is aimed at developing a method for dynamically determining the coefficient of thermal diffusivity of biological tissues in the process of laser soldering.

2. Materials and methods

2.1. Measuring setup

The measurements were carried out at a setup for laser soldering of biological tissues. The setup uses a continuous diode laser with a radiation wavelength $\lambda = 808$ nm (Figure 1). The radiation was transported to the soldering area using an optical fiber with a diameter $d = 600$ μm. Temperature measurements were carried out by a non-contact infrared sensor. The optical axes were placed in such a way that the measurement and irradiation areas intersected. A detailed description of the installation is given in [13].

It should be noted that the design features of the device used for laser welding of biological tissues can introduce a systematic error in measurements. In the system used in this work for laser welding of biological tissues, a temperature feedback is implemented. The IR sensor measures the temperature of the laser-welded area by a non-contact method. The measurement area of the IR sensor has a divergence angle. Because of this, differences emerge in the measurement area when the IR sensor is located at different distances from the measured surface. The measuring areas of the IR sensor and the laser radiation must be equal. Because a semiconductor laser is used, the radiation intensity is proportional to the normal distribution law. This leads to a non-steady temperature distribution in the laser welding area. If the measurement area of the IR sensor is larger than the area of exposure to laser radiation, then the temperature averaging will take into account the unheated tissue areas. Thus, the measured temperature will be underestimated relative to the real one. If the measurement area of the IR sensor is smaller, the area of exposure to laser radiation, then the temperature averaging will take place over the central, more heated area, which will lead to an overestimation of the measured temperature relative to the real average.

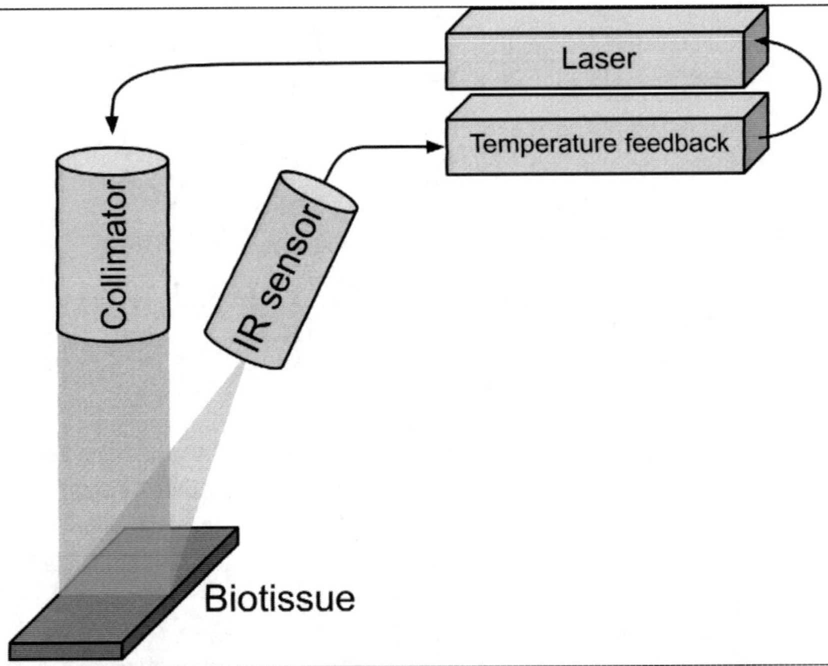

Figure 1. Scheme of the setup for laser welding of biological tissues.

2.2. Calculation of thermal diffusivity

The calculation of the thermal diffusivity a_{tissue} was carried out by minimizing a residual functional $Q(t,a)$:

$$a_{tissue} = \underset{a>0}{argmin}\big(Q(t,a)\big), \quad (1)$$

where t – time, $Q(t,a)$ - the residual functional was calculated based on the Euclidean metric:

$$Q(t,a) = \sqrt{\sum \left(T_{experimental}(t_i) - T_{simulated}(t_i,a)\right)^2}, \quad (2)$$

where t_i – i-th point in time, $T_{experimental}$ – a vector of measured temperatures, $T_{simulated}$ – a vector of simulated temperatures has been obtained solution of the heat equation in the one-dimensional case:

In general terms, the thermal diffusivity equation is:

$$\frac{dT(\vec{r},t)}{dt} - \nabla\left(a(\vec{r},t)\nabla T(\vec{r},t)\right) = f(\vec{r},t), \quad (3)$$

where T - temperature, t - time, $\vec{r}$ - coordinate vector, a - thermal diffusivity, f - heat source function. The value of the thermal diffusivity a can change over time in the case of structural changes in biological tissue. For example, with dehydration or temperature necrosis. In most of the experiments carried out, biological tissue was heated to temperatures not exceeding 42 °C, which makes it possible to exclude dehydration and thermal necrosis and to consider the coefficient of thermal diffusivity a independent of time t. The dependence of a on the spatial coordinate in the studied biological tissues (skeletal muscle, cardiac muscle and skin) is weakly expressed. Therefore, the thermal diffusivity a can be considered independent of $\vec{r}$. Heat source function $f(\vec{r},t)$ is equal to zero, since the time interval of biological tissue cooling without exposure to laser radiation is considered.

Since the biological tissue under study is isotropic and symmetric, the dependence of the temperature T can be considered on only one spatial coordinate x. Equation 3 takes the form:

$$\frac{dT(x,t)}{dt} = \Delta T(x,t), \quad (4)$$

where x - the spatial coordinate. The solution of the thermal diffusivity equation (4) was carried out by an explicit numerical method. The following boundary conditions were used:

$$\begin{cases} T_{i_0 \ldots i_{max}, k_0} = T_0 \\ T_{i_0, k_0 \ldots k_{max}} = T_0 \\ T_{i_{max}, k_0 \ldots k_{max}} = T_0 \end{cases}, \quad (5)$$

where i - coordinate index, k - time index along the time coordinate, T_0 - the temperature at the initial moment of time. Non-uniform discretization of space in time was used. Due to the peculiarities of the software and hardware solutions used, the period for recording temperature data was not constant. To calculate the error functional Q (t, a) in the simulation, the time step value was calculated as the difference in the time to obtain empirical temperature data.

2.3. Experiment protocol

Biological tissues of three types were used in the experiments: skin, skeletal and cardiac muscles. The duration of each measurement was from 10 to 20 s. Each subsequent measurement was carried out on a new, previously unused, area of biological tissue.

3. Results

The measurements were carried out for two groups of biological tissue. Group 1 used tissues exposed to prolonged (more than 3 months) freezing. Group 2 used fresh biological tissues. In group 1, measurements were carried out at temperatures of 40, 50, 60 °C, in group 2 at 60 °C. Figure 2 shows the measurement of biotissue temperature during heating by laser radiation.

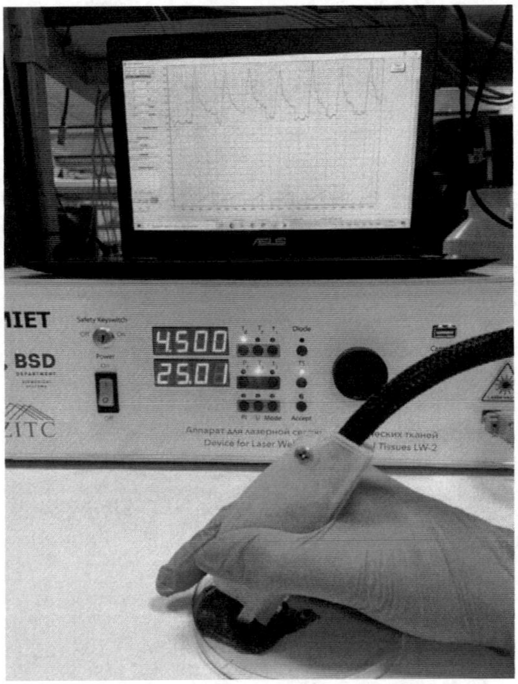

Figure 2. The measurement of biotissue temperature during heating by laser radiation.

Figure 3 shows a graph of the temperature dependence on time. The solid blue curve is the experimentally measured temperature. The red dotted line is the temperature obtained as a result of the simulation at the found thermal conductivity *a*.

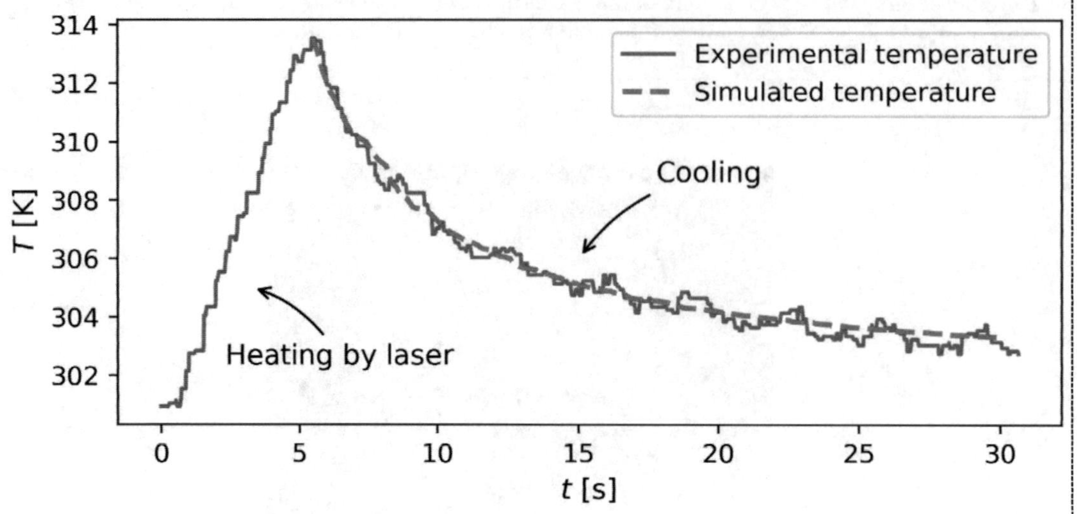

Figure 3. The dependence of biological tissue temperature T on time t during determining thermal diffusivity a.

The thermal conductivity coefficients for a cardiac muscle and skin from group 1 were calculated (Table 1). Optimization of the residual functional was carried out using the gradient descent method.

Table 1. Thermal diffusion coefficient for different types of biological tissues from group 1.

Temperature, °C	Thermal diffusivity (m^2s^{-1} 10^{-6})		
	Cardiac muscle	Laboratory bird skin	Large artiodactyl skin
40	2.3±1.0	1.6±0.4	1.7±0.2
50	2.2±1.1	1.6±0.4	1.5±0.1
60	2.7±1.2	1.6±0.4	1.6±0.4

Table 2 shows the values of thermal diffusivity for different types of biologicaltissues from group 2. The values of thermal diffusivity a for tissues from group 1 are significantly higher than for tissues from group 2. This can be explained by changes in the structure of the tissue occurring during prolonged freezing.

Table 2. Thermal diffusion coefficient for different types of biological tissues from group 2.

Temperature, °C	Thermal diffusivity, (m^2 s^{-1} 10^{-12})		
	Skeletal muscle	Cardiac muscle	Skin
40	133 ± 5	210 ± 8	118 ±5

4. Discussion

The proposed model makes it possible to determine the thermal diffusivity of biological tissues in real time during laser welding of biological tissues. According to the experimental results, the method makes it possible to determine the thermal diffusivity with an error of less than 5%. However, in the used mathematical model, a number of simplifications were allowed in order to reduce the requirements for computing power. In the event that the proposed algorithm is used with large available computing power, it is possible to increase the measurement accuracy. For example, in this article the one-dimensional case is used of the thermal diffusivity equation. For more accuracy, better to use a three-dimensional case. Discretization of both spatial and temporal coordinates also plays an important role. Downsampling any coordinate always improves the accuracy of the simulation. The proposed method can also be used in other applications in which the action of laser radiation on biological tissue is carried out. It is possible to estimate the rate and area of heating, the degree of thermal necrosis of the irradiated biological tissue. When blood vessels are irradiated through a layer of biological tissue, the degree of denaturation of blood proteins can be assessed. That will allow to assess the degree of patency of the blood vessel. In laser welding of biological tissues, a significant and undesirable side effect is thermal necrosis of the biological tissues. The method proposed in the article will allow real-time assessment of the area of temperature necrosis of tissue. Also, the method will create a spatial map of the degree of temperature necrosis. Evaluation of the degree of thermal necrosis is especially important for laser welding of optically opaque tissues, such as skin and muscles. During laser welding of optically opaque tissues, heating due to the absorption of laser radiation is carried out only in the upper layers of the tissue. The lower layers are heated by heat exchange with the warmer layers. One way to avoid thermal necrosis is to use solders with a denaturation temperature lower than denaturation temperature of the welding biological tissues.

5. Conclusion

The paper proposes the method for determining the coefficient of thermal diffusivity in the process of laser soldering of biological tissues. The values of the thermal conductivity coefficient for a cardiac muscle and skin were determined experimentally. The values of the thermal conductivity coefficient were determined for two groups of tissues: fresh and subjected to prolonged freezing. The values of the thermal diffusivity do not change significantly depending on the type of tissue. While long-term freezing increases the thermal diffusivity by orders of magnitude.

The setup for laser welding of biological tissues is controlled by a microcontroller. The PC software only allows to change a number of operating parameters, since the low speed of data exchange between the microcontroller and a PC does not allow real-time control of the laser welding process using a PC. The proposed algorithm was implemented on a PC with a processor frequency of more than 1 GHz. The calculation of one value of thermal diffusivity from experimental data took from 2 to 7 s.

The computational capabilities of the microcontroller are much less than those of a PC. It is obvious that the implementation of the algorithm on a microcontroller in the proposed form will take too much time. What is unacceptable for surgical intervention. There are several ways to solve this problem.

First, in the implementation, the temperature vectors $T_{experimental}$ and $T_{simulated}$ have a dimension of 100, which is redundant. The minimum dimension of the temperature vectors is 3, however, it is necessary to conduct additional research on the accuracy of modeling with such a small dimension of the temperature vectors.

Secondly, the algorithm uses the root-mean-square error functional. Exponentiation, and especially the extraction of the square root, are resource-intensive mathematical operations. It is possible to use the error functional of the "Manhattan length" – the sum of the modules of the vector differences.

Third, it is possible to optimize the process of searching for the minimum of the Q functional. The article proposes to optimize Q by gradient descent, however, the question of differentiability of the functional remains is open. Perhaps it is better to use methods for finding the minimum that do not require differentiability of the function. At the same time, such methods are more resource-intensive.

The article proposes an algorithm for measuring the thermal diffusivity of the biological tissue being welded. In the real process of laser welding of biological tissues, solders are used. In most cases, these are aqueous dispersions of proteins. Indocyanine green can be used as a chromophore, and single and multi-wall carbon nanotubes are added to increase the tensile strength of welded joints. In the future, it will be necessary to experimentally determine the error of the proposed algorithm under the conditions of using solder.

An important assumption of the proposed algorithm is the assumption of an isotropic distribution of thermal diffusivity in biological tissue. In a real biological tissue, there is always an inhomogeneity of the distribution of thermal diffusivity. This may be due to various reasons. For example, with a network of blood vessels or an ensemble of various biological tissues in the field of welding.
The developed algorithm will be further used in the setup for laser welding of biological tissues to determine the value of the thermal diffusivity in real time. In the future, the value will be used to simulate the heating and cooling process, to select the PID controller coefficients.

Acknowledgements
The study was supported by the Ministry of Science and Higher Education of the Russian Federation (No. 075-03-2020-216 from 27.12.2019)

References
[1] Menabuoni L, Pini R, Rossi F, Lenzetti I, Yoo S, Parel J M 2007 Laser-assisted corneal welding in cataract surgery: Retrospective study *J. Cataract Refract. Surg.* **33**(9) pp 1608–1612
[2] Schöni D, Bogni S, Bregy A, Wirth A, Raabe A, Vajtai I, Pieles U, Reinert M, Frenz M 2011 Nanoshell assisted laser soldering of vascular tissue *Lasers Surg. Med.* **43**(10) pp 975–983
[3] Bloom J D, Bleier B S, Goldstein S A, Carniol P J, Palmer J N, Cohen N A 2012 Laser facial nerve welding in a rabbit model *Archives of Facial Plastic Surgery* **14** pp 52–58
[4] Chudnovskii V M, Yusupov V I, Dydykin A V, Nevozhai V I, Kisilev A Y, Zhukov S A, Bagratashvili V N 2017 Laser-induced boiling of biological liquids in medical technologies *Quantum Electronics* **47** pp 361–370
[5] Lauto A, Hamawy A H, Phillips A B M, Petratos P B, Raman J, Felsen D, Ko W, Poppas D P 2001 Carotid artery anastomosis with albumin solder and near infrared lasers: A comparative study *Lasers in Surgery and Medicine* **28** pp 50–55
[6] Strassmann E, Livny E, Loya N, Kariv N, Ravid A, Katzir A, Gaton D D 2013 CO_2 laser welding of corneal cuts with albumin solder using radiometric temperature control *Ophthalmic Research* **50** pp 174–179
[7] Kamolz L, Andel H, Haslik W, Donner A, Winter W, Meissl G, Frey M 2003 Indocyanine green video angiographies help to identify burns requiring operation *Burns* **29** pp 785–791
[8] Saxena V, Sadoqi M, Shao J 2004 Indocyanine green-loaded biodegradable nanoparticles: Preparation, physicochemical characterization and in vitro release *International Journal of Pharmaceutics* **278** pp 293–301
[9] Holley G P, Alam A, Kiri A, Edelhauser H F 2002 Effect of indocyanine green intraocular stain on human and rabbit corneal endothelial structure and viability *Journal of Cataract and Refractive Surgery* **28** pp 1027–1033
[10] Chetoni P, Burgalassi S, Monti D, Tampucci S, Rossi F, Pini R, Menabuoni L 2007 Healing of rabbits' cornea following laser welding: Effect of solid and semisolid formulations containing indocyanine green *Journal of Drug Delivery Science and Technology* **17** pp 25–31
[11] Modaresifar K, S Azizian S, Hadjizadeh A 2016 Nano/Biomimetic Tissue Adhesives Development: From Research to Clinical Application *Polymer Reviews* **56** pp 329–361
[12] Matteini P, Ratto F, Rossi F, Pini R 2014 Laser-activated nano-biomaterials for tissue repair and controlled drug release *Quantum Electronics* **44** pp 675–682
[13] Gerasimenko A Yu, Ichkitidze L P, Palov A A, Piyankov E S, Ryabkin D I, Savelyev M S, Selishchev S V, Rimshan I B, Zhurbina N N, Podgaetskii V M 2016 Laser system with adaptive thermal stabilization for welding of biological tissues *Biomed. Eng. (NY)* **49**(6) pp. 344–348
[14] Porat Y, Gabay I, Varssano D, Barequet I, Neudorfer M, Rosner M, Katzir A 2015 Optical coherence tomography (OCT) in laser tissue bonding of incisions in the cornea *Proc. SPIE* **9317** pp 93170M/1-93170M/7

Analysis and regulation of the cardiovascular system of univentricular heart with different configurations of the total cavopulmonary connection

A A Galiastov[1, 2] and D V Telyshev[1, 2]

[1] Institute of Biomedical System, National Research University of Electronic Technology (MIET), Shokin Square, 1, Zelenograd, Moscow, 124498, Russia
[2] Institute for Bionic Technologies and Engineering, Sechenov First Moscow State Medical University, Bolshaya Pirogovskaya street, 2, bld. 4, Moscow, Russia

E-mail: galyastov@bms.zone

Abstract. The cardiovascular system (CVS) is a complex mechanism capable of reacting and regulating various changes in external and internal parameters. A particular problem is the study of CVS with univentricular heart. We studied the hemodynamic response of the CVS to the use of various configurations of bypassing the right side of the heart. During our research, we selected three different configurations of the total cavopulmonary connection. All configurations were made from Clear Flex 50 and tested in an experimental workbench. According to the research results, the TCPC-3 configuration is the most optimal. It has symmetrical output values of flows and pressures and there is no significant increase in inlet pressure. The results of this study can be used to optimize the control of the parameters of the ventricular assist device and the total artificial heart.

1. Introduction

The cardiovascular system (CVS) works continuously to maintain the required level of blood flow. it reacts to constantly changing conditions of the external environment, the needs of the body and disruptions in the work of internal organs. Sufficient cardiac output and an optimal level of mean systemic blood pressure are necessary for normal systemic circulation. Normally, these values vary widely. They can change more than five times. Since the CVS is a relatively closed system, the cardiac output with each contraction is approximately the same. The adaptation of the work of the heart to the changing parameters of the body occurs with the help of various regulatory mechanisms. The same mechanisms come into operation when the work of the heart itself or its parts is disrupted.

Fontan operation is a corrective hemodynamic operation involving the redirection of non-oxygenated venous blood from the vena cava into the pulmonary arteries. Currently, the Fontan procedure is used for many complex congenital heart defects. Due to the lack of donor hearts, this procedure gives a chance to life for thousands of patients. To form the total cavopulmonary connection (TCPC), during the Fontan procedure, the superior vena cava (SVC) and inferior vena cava (IVC) are anastomosed directly to the left pulmonary artery (LPA) and right pulmonary artery (RPA) [1].

Understanding the hemodynamics of CVS coupled with TCPC can lead to improved surgical procedures, leading to more efficient univentricular circulation. We examined the parameters of CVS with different configurations of TCPC.

2. Materials and methods

The normal cardiovascular system shown in Figure 1A is described by a parametric function of

$$f(Q_V, Q_O, Q_A, Q_{LV}, Q_{LA}, Q_{PV}, Q_{Pul}, Q_{PA}, Q_{RA}, Q_{RV}, P_V, P_O, P_A, P_{LV}, P_{LA}, P_{PV}, P_{Pul}, P_{PA}, P_{RA}, P_{RV})$$

where P_i, Q_i – pressure and flow in the corresponding section of the cardiovascular system.

During Fontan operation, the total cavopulmonary connection is formed bypassing the right side of the heart. The block diagram of the CVS is transformed into Figure 1B. This block (highlighted in red) can be characterized by the parameters: $P_{TCPC}(F)$ is the head loss when passing through the block, F is the geometric configuration of the block, R is the characteristic size of the block.

The parametric function of the CVS is transformed

$$f(Q_V, Q_O, Q_A, Q_{LV}, Q_{LA}, Q_{VP}, Q_{Pul}, Q_{PA}, Q_{TCPC}(F), Q_{IVC}(F), Q_{SVC}(F), Q_{LPA}(F), Q_{RPA}(F), P_V, P_O, P_A, P_{LV}, P_{LA}, P_{VP}, P_{Pul}, P_{PA}, P_{TCPC}(F), P_{IVC}(F), P_{SVC}(F), P_{LPA}(F), P_{RPA}(F)).$$

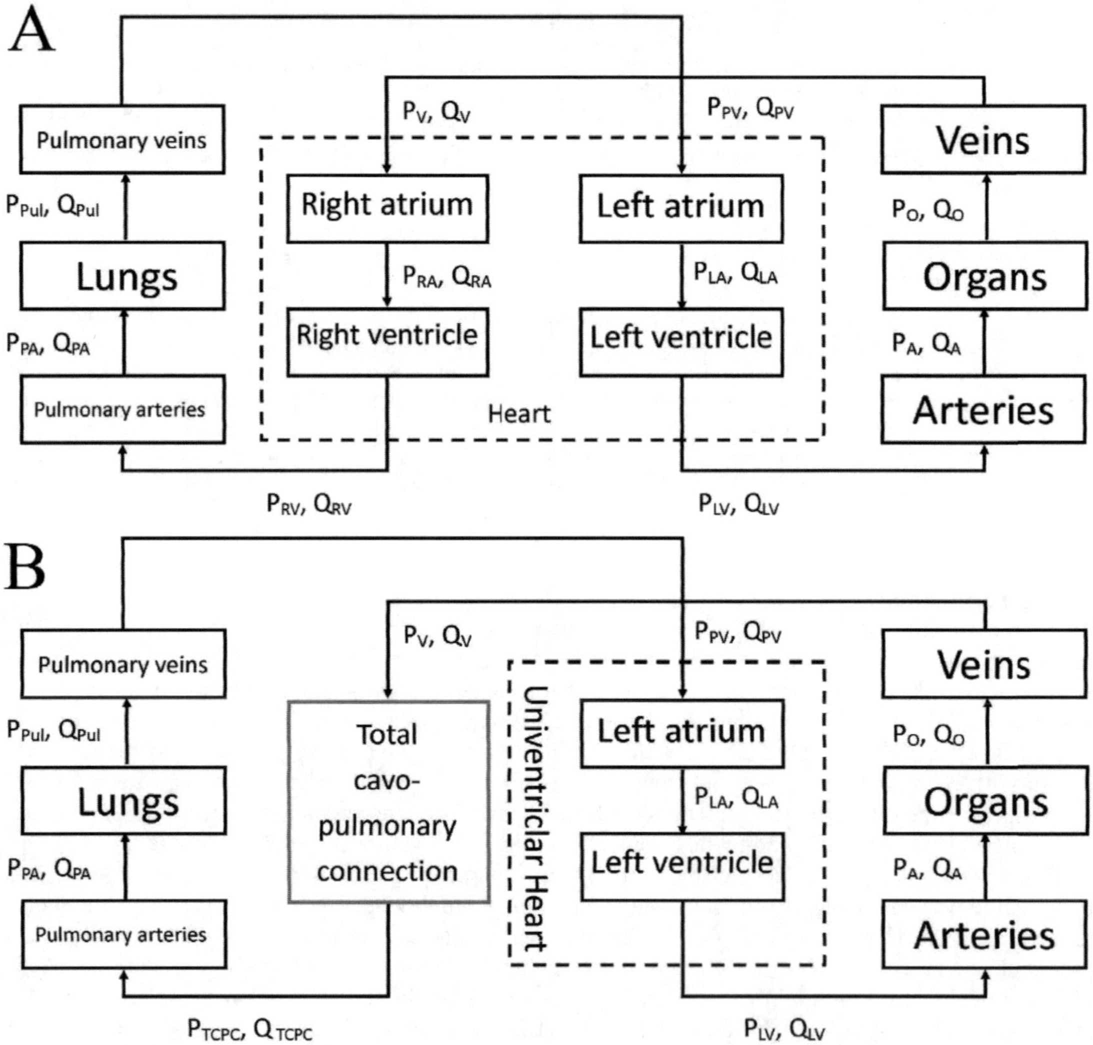

Figure 1. The block diagram of the CVS in normal (A) and Fontan operation (B)

The cardiovascular system can be described by a large number of parameters: pressures and flows at various points, cardiac output, pulsation amplitude, heart rate, stroke volume, hydraulic resistance of veins, arteries and valves.

In this work, we do not take into account the elasticity of the vessel walls, considering them absolutely inelastic, pulsations, considering the model with a constant fluid flow, and also simplify the model to the interaction of the TCPC with the CVS.

One of the most common methods for studying CVS is modeling. Reference [2, 3] demonstrates the possibilities of mathematical modeling of CVS in various diseases. Reference [4-6] demonstrates the possibilities of computer modeling and the study of the hydraulic characteristics of TCPC and an axial blood pump, without CVS. But these types of modeling have different types of limitations and may not affect the anatomical or functional features of biological parts. We applied experimental modeling to study the interaction of the TCPC with CVS.

Table 1 shows the parameters reflecting the interaction of the TCPC with the CVS in various conditions [3, 7].

Table 1. CVS parameters in various conditions.

	Cardiac output (lpm)	P_{SVC} (mm Hg)	P_{IVC} (mm Hg)	P_{LPA} (mm Hg)	P_{RPA} (mm Hg)	Q_{SVC} (lpm)	Q_{IVC} (lpm)	Q_{LPA} (lpm)	Q_{RPA} (lpm)
Normal	3.2	5	5	15	15	1.5	1.7	1.6	1.6
Fontan operation	2.2	15	15	11	11	0.81	1.38	1.1	1.1

We chose 3 different TCPC configurations with the same characteristic dimensions (Figure 2) [5]. The TCPC-1 and TCPC-2 configurations are often used in surgery. The TCPC-3 configuration is a new type. We assume that this configuration will show the best result, because there is no direct collision of fluid flows. They are made from transparent polyurethane Clear Flex 50.

These TCPC configurations were placed in a mock circulation loop (Figure 3). It simulates the work of the cardiovascular system with univentricular heart [6, 8-9]. A mock circulation loop consists of a ventricular assist device (VAD) Sputnik pump [10], an open reservoir, ultrasonic flow meters (ME-11PXL Clamp-on Tubing Flowsensors, Transonic System Inc., Ithaca, NY, USA) and a Tygon

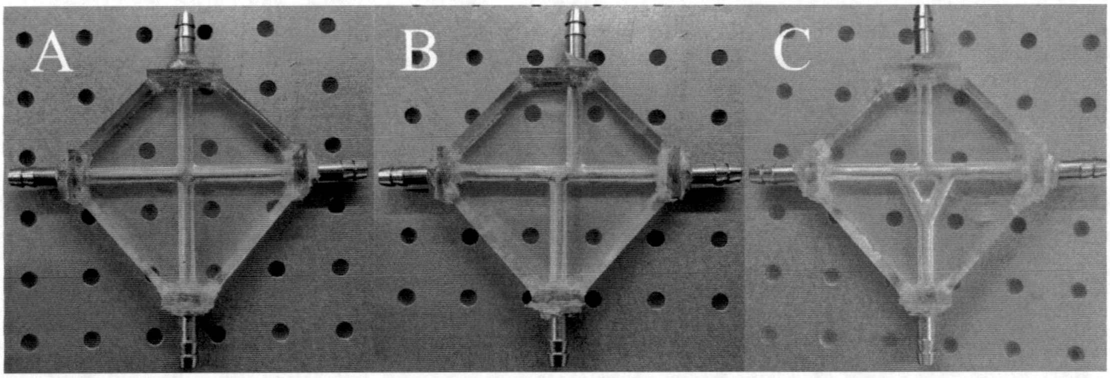

Figure 2. Three different geometry configurations of the TCPC chosen for the research in this work. (A) Simple direct joint, where the inferior vena cava (IVC) and superior vena cava (SVC) are coaxially arranged. (B) Compound where the IVC and SVC are shifted by the distance of a single diameter (6 mm). (C) More complex type of curvilinear connection, where IVC is divided into two vessels and connected to the left pulmonary artery (LPA) and right pulmonary artery (RPA) at an angle, directing the flow along a smoother trajectory.

transparent laboratory tube system (TygonS3 E-3603, Saint-Gobain Performance Plastics, Courbevoie, France) with an internal diameter of ¼ and ½ inch. VAD "Sputnik" is a continuous type rotary blood pump. An aqueous solution of glycerol was used in experiment. The working fluid density is 1104 kg/m3, the dynamic viscosity is 0.0035 Pa·s at 25°C.

3. Results and conclusion

During the experiments, we obtained the values of pressures and flows at 4 points: IVC, SVC, LPA and RPA. The data are presented in Table 2.

Table 2. CVS parameters in various conditions.

Configurations of the TCPC	P_{SVC} (mm Hg)	P_{IVC} (mm Hg)	P_{LPA} (mm Hg)	P_{RPA} (mm Hg)	Q_{SVC} (lpm)	Q_{IVC} (lpm)	Q_{LPA} (lpm)	Q_{RPA} (lpm)
TCPC-1	28	28	11	11	1,38	0,81	1,12	1,11
TCPC-2	23	23	12	10	1,40	0,81	1,16	1,06
TCPC-3	22	17	11	11	1,38	0,82	1,13	1,13

The TCPC-1 configuration is symmetric, which is reflected in the results - the flows in LPA and RPA are the same. However, in order to maintain the standard flow, the CVC responds with a significant increase in pressure in the IVC and SVC, which increases the load on the heart muscle.

The TCPC-2 configuration is asymmetric and the flow is dominated by the flow from IVC to LPA. The system cannot cope with the asymmetry of LPA and RPA flows and in the long term this can lead to pulmonary edema. The system reacts by increasing the pressure in the IVC and SVC, however, it is less than in the TCPC-1.

According to the research results, the TCPC-3 configuration is the most optimal. It has symmetrical output values of flows and pressures and there is no significant increase in inlet pressure. The pressure

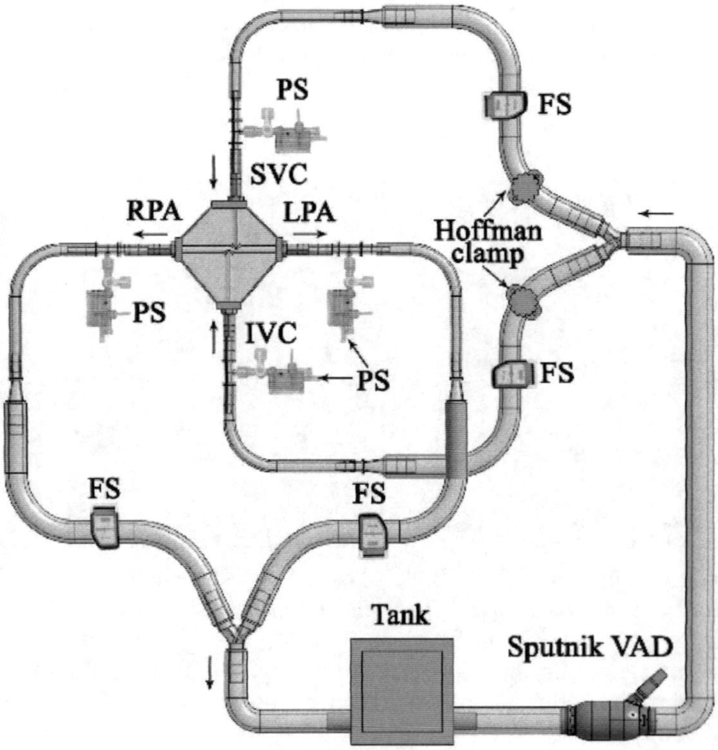

Figure 3. Mock circulation loop, where constant flow generated by Sputnik pump. Changes in pressure and flow in LPA, RPA, IVC, and SVC were detected by FSs (flow sensor) and PSs (pressure sensor). The flow direction is shown by arrows.

in the SVC remained close to ideal values.

In the course of this study, we showed the dependence of the parameters of CVS with the TCPC on configurations of the total cavopulmonary connection. Even with different types of conduits, the blood circulation parameters may not return to normal, however, they enclose the fluid flow. In the following, we will consider CVS with the total cavopulmonary connection and a rotary blood pump. We will study the reaction of the CVS to the operation of these units and show the possibility of controlling the CVS parameters using a rotary blood pump.

Acknowledgments
This study was supported by a grant from the Russian Science Foundation (No. 20-49-04404).

4. References

[1] Sprindzhuk M V 2010 Fontan operation: criteria for performance, indications and contraindications, risk factors *Sovremennye tehnologii v medicine (Modern technologies in medicine)* **3** pp 98-103 (in Russian)

[2] Petukhov D S and Telyshev D V 2016 A Mathematical Model of the Cardiovascular System of Pediatric Patients with Congenital Heart Defect *Biomedical Engineering* **50** pp 229–232

[3] Rubtsova E N, Chernovolenko A and Telyshev D 2019 A mathematical model of the univentricular Fontan *2019 IEEE Conference of Russian Young Researchers in Electrical and Electronic Engineering (EIConRus)* pp. 2337-2340.

[4] Pugovkin A A, Markov A G, Selishchev S V, Korn L, Walter M, Leonhardt S, Bockeria L A, Bockeria O L and Telyshev D V 2019 Advances in Hemodynamic Analysis in Cardiovascular Diseases Investigation of Energetic Characteristics of Adult and Pediatric Sputnik Left Ventricular Assist Devices during Mock Circulation Support *Cardiology Research and Practice* **2019** pp 1–5

[5] Telyshev D, Denisov M, Markov A, Fresiello L, Verbelen T and Selishchev S 2020 Energetics of blood flow in Fontan circulation under VAD support *Artificial Organs* **44** pp 50–57.

[6] Porfiryev A, Markov A, Galyastov A, Denisov M, Burdukova O, Gerasimenko A Y and Telyshev D 2020 Fontan hemodynamics investigation via modeling and experimental characterization of idealized pediatric total cavopulmonary connection *Appl. Sci.* **10** 6910.

[7] Rodefeld M D, Frankel S H and Giridharan G A 2011 Cavopulmonary assist: (em)powering the univentricular Fontan circulation *Semin. Thorac Cardiovasc. Surg. Pediatr. Card. Surg. Ann* **14** pp 45–54

[8] Telyshev D V, Pugovkin A A and Selishchev S V 2017 A Mock Circulatory System for Testing Pediatric Rotary Blood Pumps *Biomedical Engineering* **51** pp 83–87

[9] Galiastov A A 2021 Development of a laboratory bench for studying the work of the cardiovascular system *28-ja Vserossijskaja mezhvuzovskaja nauchno-tehnicheskaja konferencija studentov i aspirantov «Mikrojelektronika i informatika - 2021» (28th All-Russian Interuniversity Scientific and Technical Conference of Students and Postgraduates "Microelectronics and Informatics - 2021")* p 170 (in Russian)

[10] Telyshev D, Denisov M, Pugovkin A, Selishchev S and Nesterenko I 2018 The Progress in the Novel PediatricRotary Blood Pump Sputnik Development *Artif. Organs*,**42**, pp 432–443

Lumped parameter model of the cardiovascular system with baroreflex

E Rubtsova[1, 2] and D Telyshev[1, 2]

[1]Institute of Biomedical Systems, National Research University of Electronic Technology 1 Shokin Square, Zelenograd, Moscow 124498, Russia
[2]Institute for Bionic Technologies and Engineering, Sechenov University 8 Trubetskaya street, Moscow 119991, Russia

E-mail: rubtsova@bms.zone

Abstract. The patient's activity directly affects the state of the cardiovascular system (CVS). One of the mechanisms contribute to the regulation of the arterial blood pressure is the baroreflex. It is maintains an optimal level of the arterial pressure as a response to changes in the patient's activity. The model developed in this work simulates CVS at the different patient's activity levels. The model of baroreflex reproduces changes that occur during an increasing level of patient's activity. The correlation between heart rate (79 bpm –rest, 123 bpm – exercise) and cardiac output (5.1 l/min –rest, 8.2 l/min – exercise) was achieved. The developed model can be used for personal modelling of various states of the CVS. The model parameters possible to apply for individual patient characteristics. For further development, the model can be integrated with the ventricular assist device during the heart failure state in order to study interactions in the biotechnical system.

1. Introduction

The cardiovascular system (CVS) is characterized by complex interactions between various control mechanisms and physiological processes [1]. CVS is subject to quasi-static and dynamic changes. The quasi-static ones include age-related changes, under the influence of CVS diseases, etc. Dynamic changes occur in response to changes in the patient's activity, the needs of other organs and tissues, etc. [2], [3].

One of the important factors affecting the state of the cardiovascular system (CVS) is the patient's level of activity. The baroreflex is one of the mechanism of automatically regulation of arterial blood pressure. Baroreflex is a natural negative feedback control mechanism – the pressure increasing in the system acts on baroreceptors, which send a signal to the medulla oblongata about the need to reduce the vascular resistance and heart rate (HR) [2], [3].

Modeling allows us to study the natural control mechanisms in biotechnical systems (BTS) [1], [4]. The BTS models are of two types: physical and mathematical. Physical models include test benches [5], [6]. Mathematical models are systems of differential equations linking hydrodynamic parameters [1], [4], [7], [8].

The aim of this work is to developing the lumped parameter model of the CVS with the baroreflex mechanism.

2. Materials and methods

The mathematical modeling via MATLAB software (MathWorks, Natick, Massachusetts, USA) was based on the electro-hydraulic analogy method. The number of nonlinear elements (compliance of an organ or vessel, local inertia of a blood) in the system determines the number of equations. Such a model called a lumped-parameter model and allows to study the global distributions of parameters in the CVS, omitting the features of individual vessels. These models represent CVS as a set of blocks, such as ventricles, atriums, vessels and others [1], [4], [8] – [10].

2.1. Model of the cardiovascular system

The development model consists of the right and left ventricle, right and left atrium, systemic and pulmonary vessels, aorta and pulmonary artery, and also includes four heart valves (Figure 1.). The model parameters are determined for a healthy patient [11] – [13] and shown in Table 1.

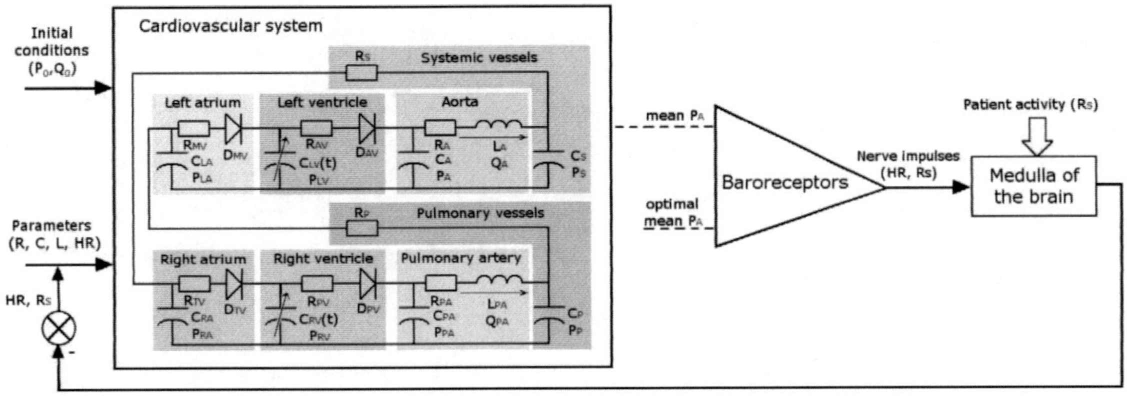

Figure 1. Illustration of a lumped parameter model of CVS coupled with baroreflex.

The method of lumped parameter modeling was described by A. Ferreira [10]. The cardiac cycle is essentially split into four phases [1], [2], [10]. The mathematical model is a system of differential equations for each phase of the cycle. In matrix form, the model is written as:

$$\dot{x} = A(t)\cdot x + B(t)\cdot b(t), \tag{1}$$

where $\dot{x}$ – the column vector of the derivatives of the column vector of variables x:

$$x = \left[P_{LV}\ P_{LA}\ P_A\ P_S\ P_{RV}\ P_{RA}\ P_{PA}\ P_P\ Q_A\ Q_{PA}\right]^T, \tag{2}$$

where $b(t)$ – the column vector governing valve opening:

$$b(t) = \begin{bmatrix} \dfrac{(P_{LV}-P_A)\cdot\eta(P_{LV}-P_A)}{R_{AV}} \\ \dfrac{(P_{RV}-P_{PA})\cdot\eta(P_{RV}-P_{PA})}{R_{PV}} \\ \dfrac{(P_{LA}-P_{LV})\cdot\eta(P_{LA}-P_{LV})}{R_{MV}} \\ \dfrac{(P_{RA}-P_{RV})\cdot\eta(P_{RA}-P_{RV})}{R_{TV}} \end{bmatrix}, \tag{3}$$

where η – the Heaviside function, which simulates heart valves; $A(t)$ and $B(t)$ – time-varying matrices, respectively, expressed as:

$$A(t)=\begin{bmatrix} \frac{\dot{E}_{LV}(t)}{E_{LV}(t)} & 0 & 0 & 0 & 0 & 0 & 0 & 0 & 0 & 0 \\ 0 & -\frac{1}{R_P C_{LA}} & 0 & 0 & 0 & \frac{1}{R_P C_{LA}} & 0 & 0 & 0 & 0 \\ 0 & 0 & -\frac{1}{R_S C_S} & 0 & \frac{1}{R_S C_S} & 0 & 0 & 0 & \frac{1}{C_S} & 0 \\ 0 & 0 & 0 & \frac{\dot{E}_{RV}(t)}{E_{RV}(t)} & 0 & 0 & 0 & 0 & 0 & 0 \\ 0 & 0 & \frac{1}{R_S C_{RA}} & 0 & -\frac{1}{R_S C_{RA}} & 0 & 0 & 0 & 0 & 0 \\ 0 & \frac{1}{R_P C_P} & 0 & 0 & 0 & -\frac{1}{R_P C_P} & 0 & 0 & 0 & \frac{1}{C_P} \\ 0 & 0 & 0 & 0 & 0 & 0 & 0 & 0 & -\frac{1}{C_A} & 0 \\ 0 & 0 & 0 & 0 & 0 & 0 & 0 & 0 & 0 & -\frac{1}{C_{PA}} \\ 0 & 0 & -\frac{1}{L_A} & 0 & 0 & 0 & \frac{1}{L_A} & 0 & -\frac{R_A}{L_A} & 0 \\ 0 & 0 & 0 & 0 & 0 & -\frac{1}{L_{PA}} & 0 & \frac{1}{L_{PA}} & 0 & -\frac{R_{PA}}{L_{PA}} \end{bmatrix}, \tag{4}$$

$$B(t)=\begin{bmatrix} -E_{LV}(t) & 0 & E_{LV}(t) & 0 \\ 0 & 0 & -\frac{1}{C_{LA}} & 0 \\ 0 & 0 & 0 & 0 \\ 0 & -E_{RV}(t) & 0 & E_{RV}(t) \\ 0 & 0 & 0 & -\frac{1}{C_{RA}} \\ 0 & 0 & 0 & 0 \\ \frac{1}{C_A} & 0 & 0 & 0 \\ 0 & \frac{1}{C_{PA}} & 0 & 0 \\ 0 & 0 & 0 & 0 \\ 0 & 0 & 0 & 0 \end{bmatrix}. \tag{5}$$

Equation (1) was transformed for numerical simulation using Euler's method generalized to systems of differential equations.

Table 1. Models parameters and variables

Parameter	Value	Physiological meaning
		Resistance (mmHg s/ml)
R_{MV}	0.005	Mitral valve
R_{AV}	0.03	Aortic valve
R_A	0.06	Aortic
R_S	1.1 / 0.5	Systemic vessels (rest / activity)
R_{TV}	0.005	Tricuspid valve
R_{PV}	0.025	Pulmonary valve
R_{PA}	0.05	Pulmonary artery
R_P	0.1	Pulmonary vessels
		Compliance (ml/mmHg)
$C_{LV}(t)$	$1/E_1(t)$	Left ventricular
C_{LA}	4.4	Left atrium
C_A	0.8	Aortic
C_S	0.35	Systemic vessels
$C_{RV}(t)$	$1/E_1(t)$	Right ventricular
C_{RA}	6.6	Right atrium
C_{PA}	2	Pulmonary artery
C_P	3	Pulmonary vessels
		The elastance (mmHg/ml)
E_{LV}^{max}	1.8	The maximum value of the left ventricle
E_{LV}^{min}	0.05	The minimum value of the left ventricle
E_{RV}^{max}	0.6	The maximum value of the right ventricle
E_{RV}^{min}	0.06	The minimum value of the right ventricle
		Other
L_A	0.0005	Inertance of blood in aorta (mmHg s^2/ml)
L_{PA}		Inertance of blood in pulmonary artery (mmHg s^2/ml)
HR	80	Heart rate (bpm)
V_0	8	Ventricular volume at zero pressure (ml)
D_{MV}	$\eta(P_{LA} - P_{LV})$	Mitral valve
D_{AV}	$\eta(P_{LV} - P_A)$	Aortic valve
D_{TV}	$\eta(P_{RA} - P_{RV})$	Tricuspid valve
D_{PV}	$\eta(P_{RV} - P_{PA})$	Pulmonary valve
		Variables
P_{LV}		Left ventricular pressure (mmHg)
P_{LA}		Left atrial pressure (mmHg)
P_A		Aortic pressure (mmHg)
P_S		Systemic vessel pressure (mmHg)
P_{RV}		Right ventricular pressure (mmHg)
P_{RA}		Right atrial pressure (mmHg)
P_{PA}		Pulmonary artery pressure (mmHg)
P_P		Pulmonary vessel pressure (mmHg)
Q_A		Aortic flow (ml / s)
Q_{PA}		Pulmonary artery flow (ml / s)

2.2. Model of the baroreflex

One of systems contribute to the regulation of arterial blood pressure is the baroreceptors system. In the wall of the aorta are many nerve receptors called baroreceptors, which are stimulated by stretch of the arterial wall. When the arterial pressure rises too high, the baroreceptors send barrages of nerve impulses to the medulla of the brain (Figure 1.). Here these impulses inhibit the vasomotor center, which in turn decreases the number of impulses transmitted from the vasomotor center through the sympathetic nervous system to the heart and blood vessels. Lack of these impulses causes diminished pumping activity by the heart and also dilation of the peripheral blood vessels, allowing increased blood flow through the vessels. Both of these effects decrease the arterial pressure back toward normal and vice versa [2], [3]. Baroreflex is a natural feedback system. The feedback of the baroreflex mechanism can be represented similarly to the principle of the PID-controller (Figure 1.).

The Poiseuille law states that the blood flow velocity (Q) in a vessel is directly proportional to the pressure drop at the ends of the vessel (ΔP) and inversely proportional to the resistance of the vessel (R) [1]:

$$Q = \frac{\Delta P}{R}. \tag{6}$$

The resistance of the vessel is described by the equation:

$$R = \frac{8\nu l}{\pi r^4}, \tag{7}$$

where ν is the blood viscosity, l is the length of the vessel, r is the radius of the vessel. The total peripheral vascular resistance is calculated by the formula:

$$R_\Sigma = N \cdot R, \tag{8}$$

where N is the number of vessels.

As physical activity increases, so does the oxygen demand. The lumen of the vessels is increased to ensure greater blood flow to organs and tissues (Figure 1.). According to equation (7), the radius of the vessel increases, and the resistance of the vessels decreases. This leads to a decrease in blood pressure in accordance with Poiseuille law (6). At this moment, the baroreflex mechanism is activated. Baroreceptors detect a decrease in blood pressure and give a command to the vasomotor center. It increases the HR and activates the vasoconstrictor effect, resulting in increased vascular resistance. Baroreflex activation takes less time than one heartbeat. Likewise, when going into a state of rest.

The baroreflex mechanism has been implemented as an add-on feature for the CVS model. Its function is similar to the task of baroreflex – to prevent pressure changes. Baroreceptors compare the effective mean aortic pressure with the optimal mean aortic pressure (Figure 1.). Under the influence of nerve impulses from baroceptors and the patient's activity, the medulla of the brain decides in which direction to change the control parameters (HR, R_S). In mathematical terms, this mechanism can be described in the form of differential equations presented in matrix form:

$$\begin{bmatrix} \dot{HR} \\ \dot{R}_S \end{bmatrix} = \begin{bmatrix} h_1 & h_2 \\ k_1 & k_2 \end{bmatrix} \cdot \begin{bmatrix} \eta\left(\left(optimal\ meanP_A - dP\right) - meanP_A\right) \\ \eta\left(meanP_A - \left(optimal\ meanP_A + dP\right)\right) \end{bmatrix}, \tag{9}$$

where k_1, k_2, h_1, h_2 – coefficients, dP – permissible pressure deviation.

3. Results

The developed model (Figure 1.) was applied to model CVS during the transition from a state of rest to physical activity. The baroreflex function monitored changes in mean aortic pressure and gave a command to change the HR and peripheral resistance, if necessary (Figure 1., equation (7)).The vascular resistance (R_S) depend on patient's activity. R_S was set as 1.1 mmHg s/ml for rest and 0.5 mmHg s/ml for exercises. Coefficients k_1, k_2, h_1, h_2 in equation (9) were determined by the fitting method and are equal to 2 mm Hg/ml; 2 mmHg/ml; 0.07 beats/s^2; 0.03 beats/s^2, respectively. The optimal mean aortic pressure was determined as the arithmetic mean between the mean pressures at rest and activity [12] and amounted to 103 mm Hg. The allowable pressure deviation from the exact value of the optimal mean aortic pressure (*dP*) was determined as the difference between the mean pressures at rest and activity [12], and amounted to 5 mm Hg.

The CVS simulation result is distribution of pressure in the left ventricle and the aorta (Figure 2 (a)), as well as PV-loops of the left ventricle (Figure 2 (b)). The *HR* and the R_S was changed under the influence of baroreflex in equation (9). The *HR* increased at 79 to 123 bpm (Table 1). The R_S increased by 0.14 mm Hg/min, and amounted to 0.64 mm Hg/min. As a result of the effect of baroreflex on CVS, cardiac output increased, which is consistent with theoretical values (Table 1) [12].

Figure 2 (b) shows PV-loops for two simulated states. In a state of activity, the stroke volume decreases due to the increased HR. The time of the cardiac cycle is shortened, so less blood volume accumulates in the ventricle during the diastolic phase, respectively, less blood is released into the systolic phase, according to the Frank-Starling law [13]. But such a slight decrease in stroke volume is compensated by an increase in HR, cardiac output increases from 5.4 to 8 l / min (Table 2) [12]. The relative modeling error is 6%, which is acceptable and satisfies the modeling goals.

Table 2. Patient data and simulated data depending on the patient's activity.

Parameters	**Heart rate, bpm**		**Mean aortic pressure, mm Hg**		**Cardiac output, l/min**	
	Theoretical value	Modeling	Theoretical value	Modeling	Theoretical value	Modeling
Rest	80	79	100	106	5	5.1
Exercise	122	123	107	108	8.2	8.2

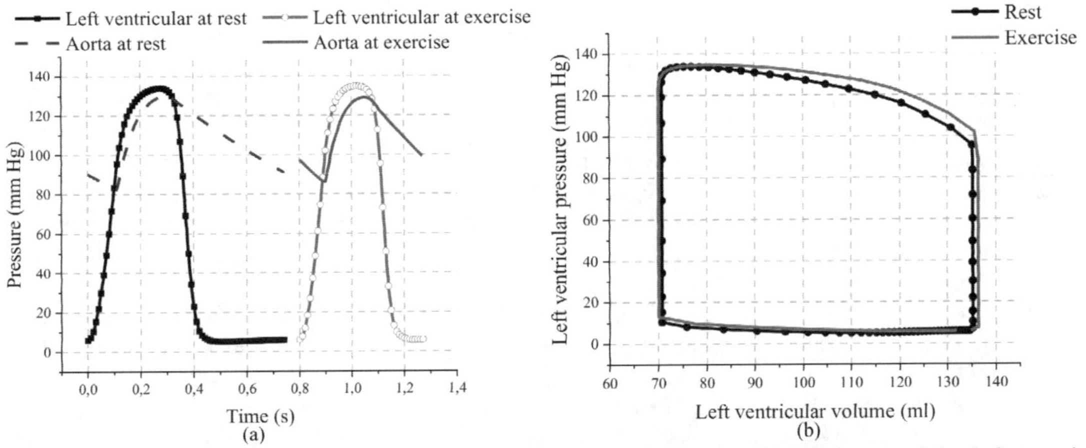

Figure 2. (a) The pressure distribution in CVS for rest and exercise (b) PV-loops of the left ventricle for rest and exercise.

4. Conclusion

This paper describes the mathematical model of the CVS with baroreflex. Baroreflex is one of the automatic control mechanisms of blood pressure regulation, which is responsible for high-speed regulation [2]. The model represents the ordinary differential equation system and is based on the electro-hydraulic analogy [7] – [10]. Comparing the condition of CVS by the rest and the physical activity, we can note increase HR (79 bpm –rest, 123 bpm – exercise) and cardiac output (5.1 l/min – rest, 8.2 l/min – exercise). The simulation results consistent with theoretical values [12].

This study presents the lumped parameter mathematical model of the CVS with the baroreflex mechanism. The work investigates distributions of pressure, volume and flow in the CVS in depending on physical activity. The integration of the baroreflex expands the scope of the model, allowing one to study not only the quasi-static state of the system, but also the dynamic one – in response to changes in the patient's activity. The developed model can be used for personal modeling of various states of the CVS. The model parameters could be personified to individual patient characteristics in order to study the blood circulation. For further development, the model can be supplemented with the ventricular assist device during the heart failure state in order to study interactions in the biotechnical system [10], [12], [14] – [17].

Acknowledgment
This study was supported by a grant from the Russian Science Foundation (No. 20-49-04404).

References

[1] Kokalari I, Karaja T and Guerrisi M 2013 Review on lumped parameter method for modeling the blood flow in systemic arteries *Biomedical Science and Engineering* (Wuhan: Scientific Research) **6**(1) pp 92–9
[2] Hall J E 2011 Guyton and Hall textbook of medical physiology Saunders Elsevier (Philadelphia) p 1112
[3] Fresiello L Zielin´ ski K, Jacobs S, Di Molfetta A, Pałko K J, Bernini F, Martin M, Claus P, Ferrari G, Trivella M G, Górczyn´ ska K, Darowski M, Meyns B and Kozarski M 2014 Reproduction of Continuous Flow Left Ventricular Assist Device Experimental Data by Means of a Hybrid Cardiovascular Model With Baroreflex Control *Artificial Organs* (Hoboken: Wiley) **38**(6) pp 456–68
[4] Capoccia M 2015 Development and Characterization of the Arterial Windkessel and Its Role During Left Ventricular Assist Device Assistance *Artificial Organs* (Hoboken: Wiley) **39**(8) pp 138–53
[5] Pugovkin A A, Markov A G, Selishchev S V, Korn L, Walter M, Leonhardt S, Bockeria L A, Bockeria O L and Telyshev D V 2019 Advances in Hemodynamic Analysis in Cardiovascular Diseases Investigation of Energetic Characteristics of Adult and Pediatric Sputnik Left Ventricular Assist Devices during Mock Circulation Support Cardiol Res Pract. (London: Hindawi) article ID 4593174 pp 1-15
[6] Telyshev D V, Pugovkin A A and Selishchev S V 2017 A Mock Circulatory System for Testing Pediatric Rotary Blood Pumps *Biomedical Engineering* (New York: Springer) **51**(2) pp 83-7
[7] Petukhov D S and Telyshev D V 2016 A Mathematical Model of the Cardiovascular System of Pediatric Patients with Congenital Heart Defect *Biomedical Engineering* (New York: Springer) **50**(4) pp. 229-32
[8] Telyshev D, Denisov M, Markov A, Fresiello L, Verbelen T and Selishchev S 2020 Energetics of blood flow in Fontan circulation under VAD support *Artificial Organs* (Hoboken: Wiley) **44**(1) pp 50-7
[9] Golov A V and Simakov S S 2020 Personalized Computational Evaluation of Physical Endurance in a Treadmill Test with Increasing Load *Lobachevskii J* (New York: Springer) **41**, pp. 2648–63
[10] Ferreira A, Chen S, Simaan M A, Boston J R and Antaki J F 2005 A Nonlinear State-Space Model of a Combined Cardiovascular System and a Rotary Pump *Proc. of the 44th IEEE Conference on Decision and Control* (New Jersey: IEEE Xplore) pp 897–902
[11] Darowski M, Kozarski M, Ferrari G, Zielinski K, Gorczynska K, Szczepanowski A, Palko K J, Fresiello L and Di Molfetta A 2013 A new hybrid (hydro-numerical) model of the circulatory system *Bulletin of the Polish Academy of Sciences. Technical Sciences* (Warsaw: Polish Academy of Sciences) **61**(4) pp 993-1003
[12] Meki M, Wang Y, Sethu P, Ghazal M, El-Baz A and Giridharan G 2020 A Sensorless Rotational Speed-based Control System for Continuous Flow Left Ventricular Assist Devices *IEEE Trans Biomed Eng* (New Jersey: IEEE Xplore) **67**(4) pp 1050–60
[13] Ribaric S and Kordas M 2012 Simulation of the Frank-Starling Law of the Heart *Computational and Mathematical Methods in Medicine* (London: Hindawi) article ID 267834 12 p.

[14] Cox L G E, Loerakker S, Rutten M C M, de Mol B A J M and van de Vosse F N 2009 A Mathematical Model to Evaluate Control Strategies for Mechanical Circulatory Support. *Artificial Organs* (Hoboken: Wiley) **33**(8) pp 593–603
[15] Petukhov D S, Selishchev S V and Telyshev D V 2015 Development of Left Ventricular Assist Devices as the Most Effective Acute Heart Failure Therapy *Biomedical Engineering* (New York: Springer) **48**(6) pp 328–30
[16] Faragallah G and Simaan M A 2013 An Engineering Analysis of the Aortic Valve Dynamics in Patients with Rotary Left Ventricular Assist Devices *Journal of Healthcare Engineering* (London: Hindawi) **4**(3) pp 307–27
[17] Martina J R, Bovendeerd P H M, Jonge N, Mol B A J M, Lahpor J R and Rutten M C M 2013 Simulation of Changes in Myocardial Tissue Properties During Left Ventricular Assistance With a Rotary Blood Pump *Artificial Organs* (Hoboken: Wiley) **37**(6) pp 531–40

Study of bioactive molecular systems by gas-phase electron diffraction: examples, limitations and possibilities

E P Altova, A N Rykov and I F Shishkov

Moscow State University, Chemistry Department, Moscow 119991, Russia

E-mail: katrin.27@inbox.ru

Abstract. The paper is aimed to the development of structural chemistry of biomolecules, which play essential role for living systems. The structures of biogenic amine, adrenaline, was determined using gas phase electron diffraction (GED) technique in comparison with tryptamine and noradrenaline investigation. Possibilities of method and its limitations are shown in application to bioactive molecular systems. The investigations of the molecules in gas phase allows us to find and describe details of their structures, molecular dynamic which leads to the behavior of the substances in the various processes.

1. Introduction

Gas phase electron diffraction (GED) is a diffraction method for analyzing the geometrical design of different free molecules. The electron beam, accelerated in the 60 kV electric field, is scattered by a vapor stream and then is recorded on a photo film. Deciphering the electron diffraction patterns provides for finding out the geometry of molecules and solving the structural chemistry problems. Structural investigations are essential in the modern chemistry, since the features of the molecular structure are responsible for the entire set of properties of substances according to the fundamental triad «structure — dynamics — property». It should be noted that the structures of free (isolated) molecules, not being affected by collective interactions occurring in the solid/liquid state, are of the most importance since they reflect their individual properties. Thus, the investigations of the substances in gas allows finding and describing details of their structures, molecular dynamics, which leads to behavior of substances in the various processes. The obtained structures of biogenic amines data is expected to provide the reference points for the assessment of some properties of other representatives of this class of hormones and neurotransmitters.

2. Analysis of the Gas-Phase Electron Diffraction Data

2.1. Gas-Phase Electron Diffraction Experiment

The GED experiment of adrenaline was carried out at the Moscow State University on the EG-100 M apparatus (figure 1). The electron wavelength was calibrated with CCl_4 scattering patterns. The diffractograms were recorded on the Maco EM-Films EMS photo films. The films were digitized and averaged to obtain mean electron diffraction intensities I(s) for each nozzle–plate distance. The optical densities were measured using a commercial Epson Perfection 4870 photo scanner. The other details of the experimental conditions are given in table 1. The UNEX [1] program was chosen to carry out the analysis of GED data, where the least-squares method was applied to the molecular intensities. The background lines were constructed by approximation with cubic splines. The main adrenaline parameters were refined in groups (table 2) because of strong correlation between independent structural parameters. The differences between bond lengths and valence angles, refined in one groups, were taken from MP2/cc-pVTZ calculations and kept fixed at the least-square procedure. It should be noted, that the vibration amplitudes and harmonic vibrational corrections to the internuclear distances (r_{h1} — r_a) were calculated from MP2/cc-pVTZ force constants using the SHRINK program [4].

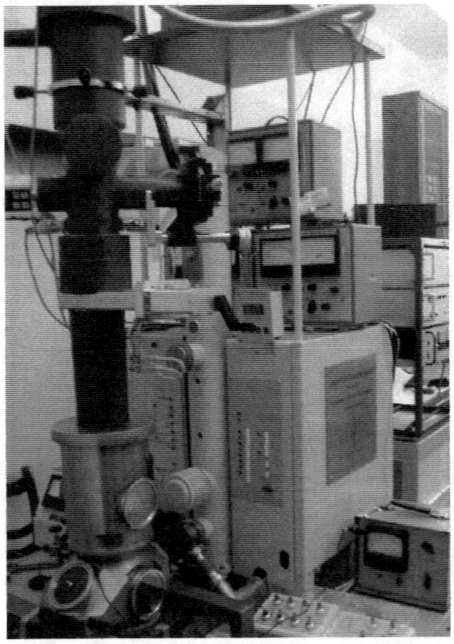

Figure 1. EG-100M device of Gas Electron Diffraction laboratory, Moscow State University.

Table 1. Experimental conditions of adrenaline GED experiment compared to noradrenaline and tryptamine structural experiments.

Compound	Camera distance (mm)	Nozzle temperature (K)	Electron wavelength (Å)	Range of s value (Å^{-1}) [a]
Adrenaline[b]	193.94	467	0,04983	6,6– 32,4
	362.28		0,04981	4,4 – 18,8
Noradrenaline [2][b]	193.94	446	0.05023	6.6–32.4
	362.28		0.04990	3.6–17.8
Tryptamine [3][c]	193.94	405	0,05008	6.6– 30.2
	362.28		0,04975	3.6– 16.8

[a] $s=4\pi\lambda^{-1}\sin(\theta/2)$, where θ is the scattering angle, λ is the electron wavelength.
[b] Commercial sample with a purity of 99% purchased from Aldrich Chemical Company.
[c] Purityof sample of 98% purchased from Acros Organics taken for experiments with no further purification.

2.2. Theoretical calculations
The geometries of the studied biogenic amine conformers were optimized with Gaussian 03 program [5] using the density functional (DFT) theory (B3LYP hybrid potential [6]) and the MP2 [7] theory.

3. Results and discussion
We decided to focus on the group of biogenic amines, like tryptamine, noradrenaline and adrenaline. These compounds are the most important hormones. Direct acting of adrenaline and noradrenaline on α - and β -adrenergic receptors is overlapping. The molecular structure of adrenaline (figure 3) was

studied by GED and quantum chemical calculations using MP2 methods with Dunning's cc-pVTZ [8] basis set. The adrenaline conformational analysis was performed previously [9]: the structures of the eight most stable conformers were found. This computational and spectroscopic study of the gas phase conformers of adrenaline [9] was able to give the explanation of the observed most stable conformer AG1a. In present GED study the fit of the experimental electron scattering intensities was obtained for the AG1a conformer model (see table 2, figure 2). The experimental structural parameters are found to be in good agreement with the results of theoretical calculations.

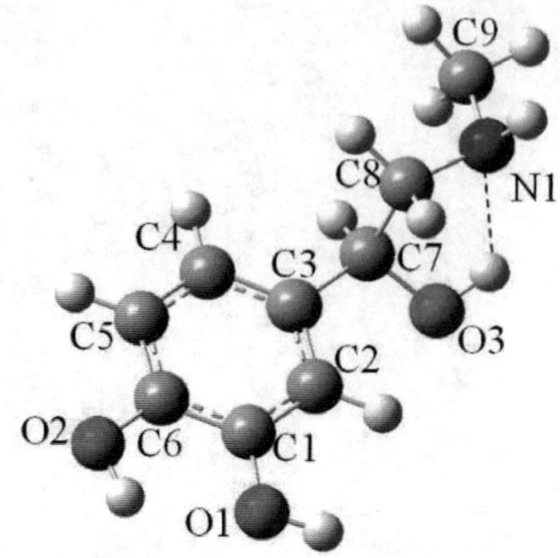

Figure 2. Molecular structure of the most abundant adrenaline conformers. AG1a conformer has atom numbering as defined in this paper.

Table 2. Molecular structure of the most abundant AG1a adrenaline conformer obtained by gas-phase electron diffraction and quantum chemical calculations

Parameter [a]	r_{h1}[GED]	MP2/cc-pVTZ	Parameter [a]	MP2/cc-pVTZ
r(C2-C1)	1,389(4) [1b]	1,392	∠(O1-C6-C5)	120,4 [c]
r(C3-C2)	1,401(4) [1b]	1,404	∠(O2-C1-C2)	124,3 [c]
r(C6-C5)	1,392(4) [1b]	1,395	∠(C7-C3-C2)	120,2 [c]
r(C6-C1)	1,405(4) [1b]	1,408	∠(C8-C7-C3)	111,2 [c]
r(O2-C1)	1,379(11) [2b]	1,369	∠(N1-C8-C7)	109,7 [c]
r(O3-C7)	1,416(11) [2b]	1,407	∠(C9-N1-C8)	114,7 [c]
r(C7-C3)	1,515(8) [3b]	1,514	∠(O3-C7-C3)	110,7 [c]
r(C8-C7)	1,548(8) [3b]	1,548	∠(O1-C6-C5-C4)	179,4 [c]
r(N1-C8)	1,467(15) [4b]	1,447	∠(C8-C7-C3-C2)	96,8 [c]
r(C9-N1)	1,463(15) [4b]	1,443	∠(N1-C8-C7-C3)	170,0 [c]

[a] Bond lengths in Å and angles in degrees.
[b] Values in parentheses are three times the standard deviations. Parameters with equal superscripts were refined in one group. Differences between parameters in the same group were fixed at the values from MP2/cc-pVTZ computations.
[c] Assumed at the values of MP2/cc-pVTZ calculations.

Thus, experimental structural determination study on adrenaline structure was carried out. The main conformations are stabilized by intramolecular hydrogen bonds. Some differences in bond lengths for gas and crystal phase [10] occur due to either neutral or zwitter-ion form presented.

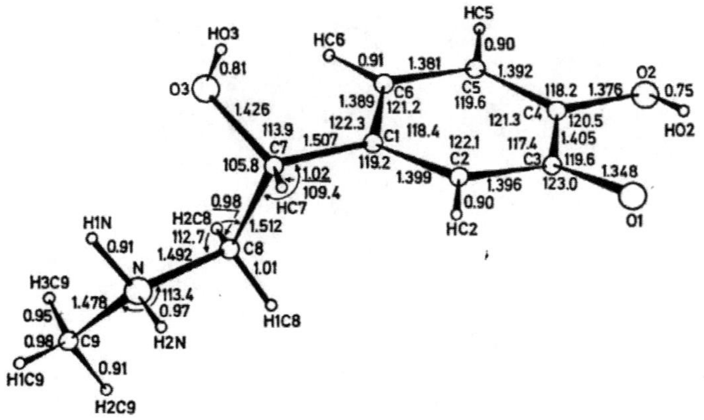

Figure 3. Structure of adrenaline according to X-ray diffraction [10]

Gas electron diffraction study of noradrenaline [2] was concluded that there were several intramolecular hydrogen bonds in noradrenaline (figure 4), namely in ethanolamine chain N···HO (2.05 Å), in catechol part of molecule O-H···O (2.05 Å), and between each other C-H···O. Thus, the ethanolamine part of noradrenaline is the most flexible part in this structure.

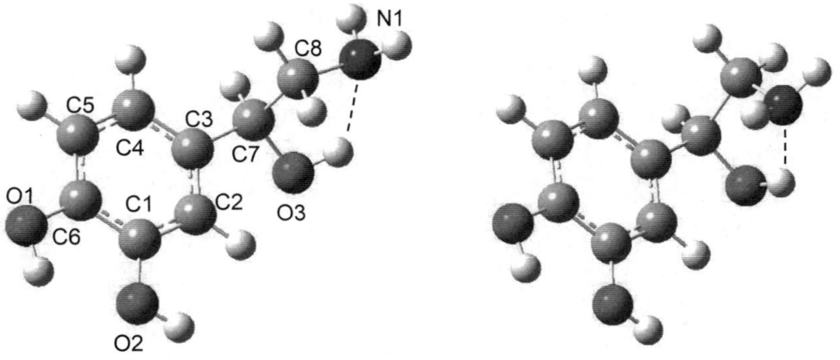

AG1a GG1a

Figure 4. The structures of conformers of noradrenaline.

Similar to adrenaline, the hydrogen bond of noradrenaline N···OH stabilizing both (Fig.4.) conformers. Intramolecular hydrogen bonds N···HO value determined from GED study (2.05 Å) was less the sum of the van der Waals radii of hydrogen and nitrogen atoms (2.6 Å). The structural parameters of two studied form noradrenaline (AG1a and GG1a) obtained from GED data were in consistent agreement with the results of MP2 and CCSD calculation [2].
Tryptamine structure has two most favorable conformers ($A^{+}G^{-}G^{+}$ and $A^{+}G^{-}G^{-}$). These forms (figure 5) can be hardly distinguished for separately detecting during gas phase electron diffraction.

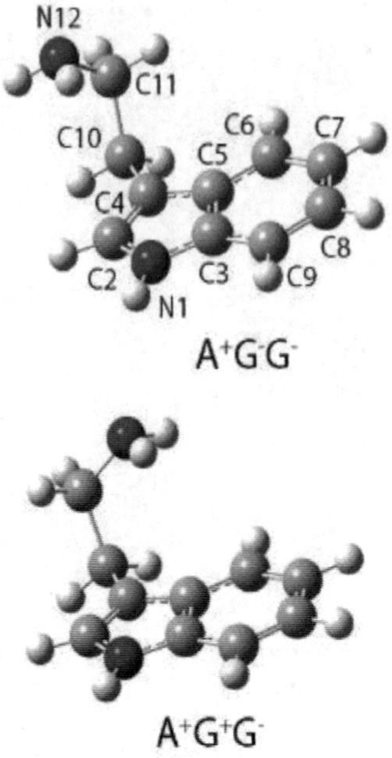

Figure 5. The structures of most favorable $A^+G^-G^+$ and $A^+G^-G^-$ conformers of tryptamine.

Despite the differing locations of amino hydrogen atoms, both conformers of tryptamine ($A^+G^-G^+$ and $A^+G^-G^-$) have the same orientation of indole moiety and flexible gauche(-) ethylamine fragment. Due to structural identity, their structures were indistinguishable for GED experiment, showing identical disagreement factors ($R_f = (\Sigma(sM(s)_{exp} - sM(s)_{theor})^2)/(\Sigma(sM(s)_{exp})^2$, $sM(s) = s \cdot I_{mol}(s)/I_{at}(s)$) equaled to 3.8%.

4. Conclusions

The structure of adrenaline was determined using GED technique along with previously determined other biogenic amines, noradrenaline and tryptamine. The most abundant conformers are found to be preferable due to stabilization by intramolecular hydrogen bonds between the nitrogen atom of amino group and hydrogen adjusted the oxygen of hydroxyl group N···HO, attached to ethylamine carbon atom. The results show that intramolecular hydrogen bonding controls both the conformational structure and the relative conformational abundance. Due to the similarity of these molecules most of the structural parameters of adrenaline are close to those of noradrenaline (table 3). The experimental structural parameters determined agree well with the results of ab initio calculations.

Table 3. Comparison of the experimental geometric parameters of adrenaline, noradrenaline and tryptamine.

Parameters [a,b]	Adrenaline AG1a[this work]	Noradrenaline AG1a / GG1a [2]	Tryptamine ($A^+G^-G^+$/ $A^+G^-G^-$)[3]
$r(C_{(cycle)}-C_{(CH2)})$	1.515 (8)	1.504(9) /1.519(9)	1.491(12)/1.489(12)
$r(C_{(H2)}-C_{(CH2)})$	1.548 (8)	1.532(9) /1.533(9)	1.527(12)/1.538(12)
$r(C_{(H2)}-N)$	1.467 (15)	1.466[c] /1.464[c]	1.475(26)/1.471(26)
$<(C_{(CH2)}-C_{(H2)}-N)$	109.7[c]	108.2[c] /107.1[c]	106.8(38)/112.5(38)
$<(C_{(cycle)}-C_{(CH2)}-C_{(H2)})$	111.2[c]	112.2(72) / 111.7(72)	112.9(28)/113.0(28)

[a] Bond lengths in Å and angles in degrees.
[b] Values in parentheses are three times the standard deviations.
[c] Assumed at the values of MP2/cc-pVTZ calculations.

Thus, the obtained structures of biogenic amines data is expected to provide the reference points for the assessment of some properties of other representatives of this class of hormones and neurotransmitters.

Both GED and rotational spectroscopy are the only experimental ways for studying the geometries of molecules with no intermolecular interactions. Theory of GED has no restrictions on molecular size, polarity or symmetry, therefore it may be preferred research method. The study of large size molecules and systems with heavy atoms is difficult for GED due to ambiguous interpreting of structural analysis data.

Acknowledgments

This study has been supported by Russian Foundation for Basic Research (project number 20-03-00747 A).

References

[1] Vishnevskiy Y V 2021 *UNEX v.1.6* http://unexprog.org
[2] Altova E, Rykov A, Abaev M, Marochkin I, Vilkova A, Popik M and Shishkov I 2019 Molecular structure of noradrenaline according to gas-phase electron diffraction data and quantum chemical calculations *Mendeleev Communications* **29** 4 pp 411-413
[3] Marochkin I, Altova E, Rykov A and Shishkov I 2017 Molecular structure of tryptamine in gas phase according to gas electron diffraction method and quantum chemistry calculations *J Mol Struct* **1148** 15 pp179-184
[4] Sipachev A V 2004 The use of quantum-mechanical third-order force constants in structural studies *J Mol Struct* **693** pp 235–240
[5] Frish M, Trucks G, Schlegel H, Scuseria G, Robb M, Cheeseman J, Zakrzewski V, Stratmann J, Burant J, Dapprich S, Millam J, Daniels A, Kudin K, Strain M, Farkas O, Tomasi J, Barone V, Cossi M, Cammi R, Mennucci B, Pomelli C, Adamo C, Clifford S, Ochterski J, Petersson G, Ayala P, Cui Q, Morokuma K, Malick D, Rabuck A, Raghavachari K, Foresman J, Cioslowski J, Ortiz J, Baboul A, Stefanov B, Liu G, Liashenko A, Piskorz P, Komazomi I, Gomperts R, Martin R, Fox D, Keith T, Al-Laham M, Peng C, Nanayakkara A, Challacombe M, Gill P, Johnson B, Chen W, Wong M, Andres J, Gonzales C, Head-Gordon M, Replogle E and Pople J 2003 Gaussian 03 (Revision D01) Gaussian Inc Pittsburgh
[6] Becke A 1988 Density-functional exchange-energy approximation with correct asymptotic-behavior *Phys Rev A* **38** pp 3098-3100
[7] Møller C and Plesset M 1934 Note on an approximation treatment for many-electron systems *S Phys Rev* **46** pp 618-622
[8] Dunning T 1989 Gaussian Basis Sets for Use in Correlated Molecular Calculations I The Atoms Boron through Neon and Hydrogen *J Chem Phys* **90** 2 pp 1007-1024
[9] Carcabal P, Snoek L, and Mourik van T 2005 A computational and spectroscopic study of the gas-phase conformers of adrenaline *Molecular Physics* **103** pp1633-1639
[10] Andersen A 1975 Structural Studies of Metabolic Products of Dopamine III Crystal and Molecular Structure of (-)-Adrenaline *Acta Chem Scand B* **29**(2) pp 239-44

Using atomic force microscopy to work with the cell surface of Candida albicans

E M Filippova[1], U V Nesvizhski [2], S A Titov [3] and A I Glukhov [4]

[1] Sechenov University (MSMU) Trubeckaya 8, Moscow, Russia, liza.filippova97@yandex.ru
[2] Sechenov University (MSMU) Trubeckaya 8, Moscow, Russia, nesviz@mail.ru
[3]Federal State Budgetary Institution of Science Scientific and Technological Centre of Unique Instrumentation of the Russian Academy of Sciences, Butlerova 13, Moscow, Russia, sergetitov@mail.ru
[4] Sechenov University (MSMU) Trubeckaya 8, Moscow, Russia, aiglukhov1958@gmail.com

Abstract. Candida albicans is a yeast-like fungus that lives on human mucous membranes and skin and does not cause infections. However, it plays a role in the development of opportunistic infections in immunocompromised people. In this work, we would like to evaluate the possibility of studying the cell wall of C. albicans by atomic force microscopy, as well as compare the operating modes of the microscope and choose optimal one for working with the fungus. Atomic force microscopy is a powerful tool for evaluating surfaces, including the cell walls of biological objects. The microscope is capable of operating in different modes, but in this study we compared two of them: contact and semi-contact. These methods are the most popular for evaluating the surfaces of biological objects. Comparison of the modes was carried out on the C. albicans strain. The surface of the strain was scanned by atomic force microscopy, and the curves of the dependence of the sensor deviation from the distance to the object were recorded. Scanning and recording of curves were carried out in two modes of operation of the microscope: contact and semi-contact, as well as three sensors: soft, medium and hard.

1. Introduction

Over the past decades, it has been shown that the number of infections caused by fungi, in particular candida, is on the rise. This is mainly due to the growing number of immunocompromised individuals, for example, people with immunodeficiency or receiving immunosuppressants after transplantation, as a consequence of autoimmune diseases or for suppression of inflammation, as well as undergoing chemotherapy for cancer. The most common cause of mycoses are yeast-like fungi of the genus Candida, which can cause both superficial and systemic processes characterized by high mortality. The most common and well-studied member of this genus of fungi is Candida albicans.

To date, more than 200 species of fungi of the genus Candida are known, the identification of which is difficult due to the peculiarities of their biology: they have both a haploid and a diploid set of chromosomes, a variable morphology, a high strength of the cell wall, are resistant to the action of immune factors, etc. For decades, a steady expansion of the pool of strains with high resistance to antimycotic drugs was clearly indicated, which complicates therapeutic measures [23].

The foregoing determined the urgency of the problem of identifying and treating infections caused by yeast-like fungi of the genus Candida, and the search for new diagnostic approaches, one of which can be atomic force microscopy (AFM). The essence of the technique is scanning the surface of an object with a sensitive probe, which makes it possible to obtain its topographic characteristics. This tool is perfect for studying biological objects, in particular cells, since it provides a higher resolution than a light microscope [10].

AFM is used for high-resolution imaging to study the topography of a cell, cell wall and its components. AFM is not limited to the study of surface topography. It is also used to study the mechanical properties of cells, the strength of adhesion and interactions between individual molecules, microorganisms on the surface of any object. AFM makes it possible to assess the elastic properties of membranes, which are closely related to the function of the object of study and its pathology [11]. This tool is a very promising research method, and also gives hope for the invention of new diagnostic methods.

The principle of operation of an atomic force microscope is based on recording the force interaction between the surface of the sample and the probe. A nanoscale tip located at the end of an sensor called a cantilever is used as a probe. The force acting on the probe from the surface causes the arm to bend. The appearance of hills or depressions under the tip leads to a change in the force acting on the probe, and hence to a change in the magnitude of the cantilever bending. Thus, by registering the amount of bending, it is possible to draw a conclusion about the surface relief. The forces acting between the probe and the sample, first of all, mean long-range van der Waals forces, which are at first attractive forces, and upon further approach they transform into repulsive forces [25].

In accordance with the above, the purpose of this work was to study the physicochemical properties of the surface of C. albicans by the AFM method: scanning the fungus surface and recording the curves of the interaction of the bioorganic structures of the fungal cell wall with the scanning sensor of the device.

2. Methods

We used the strain 30.2 C. albicans from the collection of the FBSI MNIIEM Institute named after N.G. Gabrichevsky Rospotrebnadzor. The strains were stored frozen at -70 ° C. Before the work, the strain was defrosted, recovered and sowed on a solid nutrient medium Sabouraud (agar). Petri dishes were stored under sterile conditions. A thick suspension was prepared from cultures of microbes grown on Saburo. Then it was applied to defatted glass slides for the preparation of 4 samples, which were subsequently dried in the open air under the sterile conditions of a laminar box and stored until required. The quality of the preparation was assessed by light microscopy in transmitted light.

For AFM, we used an NT-MDT microscope, model Solver PRO. The work was carried out in contact and semi-contact modes. On each sample, a study of 8-11 points was carried out, and the DFL (Height) curves were recorded at each approach and withdrawal of the cantilever. We compared three types of cantilevers with different elastic constants: "soft" CSG 11 (elastic constant 0.03N / m), "medium" HA_FM / W2C (elastic constant 3.5N / m), "hard" NSG 11 (elastic constant 5.5N / m).

To obtain an image, a 5x5 nm section of the object was scanned. To assess the image quality, two parameters were introduced - contrast and image quality. The latter was assessed by the method of expert evaluation for the absence of visual defects: stripes and gaps in individual areas of the image (Figure 1). In order to choose which of the sensors / modes is more suitable for obtaining images, we kept a record of the number of contrast / non-contrast and defective images. And then, using 2x2 contingency tables, we first compared the contact and semi-contact modes for each sensor with each other, and then the sensors between each other in different modes.

In addition to the images, the curves of the dependence of the cantilever deviation on the height were obtained. They had a typical form for such curves (Figure 2). In the figure, the 'approach' curve of the cantilever is marked in red, and the 'retraction' curve is shown in blue. The abscissa shows the Height, measured in nanometers. The ordinate is the vertical deflection of the cantilever (DFL) measured in nanoamperes.

The following points were determined, each of which has two coordinates along the abscissa and ordinate axes. In the "approach" mode (Figure 3): "1" - the beginning of the action of attractive forces on the cantilever, "2" - the change of the action of forces to repulsion, "3" - the forces arising from direct contact with the cell surface. In the "retract" mode (Figure 3): "1" - the minimum effect of the adhesion forces, "2" - the maximum manifestation of the adhesion forces, "3" - the forces arising from direct contact with the cell surface.

The research results were subjected to standard statistical processing using the STATISTICA for Windows version 6.0 program. The comparative analysis used the nonparametric Mann-Whitney and Chi-square methods. The criterion for the statistical reliability of the data obtained was considered to be the generally accepted value of $p < 0.05$ in medicine [24].

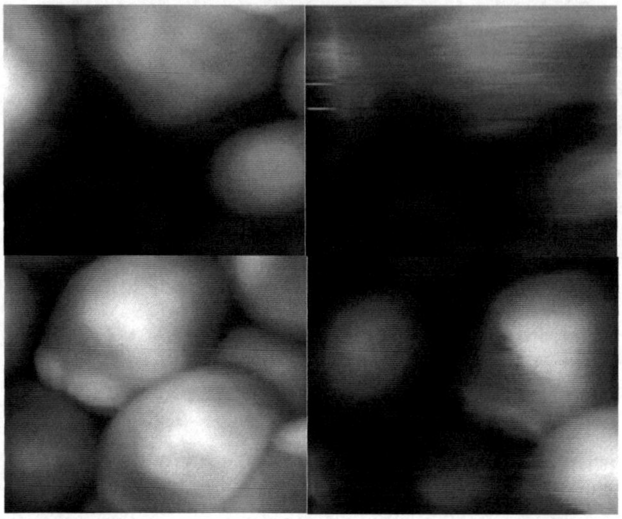

Figure 1. Images without defects are shown on the left, with defects on the right.

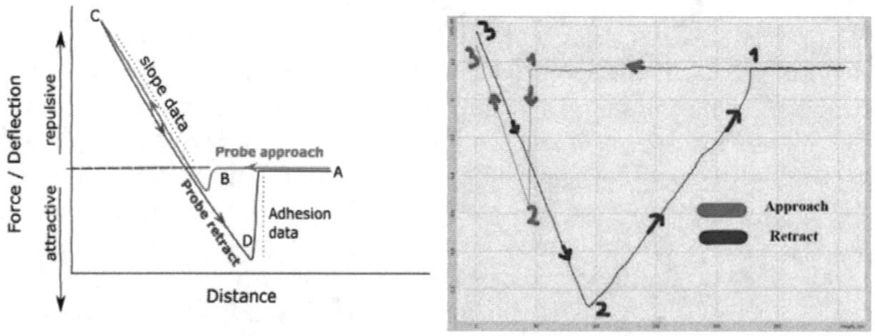

Figure 2. Example of DFL(Height) curves

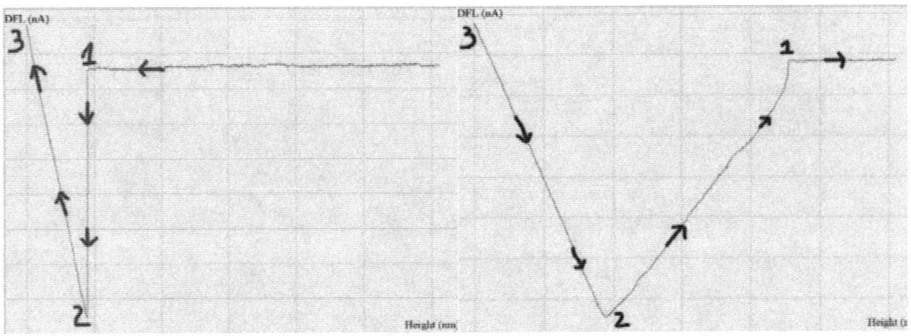

Figure 3. Examples of DFL (Height) curves in the "approach" (left) and "retract" (right) modes of the cantilever.

3. Results

As a result of the work, we managed to shoot 52 images: 18 with a "soft" cantilever, 18 with a "medium" and 16 with a "hard" cantilever. Each cantilever operated in two modes: contact and semi-contact. Accordingly, half of the images for each sensor were captured in one mode, the rest - in another.

In Figure 4, 5 and 6 show examples of the images obtained for each mode and sensor, and the results of an expert assessment of their quality are presented in Table 1. According to the comparative analysis (Table 2), it can be concluded that for obtaining images of C. albicans in AFM in the semi-contact mode, "medium" and "hard" cantilevers are better suited, and in the contact mode, "soft" cantilevers are better. In this case, the semi-contact mode seems to be more preferable.

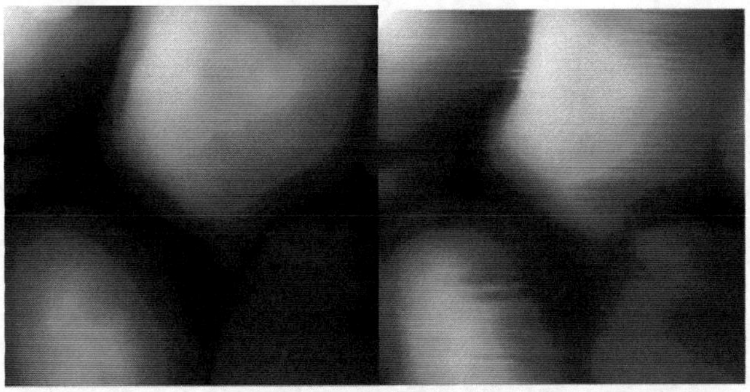

Figure 4. Image of a C. albicans cell in semi-contact (left) and contact (right) modes when working with a "hard" cantilever. The image was obtained in the Nova program.

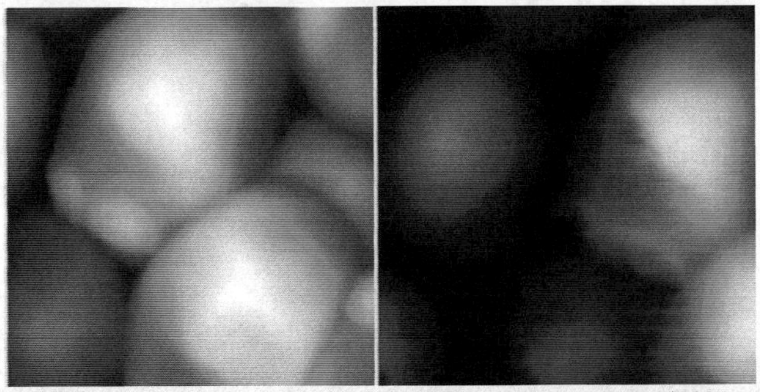

Figure 5. Image of a C. albicans cell in semi-contact (left) and contact (right) modes when operating with a “medium” cantilever. The image was obtained in the Nova program.

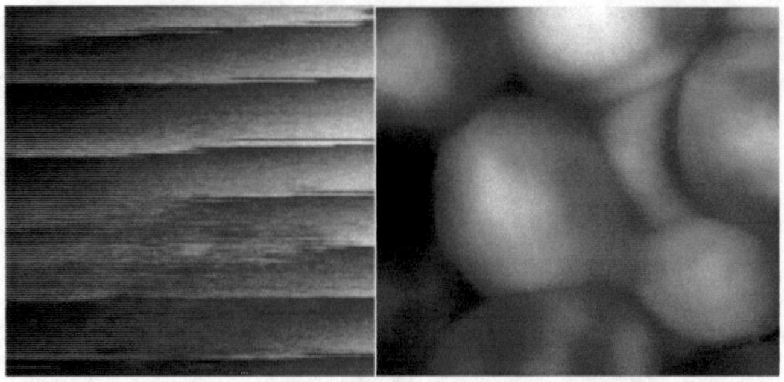

Figure 6. Image of a C. albicans cell in semi-contact (left) and contact (right) modes when working with a “soft” cantilever. The image was obtained in the Nova program.

Table 1. Results of image quality assessment

Mode		n	Contrast		Image quality	
			n	%	n	%
«soft»	Contact	9	9	100,0	8	88,9
	Semi-contact	9	0	0,0	0	0
«medium»	Contact	9	3	33,3	0	0
	Semi-contact	9	9	100,0	8	88,9
«hard»	Contact	8	5	62,5	1	12,50
	Semi-contact	8	6	75,0	5	62,50

Table 2. Comparison of the efficiency of sensors in terms of parameters for assessing image quality at different scanning modes

	Contact mode		Semi-contact mode	
	Contrast	Image Quality	Contrast	Image Quality
«Soft»/ «Medium»	<0,01	<0,001	<0,001	<0,001
«Soft»/ «Hard»	<0,05	<0,01	<0,01	>0,05
«Medium»/ «Hard»	>0,05	>0,05	>0,05	>0,05

At each of the 52 analyzed points, the interaction of the bioorganic structures of the fungal cell wall with the AFM scanning device was investigated. During this study, limitations related to the nature of the device's operation appeared: in the semi-contact mode, when approaching the object under study, the cantilevers exhibited high-frequency vibration, which did not allow fixing the control points. In addition, the "hard" cantilever did not detect these points even in the contact mode. At the same time, control points were well detected in 100% of cases with “soft” and “medium” cantilevers in the contact mode.

Table 3 shows the results of measuring the values of coordinates (Height and DFL) of the points of the investigated graphs in the mode of "approach" by various sensors. The average Height value for the “1” point measured by the soft sensor does not differ significantly from the value obtained by the “medium” sensor ($p > 0.5$). For "2" and "3" points, the difference is also insignificant ($p > 0.5$). For the DFL characteristics, the values of all three points obtained by the “soft” and “medium” sensors turned out to be significantly different ($p < 0.05$).

Table 4 contains similar information for the retraction mode. The average value for the height at the "1" point, measured by the "soft" sensor, significantly differs 1.64 ($p < 0.05$) times from the given value measured by the "medium" sensor. The value for the "2" point also differs significantly by 1.86 times ($p < 0.05$). Similar values for the "3" point do not differ significantly ($p > 0.05$). The results for DFL were as follows: "1" point differs significantly by -3.18 times ($p < 0.05$), and "2" and "3" are insignificant by 0.87 ($p > 0.05$) and 1.18 ($p > 0.05$), respectively

Table 3. Comparative efficiency of cantilevers in the "approach" mode.

Characteristic	Point	Cantilever		К	p
		«Soft»	«Medium»		
Height	«1»	208,55 ± 25,44	174,7125 ± 9,88	1,19	>0,05
	«2»	199,80625 ± 26,08	171,74375 ± 9,92	1,16	>0,05
	«3»	0	0	0	-
DFL	«1»	482,95625 ± 46,18	-144,4375 ± 28,32	-3,34	<0,05
	«2»	432,1875 ± 42,86	-195,5625 ± 26,12	-2,21	<0,05
	«3»	1496,1875 ± 80,45	- 1246,04375 ± 84,22	1,20	<0,05

To compare the effectiveness of the "soft" and "medium" sensor for fixing control points, the results were summarized in Table. 5. As you can see, the "soft" sensor more often made an error when examining an object such as Candida albicans.

Table 4. Comparative efficiency of cantilevers in the "retract" mode.

Characteristic	Point	Cantilever		К	p
		«Soft»	«Medium»		
Height	«1»	371,7625 ± 32,92	226,76875 ± 9,93	1,64	<0,05
	«2»	367,0625 ± 33,74	197,90625 ± 10,51	1,86	<0,05
	«3»	0	0	0	-
DFL	«1»	488,875 ± 47,55	-153,6875 ± 31,45	-3,18	<0,05
	«2»	-234,75 ± 56,97	-271,1875 ± 28,75	0,87	>0,05
	«3»	1540,25 ± 82,81	1309,625 ± 86,03	1,18	>0,05

Table 5. Comparison of the ability of sensors to detect differences between points

Characteristic	Mode	Cantilever	Comparable points		
			«1»/ «2»	«1»/ «3»	«2»/ «3»
Height	«approach»	«Soft»	-	<0,05	<0,05
		«Medium»	-	<0,05	<0,05
	«retraction»	«Soft»	-	<0,05	<0,05
		«Medium»	<0,05	<0,05	<0,05
DFL	«approach»	«Soft»	-	<0,05	<0,05
		«Medium»	<0,05	<0,05	<0,05
	«retraction»	«Soft»	<0,05	<0,05	<0,05
		«Medium»	<0,05	<0,05	<0,05

4. Conclusion

Thus, we have shown that AFM successfully allows one to visualize such a microscopic object as yeast-like fungi of the genus Candida, and to study the integral physicochemical characteristics of the bioorganic structures of its cell wall.

In this study, although on a limited number of samples, the test was carried out in parallel in the contact and semi-contact modes and several cantilevers during their retraction and retraction, which already at this stage makes it possible to outline the optimal mode for further studies of the study of the physicochemical properties of this object by the AFM method.

Our results were only partially consistent with the work of scientists studying C. albicans. The results obtained by us are consistent with the data obtained [11, 17] during the development of the modes of investigation of C. albicans. At the same time, they contradict the way of detecting force versus distance curves by Sofiane El-Kirat-Chatel et al. [12]. Indeed, C. albicans is better imaged in semi-contact mode when using sensors with a low elastic constant. However, in our work it was shown that the "medium" cantilever is more sensitive than the "soft" cantilever, which is usually used by scientists to study the physicochemical properties of the cell wall of C. albicans, and the "hard" one, at the same time, is of little use. It is quite obvious that in such a situation, in order to overcome the disagreements and develop effective modes of atomic force microscopy of C. albicans, it is necessary to continue our research on a larger sample.

Conclusions:

1. The "soft" cantilever proved to be effective for the visual assessment in the contact mode, and the "medium" one in the semi-contact mode.
2. The "medium" cantilever is more effective in studying the structures of C. albicans as compared to the "soft" one, as it turned out to be more effective in relation to the Height characteristic for "abduction" and DFL for "abduction".
3. In C. albicans AFM, the DFL characteristic is more sensitive and effective than Height, since it is well differentiated when "approach" and "retract".

References:

[1] Zverev V V et al. Fundamentals of microbiology and immunology 2014 *GEOTAR-Media* 368
[2] Elinov N P Toxigenic fungi in human pathology 2002 *Problems of medical mycology* **4** p37
[3] Gow N A, Brown A J, and Odds F C Fungal morphogenesis and host invasion *Curr. Opin. Microbiol* 2002 **5** p366371
[4] Gazzard B Fungal infection of the gastrointestinal tract *Principles and Practice of Clinical Mycology* 1996 p165-177
[5] Araviyskiy R A et al Candidal sepsis *Diagnosis of mycoses*
[6] Arnaldo Lopes Colombo, Thaís Guimarães et al, Brazilian guidelines for the management of candidiasis – a joint meeting report of three medical societies *The Brazilian Journal of Infectious Diseases* **17** 2013 p283–312
[7] Martha F Mushi et all High Oral Carriage of Non-albicans Candida spp. among HIV-infected individuals *International Journal of Infectious Diseases* **49** 2016 p185–188
[8] Leach M D et al. Hsp90 orchestrates transcriptional regulation by Hsf1 and cell wall remodelling by MAPK signalling during thermal adaptation in a pathogenic yeast. *PLoS Pathog* 2012
[9] Daniel Poulain Candida albicans, plasticity and pathogenesis, *Crit Rev Microbiol* 2015 p208-17.
[10] Hayden K Webb et al, Physico-mechanical characterisation of cells using atomic force microscopy - Current research and methodologies *Journal of Microbiological Methods* **86** 2011 p131–139
[11] S Kasas, et al., AFM contribution to unveil pro- and eukaryotic cell mechanical properties, *Semin Cell Dev Biol* **73** 2017 p177-187
[12] Sofiane El-Kirat-Chatel et al. Nanoscale analysis of caspofungin-induced cell surface remodelling in Candida albicans *Nanoscale* 2013 **5** 1105–1115.
[13] Chen Y-Y et al Surface rigidity change of Escherichia coli after filamentous bacteriophage infection. *Langmuir* **25** 2010 p4607–4614

[14] Vadillo-Rodriguez V et al, Surface viscoelasticity of individual Gram-negative bacterial cells measured using atomic force microscopy *J Bacteriol* **190** 2008 p4225–4232
[15] Volle C B et al Quantitative changes in the elasticity and adhesive properties of Escherichia coli ZK1056 prey cells during predation by Bdellovibrio bacteriovorus 109 *J. Langmuir* **24** 2008 p8102–8110
[16] Mei L et al Poisson analysis of streptococcal bond strengthening on stainless steel with and without a salivary conditioning film *Langmuir* **25** 2009 p6227–6231
[17] Devendra Khatiwada and shobha Kanta lamichhane A Brief overview of AFM Force distance spectroscopy, *The Himalayan Physics* **2** 2011 p80-83
[18] Boks N P et al Bond-strengthening in staphylococcal adhesion to hydrophilic and hydrophobic surfaces using atomic force microscopy *Langmuir* **24** 2008 p12990–12994
[19] Park B J et al A correlation between the virulence and the adhesion of Listeria monocytogenes to silicon nitride: an atomic force microscopy study *Colloids Surf* **73** 2009 p237–243
[20] Chandraprabha M N et al A Modeling and analysis of nanoscale interaction forces between Acidithiobacillus ferrooxidans and AFM tip *Colloids Surf* **75** 2010 p310–318
[21] *Measuring and Understanding Force Distance Curves AFMWorkshop*
[22] Garcia-Rubio R et al The Fungal Cell Wall: Candida, Cryptococcus and Aspergillus Species *Front Microbiol* 2020 **10** p2993
[23] Voropaev A D et al Mutations in the ERG11 gene of Candida albicans isolated from HIV-infected patients Problems of medical mycology 2021 p68
[24] Urbach VYu Statistical analysis in biological and medical research 1975 p295
[25] B.Cappella and G.Dietler Force-distance curves by atomic force microscopy *Surface Science Reports* **34** 1999 p1-3 p5-104

A method for presenting biomedical images based on a model of receptive fields of human visual perception

V E Antsiperov[1], V A Kershner[1]

[1] Kotelnikov Institute of Radioengineering and Electronics of Russian Academy of Sciences 11 Mokhovaya street, building 7, Moscow 125009, Russia

E-mail: antciperov@cplire.ru, vladkershner@mail.ru

Abstract. The paper is devoted to the development of a new method for presenting biomedical images based on local characteristics of the intensity of their shape. The proposed method of image processing is focused on images that have low indicators of the intensity of the recorded radiation, resolution, contrast and signal-to-noise ratio. The method is based on the principles of machine (Bayesian) learning and on samples of random photo reports. This paper presents the results of the method and its connection with modern approaches in the field of image processing.

1. Introduction

Among the well-known recognition algorithms, algorithms of biological origin are of particular interest, namely algorithms that are associated with the perception of the surrounding world by living organisms. Currently, one of the best such systems is the human visual system. Its distinctive properties are speed, consumption of a small amount of resources sufficient for fast processing of information, and obtaining the result necessary for further operations. Among other things, the process of sensory adaptation is of particular interest, as a result of which the organs of the visual system translate the disturbed information, which has unrelated components, into a new representation, which is subsequently perceived as whole data [1]. For example, an image presented in the form of a mosaic grid over time is perceived as a whole image on which certain elements can be recognized.

It is known that a huge number of cells are responsible for the perception of an image. Neural impulses from many photoreceptors pass through bipolar neurons and converge in a relatively small group of ganglion cells [2,3]. There is reason to assume that ganglion cells, which sum up information from primary receptors in some way, form a certain image, which, apparently, is close to the amplitude modulation of the original image. It is assumed that ganglion cells can be an analog of a part of the modern system of auto-coding of neural networks that are responsible for reducing the dimension of input data.

In this paper, a mathematical model of the work of cells responsible for image transformation was developed. To do this, we convert an ordinary image into a frequency-modulated one, forming a sample from random photo reports. Thus, we implement the first stage of our autoencoder, creating an intermediate layer by reducing the initial dimension of the image. Then, based on the modified expectation maximization algorithm (EM algorithm), the frequency-modulated image is transformed into some similarity of amplitude modulation, where the component nodes are in regular positions. The value of the covariance matrix in this case can be different. This algorithm is responsible for allocating input data to common features that are encoded in the values of the nodes of the neural network. Where the elements of the diagonal of the matrix have the greatest value, we have a large white point, a small value – a small white point. The scale of individual nodes varies depending on the intensity of the incident radiation, thereby simulating amplitude modulation.

2. Results of image processing by the recognition algorithm

The image processing procedure described above is shown in Figure 1. For processing, objects of a simple structure were taken from the MPEG-7 Core Experiment CE-Shape1 database [4]. This database contains 20 images from 70 categories of objects and is used to compare algorithms for recognizing the shapes of objects.

From the database [4], images of 11 categories were randomly selected as precedents: "apple-5"," bell-7", "bird-16", "camel-13", "classic-12", "crown-3", "device0-1", "device2-3", "device5-3", "frog-1", "spring-11". The images were unified in size (500x500 pixels), and the objects were centered in them. Figure 1A shows an example of the "bird-16 " precedent. The registration of the precedent data was simple: 2000 points were selected from a large number of points randomly generated by intensity and location, the corresponding registration data is shown in Figure 1B. Figure 1C shows a precedent processed by a modified expectation maximization algorithm.

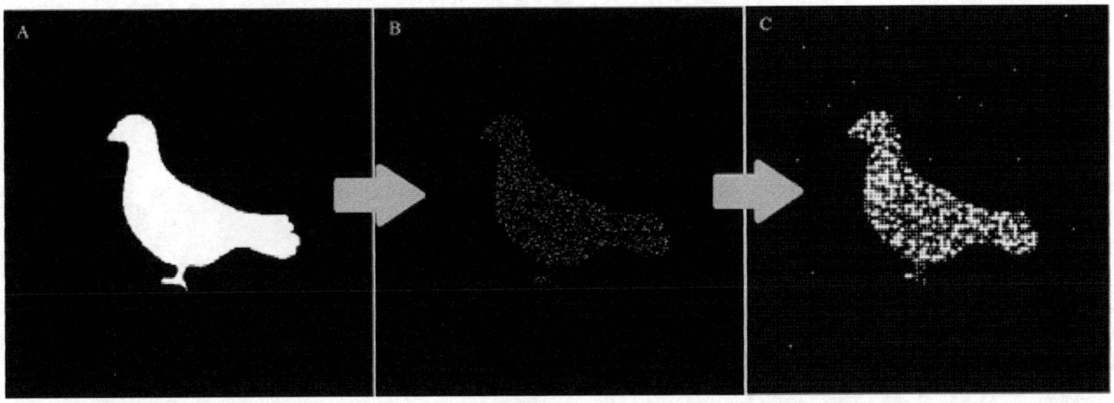

Figure 1. Image processing process: A. The original binary image of the "bird-16 " precedent from the MPEG-7 Core Experiment CE-Shape1 database, B. A model consisting of 2000 registered photo reports distributed throughout the image, C. An image obtained after processing by the algorithm.

Together with the images taken as precedents, test images from the same 11 categories were selected: "apple-6"," bell-8", "bird-17", "camel-14", "classic-16", "crown-6", "device0-4", "device2-10", "device5-5", "frog-7", "spring-12". These images were processed in the same way as the use cases, and the use cases were compared with test objects using the Kullback-Leibler divergence:

$$D_{KL}\left(P_{test}||P_{prec}\right) = \sum_{x\in X} P_{test}(x) log\left(\frac{P_{test}(x)}{P_{prec}(x)}\right) \quad (1)$$

Table 1. Comparison of use cases and test images using the Kullback-Leibler divergence

		Precedents										
		apple	bell	bird	camel	classic	crown	device0	device2	device5	frog	spring
Test objects	apple	2,9378	3,6005	4,5571	3,9956	4,4305	4,2063	4,4329	3,6306	4,0070	3,5308	5,4322
	bell	4,3818	3,7303	6,6922	5,8712	6,3172	6,6138	7,3664	5,3085	6,7210	4,1121	7,9582
	bird	14,4641	17,7542	4,8902	12,3595	12,1745	13,7451	14,0721	13,3026	11,7710	14,5675	22,0714
	camel	7,4579	10,2635	9,1607	4,2007	7,6891	8,3173	8,7186	7,6051	8,9279	9,0086	10,6597
	classic	8,8563	10,6085	7,5672	9,5916	2,2671	10,8005	10,2134	8,7780	7,5760	10,9850	18,1666
	crown	7,9141	8,4355	9,3753	9,5292	10,4030	7,2756	12,0035	8,6558	10,1897	9,4557	10,8039
	device0	10,1890	10,5740	8,4158	8,4856	8,7517	10,5789	3,7543	9,5861	8,7668	9,5780	11,9885
	device2	6,1083	5,7442	7,7601	6,0957	7,6407	6,1859	7,1800	2,8173	7,2833	7,7665	7,7695
	device5	15,6190	15,9507	13,9685	15,9346	15,6059	16,2344	15,1375	17,3147	9,0313	16,2801	14,5635
	frog	6,5004	5,4055	6,1467	7,8254	7,9172	6,7486	8,8172	5,6480	8,4859	2,8270	12,2837
	spring	21,3288	20,7868	24,3270	23,2406	23,9532	21,3058	22,1683	21,2125	22,6369	22,9784	8,2418

3. Conclusion

The method of image autocoding obtained by us is exceptionally suitable for the transition from one type of representation to another type of image representation, characterized by low contrast, resolution and signal-to-noise ratio associated with a low intensity of the recorded radiation. It is assumed that the processing of graphic information by recognition algorithms inspired by human perception will allow faster and better processing of the necessary information.

The work was carried out within the framework of the state task and partially supported by the Russian Foundation for Basic Research, grant No. 18-07-01295 A.

4. References

[1] Kharauzov A K, Vasiliev P P, Shelepin Yu E, etc. 2015 *Optical Journal. Image perception in visual search tasks under dynamic interference conditions Vol. 82*, No. 5. - pp. 42-55.2.

[2] Gusev A N 2007 *General psychology. vol. 2: Sensation and perception.* Publishing center "Academy"

[3] Alexandrov Yu I 2014 *Psychophysiology: Textbook for universities.* St. Petersburg: Peter

[4] Latecki L J, Lakmper R and Eckhardt U 2000 *Shape Descriptors for Non-rigid Shapes with a Single Closed Contour Proc.* IEEE Conf. Computer Vision and Pattern Recognition

Journal of Physics: Conference Series **2091** (2021) 012028 doi:10.1088/1742-6596/2091/1/012028

Spatio-temporal analysis the results of solving the inverse problem of electrocardiography

A I Shcherbakova, Y A Kupriyanova and G V Zhikhareva

National Research University "MPEI", 14 Krasnokazarmennaya street, Moscow 111250, Russia

Abstract. The results of the development and testing an algorithm for the physiological interpretation the results of solving the inverse problem of electrocardiography are presented. The solution to the inverse problem of electrocardiography is the distributions of equivalent current sources on the quasi-epicardium, restored from the electric potentials created by the heart on the surface of the chest. The developed algorithms are based on the space-time analysis of the distributions of equivalent current sources on the quasi-epicardium. The development and testing of the algorithm was carried out using a simulation model of the electrical activity of the heart based on cellular automata. The efficiency of the algorithm has been demonstrated when simulating the electrical activity of the heart in normal conditions, as well as in the presence of pathological changes in the myocardium in the form of areas with delayed conduction of excitation.

1. Introduction

The urgent task of electrocardiology is the development of algorithms for solving the inverse problem of electrocardiography. [1-3] Such algorithms make it possible to obtain and analyze the distribution of equivalent current sources on the surface of the quasi-epicardium (a spherical surface surrounding the heart in its immediate vicinity) and are designed to increase the sensitivity of electrocardiographic examinations, especially in the early stages of heart disease.

The results of solving the inverse problem of electrocardiography, considered in the work, represent the distributions of simple and double layers of equivalent current sources, i.e. electrodynamic characteristics of the heart. [4] To be able to use these results by cardiologists, a transition from electrodynamic to electrophysiological terms describing the state of the myocardium is necessary. In the present work, an algorithm for such a transition is proposed.

To develop this algorithm, it was necessary to solve the following tasks:

- to determine the main electrophysiological conditions of the myocardium during a single cardiac cycle;
- to simulate the ideal results of reconstruction of simple and double layers of current sources using a mathematical model of the electrical activity of the heart based on cellular automata [5];
- to conduct a spatio-temporal analysis of the distributions of simple and double layers of current sources on the quasi-epicardium (compare the values of reconstructed sources with the electrophysiological states of the myocardium) and formulate an algorithm for determining the electrophysiological states of the myocardium by the values of equivalent current sources;
- to investigate the efficiency of the developed algorithm and the relevance of the selected parameters of the algorithm in the presence of pathological changes in the myocardium.

Let's consider the solution of these tasks in more detail.

2. Physiological interpretation the results of solving the inverse problem of electrocardiography

2.1. Determination of electrophysiological states of the myocardium

To determine the electrophysiological states of the myocardium, consider the transmembrane action potential of the cells of the working myocardium [6] and distinguish four electrical states of the membrane: 0 - rest, 1 - depolarization, 2 - refractoriness, 3 - repolarization (Figure 1).

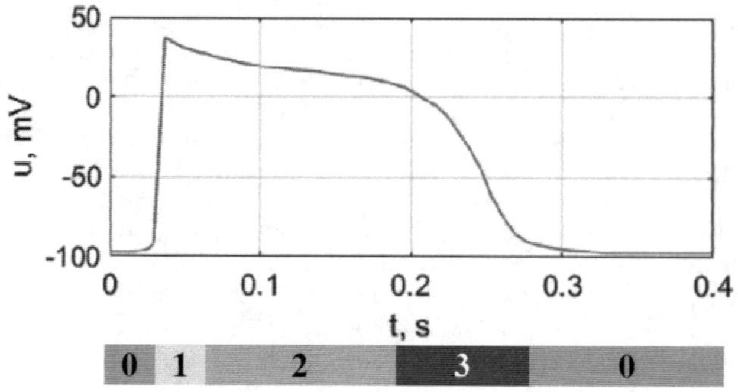

Figure 1. Transmembrane action potential of myocardial cells and electrical states of the membrane: 0 - rest, 1 - depolarization, 2 - refractoriness, 3 - repolarization.

The state of depolarization (1) corresponds to the process of excitation the myocardium and leads to contraction the heart muscle, the state of repolarization (3) corresponds to the process restoration and relaxation of the heart muscle. During these two phases, ionic currents actively flow through the membranes of myocardial cells, which are sources the electromagnetic field of the heart. And it is precisely these states that must be recognized by the time dependences the equivalent current sources at various points of the quasi-epicardium.

2.2. Modeling reconstruction results

The results of solving the inverse problem of electrocardiography are the distributions on the quasi-epicardium of simple γ and double D layers of equivalent current sources. Distributions are obtained in the form of discrete samples $\{\gamma_i\}$ or $\{D_i\}$, evenly spaced along the quasi-epicardium [7]. Figure 2 shows the distributions of a simple layer of current sources at time t_R corresponding to the apex of the R-wave (Figure 3): calculated using the mathematical model the heart generator $\{\gamma_i\}$ and obtained as a result of solving the inverse problem of electrocardiography $\{\gamma_{pi}\}$ by potentials $\{\phi_i\}$ created by the heart generator on the surface of the chest model (Figure 4). Similar distributions for the double layer $\{D_i\}$ and $\{D_{pi}\}$ are shown in Figure 5.

To formulate an algorithm for physiological interpretation using a model an electric generator of the heart, the idealized results of solving the inverse problem of electrocardiography $\{\gamma_{ij}\}$ and $\{D_{ij}\}$ were calculated at all points of the quasi-epicardium ($i = 1 \ldots 866$) for all discrete moments a time of a single cardiac cycle $t_j = (j-1)\Delta t$ where $\Delta t = 5$ ms ($j = 1\ldots1001$), i.e. on the interval from 0 to 0.5 s.

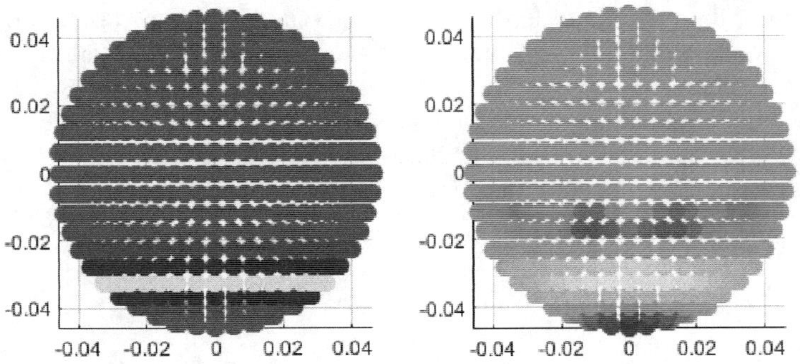

Figure 2. Distributions a simple layer of current sources at time t_R: on the left - $\{\gamma_i\}$, calculated by the model of the heart generator; on the right - $\{\gamma_{pi}\}$, obtained as a result of solving the inverse problem of electrocardiography by the simulated potentials $\{\phi_i\}$.

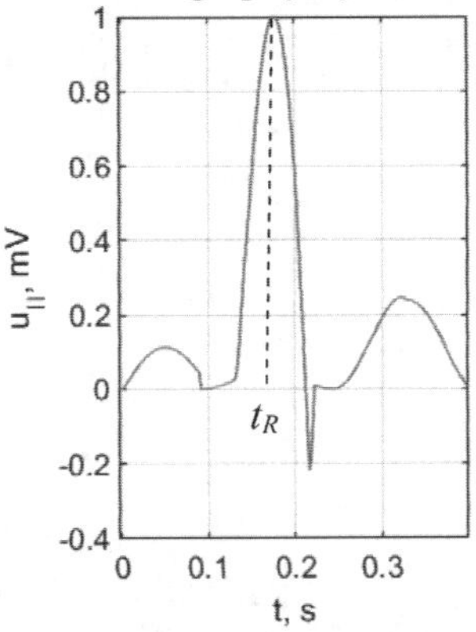

Figure 3. Electrocardiogram in standard lead II, calculated using the model the electric generator of the heart, t_R - apex of the R-wave.

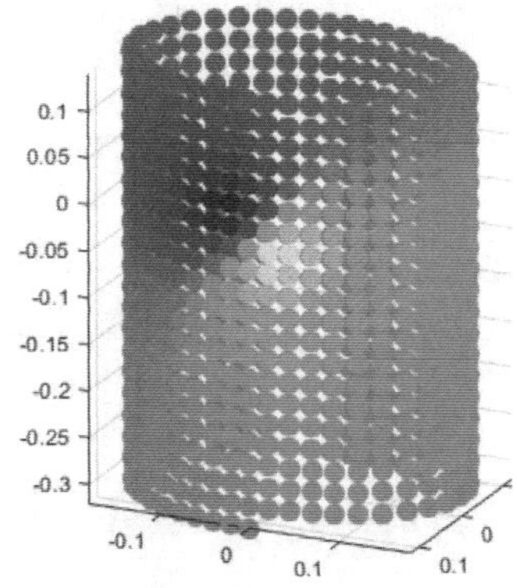

Figure 4. Potentials $\{\phi_i\}$ on the surface of the chest model at time t_R, calculated using the model the electric generator of the heart.

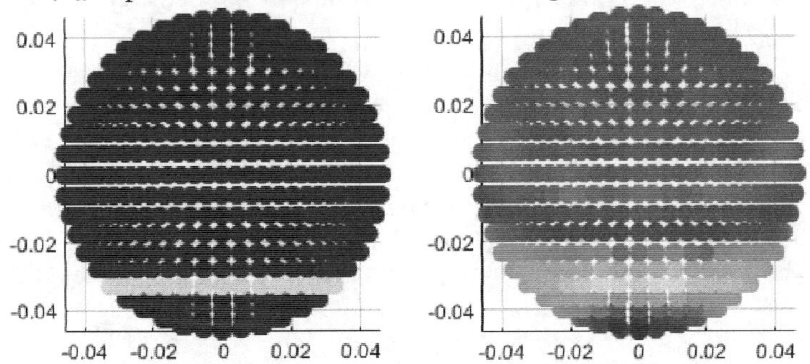

Figure 5. Distributions the double layer of current sources at the moment of time t_R: on the left - calculated by the model of the heart generator, $\{D_i\}$; on the right - the result obtained as a result of solving the inverse problem of electrocardiography by the modeled potentials $\{\phi_i\}$, $\{D_{pi}\}$.

2.3. Spatio-temporal analysis the distributions of current sources on the quasi-epicardium and the formulation of the algorithm

To formulate the algorithm, the time dependencies $\{\gamma_{ij}\}$ and $\{D_{ij}\}$ for various points of the quasi-epicardium surface were compared with the electrical states of the myocardium laid down in the model of the heart generator. Examples of comparison for one of the points the quasi-epicardium ($i = 467$) are shown in Figures 6 and 7. Dependences $\{\gamma_{467j}\}$ and $\{D_{467j}\}$ are plotted in a normalized form (red lines), the corresponding myocardial states are shown by blue lines of levels (state 1 - level 1/3, state 2 - level 2/3 and state 3 - level 1, state 0 - level 0).

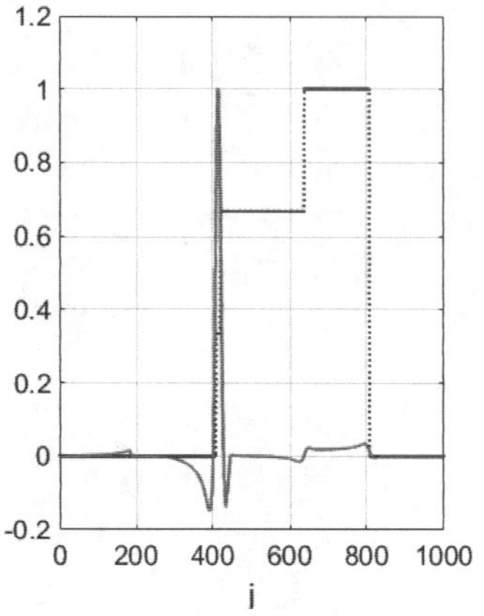

Figure 6. Time dependence $\{\gamma_{467j}\}$ (red line) and myocardial state (blue line).

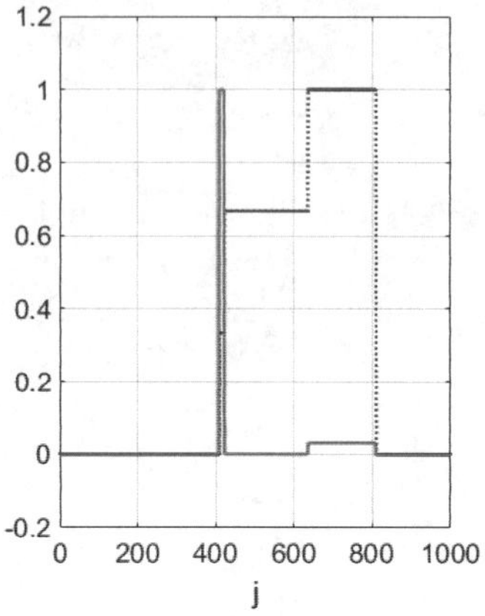

Figure 7. Time dependence $\{D_{467j}\}$ (red line) and myocardial state (blue line).

As a result of a similar analysis at various points of the quasi-epicardium, an algorithm was formulated for determining the physiological states of the myocardium, which is reduced to the following steps:

Step 1. Determination of the time intervals corresponding to the *P*-wave, according to the ECG signal in the II standard lead: $[j_{1P}, j_{2P}]$.

Step 2. Determination of the maximum values $\gamma_{i\,\max1}$ and $\gamma_{i\,\max2}$ (or $D_{i\,\max1}$ and $D_{i\,\max2}$) and the boundaries of the intervals $[j_{1\max1}, j_{2\max1}]$ and $[j_{1\max2}, j_{2\max2}]$, corresponding to the time counts of the neighborhoods of the maximum values, for each *i*-th point of the quasi-epicardium according to the dependence $\{\gamma_{ij}\}$ (or $\{D_{ij}\}$) from *j*.

Step 3. Division of all points of the quasi-epicardium into the atrial and ventricular regions according to the rule: if $j_{1P} \le j_{1\max1} \le j_{2P}$, then point *i* belongs to the atria, otherwise - to the ventricles.

Step 4. Interpolation of the values of the boundaries of the intervals of time samples $[j_{1\max1}, j_{2\max1}]$ and $[j_{1\max2}, j_{2\max2}]$ on a more frequent grid of the surface of the quasi-epicardium (points $k = 1 \ldots 31106$) by the method of inverse weighted distances [8]. Interpolation is performed separately for points belonging to the atria and points belonging to the ventricles.

Step 5. Determination of physiological states at each *k*-th point of the surface of the quasi-epicardium for each *j*-th moment in time according to the rules:

$$s_{kj} = \begin{cases} 0 \text{ if } j < j_{1\max1} \text{ or } j > j_{2\max2}, \\ 1 \text{ if } j_{1\max1} \le j \le j_{2\max1}, \\ 2 \text{ if } j_{2\max1} < j < j_{1\max2}, \\ 3 \text{ if } j_{1\max2} \le j \le j_{2\max2}. \end{cases}$$

Step 6. Result visualization: animation of the physiological states of the quasi-epicardium at all times of a single cardiac cycle.

2.4. Algorithm approbation results

Figures 8 and 9 show the pictures of the physiological states of the quasi-epicardium at time tR, determined using the above algorithm and corresponding to the distributions $\{\gamma_i\}$ (Figure 2) and $\{D_i\}$ (Figure 5). The electrophysiological states of the surface of the model of the electrical activity of the heart at time t_R are shown in Figure 10.

By comparing the states inherent in the heart generator model and reconstructed from the distributions $\{\gamma_{ij}\}$ and $\{D_{ij}\}$, it was found that the developed algorithm gives an error in determining states in 7.4% of surface points for $\{\gamma_{ij}\}$ and 6.5% of surface points for $\{D_{ij}\}$. In this case, the error is mainly due to the difference in the number of points of the quasi-epicardium when solving the inverse problem of electrocardiography (i) and when displaying myocardial states (k). When analyzing $\{\gamma_{ij}\}$, the error is slightly higher due to the more complex form of time dependences.

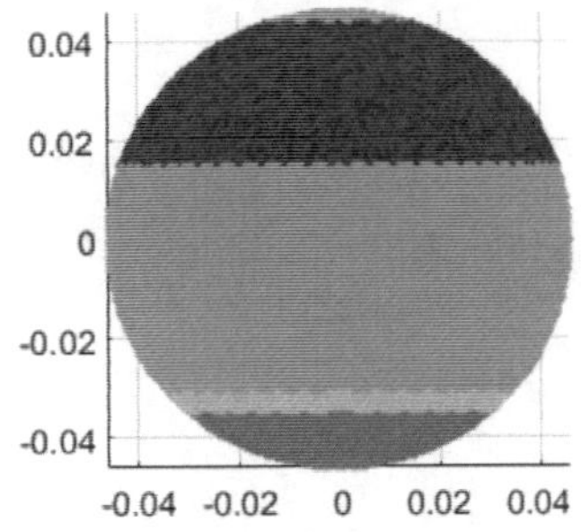

Figure 8. The states of the quasi-epicardium $\{s_k\}$ at time t_R, determined from the distributions $\{\gamma_{ij}\}$.

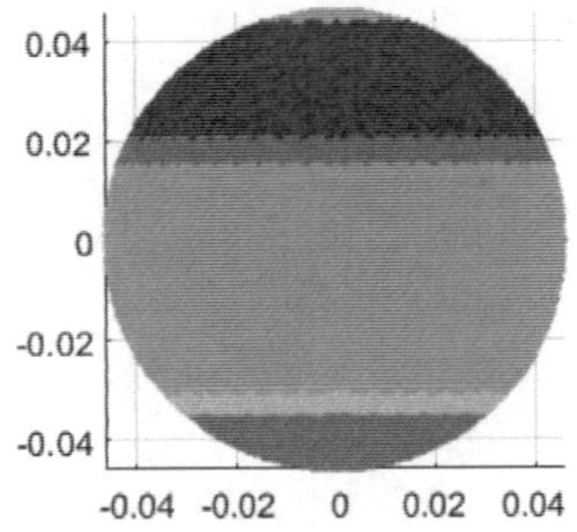

Figure 9. The states of the quasi-epicardium $\{s_k\}$ at time t_R, determined from the distributions $\{D_{ij}\}$.

The sensitivity of the algorithm to the presence of areas with delayed conduction of excitation (typical for myocardial ischemia) was investigated. In this case, the parameters of the algorithm were not changed. Figure 11 shows the state of the myocardium in the model of the heart generator at the moment of time ($j = 430$), when the excitation wave passes through the pathological area [4].

Figures 12 and 13 show the simulated results of the reconstruction of $\{\gamma_{ij}\}$ and $\{D_{ij}\}$at $j = 430$, and Figures 14 and 15 show the results of the physiological interpretation algorithm for the distributions $\{\gamma_{ij}\}$ and $\{D_{ij}\}$in the same moment of time.

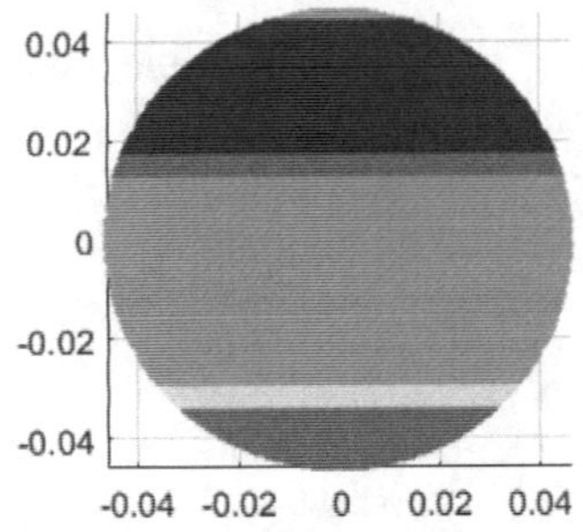

Figure 10. The states of the myocardium in the model of the heart generator at time t_R.

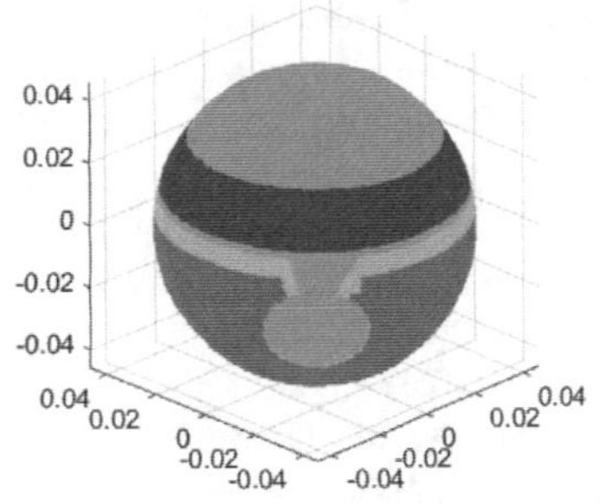

Figure 11. The states of the myocardium in the model of the heart generator in the presence of pathology ($j = 430$).

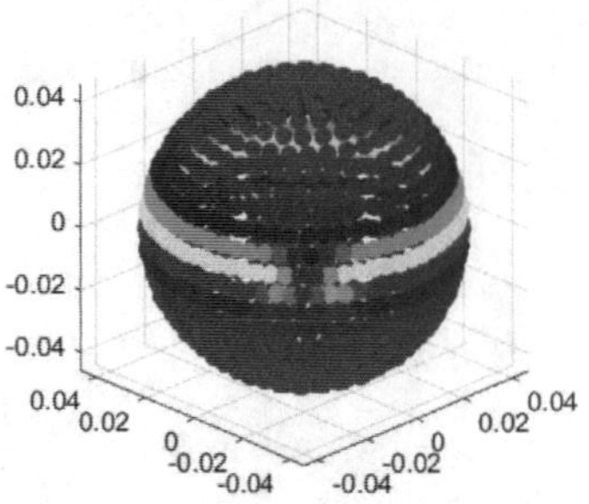

Figure 12. Distribution $\{\gamma_{i430}\}$ in the presence of pathology.

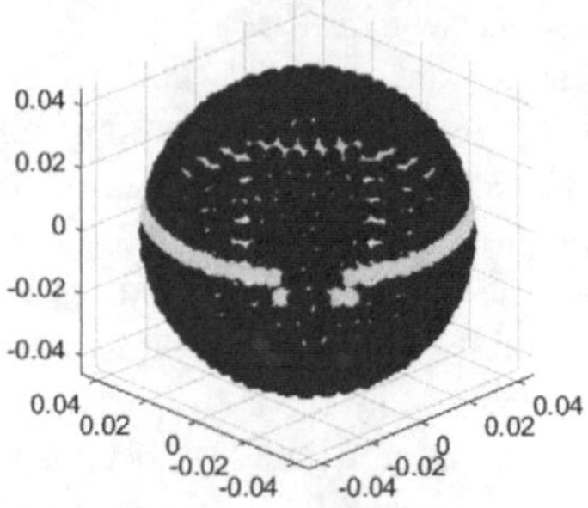

Figure 13. Distribution $\{D_{i430}\}$ in the presence of pathology.

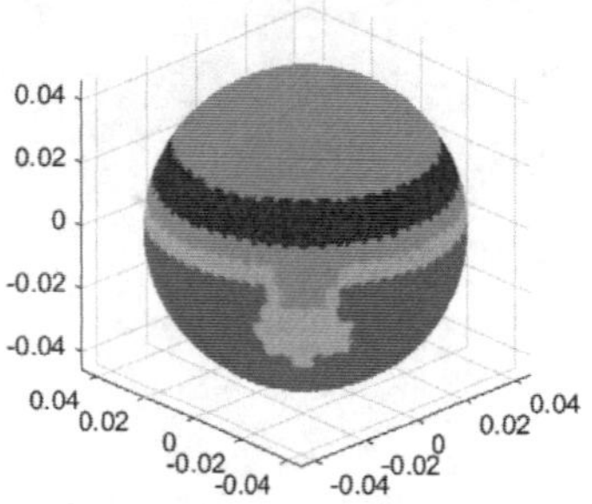

Figure 14. The states of the quasi-epicardium $\{s_{k430}\}$ determined from the distributions $\{\gamma_{ij}\}$.

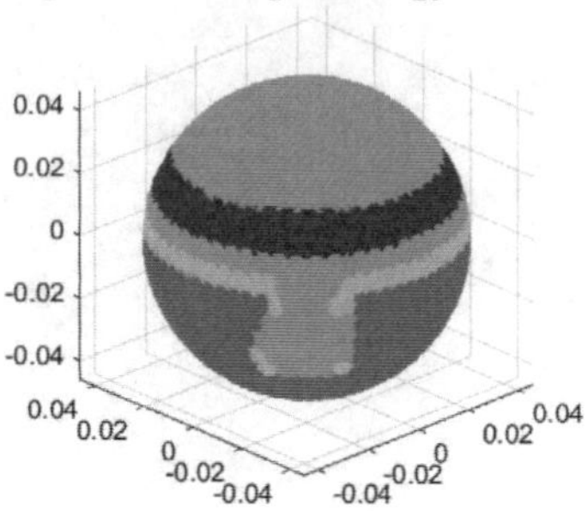

Figure 15. The states of the quasi-epicardium $\{s_{k430}\}$ determined from the distributions $\{D_{ij}\}$.

From a comparison of Figures 14 and 15 with Figure 11, it can be seen that the state in the pathological area differs from the state of normally functioning areas. The area appears more clearly when the algorithm is applied to the distributions $\{D_{ij}\}$.

3. Conclusion

The developed algorithms make it possible to obtain pictures of the physiological states of the surface of the quasi-epicardium at each moment of time of the cardiac cycle by means of spatial-temporal analysis of the results of reconstruction.

The algorithm can be used to localize areas with delayed conduction of excitation, while the parameters of the algorithm do not need to be changed.

To apply to real (not idealized) initial data, it may be necessary to refine the algorithm parameters: more fine tuning of the threshold values when determining the intervals of depolarization and repolarization.

References

[1] Titomir L I, Trunov V G and Aidu E A I 2003 Noninvasiv lectrocardiotopography (Moscow: Nauka)

[2] Macfarlane P W, van Oosterom A, Pahlm O, Kligfield P, Janse M and Camm J 2011 Comprehensive Electrocardiology (London: Springer) pp 194–225

[3] Bokeriya L A, Revishvili A S, Kalinin A V, Kalinin V V, Lyadzhina O A and Fetisova E A 2008 Hardware-software system for noninvasive electrocardiographic heart examination based on inverse problem of electrocardiography *Biomed Eng* **42(6)** 273–9

[4] Zhikhareva G V, Kramm M N, Bodin O N, Seepold R, Madrid N M, Chernikov A I, Kupriyanova Y A and Zhuravleva N A 2020 Conversion from electrocardiosignals to equivalent electrical sources on heart surface. *BMC Bioinformatics* 21(2020)87

[5] Aladiev V Z, Boyko V K and Rovba E A 2008 Classical homogeneous structures: theory and applications (Grodno: Grodno State University Press)
[6] Murashko V V and Strutynsky A V 2016 Electrocardiography (Moscow: MEDPRESS-INFORM)
[7] Kupriyanova Y A, Zhikhareva G V, Mishenina T B and Kuropatkin S A 2020 Selection of parameters algorithm for solving the inverse problem of electrocardiography *Physics and Radioelectronics in Medicine and Ecology: 14th International Scientific Conference, FREME 2020, July 1-3 2020, Suzdal, Russia, Proc. Book 1 (Biocybernetics And Mathematical Modeling, section 2)* ed L T Sushkova et al (Vladimir: Vladimir State University press) pp 123-7
[8] Pavlova A I 2017 Analysis of elevation interpolation methods for developing digital terrain models *Avtometriya* **2(53)** 86-94

The effect of queue size on the throughput, in group failure mode, for the loaded transport channel

P H Karim[1] K J Ghafoor[2] and S P Suschenko[1]

[1]National Research Tomsk State University, Lenina str., 36, 634050 Tomsk, Russia

[2]University of Halabja, Ababaile, Halabja, Kurdistan Region, Iraq

E-mail: peshangkarimov@gmail.com

Abstract: The external data flow decreases the throughput of the transport connection. The indicator of this external load is the queue size in front of the protocol data. In this article, using a mathematical model in analytical and numerical forms, the relation between the throughput of the channel and the protocol parameters are presented including the queue size parameter. In this work the effect of the queue size on time-out duration has been shown, which is one of the important parameters and it's studied weakly in researches. Also, the relation between round-trip delay, the reliability of the transmission of the information segments with queue size are also shown.

1. Introduction

Transmission connection throughput has an extremely significant feature of computer networks. This indicator determines the quality of network services for subscribers and it's determined by the values of protocol parameters such as (window size and timeout duration), and the characteristics of the data transmission path (duration round trip delay, reliability of the transmission of the information segments for both directions of the transport connection). Also, the external data flow acts to determine the throughput of the selected transport connection. This external flow decreases the throughput of the channel even if they have one common path route. The main indicator for this external load on the transport connection is the queue size in front of the protocol block data in selected sections of the transit nodes. By studying this indicator, the distribution of queue lengths in transit nodes from external network streams for analyzed connection can be estimated, and then it is used to manipulative the active characteristics of the linking and the selection of protocol parameters for the communication time between given subscribers. TCP [1] is an important transport layer protocol. Modern computer networks with the TCP protocol have a significant role to cope with a huge number of today's network problems.

As accrued before, many problems were solved by adding new algorithms like Reno, New Reno, Tahoe [17]. These modifications can be considered long-standing and the research continues to improve the speed of the transport connection. Nowadays, the external loads on the shared network resources are not be considered by known models of asynchronous control procedures of a separate data link and transport protocol [2]–[8]don't consider the external loads on the shared network resources. These loads could be on any virtual connection along the path between subscribers, which have common nodes. In [9],[10], the study of the data conversion process of the load transmission connection is carried out under the condition that the protocol parameters and channel attributes are very limited. Many modern studies on transport connection channels show that only a few factors or parameters such as timeout, round-trip delay, or congestion, have affected the speed of transport connection. However, there are no clearly expressed analytical formulas or numerical analyzes [17]. In [16], to improve the speed of transport protocols a thorough presentation of modern modifications is carried out. The study clearly showed that modern studies consider only certain factors of congestions, timeout, and round-trip delay. But the work [15] does not consider the probability of data delivery reliability in the forward and reverse channels. While in our study, the conditions for the probability of data delivery reliability in the communication channel have been considered for both directions, analytical solutions were found and their confirmation, using numerical studies. In this paper, we present the mathematical model of the transmission connection under the transmission protocol connection in the group failure mode. This model is taking to account the factors such as losing data in forward and reverse directions, the mechanism of data retransmission, which counts time-out losing acknowledgments from receiving host, and another important factor "non-zero queue" from external interconnection. The significant improvement of this study compares to the study of [13] is the solution for all states of the Markov chain have been found except one case, which is not such important for the research. The mathematical model that is shown here is the same model, which was published in [12], but here the model has been shown for group failure mode, also the tasks and conditions are dissimilar.

2. Description of the mathematical model

Let's consider data exchange between subscribers, which are connected by a multi-link data path. We suppose that the following conditions are true [11],[12]: The connection between nodes is duplex and having the same speed in both forward and reverse directions. The path length, which is expressed by hop numbers, is equal to D_f. The reverse channel, in which acknowledgments are delivered to the sender, has a length D_o. The confirmation is a receipt that contains information about whether the segment data sequence was correctly sent to the recipient. The reliability of the transmission of the information segments in the path, from sender to receiver and vice versa (the forward and reverse channel), are given as F_f and F_o respectively. The time processing of segments is equal in all nodes. Besides, the interacting subscribers have an unlimited flow of segments for transition and the length of segments is equal. Acknowledgments from the receiver are sending with their counterflow. Also, we propose that segment retransmission is organized according to group failure mode [23]. We suppose that, the loss of segments is not happening of the lack of buffer memory. The function of probability $b_n, n = \overline{0, N}$ is given. This means each segment in the flow, that we are analyzing, will meet a queue with $n \leq N$, where N is the maximum length of the queue. The maximum length of queues N defined by the buffer pools capacity of the transition nodes. We call the time t, which is needed to take the segment into the output line as a cycle. The cycle is defined as a sum of time needed to take segment into the output line, channel propagation signal, and the time for processing segment for the node side. The timeout S is expressed as the length of t. It is launched before the beginning of the first transition segment in the queue and it will be fixed for all segments within a window size. We consider that the window size of the controlling protocol, is defined by W. $S > W$ sets the timeout duration. The sum of the length of forward and reverse path $D = D_f + D_o$ can be presented as round-trip delay in the unloaded channel, which was expressed in cycles t. After sending the next segments, the protocol will copy it into the queue of transmitted but unconfirmed segments, then it will launch timeout. As soon as the queue size becomes equal to the width of the window W, the control protocol will pause transmission while waiting for the acknowledgment or the expiration of timeout S for confirmation. If the acknowledgment is positive, the segments, which sent successfully, will be deleted from the queue. If the timeout, for determining segment, is reached, that segment will be retransmitted and the timeout of confirmation will be reset and launched again. Then the time of confirmation by the source of end-to-end acknowledgment is dispersed according to geometric law with F_o parameter and the duration of sampling cycle t. The usefulness of virtual association in a loaded multilink that is controlled by the transport protocol, data transmission path with segment queues before sending data or confirmations can be described by a markovized process of the dynamics of a transmitted queue but not confirmed segments, in which the queue size of the advancing or inverse data flow of the testing connection is an additional variable of

Markov process. In the Markov chain state (i,n), a sequence of size $(i-n)$ segments that have been sent from the source, in which one of the links in the process of transfer met a queue with the length of n segments. States of Markov chain $i=\overline{0,W+n}$, $n=\overline{0,N}$, corresponding to the size of the queue which is transmitted but not yet confirmed segments in the flow source. And the time from the beginning of the transmission of the sequence, while the states $i=\overline{W+n+1,S-1}, n=\overline{0,N}$ refer to the time, during which the sender is not active and is waiting for an acknowledgment of correct reception of the sent sequence of W segments.

We define $P(i,n)$, $i=\overline{0,S-1}, n=\overline{0,N}$ as the probabilities of Markov chain states. Then the sequence of transmitted, but not confirmed data segments of considered virtual connection with a zero-length queue grows to the state of a Markov chain with coordinates $(D-1{,}0)$ with probability b_0. The further increasing size of this sequence occurs with a probability of $b_0(1-F_o)$. In the states (i,n), $i=\overline{D-1+n,S-1}, n=\overline{0,N}$, it is possible, that sender receives the acknowledgment, and depending on the acknowledgment results, the sender transmits new segments (with a positive acknowledgment), or retransmit distorted segments. Since the transmitted sequence of segments of the virtual connection, that we are analyzing, may encounter a queue of non-zero length at any moment of transferring process (on the path of the sequence to the addressee or when transferring confirmation to the sender of information flow), the transition from the state $(i,0)$, $i=\overline{0,S-2}$ to state (i,n), $i=\overline{0,S-2}, n=\overline{1,N}$ occurs with probability bn.

3. State probabilities for Markov chain

The transition probabilities of the Markov chain can be expressed as π_{in}^{jm}, where (i,n) is the coordinates of the initial state, and the resulting state (j,m) from the initial state into the chain. Then the dynamics of the process of transmitting information flow in the group failure mode for loaded data transmission channel can be set with the following values of transition probabilities:

$$\pi_{in}^{jm}=\begin{cases} b_0, \; j=i+1, m=0; i=\overline{0,D-2}, n=0;\\ b_0(1-F_0), j=i+1, m=0;\; i=\overline{D-1,S-2}, n=0;\\ b_m, \; j=i, m=\overline{1,N};\; i=\overline{0,S-2}, n=0;\\ b_0F_oF_f{}^{i-D+2}, \; j=D-1, m=0, i=\overline{D-1,W-1}, n=0,;\\ b_0F_oF_f{}^{i-D+2}, j=W+D-2-i, m=0; i=\overline{W,W+D-3}, n=0,\\ b_0F_o(1-F_f{}^{i-D+2}), \; j=0, m=0;\; i=\overline{D-1,W+D-3}, n=0,\\ b_0F_o, j=0, m=0;\; i=\overline{W+D-2,S-2}, n=0,\\ 1, \; j=0, m=0;\; i=S-1, n=\overline{0,N},\\ 1, j=i+1, m=n;\; i=\overline{0,D-2+n}, n=\overline{1,N};\\ 1-F_0, j=i+1, m=n;\; i=\overline{D-1+n,S-2}, n=\overline{1,N},\\ F_oF_f{}^{i-D+2-n}, j=D-1, m=0;\; i=\overline{D-1+n,W-1+n}, n=\overline{1,N},\\ F_oF_f{}^{i-D+2-n}, j=W+D-2-i, m=0;\; i=\overline{W+n,W+D-3+n}, n=\overline{1,N},\\ F_o(1-F_f{}^{i-D+2-n}), j=0, m=0; i=\overline{D-1+n,W+D-3+n}, n=\overline{1,N}, \end{cases} \quad (1)$$

There are different solutions for the equilibrium system of Markov chain state probabilities, and can be determined by the relationship between W,S,D, and N (window size, timeout, path length, and maximum queue length) The time-out length must be bigger or equal to the round trip-delay length ($S\geq D$) and must go beyond the width of window size, also time-out should be longer than the waiting time for the beginning data transiting in transmission nodes because of existed queues inside the virtual channel. Due to the wide variety of protocol parameter values, the system has several different solutions. For the analysis and research of the data transmission process of loading arbitrary values of protocol parameters in the channel, $b_0=0$ is an indispensable condition. In this paper, the study works to analyze

the data transmission process in the load channel with a non-zero queue length ($b_0 = 0$) and a protocol parameter with a common path and the maximum queue length $S \geq W + D + N - 1$.
For $W \geq D$, the solution of equilibrium system from system equation (1) can be written as the following:

$$P(0,0) = F_o \sum_{n=1}^{N} \sum_{i=D-1+n}^{W+D-3+n} (1 - F_f^{\,i-D+2-n}) P(i,n) + \sum_{n=1}^{N} P(S-1,n) + F_o \sum_{n=1}^{N} \sum_{i=D+W-2+n}^{S-2} P(i,n) \qquad (2)$$

$$P(i,0) = F_o \sum_{n=1}^{N} F_f^{\,W-i} P(D + W - 2 - i + n, n), i = \overline{1, D-2} \qquad (3)$$

$$P(D-1,0) = F_o \sum_{n=1}^{N} \sum_{i=D-1+n}^{W-1+n} P(i,n) F_f^{\,i-D+2-n} \qquad (4)$$

$$P(i,0) = 0, i = \overline{D, S-1} \qquad (5)$$

$$P(0,n) = b_n P(0,0), n = \overline{1,N} \qquad (6)$$

$$P(i,n) = P(i-1,n) + b_n P(i,0), i = \overline{1, D-1}, n = \overline{1,N} \qquad (7)$$

$$P(i,n) = P(i-1,n), i = \overline{D, D-1+n}, n = \overline{1,N} \qquad (8)$$

$$P(i,n) = (1 - F_0) P(i-1,n), i = \overline{D+n, S-1}, n = \overline{1,N} \qquad (9)$$

Let's start solving the equilibrium system. According to equation (8), we get : $P(i,n) = P(D-1,n), i = \overline{D, D-1+n}, n = \overline{1,N}$, and from equation (9) we have: $P(i,n) = P(D-1+n,n)\bar{F}_o^{\,i-D-n+1}, i = \overline{D+n, S-1}, n = \overline{1,N}$, taking to account these relations and from equations (4) and (5) for $i = \overline{1, D-1}$ we find:

$$P(i,0) = \frac{F_o \Phi^{W-i}}{\bar{F}_o} \sum_{m=1}^{N} P(D-1,m), i = \overline{1, D-2}, Where: \Phi = F_f(1 - F_o), \bar{F}_o = 1 - F_o \qquad (10)$$

$$P(D-1,0) = \frac{F_o F_f (1 - \Phi^{W-D+1})}{1-\Phi} \sum_{m=1}^{N} P(D-1,m) \qquad (11)$$

With equations (7), (6) and (10) we can find:

$$P(i,n) = b_n [P(0,0) + \frac{F_o \Phi^{W-i}}{\bar{F}_o} \sum_{m=1}^{N} P(D-1,m)], i = \overline{1, D-2}, n = \overline{1,N}$$

$$P(D-1,n) = b_n [P(0,0) + \frac{F_o F_f (1 - \Phi^{W-1})}{(1-\Phi)} \sum_{m=1}^{N} P(D-1,m)] \qquad (12)$$

Accordingly, from equation (12) we can find the state probabilities $P(D-1,n)$, $m = \overline{n+1, N}$ for arbitrary $n = \overline{1,N}$ through $P(D-1,m)$ and we get:

$$P(D-1,n) = \frac{b_n}{1 - \Phi - F_o F_f (1 - \Phi^{W-1}) \sum_{m=1}^{n} b_m} \left[P(0,0)(1-\Phi) + F_o F_f (1 - \Phi^{W-1}) \sum_{m=n+1}^{N} P(D-1,m) \right], n = \overline{1,N} \qquad (13)$$

When $n = N$, we get: $P(D-1,N) = b_N \frac{P(0,0)(1-\Phi)}{1-\Phi-F_oF_f(1-\Phi^{W-1})}$, substituting this relation into equation (13) for values $n = N-1$ to 1, recursively, we can find the functional expression for state probabilities $(D-1,n)$ via $P(0,0)$: $P(D-1,n) = b_n \frac{P(0,0)(1-\Phi)}{1-F_f+F_oF_f\Phi^{W-1}}, n = \overline{1,N}$. From here, up to the probability of the initial state, we obtain the probability distribution of states of the Markov chain:

$$P(i,0) = \frac{P(0,0) F_o F_f \Phi^{W-1}(1-\Phi)}{1-F_f+F_oF_f\Phi^{W-1}} \Phi^{-i}, i = \overline{1, D-2}$$

$$(D-1,0) = \frac{P(0,0) F_o F_f (1-\Phi^{W-D+1})}{1-F_f+F_oF_f\Phi^{W-1}}$$

$$P(i,n) = P(0,0) b_n \frac{1-F_f+F_oF_f\Phi^{W-1-i}}{1-F_f+F_oF_f\Phi^{W-1}}, i = \overline{0, D-2}, n = \overline{1,N}$$

$$P(i,n) = b_n \frac{P(0,0)(1-\Phi)}{1-F_f+F_oF_f\Phi^{W-1}}, i = \overline{D-1, D-1+n}, n = \overline{1,N}$$

$$P(i,n) = b_n \frac{P(0,0)(1-\Phi)}{1-F_f+F_oF_f\Phi^{W-1}} \bar{F}_o^{\,i-D-n+1}, i = \overline{D+n, S-1}, n = \overline{1,N}$$

By using the normalization condition, finally we can get the relation for initial state $P(0,0)$:

$P(0,0)=\frac{(1-F_f+F_oF_f\Phi^{W-1})(1-\Phi)F_o}{F_o(1-\Phi)[F_oF_f+D(1-F_f)+(1-\Phi)(1+\bar{N})]+F_o{}^2F_f(\Phi^{W-D+1}-\Phi^W)+(1-\Phi)^2\left(\bar{F}_o-\bar{F}_o{}^{S-D+1}\sum_{n=1}^{N}\frac{b_n}{(1-F_o)^n}\right)}$, Where $\bar{N}=\sum_{n=1}^{N}nb_n$, is the average length of the queue. Note: if we put the value of $F_f=1$, we can get the same results for selective failure mode, which are represented in [12].

Let's consider the case when the size of the window does not exceed the duration of round-trip delay $(W<D)$ and the duration of time-out is the same $S\geq W+D+N-1$. According to equation (1) the system of equilibrium will change as the followings:

The equations of (2) (6) (9) will remain the same. Equation (3) is true for $i=\overline{1,W-1}$. Equation (7) is true for $i=\overline{1,W-1},n=\overline{1,N}$. Equation (4) will change to $P(D-1,0)=0$. Equation (8) will be changed to $i=\overline{W,D-1+n},n=\overline{1,N}$. The solutions of the equilibrium system will be as the following for $W<D$:

$$P(i,0)=\frac{P(0,0)(1-\Phi)F_oF_f\Phi^{W-1-i}}{1-F_f+F_oF_f\Phi^{W-1}},i=\overline{1,W-1}$$

$$P(i,n)=b_nP(0,0)\left[\frac{1-F_f+F_oF_f\Phi^{W-1-j}}{1-F_f+F_oF_f\Phi^{W-1}}\right],i=\overline{0,W-1},n=\overline{1,N}$$

$$P(W-1,0)=\frac{P(0,0)(1-\Phi)F_oF_f}{1-F_f+F_oF_f\Phi^{W-1}}$$

$$P(i,n)=b_n\frac{P(0,0)(1-\Phi)}{1-F_f+F_oF_f\Phi^{W-1}},i=\overline{W-1,D-1+n},n=\overline{1,N}$$

$$P(i,n)=b_n\frac{P(0,0)(1-\Phi)}{1-F_f+F_oF_f\Phi^{W-1}}\bar{F}_o{}^{i-D-n+1},i=\overline{D+n,S-1},n=\overline{1,N}$$

As the way before we can find the initial state using normalization condition:

$$P(0,0)=\frac{(1-F_{\text{П}}+F_oF_{\text{П}}\Phi^{W-1})(1-\Phi)F_o}{F_o[F_oF_{\text{П}}(\Phi-\Phi^W)+F_oF_{\text{П}}(1-W)(1-\Phi)+(1-\Phi)^2(D+1+\bar{N})]+(1-\Phi)^2\left(\bar{F}_o-\bar{F}_o{}^{S-D+1}\sum_{n=1}^{N}\frac{b_n}{(1-F_o)^n}\right)}$$

Then, the transmission process with interval restrictions has been analyzed on the duration of the timeout. Let's present already found solutions. With restrictions $W\geq D,W+D-1\leq S\leq W+D+N-1,1\leq N\leq D-2$, the equations of the original equilibrium system (2-9) are transformed into:

$$P(0,0)=F_o\sum_{n=1}^{N}\sum_{i=D-1+n}^{W+D-3+n}(1-F_f{}^{i-D+2-n})P(i,n)+\sum_{n=1}^{N}P(S-1,n)$$
$$+F_o\sum_{n=1}^{S-(W+D-1)}\sum_{i=D+W+n-2}^{S-2}P(i,n)$$

$$P(i,0)=F_o\sum_{n=1}^{Vi}F_f{}^{W-i}P(D+W-2-i+n,n),i=\overline{1,D+W+N-1-S},Vi=S-D-W+i$$

$$P(i,0)=F_o\sum_{n=1}^{N}F_f{}^{W-i}P(D+W-2-i+n,n),i=\overline{D+W+N-S,D-2}$$

$$P(D-1,0)=F_o\sum_{n=1}^{N}\sum_{i=D-1+n}^{W-1+n}P(i,n)F_f{}^{i-D+2-n}$$

$$P(0,n)=b_nP(0,0),n=\overline{1,N}$$

$$P(i,n)=P(i-1,n)+b_nP(i,0),i=\overline{1,D-1},n=\overline{1,N}$$

$$P(i,n)=P(i-1,n),i=\overline{D,D-1+n},n=\overline{1,N}$$

$$P(i,n)=(1-F_0)P(i-1,n),i=\overline{D+n,S-1},n=\overline{1,N}$$

Similarly, we solve the equations and as a result we obtain the probabilities of the states of the Markov chain:

$$P(i,0)=\frac{P(0,0)(1-\Phi)F_oF_f\Phi^{W-i-1}\sum_{m=1}^{S-D-W+i}b_m}{1-F_f+F_oF_f\left[\Phi^{W-1}\sum_{m=1}^{S-D-W}b_m+\sum_{m=S-D-W+1}^{N}b_m\Phi^{S-D-m}\right]},i=\overline{1,D+W+N-1-S}$$

$$P(i,0)=\frac{P(0,0)(1-\Phi)F_oF_f\Phi^{W-i-1}}{1-F_f+F_oF_f\left[\Phi^{W-1}\sum_{m=1}^{S-D-W}b_m+\sum_{m=S-D-W+1}^{N}b_m\Phi^{S-D-m}\right]},i=\overline{D+W+N-S,D-2}$$

$$P(D-1,0)=\frac{P(0,0)F_oF_f(1-\Phi^{W-D+1})}{1-F_f+F_oF_f\left[\Phi^{W-1}\sum_{m=1}^{S-D-W}b_m+\sum_{m=S-D-W+1}^{N}b_m\Phi^{S-D-m}\right]}$$

$$P(i,n)=\frac{b_nP(0,0)\{1-F_f+F_oF_f[\Phi^{W-1-i}\sum_{m=1}^{S-D-W+i}b_m+\sum_{m=S-D-W+1+i}^{N}b_m\Phi^{S-D-m}]\}}{1-F_f+F_oF_f[\Phi^{W-1}\sum_{m=1}^{S-D-W}b_m+\sum_{m=S-D-W+1}^{N}b_m\Phi^{S-D-m}]},i=\overline{1,D+W+N-S-1}$$

$$,n=\overline{1,N}$$

$$P(i,n)=\frac{b_nP(0,0)[1-F_f+F_oF_{nf}\Phi^{W-i-1}]}{1-F_f+F_oF_f[\Phi^{W-1}\sum_{m=1}^{S-D-W}b_m+\sum_{m=S-D-W+1}^{N}b_m\Phi^{S-D-m}]},i=\overline{D+W+N-S,D-2},n=\overline{1,N}$$

$$P(i,n)=\frac{b_nP(0,0)(1-\Phi)}{1-F_f+F_oF_f[\Phi^{W-1}\sum_{m=1}^{S-D-W}b_m+\sum_{m=S-D-W+1}^{N}b_m\Phi^{S-D-m}]},i=\overline{D-1,D-1+n},n=\overline{1,N}$$

$$P(i,n)=\frac{b_nP(0,0)(1-\Phi)\bar{F}_o^{\,i-D-n+1}}{1-F_n+F_oF_f[\Phi^{W-1}\sum_{m=1}^{S-D-W}b_m+\sum_{m=S-D-W+1}^{N}b_m\Phi^{S-D-m}]},i=\overline{D+n,S-1},n=\overline{1,N}$$

As before we can find the initial state $P(0,0)$ using normalization condition:

$$P(0,0)=F_o(1-\Phi)\{1-F_f+F_oF_f[\Phi^{W-1}\sum_{m=1}^{S-D-W}b_m+\sum_{m=S-D-W+1}^{N}b_m\,\Phi^{S-D-m}]\}\{F_o{}^2F_f[\Phi^{W-D+1}+1-\Phi-\Phi^{W}\sum_{m=1}^{S-D-W}b_m-\sum_{m=S-D-W+1}^{N}b_m\,\Phi^{S-D-m+1}+(1-\Phi)\sum_{m=S-D-W+2}^{N}b_m\,\Phi^{S-D-m}(m-S+D+W-1)]+(1-\Phi)[F_o(1-F_f)D+F_o(1-\Phi)(\bar{N}+1)+(1-\Phi)\left(\bar{F}_o-\sum_{n=1}^{N}b_n\,\bar{F}_o^{\,S-D-n+1}\right)]\}^{-1}$$

Consider analyzing the process of information transfer for window size less than the time-trip delay ($W\le D$) and interval restrictions on the timeout duration $D+W-1\le S\le D+W+N-1$ and the maximum of queue size $1\le N\le W-2$, then the equilibrium equations from (2-9) take the form:

$$P(0,0)=\sum_{n=1}^{N}P(S-1,n)+F_o\sum_{n=1}^{V}\sum_{i=D+W+n-2}^{S-2}P(i,n),V=S-D-W+1$$

$$\mathrm{P}(\mathrm{i},0)=F_o\sum_{n=1}^{Vi}F_f{}^{W-i}P(D+W+n-2-i,n),i=\overline{1,D+W+N-1-S},Vi=S-D-W+i$$

$$\mathrm{P}(\mathrm{i},0)=F_o\sum_{n=1}^{N}F_f{}^{W-i}P(D+W+n-2-i,n),i=\overline{D+W+N-S,W-1}$$

$$\mathrm{P}(0,\mathrm{n})=b_n\mathrm{P}(0,0),n=\overline{1,N}$$

$$P(i,n)=P(i-1,n)+b_nP(i,0),i=\overline{1,W-1},n=\overline{1,N}$$

$$P(i,n)=P(i-1,n),i=\overline{W,D-1+n},n=\overline{1,N}$$

$$P(i,n)=P(D-1,n)=P(W-1,n)$$

$$P(i,n)=P(i-1,n)(1-F_0),i=\overline{D+n,S-1},n=\overline{1,N}$$

$$P(i,n)=P(D-1,n)\bar{F}_o^{\,i-D-n+1}=P(W-1,n)\bar{F}_o^{\,i-D-n+1}$$

Similarly, when solving equations, the probabilities of states of the Markov chain can be obtained:

$$P(i,0)=\frac{P(0,0)F_oF_f(1-\phi)\phi^{W-i-1}\sum_{m=1}^{S-D-W+i}b_m}{1-F_f+F_oF_f[\phi^{W-1}\sum_{m=1}^{S-D-W}b_m+\sum_{m=S-D-W+1}^{N}b_m\phi^{S-D-m}]},i=\overline{1,D+W+N-1-S}$$

$$P(i,0)=\frac{P(0,0)F_oF_f(1-\phi)\phi^{W-i-1}}{1-F_f+F_oF_f[\phi^{W-1}\sum_{m=1}^{S-D-W}b_m+\sum_{m=S-D-W+1}^{N}b_m\phi^{S-D-m}]},i=\overline{D+W+N-S,W-1}$$

$$P(i,n)=\frac{b_nP(0,0)\{1-F_f+F_oF_n[\phi^{W-1-i}\sum_{m=1}^{S-D-W+i}b_m+\sum_{m=S-D-W+i+1}^{N}b_m\phi^{S-D-m}]\}}{1-F_f+F_oF_f[\phi^{W-1}\sum_{m=1}^{S-D-W}b_m+\sum_{m=S-D-W+1}^{N}b_m\phi^{S-D-m}]},i=\overline{1,D+W+N-1-S}$$

$$,n=\overline{1,N}$$

$$P(i,n)=\frac{b_nP(0,0)[1-F_f+F_oF_f\phi^{W-1-i}]}{1-F_f+F_oF_f[\phi^{W-1}\sum_{m=1}^{S-D-W}b_m+\sum_{m=S-D-W+1}^{N}b_m\phi^{S-D-m}]},i=\overline{D+W+N-S,W-1},n=\overline{1,N}$$

$$P(i,n)=\frac{b_n(1-\phi)P(0,0)}{1-F_f+F_oF_f[\phi^{W-1}\sum_{m=1}^{S-D-W}b_m+\sum_{m=S-D-W+1}^{N}b_m\phi^{S-D-m}]},n=\overline{W-1,D-1+n},n=\overline{1,N}$$

$$P(i,n)=\frac{b_n(1-\phi)P(0,0)}{1-F_f+F_oF_f[\phi^{W-1}\sum_{m=1}^{S-D-W}b_m+\sum_{m=S-D-W+1}^{N}b_m\phi^{S-D-m}]}\bar{F}_o^{\,i-D-n+1},n=\overline{D+n,S-1},n=\overline{1,N}$$

Next the initial state $P(0,0)$ has been found.

$$P(0,0)=\{1-F_f+F_oF_f[\phi^{W-1}\sum_{m=1}^{S-D-W}b_m+\sum_{m=S-D-W+1}^{N}b_m\,\phi^{S-D-m}]\}F_o(1-\phi)\{F_o{}^2F_f\,[1-\phi^{W}\sum_{m=1}^{S-D-W}b_m-\sum_{m=S-D-W+1}^{N}b_m\,\phi^{S-D-m+1}+(1$$

$$-\phi)\sum_{m=S-D-W+2}^{N} b_m\,\phi^{S-D-m}(m-S+D+W-1)]+(1-\phi)^2[1+F_o(D-W+\bar{N}) \quad +F_oW(1-F_f)-\sum_{n=1}^{N} b_n\,\bar{F_o}^{\,S-D-n+1}]\}^{-1}$$

Consider another variety of interval restrictions on protocol parameters and the maximum queue size of the form $W\ge D, D+W-1\le S\le W+N+1, D-2\le N\le W-2$. Under these constraints, the equations of the original equilibrium system (2-9) will change in the following:

$$P(0,0)=F_o\sum_{n=1}^{N}\sum_{i=D-1+n}^{W+D-3+n}(1-F_f^{\,i-D+2-n})P(i,n)+\sum_{n=1}^{N}P(S-1,n)+F_o\sum_{n=1}^{N}\sum_{i=D+W-2+n}^{S-2}P(i,n)$$

$$P(i,0)=F_o\sum_{n=1}^{V_1}F_f^{\,W-i}P(D+W-2-i+n,n)\,,i=\overline{1,D-2},V_1=S-2-(D+W-2-i)$$

$$P(D-1,0)=F_o\left[\sum_{n=1}^{V_2}\sum_{i=D-1+n}^{W-1+n}P(i,n)\,F_f^{\,i-D+2-n}+\sum_{n=V_2+1}^{N}\sum_{i=D-1+n}^{S-2}P(i,n)\,F_f^{\,i-D+2-n}\right],V_2=S-W-1$$

$$P(i,0)=0, i=\overline{D,S-1}$$

$$P(0,n)=b_nP(0,0), n=\overline{1,N}$$

$$P(i,n)=P(i-1,n)+b_nP(i,0), i=\overline{1,D-1}, n=\overline{1,N}$$

$$P(i,n)=P(i-1,n), i=\overline{D,D-1+n}, n=\overline{1,N}$$

$$P(i,n)=(1-F_0)P(i-1,n), i=\overline{D+n,S-1}, n=\overline{1,N}$$

By solving the equations, the following results achieved the probabilities of the states of the Markov chain:

$$P(i,0)=\frac{P(0,0)(1-\phi)F_oF_f\phi^{W-i-1}\sum_{m=1}^{S-D-W+i}b_m}{1-F_f+F_oF_f[\phi^{W-1}\sum_{m=1}^{S-D-W}b_m+\sum_{m=S-D-W+1}^{N}b_m\phi^{S-D-m}]}, i=\overline{1,D-2}$$

$$P(D-1,0)=\frac{P(0,0)F_oF_f[1-\phi^{W-D+1}\sum_{m=1}^{S-W-1}b_m-\sum_{m=S-W}^{N}\phi^{S-D-m}b_m]}{1-F_f+F_oF_f[\phi^{W-1}\sum_{m=1}^{S-D-W}b_m+\sum_{m=S-D-W+1}^{N}b_m\phi^{S-D-m}]}$$

$$P(i,n)=b_nP(0,0)[\frac{1-F_f+F_oF_f[\phi^{W-1-i}\sum_{m=1}^{S-D-W+i}b_m+\sum_{m=S-D-W+i+1}^{N}b_m\phi^{S-D-m}]}{1-F_f+F_oF_f[\phi^{W-1}\sum_{m=1}^{S-D-W}b_m+\sum_{m=S-D-W+1}^{N}b_m\phi^{S-D-m}]}], i=\overline{0,D-2}, n=\overline{1,N}$$

$$P(D-1,n)=\frac{b_nP(0,0)(1-\phi)}{1-F_f+F_oF_f[\phi^{W-1}\sum_{m=1}^{S-D-W}b_m+\sum_{m=S-D-W+1}^{N}b_m\phi^{S-D-m}]},\ n=\overline{1,N}$$

$$P(i,n)=\frac{b_nP(0,0)(1-\phi)}{1-F_f+F_oF_f[\phi^{W-1}\sum_{m=1}^{S-D-W}b_m+\sum_{m=S-D-W+1}^{N}b_m\phi^{S-D-m}]},\ i=\overline{D-1,D-1+n}, n=\overline{1,N}$$

$$P(i,n)=\frac{b_nP(0,0)(1-\phi)\bar{F_o}^{\,i-D+1-n}}{1-F_f+F_oF_f[\phi^{W-1}\sum_{m=1}^{S-D-W}b_m+\sum_{m=S-D-W+1}^{N}b_m\phi^{S-D-m}]},\ i=\overline{D+n,S-1}, n=\overline{1,N}$$

From the normalization condition, we obtain the initial state P (0,0):

$$P(0,0)=\{1-F_f+F_oF_f[\phi^{W-1}\sum_{m=1}^{S-D-W}b_m+\sum_{m=S-D-W+1}^{N}b_m\,\phi^{S-D-m}]\}(1-\phi)\{(1-\phi)(1-F_f)F_oD+F_o^{\,2}F_f[\phi^{W-1}(2-2F_f+F_oF_f\phi(1+2\phi^{1-D}-\phi^{2-D}))\sum_{m=1}^{S-D-W}b_m$$
$$+(2-\phi)F_oF_\Pi\sum_{m=S-D-W+1}^{S-W-2}b_m\,(\phi^{W-D+1}-\phi^{S-D-m})+(1-\phi)\sum_{m=S-D-W+1}^{S-W-1}b_m\,\phi^{S-D-m}(m-S+D+W+1)+(D-1)(1-\phi)\sum_{m=S-W}^{N}b_m\,\phi^{S-D-m}+(1-\phi)-(1-\phi)\phi^{W-D+1}\sum_{m=1}^{S-W-1}b_m]+(1-\phi)^2(F_o\bar{N}$$
$$+1-\sum_{m=1}^{N}b_m\,\bar{F_o}^{\,S-D-m+1})\}^{-1}$$

Consider to present one more solution of the system of equilibrium equations for the parameters $W\ge D, D+W-1\le S\le W+N+1,\ W-2\le N$. Under these constraints, the equations of the original equilibrium system (2-9) will change as the following:

$$P(i,0)=F_o\sum_{n=1}^{V_i}F_f^{\,W-i}P(D+W-2-i+n,n)\,,i=\overline{1,D-2}$$

$$P(D-1,0)=F_o\left[\sum_{n=1}^{E}\sum_{i=D-1+n}^{W-1+n}P(i,n)\,F_f^{\,i-D+2-n}+\sum_{n=E+1}^{X}\sum_{i=D-1+n}^{S-2}P(i,n)\,F_f^{\,i-D+2-n}\right]$$

$$P(0,n)=b_nP(0,0),n=\overline{1,N}$$

$$P(i,n)=P(i-1,n)+b_nP(i,0),i=\overline{1,D-1},n=\overline{1,N}$$

$$P(i,n)=P(i-1,n),i=\overline{D,D-1+n},n=\overline{1,X}$$

$$P(i,n)=P(i-1,n),i=\overline{D,S-1},n=\overline{X+1,N}$$

$$P(i,n)=(1-F_0)P(i-1,n),i=\overline{D+n,S-1},n=\overline{1,X}$$

Where $V_i=S-2-(D+W-2-i)$, $E=S-2-(W-1)$, $X=S-2-(D-1)$.
By solving the equations, the following results are achieved:

$$P(i,0)=\frac{(1-\phi)P(0,0)F_oF_f\phi^{W-i-1}\sum_{m=1}^{S-D-W+i}b_m}{1-\phi-F_oF_f[(1-\phi^{W-1})\sum_{m=1}^{S-D-W}b_m+\sum_{m=S-D-W+1}^{S-D-1}b_m(1-\phi^{S-D-m})]},i=\overline{1,D-2}$$

$$P(D-1,0)=\frac{F_oF_fP(0,0)\left[\sum_{m=1}^{S-D-1}b_m-\phi^{W-D+1}\sum_{m=1}^{S-W-1}b_m-\sum_{m=S-W}^{S-D-1}b_m\phi^{S-D-m}\right]}{1-\phi-F_oF_f[(1-\phi^{W-1})\sum_{m=1}^{S-D-W}b_m+\sum_{m=S-D-W+1}^{S-D-1}b_m(1-\phi^{S-D-m})]}$$

$$P(i,n)=\frac{b_nP(0,0)\{1-\phi-F_oF_f\left[\sum_{m=1}^{S-D-1}b_m-\phi^{W-i-1}\sum_{m=1}^{S-D-W+i}b_m-\sum_{m=S-D-W+1+i}^{S-D-1}b_m\phi^{S-D-m}\right]\}}{1-\phi-F_oF_f[(1-\phi^{W-1})\sum_{m=1}^{S-D-W}b_m+\sum_{m=S-D-W+1}^{S-D-1}b_m(1-\phi^{S-D-m})]},i=\overline{0,D-2},n=\overline{1,N}$$

$$P(D-1,n)=\frac{b_n(1-\phi)P(0,0)}{1-\phi-F_oF_f[(1-\phi^{W-1})\sum_{m=1}^{S-D-W}b_m+\sum_{m=S-D-W+1}^{S-D-1}b_m(1-\phi^{S-D-m})]},\overline{1,S-D-1}$$

$$P(i,n)=\frac{(1-\phi)P(0,0)b_n}{1-\phi-F_oF_f[(1-\phi^{W-1})\sum_{m=1}^{S-D-W}b_m+\sum_{m=S-D-W+1}^{S-D-1}b_m(1-\phi^{S-D-m})]},i=\overline{D,D+n-1},n=\overline{1,X}$$

$$P(i,n)=\frac{(1-\phi)P(0,0)b_n}{1-\phi-F_oF_f[(1-\phi^{W-1})\sum_{m=1}^{S-D-W}b_m+\sum_{m=S-D-W+1}^{S-D-1}b_m(1-\phi^{S-D-m})]},i=\overline{D,S-1},n=\overline{X+1,N}$$

$$P(i,n)=\frac{P(0,0)(1-\phi)b_n\bar{F_o}^{\,i-D-n+1}}{1-\phi-F_oF_f[(1-\phi^{W-1})\sum_{m=1}^{S-D-W}b_m+\sum_{m=S-D-W+1}^{S-D-1}b_m(1-\phi^{S-D-m})]},i=\overline{D+n,S-1},n=\overline{1,N}$$

From here we can get $P(0,0)$:

$$\begin{aligned}P(0,0)=\{&1-\phi-F_oF_f[(1-\phi^{W-1})\sum_{m=1}^{S-D-W}b_m+\sum_{m=S-D-W+1}^{S-D-1}b_m(1-\phi^{S-D-m})]\}F_o(1-\phi)\{(1\\&-\phi)^2F_oD-(1-\phi)F_o^{\,2}F_f[(D-1)\sum_{m=1}^{S-D-1}b_m-\sum_{m=S-D-W+1}^{S-D-1}b_m\,\phi^{S-D-m}\\&+\sum_{m=S-D-W+1}^{S-W-1}b_m\,\phi^{S-D-m}+\phi^{W-1}\sum_{m=1}^{S-D-W}b_m+\sum_{m=S-D-W+1}^{S-W-1}b_m\,\phi^{S-D-m}(m-S\\&+D+W)]-F_o^{\,2}F_f[\phi^{W-D+1}\sum_{m=1}^{S-W-2}b_m-\phi^{W-1}\sum_{m=1}^{S-D-W}b_m\\&-\sum_{m=S-D-W+1}^{S-W-2}b_m\,\phi^{S-D-m}]+(1-\phi)^2[F_o\sum_{m=1}^{S-D-1}(mb_m+1)\\&+F_o\sum_{m=S-D}^{N}b_m\,(S-D)+\bar{F}_o-\bar{F}_o^{\,S-D+1}\sum_{m=1}^{N}b_m\,\bar{F}_o^{\,-m}]\}^{-1}\end{aligned}$$

The last variety of interval restriction on protocol parameters and the maximum queue size has form of $W\geq D, D<S<D+N+1$, when $F_o=1$, the equilibrium equations are like the following:

$$P(0,0)=(1-F_f)\sum_{n=1}^{V}P(D-1+n,n)+\sum_{n=V+1}^{N}P(S-1,n),\text{ where } V=S-2-(D-1)$$

$$P(i,0)=0,i=\overline{1,D-2}$$

$$P(D-1,0)=F_f\sum_{n=1}^{S-D-1}P(D-1+n,n)$$

$$P(0,n)=b_nP(0,0),n=\overline{1,N}$$

$$P(i,n)=P(i-1,n),i=\overline{1,D-2},n=\overline{1,N}$$

$$P(D-1,n)=P(D-2,n)+b_nP(D-1,0),n=\overline{1,N}$$

$$P(i,n)=P(i-1,n),i=\overline{D,D-1+n},n=\overline{1,N}$$

The initial state is obtained by solving all the equilibrium equations and from the normalization condition:

$$P(0,0)=\frac{1-F_f\sum_{m=1}^{V}b_m}{1+D-(D-1)F_f\sum_{m=1}^{V}b_m+\sum_{m=1}^{V}b_mm+(S-D)\sum_{m=V+1}^{N}b_m},\quad\text{where } V=S-D-1$$

Research in a multidimensional area (S, W, D, F_o, F_f) is a difficult task. The solution for this problem is the dimension reduction of the attribute space. Effective options for this reduction are absolute reliable reverse data transmission path $(F_o = 1)$ and the case of a uniform forward and reverse data transmission path $(F_o, = F_f = F)$ with conditions of unlimited window size $(W \to \infty)$ and, consequently, timeout duration $(S \to \infty)$. The initial probability case for $F_o = 1$ looks as the following:

$$P(0,0) = \frac{1-F_f}{1+D+\bar{N}+F_f(1-D)} \text{ for } W \geq D.$$

$$P(0,0) = \frac{1-F_f}{1+D+\bar{N}+F_f(1-W)} \text{ for } W < D.$$

For the case of a uniform forward and reverse data transmission path $(F_o, = F_f = F)$ with conditions of unlimited window size $(W \to \infty)$ and, consequently, time-out duration $(S \to \infty)$, the probability of initial states of Markov chain gets this form:

$$P(0,0) = \frac{(1-F)F}{D(1-F)F+(1-F+F^2)F(1+\bar{N})+1-2F+2F^2} \text{ for } W \geq D.$$

$$P(0,0) = \frac{(1-F+F^2\Phi^{W-1})(1-\Phi)F}{F[F^2(\Phi-\Phi^W)+F^2(1-W)(1-\Phi)+(1-\Phi)^2(D+1+\bar{N})]+(1-\Phi)^2\left((1-F)-(1-F)^{S-D+1}\sum_{m=1}^{N}\frac{b_m}{(1-F)^m}\right)} \text{ for } W < D.$$

after we found all the cases, now we can find the throughput for the selected loaded channel, which we study and to show how the throughput is behaving.

4. Throughput of the loaded channel

The capacity of the transport connection in the conditions of competition flows of various subscribers for transmission channel throughput is defined as the ratio of the average amount of data, transmitted between two consecutive acknowledgements, to the average time of getting the acknowledgements [7], [8]. The states of Markov chain, for which it is possible to get receipts, have a contribution to the throughput of the virtual connection.

The following equations can calculate the bandwidth for the upper constraint $S \geq W + D + N - 1$:

$$Zg = P(0,0)F_f\{F_o{}^2F_f(1-\Phi^{W-D+1}) + [(1+\Phi^{W+1}-\Phi^W)\sum_{n=1}^{N}\frac{b_n}{n+1}$$
$$+\frac{\bar{F}_o{}^{S-D+1}(F_f{}^W-F_f{}^{W+1}\bar{F}_o-1+\Phi)}{1-F_f}\sum_{n=1}^{N}\frac{b_n}{(n+1)\bar{F}_o{}^n}]\}\{1-F_f+F_oF_f\Phi^{W-1}\}^{-1} \text{for } W \geq D.$$

$$Zg = \frac{P(0,0)F_f}{1-F_f+F_oF_f\Phi^{W-1}}\sum_{n=1}^{N}\frac{b_n}{n+1}\{(1-\Phi^W)+\frac{(1-\Phi)(F_f{}^W-1)\bar{F}_o{}^{S-D-n+1}}{1-F_f}\} \text{ for } W < D.$$

For the cases of absolute reliable reverse data transmission path $(F_o = 1)$, for $W < D$ and the throughput depend on the closeness of the value of windows size with the value of round-trip duration, but for $W \geq D$, is invariant to D. For calculating we get the following formulas for $S \geq W + D + N - 1$:

$$Zg = \frac{F_f(F_f+\sum_{n=1}^{N}\frac{b_n}{n+1})}{1+D+\bar{N}+F_f(1-D)} \text{ for } W \geq D.$$

$$Zg = \frac{F_f\sum_{n=1}^{N}\frac{b_n}{n+1}}{1+D+\bar{N}+F_f(1-W)} \text{ for } W < D.$$

For the case of a uniform forward and reverse data transmission path $(F_o = F_f = F)$ with conditions of unlimited window size and time-out duration $(W \to \infty, S \to \infty)$, the throughput is as the followings:

$$Zg(W=\infty, S=\infty) = \frac{P(0,0)F}{1-F}(F^3+\sum_{n=1}^{N}\frac{b_n}{n+1}) \text{ for } W \geq D$$

$$Zg(W, S=\infty) = \frac{P(0,0)F(1-\Phi^W)}{1-F}\sum_{n=1}^{N}\frac{b_n}{(n+1)} \quad \text{for } W < D$$

Consider the following solution for the throughput for the interval restrictions on the duration of the timeout $W + D - 1 \leq S \leq W + D + N - 1$ and the size $1 \leq N \leq D - 2$ of the opponents' queues when window size is greater than round-trip delay $W \geq D$:

$$Zg = \frac{F_n P(0,0)}{Y(1-F_f)}[F_o{}^2F_f(1-\Phi^{W-D+1})(1-F_f)+\sum_{n=1}^{S-D-W+1}\frac{b_n}{n+1}[(1-F_f)(1-\Phi^W)+(1-F_f{}^W)(1$$
$$-F_f)(\Phi-1)\bar{F}_o{}^{S-D-n+1}]+\sum_{n=S-D-W+2}^{N}\frac{b_n}{n+1}\left(1-\bar{F}_o{}^{S-D-n+1}\right)]$$

Where $Y = 1 - F_f + F_o F_f \left[\Phi^{W-1} \sum_{m=1}^{S-D-W} b_m + \sum_{m=S-D-W+1}^{N} b_m \, \Phi^{S-D-m} \right]$

For reverse data transmission path ($F_o = 1$), it takes the form:

$$Zg = \frac{F_f\{F_f(1-F_f)+(1-F_f)\sum_{n=1}^{S-D-W+1}\frac{b_n}{n+1}+\sum_{n=S-D-W+2}^{N}\frac{b_n}{n+1}\}}{(1-F_f)\{D(1-F_f)+F_f+1+\bar{N}\}}$$

For the case of a uniform forward and reverse data transmission path ($F_o, = F_f = F$):

$$Zg(W=\infty, S=\infty) = \frac{F^5}{1+F^3+\bar{N}F(1-\Phi)+\Phi(D-1)}$$

For queue size of $D - 2 \le N \le W - 2$ we get:

$$Zg = \frac{P(0,0)F_f}{Y}\{F_o{}^2 F_f[1-\phi^{W-D+1}\sum_{m=1}^{S-W-1} b_m - \sum_{m=S-W}^{N}\phi^{S-D-m} b_m] + \frac{1-\phi}{1-F_f}[\sum_{m=1}^{S-D-W+1}\frac{b_m}{m+1}((1$$

$$-\phi^W)(1-F_f) - \bar{F}_o{}^{S-D+1-m}(1-F_f{}^W)) + \sum_{m=S-D-W+2}^{N}\frac{b_m}{m+1}(1-\bar{F}_o{}^{S-D-m+1}$$

$$-F_f(1-\phi^{S-D-m+1}))]\}^{-1}$$

And for $N \ge W - 2$ the throughput can be calculated as the followings:

$$Zg = \frac{P(0,0)F_f}{U}[F_o{}^2 F_f(\sum_{m=1}^{S-D-1} b_m - \phi^{W-D+1}\sum_{m=1}^{S-W-1} b_m) + \frac{F_f(1-\phi)}{1-F_f}\sum_{m=1}^{S-W-1}\frac{b_m}{m+1}[(1-\phi^W)(1-F_f)$$

$$-(1-F_f{}^W)\bar{F}_o{}^{S-D-n+1}]]$$

Where $U = 1 - \phi - F_o F_f[(1-\phi^{W-1})\sum_{m=1}^{S-D-W} b_m + \sum_{m=S-D-W+1}^{S-D-1} b_m (1-\phi^{S-D-m})]$.

For reverse data transmission path ($F_o = 1$), it takes the form:

$$Zg = \frac{P(0,0)F_f}{1-F_f\sum_{m=1}^{S-D-1} b_m}\{F_f\sum_{m=1}^{S-D-1} b_m + \sum_{n=1}^{S-W-1}\frac{b_m}{m+1}\},$$

For the case of a uniform forward and reverse data transmission path ($F_o, = F_f = F$):

$$Zg = \frac{P(0,0)F}{U}[F^3(\sum_{m=1}^{S-D-1} b_m - \phi^{W-D+1}\sum_{m=1}^{S-W-1} b_m) + \frac{F(1-\phi)}{1-F}\sum_{n=1}^{S-W-1}\frac{b_m}{m+1}[(1-\phi^W)(1-F)$$

$$-(1-F^W)(1-F)^{S-D-m+1}]]$$

For the interval of $W + D - 1 \le S \le W + D + N - 1, 1 \le N \le W - 2, W \le D$ the following is obtained:

$$Zg = \frac{P(0,0)F_f}{1-F_f+F_oF_f[\Phi^{W-1}\sum_{m=1}^{S-D-W} b_m+\sum_{m=S-D-W+1}^{N} b_m\Phi^{S-D-m}]}[\sum_{n=1}^{N}\frac{b_n}{n+1}(1-\Phi^W) + \frac{(1-\Phi)(F_f{}^W-1)\bar{F}_o{}^{S-D-n+1}}{1-F_f}]$$

For reverse data transmission path ($F_o = 1$) we get the followings:

$$Zg = \frac{F_f\sum_{n=1}^{N}\frac{b_n}{n+1}}{1+D+\bar{N}-F_f(W-1)}$$

For the case of a uniform forward and reverse data transmission path ($F_o = F_f = F$):

$$Zg(W, S=\infty) = \frac{F^2(1-\phi)(1-\Phi^W)\sum_{n=1}^{N}\frac{b_n}{(n+1)}}{F^3+(1-\phi)^2[1+F(D-W+\bar{N})+W\phi]}$$

The ultimate throughput solution for $W \ge D, D < S < D + N + 1$, $F_o = 1$ has the following form:

$$Zg = \frac{F_f[F_f\sum_{m=1}^{S-D-1} b_m + \sum_{m=1}^{S-W-1}\frac{b_m}{m+1}]}{1+D-(D-1)F_f\sum_{m=1}^{S-D-1} b_m + \sum_{m=1}^{S-D-1} m b_m + (S-D)\sum_{m=S-D}^{N} b_m}$$

As a result, numerical analyzes show that by increasing the queue number of rivals, the throughput of the transport channel decreases. You can easily see in (Figure 1,Figure 2), with a length of N = 8, the throughput is lower than with N = 6,4,2.

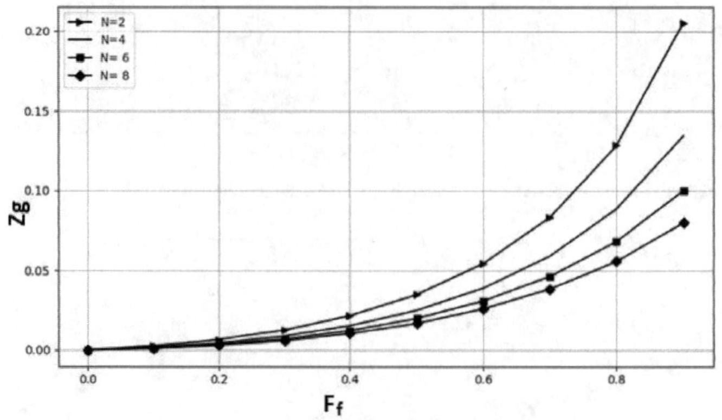

Figure 1. Dependency of the throughput on different queue sizes, for $D = 18, W = 19, b = 5, S \geq D + W + N - 1, F_o = 1$.

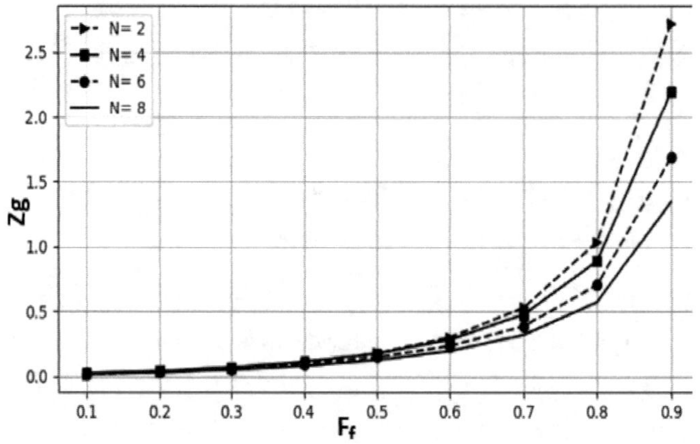

Figure 2. Dependency of the throughput on the reliability of data transmission in forwarding channel, for $D = 12, W = 15, b = 5, 1 \leq N \leq D - 2, D + W - 1 \leq S \leq D + W + N - 1, F_o = 1$.

Figure 3 shows how the throughput of transport connection changes by moving the round-trip delay with fixed values of the queue length and the geometric distribution of the queue length. a parabolic dependence is observed.

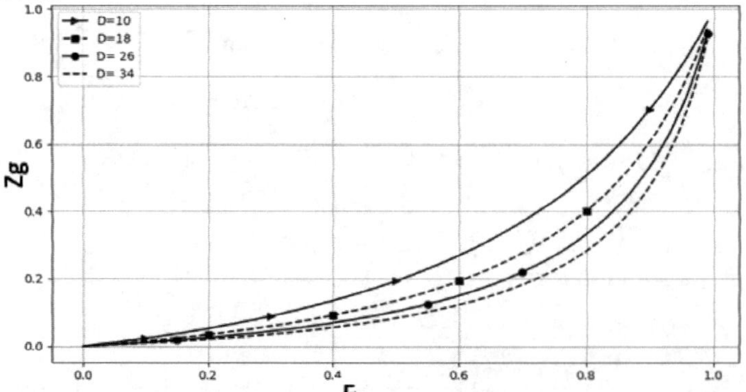

Figure 3. Dependency of the throughput on the reliability of data transmission in forwarding channel for different round-trip delay and with parameters of $N = 2, b \in [2..15]\ with\ step = 3, S \leq D + W + N - 1, F_o = 1$.

Journal of Physics: Conference Series **2091** (2021) 012029 doi:10.1088/1742-6596/2091/1/012029

5. Conclusion

In this paper, the mathematical model, in analytical and numerical forms has been proposed, and the relationship between the parameters of the loaded transmission channels of different subscribers is found. Also, the throughput of the virtual channel is described with which, we can calculate throughput of the transport connections. The analysis of equations shows that the throughput of $W \geq D$ is unchanged for the round-trip duration, but for $W < D$, the throughput of the channel depends on how close the window size value is to the round-trip duration value. From the equations, we can see that as the competition between subscribers in the transmission connection intensifies, for $W < D$, the throughput of the channel decreases. According to the results of the numerical analysis, the throughput dependency on the reliability of transmission data in the forward channel has a parabolic dependence and the form of the throughput dependency on the queue size, and on the round-trip delay is hyperbolic. Consequently, the longer the queue size, the lower the throughput. In [11], the efficiency of the FEC (Forward Error Correction) model was presented, without considering the load on the transport channel. Formerly, after the completion of this work, the research will be continued. As a next step of improving this study, the method of forward error correction considering the queue size on the transport connection will be applied. Then, identify those cases in which the FEC method has positive results in comparison with the classical transport channel (without using FEC) in conditions of an increased level of errors and at long distances between hosts.

References

[1] Fall K R and Stevens R W 2012 *TCP/IP Illustrated the Protocols 2nd Edition* (Addison-Wesley Professional Computing Series) p 1017

[2] Padhye J, Firoiu V, Towsley D and Kurose J 2000 Modeling TCP reno performance: a simple model and its empirical validation *IEEE/ACM Trans. on Net.* **8** 133–45

[3] Borodikhin E and Korotaev I 1993. Analysis of functioning and optimization of HDLC protocol *Avto. i Vychislitel. tekh.* **2** 47–51

[4] Ewald N and Kemp A 2009 Analytical Model of TCP New Reno through CTMC *Comp. Perf. Eng. EPEW* **5652** 183–96

[5] Boguslavsky L and Gelenbe E 1980. Analytical models transmission link control procedures for data computer networks with packet *Auto. and Remote Cont.* **7** 181–92

[6] Gelenbe E, Labetoulle J and Pujolle G 1978 Performance Evaluation of the HDLC Protocol *Comp. Net.* **2** 409–15

[7] Sushchenko S 1988 Analytical models of asynchronous procedures for data link control *Avto. i Vychislitel. Tekh.* 32–40

[8] Kokshenev V and Sushchenko S 2008 Performance analysis of asynchronous procedure of data link control *Vichislitel. Tech.* **13(5)**, 61–65.

[9] Kokshenev V, Mikheev P and Sushchenko S 2013 Analysis of the selective mode of the transport protocol in loaded data path *Vestnik TSU. Ser. Uprav. vichislitel. Techn. i inform* **3(24)** 78–94

[10] Ronaldo H 2017 Modeling, and comparative analysis of forward error correction in the context of multipath redundancy *Telecom. Sys. Mod. Analysis, Designand Manage.* **65** 783-94

[11] Karim P, Mikheev P, Podubny V and Suschenko S 2020 Numerical studies of transport protocol throughput with forward error correction mechanism in intersegment space *Vestnik TSU УВТиИ* **50** 89–96

[12] Mikheev P, Pristupa P and Suschenko S 2019 Performance of Transport Connection with Selective Failure Mode When Competing for Throughput of Data Transmission Path. In: Vishnevskiy V Samouylov K and Kozyrev D (eds) Distr. Com. and Commun. Net. DCCN 2019. Comm. in Com. and Inf. Sci. **1141** 89-103. *Springer, Cham*

[13] Karim P, Challoob A, Ghafoor K and Suschenko S 2020. Throughput of loaded channel for different subscribers with competition for network resources in group failure mode // *Journal of advanced research in dynamical and control Sys.* **12(8)** 403-08

[14] Olifer V and Olifer N 2016 *Computer Networks. Principles, Technologies, Protocols: Textbook for Universities* (St. Petersburg: Peter) p 862

[15] Cao Y, Balasubramanian A and Gandhi A 2017 Rethinking TCP throughput and latency modeling *SIGCOMM Posters and Demos 17 Los Angeles CA. USA*

[16] Polese M, Chiariotti F, Bonetto E, Rigotto F, Zanella A and Zorzi M 2019 A survey on recent advances in transport layer protocols *IEEE Com. Surveys & Tuto.* **21(4)** 3584-608 doi: 10.1109/COMST.2019.2932905

[17] Chaudhary P and Kumar S 2017. Comparative study of TCP variants for congestion control in wireless network *International Conf. on Com. Comm. and Auto ICCCA Greater Noida* 641–46 doi: 10.1109/CCAA.2017.8229880

[18] Gomez C, Arcia-Moret A and Crowcroft J 2018 TCP in the internet of things: from ostracism to prominence *IEEE Internet Com.* **22(1)** 29-41 doi: 10.1109/MIC.2018.112102200

[19] Sivasankari M and Hariharan S 2019 Challenges for TCP protocols in data center network: a survey *Intern. J. of engin. research & tech. IJERT* **08(06)**

[20] Lundqvist H and Karlsson G 2004 TCP with end-to-end FEC *Comm. Intern. Zurich Sem.* 152–56

[21] Barakat C and Altman E 2002 Bandwidth tradeoff between TCP and link-level FEC *Comp. Net.* 39 133–50

[22] Shalin R and Kesavaraja D 2012 Multimedia data transmission through TCP/IP using hash based FEC with auto–XOR scheme *J. on Comm. Tech.* **03**(03) 604–9

[23] L. Boguslavskii 1984 *Data flow control in computer networks* Energoatomizdat p 168

Journal of Physics: Conference Series **2091** (2021) 012030 doi:10.1088/1742-6596/2091/1/012030

The calibration method of a tandem queueing model with PH service time using NS-3 simulation of a multihop wireless network

A A Larionov[1], A A Mukhtarov[1] and A M Sokolov[1]

[1]V. A. Trapeznikov Institute of Control Sciences of Russian Academy of Sciences 65 Profsoyuznaya street, Moscow 117997, Russia

E-mail:larioandr@gmail.com,mukhtarov.amir.a@gmail.com, aleksandr.sokolov@phystech.edu

Abstract. End-to-end delay is one of the key characteristics of communication network performance. This characteristic determines the possibility of using the network for various delay-critical applications like voice or video transmission. One of the widely used approaches to estimating delays is the use of the queuing theory. According to this approach, a telecommunication network is modeled using a multiphase queuing system. Communication channels are modeled using service devices, and the incoming traffic is modeled with random distributions of the inter-arrival intervals between packets. The accuracy of this network model directly depends on how well the service time distributions are chosen. These distributions must consider the specifics of complex telecommunication protocols, size distributions of the transmitted packets, and, in case of wireless channels, the rate of collisions and retransmissions. The paper presents a study of the accuracy of estimates of end-to-end delays in a multi-hop wireless network using a queuing network with a phase-type (PH) service time distributions. To calibrate the model, PH distributions are found using the moments-matching method based on sample data on the duration of packet transmission in IEEE 802.11 channels. This sample data was obtained using a simulation model written in NS-3, taking into account the features of the IEEE 802.11 protocol and the presence of collisions in the network. To evaluate the accuracy, end-to-end delays are calculated using the queuing network and the wireless network simulation model. It is shown that it is possible to obtain reasonably accurate estimates for small networks, but with an increase in the size of the network, the accuracy decreases. In conclusion, recommendations are given to improve the accuracy of modeling.

1. Introduction

The design of communication networks and the development of new protocols include network performance evaluation, particularly end-to-end delays. These estimates allow us to conclude that the network is suitable for a variety of latency-critical applications and quickly identify performance "bottlenecks." Researchers often use the queuing theory methods to evaluate network performance characteristics. And the distribution of intervals between packets in the traffic characterizes the arrival flow transmitted by the network. It is necessary to have better distributions to get accurate estimations. However, choosing "good" distributions for channel modeling is a complex problem since telecommunication protocols, especially wireless ones, turn out to be quite complicated. In addition, the transmission time takes into account a distribution of packet sizes and errors occurring. So, it is required to select the distribution depending on the protocol used and traffic transmitted by the network.

In this paper, we consider the particular case of a multi-hop wireless network, in which base stations have the linear topology, and use the same communication channel. Examples of such networks are vehicular ad-hoc networks (VANET) along long highways. We use IEEE 802.11 protocol in our work. Base stations compete with one another to use the shared medium in such a network. We assume that the incoming traffic arrives in the network according to the simplest (Poisson) flow, in which intervals between sequential packets have an exponential distribution. We use phase-type distributions (PH) to model the packet transmission duration. It allows us to describe wired and wireless channels accurately [1].

To calibrate the model and build PH-distributions, we need to have an actual sample of transmission intervals of separate packets in the channel, or at least the first moments of this sample. It is very complicated to use actual wireless equipment for these purposes. A more straightforward method that we apply in our work is implementing a network in the NS-3 modeling system. To construct a sample of time intervals, we also consider that stations located at the beginning and end of the network are less under collisions than intermediate stations. Intermediate stations have to compete with their neighbors for channel access. We use the found PH-distributions in M/PH/1/N → •/PH/1/N →...→ •/PH/1/N queuing model to estimate end-to-end delays. Further, we will compare these estimates with end-to-end delays obtained using the simulated wireless network model implemented in NS-3. As will be shown below, the queuing system allows a fairly accurate estimate of end-to-end delays for networks with a small number of stations. It requires much less calculating time than simulating the wireless network. However, as the network size or arrival traffic rate grow, the accuracy decreases.

2. Related work

The network performance evaluation using queuing methods found a wide application in global literature. Initial research in this area contained simplified assumptions about the nature of information flows in telecommunication networks (L. Kleinrock [2], L. Kleinrock [3], V. Vishnevsky [4,5], etc.). The development of stochastic models has led to widespread use of the phase-type distribution. A tandem queueing system with PH distribution is used in [6-13, et all]. Approximating general distributions by phase-type (PH) distributions is a widespread technique in the stochastic analysis since the Markovian property of PH distributions often allows analytical tractability [14]. Researchers in [1,15-17] carried out the approximation of the time spent in the channel using the PH distribution for IEEE 802.11 wireless networks. Most of the early work used simple simulation models or other mathematical models as a data source, for example, semi-Markov random processes based on the Bianchi model. In contrast to these, in this article, we take the data to construct PH-distributions from the high-precision IEEE 802.11n simulation model in NS-3. Open-Source Project Network Simulator 3 is widely used to simulate wireless networks in many papers [18-23].

The problem of PH-distributions fitting is also presented in many papers [24-31, et all]. In this work, we will use a simple algorithm proposed by Johnson [24] to fit the PH distribution by the first three moments, but instead, you can use more complex techniques, for example, the EM procedure. The methodology of developing a wireless network model proposed in our article does not change from this.

3. Tandem queueing system

Let us consider a multi-hop wireless network with linear topology. Such a network consists of sequentially connected stations, each of which is connected only to its immediate neighbors. We will consider the case when antennas at the stations are omnidirectional, all stations use the same channel and packets enter the network only at the first station, pass all intermediate stations and leave the network at the last one. For the sake of simplicity, we will assume that the stations operate using the IEEE 802.11 DCF protocol without using RTS/CTS. Note that similar transmission protocols are used, for example, in IEEE 802.15.4 (ZigBee) sensor networks.

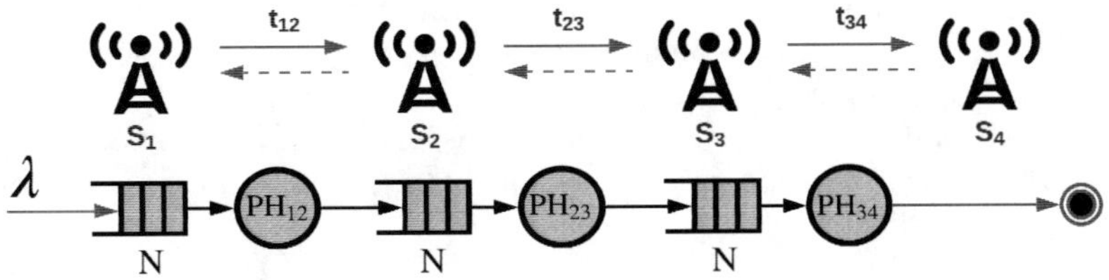

Figure 1. Queueing system model of a wireless network with linear topology.

When modeling a wireless network using the queuing system M/PH/1/N → •/PH/1/N →...→ •/PH/1/N (see Fig. 1), we will model the transmission time in channels using PH-distributions, and buffers at stations, in which packets are awaiting transmission, by queues of finite capacity N. To simplify the model, we will assume that packets have a fixed size B bits, and the intervals between their arrivals have a Poisson distribution with intensity λ. Note that the distribution of packet sizes (in our case, a constant) affects only the service time and is taken into account when constructing PH-distributions.

The PH distribution is a generalization of the exponential distribution and can be used to approximate any positive continuous distribution [24]. Random variable is phase-type distributed if its value can be expressed as the time till absorption in a continuous absorbing Markov chain with generator $\tilde{S}$. PH-distribution is defined with a subgenerator matrix $S \in \mathbb{R}^{V \times V}$ and initial probability row vector $\overline{\tau} \in \mathbb{R}^{V}$, such that:

$$\tilde{S} = \begin{bmatrix} S & -S\mathbf{1} \\ \mathbf{0} & 0 \end{bmatrix}, \forall i = \overline{1,V}: 0 \leq \tau_i \leq 1, \sum_{i=1}^{V} \tau_i = 1.$$

Since the stations on the network use the same channel, collisions can occur between them. For example, consider the three-hop network shown in Fig. 1. Simultaneous transmissions from station 1 to station 2 and from station 3 to station 4 will lead to a collision at station 2. According to CSMA/CA, on which IEEE 802.11 protocol is built, station 1 will have to retry transmitting its packet at a random time. Notice that the transmission from station 3 may still be successful, since the transmission from station 1 doesn't interfere with it at station 4. This leads to an increase in losses, especially at stations that are located closer to the data source, and increase in the duration of the transmission of each packet, which will not necessarily be the same at all stations.

In order to take into account this feature of the network operation, we will model service time in each channel with its own PH distribution. To construct these distributions, we will simulate the network shown in Fig. 1 using the NS-3 simulation system. For each channel, we collect samples of transmission durations (from the beginning of the first attempt to transmit a packet until the receipt of an ACK delivery confirmation) for each of the channels t12, t23, and t34. Based on these samples, we calculate the first three moments and, using the method described in Johnson [24], find the PH-distributions for each of the channels. This calibration method is shown in Fig. 2.

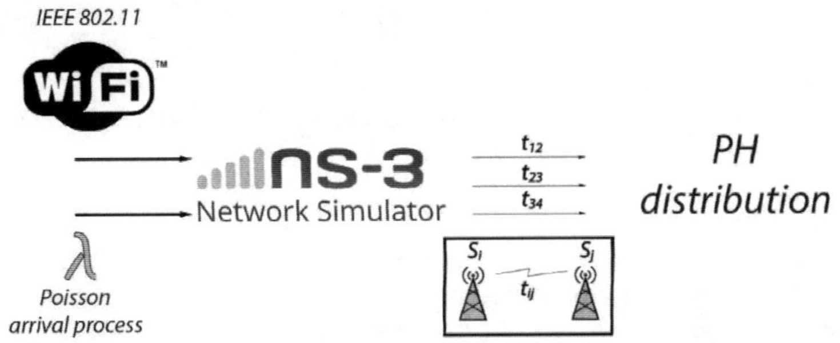

Figure 2. The calibration method of service time distribution of each station in the tandem network.

PH-distribution fitting and end-to-end delays estimations of the tandem queueing networks were implemented in the tool “pyqumo” implemented in Python 3 language. This is an open-source project available at https://github.com/ipu69/pyqumo.
Before proceeding to the numerical results and analysis of the accuracy of estimates obtained using the queuing network, we will briefly consider the main elements and features of the NS-3 simulation model, which was used in the experiment.

4. Wireless network simulation in NS-3

In this paper we implemented a wireless network simulation model using Network Simulator (NS-3). The model has three parameters: number of stations, simulation time and arrival traffic rate. Settings of the IEEE 802.11n protocol don’t change between experiments, their values are given in the numerical experiments chapter. We used IEEE 802.11n models from NS-3, slightly extended for precise transmission time estimation and random traffic generation.
To simulate network traffic we implemented a traffic generator with exponential distribution of intervals between packets. In terms of NS-3 traffic generator is just an application for scheduling events. Parameters of this application are the mean interval between the packets, the destination station address and the payload size.
Service duration is the time from the moment when the sending station senses the channel and starts DIFS waiting in the first transmission attempt, and till it finishes the reception of ACK frame from the receiving station. Any busy channel waiting intervals, packet transmission attempts and collisions are included in the service duration. If the number of retransmission attempts exceeds the given limit, the packet is considered lost, its service duration is not added to the final statistics. To collect service durations we used the NS-3 events mechanism. To record the timestamp of the first channel sensing we had to slightly modify the model implementation, because the current version doesn’t provide such means. As for ACK delivery time, we used standard event of WiFi MAC level

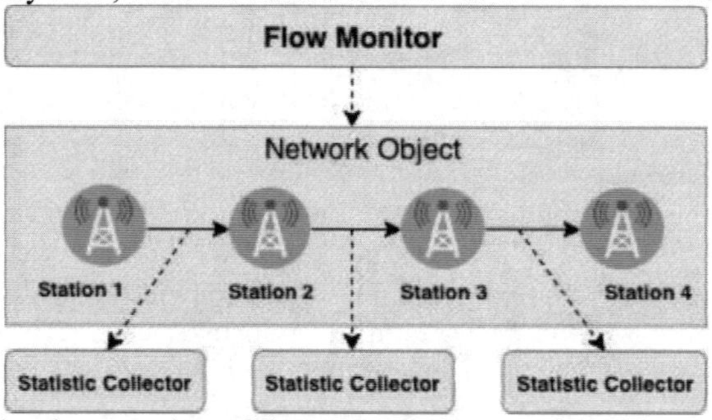

Figure 3. The implementation of the NS-3 tandem model with statistics modules.

Model architecture is shown in Fig. 3. Statistic Collector was used to collect service durations, it measured intervals between channel sensing and ACK delivery. To measure mean end-to-end delays we used a standard NS-3 class FlowMonitor. It collects network statistics and determines some network parameters like end-to-end delays, number of lost packets, number of transmitted and received packets, etc.

5. Numerical experiments

In the numerical experiment, simulated stations operated using IEEE 802.11n protocol with bit rate equal to 54 Mbps. Stations were equipped with buffers for 500 packets. Each packet carried 1400 bytes of payload, and in each simulation the traffic generator produced 500’000 packets.
We considered four distinct arrival packet rates $\lambda = 100, 200, 500$ and 1000, that corresponded to traffic bitrate of 1.2 Mbps, 2.4 Mbps, 6 Mbps and 12 Mbps, respectively. For each arrival rate we collected service time samples in each channel and computed mean end-to-end delays. Each set of service

durations contained about 500'000 samples excluding the packets lost during the transmission. After running the simulation we calculated three moments of service time durations for PH-fitting.
Besides non-trivial PH-distributions, we also considered service time approximation with exponential distribution with the mean value equal to the mean service time in the channel, and compared end-to-end estimation accuracy of M/PH/1/N → •/PH/1/N →...→ •/PH/1/N and M/M/1/N → •/M/1/N →...→ •/M/1/N queueing networks.

5.1. End-to-end delays in calibration tandem with four stations

First of all, we studied end-to-end delays in the networks with four stations and three channels like the one shown in Fig. 1. We call this a calibration tandem, since later we use distributions, fitted from samples measured in this model, for delays estimation in networks of arbitrary length.
After NS-3 simulation and service time samples collection, we fitted PH and exponential distributions of service time and measured end-to-end delays using M/PH/1/N →•/PH/1/N →...→ •/PH/1/N and M/M/1/N → •/M/1/N →...→ •/M/1/N queueing networks. For each arrival traffic rate λ=100 (1.2Mbps), λ =200 (2.4Mbps), λ =500 (6Mbps) and λ =1000 (12Mbps) we ran network simulation, service time distribution fitting and end-to-end delays estimations 10 times.

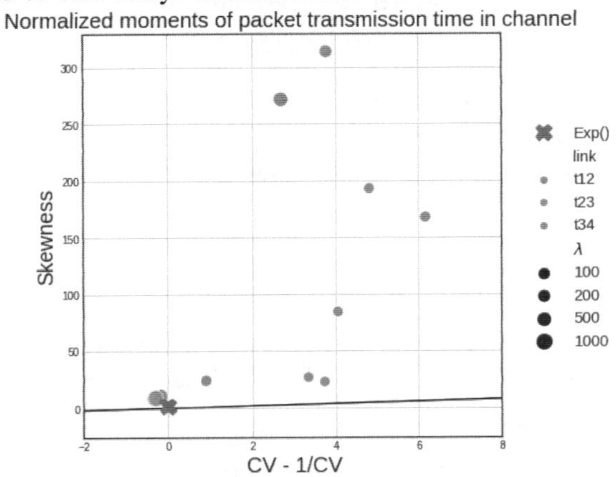

Figure 4. Coefficient of variation (CV) and skewness for each channel.

To prove that service time is better fitted with PH than with trivial exponential distribution, Fig. 4 shows the coefficients of variation and coefficients of skewness for each channel in the calibration tandem. It can be seen that most service time distributions have coefficients of variation much higher than 1, especially in cases of low arrival rates. Thus modeling service time with exponential distribution leads to error in variation, which may affect the end-to-end delays estimation precision. However, when the arrival traffic rate is high, the coefficient of variation is closer to 1, and the PH-distribution may be replaced with an exponential distribution with higher precision.
Numerical results of end-to-end delays estimations using NS-3 simulation and queueing models for the network with four nodes are shown in Fig. 5 using box plots. As it can be seen, for low intensity traffic the M/PH/1/N model provides higher accuracy than the M/M/1/N model. For high intensity traffic the M/M/1/N model suddenly provides a slightly better approximation than the M/PH/1/N model. This fact correlates with the observation that the coefficient of variation decreases when the arrival rate grows.

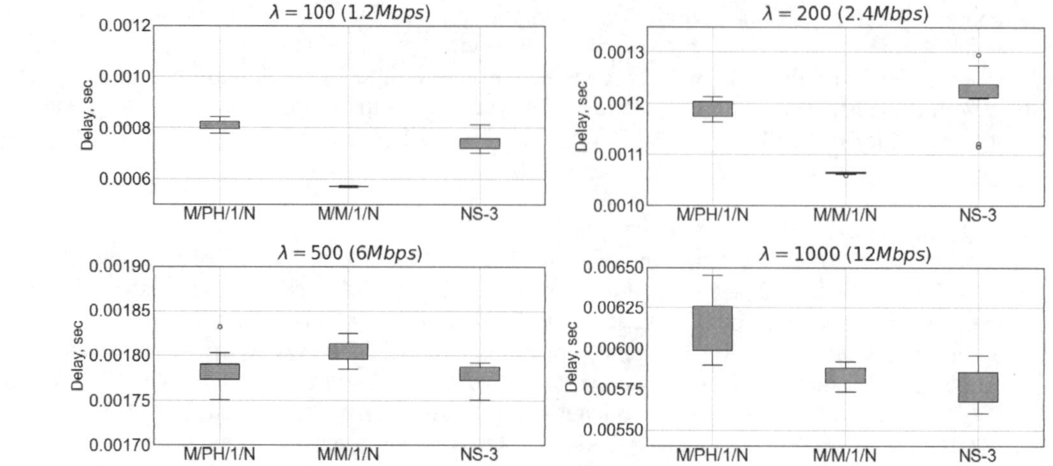

Figure 5. Boxplots for different arrival rates for NS-3 simulation and queueing models for various arrival rates.

5.2. Comparison of end-to-end delay for networks of arbitrary length.

To study the accuracy of M/PH/1/N → •/PH/1/N →…→ •/PH/1/N and M/M/1/N → •/M/1/N →…→ •/M/1/N models, we investigated the dependency of end-to-end delay on the number of stations in the tandem. Like before, we considered several packet arrival rates using both queueing networks and the NS-3 simulation model. Our key assumption was that all channels can be modeled by distributions from the calibration experiment for four stations. The first and the last channels were modeled using distributions for the first and last channels of the calibration model (1-2 and 3-4, respectively, see Fig. 1), while all the intermediate channels were modeled with the same distribution for the channel 2-3 in the calibration model. This approach allows us to model a tandem network of arbitrary length.

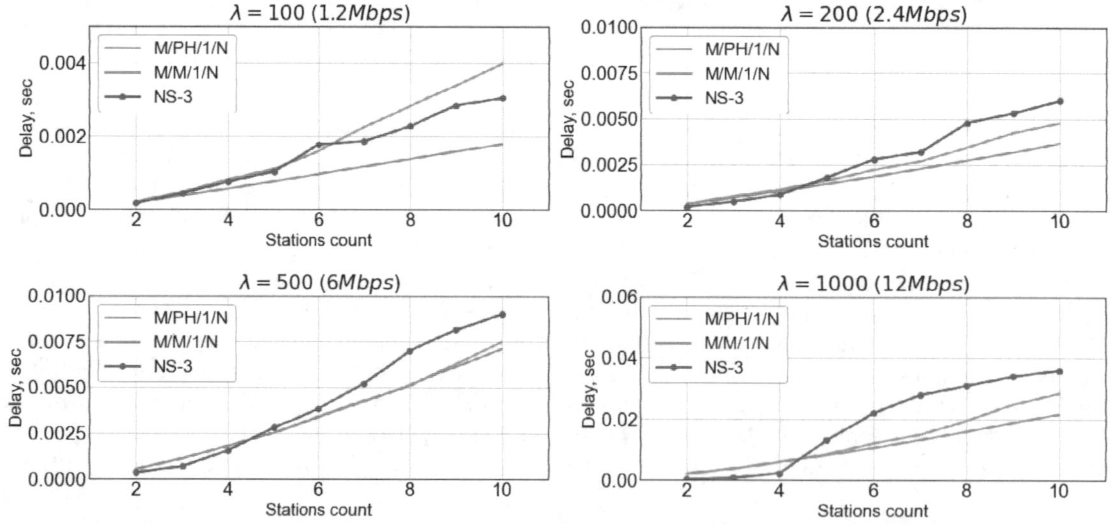

Figure 6. End-to-end delays in networks with different numbers of stations.

Plots in Fig. 6 show the dependency of end-to-end delays on the number of stations for NS-3 simulation and both queueing models. The M/PH/1/N model provides accuracy about 75% in most cases when the arrival traffic rate is low, while the M/M/1/N model has much higher relative error, up to 50%, for large networks. We should note that the accuracy difference between the models for short networks with up to 4 stations isn't high, and M/M/1/N is fairly suitable for these cases. However, as long as the arrival traffic bandwidth grows, accuracy of both models decrease dramatically.

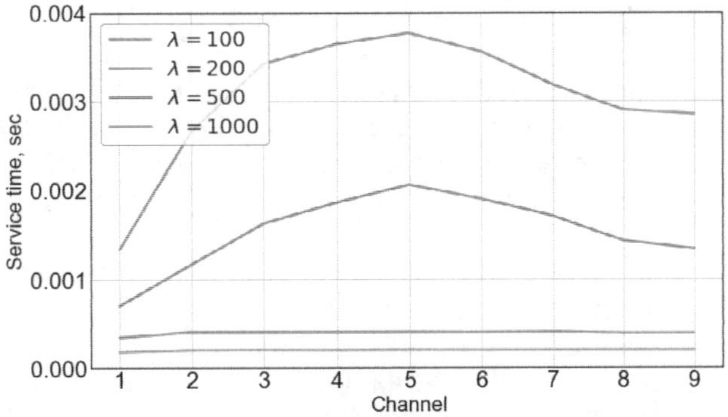

Figure 7. Mean service time of each channel in the tandem with length=10.

One of the possible reasons for accuracy degradation under the heavy traffic, is that channel 2-3 is not suitable for modeling intermediate channels in large networks. To study this effect, Fig. 7 shows the mean service times in each channel in the network with 10 stations measured in the NS-3 simulation model. The plots show that for small traffic rates most intermediate channels have very close properties. However, for larger rates the service time mean values in the intermediate channels differ: service time increases till the fifth station, and then it starts decreasing. This can be explained by the fact that most of the collisions and losses occur at the beginning of the transmission, and as the packet moves through the network, the traffic is filtered and the service becomes more deterministic.

6. Conclusion

In this paper we described an approach to multi-hop wireless networks modeling using M/PH/1/N → •/PH/1/N →…→ •/PH/1/N tandem queueing systems. We showed that calibrating PH-distributed service times using a simulation network with three hops allows us to estimate end-to-end delays with fairly good precision in large networks with up to ten stations. However, the precision depends on the arrival traffic rate, it degrades as long as the rate grows. For short networks even simple M/M/1/N models provide fairly good accuracy, but using exponential service time approximation leads to errors as long as the network becomes larger. Big networks simulation using NS-3 takes much longer time than estimation of end-to-end delays using queueing networks, so the presented approach can be used when the delays are needed as a property in some larger problem like network topology optimization.

7. References

[1] Vishnevsky V and Larionov A 2020 Stochastic Multiphase Models and Their Application for Analysis of End-to-End Delays in Wireless Multihop Networks pp 457–71

[2] Kleinrock L 1975 *Theory, Volume 1, Queueing Systems* (USA: Wiley-Interscience)

[3] James M 1972 *Systems analysis for data transmission.*

[4] Vishnevsky V 2003 *Teoreticheskie osnovi proektirovania compjuternih setej [in Russian]*

[5] Vishnevsky, V and Zhozhikashvili V 1988 *Queueing Networks. Theory and applications for computer networks*

[6] Dudin A N, Klimenok V I and Vishnevsky V M 2019 The theory of queuing systems with correlated flows *Theory Queuing Syst. with Correl. Flows* 1–410

[7] Kim C, Dudin A, Dudin S and Dudina O 2013 Tandem queueing system with impatient customers as a model of call center with Interactive Voice Response *Perform. Eval.* **70** 440–53

[8] Ayyappan G and Gowthami R 2021 Analysis of MAP,PH2OA/PH1I,PH2O/1 retrial queue with vacation, feedback, two-way communication and impatient customers *Soft Comput.* **25** 9811–38

[9] Jose J K and Deepthi V 2021 M/PH/1 queueing model with re-servicing *Commun. Stat. Simul. Comput.*
[10] Krishnamoorthy A, Joshua A N and Vishnevsky V 2021 Analysis of a k- stage bulk service queuing system with accessible batches for service *Mathematics* **9**
[11] Krishnamoorthy A, Joshua A N and Kozyrev D 2021 Analysis of a Batch Arrival, Batch Service Queuing-Inventory System with Processing of Inventory While on Vacation *Mathematics* **9** 1–29
[12] Chakravarthy S R, Ozkar S and Shruti 2021 Analysis of m/m/c retrial queue with thresholds, p h distribution of retrial times and unreliable servers *J. Appl. Math. Informatics* **39** 173–96
[14] Osogami T and Harchol-Balter M 2006 Closed form solutions for mapping general distributions to quasi-minimal PH distributions *Perform. Eval.* **63** 524–52
[15] Larionov A A, Vishnevsky V M, Ivanov R E and Semenova O V. 2017 Estimation of IEEE 802.11 DCF access performance in wireless networks with linear topology using PH service time approximations and MAP input *Proc. 11th IEEE Int. Conf. Appl. Inf. Commun. Technol. (AICT2017, Moscow)* **2** 85–9
[16] Larionov A, Vishnevsky V, Semenova O and Dudin A 2019 A Multiphase Queueing Model for Performance Analysis of a Multi-hop IEEE 802.11 Wireless Network with DCF Channel Access *Commun. Comput. Inf. Sci.* **1109** 162–76
[17] Vishnevski V, Larionov A and Ivanov R 2016 An open queueing network with a correlated input arrival process for broadband wireless network performance evaluation *Commun. Comput. Inf. Sci.* **638** 354–65
[18] Mukhtarov, A. and Sokolov M 2021 A Base Station Placement Of An Wireless Network With Linear Topology And A Network Performance Evaluation With NS-3 *Inf. Telecommun. Technol. Math. Model. High-Tech Syst.* 425–30
[19] Prakash M and Abdrabou A 2020 On the fidelity of ns-3 simulations of wireless multipath tcp connections *Sensors (Switzerland)* **20** 1–23
[20] Fontes H, Lamela V, Campos R and Ricardo M 2019 ns-3 NEXT: Towards a reference platform for offline and augmented wireless networking experimentation *ACM Int. Conf. Proceeding Ser.* 65–72
[21] Cruz R, Fontes H, Ruela J, Ricardo M and Campos R 2020 On the Reproduction of Real Wireless Channel Occupancy in ns-3 *ACM Int. Conf. Proceeding Ser.* 41–8
[22] Patidar R, Roy S, Henderson T R and Chandramohan A 2017 Link-to-system mapping for ns-3 Wi-Fi OFDM error models *ACM Int. Conf. Proceeding Ser.* **Part F128360** 31–8
[23] Afaqui M S, Cano C, Kotzschmar V, Felber C and Nitzold W 2019 Real-time operation of LTE/Wi-Fi interworking via NS-3 and SDR interfacing *INFOCOM 2019 - IEEE Conf. Comput. Commun. Work. INFOCOM WKSHPS 2019* 981–2
[24] Johnson M A and Taaffe M R 1989 Matching moments to phase distributions: Mixtures of erlang distributions of common order *Commun. Stat. Stoch. Model.* **5** 711–43

[25] Feldmann A and Whitt W 1998 Fitting mixtures of exponentials to long-tail distributions to analyze network performance models *Perform. Eval.* **31** 245–79

[26] Horváth A and Telek M 2000 Approximating heavy tailed behaviour with Phase type distributions *3rd Int. Conf. Matrix-Analytic Methods Stoch. Model. MAM3* 191–214

[27] Bobbio A, Horváth A and Telek M 2002 PhFit: A general phase-type fitting tool *Proc. 2002 Int. Conf. Dependable Syst. Networks* 543
[28] Riska A, Diev V and Smirni E 2004 An EM-based technique for approximating long-tailed data sets with PH distributions *Perform. Eval.* **55** 147–64
[29] Gu W, Fan N and Liao H 2021 Fitting aggregated phase-type distributions to the length-of-stay in intra-hospital patient transfers *Oper. Res. Heal. Care* **29**

[30] Bladt M and Rojas-Nandayapa L 2018 Fitting phase–type scale mixtures to heavy–tailed data and distributions *Extremes* **21** 285–313
[31] Bražėnas M, Horváth G and Telek M 2018 Parallel algorithms for fitting Markov arrival processes *Perform. Eval.* **123–124** 50–67

Acknowledgments
The reported study was funded by RFBR, project number 19-29-06043.

Groups of mobile cellular service stations for operational radio communication tasks in remote areas

O N Pishchin, D S Puzankov, K V Perova and I O Chulkov

Astrakhan State Technical University, Tatischeva str, 16, Astrakhan, Russia

E-mail: o.pishin@yandex.ru

Abstract. The authors of the article consider solutions needed to organize mobile cellular communication systems in remote areas. Modern mobile stations do not have the capability to overcome marshy areas, steppe, heavily rugged reliefs of the desert, water barriers and forests. Moreover, the coverage area from one base station is not highly localized (radio coverage area is not significant). Furthermore, modern mobile communication facilities have poor autonomy. The article proposes measures for an operational organization of cellular communications in large remote areas with a coverage radius of a medium-sized city. Article also considers the issues posed by a temporary organization of communication service in large sections near important locations (pipelines, riverbeds, transport and railway lines). The need for such an operational usage of mobile communication systems is most associated with geological exploration and scientific research in large remote areas and large areas with a damaged communication system due to natural emergencies or military operations.

1. Introduction

At the present moment, mobile base stations are actively used by cellular operators to implement temporary tasks to remove high traffic load in places of large social events (forums, festivals with the deployment of a field camp, large sporting events, etc.). The solution to providing radio access and phone communications is created by measures developed to organize access to necessary energy sources, connection to public cellular service and radio frequency support.

However, there is a high demand for access to the public telephone and Internet networks from large remote areas. There are several types of instances that are currently not provided with those types of necessary services: geological research sites, sites that host teams employed in exploring natural deposits and other various kinds of resources, teams providing emergency rescue operations in remote areas, places of natural disasters and in areas with destroyed infrastructure after military operations. Solving the goal of operational (up to 24 hours) installation of a mobile communication system with a radio coverage area of several thousand square kilometers (first usage option) or extended territories (second usage option) is possible when solving the following problems:

1. Determining the necessary elements of the mobile grouping composition and methods of information exchange;
2. Determining the necessary elements of an autonomous mobile base station;
3. Getting through areas impassable for conventional vehicles;
4. Autonomy and power supply for the system components;
5. Operational management of mobile cellular service stations.

2. Methods of organizing a mobile communication node grouping to group mobile base stations

Based on the tasks of the grouping, its composition should include (figure 1):

1. The control communication node (CCN), which is responsible for collecting grouping information on point-to-multipoint, point-to-point radio relay communication systems and transmitting it to the public telephone network or to the operator's system by space communication means (SC). The CCN provides continuous monitoring of the technical condition and management of autonomous mobile objects (automatic after installation).
2. Autonomous mobile objects (AMO) are mobile base stations designed to deploy a general access to the network and equipped with radio relay communication systems (point - multipoint for the first variant of the system or "point-to-point" for the second variant of use) for transmitting communication channels to the UUS.
3. Mobile maintenance and repair unit (MMaRU) - for prompt response to failure of any system components.
4. Multifunctional life support unit (MLSU) - for rest of operational and maintenance personnel.

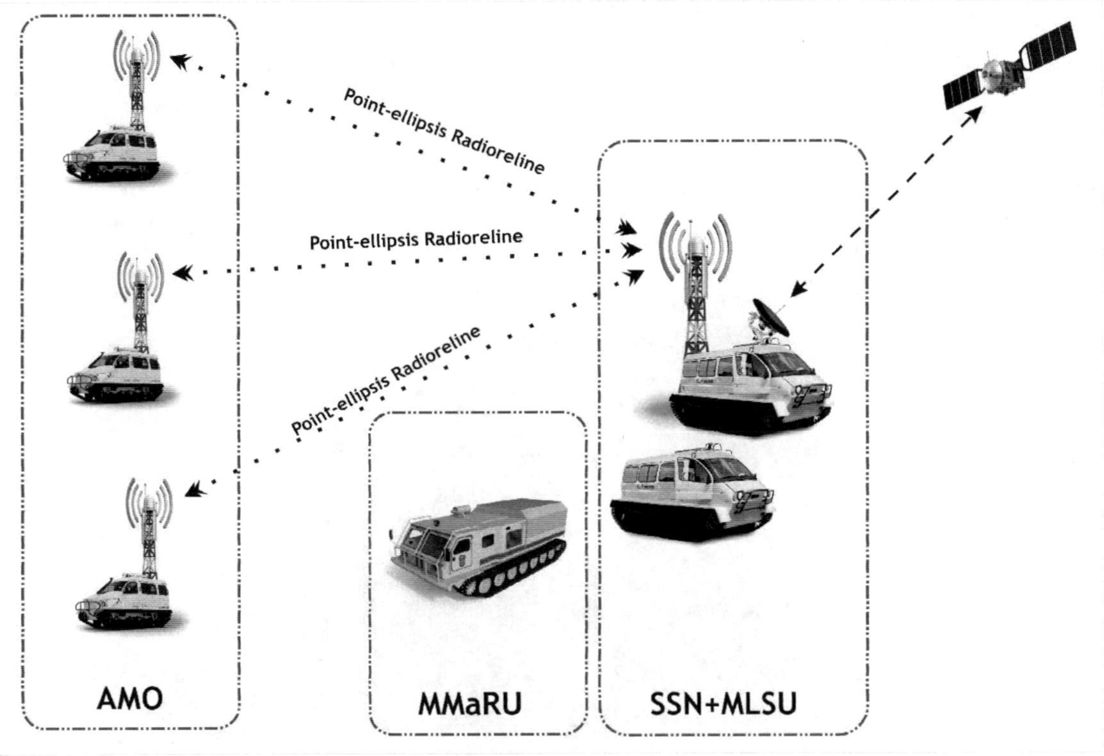

Figure 1. Mobile Base Station Mobile Grouping Functional Diagram.

Currently, technical problems of off-road access, the delivery of personnel and equipment to remote areas have several solutions, however, there are no solutions for the operation of traveling cellular stations. An effective solution can be the use of a tractor on caterpillar traction (for example, GAZ 3409 Bobr, figure 2),

Figure 2. Snow and swamp-going vehicle of GAZ-3409 "Bobr".

classified as a bog with the ability to carry up to 800 kg, which is enough for a set of equipment of a base and radio relay station with antenna equipment and associated equipment of a base station. There's also an option to use a nodal application of swamp and water-going vehicle the Premium type GAZ-34039 "Snow leopard", with a loading capacity up to 1600 kg and a possibility of transportation of the trailer up to 2000 kg (figure 3).

Figure 3. Snowshoe GAZ-34039 "Irbis" Premium.

3. Methods for implementing communication between spaced apart mobile base stations.

Based on the tasks of the mobile groups, the efficiency of the entire system can be high with the maximum possible distance between base stations to organize the widest coverage area. To organize the connectivity of a communication center, it is necessary to collect all of the base station's communication channels into a single center for further broadcasting to a mobile switching center. Let us consider the methods of implementing the tasks of the mobile grouping node, which make it possible to connect all mobile base stations promptly into a single network through radio relay communication lines.

The easiest way to implement this is by using a system of wireless broadband radio access system designed for the construction of multi-service communication and data transmission networks (MDMS - Marconi Digital Multipoint System).

Maximum distance values need to be calculated. To do this, you can use the following equations (1-3):

$$P_{вх\_пр} = P_{пер} - G_{пер} - L_{простр} + G_{пр} - L_{пер} - L_{пр}. \quad (1)$$

$$L_{простр} = 20\log(\frac{4\pi df}{c}). \quad (2)$$

$$P_{зап} = P_{вх\_пр} - Ch_{пр}, \quad (3)$$

where $P_{вх\_пр}$ is the signal level at the input to the receiver (dB), $P_{пер}$ - is the output power of the transmitter of the radio relay station (dBi), $G_{пер}$ is the antenna gain of the transmitting radio relay station (dBi), $L_{простр}$ is the loss in space between the receiving and transmitting antennas (dB), $G_{пр}$ is the antenna gain receiver RRL station (dBi), $L_{пер}$ - losses in the feeder path between the transmitter and the antenna (the feeder path is understood as a structure that consists of a cable and connectors; in the MDMS system, the transmitter's power amplifier is located in the outdoor module and is connected by a connector directly to the antenna, then only the losses occurring in this connector are taken into account, in this case the numerical value is in the range of 0.5-1 dB). $L_{пр}$ - losses in the feeder path between the transmitter and the antenna (the numerical value is within 0.5-1 dB), c - the speed of light in vacuum, $3\cdot10^8$, d - the distance between the PPC (m), f - the frequency at which the transmitter operates (Hz), $Ch_{пр}$ - the sensitivity of the radio relay station receiver (dBm), $P_{зап}$ - the margin for fading, which shows the channel's energy reserve for additional losses caused by atmospheric properties (dust, rain, fog, etc.), the numerical value of this parameter at the most distant point should be within 20-30 dB (if this parameter is below 20 dB, then this indicates low channel reliability) (dB)

As we can see in the above formulas, there is no clear dependence of the distance on the other characteristics, so we derive it from these equations 1, 2, 3.

$$\frac{L_{простр}}{20} = \frac{4\pi df}{c} => d = \frac{10^{\frac{L_{простр}}{20}}\cdot c}{4\pi f}.$$

$$P_{зап} = P_{вх\_пр} - Ch_{пр} => P_{вх\_пр} = P_{зап} + Ch_{пр},$$

$$P_{зап} + Ch_{пр} = P_{пер} + G_{пер} - L_{простр} + G_{пр} - L_{пер} - L_{пр}, \quad (4)$$

$$L_{простр} = P_{пер} + G_{пер} + G_{пр} - L_{пер} - L_{пр} - P_{зап} - Ch_{пр},$$

$$d = \frac{10^{(\frac{P_{пер}+G_{пер}+G_{пр}-L_{пер}-L_{пр}-P_{зап}-Ch_{пр}}{20})}\cdot c}{4\pi f}.$$

Let us find the values of the distances between the transmitting and receiving radio relay stations according to formula 4 in two directions (BS → TS and TS → BS), with a fading margin equal to 20 dB. The smallest value of these 2 will be the maximum distance providing stable communication. If we take values, then the connection between them will be either simplex or not reliable in one direction. The characteristics of the transceivers are shown in tables 1 and 2.

Table 1. Base station transceiver characteristics.

Name	DMS 3.5	DMS 10.5			DMS 26		
Frequency range, GHz	3,4-3,6	10,15-10,65			24,5-26,5		
Directional pattern width, degrees	60	15	45	90	15	45	90
Transmitter power, dBm	23	24			18		
Antenna gain, dBi	12	19	17	12.5	21.5	17	15
Receiver sensitivity, dBm	-91.5						
Dimensions, mm	480x340x110	450x275x117			390x212x142		
Weight, kg	7	9			8		

Table 2. Terminal station transceiver characteristics.

Name	DMS 3.5	DMS 10.5	DMS 26		
Frequency range, GHz	3,4-3,6	10,15-10,65	24,5-26,5		
Directional pattern width, degrees	20	9	5	2,5	1,2
Transmitter power, dBm	11	24	8		
Antenna gain, dBi	17	25	29	34	40
Receiver sensitivity, dBm	-91,5				
Dimensions, mm	340x320x110	450x275x117	417x212x 106	30 см diameter	60 см diameter
Weight, kg	5	5	6		

Since we are looking for equipment with the longest communication range, to reduce the time, we take the value of the center frequency of the range:

$$f = \frac{f_1 + f_2}{2}. \tag{5}$$

For example, DMS 3.5 equipment operates in the 3.4-3.6 GHz frequency range, then

$$f = \frac{(3{,}4 + 3{,}6)\cdot 10^9}{2} = 3{,}5 \cdot 10^9.$$

The maximum distance between the transmitting and receiving radio relay stations, created on DMS 3.5 equipment with a base station antenna pattern of 60 degrees:

$$d_1 = \frac{10^{(\frac{23+12+25-1-1-20-(-91,5)}{20})} \cdot c}{4\pi \cdot 3,5 \cdot 10^9} \approx 8100.$$

$$d_2 = \frac{10^{(\frac{24+25+12,5+-1-1-20-(-91,5)}{20})} \cdot c}{4\pi \cdot 3,5 \cdot 10^9} \approx 24200.$$

$$d = 8100$$

The maximum distance between the transmitting and receiving radio relay stations, created on DMS 10.5 equipment with a base station antenna pattern of 90 degrees:

$$d_1 = \frac{10^{(\frac{24+12,5+25-1-1-20-(-91,5)}{20})} \cdot c}{4\pi \cdot 10,5 \cdot 10^9} \approx 8140.$$

$$d_2 = \frac{10^{(\frac{24+12,5+25-1-1-20-(-91,5)}{20})} \cdot c}{4\pi \cdot 10,5 \cdot 10^9} \approx 8140.$$

$$d = 8140$$

The maximum distance between the transmitting and receiving radio relay stations, created on DMS 10.5 equipment with a base station antenna pattern of 45 degrees:

$$d_1 = \frac{10^{(\frac{24+25+17-1-1-20-(-91,5)}{20})} \cdot c}{4\pi \cdot 10,5 \cdot 10^9} \approx 13670.$$

$$d_2 = \frac{10^{(\frac{24+17+25-1-1-20-(-91,5)}{20})} \cdot c}{4\pi \cdot 10,5 \cdot 10^9} \approx 13670.$$

$$d = 13670$$

The maximum distance between the transmitting and receiving radio relay stations, created on DMS 10.5 equipment with a base station antenna pattern of 15 degrees:

$$d_1 = \frac{10^{(\frac{24+19+25-1-1-20-(-91,5)}{20})} \cdot c}{4\pi \cdot 10,5 \cdot 10^9} \approx 17210.$$

$$d_2 = \frac{10^{(\frac{24+19+25-1-1-20-(-91,5)}{20})} \cdot c}{4\pi \cdot 10,5 \cdot 10^9} \approx 17210.$$

$$d = 17210$$

The DMS 26 equipment has 3 terminal antennas with different gains: 29, 34, 40 dBi. In formula 4, it can be seen that the larger the gain, the greater the value of d. Therefore, we take for the calculation a terminal antenna with a value of 40.

The maximum distance between the transmitting and receiving radio relay stations, created on the DMS 26 equipment with the antenna pattern of the base station 90 degrees:

$$d_1 = \frac{10^{(\frac{18+15+40-1-1-20-(-91,5)}{20})} \cdot c}{4\pi \cdot 25,5 \cdot 10^9} \approx 3950.$$

$$d_2 = \frac{10^{(\frac{8+15+40-1-1-20-(-91,5)}{20})} \cdot c}{4\pi \cdot 25,5 \cdot 10^9} \approx 12484.$$

$$d = 3950$$

The maximum distance between the transmitting and receiving radio relay stations, created on DMS 26 equipment with a base station antenna pattern of 45 degrees:

$$d_1 = \frac{10^{(\frac{18+40+70-1-1-20-(-91,5)}{20})} \cdot c}{4\pi \cdot 25,5 \cdot 10^9} \approx 15715.$$

$$d_2 = \frac{10^{(\frac{8+70+40-1-1-20-(-91,5)}{20})} \cdot c}{4\pi \cdot 25,5 \cdot 10^9} \approx 4970.$$

$$d = 4970$$

The maximum distance between the transmitting and receiving radio relay stations, created on DMS 26 equipment with a base station antenna pattern of 15 degrees:

$$d_1 = \frac{10^{(\frac{18+15+40-1-1-20-(-91,5)}{20})} \cdot c}{4\pi \cdot 25,5 \cdot 10^9} \approx 26985.$$

$$d_2 = \frac{10^{(\frac{18+40+15-1-1-20-(-91,5)}{20})} \cdot c}{4\pi \cdot 25,5 \cdot 10^9} \approx 8343.$$

$$d = 8343.$$

For convenience, we will place the calculation results in table 3. [5]

Table 3. The calculation results.

Name	Base station antenna pattern width, degrees	Distance between stations of the BS → TS channel, km	Distance between stations of the TS → BS channel, km	Maximum distance b{etween stations
DMS 3.5	60	8,1	24.2	8,1
	90	8,14	8,14	8,14
DMS 10.5	45	13,67	13,67	13,67
	15	17,21	17,21	17,21
	90	12,48	3,95	3,95
DMS 26	45	15,71	4,97	4,97
	15	26,39	8,34	8,34

As we can see in table 3, the DMS 10.5 equipment with a base station antenna has the greatest range, the AP width is 15, but the base station is capable of supporting up to 8 sectors, therefore, the base station provided with this equipment will be able to provide the beam width up to 120 degrees. Therefore, we will choose the DMS 10.5 system with a base station antenna, whose antenna pattern is 45 degrees, to ensure a circular coverage area.

4. Calculation of RRL system MDMS in the range of 10.5 GHz

Based on the need to organize the maximum RRS range, frequencies in the 10 GHz range will be selected (figure 4, 5). Let us construct graphs of the dependence of losses in space on the distance according to the formula (2).

Since the MDMS system uses several modulation schemes, we will use the scheme with the maximum performance, namely 16 TCM 3/4, which will maintain the maximum channel rate. To ensure reliable communication, you need to ensure the lowest BER ((Bit Error Rate) - the ratio of the number of erroneously received bits to the total number of bits received.).

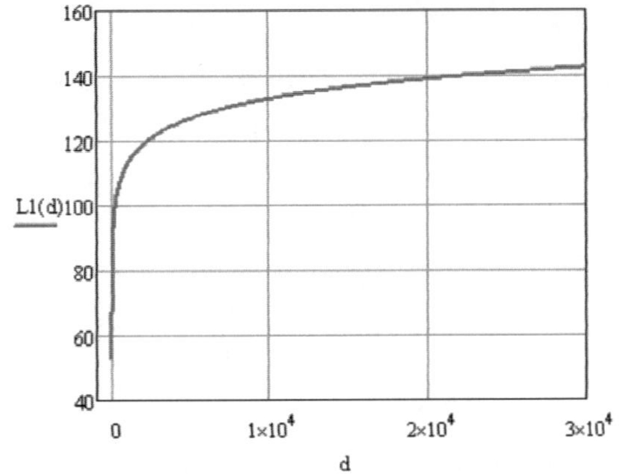

Figure 4. Dependence of the value of losses (signal attenuation) depending on the distance for the frequency f = 10.3 GHz.

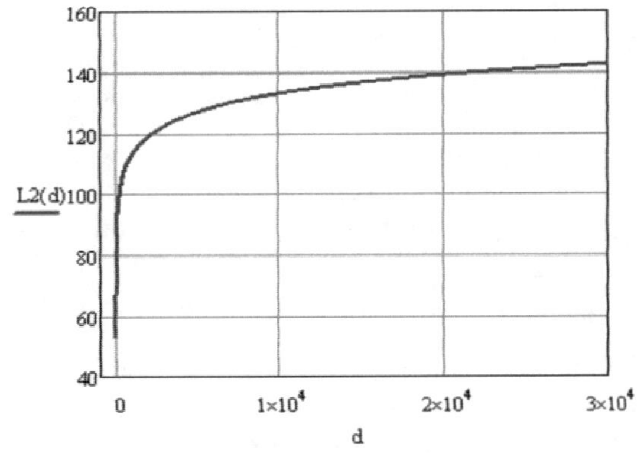

Figure 5. Dependence of the value of losses (signal attenuation) depending on the distance for the frequency f = 10.65 GHz.

Margin values have been obtained from Ericsson that can provide specific BERs for various modulation schemes for MDMS equipment.

The margin value = 24 dB will provide the BER value = and the margin for fading required for RRL systems [4]

Using formula (4), we find the maximum distance between stations

$$d_1 = \frac{10^{(\frac{24+17+25-1-1-20-(-91,5)}{20})} \cdot c}{4\pi \cdot 10,5 \cdot 10^9} \approx 8711.$$

$$d_2 = \frac{10^{(\frac{24+25+17-1-1-20-(-91,5)}{20})} \cdot c}{4\pi \cdot 10,5 \cdot 10^9} \approx 8425.$$

$$d = 8425.$$

Check the resulting value using formulas 1, 2, 3 for the frequency f_1

$$L_{простр} = 20\log(\frac{4\pi \cdot 8425 \cdot 10,3 \cdot 10^9}{c}) = 131,21.$$

$$P_{вх\_пр} = 24 + 17 - 131,21 + 25 - 1 - 1 \approx -67,5.$$

$$P_{зап} = -67,5 + 91,5 = 24.$$

And for f_2:

$$L_{простр} = 20\log(\frac{4\pi \cdot 8425 \cdot 10,65 \cdot 10^9}{c}) = 131,5.$$

$$P_{вх\_пр} = 24 + 17 - 131,21 + 25 - 1 - 1 \approx -67,5.$$

$$P_{зап} = -67,5 + 91,5 = 24.$$

Another important step during calculations is to determine the height of the antenna suspension. The main criterion for the correct choice of the desired suspension height is the absence of obstacles both directly in the line of sight, and in the signal energy propagation zone, namely in the first Fresnel zone), while the excess suspension height leads to an unjustified increase in the mast height.

Calculation of the antenna suspension height consists of the following steps:

1. Determination of the curvature of the earth by the formula:

$$y = (\frac{R_0^2}{2a}) \cdot k_i(1 - k_i) \cdot 10^3, \quad (6)$$

where R_0 is the distance between radio relay stations (km), $k_i = \frac{R_i}{R_0}$ is the coefficient characterizing the ratio of the distance to a given point R_i to the distance, R_0, a is the radius of the earth 6.371 km

2. Determination of the size of the first Fresnel zone by the formula:

$$H_i = \sqrt{\frac{R_0 \cdot 1000 \cdot \lambda \cdot k_i(1 - k_i)}{3}}, \quad (7)$$

where λ is the wavelength in meters

3. Construction of a span profile, taking into account the terrain, is carried out according to a topographic map or special geoinformation systems. Wireless network design systems such as Oneplan RPLS-Link use special cards. It is in it that this calculation will be carried out.

On an arbitrary open area of the Astrakhan region in the software package Oneplan RPLS-Link, a profile (figure 6) of the span was built taking into account the terrain. The figure shows that obstacles do not fall into 1 Fresnel zone when the antennas are suspended 8 m above the ground.

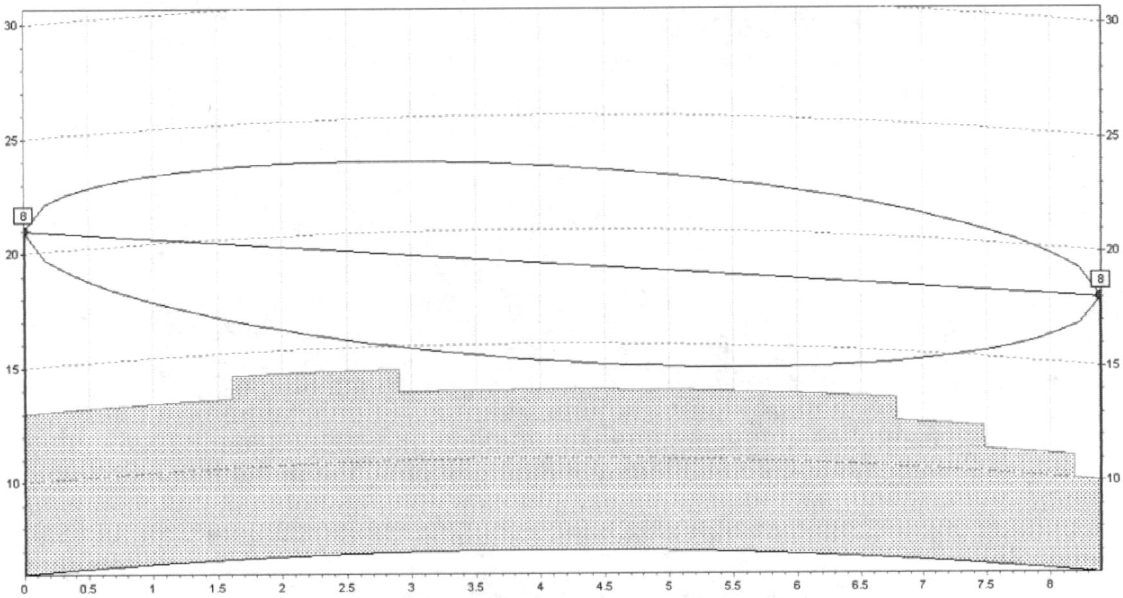

Figure 6. Span profile taking into account the terrain.

5. Automated calculation of radio coverage and energy characteristics

The calculation of the range of the "point-to-multipoint" system can be carried out using the RPLS ONEGA computer-aided design system. For the calculation, a digital map of the Astrakhan region was used.

The calculation was made by sector. The sector is discretized into areas (squares) with the side size "a".

The distance between the center of the square and the BS (b) and the angle between the direction of the BS antenna and the resulting segment (α) are determined. The value of the gain (G) varies with the angle α according to the dependence shown in the antenna radiation pattern. The value b is substituted instead of d in the formula (2), and the value of G instead of G_1 in the formula 1 and the value $P_{\text{вх\_пр}}$ in each such square is found. These squares are marked with different colors depending on the value $P_{\text{вх\_пр}}$. The resulting map will be called a field level map. Oneplan RPLS-Link software complex performs the above operations. The result is shown in figures 7 and 8.

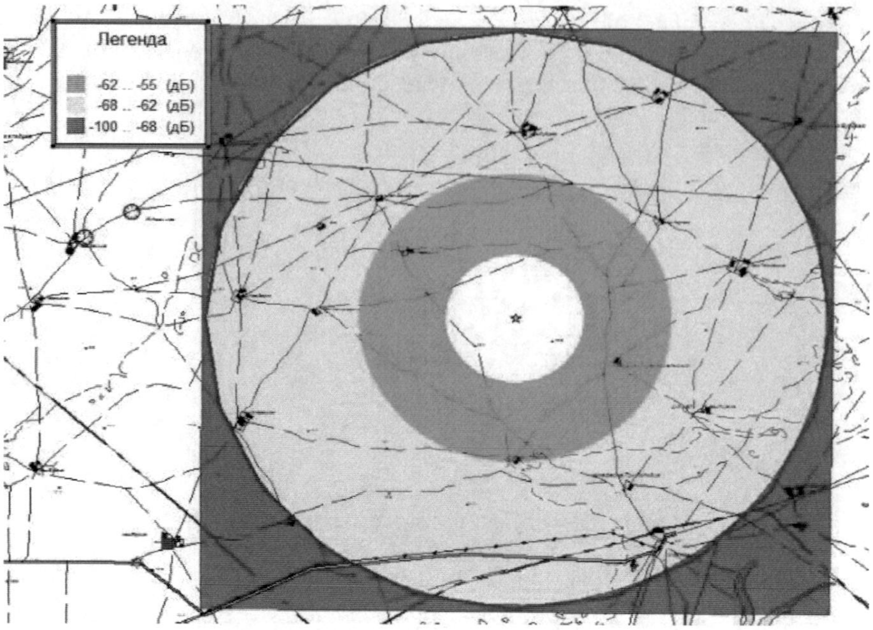

Figure 7. Field level map.

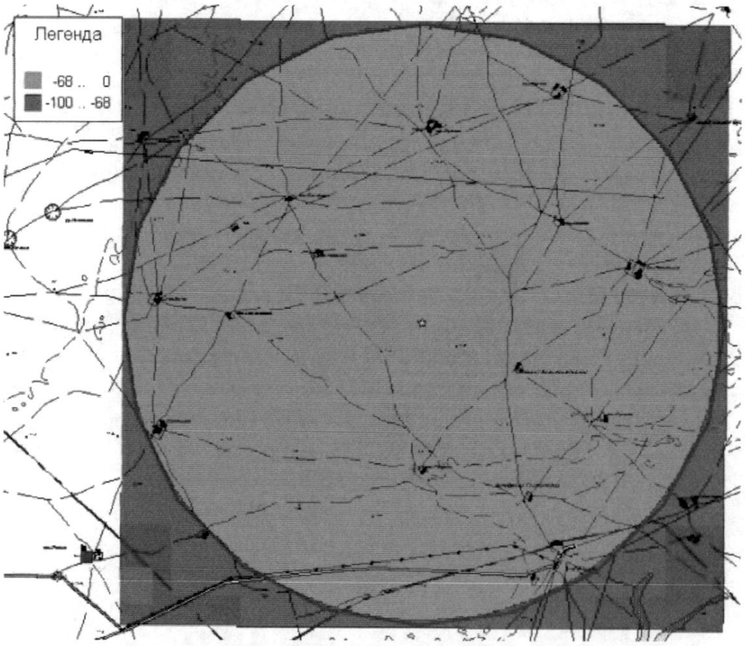

Figure 8. Coverage map.

Using the most efficient equipment configuration, MDMS equipment can provide communication at a speed of up to 8 Mbit / s to a terminal with a coverage radius of 8.4 km. In this configuration, it is possible to install up to 20 terminals per sector, since the maximum speed is achieved using 4 E1 channels.

6. Conclusions

As a result of the study, the following results were obtained: new functioning equipment scheme was selected for organizing a new mobile communication system (mobile city). The selected equipment contains such elements as: solar panels, diesel generator units mounted on the chassis, mobile antenna-mast devices, radio relay communication stations, mobile base stations of space communication. Calculations were made by an automated software program on a digital map of the Astrakhan region. As a result of these findings and calculations we can confirm a possibility of a fast-functioning system of cellular communication in remote areas with radius coverage of up to 25 kilometers on open (semi-open) surface with consideration of the reach attainable by basic and radio-relaying stations.

References

[1] *Snowshoe GAZ-34039 Irbis Premium* https://rus-ap.ru/umg/snegobolotohody/snegobolotohod-gaz-34039-irbis.html?region =/region/ulyanovskaya-oblast

[2] Pishchin O N and Perova K V 2021 New possibilities in modern radio planning *Proc. of the IV interregional webinar "Radio frequency planning and electromagnetic compatibility"* pp 18-23

[3] Pishchin O N, Shavandasht A D and Nikulin V V 2020 Frequency support of mobile mobile communication facilities *"Ogarev-on-line" Electronic periodical for students and graduate students* **15** p 10

[4] *Broadband Wireless Subscriber Radio Access System Marconi DMS* http://telecomnetworks.ru/datadocs/doc_191hi.pdf

[5] MDMS V4.0 - Technical Overview. *Customer Information*

[6] Pishchin O N, Bestaeva N V, Zubova A D and Orlova A A 2017 Analysis of the model of propagation of radio waves over the water surface and their use in calculating electromagnetic field levels in mobile radio communication systems. *Caspian magazine. Management and high technology* **3** pp 121-130

[7] Naryshkin M I 2017 Antennas for transportable base stations for mobile radio based on structures with complex configuration *Antennas* **2** pp 14-21

[8] Vishnevsky V M, Vytovtov K A and Barabanova EA 2020 Model of Navigation and Control System of an Airborne Mobile Station *Lecture Notes in Computer Science* LNCS **12563** pp 643–57

[9] Vasilenko G O 2008 Assessment of attenuation of signals of mobile communication networks on short lines of sight *Electronics: Science, Technology, Business* **4** p 72-74

[10] Mogensen P E, Eggers P, Jensen C and Andersen J B 1991 Urban Area Radio Propagation Measurements at 955 and 1845 MHz for Small and Micro Cells *Proc. of IEEE Global Commun. Conf. (GLOBECOM)* pp 1297-1302

[11] Popov V 2003 UNF radio wave propagation through woodlands in cellular mobile communication systems *44nd. International Scientific Conf.*

[12] Popov V, Chaiko A, Homicky V and Mogorit N 2001 Effective complex dielectric permittivity of forest media for Radio waves *Latvian Journal of Physics and Tehnical Sciences* **2** p 46-50

The method for constructing fault-tolerant photonic switches for high-performance computing systems

E A Barabanova, K A Vytovtov, V M Vishnevsky and V S Podlazov

V. A. Trapeznikov Institute of Control Sciences of Russian Academy of Sciences
65 Profsoyuznaya street, Moscow 117997, Russia

E-mail: elizavetaalexb@yandex.ru

Abstract. In this paper the new type of fault-tolerant non-blocking photonic switch is presented for the first time. The proposed switch architecture is based on quasi-complete graph topology which use provides non-blocking and fault-tolerant switching process. The new two-stage switch architecture uses the stage of dual photonic switches and pairs of photonic demultiplexers and multiplexers which have been described in detail by authors in their previous works. Depending on the number of different backup connections, the two types of fault-tolerant pho-tonic switches are considered in this paper: single-channel fault-tolerant photonic switch and dual-channel fault-tolerant photonic switch. The mathematical expressions for calculating the switching and fiber complexities of these two types of fault-tolerant photonic switches are also presented here for the first time. The numerical calculations shown that the increasing the reliability of the fault-tolerant photonic switches twice leads to an increasing their switching complexity in 1.4 times and fiber complexity in 1.8 times.

1. Introduction

Scalable photonic switching systems are the perspective direction of high-performance interconnection development for Exascale supercomputer systems [1-13]. This is because that photonic switches have a lot of advantages over electronic ones. They provide high speed, low latency and low-power communication between optical [14] or electron microprocessors [15-16] and must have non-blocking, reliable and scalable architecture. In a number of recent works, various types of photonic switches have been presented. For example, a single-stage photonic switching architecture which provides nanosecond switching process has been proposed in [2]. Lee et al. propose the new design of switching fabric that decreases the crosstalk effect [4]. Most research works are focused on conventional switch topologies such as matrix scheme, Clos network, Banyan switch, etc. However, these topologies are not suitable for constructing non-blocking scalable photonic switching systems.

The method for constructing non-blocking photonic switches for optical super-computer and communication systems has been proposed by the authors of this paper in [9-13]. The main functioning principle of these switches is the dual method of conflict resolution based on the combination of time and space optical signal division combination. Based on the authors' proposals, dual photonic switches based on 4×4 [9] and 8×8 [10] elements have been realized in low complexity. These only contain one physical channel between the input and the output of the switch. Besides that, the proposed photonic switches use the method of parallel transmission of in-formation and control signals at different frequencies [13]. Note that combinations of information and control optical signals are separated in time. So, these methods provide non-blocking switching process without an external control [11]. However, the use of optical signal division and parallel transmission of optical signals for design of high capacity photonic switches led to a decrease in their speed [9-11]. In results, the performance of 256×256 photonic switches photonic switches is twelve times faster than the 16×16 ones. To solve this problem the authors proposed to use the method of invariant expansion of system networks which have been used for design electron-ic switching systems [12,17]. This approach is based on the use of quasi-complete digraph topology [12,17] and on an extending scheme base consisting of p×p switch-es, p×1 multiplexers and p×1 demultiplexers. So, this method has provided the balance of complexity and performance of scalable photonic switches. Our last development in this direction is a two-stage photonic switch based on quasi-complete di-graph topology and provided higher performance and less number of stages with respect to [12] with our own switching architectures [9-11].

The first scheme contains one additional physical communicating channel and is called fault-tolerant single-channel. The other one contains two backup physical channels, and is called fault-tolerant dual-channel switches.

Besides low complexity and high performance a very important problem for practical use of photonic switching systems is their reliability that has not been considered yet. Here we propose a new method for fault-tolerant photonic switch design based on increasing number of communicating channels between each pair of inputs and outputs. We consider two types of fault-tolerant photonic switches. The first scheme contains one additional physical communicating channel and is called fault-tolerant single-channel. The other one contains two backup physical channels, and is called fault-tolerant dual-channel switches. For each of the schemes we computed the switching and the fiber complexities and the results of the numerical calculations showed that increasing the number of backup channels increases the complexity of photonic switches. To the best of our knowledge, this is the first time that the switching and the fiber complexities of new fault-tolerant photonic switches and photonic switches without backup communicating channels are compared.

The principle of constructing of the new non-blocking photonic switches based on the quasi-complete graph topology is described in Section 2. The schemes and formulas for calculation of switching and fiber complexities of non-blocking two-stage single-channel and dual-channel fault-tolerant photonic switches are presented in Sections 3 and 4. Section 5 reports on the comparison of proposed schemes.

2. The non-blocking photonic switch based on the quasi-complete graph topology

The non-blocking photonic switch based on quasi-complete graph topology QCG (N_1, p, σ) [12,17] consists of the $p \times p$ dual switches (DS), the $1 \times p$ demultiplexers (D) and the $p \times 1$ multiplexers (M). Here $N_1 = p(p - 1) / \sigma + 1$ is the number of inputs of the non-blocking photonic switch, p is the number of inputs of the basic $p \times p$ dual switch DS and σ is the number of duplex channels between any two processors through different $p \times p$ dual switches DS.

The QCG-graph (N, p, σ) is isomorphic to such a mathematical object as an incomplete balanced block diagram B (N, p, σ). It contains N blocks and N elements, which are arranged in blocks so that each block contains p different elements, each element is included in exactly different blocks, and each pair of elements is included in exactly σ blocks. Moreover, B (N, p, σ) defines the maximum N for given p and σ.

The scheme of photonic switch based on quasi-complete graph QCG (7, 4, 2) with seven duplex inputs containing seven 4×4 dual switches DS4, seven pairs of 1×4 demultiplexers DM4 and 4×1 multiplexers M4 is presented in figure 1.

In this case, single-channel fault tolerance is provided by the presence of two different optical paths through the different switches 4×4 DS4 between the input 1×4 demultiplexers D4 and the output 4×1 multiplexers M4.

In general case, (σ - 1) -channel fault tolerance is provided by the presence of σ different optical paths through the different $p \times p$ dual switches between any input $1 \times p$ demultiplexers and any output $p \times 1$ multiplexers.

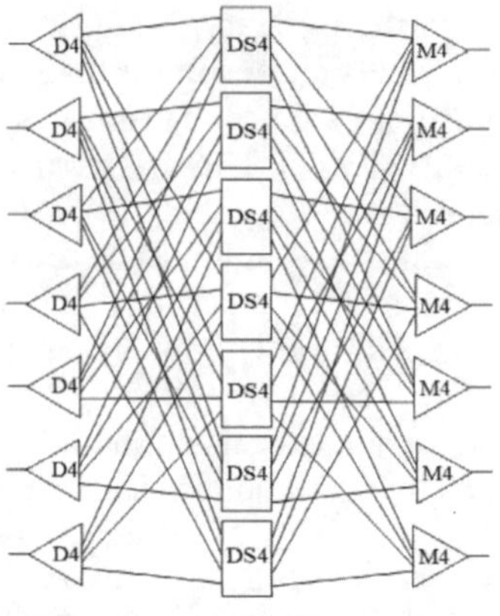

Figure 1. The non-blocking switch based on QCG (7, 4, 2).

3. The non-blocking two-stage single-channel fault-tolerant photonic switch

Let us consider the principle of constructing single-channel fault tolerance photonic switches using the example of the simplest scheme of quasi-complete *2×2* photonic switch (PS_12) which can be described by the model of quasi-complete graph QCG (2,2,2) presented in figure 2.

First of all, it must be constructed the *4×4* two-stage blocking switching system BS_24 on based of four QCG (2,2,2) schemes presented in figure 3. Note that this system is blocking because there are conflicts in the *2×1* multiplexers of the first stage highlighted by filling. Besides this system is not single-channel fault tolerance. Then the switching system BS_24 must be transformed into the non-blocking switch NS_24 by using two copies of the second stage and applying the method of internal parallelization considered in Figure 4. This transformation is carried out by removing all *2×1* multiplexers from the first stage and distributing their input channels into two copies of the second stage. Cut out *2×1* multiplexers combine the output channels of two QCG (2,2,2) schemes of the second stage and corresponding channels of their two copies. In this case, any optical signal arrived to the input of the system can pass to its output through two different paths, which provide single-channel fault tolerance. Besides the optical signals going to the inputs of the PS_12 are time divided so there no any conflicts in the system and the non-blocking property of the system is persisted.

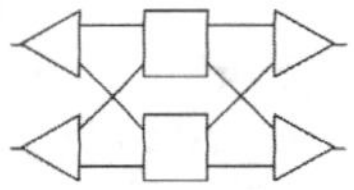

Figure 2. The *2×2* photonic switch (PS_12).

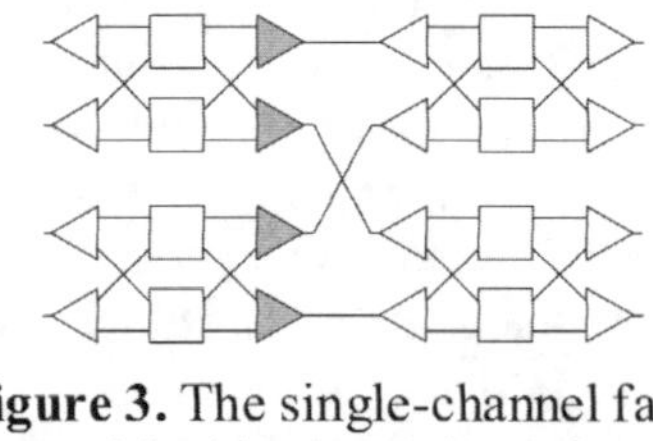

Figure 3. The single-channel fault-tolerant blocking *4×4* photonic switch (BS_24).

In the general case, $p > 2$, $\sigma = 2$ and the single-channel fault-tolerant photonic switching systems with N_2 inputs can be constructed on the base of graph topology QCG (N_1, p, 2). Let us consider the *4×4* photonic switching system which can be described by the minimal non-trivial quasi-complete graph QCG (4, 3, 2) presented in figure 5. It consists of four *3×3* dual photon switches in the central stage, four *1×3* input multiplexers and four *3×1* output demultiplexers.

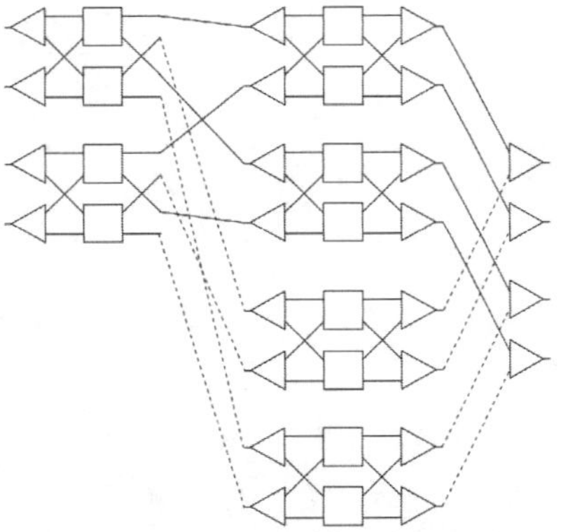

Figure 4. The single-channel fault-tolerant non-blocking 4×4 photonic switch (NS_24).

The two-stage non-blocking switching system NS_216 with number of inputs $N_2 = N_1^2 = 16$ is based on the 4×4 switches NS_14. The NS_216 is constructed in an analogous above principle and presented in figure 6. The first stage of the NS_216 consists of four NS_14 without output 3×1 multiplexers. The second stage of NS_216 consists of $p = 3$ copies of switch groups of four NS_14 in each group. And the output stage of NS_216 consists of N_2=16 output multiplexers.

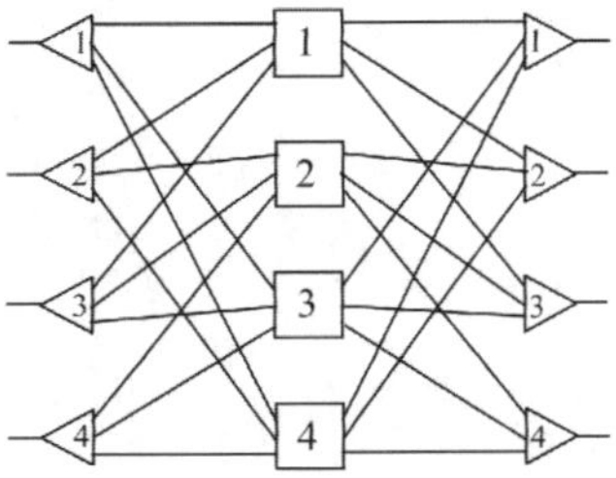

Figure 5. The single-channel fault-tolerant non-blocking 4×4 photonic switch (NS_14).

To determine how an increase in the reliability of the new type of switches will affect their complexity, let us define mathematical relations for calculating their switching and fiber complexities. The concepts of switching and fiber complexity for photonic switches were introduced by the authors earlier [10,12]. The switching complexity is equal to the total number of switching components of the scheme and the fiber complexity is equal to the total number of fiber connections between stages of the system.

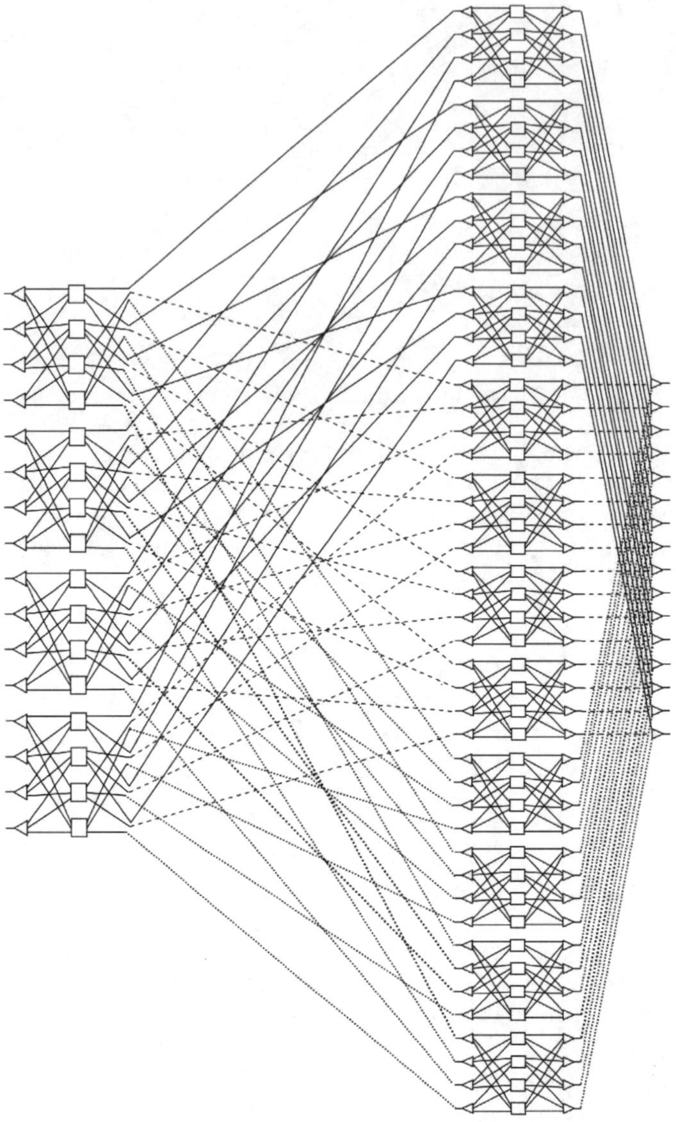

Figure 6. The single-channel fault-tolerant 16×16 photonic switch ($NS_2 16$).

The switching complexity of the $p\times p$ dual switch is determined by the formula $S_0 = 2p^2$ [12]. Then the expression for the switching complexity of the photonic switch based on the QCG (N_1, p, σ) has the form $S_1 = N_1S_0 + 2pN_1$, while the complexity of the quasi-complete switch based on QCG (N_1, p, σ) without external multiplexers is determined by the formula $S_1' = N_1S_0 + pN_1$. Thus, the complexity of the two-stage photonic switch based on the quasi-complete graph will be $S_2 = S_1' N_1 + pS_1N_1 + pN_1^2$. Since $N_2 = N_1^2$, then $S_2 = 2N_2 (p^3 + 2p^2 + p)$. It has been indicated above that for the considered type of photonic switches $N_1 = p(p-1)/\sigma + 1$. For this case, $\sigma = 2$ and $N_1 = p(p-1)/2 + 1$. Expressing from the last formula p, we obtain the final expression for the switching complexity of the single-channel fault tolerance non-blocking photonic switch depending on the number of inputs N_2:

$$S_2 = 2N_2(\sqrt{N_2}(8\sqrt{N_2}-7)^{1/2} + 7\sqrt{N_2} + (8\sqrt{N_2}-7)^{1/2} - 5) \qquad (1)$$

Taking into account that the expression for the fiber complexity of the photonic switch based on QCG (N_1, p, σ) has the form $L_1 = 2pN_1$, the fiber complexity of the photonic switch based on QCG (N_1, p, σ) without external multiplexers $L_1' = 2pN_1$ and the total fiber complexity of the two-stage switch $L_2 = L_1' N_1 + pL_1N_1 + pN_1^2$, we obtain the fiber complexity of the single-channel fault tolerance photonic switch depending on the number of inputs N_2:

$$L_2 = 0.5N_2(1+(8\sqrt{N_2}-7)^{1/2})(4+(8\sqrt{N_2}-7)^{1/2}) \quad (2)$$

4. The non-blocking two-stage dual-channel fault-tolerant photonic switch

For constructing the dual-channel fault tolerance photonic switches it is necessary to use the dual switches based on the QCG ($N_1, p, 3$). The smallest type of the QCG ($N_1, p, 3$) is QCG (3, 3, 3). Let us consider the principle of constructing the dual-channel fault-tolerant photonic switches based on the graph QCG (5, 4, 3) presented in figure 7.

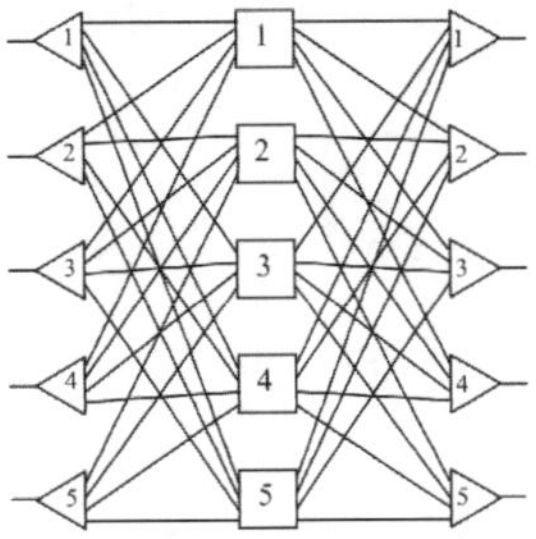

Figure 7. The dual-channel fault-tolerant *5×5* photonic switch (NS_15).

The dual-channel fault-tolerant photonic switching system NS_225 with number of inputs $N_2 = N_1^2 = 25$ is based on the *5×5* switches NS_15. In this case the first stage of NS_225 consists of five *5×5* switches NS_15 without output *4×1* multiplexers. The second stage of NS_225 consists of $p = 4$ copies of switch groups of $N_1 = 5$ complete NS_15 in each group. The output stage consists of N_2=25 output *4×1* multiplexers.

In the general case, for $p \geq 3$ and $\sigma = 3$, the two-stage dual-channel fault-tolerant photonic switch with number of inputs $N_2 = N_1^2$ has analogous constructing principle. So, in the above described way, by means of internal parallelization using p copies of the second stage, NS_1N_1 is converted into the two-stage photonic switch NS_2N_2, which has two additional channels for fault tolerance.

The property of (σ - 1) -channel fault tolerance of the proposed switches is provided by $p \geq \sigma$ copies of two-stage dual photonic switch groups in the second stage. So, in other words fault tolerance is provided by the additional optical paths between input demultiplexers and output multiplexers.

Considering that for $\sigma = 3$ $N_1 = p(p-1)/3 + 1$ the expressions for the switching and fiber complexities of this type of photonic switches are as follows:

$$S_2 = 0.5 \cdot N_2(1+(12\sqrt{N_2}-11)^{1/2})(6\sqrt{N_2}-1+3(12\sqrt{N_2}-11)^{1/2}) \quad (3)$$

$$L_2 = 0.5 \cdot N_2(1+(12\sqrt{N_2}-11)^{1/2})(4+(12\sqrt{N_2}-11)^{1/2}) \quad (4)$$

5. Comparative analyses

This section provides a comparative analysis of the three types of dual photonic switches. The first type is the two-stage dual photonic switch [12]; the second type is single-channel fault tolerance photonic switch and the third one is the dual-channel fault tolerance photonic switch. The expressions for switching and fiber complexities of the two types fault tolerance photon switches are presented above.

The corresponding expressions for the two-stage dual photonic switches have been presented in [12]. Here they are written in the following form

$$S_2 = 2N_2^{4/5}(N_2^{1/4}+1)^2 \quad (5)$$

$$L_2 = N_2^{4/5}(2N_2^{1/4}+3) \quad 6)$$

Figures 8a and 8b show the dependences of switching and fiber complexities on the number of inputs for three types of switches.

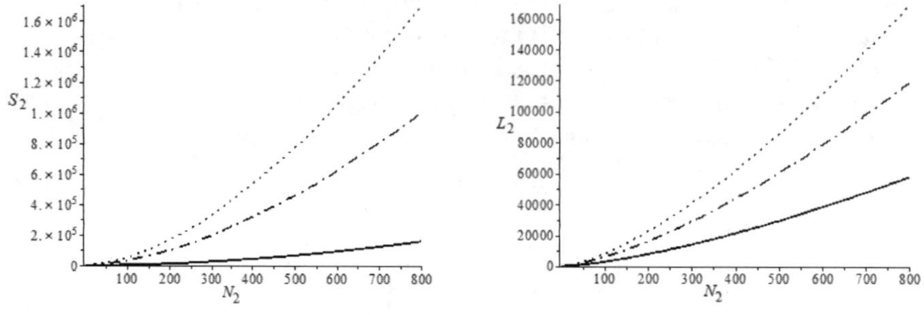

a) b)

Figure 8. The dependences of a) switching and b) fiber complexities of the dual photonic switches on the number of inputs.

The first type is the dual two-stage photonic switch (solid line); the second type is the single-channel fault tolerance dual photonic switch (dash-dotted line). The third type is the dual-channel fault tolerance dual photonic switch (dashed line). The graphs are built according to formulas (1)-(6) respectively.

It should be noted that according to the graphs in figure 8a and figure 8b, fault tolerance two-stage non-blocking switches have higher switching and fiber complexities in comparison with two-stage photonic switches which do not have the fault tolerance property [12]. This difference becomes noticeable when the number of inputs is more than 144. For example, for $N_2 = 800$ the switching complexity of the dual-channel fault tolerance photonic switches in 1.8 time more than switching complexity of the single-channel fault tolerance photonic switches and in 11 time more than switching complexity of the photonic switches which do not have the fault tolerance property. And for this number of inputs the fiber complexity of the dual-channel fault tolerance photonic switches in 1.4 time more than switching complexity of the single-channel fault tolerance photonic switches and in 3 time more than switching complexity of the photonic switches which do not have the fault tolerance property.

Figure 9 shows the dependences of the switching and fiber complexities on the parameters and p and σ.

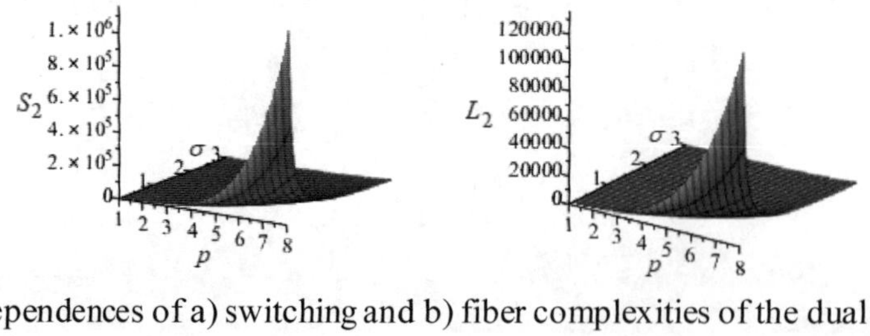

Figure 9. The dependences of a) switching and b) fiber complexities of the dual photonic switches on the values of p and σ.

Based on the results of the analysis it can be concluded that both the switching and fiber complexities increase at $p > 4$, and an increase in σ does not significantly affect the complexity of the proposed fault-tolerance switching systems.

6. Conclusion

We have proposed a new type of non-blocking two-stage photonic switches which have $(\sigma - 1)$-channel fault tolerances property. Replacing the basic dual switch based on the quasi-complete digraph topology [5] with the dual switch based on the quasi-complete graph topology increases the efficiency of these switches. In particular, the new types of fault tolerant photonic switches have more than one physical communicating channels between each input and each output, which is not the case with the previous types of photonic switches [9-13].

Each additional channel can be used when the main channel fails. As expected, in-creasing the number of backup channels increases the fiber and the switching complexities of the new types of switches. We have also presented for the first time formulas for calculating these characteristics for the new photonic switches and their graphical dependences on the number of inputs.

7. References

[1] Cheng Q, Glick M and Bergman K 2019 Optical Interconnection Networks for High Performance Systems *Optical Fiber Telecommunications* **7** pp 785–825

[2] Maniotis P, Dupuis N, Kuchta D M, Taubenblatt M A and Lee B G 2020 Intra-node high-performance computing network architecture with nanosecond-scale photonic switches *Journal of Optical Communications and Networking* **12** 12 pp 367-77

[3] Duro J, Petit S, Gómez M E and Sahuquillo J 2021 Segment Switching: A New Switching Strategy for Optical HPC Networks *IEEE Access* **9** pp 43095-106

[4] Lee B G, Dupuis N, Pepeljugoski P, Schares L, Budd R, Bickford J R and Schow C L 2015 Silicon photonic switch fabrics in computer communications systems *Journal of Lightwave Technology* **33** 4 pp 768-77

[5] Krishnamoorthy A V, Ho R, Zheng X, Schwetman H, Lexau J, Koka P, Li G, Shubin I and Cunningham J E 2009 Computer systems based on silicon photonic interconnects *Proceedings of the IEEE* **97** 7 pp 1337-61

[6] Cheng Q, Rumley S, Bahadori M and Bergman K 2018 Photonic switching in high performance datacenters *Optics express* **26** 12 16022

[7] Meyer H, Sancho J C, Mrdakovic M, Miao W and Calabretta N 2018 Optical packet switching in HPC: an analysis of applications perfomance *Future Generation Computer Systems* **82** pp 606-16

[8] Seok T J, Kwon K, Henriksson J, Luo J and Wu M C 2019 Wafer-scale silicon photonic switches beyond die size limit *Optica* **6** 4 pp 490-94

[9] Barabanova E A, Vytovtov K A, Maltseva N S, Kravchenko O V and Kravchenko V F 2019 Models and algorithms of optical switching systems with decentralized control *2019 IEEE Conference of Russian Young Researchers in Electrical and Electronic Engineering* 8657063 pp 64-8

[10] Barabanova E, Vytovtov K and Podlazov V 2019 Model and algorithm of next generation optical switching systems based on 8×8 elements *LNCS* **11965** pp. 58-70

[11] Barabanova E A, Vytovtov K A and Nguyen T T 2019 The control system elements of the new generation optical switching cell *Journal of Physics: Conference Series* **1368** 2 022002

[12] Barabanova E A, Vytovtov K A, Podlazov V S and Vishnevsky V M 2020 High-capacity photon switching systems based on the two-stage 256 × 256 switch *CCIS* **1337** pp 57-69

[13] Vytovtov K A, Barabanova E A and Zouhdi S 2018 Optical Switching Cell Based On Metamaterials And Ferrite Films. *Proc. 12th International Congress on Artificial Materials for Novel Wave Phenomena – Metamaterials* (Espoo) pp 424-26

[14] Stepanenko S 2019 Structure and Implementation Principles of a Photonic Computer *EPJ Web of Conferences* **224**, 04002
[15] Kutuzov D, Stukach O, Starov D and Osovsky A 2020 Decentralized Control of Switching Elements in the Intra-Chip Network-on-Chip Routers *Proc. 2020 Moscow Workshop on Electronic and Networking Technologies* (Moscow) pp 1-4
[16] Kutuzov D 2009 Parallel switching on the coordinate switchboard IMS C004 *Proc. 2009 International Siberian Conference on Control and Communications* (Moscow) pp 255-58,
[17] Karavai M F and Podlazov M F 2010 An invariant extension method for system area networks of multicore computational systems An ideal system network *Automation Remote Control*, **71** 12, 2644–54

Mathematical model for reliability indicators calculation of tethered UAV hybrid navigation system

V M Vishnevsky, K A Vytovtov, E A Barabanova and V E Buzdin

V. A. Trapeznikov Institute of Control Sciences of Russian Academy of Sciences 65 Profsoyuznaya street, Moscow 117997, Russia

E-mail: vytovtov_konstan@mail.ru

Abstract. The mathematical model for reliability indicators calculation of the hybrid navigation system containing microwave and technical vision subsystems is proposed in this paper for the first time. The proposed method is based on the translation matrix concept of solutions to the Kolmogorov equation system and it allows us to obtain the mathematical expression of availability factor, downtime ratio, and other reliability indicators. Also the presented approach allows finding the reliability indicators for the cases of jump change of transition intensities caused by external influences. Besides the analytical method can be used for investigation of hybrid navigation system transient mode functioning. The results of the numerical calculations clearly demonstrated correctness of the proposed approach.

1. Introduction

Unmanned aerial vehicles (UAVs) are widely used in various fields of economics, industry and military [1]. The disadvantage of most UAVs using is necessarily of recharging of a rechargeable battery, which is not always which does not always meet the requirements of providing uninterrupted high-quality communication. For example, in the case of placing a mobile base station or an equipment of video surveillance system on UAV it is important problem to provide long working time of it. In such cases, the so-called tethered UAVs are widely used, which receive power from a generator located at the ground station via an electric cable [1-3]. At the present time the papers describing the features of building communication systems on base of tethered UAVs, the features of power supply of such a system, the functioning principles of autonomous navigation system are presented in scientific literature [1-5]. For example, the paper [2] presents the features of building an autonomous navigation system based on radar methods. The proposed navigation method [2] allows us to accurately determine the coordinates of UAV at an altitude of 10 meters and above. But at altitudes up to 10 meters, the accurate of the navigation system is insufficient and vision-based navigation systems are more are more suitable for solving this problem [6]. So, a hybrid navigation system based on radar and vision-based navigation systems is the most effective solution. But there are some cases when such a hybrid navigation system may be in unsteady state. For example, the process of take-off or landing of an UAV leads to switching from one navigation system to another. In such cases the system is in a transient mode. During this period, the reliability indicators of the hybrid navigation system can differ significantly from the parameters in the stationary mode. Therefore, the actual problem is calculating the reliability indicators of a hybrid navigation system in steady and transient mode.

2. The statement of the problem

Let us consider the hybrid navigation system consists of the two subsystems based on different principals. The first one is the radar subsystem of the terahertz domain and the second one is the vision-based navigation subsystem. Such the system has been proposed and described recently in [7]. However, the problem of reliability of its functioning has not been studied in [7]. First of all the probability of operation of each of the subsystems depends on the height. This probability is also determined by the technical characteristics of the noted above subsystems. The main navigation system is the radar subsystem at an altitude above 10m, and the main navigation system is the technical vision subsystem at an altitude of less than 10m. It should be noted that when the UAV is raised above 10 meters, the vision-based navigation subsystem does not turn off, it is switched in a cold standby mode. Analogously, the radar navigation subsystem transits on in the reserve mode after the UAV is descended below the level of 10 meters. It is obvious that the conditions for switching on one or the other subsystem are very important both theoretical and practical problems.

The aim of this treatment is investigation of reliability indicators in transient mode which accrues after switching from one subsystem to another. Such hybrid system can be described by using the theory of Markov processes [8,9]. Figure 1 shows the transition graph of the corresponding Markov chain. In our case the Markov chain consists of the four states. The first state $S_1(t)$ is the one of functioning of the radar navigation subsystem only [2], the second state $S_2(t)$ is the one of switching from the radar navigation subsystem to the technical vision navigation subsystem, $S_3(t)$ is the state of functioning only of the technical vision subsystem [5], and $S_4(t)$ is the state of switching from the technical vision navigation subsystem to the radar navigation subsystem. The intensities of transitions $\lambda_{ij}(t)$ are generally functions of altitude, time, and many other parameters characterizing the navigation system. A change of transition intensities is associated with a change of external weather conditions during take-off or landing of the UAV. Moreover an altitude is also function of t. For the possibility of obtaining the analytical expressions in this work, a piecewise-constant approximation of the original dependencies is used. Then we can write

$$\lambda_{ij}(h,t)=\begin{cases}\lambda_{ij}^{1}, & 0<t<t_1\\ \lambda_{ij}^{(2)}, & t_1<t<t_2\\ \dots\dots\dots\dots & \\ \lambda_{ij}^{(N)}, & t_2<t<t_N\end{cases} \tag{1}$$

Thus, first of all we must find the dependences of the state probabilities on the time t and obtain the the mathematical expression of availability factor, downtime ratio, and other reliability indicators by using the state probabilities.

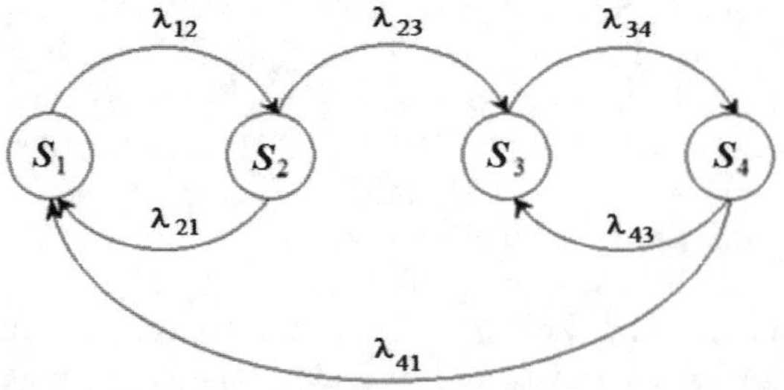

Figure 1. Transition graph of a Markov stochastic process.

We also intend to numerically investigate the behavior of the system with large fluctuations in altitude and the effect of small fluctuations in altitude when landing the drone on the system reliability indicators.

2.1. Translation matrix

At first let us to write the Kolmogorov equation system corresponding to the transition diagram presented in figure 1:

$$\begin{cases} \frac{dp_1}{dt} = -p_1(t)\lambda_{12}(t) + p_2(t)\lambda_{21}(t) + p_4(t)\lambda_{41}(t) \\ \frac{dp_2}{dt} = p_1(t)\lambda_{12}(t) - p_2(t)[\lambda_{21}(t) + \lambda_{23}(t)] \\ \frac{dp_3}{dt} = p_2(t)\lambda_{23}(t) - p_3(t)\lambda_{34}(t) + p_4(t)\lambda_{43}(t) \\ \frac{dp_4}{dt} = p_3(t)\lambda_{34}(t) - p_4(t)[\lambda_{43}(t) + \lambda_{41}(t)] \end{cases} \quad (2)$$

Here p_i is the probability of the system being in state i.

To find the reliability indicators in this work, it is proposed to use the method of the so-call translation matrix of solutions to the Kolmogorov system (2). This matrix $\mathbf{L}(t)$ relates the probabilities of the states of the system at an arbitrary time t with these probabilities at the initial time.

$$\mathbf{P}(t) = \mathbf{L}(t)\mathbf{P}(t_0) \ . \quad (3)$$

The various methods for finding the translation matrix $\mathbf{L}(t)$ are described in mathematics, radio engineering and microwave theory [7] in detail, therefore, we do not consider this issue in detail here. It is important that the translation matrix has the same order as the Kolmogorov system. In this case it is a 4x4 matrix.

In the case of the abrupt change transition intensities, the translation matrix after the jump is found as the multiplication of the matrix of the first interval and the translation matrix of the second interval

$$\mathbf{L}(t) = \mathbf{L}_2(t)\mathbf{L}_1(t_1) \ . \quad (4)$$

Analogously, in the case of N jumps the behavior of the system on the i-th time interval is described by the matrix

$$\mathbf{L}(t) = \mathbf{L}_{N+1}(t - \sum_{i=1}^{N} t_i)\prod_{i=N}^{1}\mathbf{L}_i(t_i) \quad (5)$$

The matrix (5) allows us to determine the state probabilities (figure 1) at any time and it can be used for obtaining the reliability parameters of the hybrid navigation system.

2.2. Reliability parameters

Since the availability factor at an arbitrary time moment t is equal to the sum of the probabilities that either the radar navigation subsystem or technical vision navigation subsystem is working, then we can write $R(t) = p_1(t) + p_3(t)$. And taking into account the kind of the fundamental matrix, we can write the following expression

$$R(t) = \sum_{i=1}^{4}[L_{1i}(t) + L_{3i}(t)]p_i(t_0) \ , \quad (6)$$

for the availability factor, where $L_{1i}(t)$ и $L_{3i}(t)$ are the elements of the fundamental decision matrix.

Analogously the downtime ratio at an arbitrary time moment t is equal to the sum of probabilities that both subsystems are not functioning and the system is in switching mode. Taking into account this fact we can write the expression for this parameter

$$Q(t) = \sum_{i=1}^{4} \left[L_{2i}(t) + L_{4i}(t)\right] p_i(t_0) \tag{7}$$

The important problem for investigation of hybrid navigation system reliability in transient mode is estimation of transient time. The transient time is the time period from the beginning of the system functioning to the transition it in the stationary mode when the probabilities of system states do not depend on time. Now, the general solution of (2) can be written in the form

$$p_i(t) = \sum_{k=1}^{4} A_{ik} \xi_{ik} \exp(\gamma_k t) \tag{8}$$

As we can see from (8) the transient time is determined by γ_k. Let us define the transient time as $\tau_{tr} = 5/\gamma_{k\min}$. Indeed, the probability $p_i(t)$ changes e times for the time $1/\gamma_k$, and this probability changes $5e$ times for the time τ_{tr}. Moreover, we have four values γ_k, and the maximal transient time is determined by the minimal value of γ_k.

3. Numerical results

In this section the numerical results of the hybrid navigation system reliability investigation are presented. Here we consider two cases of switching from one navigation system to another. The first one occurs during take-off or landing of UAV and the second one is the result of its height oscillation process.

3.1. The periodically jumps of transient intensities

The periodically jumps of transient intensities occurs when the UAV is on the determinate altitude but its position is changed in the result of wind or other external factor influences. For example, the case of large height oscillations is considered in figure 2. Here the altitude of UAV position can change within one meter. In other words, we are studying the fluctuations in the height of the drone.

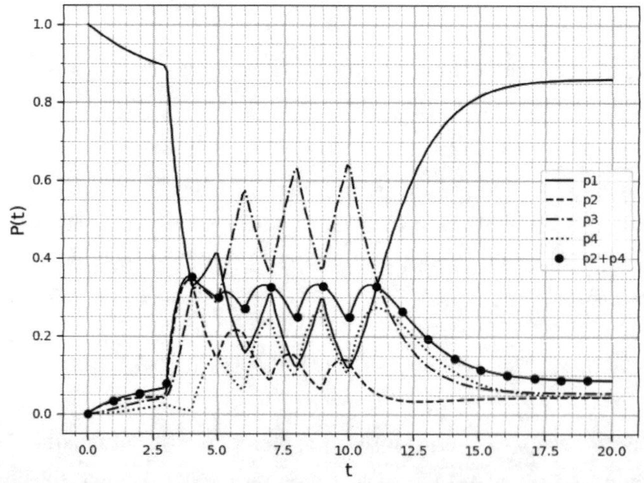

Figure 2. Dependencies of hybrid navigation system state probabilities.

The values of transient intensities before the jump are taken as $\lambda_{12} = 0.05$, $\lambda_{13} = 0$, $\lambda_{14} = 0$, $\lambda_{21} = 0.1$, $\lambda_{23} = 0.9$, $\lambda_{24} = 0$, $\lambda_{31} = 0$, $\lambda_{32} = 0$, $\lambda_{34} = 0.8$, $\lambda_{41} = 0.9$, $\lambda_{42} = 0$, $\lambda_{43} = 0.1$. The values of transient intensities after the jump are $\lambda_{12}=1$, $\lambda_{13}=0$, $\lambda_{14}=0$, $\lambda_{21}=0$, $\lambda_{23}=0.9$, $\lambda_{24}=0$, $\lambda_{31}=0$, $\lambda_{32}=0$, $\lambda_{34}=0.00002$, $\lambda_{41}=0.9$, $\lambda_{42}=0$, $\lambda_{43}=0.1$. Here we consider the particular case of the eight jumps. The numerical results show that the probability of failure or another words downtime ratio of the hybrid navigation system is changed from 0.25 to 0.35, the availability factor of radar navigation subsystem is changed from 0.08 to 0.35 and the availability factor of technical vision-based navigation subsystem is changed from 0.05 to 0.65. Thus the fluctuations in the height of the drone give the significant decrease in the reliability of the system under consideration. At the same time, the probability of the functioning of the technical vision subsystem increases. The transient time in accordance with the figure 1 is 19 second.

3.2. Investigation of hybrid navigation system reliability in transient mode

The transient mode of UAV hybrid navigation system functioning occurs during its take-off or landing. During landing of UAV the radar navigation subsystem is switching on the vision-based navigation one. And vice versa during take-off the vision-based navigation subsystem of UAV is switching on the radar navigation subsystem. The dependencies of hybrid navigation system state probabilities in transient mode are presented in figure 3. Here the case of transient mode setting in the hybrid navigation system is considered.

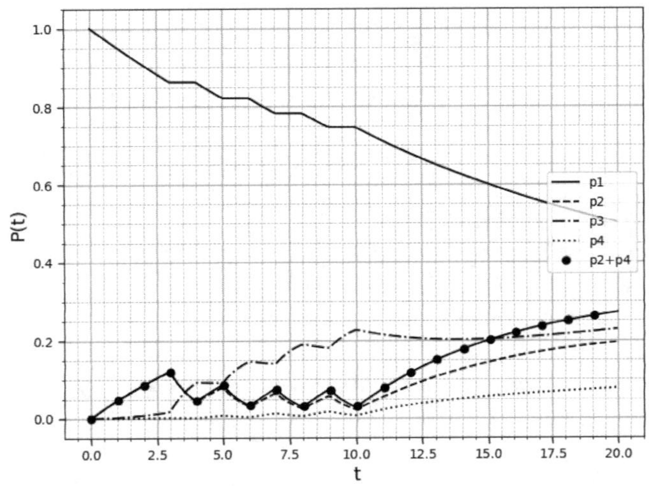

Figure 3. Dependencies of the hybrid navigation system state probabilities during the UAV landing.

In this mode small jumps of transition intensities occur until the system is in stationary mode. For example we consider the transient mode for following transition intensities before the jump $\lambda_{12}=0.05$, $\lambda_{13}=0$, $\lambda_{14}=0$, $\lambda_{21}=0.01$, $\lambda_{23}=0.9$, $\lambda_{24}=0$, $\lambda_{31}=0$, $\lambda_{32}=0$, $\lambda_{34}=0.08$, $\lambda_{41}=0.09$, $\lambda_{42}=0$, $\lambda_{43}=0.1$ and after the jump $\lambda_{12}=0.0001$, $\lambda_{13}=0$, $\lambda_{14}=0$, $\lambda_{21}=0$, $\lambda_{23}=1$, $\lambda_{24}=0$, $\lambda_{31}=0$, $\lambda_{32}=0$, $\lambda_{34}=0.00002$, $\lambda_{41}=0.00001$, $\lambda_{42}=0$, $\lambda_{43}=0.99999$. The results show that the probability of radar subsystem functioning is gradually decreasing from 1 to 0.5 and the probability of vision-based navigation subsystem functioning is gradually increasing from 0 to 0.23. The transient time is approximately 20 second.

4. Conclusion

Accurate analytical investigation of tethered UAV hybrid navigation system reliability is presented in this paper for the first time. The consider navigation system consists of radar subsystem and vision-based navigation subsystem. The mathematical method used for investigation is based on the translation matrix of the Kolmogorov equation system and it allows us to obtain the mathematical expression for the availability factor and for the downtime ratio for cases of abruptly changing of transition intensitiesis. Besides the method can be used for investigation of transient mode of the UAV hybrid navigation system reliability. The numerical calculations clearly demonstrate the proposed approach.

Acknowledgments
The reported study was funded by RFBR, project numbers 19-29-06043, 20-37-70059.

References

[1] Boukoberinea M N, Zhoub Z and Benbouzid M 2019 A Critical Review on Unmanned Aerial Vehicles Power Supply and Energy Management: Solutions, Strategies, and Prospects *Applied Energy* **255** 113823

[2] Vishnevsky V M, Vytovtov K A and Barabanova E A 2020 Model of Navigation and Control System of an Airborne Mobile Station *LNCS* **12563** pp 643–657

[3] Vishnevsky V M, Tereschenko B N, Tumchenok D A, Shirvanyan A M and Sokolov A M 2019 Principles of building a power transmission system for tethered unmanned telecommunication platforms *LNCS* **11965** pp 94-110

[4] Petritoli E, Leccese F and Ciani L 2018 Reliability and maintenance analysis of unmanned aerial vehicles *Sensors* **18(9)** 3171

[5] Valavanis K P 2007 Advances in unmanned aerial vehicles (Springer Netherlands) **33** p 552

[6] Al-Kaff A, Armingol J M and Escalera A de La 2019 A vision-based navigation system for Unmanned Aerial Vehicles (UAVs) *Integrated Computer-Aided Engineering* **26(3)** pp 297-310

[7] Vytovtov K A, Barabanova E A, Vishnevsky V M and Kvyatkovskaya I Y 2020 The Analytical Model of Six-Dimensional Linear Dynamic Systems with Arbitrary Piecewise-Constant Parameters *LNCS* **12563** pp 617–30

[8] Rykov V, Kozyrev D, Filimonov A and Ivanova N 2020 On reliability function of a k-out-of-n system with general repair time distribution *Probability in the Engineering and Informational Sciences* **35(4)** pp 1-18

[9] Ivanova N 2020 Modeling and Simulation of Reliability Function of a k-out-of-n *Communications in Computer and Information Science* **1337** pp 271–85

Challenges in design of an on-board power supply system for a tethered unmanned aerial telecommunication platform

V M Vishnevsky, V P Morozov, K A Alikin

V. A. Trapeznikov Institute of Control Sciences of Russian Academy of Sciences
65 Profsoyuznaya street, Moscow 117997, Russia

E-mail: ak-evmt@yandex.ru

The increasing demand for tethered unmanned aerial platforms in various fields and applications, particularly in telecommunications, leads to growing requirements for the amount and quality of electrical power supplied on board. The paper explores the challenges in hardware, electronic and circuit design of the systems that supply power to the equipment installed on tethered aerial platforms. The arguments for implementing a two-stage DC voltage conversion are provided. In the study the expressions for static and dynamic parameters of the on-board bus voltage regulator and stability conditions for the first stage stabilizer with multiple feedback loops were determined. Additionally, the design of the error signal generation circuit is provided.

1. Introduction

The amount of energy available on-board is a crucial parameter for a variety of unmanned aerial vehicles (UAV), that defines their payload capacity and subsequently the achievable functionality. This is the reason for the growing number of publications dedicated to the challenges in supplying power to the on-board equipment of multicopter drones.

In [1] the authors analyze various ways of generating electrical power on-board and provide arguments for the use of on-board electrical power sources for supplying the electrical energy to the on-board electronic equipment and the motors of light UAVs. A similar conclusion is reached as a result of the computer modeling and the analysis of experimental data on certain civilian UAVs in [2]. The recommendations on the design and structures of on-board power sources for drones provided in [3] are hard to implement in practice without accounting for the issue of converting the voltage supplied on board.

The growing use of the tethered aerial platforms (TAP) in such fields as area surveillance or the deployment of local mobile communication networks requires improvements in the quality of the power supplied on board. But neither the aforementioned papers, nor an in-depth analysis of the TAP power supply structures [4] do not address the issue of stabilizing the voltage that powers the on-board equipment. Meanwhile, powering each motor from a dedicated power supply module is examined as a way of reducing the voltage ripple in response to the load transients.

Additionally, there is a lack of research on the issue of minimizing the weight of the tether, used to supply the electrical power on board, which directly affects the payload capacity of TAPs.

Therefore, insufficient attention is given to some of the crucial challenges in the design of power supply systems for TAPs.

In this paper we provide the possible approaches to solving the issues that arise during the design and realization of the electronic modules of direct current power supply systems that power the motors and electronic telecommunication equipment on-board of the tethered aerial platforms.

2. Power transmission methods

Currently there is a lot of on-going research and experiments dedicated to transferring the power on board of an UAV wirelessly, using laser, focused beams of electromagnetic radiation, etc. But to this day only wired power transfer has been implemented in practice.

Power supply systems designed to transfer electrical energy on board of a TAP may use direct current (DC) or alternating current (AC). Systems utilizing DC are a lot more common in practice. With either approach the parameters of the on-board power supply unit significantly affect the weight of the power line, accounting for a noticeable portion of the total flight weight of a TAP. A technique for the calculation of the power line weight for AC power supply systems using telegraph equations is provided in [5].

For DC power supply systems, the weight of the tether is

$$F = \frac{2P \cdot h_{\max} d_{\mathrm{M}}}{V_{\mathrm{LINE}} J}(1 + k_{\mathrm{ins}}), \tag{1}$$

where P is power supplied on board, $h_{\max}$ is maximum altitude of the TAP, d_{M} is the density of the cable material, V_{LINE} is the line voltage, J is the current density in the cable, and k_{ins} is a coefficient that accounts for the weight of the insulation.

The comparative analysis provided in [5] shows that above a certain power level high frequency AC power supply system gives an advantage in the tether weight over the DC one. But in practice the choice of the current type is affected by additional criteria, such as the weight of the on-board voltage conversion equipment, the limits on cable stretch and temperature. Additionally, taking into account the accessibility of electronic components required for the implementation of the specific power supply system makes the decision over the current type even less obvious.

3. Power supply system structure

The main components of the discussed systems are the ground power supply unit, the power line (tether) and the on-board power conversion modules.

From the relationship (1) it is obvious that given fixed supplied power, it is necessary to maximize the line voltage V_{LINE} in order to decrease the weight of the power line. At the same time, the DC voltage V_{BUS} used to supply the on-board equipment and drive the electrical motors is relatively low and usually does not exceed tens of volts. Thus, regardless of the chosen current type an on-board step-down voltage converter with conversion ratio $K = V_{\mathrm{BUS}} / V_{\mathrm{LINE}} \ll 1$ is required.

A high-level block diagram of the tethered aerial platform power supply system is shown in figure 1. The ground power supply unit converts the power supplied either from the industrial power grid or from a three-phase generator into the chosen power line current type and voltage. The power line is commonly realized as a flexible cable rated for at least 1 kV nominal voltage.

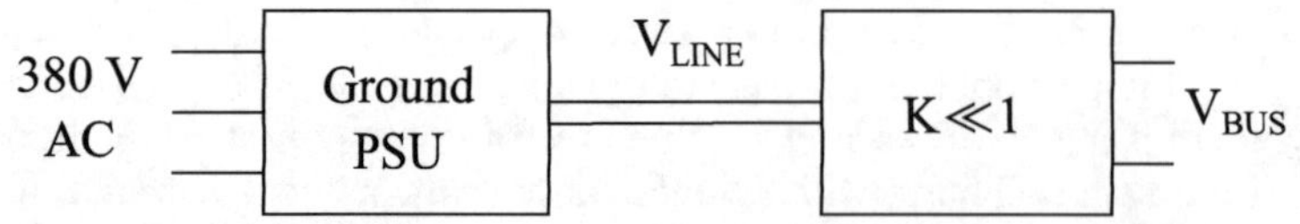

Figure 1. TAP power supply system block diagram.

For the AC power systems, the choice of the operating frequency is also important. The use of frequencies of 100 kHz and above [5] provides a significant reduction in the size and weight of the on-board step-down transformer. But there is a considerable power loss inherent to this design despite its simplicity. Notably, the load current transients cause changes in the input impedance of the on-board transformer. Considering the relatively large length of the power line, these changes may lead to the impedance mismatch with the ground PSU and cause reflected waves to be generated. Another source of the power loss is the on-board rectification of the supplied high frequency power. Cooling the rectifier diodes working at high currents requires installing large heat sinks that reduce the maximum achievable payload. Additionally, high current on the secondary side of the transformer, that may reach hundreds of amperes at a few kilowatts power levels, significantly complicates the winding design. The issue of stabilizing the on-board bus voltage in response to load current changes remains unsolved for such power systems as well.

In the DC power supply systems, it is feasible to implement the voltage conversion using PWM-controlled switching step-down converters [6], as it would allow the output voltage to be stabilized relatively easily. However, achieving the necessary conversion ratio of $K = V_{BUS}/V_{LINE}$ would require the convertor to operate at very low duty cycle values, limiting its voltage control range and decreasing stability. Thus, achieving the necessary conversion ratio at the required power levels within a single device in practice is difficult regardless of the chosen current type.

Implementing a two-stage voltage conversion on-board would make it possible to realize the TAP power supply system that fulfills the conflicting requirements defined above. It should be noted, that this design is not related to the multielement power supply system described in [7].

The primary stage of the proposed system is a PWM controlled non-isolated regulated DC-DC step-down converter operating at convenient duty cycle ranges of $D = 0.3\ldots0.5$. The secondary stage is an array of isolated non-regulated fixed ratio DC/DC converter modules connected in parallel to achieve the required power levels. Such converter modules are produced by various power technology companies and provide high efficiency conversion in compact and lightweight formfactors. For example, the BCM1441VDI1E5 module is rated for the maximum continuous power of 1.6 kW and nominal input voltage of 400 V, has the conversion ratio $K = 0.125$, and achieves over 97% efficiency. But the lack of output voltage stabilization in these compact voltage conversion modules requires additional measures to be implemented, such as a dedicated feedback channel from the on-board power supply system to the regulated ground power supply unit that stabilizes the on-board bus voltage [8].

Two-stage voltage conversion system allows both stages to operate in close to optimal modes. Additionally, it allows the power line voltage to be selected independently from the input voltage of the on-board conversion modules.

An example of the on-board power supply unit with a two-stage voltage conversion is shown in figure 2. VC1 is the primary stage regulated switching DC-DC converter that converts the power line voltage V_{LINE} to the nominal input voltage of the secondary stage converter modules VC2–VC5. Consequently, the with the power line voltage decoupled from the relatively low nominal input voltage of the on-board converters it may be significantly increased leading to a noticeable reduction in the tether weight.

An important requirement for the TAP power supply systems is the ability to power the on-board bus from a reserve power source in the event of power delivery from the ground being interrupted. At the same time measures should be implemented to prevent the reserve power source from being discharge during normal operation. Another important function of the reserve power source is providing energy to the control circuitry of the primary conversion stage during the system startup.

For fulfill these requirements the reserve power source is connected to the on-board bus through a diode, during normal operation this diode is reverse biased and prevents the battery discharge. If the on-board bus voltage drops and the diode becomes forward biased, the on-board equipment is supplied from the battery. An additional antiparallel diode is added to enable the battery to be charged during normal operation if its voltage drops more than about 1 V below the bus voltage V_{BUS}. This additional diode also allows the battery to absorb the current generated by the electrical motors during the TAP descent without it damaging the on-board electronic equipment.

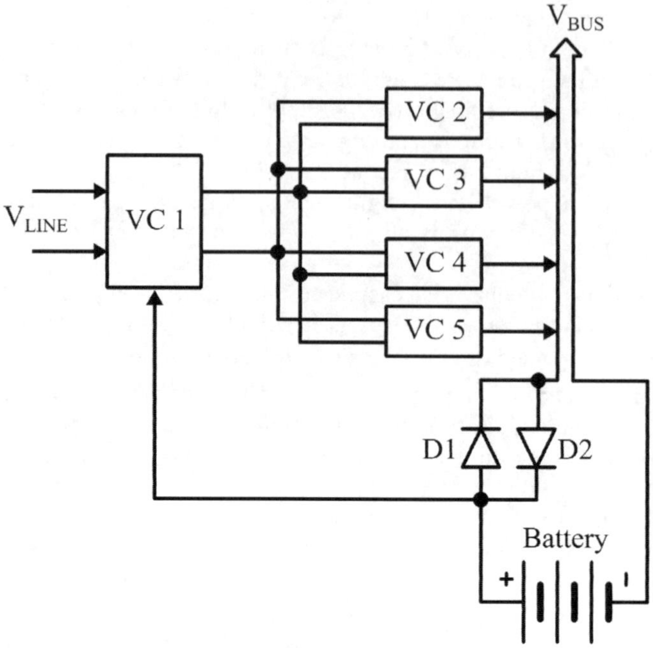

Figure 2. On-board power supply with a backup battery unit.

4. Stabilization of the on-board bus voltage

The power supplied to the tethered aerial platform is used to power on the on-board electronic equipment, as well as electric motors that drive the rotors of the platform. Electric motors usually account for more than 90% of the total TAP power requirements. Furthermore, this portion of the power consumption is heavily affected by external factors, such as wind gusts, and may vary in a wide range. In the absence of on-board voltage stabilization, these changes in power consumption frequently cause the on-board bus being supplied by the reserve battery and may lead to it being discharged uncontrollably. This necessitates the static and dynamic stabilization of the on-board bus voltage.

4.1. Static control

The primary kind of disturbances that should be corrected by the first-stage voltage stabilization system are the power line voltage fluctuations caused by the variations in the on-board current consumption. These fluctuations depend on the power line resistance, which can be calculated as

$$R_{\text{LINE}} = \frac{2\rho \cdot h_{\max}}{S}\left[1 + \alpha \cdot \Delta T(I)\right], \tag{2}$$

where $h_{\max}$ is maximum altitude of the TAP, ρ, α and S are the electrical resistivity, temperature coefficient of resistivity and cross-section area of the power line conductor respectively, and $\Delta T(I)$ is the change in the power line conductor temperature caused by the passing current I.

Computing the value of $\Delta T(I)$ for real world cable configurations in which a section of the cable extends between the ground station and the TAP and another section is reeled on the cable reel would require complex thermophysical calculations. In practice it is much easier to determine this value experimentally [9].

Considering (2) the relationship between the amplitude of the power line voltage difference ΔV_{LINE} caused by the change in the load current, the power line resistance, and the maximum power supplied on board $P_{\max}$ may be expressed as

$$\Delta V_{\text{LINE}} = \frac{2P_{\max} \cdot \rho \cdot h_{\max}}{V_{\text{LINE}} S}\left[1 + \alpha \cdot \Delta T(I)\right]. \tag{3}$$

To counteract the slow voltage changes in the on-board power supply systems, the VC1 converter module has to include a regulator. The feedback loop in the switching DC-DC converters is comprised of the PWM controller and the error signal generation (ESG) circuitry connected to its input. The PWM controller varies the duty cycle of pulses that open the electronic switch (SW) of the voltage converter shown in figure 3. Most often the electronic switch is either a MOSFET or an IGBT.

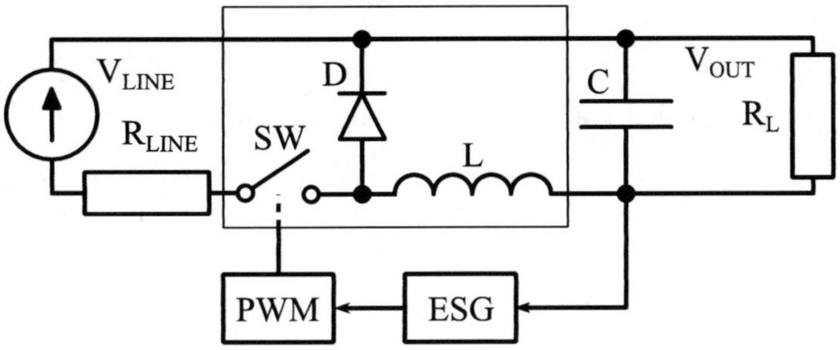

Figure 3. Voltage feedback loop.

Assuming that the secondary-stage convertor module array VC2–VC5 constitutes an ideal proportional element with a gain of $V_{\text{BUS}}/V_{\text{OUT}}$, let us assess the required static parameters of the regulator controlling the voltage V_{OUT} (figure 3). Given the allowed relative instability of the on-board bus voltage is δV_{OUT} and considering that fixed ratio conversion from V_{OUT} to V_{BUS} does not change the relative instability value, the required gain of the VC1 proportional feedback element is

$$K_{\text{P}} = \frac{\Delta V_{\text{LINE}}}{V_{\text{LINE}} \cdot \delta V_{\text{OUT}}}. \tag{4}$$

The voltage feedback loop in VC1 consists of the MOSFET electronic switch (SW), inductor L, output filter capacitor C, error signal generation circuit (ESG) and PWM controller. Considering (2), the product of the gains of the feedback loop components should satisfy the inequality

$$K_{\text{ESG}} \cdot K_{\text{PWM}} \cdot D \geq \frac{\Delta V_{\text{LINE}}}{V_{\text{LINE}} \cdot \delta V_{\text{OUT}}}, \tag{5}$$

where D – is the duty cycle.

In practice as the assumption that VC2–VC5 convertor modules are an ideal proportional element is not universally correct, a single feedback loop is insufficient to ensure the stability of the on-board bus voltage V_{BUS}. The convertor modules have a non-zero output resistance and feature internal control circuitry that can alter their operation mode in response to outside events, such as overheating. This necessitates the addition of a second feedback loop generating the error signal from the on-board bus voltage V_{BUS}.

A schematic of an error signal generation circuit with multiple feedback loops designed for one of the DC-DC converter variants is shown in figure 4.

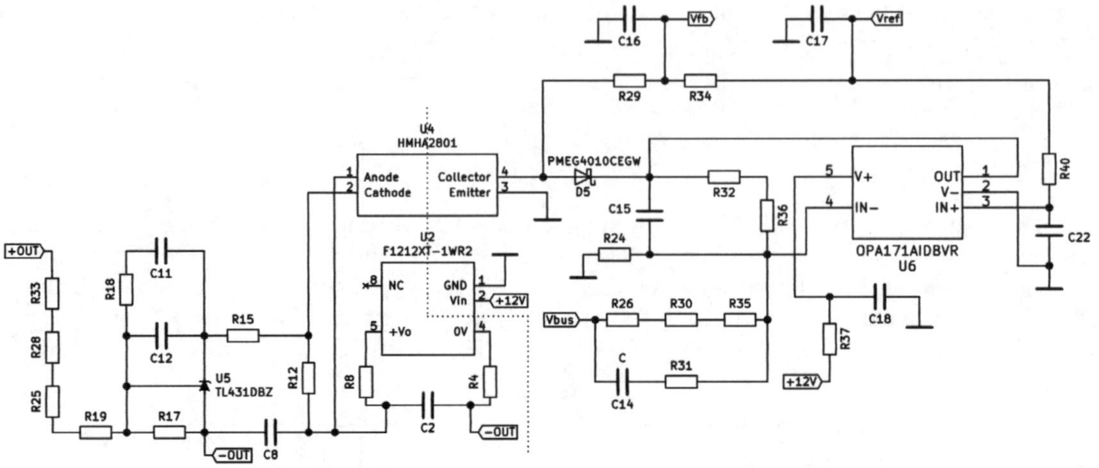

Figure 4. Error signal generation circuit schematic.

A circuit featuring a TL431 voltage reference (U5) is used to generate the error signal from the first stage output voltage V_{OUT}. Through the optocoupler U4 and the error signal is level shifted with a voltage divider R29–R34 and passed to the feedback input of the PWM controller. An additional feedback circuit containing an operational amplifier U6 extracts the error signal from the on-board bus voltage V_{BUS}. This secondary error signal only passes to the input of the PWM controller through the diode D5 if the relative error for the on-board bus voltage is higher than the relative error for the first stage output voltage.

In the PWM controller the error signal is transformed into a pulse sequence with a coefficient of dD/dV_{ESG}. The output signal is then recovered by the first stage output LC-filter with a coefficient of dV_{OUT}/dD. Therefore, the static gain of the PWM controller is $K_{\mathrm{PWM}} = dV_{\mathrm{OUT}}/dV_{\mathrm{ESG}}$. The second unknown component of (5), K_{ESG} may be determined knowing the parameters of feedback circuit components and the gain of the optocoupler.

4.2. Dynamic control

The output LC-filter of the first stage converter is component with the largest time constant in the feedback loop. The transfer function of the first stage converter dependent on the parameters of the LC-filter and the load resistance $R_{\mathrm{L}} = P/V_{\mathrm{OUT}}$ is

$$K_{\mathrm{SW}} = \frac{D}{T_1^2 p^2 + T_2 p + 1}, \tag{6}$$

where $T_1 = \sqrt{LC}$, $T_2 = L/R_{\mathrm{L}}$ are the time constants.

It is evident that the system with proportional regulator transfer function (4) is unstable. As shown in figure 3, components L, C, R_{L} and R_{LINE} form an RLC-circuit with the resonant frequency $\omega = \sqrt{LC}$ and Q-factor $Q = \omega L/(R_{\mathrm{L}} + R_{\mathrm{LOAD}})$. If the parameters of the switching regulator are chosen to ensure that $Q \le 1$ then the resonance in this structure would not impede the dynamic control [10].

However, in power converter topologies similar to the one shown in figure 3 and additional feedback loop may be added to generate the PWM controller ramp voltage directly from the inductor current. In current control mode this inner feedback loop turns the inductor into a voltage-controlled current source and basically removes the inductor from the voltage feedback loop at low frequencies [11].

To counteract the rapid changes in the on-board bus voltage, the regulator has to have sufficient dynamic parameters. An experimental measurement of the on-board bus voltage ripple recorded during flight of the TAP powered by a power supply system without voltage stabilization is shown in figure 5. Even though in this mode of operation the on-board equipment is partially supplied by the reserve battery, the voltage ripple exceeds 5% of the nominal bus voltage.

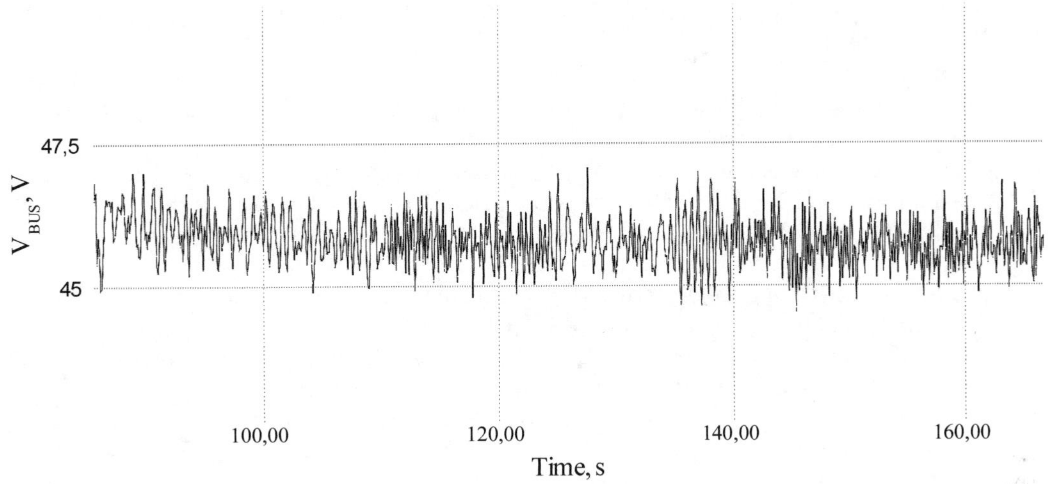

Figure 5. On-board bus voltage ripple in flight.

Dynamic parameters of the designed regulator should be sufficient to suppress the on-board bus voltage ripple. To evaluate required gain of the differential term of the regulator the following expression may be used:

$$K_{\mathrm{D}} = \frac{V_{f1}}{\Delta V_{\mathrm{BUS}}}, \qquad (7)$$

where V_{f1} is the amplitude of main frequency oscillations and ΔV_{BUS} is the acceptable amplitude of the on-board bus voltage ripple.

5. Conclusion

This paper demonstrates the feasibility of implementing two-stage on-board voltage conversion to solve the challenges in designing the power supply system for a contemporary TAP. Presented results demonstrate the possibility of the elementwise analysis of the on-board bus voltage stabilization system with multiple feedback loops.

However, it should be mentioned that the difficulty in defining the transfer functions of individual feedback loops containing non-linear elements complicates determining the exact relationships between proportional and differential components of the error signal. Because of that, the fine tuning the stabilization system by applying the test signals and measuring the system response may be necessary [6].

Acknowledgments

The reported study was funded by RFBR, project number 19-29-06043.

References

[1] Cardone M, Bonaventura G and Fornaro E 2021 Modelling and experimental validation of a hybrid electric propulsion system for light aircraft and unmanned aerial vehicles *Energies* **14(13)** p 3969

[2] Bongermino E, Mastrorocco F, Tomaselli M, Monopoi V G and Naso D 2017 Model and energy management system for a parallel hybrid electric unmanned aerial vehicle *IEEE 26th Int. Symp. on Industrial Electronics (Edinburgh, UK)* (IEEE) pp 1868–73
[3] Boukoberine M, Zhou Z and Benbouzid M 2019 Power supply architectures for drones - a review *IECON 2019 - 45th Annual Conf. of the IEEE Industrial Electronics Society (Lisbon, Portugal)* (IEEE) pp 5826–31
[4] Walendziuk V, Falkowski P and Kulikowski K 2020 The analysis of power supply topologies for tethered drone *applications Proc. of the 9th Innovations-Sustainability-Modernity-Openness Conf. (Bialystok, Poland)* (MDPI) p 25
[5] Vishnevsky V M, Tereschenko B N, Tumchenok D A, Shirvanyan A M and Sokolov A 2019 Principles of building a power transmission system for tethered unmanned telecommunication platforms *Int. Conf. on Distributed Computer and Communication Networks (Moscow, Russia)* (Springer, Cham) pp 94–110
[6] Williams B W 2012 DC-to-DC converters with continuous input and output power IEEE *Transactions on Power Electronics* **28(5)** pp 2307–16
[7] Yu X, Jin K and Liu Z 2013 Capacitor voltage control strategy for half-bridge three-level DC/DC converter *IEEE Transactions on Power Electronics* **29(4)** pp 1557-61
[8] Alikin K A and Morozov V P 2019 Fiber optical feedback system with voltage-to-frequency conversion for a remote power supply *Proc. of the 22nd Int. Scientific Conf. on Distributed Computer and Communication Networks: Control, Computation, Communications (Moscow, Russia)* (Moscow: RUDN) pp 237-43
[9] Morozov V P and Osipov A B 2017 Experimental stand for the estimation of load ratings of the electrical line feeding on-board equipment in tethered telecommunication platform *Proc. of the 20th Int. Scientific Conf. on Distributed Computer and Communication Networks: Control, Computation, Communications (Moscow, Russia)* (Moscow: Technosphera) pp 641–6
[10] Goodwin G C, Graebe S F and Salgado M E 2000 *Control System Design* (Pearson) p 911
[11] Ridley R 2001 *A more accurate current-mode control model* (Ridley Engineering Inc.) p 225

Calculation of the power required to operate a tethered unmanned airborne platform in gust conditions

V M Vishnevsky and A M Shirvanyan

V. A. Trapeznikov Institute of Control Sciences of Russian Academy of Sciences 65 Profsoyuznaya street, Moscow 117997, Russia

E-mail: vishn@inbox.ru

Abstract. This work aims to assess the necessary power for a tethered high-altitude unmanned airborne platform, in which the propulsion system and payload are powered from a ground-based cable power source. A summary and solution to a system of differential equations are given to describe the spatial position of the cable and its action on the high-altitude unmanned platform in gust conditions. The mass of the airborne platform including payload, the effect of the cable on the airborne platform, and its wind resistance are taken into account when calculating the power. The paper describes the system architecture, consisting of a ground station, a cable, and an airborne platform, and presents the results of field tests, which showed good agreement with the results of theoretical studies.

1. Introduction

High-altitude telecommunication platforms are widely used at present. They are based on autonomous unmanned aerial vehicles (UAVs) [1, 2]. Limited operating time is the main disadvantage of UAVs. This is due to the low resource of batteries for UAVs equipped with electric engines or limited fuel supply for internal combustion engines. In this regard, such UAVs cannot be effectively used in systems where a long operating time is required, for example, in safety management systems and protection of critical objects (nuclear power plants, airfields, long bridges, border sections) from terrorist threats. Tethered high-altitude unmanned platforms provide long-term operation if they are powered from a ground-based energy source [3-5]. Transmitting high-power energy (10-15 kW) from the ground to a UAV via cable allows lifting and holding a useful telecommunication load at a height of 100-200 m for a long time, limited only by the reliability characteristics of the UAV [6-9, 22].

Tethered high-altitude platforms hold an intermediate position between satellite systems and ground-based systems, the equipment of which (cellular base stations, radio-relay and radar equipment, etc.) is located on high-rise structures [23]. Compared to expensive satellite systems, tethered high-altitude platforms are cost-effective, and they outperform terrestrial telecommunication systems in vast areas of telecommunication and video coverage. Research centres of world-leading countries design and use tethered high-altitude unmanned platforms, both in the civilian and defence sectors, given the potential for the practical application of such platforms.

It should be noted that the study of autonomous UAVs and the deployment of UAV-based broadband wireless networks have been the subject of numerous publications, described, for example, in the reviews [1, 2, 24]. At the same time, methods for designing tethered high-altitude unmanned platforms with extensive applications are poorly described in modern literature.

The take-off weight of an unmanned module, including the weight of a useful telecommunication load; the climbing and hovering height of the unmanned module; the ultimate force of wind loads, which the UAV is able to withstand; and the time of continuous operation without landing are the main initial parameters when designing tethered unmanned high-altitude platforms. A cellular base station can be used as a payload to quickly create a modern telecommunications infrastructure in emergency conditions, as well as radio-relay and radar equipment to organise long-distance communications in line-of-sight conditions or radar coverage of a vast territory or equipment for video surveillance, environmental monitoring, etc.

Calculations of the required power transmitted from the ground to a high-altitude unmanned module, depending on the initial data listed above, determine the choice of the main parameters of all components of the tethered high-altitude platform. First, these are the parameters of a land-to-board energy transmission system, including the energy transmission method (direct or alternate current [10, 11]). The magnitude of the transmitted power determines the characteristics of propulsion systems and the architecture of the multi-rotor high-altitude module [12], as well as the choice of parameters and the structure of the high-voltage cable, including copper wires, optical fibre, and Kevlar thread [13]. In this regard, the theoretical and experimental methods proposed in this article for studying tethered unmanned platforms operating in a turbulent atmosphere are important when designing such robotic systems.

The article is organised as follows. The second section describes the method for calculating the required power transmitted from the ground to the UAV, depending on the climbing height, wind load speed, and take-off weight of the high-altitude unmanned module. The third section discusses the mathematical model of how a tethered platform operates in a turbulent atmosphere and the solution to a system of differential equations to calculate the cable forces affecting the UAV under wind loads. The fourth section contains the numerical results of calculating the required power depending on the climbing height, cable tension, and wind speed. The fifth section describes the architecture of the Albatross tethered high-altitude unmanned platform, developed under the guidance and with the participation of the authors of the article, and the experimental studies that confirm the high accuracy of the theoretical results.

2. Method for calculating the required power transmitted from the ground to the UAV

Let us consider a UAV and effective wind forces.Figure 1 shows the forces affecting an unmanned aerial vehicle in a turbulent atmosphere: gravity force $\vec{P}$is the take-off weight of a UAV; UAV wind resistance force ($\overrightarrow{F_w}$); cable pull force ($\vec{T}$), taking into account the weight of the cable.

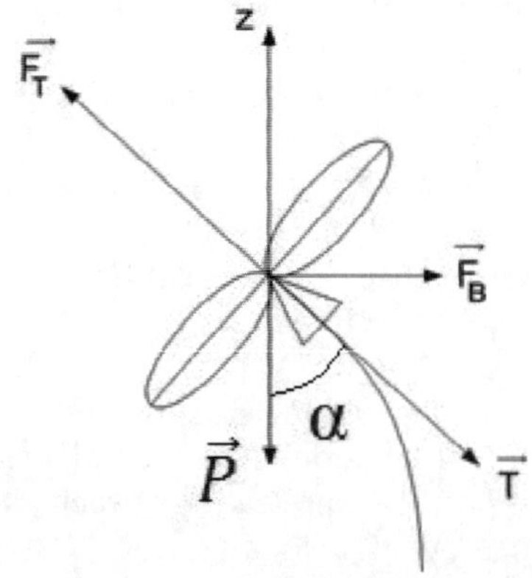

Figure 1. Forces that affect an unmanned aerial vehicle.

The sum of these forces determines the absolute value and the vector of the necessary engine propulsive thrust ($\overrightarrow{F_T}$) to create the climbing power that holds the UAV at a given point:

$$-\vec{F_T} = \vec{F_в} + \vec{P} + \vec{T} \tag{1}$$

The absolute value of the required engine thrust is calculated by the following formula:

$$\left|\vec{F_T}\right| = \sqrt{(\left|\vec{P}\right| + \left|\vec{T}\right| \cdot \sin\alpha)^2 + (\left|\vec{F_w}\right| + \left|\vec{T}\right| \cdot \cos\alpha)^2} \tag{2}$$

where α is the angle of deviation of the tension force from the vertical.

After the absolute value of the resulting force, the engine power can be calculated by the formula:

$$W = \left|\frac{\vec{F_T}}{k \cdot \vec{g}}\right|, \tag{3}$$

where W is the required engine power for hovering at a given point, kW; $\left|\frac{\vec{F_T}}{\vec{g}}\right|$ is the total thrust of UAV engines, kg; $\vec{g}$ is the gravity factor, m/s^2; k is the performance of the propulsion system, characterising the ratio of engine power to the total engine thrust, kg/kW.

Gravity $\vec{P}$ depends on the take-off weight, including the weight of the unmanned aerial vehicle and the payload weight

$$\vec{P} = m_1\vec{g} + m_2\vec{g}, \tag{4}$$

where m_1 is the own weight of the UAV; m_2 is the payload weight.

The wind resistance force ($\vec{F_w}$) of a UAV can be calculated as follows:

$$\left|\vec{F_w}\right| = C_x \cdot w \cdot S, \tag{5}$$

where C_x is the aerodynamic drag coefficient (nondimensional factor); w is wind pressure; S is the cross-sectional area of the UAV.

The factor C_x and the UAV cross-sectional area change their values depending on the angle of inclination α. For the calculations, the following average values were taken: $C_x = 1.5$ and $S = 1m^2$ (an estimated UAV cross-section area) [14].

The approximate nominal wind pressure on an obstacle is calculated by the formula [15]:

$$w = 0.5 \cdot \rho \cdot v^2,$$

where ρ is air density, depending on its humidity, temperature and atmospheric pressure ($1.18415\ kg/m^3$); v is the wind speed, m/s.Then

$$\left|\vec{F_w}\right| \approx 0.9 \cdot v^2 \tag{6}$$

The performance of the propulsion system $k$is calculated as follows.

It is known from the practice of designing UAVs and selecting a propulsion system that the best value for k is 10 kg/kW and higher. This means that every kW of energy can lift 10 kg of load. The value of performance k largely depends on the engine type, the range of length, and the pitch selected, which depends, in turn, on the manufacturers.

The performance k remains constant and does not depend on the climbing height for autonomous UAVs at constant wind speed and payload. The load weight increases when climbing higher (because the weight of the cable increases), and, accordingly, the performance of the propulsion system is reduced in a tethered unmanned platform. A graph is shown below illustrating the underperformance of an 8-motor high-altitude unmanned module equipped with HobbyWingXRotor 8 Series engines [1] and 3010 propellers for a comparative analysis of increasing energy consumption with increasing load.

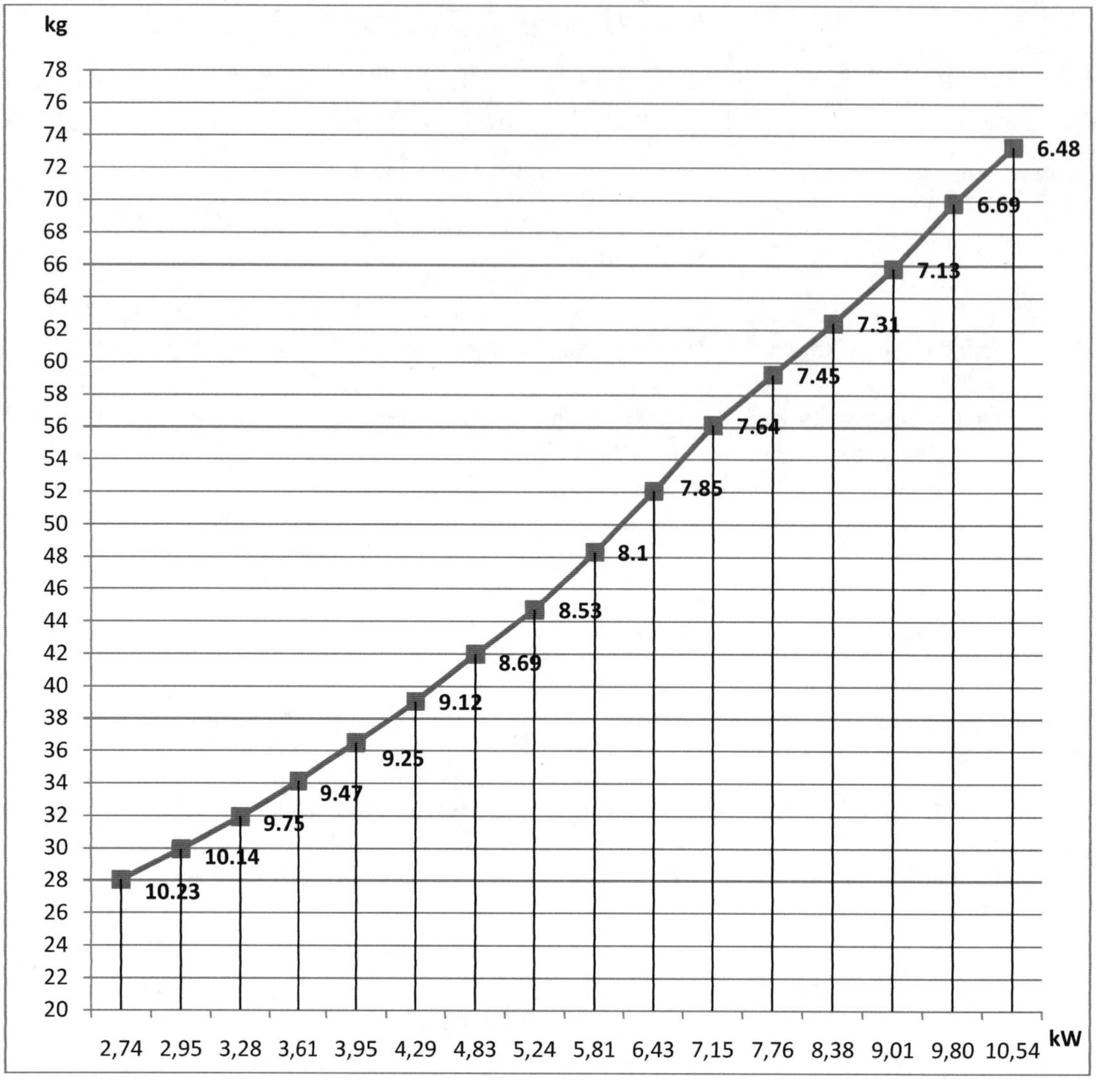

Figure 2. Performance of an 8-rotor coaxial UAV P80III 100kV + MF3218.

To calculate the coefficient k, let us construct an approximate function based on the values in figure 2:

$$\left|\overrightarrow{F_w}\right| \approx 0.9 \cdot v^2. \tag{7}$$

Similar dependencies can be obtained using a T-MOTOR propulsion system [17].

The power necessary for a UAV can be calculated as follows, taking into account (1, 2, 3, 4):

$$|W| = \frac{\sqrt{(|m_1\vec{g}| + |m_2\vec{g}| + |\vec{T}| \cdot \sin\alpha)^2 + (|\vec{F_в}| + |\vec{T}| \cdot \cos\alpha)^2}}{k \cdot g} \tag{8}$$

Thus, all the forces necessary to calculate the power are determined, except for the tension force acting from the cable to the platform and its angle of inclination. To calculate them, it is necessary to describe a mathematical model determining the position of the cable in space.

3. Mathematical model of cable operation in a turbulent atmosphere

As previously stated, power is supplied to propulsion systems and telecommunication payloads of a high-altitude unmanned module from the ground via a cable, which allows a long-term operation of the tethered high-altitude platform. The cable includes copper wires, optical fibre, and Kevlar threads and is a mechanical, electrical, and information connection between a ground module and the high-altitude unmanned module. The outer diameter of the cable determines its aerodynamic resistance to the wind flow, and thereby affects significantly the nature and magnitude of the disturbing effect on the unmanned module at various wind loads.

The differential equations for the tether equilibrium and the acting forces have been calculated previously for various tethered objects. A number of articles discuss different tasks with tethered UAVs: how to control the cable tension, when following a moving vehicle, etc. [1, 4-5]. However, the aforementioned works did not consider how cables affected unmanned vehicles under wind loads, although they have a meaningful effect on how tethered unmanned platforms operate.

The gravity determined by the cable weight, the take-off weight of a UAV, and the force of the wind load must be taken into account when calculating the forces affecting the UAV in a turbulent atmosphere. Let us consider an unmanned aerial vehicle, which is supplied with electricity using a cable from a ground power point. It is assumed that the lower end of the cable is secured with a winch providing constant tension to the lower end of the cable T_0 (figure 3), and the upper end is pivotally attached to the platform. The platform provides cable tension along its entire length up to the winch.

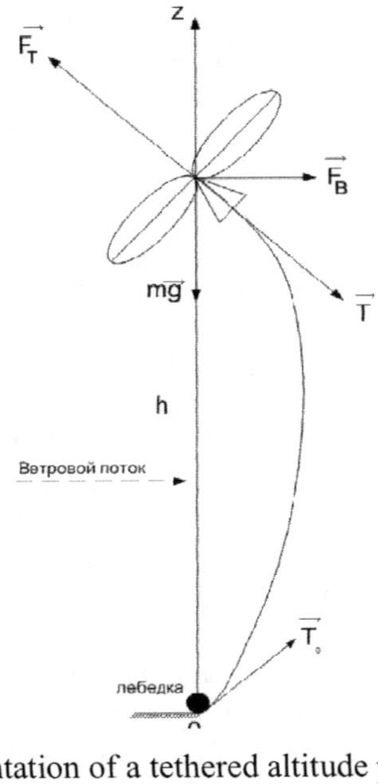

Figure 3. Schematic representation of a tethered altitude platform and the effective forces.

The position of the cable in space will be set in the rectangular coordinate system (x, y, z). Two equations set the equilibrium of the cable in the OXYZ coordinate system:

$$\begin{cases} x = x(z); \\ y = y(z) \end{cases}, \tag{9}$$

where $z \in [0; H], H$ is the climbing height of the platform.

To calculate the wind load on the cable, suppose that the wind speed is specified by two deterministic projections $v_x = v_x(z)$ and $v_y = v_y(z)$ on the axes *OX* and *OY*, respectively. The projection of wind speed is identically zero on the vertical axis *OZ*. The wind speed does not change over time. When calculating the load, only the wind load that is normal in relation to the cable axis is taken into account. The tangential load will be considered equal to zero.

The cable length L is described by the following first-order differential equation:

$$\frac{dL}{dz}=\sqrt{1+\left(\frac{dx}{dz}\right)^2+\left(\frac{dy}{dz}\right)^2}, \tag{10}$$

with the initial condition that $L(0) = 0$.

To simplify the calculations, it can be assumed that the wind direction coincides with the direction of the X axis, which allows considering the position of the cable in the OXZ plane. Earlier in [21], the authors presented a detailed derivation of the system of differential equations describing the position of the cable in space:

$$\begin{cases} \dfrac{dx_1}{dz}=\rho\cdot g; \\ \dfrac{dx_2}{dz}=x_3; \\ \dfrac{dx_3}{dz}=\dfrac{-\rho\cdot g\cdot x_3\cdot(1+x_3^2)-A\cdot v^2\cdot\sqrt{1+x_3^2}}{x_1}; \\ \dfrac{dx_4}{dz}=\sqrt{1+x_3^2}. \end{cases} \tag{11}$$

where z is the current height (m) of the cable element dx_2 above ground; x_1 is the cable tension force (N); $\rho = 0.05$ is the linear density of the cable (kg/m); $g = 9.8$ is the gravity factor (m/s^2); x_2 is the cable vertical deviation (m); x_3 is the cable tangent vertical deviation; $A = 0.003$ is the aerodynamic drag coefficient (kg/m^2) [16];v is the wind speed (m/s); x_4 is the cable length from the winch to the actual height (m).

Based on the first equation of the system, it follows that the cable tension depends only on the altitude of the UAV and the cable tension on the winch. The solution to the system of equations will allow finding the cable action force affecting the UAV and the angle of deviation from the vertical of the cable, depending on various source data. This remains the last unknown in the equation (8).

The complexity of the problem under consideration lies in that not the Cauchy problem must be solved, but the boundary value problem, when the conditions are specified for various values of the argument z. The article [18] gives the derivation of the system of differential equations, but not its algorithm and, accordingly, numerical results. This gap is filled in this paper.

For calculations, let us bring the lower end of the cable into coincidence with the origin of the $OXYZ$ coordinate system. It can be assumed that the winch provides constant tension to the lower end of the cable T_0. Using the control system, the drone is held near a point located vertically N meters above the lower end of the cable.

The above mathematical model allows taking into account different wind speeds on each section of the cable and over the entire length of the cable, so the wind speed at each point will be calculated based on the specified wind speed at a hover point.

Numerous experiments on measuring the wind velocity profile to heights of about 200 m have shown that this value depends logarithmically on height. Rough ground creates friction, which slows airflows down near the surface. The effect of surface heterogeneity is taken into account when modelling by introducing a roughness factor calculated based on wind speeds measured at different heights. The higher a UAV flies, the lower is the friction effect and the higher is the wind speed.

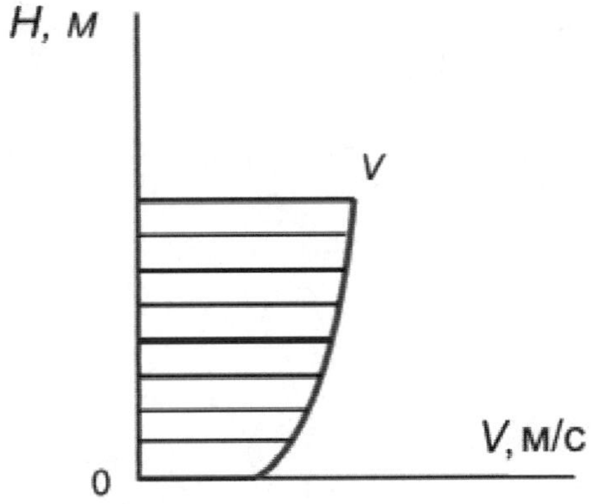

Figure 4. Speed profile near the surface.

It was shown in [17] that the following logarithmic dependence described the wind profile with good accuracy in the surface layer:

$$V_1 = V_0 \cdot \left(\frac{h_1}{h_0}\right)^{kx}, \tag{12}$$

where V_1 is the wind speed at a given height; V_0 is the wind speed at a certain height; h_1 is the given height; h_0 is the height of measurement; kx is the empirical roughness value of the underlying surface. This coefficient takes different values for different landscapes and varies from 0 for a water surface to 0.44 for large cities with skyscrapers. In this case, the average value of the coefficient is taken equal to 0.2.

Then the wind speed in the calculations will depend on the height and have a non-linear character.

Thus, the speed at a given height will be calculated by the expression:

$$V_{act} = V_{tp} \cdot \left(\frac{h_{act}}{h_{tp}}\right)^{0.2} \tag{13}$$

where V_{act} is the wind speed at a given height; V_{tp} is the wind speed at the top point; h_{act} is the actual height; h_{tp} is the top point of measurement.

The following initial conditions are given for the system of differential equations (11): $z = 0, x_1 = T_0, x_2 = 0, x_4 = 0$; and $z = h_{vert}, x_2 = x_{zadan}$. Since the value of x_3 is not defined at $z = 0$, it is better to solve the first and the third equation of the system with an additional condition, taking into account the structure of the system (11):

$$\int_0^{z_{max}} x_3(z) \cdot dz = x_{zadan} \tag{14}$$

After that, the cable length can be found by the formula:

$$L = \int_0^{h_{vert}} \sqrt{1 + (x_3(z))^2} \cdot dz \tag{15}$$

The solution method used is, in fact, one of the variations of the shooting method [18].

The function $x_1(z)$ can be solved analytically by the formula:

$$x_1(z) = x_1(0) + \rho \cdot g \cdot z \tag{16}$$

To find $x_3(z)$, let us divide the interval $0 < z < h_{vert}$ into N length intervals $\Delta z = h_{vert}/N$. Let us look for a numerical solution at the points $z_i = i\Delta z,\ i = 0, 1, \ldots, N$.

The value $(x_3)_0= B$varied and was taken as the initial value of$x_3$at$z_0 = 0$. For the solution, the quartic Runge–Kutta method is used [18]. At each i-th step ($i =0,1,...,N$), the following coefficients are calculated:

$$k_1 = \frac{-\rho \cdot g \cdot (x_3)_{(i-1)}(1+((x_3)_{(i-1)})^2) - A \cdot v^2 \cdot \sqrt{1+((x_3)_{(i-1)})^2}}{(x_1)_{(i-1)}}, \tag{17}$$

$$k_2 = -\frac{-\rho g[(x_3)_{(i-1)} + k_1\Delta z/2](1+((x_3)_{(i-1)} + k_1\Delta z/2)^2) - Av^2\sqrt{(1+((x_3)_{(i-1)} + k_1\Delta z/2)^2}}{(x_3)_{(i-1)} + \rho g\Delta z/2}, \tag{18}$$

$$k_3 = -\frac{-\rho g[(x_3)_{(i-1)} + k_2\Delta z/2](1+((x_3)_{(i-1)} + k_2\Delta z/2)^2) - Av^2\sqrt{(1+((x_3)_{(i-1)} + k_2\Delta z/2)^2}}{(x_3)_{(i-1)} + \rho g\Delta z/2}, \tag{19}$$

$$k_4 = -\frac{-\rho g[(x_3)_{(i-1)} + k_3\Delta z](1+((x_3)_{(i-1)} + k_3\Delta z)^2) - Av^2\sqrt{(1+((x_3)_{(i-1)} + k_3\Delta z)^2}}{(x_3)_{(i-1)} + \rho g\Delta z}. \tag{20}$$

Then

$$(x_3)_i = (x_3)_{i-1} + \frac{1}{6}(k_1 + 2k_2 + 2k_3 + k_4). \tag{21}$$

The integration on the right-hand side of (14) is expediently carried out using the Simpson method, which, on the one hand, is quite simple (being a modification of the trapezoidal method), and on the other hand, allows achieving high accuracy of the result:

$$I = \frac{\Delta z}{3}\left((x_3)_1 + 4(x_3)_2 + 2(x_3)_3 + \ldots + 4(x_3)_{N-1} + (x_3)_N\right). \tag{22}$$

Regarding the choice of the coefficient B, its best value can be calculated using a dichotomy as follows. From physical considerations, let us determine the values of the coefficients P_0 and Q_0 such that at $x_3(0)=P_0$ one can guarantee that $x_2(h_{vert})<x_{zadan}$, and at $x_2(0) =Q_0$, $x_2(h_{vert})>x_{zadan}$.

Let us sequentially solve the equations according to the above principle for x_2 and x_3 with the initial condition than $B = (P_j+Q_j)/2$ where $j=0,1,2,...$. If $I >x_{zadan}$, then $P_{j+1} =P_j$, $Q_{j+1} =B$must be taken in the next step. If $I <x_{zadan}$, then $P_{j+1} =B$, $Q_{j+1} =Q_j$. The sequence continues until $|(x_3)_N- x_{zadan}| < \delta$, where δ is the required accuracy of the solution. In this case,$\delta= 10^{-7}$, which is more than enough in the face of restrictions imposed on this model of cable.

After that, the cable length was calculated, the integral (14) for which can also be found by the Simpson method:

$$L = \frac{\Delta z}{3}[\sqrt{(1+((x_3)_1)^2} + 4\sqrt{1+((x_3)_2)^2} + 2\sqrt{1+((x_3)_3)^2} + 4\sqrt{1+((x_3)_{(N-1)})^2} + \sqrt{1+((x_3)_N)^2}]. \tag{23}$$

4. Calculation results of the required power depending on the climbing height, the cable tension, and the wind speed

The problem was solved on a grid consisting of $N = 3000$ points. Both the quartic Runge–Kutta method and the Simpson method provide the quartic error [19–20], giving an error that can be estimated as $O(\Delta z^4)$. Thus, one can be sure that the numerical solution differs from the exact solution of the original problem by no more than about 10^{-4}.

Using C#, a programme was developed to solve the system of equations. This programme allows obtaining the required power for various sets of input data. Users can enter linear cable density ρ into the programme, wind speed $v$at the maximum climbing height z_{max}, which slows down in accordance with a velocity profile near the surface, UAV climbing height, and the weight of the UAV with the load.

The system (11) was solved for various values of wind speed and UAV hovering height. Tables 1-3 contain the results of calculations. The estimated weight of the UAV was 30 kg including a payload. The necessary power was calculated using the formula (8).

Table 1. Calculation of the power required when T_0 is 15 N, cruising altitude is 50 m

Wind speed, m/s	Cable deflection angle at the top point, °	Cable length, m	Engine thrust required, N	Power required, kW
0	0	50	338.39	3.64
3	-0.8	50.003	338.49	3.65
6	-3.3	50.041	340.16	3.67
9	-7.5	50.209	347.02	3.78
12	-13.2	50.663	365.05	4.06
15	-20.4	51.636	400.13	4.63
18	-28.5	53.463	456.88	5.6

Table 2. Calculation of the power required when T_0 is 15 N, cruising altitude is 75 m

Wind speed, m/s	Cable deflection angle at the top point, °	Cable length, m	Engine thrust required, N	Power required, kW
0	0	75	353.094	3.87
3	-1	75.006	353.192	3.87
6	-3.9	75.097	354.858	3.9
9	-8.7	75.494	361.816	4.01
12	-15.3	76.578	379.848	4.29
15	-23.5	78.95	415.128	4.88
18	-32.7	83.628	472.066	5.87

Table 3. Calculation of the power required when $T_0$is 10 N, cruising altitude is 50 m

Wind speed, m/s	Cable deflection angle at the top point, °	Cable length, m	Engine thrust required, N	Power required, kW
0	0	50	333.40	3.57
3	-1	50.004	333.49	3.57
6	-3.9	50.065	335.16	3.59
9	-8.7	50.329	342.12	3.7
12	-15.3	51.052	360.35	3.98
15	-23.5	52.633	395.72	4.55
18	-32.7	55.752	452.66	5.52

Table 4. Calculation of the power required when T_0 is 10 N, cruising altitude is 75 m

Wind speed, m/s	Cable deflection angle at the top point, °	Cable length, m	Engine thrust required, N	Power required, kW
0	0	75	348.1	3.79
3	-1.1	75.009	348.2	3.79
6	-4.4	75.146	349.86	3.82
9	-9.8	75.745	356.92	3.93
12	-17.2	77.416	375.05	4.22
15	-26.3	81.284	410.62	4.8
18	-36.8	90.836	467.75	5.79

Based on the data from the tables, it can be seen that the high wind speed has a significant effect on the required power, because the higher the wind speed, the bigger the angle of inclination of the cable resulted from an increased tension force affecting the UAV, which, in turn, increases the horizontal component of the tension force. In addition, it can be concluded that changing the tension force of the winch significantly changes the cable profile and its length. It is important for the control system to find a balance between the pull force of the winch and the power expended. A low winch tension provides a large cable outlet, which can cause problems in the case of tangling. A strong winch tension results in an almost vertical cable profile, but at the same time, it significantly increases the power, which the UAV consumes.

The data obtained are important to assess the payload and the necessary power transmitted from the ground to the board when flying in a turbulent atmosphere, as well as in when designing the UAV control system.

5. Field studies of a tethered high-altitude unmanned platform

Field studies confirming the above theoretical results were carried out using an Albatross tethered high-altitude unmanned platform, developed at the Russian Academy of Sciences under the guidance and with the participation of the authors of this article. This tethered platform provides long-term functioning (up to 50 hours) with a payload up to 15 kg (UAV take-off weight up to 35 kg) at altitudes up to 100 m in a turbulent atmosphere (wind speed up to 15 m/s).

The tethered high-altitude unmanned platform includes the following main components:

1. A durable heavy unmanned multi-rotor vehicle is intended to lift and hold payloads, video surveillance equipment, etc. up to 100 m.
2. A high-power ground-to-board energy transmission system (up to 9 kW) supplies the propulsion systems of the unmanned multi-rotor vehicle and payload equipment with power.
3. A control and stabilisation system for the high-altitude platform, including a backup local navigation subsystem with ground-based radio beacons to improve positioning accuracy and noise immunity when satellite navigation signals are weakened or absent.
4. On-board payload equipment comprising: a 4G cellular network base station (LTE); radar and relay equipment; video surveillance equipment, etc.
5. A Kevlar-based strength-power communication cable, including small copper wires to transmit high-voltage (up to 1000 V) signals and optical fibre to transmit digital information at up to 10 Gb/s.
6. A ground control system, which includes a DC or AC voltage converter, a diagnostic system for the high-altitude platform and an intelligent microprocessor-based winch to control the tension of the cable during take-offs, landings and wind loads. A mobile version of the ground control system is located on a mobile platform with an electric generator installed on it; the output power of the generator is at least 10 kW (figure 5).

Figure 5. Mobile version of the Albatross tethered high-altitude unmanned platform.

A coaxial octocopter (8 propellers) is equipped with T-MOTOR engines and used as an aircraft. Its frame, propulsion system (engines, travel controllers, propellers) and flight controller weight 13 kg. A DC-DC voltage converter is installed on the frame of the copter and weights 3 kg. A backup battery provides 8 minutes of flight in the case of failure of the ground power source and weights 4 kg. The payload mass depends on the need and can be 5, 10, or 15 kg. The land-to-board power supply system was developed as part of the Albatross project and supplies the UAV propulsion system and the payload with up to 9 kW via a cable of 6.7 g per 1 meter.

The UAV electronic control is based on a PixHawk 2.1 Cube open-source flight controller (hardware), with an ArduPilot software flight stack (software). The flight controller determines the altitude based on the data received from an accelerometer, a GPS module, and a barometer. A LightWare SF11/C Lidar altimeter is used for more accurate height measurement. It is connected to a serial port of the flight controller and provides accurate height measurement up to 120 m. A GPS module M8N provides GPS and GLONASS communications and navigates the UAV. Two HMC5983 MAG and LIS3MDL Mag magnetometers are mounted on an external GPS module and provide directional stability. A MAUCH PL4-14S/1 primary and a DC-DC backup converter convert the on-board operating voltage of 50V to 5.3V and are used to power the flight controller. The converters are equipped with current and voltage sensors, which measure the current and voltage of the backup battery and of the converter from the ground power source. Unlike traditional UAVs, the high-altitude unmanned module of the tethered platform is additionally equipped with an emergency landing system (which includes both hardware and software) to increase reliability when the power supply is cut off from the ground, as well as an engine temperature automatic control system, a backup local navigation system, etc., taking into account the need for long-term functioning.

UAV power consumption was measured in different weather conditions, with different payload weights (from 5 kg to 15kg) and different climbing (hovering) heights of the module (from 50 m to 100 m) during lengthy field trials of the Albatross tethered high-altitude unmanned platform. Each test (without landing the module) ranged from 8 to 12 hours. The test results below were obtained under calm conditions, and with gusts of wind up to 8 m/s and lifting to a height of 75 m with 5 and 10 kg of payload.

An unmanned module was tested in calm weather with a take-off weight of 30 kg, including 10 kg of payload. A 75-m-long cable weighed 5 kg. A diesel generator was installed on a car and powered the UAV propulsion system and the payload from the ground via a cable. The average measured and consumed power was 90 A at 47 V. Accordingly, the average power consumption was 4.2 kW, and the average efficiency was k = 8.3 g/W. The current increased to 105 A with a short-term increased wind load (Figure 6). Accordingly, the power consumption increased to 5 kW. Thus, a ground source must supply a tethered high-altitude unmanned platform with at least 5 kW if the platform has a payload of 10 kg and a total weight load of 35 kg (including the weight of a 75-m cable) in calm weather.

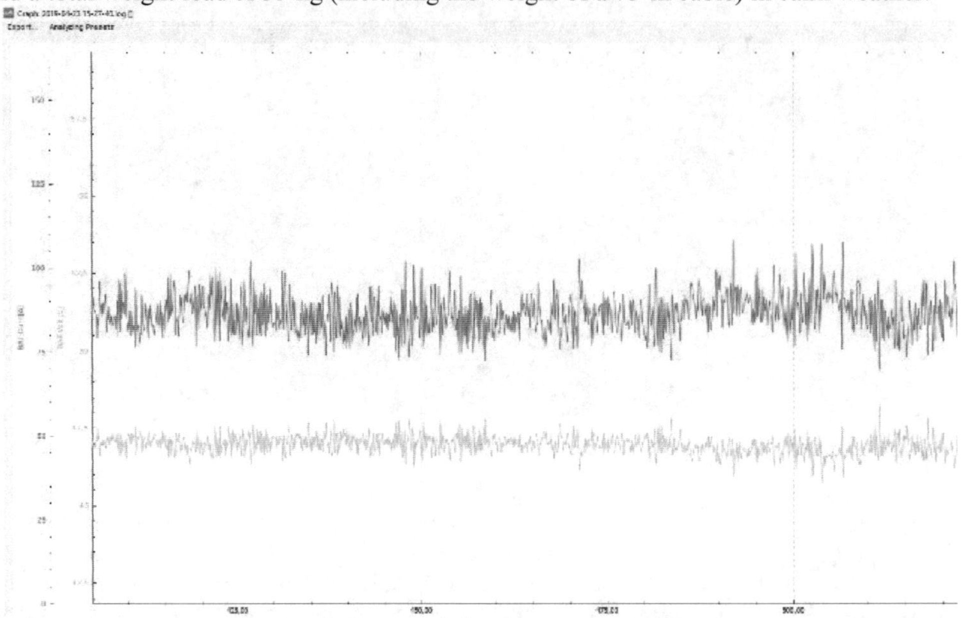

Figure 6. Current and voltage consumption.

The tests were also carried out in windy weather with a sustained wind of 5 m/s and with gusts of up to 8 m/s in order to determine the maximum energy consumed and transmitted from the ground to the UAV. Note that measuring the wind flow is a sophisticated problem, because using traditional wind speed sensors installed on UAVs results in low measurement accuracy because of disturbing effects created by UAV propulsion systems. In this regard, wind speed V_1 is calculated by the formula (12) at h_1 using the meteorological complex [24] for the high-precision measurement of the V_0 initial values at the fixed height h_0.

The take-off weight of the high-altitude unmanned module during these tests was 25 kg, including 5 kg of payload. The total weight load was 30 kg, including the 75-m cable. The measured average current was 82 A at 49.5 V, when testing in windy weather with a sustained wind of 5 m/s. Accordingly, the average power consumed by the UAV was 4 kW. The values of current consumption ranged from 60 A to 103 A, while the values of power consumption varied from 3 kW to 5 kW. The peak value of the current reached 120 A at 47 V with gusts of 8 m/s, which corresponds to the maximum energy consumption of 5.6 kW (figure 7).

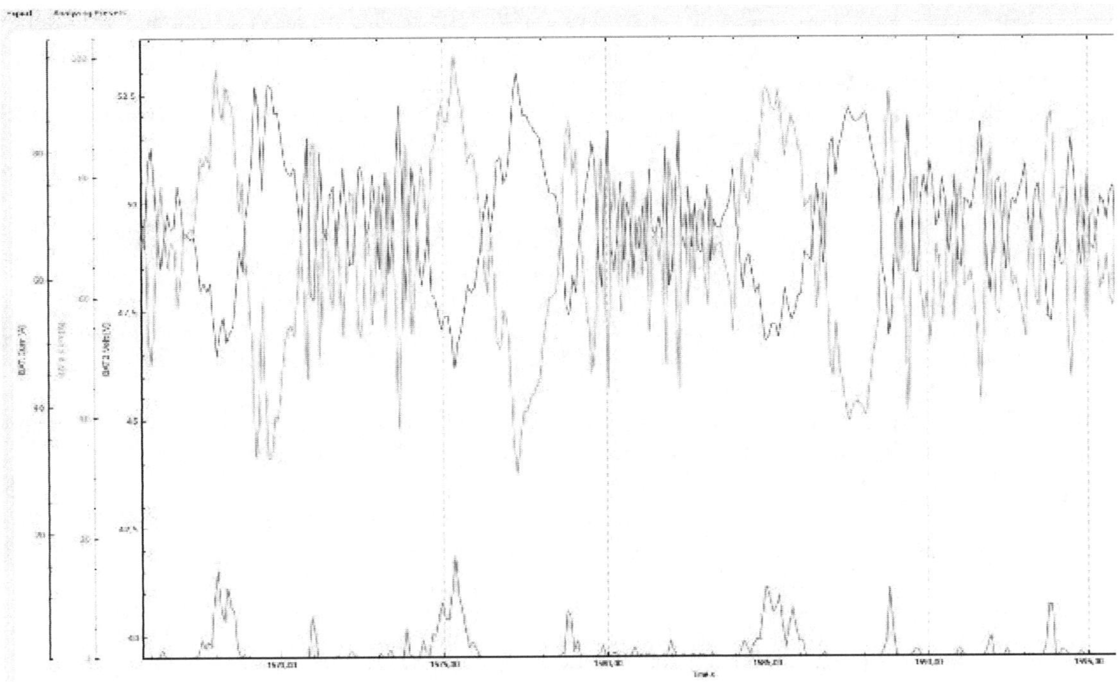

Figure 7. Current and voltage consumption.

Thus, up to 5.6 kW must be transmitted from the ground to the UAV when operating in a turbulent atmosphere (wind speed up to 8 m/s). It should also be noted that the obtained experimental results coincide with the theoretical results, described in the previous sections, with a reasonable practice accuracy. In this regard, the theoretical results can be effectively used when designing tethered high-altitude unmanned platforms.

6. Conclusion

The article poses a problem of determining the power necessary to operate a tethered high-altitude platform in a turbulent atmosphere. Its second section describes a method for calculating the required power transmitted from the ground to the UAV, depending on the climbing height, wind load speed, and the take-off weight of the high-altitude unmanned module. A mathematical model is described, built to calculate the tensile force acting from the cable on the UAV. The results of the calculation of the required power depending on the climbing height, the tension, and the wind speed are presented. The fifth section of the paper describes the architecture of a tethered high-altitude unmanned platform, developed under the guidance and with the participation of the authors of the article, and the experimental studies that confirm the high accuracy of the theoretical results.

Acknowledgments

The reported study was funded by RFBR, project numbers 19-29-06043, 20-37-70059.

References

[1] Mozaffari M, Saad W, Bennis M, Nam Y-H and Debbah M 2019 A Tutorial on UAVs for Wireless Networks: Applications, Challenges, and Open Problems *IEEE Communications Surveys & Tutorials* pp 410–38

[2] Khan M, Hamila R, Kiranyaz M and Gabbou A 2019 A Novel UAV – Aided NetWork Architecture Using Wi Fi Derect *IEEE Access* **7** pp 67305–18

[3] Kiribayashi S, Yakushigawa K and Nagatani K 2018 Design and Development of Tether-Powered Multirotor Micro Unmanned Aerial Vehicle System for Remote-Controlled Construction Machine *Field and Service Robotics* pp 637–48
[4] Kiribayashi S, Ashizawa J and Nagatani K 2015 Modeling and Design of Tether Powered Multicopter *Proceeding IEEE International Symposium on Safety, Security, and Rescue Robotics (SSRR)* pp 1–7
[5] Raj V, Raj N and Kumar J 2016 An Approach for Power Optimization of Tethered UAV *Journal of Electrical and Electronics Engineering* **11(5)** pp 23–25
[6] Tognon M and Franchi A 2017 Position tracking control for an aerial robot passively tethered to an independently moving platform *IFAC Papers OnLine* **50(1)** pp 1069–74
[7] Vishnevsky V and Tereschenko B 2017 Development and research of a new generation of high-altitude tethered telecommunications platforms *T-Comm: Telecommunications and transport* **7** pp 20–24
[8] Wang G, Samarathunga W and Wang S 2016 Uninterruptible Power Supply Design for Heavy Payload Tethered Hexaroters *International Journal of Emerging Engineering Research and Technology* **4(2)** pp 16–21
[9] Wasantha, G. Wang, S. Wang 2015 Heavy Payload Tethered Hexaroters for Agricultural Applications: Power Supply Design *International Research Journal of Engineering and Technology* **2(5)** pp 641–45
[10] VishnevskyV, Tereschenko, Tumchenok D and Shirvanyan A 2017 Optimal Method for Uplink Transfer of Power and the Design of High-Voltage Cable for Tethered High-Altitude Unmanned Telecommunication Platforms *Communications in Computer and Information Science* **700** pp 240–47
[11] Vishnevsky V and Tereschenko B 2015 Method of Remote Wired Power Supply of Objects *Patent for Invention* 2572822 RF; Dec. 16
[12] Vishnevsky V and Meshcheryakov R 2019 Experience of Developing a Multifunctional Tethered High-Altitude Unmanned Platform of Long-Term Operation *Lecture Notes in Computer Science* **11659** pp 236–44
[13] VishnevskyV, Tereschenko, Tumchenok D and Shirvanyan A 2019 Cable Design and Requirements to Power the Unmanned Tethered Telecommunications Platform *Proceedings of the 22nd International Scientific Conference on Distributed Computer and Communication Networks: Control, Computation, Communications* pp 12–18
[14] Albisser M2015 Identification of Aerodynamic Coefficients from Free-Flight Data *Ph.D. thesis*
[15] Savicky G 1972 Wind load on structures *Build public*
[16] Ebert G 1963 Short Handbook of Physics: Reference Book/Per *Fizmatgiz* pp 552
[17] Touma J 1977 Dependence of the wind profile power law on stability for various locations *Journal of the Air Pollution Control Association* **27(9)** pp 863–66
[18] Süli E and Mayers D 2003 An Introduction to Numerical Analysis *Cambridge University Press*
[19] Hazewinkel M 1994 Simpson formula *Encyclopedia of Mathematics*
[20] Press W,Teukolsky S, Vetterling W and Flannery B 2007 Section 18.1. The Shooting Method *Numerical Recipes: The Art of Scientific Computing*
[21] Vishnevskiy V, Shirvanyan A and Tumchenok D 2019 Mathematical Model of the Dynamics of Operation of the Tethered High-Altitude Telecommunication Platform in the Turbulent Atmosphere *Proceedings of International Scientific Conference «2019 systems of signals generating and processing in the field of on board communications»* pp 1–7.
[22] Tognon M and Franchi A 2020 Theory and Applications for Control of Aerial Robots in Physical Interaction Through Tethers *Tracts Adv. Robot* pp 156
[23] Xiao X, Dufek J and Murphy R 2019 Benchmarking Tether-based UAV Motion Primitives *IEEE International Symposium on Safety, Security, and Rescue Robotics (SSRR)* pp 51–55
[24] Xiao X, Fan Y, Dufek J and Murphy R 2018 Indoor UAV Localization Using a Tether*IEEE International Symposium on Safety, Security, and Rescue Robotics (SSRR)* pp 1–6

Journal of Physics: Conference Series **2091** (2021) 012036 doi:10.1088/1742-6596/2091/1/012036

Development of a methodology for migration of monolithic systems to microservice architecture using cloud technologies

A V Balzamov, D S Fomin, A V Savkina, V V Nikulin and S A Fedosin

NI MSU named after N. P. Ogarev, Bolshevistskaya str., 68, Saransk, Rep. Mordovia, 430005, Russia

E-mail: balzamovav@yandex.ru

Abstract. The long-term use of the classical monolithic architecture for the development of web applications has led to the accumulation of a voluminous knowledge base for analyzing the positive and negative aspects, as well as finding ways to solve the problems that arise when using it at the present stage. Thanks to the rapid development of the computing power of personal computers, network data exchange mechanisms, cloud technologies and Open Source software, it has become possible to use more complex and efficient approaches to architecture design and deployment of web applications, such as a service-oriented, microservice model and deployment on a virtual machine or cloud. Due to the emergence of a large number of new technologies described above, it is quite difficult to apply them effectively enough without proper experience, as well as to switch to new models of architecture design.

1. Introduction

Many software developers of different levels around the world at the stage of software design are faced with the question of the correctness of the application architecture used. Similar questions arise among specialists involved in maintaining existing software, since sooner or later a poorly designed application architecture will not allow making any changes in the program code. In other words, software is no longer flexible, changeable, and generally maintainable.

It is the archaic monolithic architecture that is susceptible to such problems, which was replaced by service-oriented and then microservice architectures. These approaches take into account and correct the shortcomings of their descendants, but they also have problem areas that will have to be faced at the initial stage of design.

Since in modern times most systems, programs and services are moving to web applications, the platform used to deploy software plays an important role. Whether it will be physical machines, virtual machines, or modern cloud services. It is the technique described in the article that will make it possible to make a decision on the transition to a new architecture, get acquainted with the options for deploying the final system, as well as the containerized method of software delivery. The described technique can be useful both for software developers and for students and engineers of technical specialties.

2. Comparision of existing architectures

2.1 Monolithic architecture

In software engineering, a monolithic model refers to a single, indivisible unit. The concept of monolithic software is that the various components of an application are combined into one program on one platform. Typically, a monolithic application consists of a database, a client user interface, and a server application. All parts of the software are unified and all of its functions are managed in one place. Like every programming language, every approach to building an application architecture has its own merits and demerits to a greater or lesser extent.

Monolithic architecture is great for small teams because of its ease of development. Everyone from a small group of developers will be able to make changes to the application or create something new, since all the necessary data and principles are concentrated within one workspace.

The components of monolithic software are interconnected and interdependent, which helps the software to be self-contained. As mentioned earlier, this architecture is "traditional", but some developers consider it obsolete.

Because the application is a “monolith,” it often contains only one directory, making it easier to deploy or upgrade further. Also, since the application is built on the basis of a monolithic kernel, developers do not need to make changes and updates separately, as they can do it right away and save time. Most applications rely on many cross-component tasks such as audit trails, logging, rate limiting, and so on. Monolithic applications are much easier to address these issues because of their common codebase. It is easier to connect components to these tasks when everything works in one application.

In a clean build, small monolithic applications tend to perform better than microservices-based applications. This is explained by the fact that monolithic applications have a common code and used memory.

Since any developing application sooner or later begins to grow, it is necessary to periodically revise its architecture. At first, it is still possible to control the process of expanding the monolithic architecture, but at a certain moment it may turn out that in order to add quite insignificant functionality, it will be necessary to update many components at once. In addition, the codebase becomes unwieldy, unreadable, and impossible to understand and change.

Also a monolithic approach to building software, imposes a lot of cohesion on the components within the application. Especially if the user interface and the server application are connected.

An important feature of monolithic applications in a production environment is the impossibility of running the application if one of the links in the system fails.

In monolithic applications, each small update requires a complete redeployment.

When multiple teams are working on the same project, flexibility can be greatly reduced.

2.2 Service-oriented architecture

Service-oriented architecture (SOA) is a style of software architecture that assumes a modular application consisting of discrete and loosely coupled software agents that perform specific functions. SOA divides components into two main roles: service provider and service consumer. These two roles can be played by software agents. The main feature of SOA is this: an application can be designed and built in such a way that its modules can be easily designed and reused later.

Because of the self-contained and loosely coupled nature of functional components in service-oriented applications, these components can be reused across multiple applications without affecting other services.

Since each software service is an independent unit, it is easy to update and maintain without affecting other services already running. For example, large corporate applications where uninterrupted and stable operation is important.

Services are much easier to debug and test than large chunks of code, as in monolithic systems. This, in turn, makes the end products based on SOA more stable and reliable.

Because service-oriented architecture is layered (middleware), it maintains concurrency throughout the development process. Independent services can be developed in parallel and completed at the same time.

The main disadvantage of a service-oriented architecture is its increased complexity. Each service must ensure that messages are delivered on time, which is achieved by connecting message brokers. The number of these messages can exceed a million at a time, making it difficult to manage all services.

SOA development requires a fairly large upfront investment in human resources, technology, and development.

In SOA, all inputs are validated before one service interacts with another service. When using multiple services, this increases the response time and increases the load on the system, which in turn reduces the overall performance.

The best application of SOA architecture is complex corporate and banking systems, due to the requirements of stability and reliability of the final product.

2.3 Microservice architecture

Microservices are a type of service-oriented software architecture focused on creating a set of self-contained components that make up an application. Microservice architecture, in contrast to monolithic, involves the use of many independent components that interact via API.

The biggest advantage of microservices over other architectures is that small, separate services can be developed, tested, and deployed independently. Since the deployment unit is small, it makes product development and release easier and faster. In addition, the release of one unit is not limited to the release of another that has not yet been developed.

With microservices, multiple development teams can work independently and quickly on their services. Each separate part of the application can be built in isolation due to the isolation of the microservice components.

The biggest disadvantage of microservices is their complexity. Dividing an application into independent microservices introduces more management artifacts. This type of architecture requires careful planning, tremendous effort, team resources, and skills.

In microservice applications, each functionality that interacts through the API from the outside increases the likelihood of network attacks.

A plus and at the same time a minus is a large number of different programming languages. On the one hand, the application will be very flexible in terms of the implementation of functionality tied to a specific programming language. On the other hand, the need for knowledge of these languages by the development team, which increases the money costs for specialists.

Microservices are great, but not for every type of application. This template is great for developing applications and complex systems. When the application is large and needs to be flexible and scalable, microservice architecture is beneficial [1].

3. Platforms for deploying

When developing software, regardless of its architectural style, developers are always faced with the acute question of choosing a platform on which this application will run and will provide its capabilities to the end user. Today, there are two main directions in the topic of software deployment - the use of cloud technologies or the preparation of your own server.

3.1 Bare metal

A separate physical server is a more traditional solution that is gradually becoming a thing of the past. This is due to the fact that the cost of using it is almost 2 times higher than the cost of cloud storage.

A separate physical server can be called the most flexible solution for organizing your own IT infrastructure, but it involves high costs. These include deployment on servers, followed by administration and maintenance of the equipment by system engineers. Only in this case can we talk about maintaining the trouble-free operation of the system.

At the same time, there are two options for using the server - this is the construction of your own server node (that is, the purchase and installation of machines in the company's office) or the lease of equipment in the data center (that is, the provision of remote access to the server based on the data center).

The fundamental advantage of your own server is the maximum freedom in choosing methods for integrating the infrastructure with other systems. However, this assumes that you are incurring additional costs for the following:

- purchase / rent of equipment and its maintenance;
- maintaining the security of the server node;
- redundancy of additional servers;
- maintaining the smooth operation of machines (internet, power supply, etc.);
- software update.

Using your own server is not recommended for companies that do not have qualified IT staff or that do not have a reliable server rental contractor. But in the event that you have the right employee, then you can ensure greater system reliability than using third-party cloud storage.

3.2 Cloud storage

The use of cloud storage means that you store all information on the supplier's server, and you get access to files using a browser or through a special program.

Cloud technology assumes two solutions: on-demand (the cloud of the CRM system manufacturer) and private cloud (any private cloud). Each of the options has features:

- on-demand. Used by those companies that are prohibited from storing customer / user data outside the country. This solution helps to reduce capital costs, provides reliable backup of data and capacities, allows you to monitor the security of databases and control access to information. But there are also disadvantages of such a cloud. For example, a number of system settings will need to be performed exclusively through technical support. When integrating CRM with other systems, you will need to use special protocols. This solution is not recommended for corporations for which it is critical to update huge amounts of customer data and do it in the shortest possible time.
- private cloud. This is another option for cloud storage, but it is not the manufacturer of the CRM you use that is responsible for it, but a third-party provider. This solution has the advantages of on-demand, including ensuring uptime of services and constant access to information. However, in this case, you will have to pay separately for the used cloud capacities. This solution is recommended for companies that have a complicated CRM system and therefore cannot always provide reliable access to databases for all employees. Not recommended for those organizations that need to quickly update huge databases.

The main advantage of cloud solutions is said to save money and time. When using this technology, you do not need to purchase equipment, maintain or replace it. Also, no maintenance is required on the staff of the system administrator. That is, your costs are only in the payment for the used cloud resources.

Among other advantages of such a solution are noted:

- data availability. With cloud technology, you have access to data anywhere you have an internet connection. This is very convenient if you are on frequent business trips or have to deal with urgent matters on the road.
- information security. All data is automatically backed up to the cloud. In the event of a failure, you can restore the lost information, and, therefore, there is no risk of its complete loss.
- lack of additional settings. You get a ready-to-use system. You do not need to make complex settings, perform updates or reinstall the platform.
- independence from hardware failures. The cloud server has a multi-level redundancy system, so the problem of equipment breakdown is not relevant for it, while the use of its own server by a company is associated with great risks.
- flexibility. Hardware and software can be reconfigured into new information systems in just a few days.

Because of these benefits, organizations are leveraging cloud services in one way or another in their business processes. It is really simple and cheap, does not require special skills and knowledge from employees.

But this decision also has its negative sides. The cloud implies working with files over an Internet connection, that is, to view data, you need access to the network. This can be a disadvantage if there are interruptions in the connection. Consequently, access to information is lost and work becomes impossible.

4. Migration criterias

When deciding whether to migrate from a monolithic architecture, there are many factors to consider, including the existing solution, if any. Also, one should not forget about the available material, human and time resources, since it is the resources that often become key when making a decision. In order to comprehensively assess the value of switching to a microservice architecture and make a final decision, you can check the list of criteria below:

- the current application codebase has become large and almost unsupported;
- new specialists who are just starting work on the application spend a lot of time getting to know it;
- impossibility of introducing new technologies;
- impossibility of prompt delivery of minor updates and changes;
- lack of the ability to reuse ready-made services, tools and solutions;
- strong coupling between the classes of a monolithic application, which makes it difficult to make changes;
- automated testing is difficult or impossible;
- attachment to one platform for deployment;
- problematic integration of modules and solutions in other programming languages;
- inability to coordinate the development process [2].

These problems are found in a lot of software. They have become especially acute at the present stage, when most software solutions are transferred to web resources for open access to them, which increases the requirements for final products.

5. Basic principles of microservices

Developing software using a microservices architecture brings more value and productivity while adhering to the basic principles of a microservices approach.

The main principle is that the smaller the microservice you create, the better. This is due to greater immutability and independence. The concept of detailing a unit of work is very important in many contexts of modern software development. Also, reducing the scope of the problem also means reducing the time it takes to complete the task, reducing the time it takes to get feedback, and reducing the size of the code to deploy [3, 4].

If the software involves the use of a user interface to interact with the user, then the principle of Interface Segregation will be useful. It consists in dividing the cumbersome service responsible for transferring data about how the final interface will look and function into smaller ones that implement the Backend for frontends (BFF) pattern. This pattern assumes the segregation of services by the types of devices used from a personal computer to a mobile phone and routing requests to a common API service. Figure 1 shows a diagram of the BFF pattern.

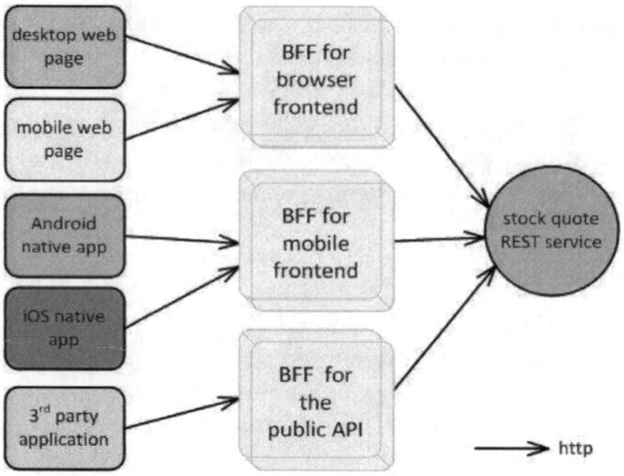

Figure 1. Pattern BFF scheme.

The development of virtualization, as well as the emergence of containerization tools, made it possible to test and deliver the developed software faster. From this follows the third principle (Deployability) of the possibility of deploying microservice applications. This principle includes:

- setting up a runtime infrastructure that includes containers, modules, clusters, persistence, security, and networking;
- expanding and shrinking microservices or porting them from one runtime to another;
- acceleration of the process of fixing, assembling, testing, deploying;
- minimization of downtime when replacing the current version;
- synchronization of changes to versions of related software;
- monitoring the health of microservices for quick troubleshooting.

Since communication of microservices occurs only at the network level, and therefore can be implemented only using one of the following methods: HTTP request, gRPC technology or GraphQL, or through asynchronous messages, which are a queue processed by a special message broker. The first two methods are mostly synchronous, but asynchronous message brokers eliminate the problem of query race. This leads to the fourth principle of the event-driven architecture of communication with the help of brokers.

Availability over Consistency is the fifth principle and is often implemented through the Command Query Responsibility Segregation (CQRS) pattern. The pattern allows you to decouple the design and implementation of operations that modify data (commands) from operations that only read data (requests). CQRS typically relies on replication of service data for queries to improve performance and autonomy.

One of the advantages and the sixth principle is Loose-coupling. It is Loose-coupling that allows you to update, maintain and rebuild even outdated software [5].

The seventh principle of microservices is Single Responsibility. The approach that implements this principle, which has become popular in the industry for empowering microservices, is to follow the principles of Domain-Driven Design (DDD) [6].

6. Containers and orchestration

At the initial stage of microservice architecture development, working with services was inconvenient and inefficient. With the advent of software for automating the deployment and management of applications in containerized environments, it has become more convenient and faster to deliver and develop software on a microservice architecture. Docker is still one of the representatives and flagship in containerization. Later, problems arose of managing and maintaining containers in a constant working state, which led to the creation of software for orchestrating containerized applications such as Kubernetes and Jenkins. With the advent of cloud platforms for deploying applications, the use of orchestration tools has become even more relevant, since for a full-fledged automated deployment and operation of an application, it is enough to define one configuration file, which will include: a description of the interaction of services with each other, the life cycle of each service, backup, recovery in case of failure, as well as load balancing between child nodes and idle physical machines [7].

Nodes are a replaceable unit that is created when there is a lack of performance or exit from other nodes. A simplified scheme of Kubernetes operation is shown in Figure 2. The user makes a request to the master node, then the user is redirected through the API Server to one of the nodes, which is currently more preferable in terms of performance. In turn, each node has one or more pods that store the Docker container image specified in the configuration. The master node also includes a low-level and highly reliable storage of key-value pairs, for finding and working with all clusters, and a scheduler responsible for the correct creation and tracking of the pod life cycle. Communication between nodes and masters occurs through kube-proxy, by proxying the internal network. kubelet. An agent running on every node in the cluster. He makes sure that the containers are running in the pod. The kubelet utility takes a set of PodSpecs and ensures that the containers defined in them are up and running.

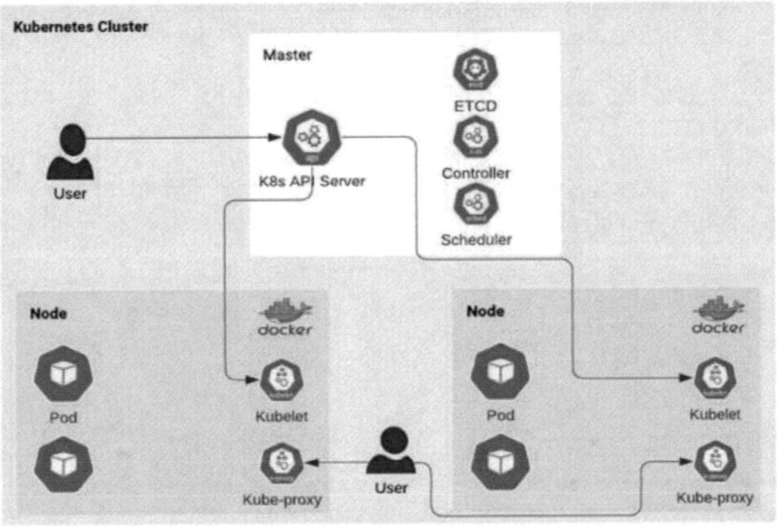

Figure 2. Kubernetes working scheme.

Thanks to the extensive functionality of configuring orchestration and containers, you can design and configure the security of services already at the level of the configuration file. The access of each service from outside and inside the system is configured through the network configuration. If you need to set security on keys, security files, authorization through tokens and databases, Secrets are used. Secrets contain a small amount of sensitive data such as a password, token, or key. Otherwise, such information could be placed in the Pod spec or in the container image [8].

To familiarize yourself with containerization and orchestration technologies, it is recommended that you first create your first cluster using virtual machines. Virtual machines also support isolated orchestration systems, which allows you to set up and debug a future system deployed in the cloud without harming a productive project. The Docker Developers offer a public repository of Docker Hub containers to use off the shelf or create your own. Without much experience in building and assembling containers, you can use ready-made ones, since they already have a starting configuration to run on any platform.

7. Conclusion

The described methodology and aspects of the transition from a monolithic architecture to a microservice architecture are generalized and applicable to a variety of subject areas. After examining the presented data, specialists will be able to objectively assess the advantages of the microservice approach, decide on the use of the platform for deploying the future system, and analyze the existing software according to the given criteria. After deciding on a future migration, he will get acquainted with the basic principles and patterns used in the design and development of microservice applications. Since the existence of microservices without an ecosystem of containerization and orchestration, in general, is impossible, a specialist without certain basic knowledge will be able to understand and apply in practice the described Docker and Kubernetes technologies.

References

[1] Fomin D S and Savkina A V 2021 *Proc. Conf. XLIX Ogarev Readings* vol 1 (Saransk: Ogarev Mordovia State University Press) p 257
[2] Fowler M 2019 *Enterprise Application Templates* (Moscow: Dialectica-Williams) p 546
[3] Lane K and Cherity M 2020 *Database. Reliability engineering* (Saint-Petersburg: Piter) p 304
[4] Bajenova I U 2006 *Developing Distributed Database Applications: A Course of Lectures* (Moscow: Moscow State University M.V. Lomonosov) p 203
[5] Martin R 2018 *Clean architecture. The art of software development* (Saint-Petersburg: Piter) p 352
[6] Ilyin V V 2006 *Business process modeling. Practical use of ARIS* (Moscow: Dialectica-Williams) p 176
[7] Goetsch K 2017 *Microservices for Modern Commerce: Report* (CA: O'Reilly Media) pp 25–47
[8] Nadareishvili I and Mitra R and McLarty M and Amundsen M 2016 *Microservice architecture* (CA: O'Reilly Media) pp 25–38

Research and Modeling of Digital Antenna Arrays with Directional Elements on Azimuth-Elevation in VHF Terrain and Vegetation Multipath Propagation Situations

Yu B Nechaev, I W Peshkov, N A Fortunova, I N Zaitseva and V A Zhigulin

Bunin Yelets State University, 28 Kommunarov street, Yelets 399770, Russia

E-mail: ilvpeshkov@gmail.com

Abstract. In the paper the problem of using digital antenna arrays equipped with dipole and directive elements simultaneously in the azimuthal and elevation planes is discussed. In particular, they are investigated for communication channels of the VHF range below 1 GHz. This range is characterized by the fact that it has found wide distribution. However, its application is fraught with many difficulties, such as attenuation, reflection from rough surfaces, etc. One of the solutions to these problems can be azimuth-elevation digital beamforming. The paper presents the simulation results for the propagation of signals in the VHF range in free space, two-ray ground-reflection and foliage models. It is assumed that a symmetrical dipole is used on the transmitting side, and a directional Yagi antenna on the receiving digital antenna array (DAA) also equipped with a dipole. The circular (CAA) and hemi-dodecahedron shapes of the arrangement of elements of the receiving digital antenna array are simulated. The bit error rates (BER) depending on the distance between the transmitter and the receiver are estimated for the three above mentioned cases.

1. Introduction

At present, a lot of wireless networks are being deployed, requiring both azimuthal and elevation beamforming. Wireless communication using ultra-short waves also presents difficulties such as high attenuation, reflections from the earth's surface, including rough terrain with multiple reflections and diffraction, and propagation in urban environments. The azimuth-elevation beamforming using a digital antenna array is also especially important to overcome the low power of the useful VHF signal, as well as to eliminate the influence of multiple reflections from the different kinds of surfaces such as rough terrain, foliage etc. [1-3].

There are many publications devoted to the problem of ultra-shortwave communication, as well as digital beamforming. Most of them consider only one side of the problem, i.e. either VHF antennas, or beamforming, or VHF transmission in any application. Therefore, it is important to examine more complex cases when the goal is to determine the spatial coordinates in azimuth and elevation using superresolution algorithms and further optimal digital beamforming via the VHF communication channel, depending on the state of the atmosphere, the type of underlying surface, forests, ravines, etc. This approach helps in obtain an optimal system of antenna elements, antenna array geometry, a spectral estimation algorithm with superresolution and subsequent digital beamforming. Hence high-quality communication (minimum bit error rate) over long distances using radio waves of the meter and decimeter ranges will be achieved.

2. Problem Formulation

2.1. Free space propagation model

Consider a few common models of radio signal propagation outside the city, as shown in figure 1. Among them we can mention such as: signal transmission in free space without obstacles in the air (figure 1a); ground wave propagation taking into account surface reflection (figure 1b), as well as inside foliage (figure 1c). The last two scenarios can be related to the multipath fading.

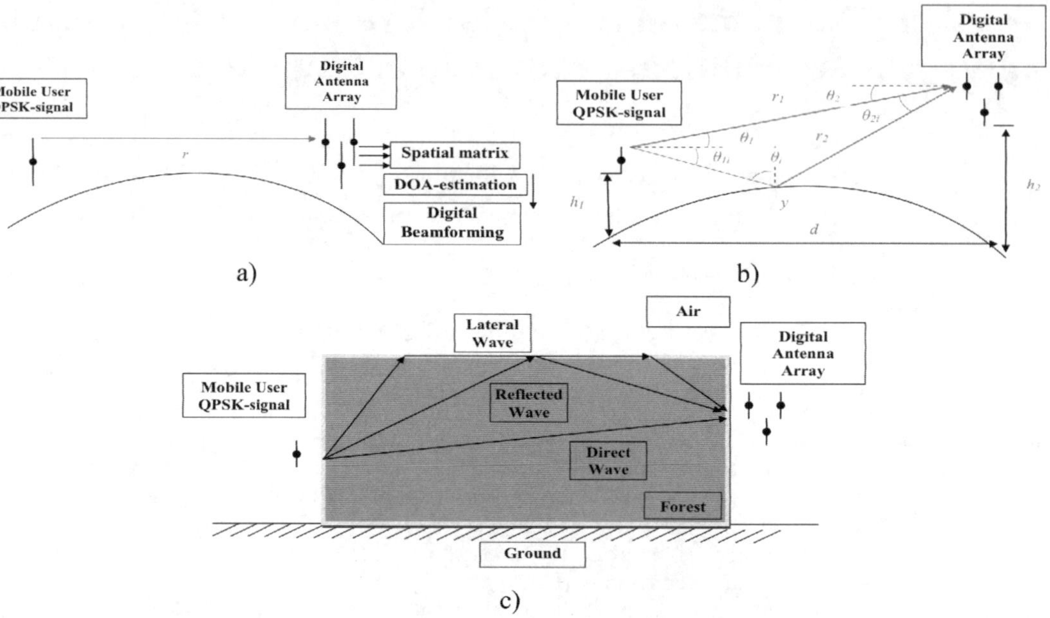

Figure 1. Digital antenna array in models: a) free space, b) two-ray ground reflection propagation and c) foliage.

Consider digital beamforming applied to VHF communication channels. The simplest situation is the direct transmission of the signal from the transmitter (mobile subscriber) to the receiver (digital antenna array) depicted in figure 1a. In addition to the useful signal, noise is superimposed here, and active interference can also affect. Often one of the common forms of antenna arrays that are used for direction-of-arrival estimation and digital beamforming is circular. These arrays can be equipped with a half-wave dipole antenna [4]:

$$E(\phi,\theta) \cong \frac{60\pi I_s l_d}{\lambda r}\left[\frac{\cos\left(\frac{\pi}{2}\cos\theta\right)}{\sin\theta}\right]. \tag{1}$$

On the receiving side a digital antenna array is deployed and also equipped with dipole elements located in space along a circle [5]:

$$\mathbf{a}(\phi,\theta)=\begin{bmatrix} g(\phi_0,\theta_0)e^{j[-kR\cos(\theta-\gamma_0)\sin(\phi)]} \\ g(\phi_1,\theta_1)e^{j[-kR\cos(\theta-\gamma_1)\sin(\phi)]} \\ \dots \\ g(\phi_{N-1},\theta_{N-1})e^{j[-R\cos(\theta-\gamma_{N-1})\sin(\phi)]} \end{bmatrix}, \quad (2)$$

or hemi-dodecahedron [6]:

$$\mathbf{a}(\theta,\varphi)=\begin{bmatrix} g_1(\theta,\varphi+\frac{\pi}{2})e^{jk\mathbf{R}_z(0)[0,0,r_{in}]^T} \\ g_2(\theta+\frac{\pi}{2.5},\varphi+\frac{\pi}{6.8})e^{jk\mathbf{R}_z\left(\frac{\pi}{2.5}\right)[0,\frac{\lambda}{2},\frac{r_{in}}{2}]^T} \\ g_3(\theta+2\frac{\pi}{2.5},\varphi+\frac{\pi}{6.8})e^{jk\mathbf{R}_z\left(2\frac{\pi}{2.5}\right)[0,\frac{\lambda}{2},\frac{r_{in}}{2}]^T} \\ g_4(\theta+2\frac{\pi}{2.5},\varphi+\frac{\pi}{6.8})e^{jk\mathbf{R}_z\left(3\frac{\pi}{2.5}\right)[0,\frac{\lambda}{2},\frac{r_{in}}{2}]^T} \\ g_5(\theta+4\frac{\pi}{2.5},\varphi+\frac{\pi}{6.8})e^{jk\mathbf{R}_z\left(4\frac{\pi}{2.5}\right)[0,\frac{\lambda}{2},\frac{r_{in}}{2}]^T} \\ g_1(\theta+5\frac{\pi}{2.5},\varphi+\frac{\pi}{6.8})e^{jk\mathbf{R}_z\left(5\frac{\pi}{2.5}\right)[0,\frac{\lambda}{2},\frac{r_{in}}{2}]^T} \end{bmatrix}, \quad (3)$$

where $\beta=\frac{2\pi}{\lambda}$ is the wavenumber or wave propagation coefficient (***λ*** is the wavelength), ***R*** is the circular antenna radius, $\mathbf{R}_z(\theta)$ is the rotation matrix around the *z* axis, $\boldsymbol{r_{in}}$ is the inner radius of hemi-dodecahedron, $\boldsymbol{g_0(\theta)}, \boldsymbol{g_1(\theta)},\ldots, \boldsymbol{g_{N-1}(\theta)}$ is the real numbers representing the amplitude of the response of the antenna element in the directions ***θ*** and , as in the expression for the field strength $E(\phi,\theta)$ (2). In addition, the paper considers the case when the Uda-Yagi antennas is simulated as the elements of the DAA as well [6]:

$$F_{yag}(\phi,\theta)\cong\left|\frac{\sin\left[\frac{N_{yag}+2}{2}kd_{yag}(1-\cos(\theta))\right]}{\sin\left[\frac{1}{2}kd_{yag}(1-\cos(\theta))\right]}\right|. \quad (4)$$

2.2. Two-ray ground reflection propagation model

Let us consider the case when the signal is reflected from the earth's surface (figure 1b). In the literature, this case is known as two-path propagation.

$$\mathbf{e}_{EMF}\cong\sqrt{A_e}\left(\mathbf{a}(\phi,\theta_2)E(\phi,\theta_1,r_1)+Rc(\theta_i)\mathbf{a}(\phi,\theta_{2i})E(\phi,\theta_{1i},r_2)e^{jk(r_1-r_2)}\right), \quad (5)$$

where $\boldsymbol{r_1}$ is the distance of the direct ray, $\boldsymbol{r_2}$ is the Reflected beam path distance, $\boldsymbol{A_e}$ is the antenna aperture, $\boldsymbol{R_c}$ is the reflection coefficient.

The distance of the direct ray can be determinedd [10]:

$$r_1=\sqrt{d^2+(h_2-h_1)}, \quad (6)$$

where d is the distance between a user and DAA, h_1 is the user antenna height, h_2 is the DAA height.

Reflected beam path length is:

$$r_2 = \sqrt{d^2 + (h_2 + h_1)}. \tag{7}$$

Distance to the point of reflection from the surface of the Earth is [11]:

$$y = \frac{dh_1}{(h_2 + h_1)}. \tag{8}$$

It is important to determine the angles of arrival at the DAA with respect to the mobile user of the direct and reflected beams to assess the accuracy of spatial filtering using non-isotopic emitters:

$$\theta_1 = \theta_2 = \tan^{-1}\frac{h_2 - h_1}{d}. \tag{9}$$

Calculating the reflected beam is much more difficult.:

$$\theta_i = 0.5\pi - \sin^{-1}\frac{h_1 + h_2}{d}, \tag{10}$$

$$\theta_{1i} = \theta_{2i} = 0.5\pi - \theta_i. \tag{11}$$

2.3. Spatial Processing

In practice, the spatial covariance matrix $\boldsymbol{R}$ is obtained from a set of K_{Sample} time samples [7]:

$$\hat{\mathbf{R}} = \frac{1}{K_{\text{Sample}}} \sum_{k=1}^{K_{\text{Sampl}}} \vec{\mathbf{x}}(k)^H \vec{\mathbf{x}}(k), \tag{12}$$

where $\vec{\mathbf{x}}(k)$ is the sample of the signal from the DAA elements at the k-th moment of time.

The direction-of-arriva estimation in the simplest case is carried out by electronically steering the main lobe of the pattern. This method can be performed mechanically, based on analog phase shifters. However, since the work examines the DAA, we will consider digital phased array [13]:

$$P_{FAA}(\varphi,\theta) = \frac{\mathbf{a}(\varphi,\theta)^H \hat{\mathbf{R}}\mathbf{a}(\varphi,\theta)}{\mathbf{a}(\varphi,\theta)^H \mathbf{a}(\varphi,\theta)}. \tag{13}$$

The spatial spectrum $P_{FAA}(\varphi,\theta)$ has low resolution, however easy to implement. The peak of the $P_{FAA}(\varphi,\theta)$ will point to the coordinates of the signal. In addition, the digital swing of the maximum radiation pattern for receiving a useful signal can be obtained after calculating the vector of weight coefficients:

$$\vec{\mathbf{w}}_{FAA} = \frac{\vec{\mathbf{a}}(\theta_0,\varphi_0)}{\sqrt{\vec{\mathbf{a}}(\theta_0,\varphi_0)^H \vec{\mathbf{a}}(\theta_0,\varphi_0)}}. \tag{14}$$

In order to estimate the coordinates of signals with a higher accuracy and resolution, the MUSIC method is used [14]:

$$P_{MUSIC}(\theta,\varphi) = \frac{1}{\vec{\mathbf{a}}^H(\theta,\varphi)\mathbf{E}_n\mathbf{E}_n^H\vec{\mathbf{a}}(\theta,\varphi)}. \tag{15}$$

where $\boldsymbol{E}_\mathbf{n}$ is the matrix A special vector of weighting coefficients is required also in order to implement the suppression of both interference and noise, as well as amplification of the useful signal, i.e. realization of the full functionality of the DAA. The weight vector that maximizes the signal-to-noise + noise ratio can be obtained [14]:

$$\vec{\mathbf{w}} = \frac{\hat{\mathbf{R}}^{-1}\vec{\mathbf{a}}(\theta_0,\varphi_0)}{\vec{\mathbf{a}}(\theta_0,\varphi_0)^H \hat{\mathbf{R}}^{-1}\vec{\mathbf{a}}(\theta_0,\varphi_0)}, \quad (16)$$

where $\vec{\mathbf{a}}(\theta_0,\varphi_0)$ is the N-dimensional vector of the useful signal after evaluating the spatial spectrum $P_{MUSIC}(\varphi,\theta)$.

2.4. Calculation of the electric field at the receiving digital antenna array in the foliage

Let's consider the last model of signal propagation outside the settlement. Here, the signal source and receiver are assumed to be in a forest. In the used model, the signal will propagate along a directed, path reflected from the crown and over the foliage as shown in figure 1c.The electric field at the end point is the sum of the field of the geometric component $\boldsymbol{E_F}$ (including direct and reflected waves) and the field of the lateral wave $\boldsymbol{E_L}$ (figure 1c):

$$E = E_F + E_L. \quad (17)$$

Electric field with a geometric component (including direct and reflected waves) is calculated using the following formula [3]:

$$E_F = 30Il\left(f_d \frac{\exp\left(-jk_0\sqrt{\xi_{rc}r_d}\right)}{r_d} + f_r \frac{\exp\left(-jk_0\sqrt{\xi_{rc}r_r}\right)}{r_r} \right). \quad (18)$$

where $\boldsymbol{I}$ is the current in the transmitting antenna, $\boldsymbol{l}$ is the length of the transmitting antenna, $\boldsymbol{r_d}$ is the distance traveled by the direct wave, $\boldsymbol{r_r}$ is the distance traveled by the reflected wave, $\boldsymbol{k_0}$ is the wave number of air, which can be expressed as follows:

$$k_0 = \omega\sqrt{\mu_0\xi_0}, \quad (19)$$

$\boldsymbol{\varepsilon_r}$ is the complex permittivity of the forest, calculated by the following formula:

$$\xi_{rc} = \xi_r - j\frac{\sigma}{\omega\xi_0} = \xi_r - j60\sigma\lambda_0, \quad (20)$$

$\boldsymbol{f_d}$ and $\boldsymbol{f_r}$ are antenna radiation patterns. In this paper, a half-wave dipole is used as an antenna. The formulas of its radiation pattern for direct and reflected waves are presentedы below:

$$f_d = \frac{\cos(\pi\cos(\alpha)/2)}{\sin(\alpha)}, \quad (21)$$

$$f_r = \frac{\cos(\pi\cos(\beta)/2)}{\sin(\beta)}. \quad (22)$$

The path length of a direct ray $\boldsymbol{r_d}$ is equal to the length of the segment ***DC*** (figure 2). We find it from the right-angled triangle ***DCG***. The path length of the reflected wave is equal to the sum of the lengths of the segments ***DE*** and ***EC***, which in turn are equal to the segment ***DB***. We find it by the property of the isosceles trapezoid ***ABCD***. The final formulas are presented below:

$$r_d = \sqrt{r_d^{\,2} + 4(h-h_d)(h-h_r)}, \quad (23)$$

$$r_d = \sqrt{d^2 + (h_r - h_t)^2}, \quad (24)$$

where $\boldsymbol{h_r}$ is the height of the receiving antenna, $\boldsymbol{h_t}$ is the height of the transmitting antenna, $\boldsymbol{d}$ is the distance between the antennas, and $\boldsymbol{h}$ is the average height of the forest. The values of the elevation angles α and β required in the radiation pattern formula (20-21) are obtained from the right triangles ***GDC*** and ***GDB*** using the following formulas:

$$\alpha = 90^0 - \arctan\left(\frac{h_r + h_t}{d}\right), \tag{25}$$

$$\beta = \arctan\left(\frac{d}{2h - h_r - h_t}\right). \tag{26}$$

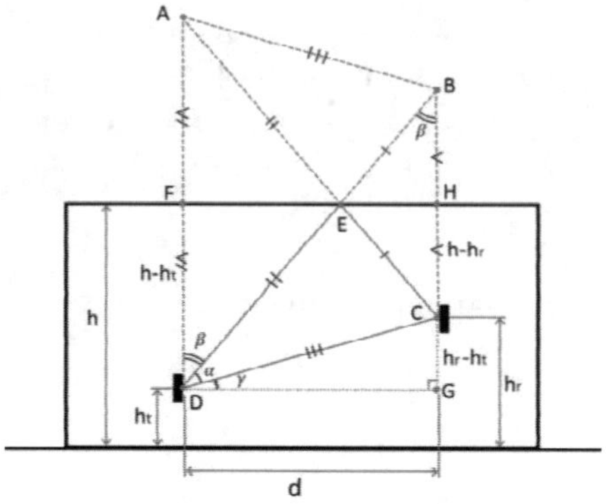

Figure 2. Constructions for calculating the path lengths of direct and reflected waves and elevation angles α and β.

The electric field of the lateral wave E_L can be calculated by the following formula:

$$E_L = \frac{60Il}{\xi_r - 1}\frac{\exp\left(-jk_0\left[r_L + s\left(\sqrt{\xi_r} - 1\right)^{0.5}\right]\right)}{r_L^2}, \tag{27}$$

where r_L is the distance traveled by a lateral wave, and s is the total distance between the antennas and the upper edge of the forest, given by the formula:

$$s = 2h - h_r - h_t\,. \tag{28}$$

For the appearance of a lateral wave, the following condition must be satisfied: the value of the angle of incidence of the ray at the forest - air boundary is equal to the angle of the limiting total reflection θ_C. The value of this angle is calculated using the following formula:

$$\theta_c = \arcsin\left(1/\sqrt{\xi_{rc}}\right). \tag{29}$$

The length of the lateral wave path r_{Rl} is equal to the sum of the lengths of the segments ***AB***, ***BC*** and ***CD*** (figure 3). The lengths of the segments ***AB*** and ***CD*** are found from the right triangles ***ABE*** and ***CDF***. The length ***BC*** is equal to the difference between the distance d between the antennas and the segments ***AE*** and ***FD***. We also find their length from the right triangles ***ABE*** and ***CDF***. The final formula for calculating the length of the lateral wave path traveled is presented below:

$$r_L = \frac{h - h_r}{\cos(\theta_C)} + \frac{h - h_t}{\cos(\theta_C)} + d - tg(\theta_C)(2h - h_r - h_t). \tag{30}$$

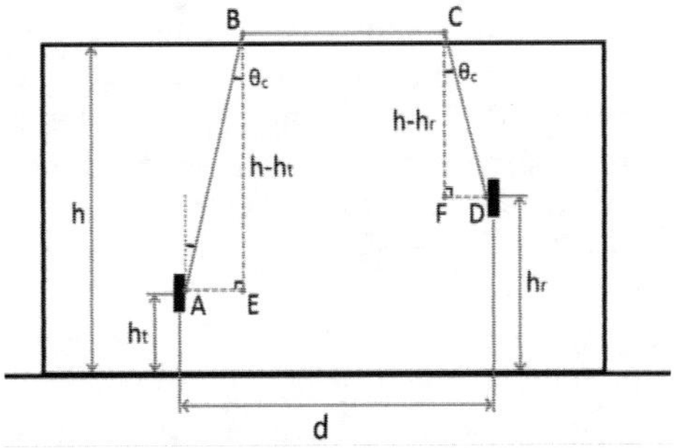

Figure 3. Constructions for calculating the length of the lateral side wave path traveled.

Thus the electric field of the geometric component (including direct, reflected and lateral waves on figure 1c) is calculated using the following formula [3]:

$$E(\phi,\theta) = 30IL\left(f_d \frac{e^{-jk\sqrt{\varepsilon_{rc}}r_d}}{r_d} + f_r \frac{e^{-jk\sqrt{\varepsilon_{rc}}r_r}}{r_r} \right) + \frac{60IL}{\varepsilon_r} \frac{e^{-jk\left(r_L + s\sqrt{\varepsilon_r - 1}\right)}}{r_L^2} \qquad (31)$$

3. Simulation

In the chapter the value of the probability of a bit error (figure 4) is estimated depending on the distance using free space, two-ray ground reflection and foliage propagation models shown on figure 1. QPSK is modeled as a signal source, on which noise is superimposed, whose power is equal to $1.53 \cdot 10^{-13}$ W. The antenna on the transmitter is described by (1), an antenna array (2) and (3) is simulated on the receiver. A dipole (1) or Uda-Yagi (4) can be modeled as antenna elements of a digital antenna array. Then the signal enters the digital spatial processing unit according to the expressions (12-15). The main lobe of the radiation pattern is controlled according to the equations (13-14) («FAA» in figure 4) and using the digital beamforming (15-16) («DAA» in figure 4).

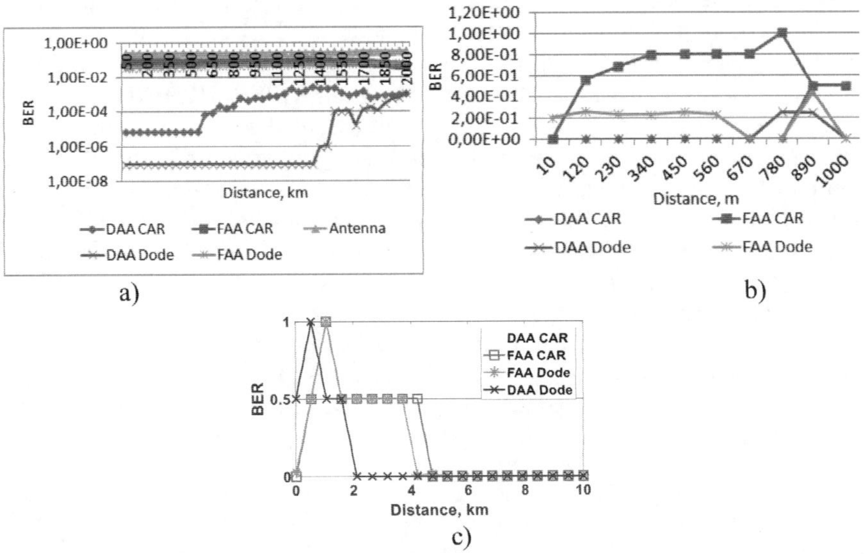

Figure 4. Simulation results: a) free space, b) using two-ray signal model, c) foliage model.

As can be seen from figure 4a, digital antenna array equipped with a hemispherical array, allows increasing the range and reliability of communication in comparison with the circular array almost twice, if we take the bit error rate on 10^{-4}. The two-ray model of signal propagation shows (figure4b) that phasing does not allow reliable communication, since there is a reflected beam, which can only be suppressed through the formation of a null. As shown in figure 4c, a DAA equipped with the hemisphere array allows reducing errors in signal propagation through the forest.

4. Conclusion

In the paper the problem of using digital antenna arrays in conjunction with half-wave dipoles, as well as directional Uda-Yagi elements simultaneously in the azimuthal and elevation planes in free space, ground reflection and foliage scenarios is researched. In particular, a study is fulfilled for VHF communication channels at a carrier frequency of 300 MHz with quadrature phase shift keying. The digital processing unit used the phase space scanning method, as well as the optimal beamforming algorithm. Thus, the calculated bit error rate is significantly reduced using digital spatial filtering and directional half-dodehedron of the receiving digital antenna array.

Acknowledgments

The research was funded by RFBR and Lipetsk Region, project number 20-47-480002.

References

[1] Terrestrial Land Mobile Radiowave Propagation in the VHF/UHF Bands 2002 *Recommendation ITU-R* http://www.itu.int/rec/R-REC-P/en

[2] Chen J 2017 Sea surface observations using a VHF radar with multireceiver beamformingtechnique *OCEANS 2017* pp 1-4

[3] Chan Chung Wei 2011 Investigation of propagation in foliage using simulation techniques *master's dissertation. Naval Postgraduate School. Monterey, USA*

[4] Balanis C A 2005 *Antenna Theory: Analysis and Design* (New York, USA: Wiley-Interscience) p 1104

[5] Nechaev Yu B, Peshkov I W, Fortunova N A and Zaitseva I.N. 2020 The Research of theDigital Beamforming Algorithm for Optimal Noise Reduction in a Cylindrical Antenna Array with Directive Radiator IEEE 2020 Systems of Signal Synchronization, Generating and Processing in Telecommunications (SYNCHROINFO) pp 1-4

[6] Chou H T, Hung K L and Chen C Y 2009 Utilization of a Yagi antenna director array to synthesize a shaped radiation pattern foroptimum coverage in wireless communications*Journal of Electromagnetic Waves and Applications* **23** pp 851-861

[7] Jun Liu, Weijian Liu, Hongwei Liu, Zi-Jing Zhang and Bo Chen 2016 Performance of the SMIbeamformer with signal steering vector errors in heterogeneous environments *Signal Processin* **123** pp 22-29

[8] Afonso D Ribas, Antonio R Carvalho, Carlos M S Figueiredo, Eduardo F. Nakamura 2013 Characterizing the communication in the Amazon rainforest: towards a realistic simulation *Journal of the Brazilian Computer Society* **19** pp 383-393

[9] Kin Lien Chee, Ferran Catalan, Saul A Torrico, Thomas Kürner 2014 Modeling tree scattering in rural residential areas at 3.5 GHz *Radio Science* **49** pp 44-52

Design of local heterogeneous system control networks of a new generation with the preservation of the optimality of the main topological functionals of the network.

M F Karavay[1] and A M Mikhailov[2]

V. A. Trapeznikov Institute of Control Sciences of Russian Academy of Sciences
65 Profsoyuznaya street, Moscow 117997, Russia

E-mail: [1]mkaravay@yandex.ru; [2]alxmikh@gmail.com

Abstract. The paper discusses On-Board Computing Control Systems (OBCS) in astronautics, avionics, autonomous mobile devices, robotics, weapons control and multi-core microprocessors. This is sort of a "backbone", which unites many sensors, calculators, control and executive devices. The architecture of these networks was developed some 30-40 years ago. At that time, these systems met the technical conditions in terms of dynamics and reliability. Nowadays, these systems must perform their functions for 10 to 15 years without maintenance. The performance of system networks must be high enough to solve such tasks as monitoring "swarms" that comprise hundreds of objects or work as a "garbage collectors" in space orbits. Nevertheless modern system networks continue to be based on bus or multi-bus architectures. Since these systems are serial for active nodes, a multi-bus solution is a main way to increase the performance of networks by using very high frequencies that amount to 2 ÷ 4 GHz. It's an extensive path of development, which is problematic. More acceptable would be an intensive path of development, which, in electronics and computer engineering, is associated with the parallelism of task execution. It means that the operating frequencies may not be ultra-high, not exceeding that of modern devices for frequencies of 10 – 600 MHz. However, such devices should work in a parallel mode. The paper proposes a new approach to designing of heterogeneous parallel control system networks, solving parallel tasks, and a conflict-free management of "passive" nodes. To the best of our knowledge, such control system networks are not available as yet.

1. Introduction

The article examines the issues with one of the most important components of almost any CONTROL SYSTEM (especially, and in particular, On-Board Computing Control Systems (OBCS)) in astronautics, avionics, autonomous mobile devices, robotics, weapons control and multi-core microprocessors. This is not a separate appliance or device, but a "system network", that is, a sort of a "backbone", which unites many sensors, calculators, control and executive devices. Failure of the "system network" causes the entire object to fail. Naturally, all the mentioned objects already have their "system networks". In most cases, the architecture of these networks was developed 30-40 years ago, when the requirements for the dynamics (speed of decision-making) and reliability of the system network met the technical conditions of that time. The requirements for OBCS are constantly growing. In terms of reliability, they must perform their functions for 10 to 15 years without maintenance, i.e., they must be both "reliable" and "fault-tolerant". In terms of dynamics, modern objects move at speeds of 6-7 km /s (or 6-7 m/ms). That is, the time of approach and guidance amounts to hundreds of microseconds. Therefore, the performance of the system network must be very high. In addition, it is often necessary to "monitor" a "swarm" of objects, numbering from a few instances to dozens and hundreds of them. A future "garbage collector" will face such tasks, for instance, in space orbits or when tracking a swarm of drones.

In terms of technology and versatility, "system networks" should allow free "scaling" within certain limits, i.e. adding or removing network nodes without changing the network architecture and its software. There are also requirements to production aimed to simplifying network diagnostics and debugging, which depend on the selected network architecture.

Modern system networks continue to be based on architectures such as National Industrial Standard in space avionics and represent a domestic analog of the firm ARINC avionics (for example, in aviation for MC-21). All such systems are bus- or multibus-based. The bus owner is the bus "master" or "controller". Therefore, access to the bus for transmitting information is serial. In case of a single controller, it determines what system nodes and when are to access the bus. Bus hijacking with multiple active devices (Ethernet) is allowed in accordance with different protocols. All these networks are serial for active nodes. It implies a simplicity and lowest performance. In these cases, a multi-bus solution is a main way to increase the performance of such networks. In order to meet the requirements, ever higher frequencies are used in microprocessors, memory systems, interfaces and in the bus system networks. But, there exist a limit. Today's avionics frequencies amount to 2 ÷ 4 GHz. It's an extensive path of development, which is problematic. More acceptable would be an intensive path of development, which, in electronics and computer engineering, is associated with the parallelism of task execution. It means that the operating frequencies may not be ultra-high not exceeding that of modern devices for frequencies of 10 – 600 MHz. However, such devices work a parallel mode. Besides, this frequency range reduces the power consumption, the level of radiated interference, sensitivity to external interference and increases a radiation resistance.

2. What are promising directions for next-generation system networks?

Another class of control networks are parallel networks, where more than one node, even all nodes can be active (transmitting) simultaneously. Examples of such networks are well-known network topologies: complete graph, Omega networks, Klos-Benesh networks, direct and reverse cube, ring, PCI-express network (interface), "thick tree", Marynet, Infiniband, relatively new SpaceWire network, etc [1]. A parallel network constitutes another main road to performance improvement. Actually, having reached the limit of the frequencies used, there remain no extensive ways of development. The alternative is only a parallelism. Supercomputers demonstrate this convincingly. They run hundreds and thousands of microprocessors in parallel mode. A list of main functional properties of new generation networks is as follows.

1) The possibility of autonomous parallel operation of the nodes of the computing core to achieve maximum performance.

2) Conflict-free parallel information transfer from any core node to any network node.

3) Conflict-free self-routing of any core node.

These points justify a need for development of conflict-free parallel system networks.

Such control system networks are not available as yet.

The paper proposes a new approach to the design of heterogeneous parallel control system networks, that contain many passive nodes and an active computing and control core, the volume of which is determined by the required maximum performance. The core should contain from 2 up to 4-6-10 or more microprocessors, partially or completely working in synchronous mode to solve the most critical tasks, or work independently, thus solving parallel tasks and providing conflict-free management of "passive" nodes. The main directions of electronics development for the coming decades were identified as parallelism and reconfiguration. These two directions are actually closely related, since they should be based on one special (minimal!) topology that is *logically equivalent* to a complete graph. Such a minimal static topology in the form of incomplete symmetric balanced block-designs *B(N, s, σ)*, is equivalent to bipartite graphs of special topology, - the graphs found in algebraic combinatorics [2]. Here *N* is the number of blocks (nodes) of the flowchart, s is the degree (valence) of the blocks, *σ* is the number of different (direct) routes between any pair of blocks. Such graphs were called quasi-complete graphs [3]. In this static topology, conflict-free connections are dynamically implemented as in a complete graph. Reconfiguration, which is a basis of diagnostics and fault tolerance is closely related to the theory of parallel systems of special topology. This bipartite topology exists extremely rarely and only for special value couples (*N, s*). That's the reason it is almost not known in computer community. In recent years, the IPU RAS is developing a theory of structural parallelism that will be a basis for building parallel system networks for high-performance OBCS.

3. Main topological functions of the network

Clearly, new topologies by itself do not solve problems, but open up a fundamental possibility for software to achieve optimal basic functional characteristics of the network:

* the highest possible network performance with autonomous parallel operation of the core. Synchronicity of operation for the sake of a highest reliability. An ability to work in any partial synchronicity-confidence relations;

* full parallel direct (i.e., without "entering" other nodes) access between active network nodes and the rest of the nodes;

* no conflicts in parallel routing;

* possibility of parallel independent self-routing;

* dynamic reconfiguration capabilities for fault tolerance;

• availability of alternative access routes in the network (for $\sigma > 1$);

* scaling (within the specified limits).

Each of these problems essentially characterizes the corresponding functionality of the system network. And new developments of control system networks can no longer be based on gradual improvements of existing developments – bus and multi-bus networks, since the latter have been exhausted to the limit over the past 30-40 years. Networks in which the specified functions are optimal in aggregate have to be fundamentally new. Back in the late twentieth century, a special American magazine called such "hypothetical" networks ideal ("*ideal system networks*") [4]. It was clear what to strive for, but it remained unknown how to implement it. We appreciated this definition and introduced it into everyday use.

Nonetheless, parallelism was fraught with several serious difficulties. First: it was not conceptually clear how to build a multiprocessor network for such tasks? This issue was a problem in telephony. How can multiple users talk at the same time, when the frequency separation was still poorly developed? For one, there is an elementary simple solution - connect everyone with everyone else and let them speak directly as much as they want. But this is a very expensive way, since the complexity increases as a square with the number of nodes. Is there a simpler solution? Strictly speaking, no. To put it briefly, the following issues are still lingering.

- So far, no algorithm was found for laying information paths in the network using intermediate switches (as in the Klos network), such that they do not conflict.
- If there are several conflict-free tracks (parallel in time), then how to draw the rest of the tracks without destroying the original ones?

These first two problems essentially boil down to one mathematical problem: given an N-nodes network, is it possible to implement a free of conflicts arbitrary algebraic permutation $N \leftrightarrow N$ (all-to-one)? Obviously, there exist N! (factorial) different transfers, which, being control signals, are neither predictable nor supposed to have conflicts. Although Klos proved that it is theoretically possible to lay down these routs in a special network, he left no clue as to how to achieve it. After 50 years of investigation, the required algorithm is not in sight as yet. The only available solution so far is a complete search. For the sake of reliability, networks' computing cores comprise 3-4-6 processors, whose connectivity pattern follows a topology of the complete graph (each-to-each). In parallel networks, such as PCI-express or SW, there is a need to analyze all information transactions of the problem being solved to ensure the absence of deadlocks and conflicts of parallel transfers.

4. Example

The proposed approach is illustrated with a small heterogeneous system network. Initial requirements are as follows.

1. Design a parallel conflict-free control network that contains $N = 21$ nodes, of which 4 nodes are microprocessors for network management and redundancy, whereas the remaining 17 nodes are passive.

2. The core of the network should be (quasi-) fully connected. Each passive node should be available to four active nodes on different routes.

3. Minimality (or non-redundancy) of the constructed network should be justified.

In this paper, the actual design of the given block diagram $B(N, s, \sigma)$ is not considered, since this had been done earlier. For mathematical details refer to [5, 6].

Figure 1 below provides an isomorphous representation of block-design *B(N, s, 1)*. Mathematical justification for the minimality of the table B(21,5,1) is omitted because it is evident from definition of block-designs. Note that it is an incidence table that represents the required block-design. Network complying with the Figure 1 possesses all mentioned above functionalities. This conclusion follows directly from the BRC-theorem by Brook, Ryser and Chowla [7].

21 Local switches

21 Local switches

	1	2	3	4	5	6	7	8	9	10	11	12	13	14	15	16	17	18	19	20	21
1	I	I	I	I	I																
2	I					I	I	I	I												
3	I									I	I	I	I								
4	I													I	I	I	I				
5	I																	I	I	I	I
6		I				I				I				I				I			
7		I					I				I				I				I		
8		I						I				I				I				I	
9		I							I				I				I				I
10			I			I							I		I					I	
11			I				I					I		I							I
12			I					I			I						I	I			
13			I						I	I						I			I		
14				I		I					I					I					I
15				I			I			I							I			I	
16				I				I					I	I					I		
17				I					I			I			I			I			
18					I	I						I					I		I		
19					I		I						I			I		I			
20					I			I		I					I						I
21					I				I		I			I						I	

Figure 1. Connection table of network with topology of special bipartite 21-node graph [B(21, 5, 1)] (all nodes are active)

This form is isomorphic to the minimal (!) quasi-complete distributed switch for 21 users. It opens a possibility of parallel conflict-free transmission of information (packets) over the network by way of arbitrary algebraic permutations. In this case, the rows of Figure 1 are interpreted as nodes (control and executive devices like processors, converters, motors, etc.), and the columns are interpreted as local switches that implement the connections shown in the table.

Figure 1 can be transformed into Figure 2 by selecting 4 arbitrary lines (highlighted in the Figure 1 with light grey color). They represent 4 active network nodes. Next, find all the columns of the table, the intersections of which with the selected rows do not contain cells marked with 1 (these columns are also marked with light-grey). The identified columns contain five marked cells, each characterizing the connections of passive nodes and only passive nodes with each other. These cells will be deleted during the synthesis of the network, see Figure 2.

These 6 local switches will be removed

4 active nodes (will be removed)

	1	2	3	4	5	6	7	8	9	10	11	12	13	14	15	16	17	18	19	20	21
1	I	I	I	I	I																
2	I					I	I	I	I												
3	I									I	I	I	I								
4	I													I	I	I	I				
5	I																	I	I	I	I
6		I				I				I				I				I			
7		I					I				I				I				I		
8		I						I				I				I				I	
9		I							I				I				I				I
10			I			I							I		I					I	
11			I				I					I		I							I
12			I					I			I						I	I			
13			I						I	I						I			I		
14				I		I					I					I					I
15				I			I			I							I			I	
16				I				I					I	I					I		
17				I					I			I			I			I			
18					I	I						I					I		I		
19					I		I						I			I		I			
20					I			I		I					I						I
21					I				I		I			I						I	

Figure 2. Connection table with 4-node computing core and 17 passive nodes on the base of block-designs [B(21, 5, 1)]

The above example applies to networks with 13 to 21 nodes. This is because of the ports of network nodes, that is, switches and microprocessors. Each microprocessor port is a relatively expensive resource. Switch ports are much simpler. They route the packet towards the destination address. What would be the best way to increase the number of ports in microprocessors? This can be achieved by setting controlled "hubs" or bifurcation taps on the I/O port of microprocessors, thus implementing a virtual multi-porting.

Lines (nodes) 3, 8, 11, 19 are selected (arbitrarily) as nodes of the **computing core** of the network. The marked cells in the columns show the numbers of nodes (lines) connected to the local switch on the top line. For the given network, where each node, including the cores, is independently available to any element of the core, the minimum number of ports of the microprocessor core must be 5. In this case, each microprocessor can connect through its "hubs" to five network 5-port routers (using one port in each router) and reach any of the other (4x5) = 20 nodes - passive or belonging to the core. Real network connections are shown in Figure 1.

In this example a further reduction of distributed connections is also a possibility, - the possibility that diminishes the core volume. Suppose, only 3 active nodes are available, which are 8, 11, and 19 (see Figure 3).

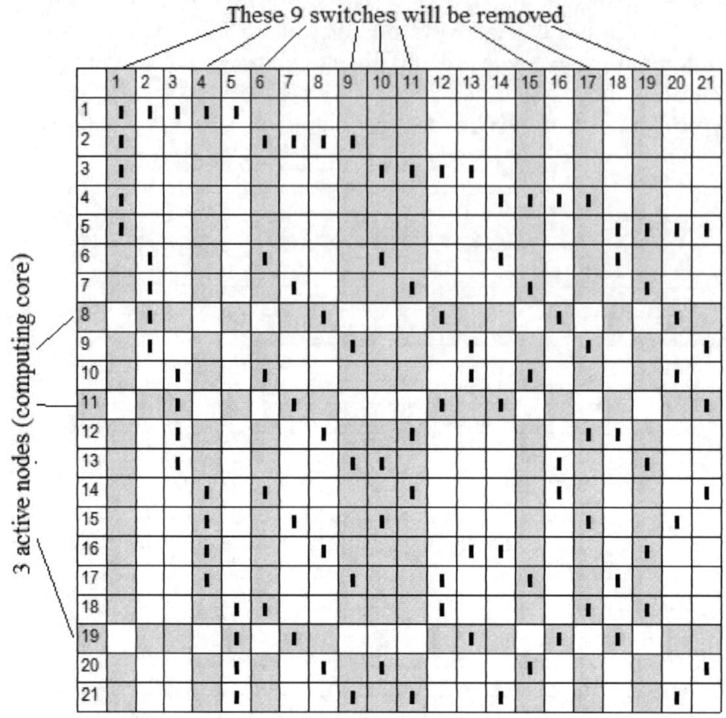

Figure 3. Connection table with 3-node computing core and 18 passive nodes on the base of block-design [B(21, 5, 1)]

Note that 4-th active core with the former number 3 was deleted. In this case, three new additional columns arise that do not have the mark 1 on intersections with rows 8, 11 and 19. These columns' numbers are 1, 9, 10. This opens up a possibility to further reduce the number of local switches by nine. Certainly, reduction of the core volume leads to a deterioration of the functional parallelism. Nevertheless, all remaining active nodes preserve fully parallel connectivity with other nodes, self-routing and conflict-free parallel transmission of packets.

A similar reduction can be applied to any other reasonable number of active nodes on a basis of any block-design construction B(*N*, *s*, 1).

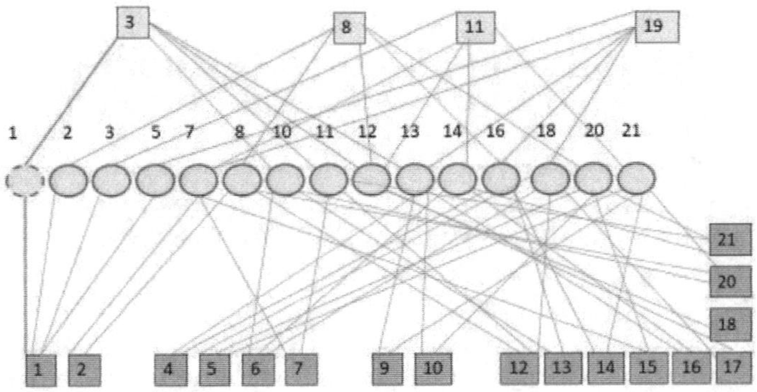

Figure 4. Minimal topology of "ideal" heterogeneous network for 21 nodes, reduced from Figure 2, containing 4 active nodes of computing core

5. Conclusion

The methodology for designing local heterogeneous system control networks based on quasi-complete graphs is offered. The systems feature such major topological characteristics of the network as parallelism of computation, the absence of "deadlocks", conflict-free parallel routing, parallel availability of all nodes for the network core and network scaling.

Although the new topologies by itself do not solve anything, they open up a fundamental possibility for software to achieve optimal basic functional characteristics of the network. With full functional symmetry of all active core nodes, the proposed network topology has the best fault tolerance. A computing core of 5 ÷ 20 microprocessors can "survive" 15 or more years without maintenance. Besides, it easier lends itself to a restoration after irradiation.

New developments of control system networks can no longer be based on gradual improvements of existing ones, that is, bus and multi-bus networks, because these possibilities are completely exhausted over the past 30-40 years. It would take fundamentally new networks to achieve functioning that is optimal in the aggregate.

References

[1] Barabanova E A, Maltseva N S and Barabanov I O 2014 Electronic components for switching chips with parallel adjustment *Proc. 2014 International Conference on Actual Problems of Electron Devices Engineering* (Saratov) **2** 6958245 pp. 204-11.

[2] FINN A and Decker R 1988 A network architecture for radar signal processing *IEEE 8-Th Digital Avionic Systems Conference* pp 614-21

[2] Karavay M, Parhomenko P and Podlavoz V 2009 Combinatorial Methods for constructing bipartite homogeneous minimum quasi-complete graphs (symmetric block-designs) *Automation.&T* **2** pp 153-70

[3] Karavay M and Podlavoz V 2010 The method of invariant expansion of system networks for multiprocessor computing systems *Automation. &T* **12** pp 166-76

[4] Podalsov V and Sokolov V 2006 Single-stage switches of big dimension for multiprocessor and multicomputer systems *Problems of Control* **6** pp 19-24.

[5] Karavay M, Parkhomenko P and Podlazov V 2010 Universal network structure for fault tolerant multiprocessor real-time systems *Proceedings conference "Technical and software management systems, control and measurements* pp 583-97

[6] Holl M 1970 Combinatorics *Mir.*

Investigating the reliability of all-optical switches in transient mode

E A Barabanova[1,2,3], K A Vytovtov[1,2,3], V M Vishnevsky[1], A V Dvorkovich[4] and V F Shurshev[2]

[1]V.A. Trapeznikov Institute of Control Sciences of Russian Academy of Sciences 65 Profsoyuznaya street, Moscow 117997, Russia
[2]Astrakhan State Technical University, 16 Tatishcheva street, Astrakhan 414056, Russia
[3]National Research University "Moscow Power Engineering Institute", 14 Krasnokazarmennaya street, Moscow 111250, Russia
[4]Moscow Institute of Physics and Technology, 9 Institutskiy per., Dolgoprudny, Moscow 41701, Russia

Abstract. In this paper the scheme of 8×8 all-optical switch with two duplicating switching elements and the algorithm of its functioning are presented. For calculating the reliability of all-optical switches, the analytical method based on the Markov process theory is presented. This method allows us to calculate reliability indicators of the all-optical switches as downtime coefficient and availability factor in stationary and transient modes and time of transient mode. The results of numerical calculations show that an increase in the number of duplicating elements by one gives decrease in the downtime coefficient by 6 percent, and the time of the transient process is halved in this case.

1. Introduction

Nowadays a development of all-optical communication elements is a very relevant issue. This is because all-optical communication networks have a lot of advantages such as high throughput and reliability and lower energy efficiency than electron telecommunication networks. To date, the scientific literature comprises a sufficient number of experimental and theoretical works on study of switching elements for optical systems, some of them have been presented in [1-5].

The advantage of our all-optical switches based on the 4×4 and 8×8 basic elements [5-7] is decentralized control of switching process providing higher switching speed. For these schemes the mathematical models for calculation of complexity and performance have been proposed in papers [5-7]. The results have been shown that the complexity of the next-generation all-optical switching system based on the 4×4 elements is in hundred times less than the complexity of crossbar all-optical switches and its performance is in 10 times more than performance of existing ones [5,6]. And the complexity of the system based on the 8×8 elements is in twenty times less than complexity of other well-known schemes [7].

Beyond calculating complexity and performance the study of switch reliability is also important problem [8-10]. The paper [8] presents the method for calculating the reliability of multistage switches using the probabilistic approach. Here are expressions for calculating the uptime probability of well-known Benes switches and multistage switches based on crossbar schemes. In paper [9] a comparison of two types of torus networks using various routing algorithms is presented. The reliability of direct torus topology and Benes networks have been analysed and compared in [10]. Another approach of switching systems reliability calculation which is based on Markov chains has been considered in [11]. Besides network topology investigation of optical switching system reliability includes analyse of optical elements. The reliability of Bragg filters, delay lines, modulators and other standard optical elements has been considered, for example, in [12].

One of widely-used methods of improving reliability of telecommunications components is duplication [13]. The authors have used this approach for design of new 8×8 all-optical switch [7]. It contains the two duplicating switching elements compared to the previous one.

In this paper we present the scheme of new 8×8 all-optical switch (Section 2) and the algorithm of its functioning (Section 3). The analytical method for reliability calculation based on Markov chains is considered in Section 4. This method allows us to estimate the all-optical switching systems reliability both in stationary and transitional modes. The results of numerical calculations are presented in Section 5.

2. The scheme of the 8×8 all-optical switch with two duplicate switching elements

The scheme of the 8×8 all-optical switch with the two duplicate switching elements is presented in figure 1. It consists of the one basic switching element SE1 and the two duplicate switching elements SE2 and SE3. Each of the switching elements consists of the Bragg filter (BF), the displacement system (DS), the frequency Detector (FD) and the control signal converter (CSC). The operation principle of the 8×8 all-optical switch with the two duplicate switching elements is similar to the operational principal of the 8×8 all-optical switch with the one duplicate switching element [7].

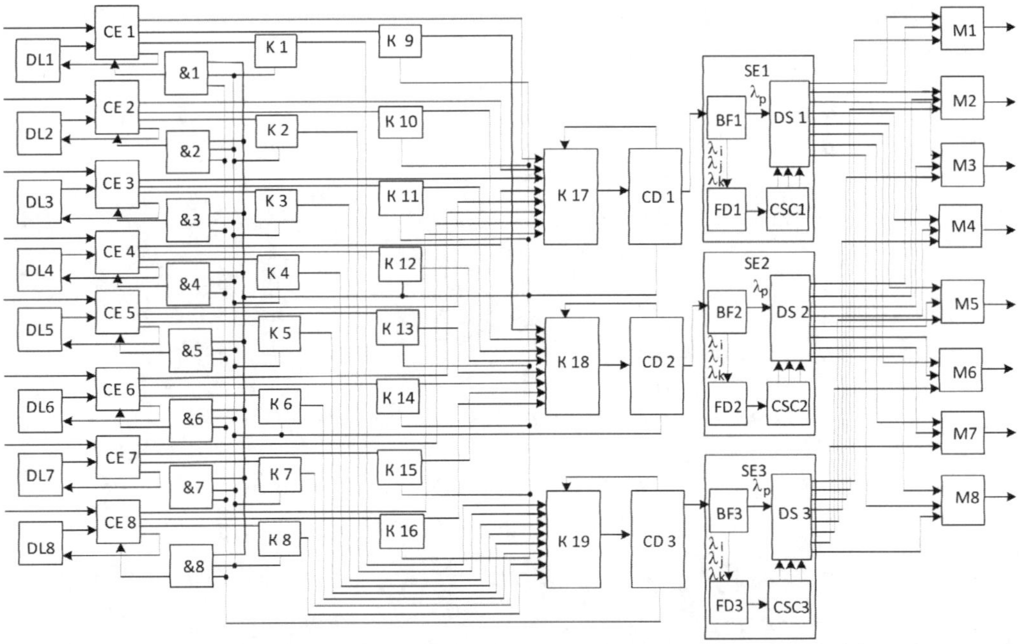

Figure 1. The diagram of the 8×8 all-optical switch with two duplicate switching elements.

The main feature of the 8×8 all-optical switch with the two duplicate switching elements is the operation of the one more switching element. So, if the basic SE1 and the first duplicate element SE2 are occupied or they are faulty information is transmitted through the second duplicate element SE3. Each switching element is controlled by the control device (CD). If the key (K17- K19) is in the on state than the control device is turned on, and the combination of control and information signals arrives to the switching element. The information signal with the wavelength λ_i is transmitted and the control signal with the wavelength λ_c is transmitted. The control signal arrives to FD and then to CSC. In the CSC optical signal is transformed to the electron one and arrived to the control inputs of the deflection system DS. Simultaneously the information signal arrives to the information input of the displacement system DS. So, the information signal from the input is transmitted to one of the eight outputs. The outputs of SE1-SE3 are paralleled and can be transmitted on the 8×8 switching system outputs through the optical multiplexers M1-M8. The input optical buffer of the 8×8 optical switching system consists of the delay lines DL1-DL8, control elements CE1-CE8, optical keys K1-K8 and K9-K16. The optical signal arrives to the buffer when the signal «one» from the END elements &1-&8 shut the control elements CE1-CE8. This case corresponds to the state of the 8×8 optical switching system when all of the switching elements are blocking and signals from three control devices CD1-CD3 arrived to the three inputs of the END elements &1-&8. If one of the switching elements is blocking one of the optical keys K17-K19 is switched to the off state and the optical signal combination arrives to the another switching element directly (on the K 17) or through two groups of the optical key K1-K8 (on the K 19) and K9-K16 (K 18).

3. The algorithm of the 8×8 all-optical switch with two duplicate switching elements

Now let us consider the 8×8 all-optical switch with the two duplicate switching elements as bipartite graph. It is presented in figure 2.

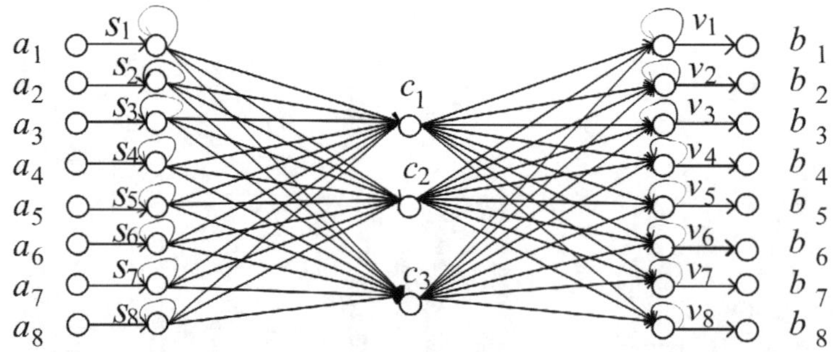

Figure 2. The bipartite graph of the 8×8 all-optical switch.

Let us assume that $A=\{a_1,a_2,a_3,a_4,a_5,a_6,a_7,a_8\}$ is the set inputs of 8×8-element; $B=\{b_1,b_2,b_3,b_4,b_5,b_6,b_7,b_8\}$ is the set of outputs of 8×8-element; $S=\{s_1,s_2,s_3,s_4,s_5,s_6,s_7,s_8\}$ is the set of input delay lines of 8×8-element; $V=\{v_1,v_2,v_3,v_4,v_5,v_6,v_7,v_8\}$ is the set of output delay lines of 8×8-element; $C=\{c_1,c_2,c_3\}$ is the set of switching elements; $Q_1=\{1,0\}$ is the set of states of the first switching element (1-busy; 0-free); $Q_2=\{1,0\}$ is the set of states of the second switching element (1-busy; 0-free); $Q_3 = \{1,0\}$ is the set of states of the third switching element (1-busy; 0-free), $Q_4 = \{1,0\}$ is the set of states of delay lines at the output of the switching element (1-busy; 0-free). These delay lines are included in the schemes of output multiplexers (M1-M8).

Let the set of variables $X = \{x_k\}$ arrives at the input nodes of the graph, where $x_k = \{\lambda_i, \lambda_j, \lambda_l, \lambda_p\}$, i = 1,2, j = 1,2, k = 1,2. Let us introduce functions that determine the states of the communication channel, the states of inputs and outputs of the element, the location of signals in the delay lines, as well as the function of the existence of the communication channel.

The function describing the states of inputs a_i of the switch at time t can be determined by the formula:

$$g(x_k(t), a_i, t) = \begin{cases} x_k(a_i, t), & \text{if a signal arrives at the input } a_i \\ 0, & \text{if a signal doesn't arrive at the input } a_i \end{cases} \quad (1)$$

where $i = \overline{1,8}$. Let at a given time t the signal arrives at the input a_i of the switch, in this case the function $g(x_k(t), a_i, t)$ will be equal to the signal value, otherwise it will be equal to zero.

We define the state functions of the first, second and third switching elements c_1, c_2, c_3, $f_1(t)$, $f_2(t)$, $f_3(t)$ as follows:

$$f_1(t) = \begin{cases} 1, & \text{if } c_1 \text{ is busy at time } t \\ 0, & \text{if the } c_1 \text{ is free at time } t \end{cases} \quad (2)$$

$$f_2(t) = \begin{cases} 1, & \text{if } c_2 \text{ is busy at time } t \\ 0, & \text{if the } c_2 \text{ is free at time } t \end{cases} \quad (3)$$

$$f_3(t) = \begin{cases} 1, & \text{if } c_3 \text{ is busy at time } t \\ 0, & \text{if the } c_3 \text{ is free at time } t \end{cases} \quad (4)$$

Let us introduce the state function of the output optical multiplexers (M1-M8):

$$f_4(t) = \begin{cases} 1, & \text{if } v_i \text{ is busy at time } t \\ 0, & \text{if the } v_i \text{ is free at time } t \end{cases} \quad (5)$$

Then the distribution functions of signals over the input $p_1(t)$ and output $p_2(t)$ delay lines with number i can be determined by the expressions

$$p_{1i}(t)=\begin{cases} x_k(a_i,t),\ \text{if} \begin{cases} f_1(t)=1 \\ f_2(t)=1 \\ f_3(t)=1 \\ g(x_k(t),a_i,t)=1 \end{cases} \\ 0,\ \text{if} \left[\begin{array}{l} f_1(t)=0 \\ f_2(t)=0 \\ f_3(t)=0 \\ g(x_k(t),a_i,t)=0 \end{array} \right. \end{cases} \tag{6}$$

$$p_{2i}(t)=\begin{cases} x_k(a_i,t),\ \text{if} \begin{cases} f_4(t)=1 \\ g(x_k(t),a_i,t)=1 \end{cases} \\ 0,\ \text{if} \left[\begin{array}{l} f_4(t)=0 \\ g(x_k(t),a_i,t)=0 \end{array} \right. \end{cases} \tag{7}$$

In other words, the function $p_1(t)$ will be equal to the value of the signal $x_k(a_i,t)$, if at a given time t the input signal arrives at the input a_i, and all three switching elements c_1, c_2, and c_3 are busy for transmitting the previous signals. And this function will be equal to zero if at a given time moment t no input signal arrives at the input a_i, or at least one of the three switching elements c_1, c_2, and c_3 is not occupied. Thus, the function (6) determines the conditions for the input signal to the corresponding delay line. The function $p_2(t)$ will be equal to the value of the signal $x_k(a_i,t)$ if at a given time t the input signal arrives at the input a_i, and the optical output multiplexer is already occupied. This function is equal to zero if no input signal arrives at the input a_i at a given time t or the optical output multiplexer is idle. Thus, the function of forming the communication channel at time t is given by the formula

$$S_i(t)=\begin{cases} 1,\ \text{if} \begin{cases} p_{1i}(t)=0 \\ p_{2i}(t)=0 \\ g(x_k(t),a_i,t)=1 \end{cases} \\ 0,\ \text{if} \begin{cases} p_{1i}(t)=1 \\ p_{2i}(t)=1 \end{cases} \end{cases} \tag{8}$$

That is, $S_i(t)$ will be equal to one if at a given time moment t the n-th channel from input a_i to output b_j turns out to be unoccupied and the input signal arrives at input a_i. If the n-th channel is busy, then $S_i(t)$ will be equal to zero. Consequently, the function $S_i(t)$ describes the condition for the formation of a new communication channel.

Using formulas (1) and (8), let us describe the operation algorithm of the 8×8 switch with duplication of the switching element which is presented in the figure 3. Let at time t the input signal arrives at one of the inputs a_i, and the control signal contains information about the output b_j of the switch from which it is removed. In this case, the function $g(x_k(t), a_i, t)$ will take one of two possible values: $x_k(a_i, t)$ or 0 in accordance with (1). Then the occupancy of the first, then the second and finally the third switching elements is determined by finding the functions $f_1(t)$, $f_2(t)$ and $f_3(t)$ at a given time in accordance with (2) - (4). In the case when all three switching elements are occupied $f_1(t)=1$, $f_2(t)=1$ and $f_3(t)=1$, then the input signal will go to the corresponding delay line $c_i(t)$, and the function $p_{1i}(t)$ will be assigned the value $x_k(a_i, t)$ (6). In the case when the channel is free, then $p_{1i}(t)=0$. If one of the switching elements is free, that is, $f_1(t)=0$ or $f_2(t)=0$ or $f_3(t)=0$, then the set of values of the control signals $\lambda_i, \lambda_j, \lambda_k$ is used to determine the output number switch b_k.

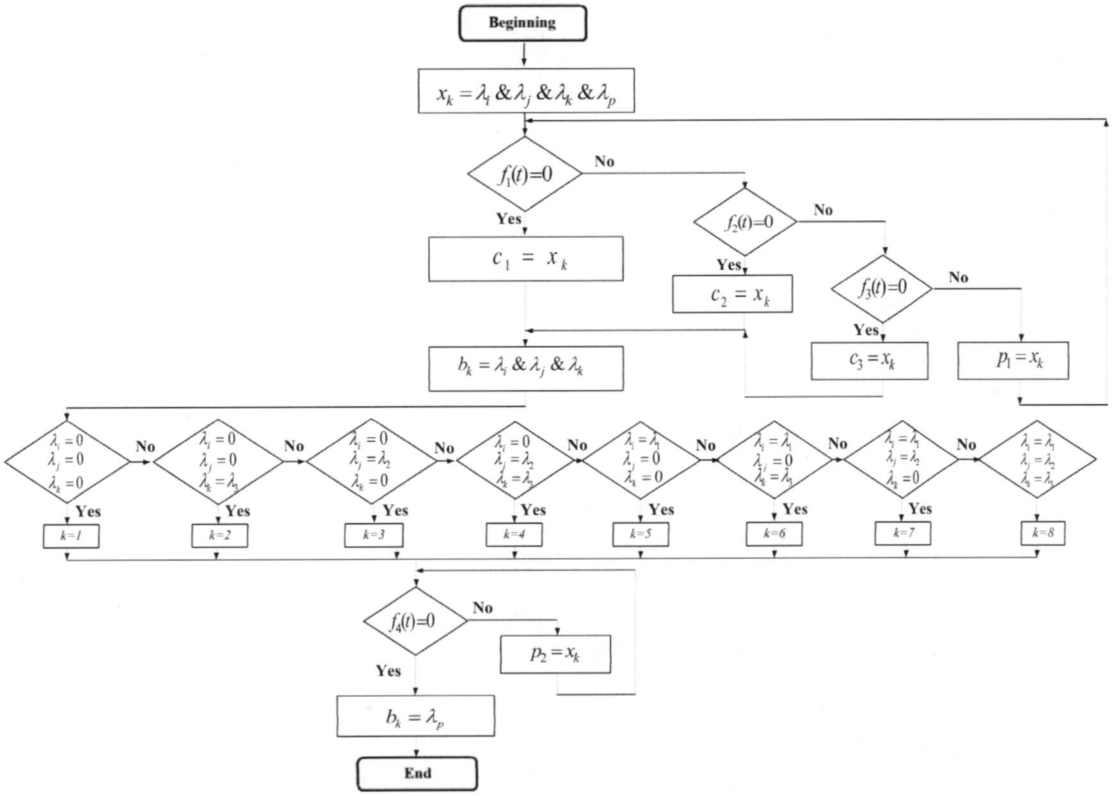

Figure 3. The algorithm of functioning of the 8×8 all-optical switch.

Next, a communication channel is established between the input a_i and the output b_j, and the function $S_i(t)$ (8) becomes equal to one. After that, useful information enters the communication channel.

4. Analytical method of reliability calculation of all-optical switching systems

This section presents the analytical method for calculating the reliability of the 4×4 [6], 8×8 all-optical switches with one duplicate element [7] and the 8×8 all-optical switch with two duplicate elements that are described in the previous section.

4.1. The state graphs of optical switching systems

According to the [6], the two states are inherent in the the 4×4 optical switch. The first state corresponds to the normal functioning of the 4×4 switch and the second state corresponds to its blocking state (figure 4a). Considered [7] the 8×8 switch with one duplicating switching element can be in three states (figure 4b). The first state is the state of both switching elements operation, the second state corresponds to the case when one switching element is working, the second switching element is blocked, and the third state is the state when both switching elements are blocked. Analogously the 8×8 switching system with two duplicating switching elements can be in four states (figure 4c). Here it will be considered the case when flows of processes of switch blocking and restoring their work are the Poisson ones with blocking rate λ and recovery rate μ. The problem is calculating probabilities of finding an optical switch in each of states, as well as an availability and downtime ratios of the systems.

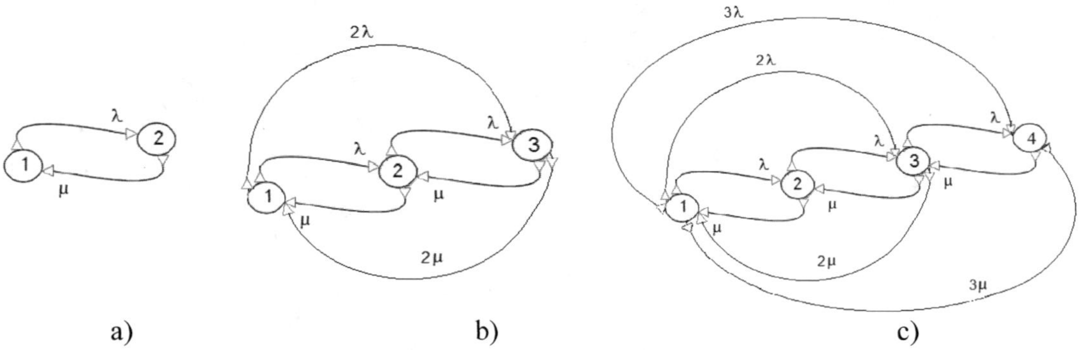

Figure 4. The state graphs of the Markov chain describing a) 4×4 switch b) 8×8 switch with one duplicating switching element c) 8×8 switch with two duplicating switching elements.

4.2. Kolmogorov equations

The systems of Kolmogorov equations describing the process of transitions of Markov chains presented in Fig.4a can be written in the form:

$$\begin{cases} \dfrac{dp_1(t)}{dt} = p_2(t)\mu - p_1(t)\lambda \\ \dfrac{dp_2(t)}{dt} = p_1(t)\lambda - p_2(t)\mu \end{cases} \tag{9}$$

where $p_1(t)$, $p_2(t)$, $p_3(t)$, $p_4(t)$ are probabilities of finding the system in the states 1,2,3,4, respectively. The solutions of the (9) are the following:

$$P_1(t)=\frac{\mu}{\lambda+\mu}+\frac{\lambda}{\lambda+\mu}e^{-(\lambda+\mu)t} \tag{10}$$

$$P_2(t)=\frac{\lambda}{\lambda+\mu}-\frac{\lambda}{\lambda+\mu}e^{-(\lambda+\mu)t} \tag{11}$$

When $t\to\infty$ the system is in stationary mode and expressions (10) and (11) for this mode are

$$\pi_1=\frac{\mu}{\lambda+\mu}, \tag{12}$$

$$\pi_2=\frac{\lambda}{\lambda+\mu}. \tag{13}$$

The availability and downtime ratios of the 4×4 optical switching system can be calculated as

$$K_a(t)=p_1(t), \tag{14}$$

$$K_d(t)=p_2(t). \tag{15}$$

Let us consider the 8×8 optical switching system with one duplicating switching element and find the analytical expressions for availability and downtime ratios. For this, solutions of the system

$$\begin{cases}\dfrac{dp_1(t)}{dt}=-3\lambda p_1(t)+\mu p_2(t)+2\mu p_3(t)\\ \dfrac{dp_2(t)}{dt}=\lambda p_1(t)-(\lambda+\mu)p_2(t)+\mu p_3(t)\\ \dfrac{dp_3(t)}{dt}=2\lambda p_1(t)+\lambda p_2(t)-3\mu p_3(t)\end{cases} \tag{16}$$

must be found. The characteristic equation of the system (16) is

$$\gamma^3+(4\lambda+4\mu)\gamma^2+(3\lambda^2+9\lambda\mu+3\mu^2)\gamma=0. \tag{17}$$

And the roots of (17) are

$$\gamma_1=0,\ \gamma_2=-2\lambda-2\mu+\sqrt{\lambda^2+\lambda\mu+\mu^2},\ \gamma_3=-2\lambda-2\mu-\sqrt{\lambda^2+\lambda\mu+\mu^2} \tag{18}$$

The solution of (16) can be obtained by using so-call translation matrix **M** that relates the state probabilities at the time t to the state probabilities at $t = 0$:

$$\mathbf{P}(t)=\mathbf{M}\cdot\mathbf{P}(0), \tag{19}$$

where $m_{ij}=\sum_{l=1}^{3}\xi_{il}A_{jl}e^{\gamma_l t}$ ($i=\overline{1,3}$, $j=\overline{1,3}$) are the elements of the matrix **M**, A_{jl} are the integration constants determined by the initial conditions, and ξ_{il} are the coefficients expressing probabilities $p_2(t)$ and $p_3(t)$ in terms of $p_1(t)$.

For the 8×8-system with the one duplicating switching element an availability ratio at any time can be calculated as

$$K_a(t)=\sum_{i=1}^{2}p_i(t), \tag{20}$$

and a downtime ratio at any time can be calculated as

$$K_d(t)=p_3(t). \tag{21}$$

Considering the 8×8 optical system with two duplicating elements it is possible to write the system of Kolmogorov equations in the form:

$$\begin{cases}\dfrac{dp_1(t)}{dt}=-6\lambda p_1(t)+\mu p_2(t)+2\mu p_3(t)+3\mu p_4(t)\\ \dfrac{dp_2(t)}{dt}=\lambda p_1(t)-(3\lambda+\mu)p_2(t)+\mu p_3(t)+2\mu p_4(t)\\ \dfrac{dp_3(t)}{dt}=2\lambda p_1(t)+\lambda p_2(t)-(3\mu+\lambda)p_3(t)+\mu p_4(t)\\ \dfrac{dp_4(t)}{dt}=3\lambda p_1(t)+2\lambda p_2(t)+\lambda p_3(t)-6\mu p_4(t)\end{cases} \tag{22}$$

The characteristic equation of the system (22) has the form:

$$\gamma^4+(10\lambda+10\mu)\gamma^3+(27\lambda^2+74\lambda\mu+27\mu^2)\gamma^2+(18\lambda^3+114\lambda^2\mu+114\lambda\mu^2+18\mu^3)\gamma=0 \tag{23}$$

And the roots of (23) are $\gamma_1=0,\ \gamma_2=-3\mu-3\lambda,\ \gamma_3=-\frac{7}{2}\lambda-\frac{7}{2}\mu+\frac{1}{2}\sqrt{27\lambda^2-30\lambda\mu+25\mu^2}$, $\gamma_4=-\frac{7}{2}\lambda-\frac{7}{2}\mu-\frac{1}{2}\sqrt{27\lambda^2-30\lambda\mu+25\mu^2}$. Now, the solutions of (22) can be found analogously in the form (19), where $m_{ij}=\sum_{l=1}^{4}\xi_{il}A_{jl}e^{\gamma_l t}$ ($i=\overline{1,4}$, $j=\overline{1,4}$). For the 8×8-system with the two duplicating switching element an availability ratio at any time can be calculated as

$$K_a(t)=\sum_{i=1}^{3}p_i(t), \tag{24}$$

and a downtime ratio at any time is

$$K_d(t)=p_4(t). \tag{25}$$

5. Numerical results

In this section the numerical calculations for the case of λ=0,4 and μ=0,6 is considered. The dependences of the state probabilities on the time are shown in figure 5. According to the definitions of the downtime ratio for different type of systems (15), (21) and (25) the downtime ratio of the 4×4 system in stationary mode is 0,4, the downtime ratio of the 8×8 system with one switching element in stationary mode is 0,26 and the downtime ratio of the 8×8 system with two switching elements in stationary mode is 0,19. Thus, the results of numerical calculations have been shown that the downtime ratio of the 8×8 switching system with one duplicating element in stationary mode less than the one of the 4×4 switching system by 14 percent. And an increase in the number of duplicate elements of the 8×8 switching system by one result to decreases of downtime ratio by 7 percent.

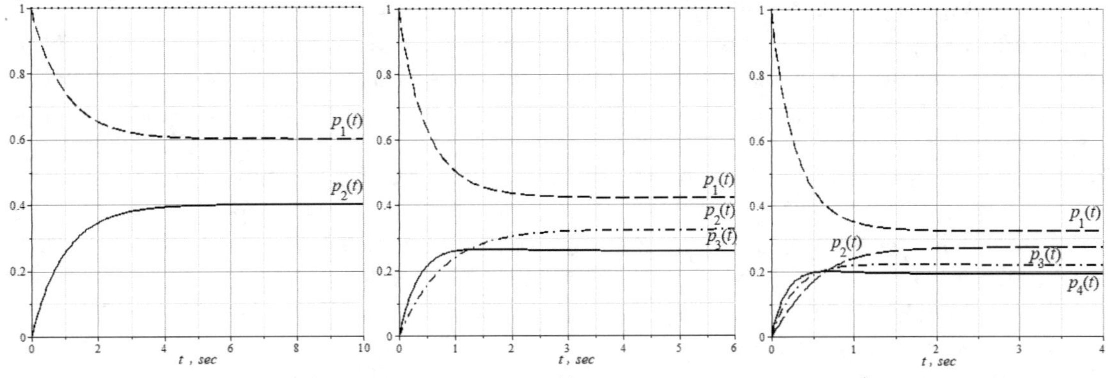

a) b) c)

Figure 5. System state probabilities: a) for the 4×4 switch; b) for the 8×8 switch with one duplicate switching element c) for the 8×8 switch with two duplicate switching elements.

Based on the numerical results it is shown that the time of the transient mode t_{tr3} = 2s for the third case (figure 5c) is in 4 time less than the time of the transient mode t_{tr1} = 8s for the first case (figure 5a) and in two time less than the one (t_{tr2} = 4s) for the second case (figure 5b).

6. Conclusion

In this paper the new scheme of the 8×8 optical switching system with two duplicating elements is presented for the first time. The model and algorithm of function are described also. The 8×8 all-optical switch model is a bipartite graph which is presented and described in Section 3.

The analytical method of the 4×4 and 8×8 optical switching elements reliability investigation is proposed. It is based on Markov random processes theory and on solving systems of Kolmogorov equations. The method allows to investigate the reliability metrics not only in stationary but also in transient mode and to determine the time of transient time.

Numerical calculations showed that the downtime ratio of 8×8 system with two switching elements in stationary mode for the case of λ = 0,4 and μ = 0,6 on 7 percent less than the downtime ratio of 8×8 system with one switching element and on 21 percent less than the downtime ratio of 4×4 system. According to the calculation results, the transient time of functioning of the 8×8 optical switching system with two duplicating elements in 2 time less than transient time of functioning of the 8×8 optical switching system with one duplicating element and in 4 time less than transient time of functioning of the 4×4 optical switching system.

References

[1] Cheng Q, Glick M and Bergman K 2019 Optical Interconnection Networks for High Performance Systems *Optical Fiber Telecommunications* **7** pp 785–825

[2] Maniotis P, Dupuis N, Kuchta D M, Taubenblatt M A and Lee B G 2020 Intra-node high-performance computing network architecture with nanosecond-scale photonic switches *Journal of Optical Communications and Networking* **12**(12) pp 367-77

[3] Seok T J, Kwon K, Henriksson J, Luo J and Wu M C 2019 Wafer-scale silicon photonic switches beyond die size limit *Optica* **6**(4) pp 490-94
[4] Cheng Q, Rumley S, Bahadori M and Bergman K 2018 Photonic switching in high performance datacenters *Optics express* **26**(12) 16022
[5] Barabanova E A, Vytovtov K A and Vishnevsky V M 2020 Method for calculating performance of all-optical linear system of transmitting and processing information *Proc. XXII International Conference Digital Signal Processing and Its Application,* 9213274
[6] Barabanova E A, Vytovtov K A, Podlazov V S and Vishnevsky V.M 2020 High-capacity photon switching systems based on the two-stage 256 × 256 switch *Communications in Computer and Information Science* **1337** pp 57-69
[7] Barabanova E A, Vytovtov K A and Podlazov V S 2019 Model and algorithm of next generation optical switching systems based on 8×8 elements *Lecture Notes in Computer Science* **11965** pp 58-70
[8] Bistouni F and Jahanshahi M 2014 Scalable crossbar network: a non-blocking interconnection network for large-scale systems *Journal of Supercomputing* **71**(2) pp 697-728
[9] Abhilasha S and Sangeetha R G 2021 Terminal and broadcast reliability analysis of direct 2-D symmetric torus network *The Journal of Supercomputing* **77** pp 1517–36
[10] Sharma A and Sangeetha R G 2018 Reliability analysis of data center network *Optical and microwave technologies* pp 71–80
[11] Bistouni F and Jahanshahi M 2020 Impact of Raising Switching Stages on the Reliability of Interconnection Networks *Journal of the Institute of Electronics and Computer* **2** pp 99-126
[12] Berghmans F, Eve S and Held M 2008 An Introduction to Reliability of Optical Components and Fiber Optic Sensors *Optical Waveguide Sensing and Imaging* ed Bock W J, Gannot I and Tanev S (NATO Science for Peace and Security Series, Springer Dordrecht) pp 73-100
[13] Gnanasivam P 2006 *Telecommunication switching and networks* (New Age International (P) Ltd Publishers) 333p.

Journal of Physics: Conference Series **2091** (2021) 012040 doi:10.1088/1742-6596/2091/1/012040

High-capacity strictly non-blocking optical switches based on new dual principle

E A Barabanova, K A Vytovtov, V M Vishnevsky and V S Podlazov

V. A. Trapeznikov Institute of Control Sciences of Russian Academy of Sciences
65 Profsoyuznaya street, Moscow 117997, Russia

Abstract. — In this paper the new dual principle of designing high-capacity strictly non-blocking optical switches is presented for the first time. The new type of switches with decentralized control based on the dual principle has been developed for the first time too. In accordance with the scheme topology this type of switching system was called as the quasi-complete dual switch. It also is analysed the optical signals minimal transmission time that provide the non-blocking property of the new schemes. For the circuit and link complexity calculations the accurate analytical expressions are obtained for the first time. The numerical investigation shows that the proposed schemes significantly benefit in comparison with the well-known switches, for example, the crossbar and Clos schemes. The results of comparing of the dual switch and other types of self-routing switches throughput show an obvious advantage of the proposed scheme. It also is shown that the offered switch type can be considered as a perspective for high-capacity strictly non-blocking optical systems. Indeed, the strictly non-blocking property, the scalability, high throughput, and the decentralized control are the main advantages of offered switches.

1. Introduction

The design of 5G / 6G communication systems, the development of cloud technologies and the Internet of Things are currently in the focus of researchers around the world [1-22]. Well-known basic concepts of next generation networks are already presented in the literature [3,4]. According to reference books, the data transfer rate in new generation mobile networks should be hundreds of Mbps. To ensure such a data transmission rate, it is necessary to increase the throughput of optical telecommunication systems to 1 Tbit/s and more.

This work is devoted to the development of high-speed optical switches that meet the requirements of both modern WDM (Wavelength Division Multiplexing) and SCN (Spatial Channel Network) systems, and future optical packet networks OPS (Optical Package Switching) [4].

It is important to note that, despite the large number of developments in the field of optical switching [1-22], the problem of insufficient switching speed has not been resolved, since all previously presented systems are actually electro-optical [9-13]. According to theoretical sources [11,21], their speed does not exceed 10^{-11} s, since electronic circuits are used in their control systems. Indeed, the experimental results demonstrate [9, 10, 20] a switching speed of the order of 10^{-9} s. As a rule, WSS (Wave-length Selective Switching) systems are based on micromechanical devices MEMS (Micro-Electro Mechanical System) [5] or liquid crystals on silicon LCoS (Liquid Crystal on Silicon) [6]. The performance of this type of switch is also limited by the physical principles of their operation and control.

It should be noted that today there are no high-speed optical control devices that could solve the switching problems, since they are based on slow nonlinear optical effects [21]. As a rule, these devices have no advantages over electronic ones. The exception is linear devices described in [2,22,27]. One of the possible ways to solve the performance problem can be the use of both decentralized control schemes [28] and high-speed linear optical effects [21].

The problem of non-blocking of the space switching system is one of the most important problems, which cannot be considered finally solved even today [11]. Indeed, it is known that such schemes as switches Spanke, Benes, Spanke-Benes, Clos, Banyan are not strictly non-blocking [12,17,18,24], and the problem of finding a free communication channel was solved using routing algorithms and control. The exception to all these schemes is the toggle switch, but its complexity is so great that they cannot be used in high-bandwidth systems [11]. There is also an article [13], which describes a strictly non-blocking 4 × 4 optical switches based on a two-layer network architecture. However, its complexity is greater than that of the switching bar if the capacity is greater than 4×4. Thus, the existing principles of constructing a non-blocking electronic switch, known electronic switching circuits and devices do not allow solving the problem of non-blocking in the optical range for large-capacity switching systems.

In this article, we propose a new principle for constructing optical switches, a new class of switching circuits based on the proposed principle, an analytical method for investigating the complexity, and numerical calculations confirming the advantages of the proposed circuits. compared to existing ones.

In Section 2 of this work the so-called dual principle of high capacity non-blocking optical switch designing is described in detail for the first time. This principle is based on the simultaneous application of the two methods: the method of parallel transmission of information and control signals and the method of time diversity of transmitted optical signals.

The new class of switch systems based on the developed dual principle is presented in Section 3. These high-capacity switches are called here as quasi-complete dual switches in accordance with scheme design. It is very important result for practical applications in high-speed systems that they are strictly non-blocking high-speed devices. Possibility of scale expansion is the important property of these schemes.

The new calculation method of the circuit and link complexity of the quasi-complete dual switches is developed and the corresponding expressions are obtained in Section 4 for the first time. The numerical calculations and comparison of the developed switches with the well-known ones are given in this section also.

The throughput of the proposed switching systems is estimated in Section 5. The numerical calculations and comparison of the quasi-complete dual switches with the blocking schemes are carried out in this section.

2. The new dual principal of optical switching

The new dual principle implies the simultaneous use of the two methods of resolving blockings: the method of parallel transmission of information and control signals and the method of time division of transmitted optical signals.

2.1. The method of information and control signal parallel transmission

First of all, to describe the method, we are introducing the conception of an optical group as a combination of information and control signals transmitted at the same moment of time on the different wavelengths (figure 1). Such an optical group consists of one information signal and a set of control signals. Besides phase and amplitude of the signal can be used as a distinguishing property of control signals within this optical group. For the example the case of frequency division of information signal and two control signals in a switching system is shown in figure 1. In this case the system is based on the 2×2 elements. In figure 1 $S_{inf}(t)$ is the information signal, $S_{c1}(t)$ is the first control signal, $S_{c2}(t)$ is the second control signal. The signals $S_{inf}(t)$, $S_{c1}(t)$, $S_{c2}(t)$ together give us the introduced optical group, $T_0 = \lambda_0/c$, $T_1 = \lambda_1/c$, $T_2 = \lambda_2/c$ are the periods of the high frequency oscillations, λ_0, λ_1, λ_2 are the wavelengths of information signal $S_{inf}(t)$ and two control signals $S_{c1}(t)$

and $S_{c2}(t)$ correspondingly, c is the light velocity, T is signal period determined by technical characteristics of the scheme.

The signals $S_{c1}(t)$ and $S_{c2}(t)$ with the different frequencies arrive simultaneously with the information and establish the necessary state of the system. The signal $S_{c1}(t)$ establishes the state of the switching element of the first stage, and signal $S_{c2}(t)$ establishes the state of the switching elements of the second stage. It is obviously that the number of control signals increases with an increase in the system capacity. As a result, the bandwidth of high-capacity optical switches will be limited by the bandwidth of the optical communication channel. To enlarge in the system capacity with a fixed bandwidth is possible due to the use of additional phase or amplitude division of the signals.

The described method can be applied to the well-known Banyan switches [16,18] (figure 2). Here each output (o_1 of the stage B, *o2* of the stage B, *o*3 of the stage B, *o4* of the stage B) is associated with the certain combination of control signals: 00, 01, 10, 11. The first digits in these combinations correspond to the control signal $S_{c1}(t)$ and the second digits correspond to the control signal $S_{c2}(t)$. If the digit is equal to zero than the control signal at the given frequency is absent ($S_{c1}(t)$=0 for 00 or 01, and $S_{c2}(t)$=0 for 00 or 10). If the digit is equal to unit than the control signal at a given frequency is presented ($S_{c1}(t)$=1 for 10 or 11, and $S_{c2}(t)$=1 for 01 or 11).

Two control signals are used to establish a path in the scheme in accordance with the proposed principle of parallel transmission of information and control signals (figure 2). The control signal S_{c1} (t) is used to determine one of four outputs (*o*1, *o*2, *o*3, *o*4) of the first stage (stage A in figure 2) and the signal S_{c2} (t) is used to determine one of four outputs (*o*1, *o*2, *o*3, *o*4) of the second stage (stage B in figure 2). Obviously, the number of control signals depends on the number of switching system inputs. So we need two control signals for 1×4, 2×4, and 4×4 switches, four control signals for 1×16, 2×16, ..., 16×16 switches, etc.

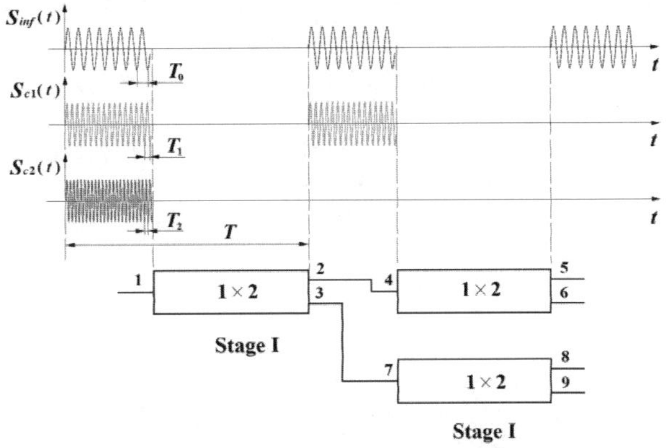

Figure 1. The illustration of parallel bitwise information and control signals transmission with frequency division.

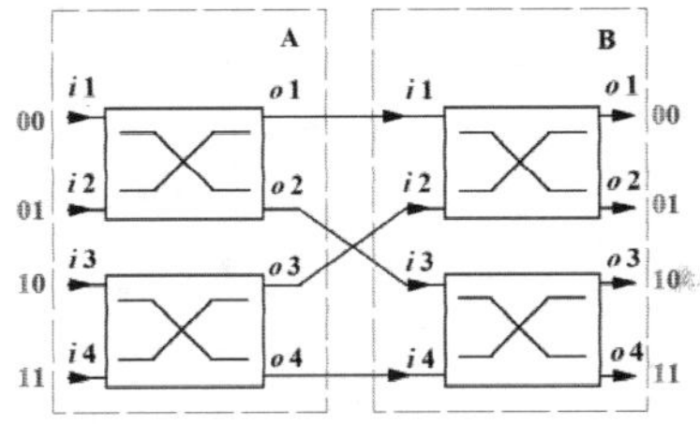

Figure 2. The typical 4×4 Banyan Switch.

It should be necessary noted that using only this method is not enough to design strictly non-blocking switching systems, and obviously there are some cases when the switching system will be blocked. For example, when signals arrive at the first (*i*1 of the stage A in figure 2) and second (*i2* of the stage A in figure 2) inputs, blocking occurs already in the first stage, and when signals arrive at the first (*i*1 of the stage A in figure 2) and third (*i*3 of the stage A in figure 2) inputs and transmit to the first (o1 of the stage B in figure 2) and second (*o*1 of the stage B in figure 2) outputs correspondently, blocking occurs in the second stage (the stage B), etc. Moreover, in the scientific literature, the problem of decentralized control in Banyan switches, in particular for the optical range, has not been studied sufficiently deeply [17,18]. Therefore, it must be applied the additional method described below for designing high-capacity strictly non-blocking optical switching systems.

2.2. The method of time division of transmitted signals

In this subsection the method of time division of transmitted optical signals that provides the non-blocking property of an optical switching system is present. In accordance with the method it is offered to use the time division of the transmitted optical groups so that simultaneous arriving of input signals at one output is impossible. To describe the method, we introduce the conception of a so-called guard interval T_g as the minimal transmission time of an optical group that provide non-blocking property of the system. For example, to provide non-blocking property of the 4×4 Banyan system based on the 2×2 elements (figure 2) the guard interval must be equal to $2T$.

Let us demonstrate this statement by the example of the time diagram (figure 3). Here figure 3a, figure 3b, figure 3c, figure 3d are the diagrams of the input optical groups for Banyan switch (figure 2), *i*1A is the optical group at the first input of the stage A, *i*2A is the optical group at the second input of the stage A, i3A is the time diagram at the third input of the stage A, i4A is the time diagram at the fourth input of the stage A. Analogously figure 3e, figure 3f, figure 3g, figure 3h are the time diagrams at the outputs of the stage A. These diagrams correspond to states of the inputs of the stage B. Figure 3j, figure 3k, figure 3l, figure 3m are the time diagrams at the outputs of the stage B.

The superscriptions of the optical group in the time diagrams correspond to the input signals in figure 3a, figure 3b, figure 3c, figure 3d. Now suppose that four optical signals arrive simultaneously at the inputs *i*1A, *i*2A, *i*3A, *i*4A of the first stage (stage A) and they must be transmitted to the outputs *o*1B, *o*2B, *o*3B, *o*4B (stage B) of the switch correspondingly. First, the signals from the inputs *i*1A and *i*3A of the stage A arrive at the outputs *o*1A, *o*3A of the stage A without delay, and the signals from the inputs *i*2A and *i*4A must be delayed for the time period T at the stage A. Therefore, they arrive at the outputs *o*2A, *o*4A of the stage A at the time moment $t = T$. Thus, the new input optical group can arrive at the inputs *i*1A, *i*2A, *i*3A, *i*4A (stage A) of the switch only after two time periods $t = T_g = 2T$.

Analogously can be considered the next optical groups that arrive at $t = T_g = 2T$. It is assumed that these signals should arrive at the outputs in accordance to the following rule *i*1A → *o*3B, *i*2A → *o*1B, *i*3A → *o*2B, *i*4A → *o*4B. It is seen that in this case the guard interval is also equal to $2T$ (figure 3).

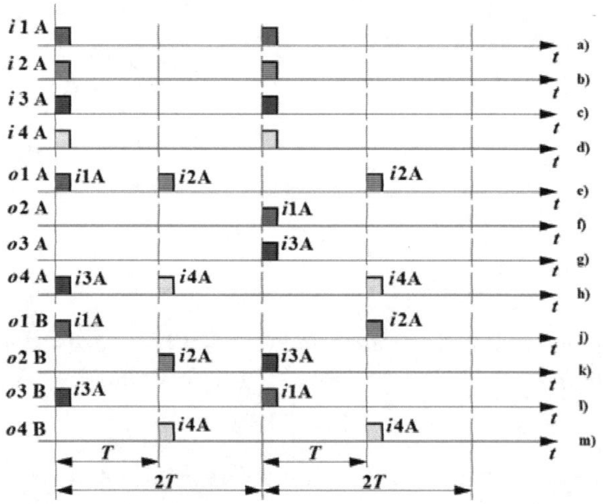

Figure 3. The timing diagram of the signal transmission in a 4 × 4 optical switches.

Numerical calculations show that the guard interval is equal to $2T$ for the two-stage 4×4 switch, and it is $4T$ for the four-stage 16×16 switch, it is $8T$ for the eight-stage 256×256 switch, it is $16T$ for the sixteen-stage 65536×65536 switch i.e. In general case the guard interval T_g is equal to $k \times T$, where k is the number of scheme stages. As the result typical Banyan schemes give unacceptably large guard interval T_g for the large number of inputs/outputs. Therefore, it is required additional fundamental changes in the design principles of high-capacity switches. Moreover, this method has not been used previously for Banyan systems with large amount of inputs. However, in some papers, for example in

[16], authors have tried to applied time diversity and for this they have used buffer devices at the inputs of each 2 × 2 elements. But in that treatment, it has not been solved the problem of strictly non-blocking on this way. Moreover, they have not even considered high-capacity switching systems.

As the conclusion we must note that despite the methods analogous to the described ones have been used separately in some types of multistage switching system [11,19,24], only the proposed dual principle allows us to solve the problem of strictly non-blocking in distributed control systems for the first time. It is also important that in order to implement the new dual principle, we propose new switching schemes.

3. The new dual principal of optical switching

To construct strictly non-blocking high-capacity optical switching systems with a small guard interval, the 4×4 all-optical decentralized element [27,30] has been proposed by authors as the basic device for the first time. This element (figure 4a) operates in accordance with the dual principle described above (Section 2). It contains four optical demultiplexers D1-D4 and four optical multiplexers M1-M4 that is connected in accordance with the principle of a complete bipartite graph [26]. The delay devices B1-B4 are installed at the inputs of each multiplexer. These devices provide a signal delay for preventing switch blocking. Note that the real optical delay devices have already been described for example in [23].

Let us consider the process of signals transmission in the proposed dual 4×4 switch (figure 4a). First of all, note that a permutation is understood in the switching theory [24] as the ordinary mathematical concept says that all input signals must be distributed between all outputs, and two or more input signals cannot arrive at one output simultaneously. Now suppose that input signals arrive at all four inputs (*i*1, *i*2, *i*3, *i*4) simultaneously and they should be transmitted to the four different outputs (*o*1, *o*2, *o*3, *o*4) in the circuit (figure 4a).

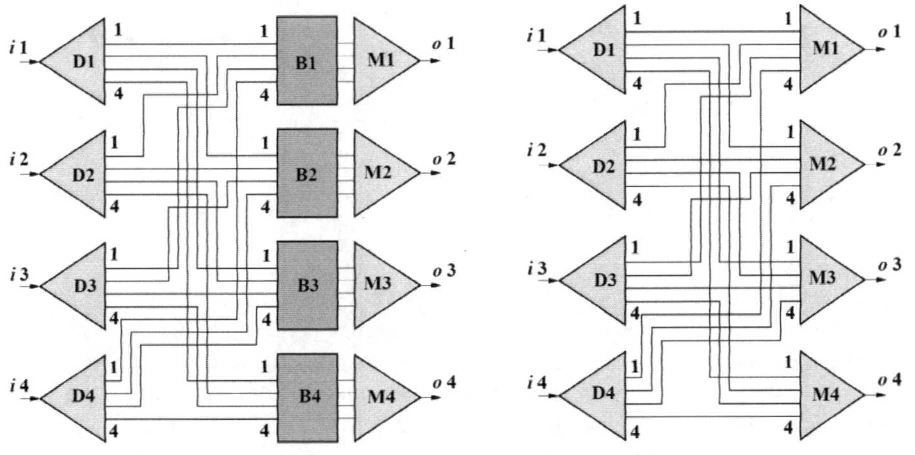

a) b)

Figure 4. The 4×4 switches built in accordance with a) the dual principle, b) the cross-connect principle.

For example, a signal from the input *i*1 should go to the output *o*1 (*i*1 → *o*1), analogously for other inputs *i*2 → *o*2, *i*3 → *o*3, *i*4 → *o*4. In accordance with the scheme, a signal from the input *i*1 goes to the first output of the demultiplexer D1, then it goes to the first input of the delay line B1, and then it goes to the first input of the multiplexor M1, from this input it goes to the first output o1. So we can write this path as the sequence *i*1 →*o*1 (D1) →*i*1 (B1) →*o*1 (B1) →*i*1 (D1) →*o*1. Similarly, for a signal from input *i*2 we can write *i*2 →*o*2 (D2) →*i*2 (B2) →*o*2 (B2) →*i*2 (D2) →*o*2 etc. Therefore, they are distributed between the outputs without delays. Thus, the delay devices B1-B4 are not required in the considered 4×4 scheme, and we can use the well-known optical cross-connect (OXC) scheme (figure 4b) for non-blocking transmission [25].

Now let us consider the well-known 16×16 OXC (figure 5a) and the developed 16×16 dual switch (figure 5b). The topology of OXC switch is a complete graph. There are the direct connections between all inputs and all outputs within this switch (figure 5a). This condition requires using 16×1 multiplexers and 1×16 demultiplexers in the scheme. As the result we have the large number of the interstage connections that gives increase of the complexity ($s = N^2 = 256$, N is the number of inputs [33]). Moreover, the schemes of 16×1 optical multiplexers and 1×16 optical demultiplexers are quite complexity. It can be used more simple 4×1 multiplexers (1×4 demultiplexers) or 2×1 multiplexers (1×2 demultiplexers), but in this case the complexity of the whole scheme will be even greater. Note that future increase in the number of inputs leads to significant complication of the OXC switch, and therefore, it leads to reliability degradation and control system complication. Thus, using OXC switches for processing large amounts of information is seen inappropriate.

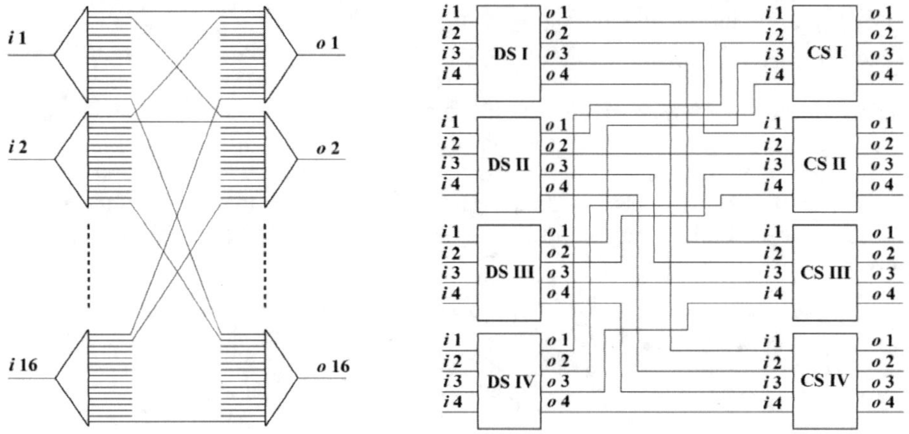

a) b)

Figure 5. The 16×16 switches built in accordance with a) the cross-connect principle, b) the dual principle.

Next we consider the operation principle of the new 16×16 dual switch (figure 5b). It consists of the four 4×4 dual switches DSI–DSIV (figure 4a) in the first stage and the four 4×4 OXC switches CSI–CSIV (figure 4b) in the second stage. Let us assume that input signals arrive at all sixteen scheme inputs simultaneously ($i1$-$i16$ of the first stage) and they should be transmitted to the sixteen different scheme outputs ($o1$-$o16$ of the second stage), so that $\boldsymbol{i}1$(DSI) → $o1$(CSI), $i2$(DSI) → $o2$(CSI), $i3$(DSI) → $o3$(CSI), $i4$(DSI) → $o4$(CSI), $i1$(DSII) → $o1$(CSII), ..., $i4$(DSII) → $o4$(CSII), $i1$(DSIII) → $o1$(CSIII), ..., $i4$(DSIII) → $o4$(CSIII), $i1$(DSIV) → $o1$(CSIV), ..., $i4$(DSIV) → $o4$(CSIV). At the first step signals from $i1$(DSI), $i2$(DSI), $i3$(DSI), $i4$(DSI) must be arrived at $o1$(DSI) consistently in according to following time diversity (figure 5b): a signal from $i1$(DSI) must be arrived at $o1$(DSI) without delay, then a signal from $i2$(DSI) delayed in the line B2 (figure 4a) on the time period T must be arrived at the output $o1$(DSI), then a signal from $i3$ delayed in B3 on the two time periods $2T$ must be arrived at this output, and then a signal from $i4$(DSI) delayed in B4 on the three time periods $3T$ must be arrived at this output too. At the second step these signals will be transmitted consistently to the outputs $o1$(CSI), $o2$(CSI), $o3$(CSI), $o4$(CSI) without additional delay. Analogously, the signals from other inputs $i1$(DSII)–$i4$(DSIV) must be transmitted to the outputs $o1$(CSII)–$o4$(CSIV) through the points $o2$(DSII), $o3$(DSIII), i4(DSIV) and the points $i2$(DSII), $i3$(DSIII), $i4$(DSIV). So, the maximum time period is also $3T$. Thus, a new input optical group can arrive at the switch inputs only after four time periods, and $T_g = 4T$. Therefore, the guard interval of our strictly non-blocking 16×16 dual switch is equal to the blocking 16×16 Banyan switch (see Subsection 2.2).

Thus, the presented 16×16 dual optical switch can be considered as the compromise solution. Indeed, its complexity is significantly less than the analogously parameter of non-blocking complete switch and its performance is equal to the performance of blocking Banyan scheme.

Further increase of outputs of the dual switch leads increase of guard interval. This property leads to a decrease in the performance of the switch. Thus, for constructing high-capacity non-blocking switch based on the dual principle, the authors propose the so-called dual quasi-complete scheme. The dual quasi-complete scheme combines the advantages of OXC switch (high performance) and a dual switch (low complexity).

The scheme of the 16×16 - dual quasi-complete switch is shown in figure 6a. This switch consists of the input, central and output stages. The central stage contains the sixteen dual switches DS1–DS16 described above (figure 5b) with the guard interval $T_g = 4T$. The input stage contains the sixty-four 1×4-demultiplexers D1–D64 that are combined in the sixteen input groups DG1-DG16 including the four ones. The output stage contains the sixty-four 4×1-multiplexers M1–M64 that are combined in the four groups MG1-MG4 including the sixteen ones.

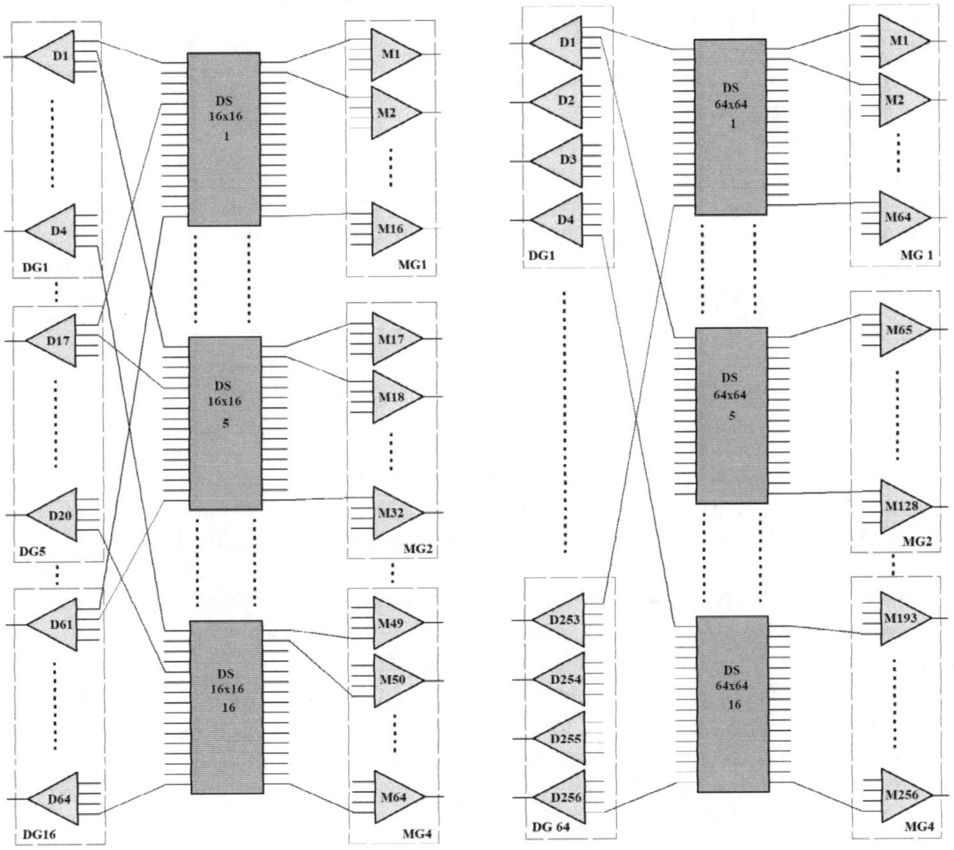

a) b)

Figure 6. The dual switch based on a quasi-complete graph a) N=64, b) N=256.

The multiplexers and demultiplexers are connected with the dual switches DS1×DS16 of the central stage in according to the principle of a quasi-complete graph [26]. The outputs of the first input group DG1 are connected with the first inputs of the central dual quasi-complete switches DS1-DS16. The outputs of the second input group DG2 are connected with the second inputs of the central dual quasi-complete switches DS1-DS16. Analogously the outputs of the sixteenth input group DG16 are connected with the sixteenth inputs of the central dual quasi-complete switches DS1-DS16. The output multiplexors M1-M64 are connected with the central stage in the following way: the first inputs of the first output group MG1 are connected with the fist dual quasi-complete switch DS1. The second inputs of the first output group MG1 are connected with the second dual quasi-complete switch DS2. The first inputs of the second output group MG2 are connected with the fifth dual quasi-complete switch DS5. The fourth inputs of the fourth output group MG4 are connected with the sixteenth dual quasi-complete switch DS16. It is very important for design of high-capacity scheme that the guard interval of such the switch is also $4T$ and it doesn't increase in comparison with the 16×16 switch (figure 5b).

The 256×256 quasi-complete dual switch (figure 6b) also contains the three stages. The input stage includes two hundred and fifty-six 1×4-demultiplexers, and the output stage includes two hundred and fifty-six 4×1-multiplexers. Analogously to the 64×64-switch the input demultiplexers are combined in sixty-four groups of four; the output multiplexors are combined in four groups of sixty-four; the central stage of the switch includes sixteen 64×64-dual switches.

For the schemes of the 1024×1024 -switch the number of the 256×256-switch of the central cascade is sixteen also, the number of the 4×1-multiplexer (4×1-demultiplexer) is one thousand and twenty-four. Analogously, the dual quasi-complete switch contains sixteen 1024×1024-dual switches, four thousand and ninety-six 4×1-multiplexer (1×4-demultiplexer).

Therefore, the very important advantage of the developed schemes is their scalability. It is convenient property for practical implementation of switches in accordance with duality principle.

We can also formulate the second main advantage of the design principle of the quasi-complete dual switches as following: the guard interval of a quasi-complete dual switch is equal to four time periods for any capacity of a scheme. For comparison the guard interval is the six time periods for 64×64 Banyan switch, and it is the eight time periods for a 256×256 one.

4. Calculation of the dual switch complexity and comparison with well-known schemes

4.1. The calculation of the circuit complexity

It is obviously that cost, reliability, and energy consumption of a switching system are depended on its hardware (circuit) complexity. The circuit complexity is a value that is equal to the total number of elements in the switch scheme. In this subsection the mathematical method for the complexity calculation and analysis of the $N\times N$ quasi-complete dual switch is presented for the first time. In the presented method the 16×16 dual switch (figure 5b) is chosen as the basic element. It is followed from figure 4a, figure 4b and figure 5b that its complexity is 32. Indeed, the 4×4 dual switch consists of the four demultiplexer-multiplexer pairs (Appendix A), each of the pair can be implemented as the separate circuit (figure 4a), so its complexity is four [27]. And the 16×16 dual switch (figure 5b) contains the eight 4×4 switches. As the result its circuit complexity is 4×8=32. It is shown above that by using the 16×16 basic element it is possible to construct the 64×64, 256×256, 1024×1024 switches and etc. Considering the schemes of these ones we can see that the capacity of the quasi-complete dual switch can be write as the following expression

$$N\times N\cong 16\cdot 2^{2n-2}\times 16\cdot 2^{2n-2} \quad , \tag{1}$$

where N is the number of inputs/outputs of the switches, 16 is the number of the basic element inputs, n = 1,2, ... is called as the serial number of a switch. Now let us denote the complexity of the considered scheme as S_n. It is seen from figure 6a and figure 6b that the circuit complexity of the 64×64 switch is 640, analogously the circuit complexity of the 256×256 switch is 10752, the 1024×1024 switch circuit complexity is 174080 etc. Analyzing the switching schemes we have obtained the expression of circuit complexity for other serial numbers *n*.

$$S_n = 2^{4n+1}\sum_{i=0}^{n-1}\frac{1}{2^{2i}} \quad . \tag{2}$$

It is followed from (1) that $N = 16\cdot 2^{2n-2}$. Thus, we can express the circuit complexity as the function of the number of the inputs *N*, and write the final expression of the circuit complexity

$$S_N = \frac{N^2}{2}\sum_{i=1}^{\log_2(\sqrt{N})-1}\frac{1}{2^{2i}}, \tag{3}$$

for the first time.

Now to evaluate the benefit in the complexity of the developed switching system in comparison with the existing ones, we propose the new concept of the absolute circuit complexity difference as the difference between the circuit complexities of some comparing switch S_{CS} and the developed quasi-complete dual switch S_D

$$\Delta S = S_{CS} - S_D . \tag{4}$$

Thus, the circuit complexity of the quasi-complete dual switch is less than the complexity of a some comparing switch if $\Delta S > 0$, and this complexity is more than the one of a some comparing switch if $\Delta S < 0$.

Additionally, we introduce the concept of the relative circuit complexity difference as

$$\Delta s = \left(\left|S_{CS} - S_D\right| / S_D\right)\cdot 100\% . \tag{5}$$

This value shows a difference between circuit complexities of some comparing switch and the developed quasi-complete dual switch in percent. Such an evaluation is advisable in many practical cases [31,32].

Now let us carry out comparison of the circuit complexity of Banyan switch, Clos switch, a crossbar switch, and the offered dual quasi-complete switches. Additionally, note again that only a crossbar switch and the presented quasi-complete dual switch are strictly non-blocking.

It is well-known that the circuit complexity of Banyan switch must be calculated by the formula $S_B = (N/4)\log_4 N$ [11,16], Clos switch must be calculated as $S_C = 3N(2N^{1/2} - 1)$ [11], the circuit complexity of a crossbar switch is $S_{CD} = N^2$ [11,24].

The results of the numerical calculations by using these formulae and the expression (3) are presented in figure 7a. It is seen in figure 7a that the complexity of the non-blocking quasi-complete dual switches is greater than the complexity of Banyan scheme, but less than the complexity of Clos switch and significantly less than the complexity of a non-blocking crossbar switch with a small number of inputs. It can also be seen in figure 7b that the complexity of the quasi-complete dual switch with the number of inputs 4096 and more will be greater than the complexity of Clos switch.

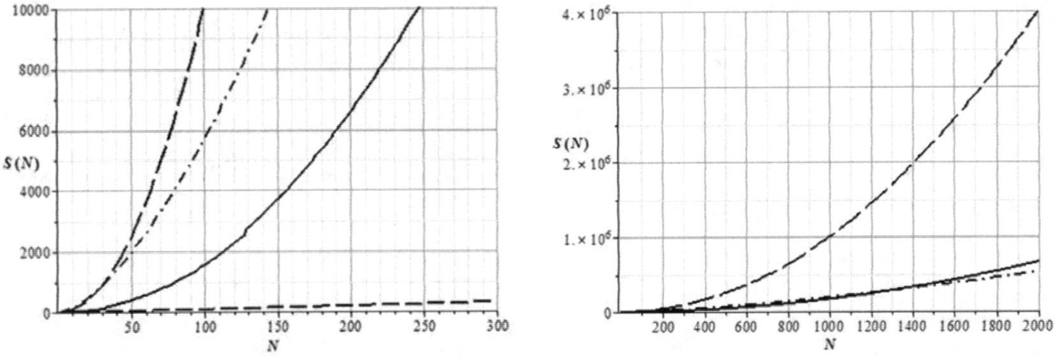

a) b)

Figure 7. The comparison of the circuit complexity of multi-stage switches. The solid line corresponds to quasi-complete dual switch, the long-dash line - to the crossbar switch, the dash-dot line - to Clos switch, and dash line - to Banyan switch: a) for $4<N<300$, b) for $4<N<2000$

But at the same time Close scheme is not strictly non-blocking. And, in comparison with a strictly non-blocking crossbar switch, the proposed schemes have the clear advantage. Moreover, as follows from the expression below the complexity difference ΔS_{CD} between proposed and crossbar switches increases with the number of inputs:

$$\Delta S_{CD} = \frac{N^2}{2}\left(2 - \sum_{i=1}^{\log_2(\sqrt{N})-1} \frac{1}{2^{2i}}\right). \quad (6)$$

The numerical calculation results of ΔS_{CD} are presented in figure 8. For the number of inputs $N = 64$ we have ΔS_{CD} =3456, for $N = 256$ we have ΔS_{CD} =54784, for $N = 1024$ we have ΔS_{CD} =874496, and for $N = 4096$ we already have ΔS_{CD} =13983744. These calculations showed that the proposed schemes significantly benefit in complexity with increase in the number of inputs. Therefore, they have higher reliability and smaller space dimensions, but their performance is four time less in comparison with a crossbar switch. At the same time the signal delay is constant for any number of switch inputs in the offered switch. Moreover, the important feature of the presented scheme is the use of the self-tuning all-optical elements [2,20,27] in the switch. Obviously, their performance is an order of magnitude higher than that of schemes with external control [10] as it determined primarily by transients in the control circuit.

As for Banyan switch, we can see that the complexity of the quasi-complete dual switch is significantly more. For example, ΔS_{BD} = 10752-256 = 10496 for N=256. However, recall again that the Banyan switch is well-known as a blocking scheme.

4.2. The calculation of the link complexity

To optimization switch construction it is important to minimize not only switch complexity but so-called link complexity too. Link complexity C is the total number of interconnections between all stages of a switch [17]. In according to this definition and the scheme figure 4a, the link complexity of the 4×4 switch is 16. Taking into account that the 16×16-switch consists of the eight 4×4 switches and there are sixteen connections between its two stages (figure 5b), the link complexity of the 16×16 switch is $C_{16} = 144$.

It is seen from the figure 6 that the 64×64-switch consists of the sixteen 16×16 switches, the sixty-four 1×4 demultiplexers and the sixty- four 4×1 multiplexers. Let us denote N_{64} as the number of the inputs that is equal to sixty-four in this case. It also is seen that the 4 x N_{64} links connect the outputs of the input stage (D1-D64) and the inputs of the central stage (DS1-DS16), and the 4 x N_{64} links connect the outputs of the central stage (DS1-DS16) and the inputs of the output stage (M1-M16). Therefore, the link complexity of this 64×64-switch is $C_{64} = 16 \cdot C_{16} + 2 \cdot 4 \cdot N_{64}$. Reasoning similarly, we can see that the link complexity of this 256×256-switch is $C_{256} = 16 \cdot C_{64} + 2 \cdot 4 \cdot N_{256}$.

In accordance to the design principle of quasi-complete dual switches we can write $N_{256} = 4 \cdot N_{64}$, $N_{1024} = 4 \cdot N_{256} = 16 \cdot N_{64}$, $N_{4096} = 4 \cdot N_{1024} = 16 \cdot N_{256} = 64 \cdot N_{64}$ etc. Thus, the expression for the link complexity of the $N \times N$ switch can be written as following

$$C_N = 16^{n-1} \cdot C_{16} + 2N\sum_{i=1}^{n-1} 4^i . \quad (7)$$

Then taking into account the relation (1) and the value $C_{16} = 144$ we obtain the final expression of the link complexity of the quasi-complete dual switch

$$C_N = \left(\frac{3N}{4}\right)^2 + 2N \sum_{i=1}^{\log_2 \sqrt{N}-2} 2^{2i} , \quad (8)$$

for the first time.

Analogously to (4) let us we introduce the concept of the absolute link complexity difference as the difference between the link complexities of some comparing switch C_{CS} and the developed quasi-complete dual switch C_D as

$$\Delta C = C_{CS} - C_D . \quad (9)$$

Thus, the link complexity of the quasi-complete dual switch is less than the complexity of a some comparing switch if $\Delta C > 0$ and this complexity is more than the one of a some comparing switch if $\Delta C < 0$.

And analogously to (5) we introduce the concept of the relative link complexity difference as

$$\Delta c = \left(\left|C_{CS} - C_D\right| / C_D\right) \cdot 100\% \tag{10}$$

for the first time.

Analogously to (4) let us we introduce the concept of the absolute link complexity difference as the difference between the link complexities of some comparing switch C_{CS} and the developed quasi-complete dual switch C_D as

$$\Delta C = C_{CS} - C_D. \tag{11}$$

Thus, the link complexity of the quasi-complete dual switch is less than the complexity of a some comparing switch if $\Delta C > 0$ and this complexity is more than the one of a some comparing switch if $\Delta C < 0$.

And analogously to (5) we introduce the concept of the relative link complexity difference as

$$\Delta c = \left(\left|C_{CS} - C_D\right| / C_D\right) \cdot 100\% \ . \tag{12}$$

Taking into account the expressions for the crossbar switch complexity $C_C = N^2$ [24] and for Banyan scheme complexity $C_B = N \cdot (\log_2 N - 1)$ [16] we can write the absolute link complexity differences as

$$\Delta C_{CD} = C_C - C_D = N^2 \left(\frac{7}{16} - \frac{1}{N} \sum_{i=1}^{\log_2 \sqrt{N} - 2} 2^{2i+1} \right), \tag{13}$$

$$\Delta C_{BD} = C_B - C_D = N \left(\log_2 N - 1 - \frac{9}{16} N - \sum_{i=1}^{\log_2 \sqrt{N} - 2} 2^{2i+1} \right), \tag{14}$$

for the crossbar and Banyan switches, correspondingly. Here C_C is the link complexity of the crossbar switch, C_B is the link complexity of Banyan switch, C_D is the link complexity of the quasi-complete dual switch.

Analogously the relative link complexity differences for the crossbar and Banyan switches are

$$\Delta c_{CD} = \left| 7N^2 - 16N \sum_{i=1}^{\log_2 \sqrt{N} - 2} 2^{2i+1} \right| \Big/ \left(9N^2 + 16N \sum_{i=1}^{\log_4 N - 2} 2^{2i+1} \right), \tag{15}$$

$$\Delta c_{BD} = \frac{\left| 16 \log_2 N - 16 - 9N - 32 \sum_{i=1}^{\log_2 \sqrt{N} - 2} 2^{2i} \right|}{9N + 32 \sum_{i=1}^{\log_2 \sqrt{N} - 2} 2^{2i}} \ . \tag{16}$$

The results of the numerical calculation of the link complexities are presented in figure 9. It shows that the link complexity of the proposed quasi-complete dual switches calculated by using (8) is much more than the one of the blocking Banyan scheme ($C_B = N \cdot (\log_2 N - 1)$ [16]) but this complexity is less than the one of the non-blocking OXC switch ($C_C = N^2$ [25]).

In Figure 10 the results of the numerical calculation of the absolute link complexity difference are presented. It is seen that the absolute link complexity difference between the OXC switch and quasi-complete dual switches ΔC_{CD} (line 1) is positive and it increases with increasing in the number of inputs. So, the benefit in the complexity of the proposed scheme type in comparison with the OXC switch is great for high-capacity switches.

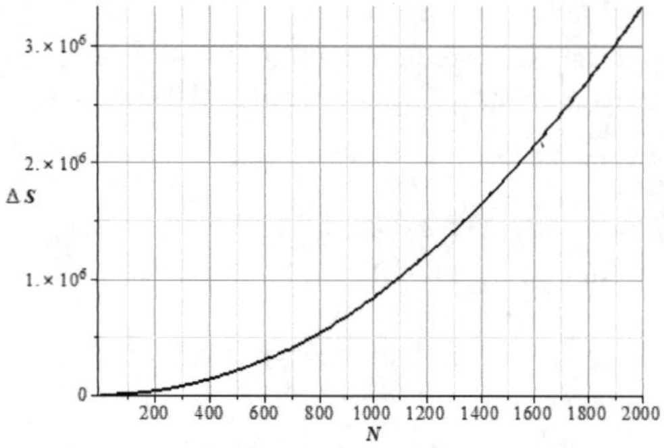

Figure 8. The absolute circuit complexity difference Δ S between quasi-complete dual switch and crossbar switches.

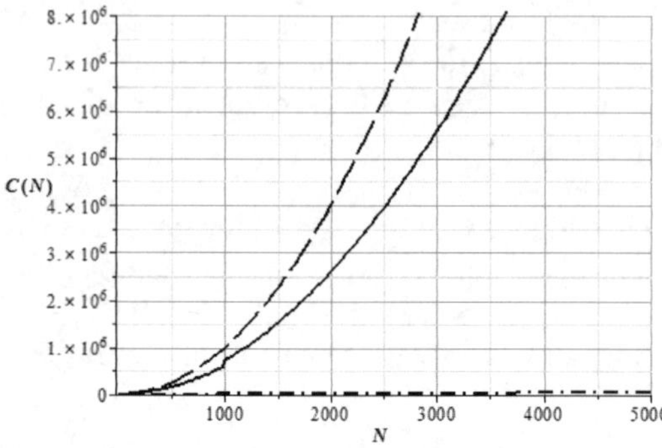

Figure 9. The comparison of the link complexity of the multi-stage switching systems. The solid line corresponds to the quasi-complete dual switch, the long-dash line corresponds to the OXC switch, and the dash line corresponds to Banyan switch.

For example, ΔC_{CD}=4554752 and Δc_{CD}=37.27% for N=4096. It also is seen in figure 9 that the absolute link complexity difference ΔC_{BD} (line 2) is negative and it increases in the absolute value with increasing the number of inputs.

For example, ΔC_{BD} = 121777408 and Δc_{BD} = 99.63% for N=4096. Thus, the quasi-complete dual switch gives us the loss in the complexity in comparison with the Banyan scheme. This is the obvious result as it is well-known in the switching theory [11,24] that the complexities of blocking schemes are less than non-blocking ones.

5. Calculation of the dual switch complexity and comparison with well-known schemes

As the proposed switches are based on the method of time division of transmitted optical signals, it is important to show that this method does not provide negative affect on their throughput. It is well-known that the throughput depends on many factors, the main of them are the switch architecture, the type of control and the switch-fabric technology [19]. These factors influence on switching time and on time that is required to provide non-blocking switching. The value of switching time of the single switching element of the dual switch has been calculated by author in [30] (T_s =1.45908·10^{-12} s). Taking into account the fact that the signal delay in the proposed switches is depended on the number of stages D=log$_2N$, it is obviously following that the guard time period can be calculated by using the following formula:

$$T = T_s D = T_s \log_2 N . \tag{17}$$

Indeed, the delay T_s is exited in each stage of the system, and the guard interval of proposed dual switch is equal to 4T (see Section 3), we can write the expression for the throughput of the N×N dual switches for the 100 percent system load

$$P_D = N / \left(4 \log_2 N \cdot T_s \right) . \tag{18}$$

According to the function principal of self-routing Banyan scheme, the connection path is established consistently stage by stage analogously the dual switch. Thus, the time period for Banyan schemes can be calculated by using (15). As it is shown in Subsection II B, the guard interval T_g for Banyan schemes is equal to $D \cdot T$, where $D = \log_2 N$ [29]. Therefore, the throughput of the $N \times N$ Banyan switch for the 100 percent system load can be written in the following form

$$P_D = N / \left((\log_2 N)^2 \cdot T_s \right). \quad (19)$$

The results of the numerical calculations of the throughputs are presented in figure 11.

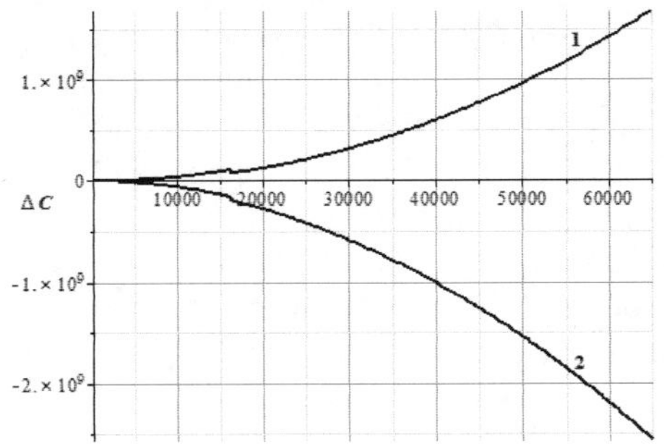

Figure 10. The absolute link complexity difference of the switching systems: 1) the absolute link complexity difference ΔC_{CD}, 2) the absolute link complexity difference ΔC_{BD}.

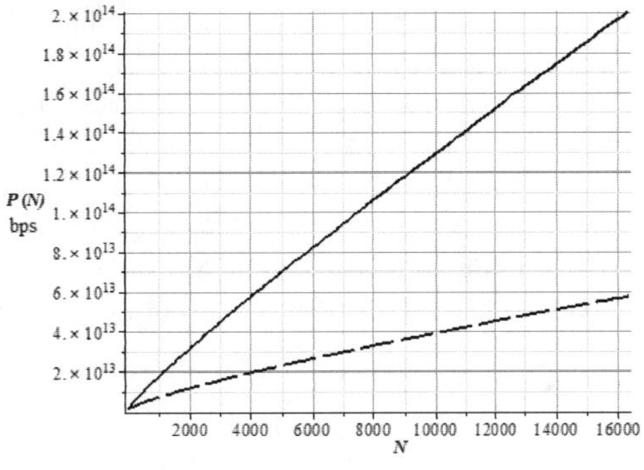

Figure 11. The throughput of the switching systems: the solid line corresponds to quasi-complete dual switch and the long-dash line corresponds to the Banyan switch.

As we can see from figure 11 the throughput of the dual optical switch is substantially more than the Banyan ones and the difference increases significantly with the number of inputs.

6. Conclusions

The main aim of this work is development of new class of strictly non-blocking optical switches that provide a tradeoff between high throughput and low hardware cost. Obviously, such switching systems are of great importance for high-speed channels of 5G/6G high-capacity communication systems.

In this paper the authors present the new dual principle for designing these optical switches for the first time. It is based on the combine using the two methods: 1) the method of optical information and control signals simultaneously transmission; 2) the method of time diversity of optical signals to ensure strictly non-blocking property. Despite the fact that the similar methods have already been used, only the new concept of their simultaneous application allows us to solve the stated problem.

For this purpose, the authors proposed the new type of the 4×4 switching element. It is important that this element is based on decentralized control principle, which obviously allows us to increase a speed of optical switching systems by an order of magnitude.

Utilizing this cell allows us to design the strictly non-blocking 16×16 all-optical element that implements the dual principle of information transfer. Moreover, it is shown that the circuit complexity of the developed scheme is less than the complexity of the crossbar 16×16-switch. The calculated delay time of the proposed element is four time periods ($4T$) and this time is equal to the time delay for the Banyan switch.

The developed 16×16 optical element allows us to design the new type of the so-called quasi-complete dual switches. It is strictly non-blocking, high-speed and combines the advantages of the crossbar and dual switches. The next important advantage of the proposed switches is the constant signal diversity period $4T$ which does not increase with increasing the number of inputs. For example, for Banyan scheme this period must necessarily increase with increasing switch capacity so that this scheme to be non-blocking. The important advantage of the developed switch is their scalability. Indeed, it is possible to build higher capacity switches by using smaller capacity switches without changing the principles of circuit design. So we proposed the following series of switches: 16, 64, 256, 1024,4096, …, $16\cdot 2^{2n-2}$, … , where n is the serial number of the scheme.

The complexity of the developed schemes is one of their most important parameters. Here the analytical study methods of the circuit and link complexities are presented for the first time. The results of the numerical calculations show that the circuit complexity of the proposed quasi-complete dual switch is less than the complexity of the crossbar one for the same number of inputs and it is less than the complexity of Clos scheme with the number of inputs up to 1024 inclusive, but it is more than this complexity of Banyan scheme. The link complexity of the proposed quasi-complete dual switches is much more than the blocking Banyan scheme but it is less than the OXC switch.

Despite the fact that the proposed switches are based on the method of time division of transmitted signals, it does not significantly negative affect their throughput. As it is shown in the separate section the proposed dual switches are several times superior in throughput Banyan schemes.

Appendix A. The connection table of 64×64 quasi-complete dual switch

In table A1 the connection rule of the central stage inputs with the outputs of the demultiplexers of the input stage in accordance with the quasi-complete dual switch is given. Here N_{inp} is the number of the input of 16×16 switch, N is the number of 16×16 switch. A numeric is a number of demultiplexer and an index is a number of the output of the demultiplexer. For example, 50_1 means that thirteen input of the second 16×16 switch connects with first output of the fifteen demultiplexer.

In table A2 the connection rule of the central stage outputs with the inputs of the multiplexers of the output stage in accordance with the quasi-complete dual switch is given. In the table A2 N_{out} is the number of the output of 16×16 switch, N is the number of 16×16 switch. A numeric is a number of multiplexer and an index is a number of the input of multiplexer. For example, 21_3 means that fifth output of the seven 16×16 switch connects with the third input of twenty-one multiplexer.

Table A1. The connection of inputs of 64×64 quasi-complete dual switch.

	N_{INP}															
N	1	2	3	4	5	6	7	8	9	10	11	12	13	14	15	16
1	1_1	5_1	9_1	13_1	17_1	21_1	25_1	29_1	33_1	37_1	41_1	45_1	49_1	53_1	57_1	61_1
2	2_1	6_1	10_1	14_1	18_1	22_1	26_1	30_1	34_1	38_1	42_1	46_1	50_1	54_1	57_1	62_1
3	3_1	7_1	11_1	15_1	19_1	23_1	27_1	31_1	35_1	39_1	43_1	47_1	51_1	55_1	58_1	63_1
4	4_1	8_1	12_1	16_1	20_1	24_1	28_1	32_1	36_1	40_1	44_1	48_1	52_1	56_1	59_1	64_1
5	1_2	5_2	9_2	13_2	17_2	21_2	25_2	29_2	33_2	37_2	41_2	45_2	49_2	53_2	57_2	61_2
6	2_2	6_2	10_2	14_2	18_2	22_2	26_2	30_2	34_2	38_2	42_2	46_2	50_2	54_2	57_2	62_2
7	3_2	7_2	11_2	15_2	19_2	23_2	27_2	31_2	35_2	39_2	43_2	47_2	51_2	55_2	58_2	63_2
8	4_2	8_2	12_2	16_2	20_2	24_2	28_2	32_2	36_2	40_2	44_2	48_2	52_2	56_2	59_2	64_2
9	1_3	5_3	9_3	13_3	17_3	21_3	25_3	29_3	33_3	37_3	41_3	45_3	49_3	53_3	57_3	61_3
10	2_3	6_3	10_3	14_3	18_3	22_3	26_3	30_3	34_3	38_3	42_3	46_3	50_3	54_3	57_3	62_3
11	3_3	7_3	11_3	15_3	19_3	23_3	27_3	31_3	35_3	39_3	43_3	47_3	51_3	55_3	58_3	63_3
12	4_3	8_3	12_3	16_3	20_3	24_3	28_3	32_3	36_3	40_3	44_3	48_3	52_3	56_3	59_3	64_3
13	1_4	5_4	9_4	13_4	17_4	21_4	25_4	29_4	33_4	37_4	41_4	45_4	49_4	53_4	57_4	61_4
14	2_4	6_4	10_4	14_4	18_4	22_4	26_4	30_4	34_4	38_4	42_4	46_4	50_4	54_4	57_4	62_4
15	3_4	7_4	11_4	15_4	19_4	23_4	27_4	31_4	35_4	39_4	43_4	47_4	51_4	55_4	58_4	63_4
16	4_4	8_4	12_4	16_4	20_4	24_4	28_4	32_4	36_4	40_4	44_4	48_4	52_4	56_4	59_4	64_4

Table A2. The connection of outputs of 64×64 quasi-complete dual switch.

N	N_{INP}															
	1	2	3	4	5	6	7	8	9	10	11	12	13	14	15	16
1	1_1	2_1	3_1	4_1	5_1	6_1	7_1	8_1	9_1	10_1	11_1	12_1	13_1	14_1	15_1	16_1
2	1_2	2_2	3_2	4_2	5_2	6_2	7_2	8_2	9_2	10_2	11_2	12_2	13_2	14_2	15_2	16_2
3	1_3	2_3	3_3	4_3	5_3	6_3	7_3	8_3	9_3	10_3	11_3	12_3	13_3	14_3	15_3	16_3
4	1_4	2_4	3_4	4_4	5_4	6_4	7_4	8_4	9_4	10_4	11_4	12_4	13_4	14_4	15_4	16_4
5	17_1	18_1	19_1	20_1	21_1	22_1	23_1	24_1	25_1	26_1	27_1	28_1	29_1	30_1	31_1	32_1
6	17_2	18_2	19_2	20_2	21_2	22_2	23_2	24_2	25_2	26_2	27_2	28_2	29_2	30_2	31_2	32_2
7	17_3	18_3	19_3	20_3	21_3	22_3	23_3	24_3	25_3	26_3	27_3	28_3	29_3	30_3	31_3	32_3
8	17_4	18_4	19_4	20_4	21_4	22_4	23_4	24_4	25_4	26_4	27_4	28_4	29_4	30_4	31_4	32_4
9	33_1	34_1	35_1	36_1	37_1	38_1	39_1	40_1	41_1	42_1	43_1	44_1	45_1	46_1	47_1	48_1
10	33_2	34_2	35_2	36_2	37_2	38_2	39_2	40_2	41_2	42_2	43_2	44_2	45_2	46_2	47_2	48_2
11	33_3	34_3	35_3	36_3	37_3	38_3	39_3	40_3	41_3	42_3	43_3	44_3	45_3	46_3	47_3	48_3
12	33_4	34_4	35_4	36_4	37_4	38_4	39_4	40_4	41_4	42_4	43_4	44_4	45_4	46_4	47_4	48_4
13	49_1	50_1	51_1	52_1	53_1	54_1	55_1	56_1	57_1	58_1	59_1	60_1	61_1	62_1	63_1	64_1
14	49_2	50_2	51_2	52_2	53_2	54_2	55_2	56_2	57_2	58_2	59_2	60_2	61_2	62_2	63_2	64_2
15	49_3	50_3	51_3	52_3	53_3	54_3	55_3	56_3	57_3	58_3	59_3	60_3	61_3	62_3	63_3	64_3
16	49_4	50_4	51_4	52_4	53_4	54_4	55_4	56_4	57_4	58_4	59_4	60_4	61_4	62_4	63_4	64_4

Appendix B. The diagrams of the 4×4 quasi-complete dual switch elements

In figure B1 the diagrams of the 1×4 demultiplexer (figure B1a) and the 4×1 multiplexer (figure B1b) are presented. In figure B2 the structure of the switching cell based on metamaterial and ferrite-garnet films is shown. 1,2,3,4 are the numbers of outputs of the 1×4 demultiplexer and the numbers of inputs of 1×4 multiplexer, i1 is one of the inputs of the 4×4 dual switch, o1 is one of the outputs of the 4×4 dual switch.

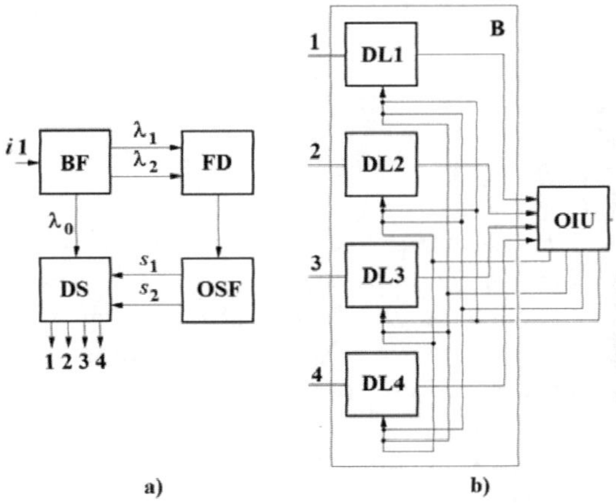

Figure B1. The diagrams of the 4×4 dual switch elements a) the 1x4 demultiplexer, b) the 4×1 multiplexer. BF- Bragg filter; FD- frequency detector; DS - deflection system; OSF -optical signal former; DL1-DL4 - delay lines; OIU - optical integrated unit, B-delay device.

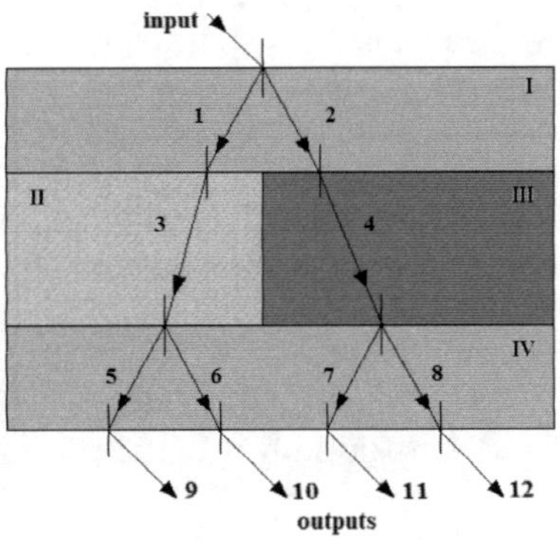

Figure B2. The structure of the switching cell based on metamaterial and ferrite-garnet films.

The information signal with the wavelength λ_0 and two control signals with the wavelengths λ_1, λ_2 is separated in the Bragg filter (BF). The information signal goes to the deflection system (DS), the control signals go to the frequency (FD) detector, and then to the optical signal former (OSF). These signals (s_1, s_2) determine the path of the information signal in the DS [2]. Four optical groups arrive to four delay lines (DL1-DL4) (figure B1b) controlled by the optical integration unit (OIU). These delay lines allow signals to pass from four inputs to one output one after the other. Detail description of such a multiplexer as an element of 4×4 switch has been given in [30].

The structure consists of the four layers (figure B2): The layers I and IV are the controlled optically transparent ferrite-garnet films. The layer II is the metamaterial one, the layer III is the ordinary dielectric one. The information beam can transmit through one of four paths: 1-3-5-9; 1-3-6-10, 2-4-7-11, 2-4-8-12 depending on the combination of control signals s_1, s_2. Detail description of this optical cell has been presented in [2].

References

[1] Tu Xin 2019 State of the art and perspectives on silicon photonic switches *Micromachines (Basel)* 10(1)

[2] Vytovtov K A and Barabanova E A 2018 Optical switching cell based on metamaterials and ferrite Films *Proc. 12th International Congress on Artificial Materials for Novel Wave Phenomena – Metamaterials* (Espoo, Finland) pp 424-26

[3] Paudel P and Bhattarai A 2018 5G Telecommunication Technology: History, Overview, Requirements and Use Case Scenario in Context of Nepal *Proc. in Conference* IT4D

[4] Argibay P J, Chiaroni L D and Qiao C 2020 Optical Packet Switching and Optical Burst Switching *Springer Handbook of Optical Networks* pp 665-701

[5] Sorimoto K, Kawashima H, Mori M, Hasama T, Ishikawa H, Tsuda H and Uetsuka H 2013 MEMS mirror with slot structures suitable for flexible-grid WSS *IEICE Electron. Express* **10**(3) 20120924

[6] Yang H, Wilkinson P, Robertson B, Giltrap S, Snowdon O, Prudden H and Chu D.P 2020 24 [1x12] Wavelength Selective Switches Integrated on a Single 4k LCoS Device *Journal of Lightwave Technology*

[7] Xie D, Wang D, Zhang M, Liu Z, You Q, Yang Q and Yu S 2017 LCoS-Based Wavelength-Selective Switch for Future Finer-Grid Elastic Optical Networks Capable of All-Optical Wavelength Conversion *IEEE Photonics Journal* **9**(2)

[8] Jinno M, Kodama T and Ishikawa T 2020 Feasibility Demonstration of Spatial Channel Networking Using SDM/WDM Hierarchical Approach for Peta-b/s Optical Transport *Journal of Lightwave technology* **38**(9)

[9] Qiao L, Tang W and Chu T 2017 32 × 32 silicon electro-optic switch with built-in monitors and balanced-status units *Scientific Reports* **7** pp 42306

[10] Seok T J, Kwon K, Henriksson J, Luo J and Wu M C 2019 Wafer-scale silicon photonic switches beyond die size limit *Optica* **6**(4) pp 490-94

[11] Kabacinski W 2005 Nonblocking electronic and photonic switching fabrics (Springer) 282p

[12] Esmaeel K H, Zeinali and Fesharaki M N 2008 NMR: A New Approach for Optimal Design of Strictly Non-Blocking Multistage Interconnection Networks *Journal of Applied Sciences* **8**(19) pp 3453-59

[13] Lu L, Li X, Gao W, Li X, Zhou L and Chen J 2019 Silicon non-blocking 4 × 4 optical switch chip integrated with both thermal and electro-optic tuners *IEEE Photonics Journal* **11**(6) 6603209

[14] Ramaswami R 2006 Optical networking technologies: what worked and what didn't *IEEE Communications Magazine* **44**(9) pp 132-39

[15] Chao H J and Bin L 2007 High Performance Switches and Routers (Published by John & Sons, Inc., Hoboken, New Jersey) p 607

[16] Vasiliadis D C, Rizos G E, Vassilakis C 2007 Performance Analysis of blocking Banyan Switches *Innovative Algorithms and Techniques in Automation, Industrial Electronics and Telecommunications* pp 107–11.

[17] Danilewicz G, Kabaci´nski W, Michalski M and Zal M 2007 Wide-Sense Nonblocking Multiplane Baseline Switching Networks Composed of d × d Switches *The IEEE International Conference on Communications* **14** pp 6386-91

[18] Suherman M Z, Pinem M, Fauzi R and Razali M 2019 Reducing Crosspoints on Multistage Switching by Using Batcher Banyan Switches *Proc. ICASI* (Banda Aceh, Indonesia)

[19] Dally W, Towles B Principles and practices of interconnection networks 2003 p 550

[20] Chi N, Olmos J J, Thakulsukanant K, Wang Z, Ansell O, Yu S and Huang D 2006 Experimental characteristics of optical crosspoint switch *Journal of Lightwave Technology* **24**(10) pp 3646-53

[21] Saleh B E A and Teich M C 2007 Fundamentals of Photonics (2nd Edition) p 1200

[22] Barabanova E A, Vytovtov K A, Vishnevskiy V M and Podlazov V.S 2019 Model of optical non-blocking information processing system for next-generation telecommunication networks *Communications in Computer and Information Science* pp 188-98

[23] Dmitriev A V, Toropov N A and Sumetsky M 2016 Miniature Optical Delay Lines and Buffers *Proc. 18th International Conference on Transparent Optical Networks (ICTON)* 16281087

[24] Hwang F. K 2004 The mathematical theory of nonblocking switching networks (2nd Edition) Series on Applied Mathematics, vol.15, Published by World Scientific Publishing Co. Pte. Ltd

[25] Nikolova D, Calhoun D M, Liu Y, Rumley S, Novack A, Baehr-Jones T, Hochberg M and Bergman K 2017 Modular architecture for fully non-blocking silicon photonic switch fabric *Microsystems & Nanoengineering* **3** 16071 pp 1-9

[26] Karavai M F, Parkhomenko P P and Podlazov V S 2009 Combinatorial methods for constructing bipartite uniform minimal quasicomplete graphs (symmetrical block designs) *Automation and Remote control* **70**(2) pp 312-27.

[27] Barabanova E A , Vytovtov K A, Maltseva N S, Kravchenko O V and Kravchenko V F 2019 Models and algorithms of optical switching systems with decentralized control *Proc. IEEE Conference of Russian Young Researchers in Electrical and Electronic Engineering* (Saint-Petersburg) **8657063** pp 64-8

[28] Barabanova E A, Vytovtov K A and Podlazov V S 2019 Model and algorithm of next generation optical switching systems based on 8×8 elements *Lecture Notes in Computer Science* **11965** pp 58-70

[29] Yang E Lu M, Yang B, Feng X and Zheng S Q 2017 A Novel Design of Self-Routing Strictly Nonblocking Switching Networks *International Journal of Computers and Applications* pp 44-50

[30] Barabanova E A, Vytovtov K A and Vishnevskiy V M 2020 Method for calculating performance of all-optical linear system of transmitting and processing information *The 22th International Conference "Digital signal processing and its applications* 9213274

[31] Kutuzov D, Osovsky A, Stukach O and Starov D 2021 Modeling of IIoT Traffic Processing by Intra-Chip NoC Routers of 5G/6G Networks *Proc. 2021 International Siberian Conference on Control and Communications (SIBCON)* pp 1-5

[32] Kutuzov D, Stukach O, Starov D and Osovsky A 2020 Decentralized Control of Switching Elements in the Intra-Chip Network-on-Chip Routers *Proc. 2020 Moscow Workshop on Electronic and Networking Technologies (MWENT)* pp 1-4

Application of machine learning methods for automated classification and routing in ITIL

V V Nikulin, S D Shibaikin, A N Vishnyakov

National Research Ogarev Mordovia State University
68 Bol'shevistskaya street, Saransk, 430005, Republic of Mordovia, Russia

E-mail: nikulinvv@mail.ru

Abstract. The article analyzes the application of machine learning methods for automated classification and routing in ITIL library. ITSM technology and ITIL are considered. The definitions of the incident and IT services are given. Then, the vectorization and extraction of keywords in the information written in natural language is carried out and lemmatization and TF-IDF measure will be used. A comparative analysis of the application of machine learning methods is given as well as a comparison of the results of automatic classification of text information using gradient boosting and a convolutional neural network. Various parameters of these methods are considered and the most effective method of machine learning is determined. The results of using machine learning methods for automated classification of incidents allows high-precision routing of requests for restoring the operability of IT services, reducing response time and errors associated with the human factor.

1. Introduction

ITSM (Information Technology Service Management) is an ideology of managing IT as a set of services. Within this framework, the IT department or outsourcer provides services to the business with measurable characteristics. All parameters of this service are described in detail in a special agreement - SLA, which is made by both interested parties - IT and business. ITSM is based on ITIL, the library of best practices. ITIL describes a business process that allows you to deliver services that meet the requirements of customers (enterprises) based on real business experience. However, ITSM and ITIL are different concepts. ITSM is an approach, and ITIL is a practical guide [1,2,3,4].

Various problems can arise in the process of providing IT services: unavailable service, execution with errors, an ability to get unauthorized access to information, and so on. Negative deviations from regular service provision are possible. ITIL defines such deviations as incidents. Thus, an incident is an unplanned interruption or decrease in the quality of an IT service. A configuration unit failure that has not yet affected the service is also an incident.

Typically, actions to eliminate IT incidents have a significant impact on the overall perception of IT from the user. To effectively manage these activities, an action plan must be established. In accordance with ITIL recommendations, the incident management process should be built for this.

2. Analysis of the application of machine learning methods

For the correct resolution of the incident, the task of its classifying [6] is critical. Classification is the assignment of an appropriate incident category for its further routing. In this work, machine learning algorithms are used for this.

Machine learning is an extensive area that uses mathematical, statistical, and computational methods to solve problems using a specific algorithm. There are many methods of machine learning, such as: the nearest neighbour method, a decision tree, a random forest, Bayesian classifier, neural networks, convolutional neural networks, adaptive boosting, gradient boosting and others. Consider each method in more detail. Neural networks are simulations of brain function. It consists of a sequence of neurons connected to each other by synapses. They consist of an input layer, a hidden layer, and an output layer. Each of the neurons receives and processes data that is transmitted to neurons at the next level. Since their signal processing is identical in order to get the right solution, we need to correctly select the parameters of the synapses that bind the neurons.

There are many varieties of neural networks, such as: convolutional neural network, impulse neural network, chaotic neural network and others. An example of a neural network is shown in figure 1.

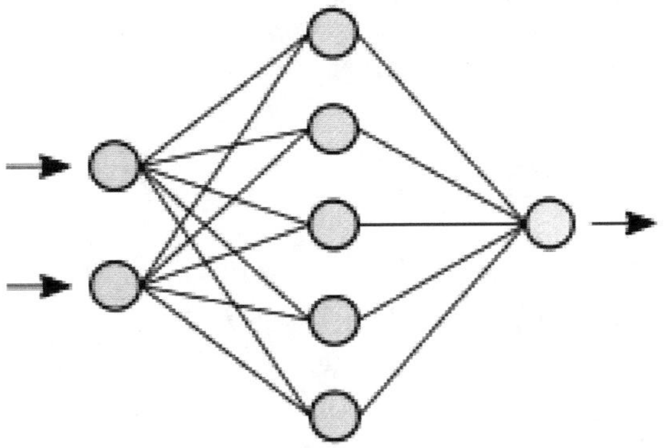

Figure 1. Example of neural network.

In this example, the first layer (marked green) is the input layer, the second layer (blue) is the hidden (computational) layer, and the third layer (yellow) is the output layer. One of the private cases of a neural network is a convolutional neural network. The difference in the methods of operation of the neural network and the convolutional neural network is that in the convolutional operation, a limited matrix is used, moving along the input layer and forming a neuron activation signal of the next layer after each movement. For all neurons, the same matrix called the convolution nucleus is used.

Boosting is a method that consists of training weak classifiers to obtain a stronger classifier. Usually, with this method, each object is assigned a certain weight, which is associated with the accuracy of the training. Then, when the weak classifier is added, the weighting factors are recalculated - the weight of each object is recalculated so that objects classified incorrectly receive more weight, and those classified correctly - less weight. Then the process of normalization takes place, so that all weights in total are equal to one, and the process starts over. Thus, in the next iteration, learning will focus more on objects that have been misclassified.

This work will compare different machine learning methods to find out the most effective approach to the task of automatic classification of incidents. Each incident consists of a list of attributes, each of which has the form "key-value." Since the list of attributes is very large, only the most informative attributes are selected. Their list with sample values is shown in Table 1.

Take a closer look at the attributes. The status indicates whether the incident is relevant at the moment, or if it has already been eliminated (OPEN, CLOSE statuses). Criticality shows how much this incident affects the service provided or the IT service in general. The object is configured manually and is required to clarify the type of parameter by which the incident occurred, and the object type shows the system module by which the metric is checked. The indicator is the specific metric by which the incident occurred. MSG (message) is displayed in the system and is required for a brief description of the incident for better understanding.

Each incident can be classified by type (exceeding acceptable metrics, equipment inoperability, etc.), as well as by categorization by the group of those assigned to eliminate the incident and by categorization by belonging to the IT service. Let us categorize services by their role. To do this, a hierarchical classification is introduced that allows you to clearly understand the nature of the incident (figure 2).

Table 1. List of selected attributes with sample values.

Status	Criticality	Object	Object class	Indicator	MSG
OPEN	MINOR	server.ru@vm-16401_server@Processor	VSM_CPU_CNTR	TotalCPU Utilization	VMware CPU Container Total CPU Utilization >= 80% for 10 min.
OPEN	MAJOR	service:[service]: Errors requests	ZabbixDefaultClass		Apiman: Errors request in API service more than 5 on server.ru
OPEN	MAJOR	service:[service]: Errors requests	ZabbixDefaultClass		Apiman: Errors request in API service/account-charges more than 5 on server.ru
OPEN	MINOR	server.ru@vm-3801_server@Processor	VSM_CPU_CNTR	TotalCPU Utilization	VMware CPU Container Total CPU Utilization >= 80% for 10 min.
OPEN	CRITICAL	IS:			Integration Service on ip:port is disconnected.
OPEN	CRITICAL	IS:			Integration Service on ip:port is disconnected.
OPEN	MAJOR	ip|Application27	PIM_Port	Port status	Application ip|port ne dostupen na servere
OPEN	MAJOR	service:[service]: Errors requests	ZabbixDefaultClass		Apiman: Errors request in API service more than 5 on server.ru
OPEN	MAJOR	ip|Application31	PIM_Port	Port status	Application ip|Application31 ne dostupen na servere ip
OPEN	MAJOR	service:[service]: Errors requests	ZabbixDefaultClass		Apiman: Errors request in API service more than 5 on server.ru
OPEN	MAJOR	server.ru@vm-8790_vm-ora-odb-22@Processor	VSM_CPU_CNTR	TotalCPU Utilization	VMware CPU Container Total CPU Utilization >= 90% for 10 min.
OPEN	MINOR	server.ru@vm-2659_infra@Processor	VSM_CPU_CNTR	TotalCPU Utilization	VMware CPU Container Total CPU Utilization >= 80% for 10 min.

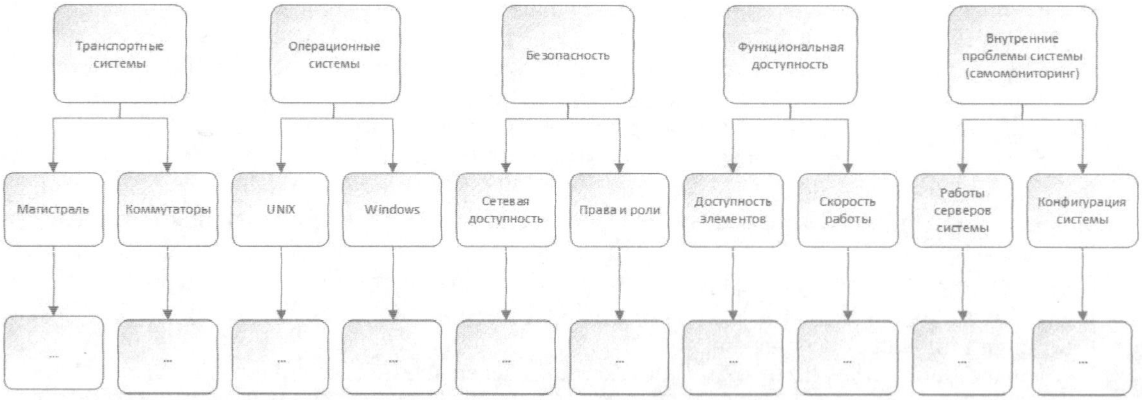

Figure 2. Hierarchical view of services.

The first level is the main division of services by the roles they perform: transport systems, operating systems, security, functional availability, internal system problems (self-monitoring). At the second level, problems become specific - for the operating system, for example, the incident is specified to relate to a UNIX or Windows system. The third level is optional for all systems. Usually, it is necessary to pinpoint the source of the incident, if possible. For example, if there is a problem in the sw itch in the transport system, layer 3 will indicate which particular switch turned out to be the source of the problem. For convenience of presentation, the figure is shown schematically. In order for the initial data to be used in machine learning, let us conduct a three-level classification of incidents. An example is shown in the following table.

Table 2. Three-level classification of incidents.

Incident description	Web-stranica url.ru nedostupna.(Ob"ekt: index.html)
Operational categorization (Level 1)	Transport systems
Operational categorization (Level 2)	Switch
Operational categorization (Level 3)	Switch (address) not available

The operational categorization of the first level is the purpose of the system, which gives a general idea of the incident - transport systems. The second layer, which is more accurate for classification, is the attribute "switch" which indicates problems with the network equipment. And the third layer is the attribute "cause and address of the switch," which allows you to pinpoint the source of the problem. Thus, to solve the problem of classifying incidents, 5 unique attributes are allocated for the first level, for the second - 10, and for the third - from 3 to 20.

Also, to categorize incidents, let us introduce an additional field "IT service." Incidents will be classified by the service provision or by belonging to a system, for example: Rostelecom's Personal Account, SystemName Database, Accounting System, etc.

BMC ProactiveNet 9.6 was selected as the system in which the classification will be carried out. It has a built-in event collector that displays open and closed incidents in real time. Figure 3 shows the event collector with examples of incidents coming into the system.

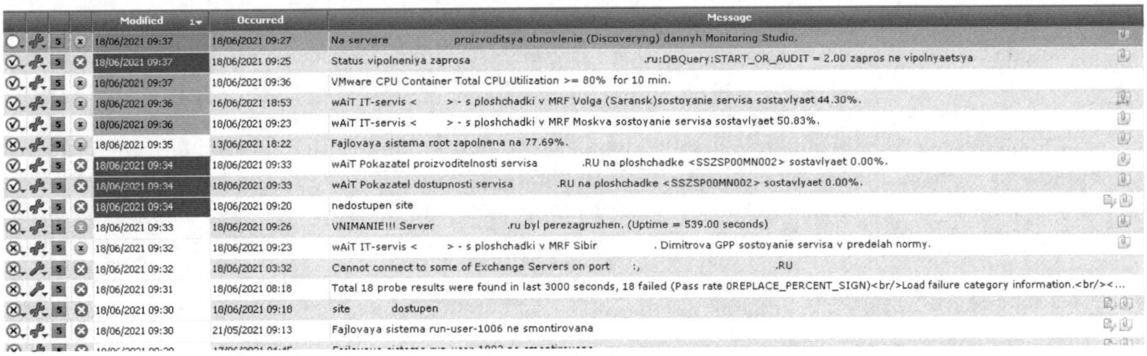

Figure 3. Event collector with incident examples.

The training sample is unloading from this system, which contains information about 10,000 incidents recorded by the system, fully or partially classified. They all have the form shown in Table 1. Before compiling the training sample, it is necessary to prepare the initial data. For this, the TF-IDF [11, 12] measure is used, which allows you to evaluate the importance of a single word in the context of one site (in this case, an incident) for the rest of the incident flow. This measure works according to the following principle: the weight of a word that carries a semantic load is proportional to the frequency of use of this word in the incident, and inversely proportional to the frequency of use in other documents. Such words are all words that are not prepositions, numbers, pronouns, interjections, particles, numerals, etc. Using this approach, you can convert the original data into a vector of fixed length. Each word that has a weight will be assigned its own number, which ultimately will allow you to get the length of the vector.

$$idf(text, D) = \log\left(\frac{D}{n}\right), \tag{1}$$

$$tf(text, D_i) = \frac{n}{D_i}, \tag{2}$$

$$tfidf() = tf(text, D_i) * idf * (text, D), \tag{3}$$

where D – list of descriptions of incidents or reference data; Di – description of the i-th incident or reference data; n – number of occurrences of a word with the weight.
Table 3 shows examples of calculating weights for some words.

Table 3. Examples of calculating the weight of words.

Word	IDF	Number of occurrences
web-stranica	1,66	65
VMware	1,81	46
Zapros	1,5	96
Status	1,46	104
Server	0,86	413
Available	2,64	7
Cluster	2,52	8
kod	2,5	9
System	1,88	39
Container	2.9	4
Adapter	2,27	16

After calculating the weight of the words and subsequent vectorization, we obtain a vector with a length of 250. This is due to the number of IT services (140 services), as well as several integrated systems that also send incidents to this system.

Once the initial data preparation has been performed, it can be used to teach machine learning algorithms. They will be the previously described neural network, convolutional neural network (CNN) and gradient boosting [8, 9, 10].

The neural network is built in 5 layers. There are 250 neurons in the input layer, the second and third hidden layers have 30 neurons each, and the number of neurons equal to the number of attributes at the classification level in the output layer. The following architecture was chosen to construct a convolutional neural network:

- input layer;
- convolution layer;
- subsampling layer;
- a fully connected hidden layer;
- output layer.

The convolution operation implies that each fragment of the input is multiplied by the matrix of weights in an element-by-element manner, and the results are summed up. This sum is an output element called a feature map. The weighted sum of inputs is passed through the activation function.

The subsampling layer (pooling) is a nonlinear feature map compaction through nonlinear transformation. Pooling is interpreted as splitting the feature map into smaller matrices, finding from maximum elements, that is, increasing the depth of values.

Also, we apply the category crossentropy parameter, which is used for multiclass labels and calculates the cross-entropy loss between labels and predictions. It is necessary to measure the similarity between the predicted attribute and the true one. To evaluate the accuracy of the classifier, the metric "F-measure" is applied. It is the harmonic average between accuracy and completeness. The system accuracy within a class is the proportion of incidents that actually belong to a given class relative to all incidents that the system has assigned to that class. Completeness of the system is the proportion of incidents found by the classifier that belong to the class relative to all incidents of this class in the test sample.

For gradient boosting [5], we apply the following parameters: maximum tree length – 10; objective = 'multi: softmax' – setting XGBoost to perform multiclass classification, for which the number of classes must also be specified (depending on the classification level specified in Table 2); num_parallel_tree = 1 (increase of random forest); subsample = 0.5 is the subsample ratio of the training sample.

After processing 10,000 incidents, performance graphs of all of the above methods were generated. Also, the accuracy of the handwritten classifier currently used in the system was calculated. The results are shown in figure 4.

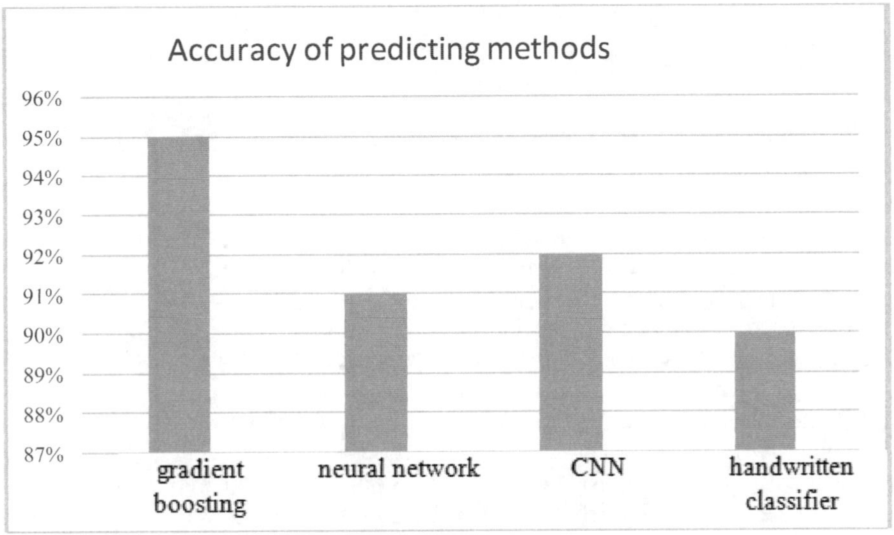

Figure 4. The accuracy of predicting the classification of incidents.

Gradient boosting showed the best result – 95% of correctly classified incidents. In the case of the neural network it was 91%, and convolutional neural network – 92%.

The 90% accuracy of the handwritten classifier is due to the fact that some of the incidents do not fall under its conditions and remain unclassified. These are, for example, metrics incidents that are currently at the stage of setting, or system notifications that automatically come to the system.

After all the above work, the question remains about the feasibility of introducing this approach into ITIL systems. To do this, we will consider the structure of the department and the expected effectiveness of the implementation of this solution. A schematic structure of the department is shown in figure 5. Systems specialists act as a router for incidents and redirect them to the specialists of the FG (functional groups) responsible for the functioning of the services.

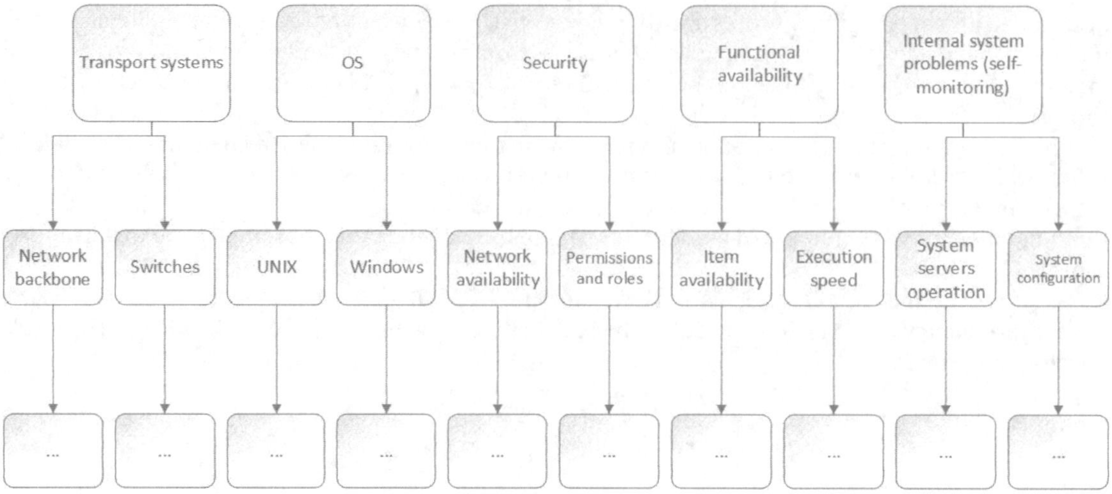

Figure 5. Schematic representation of the informatization department.

3. Conclusion

The proposed solution assumes the automation of incident routing, which will reduce the response time and eliminate the inaccuracy associated with the human factor. Also, this will optimize the workplaces in the departments responsible for the continuous operation of IT services, since the role of team specialists will be performed by a machine learning algorithm.

In conclusion, it should be said that the use of the developed classifier based on the gradient boosting approach will reduce the number of errors in the classification and routing of applications and will allow predicting the categories for incidents that did not fall under the handwritten classifier. Also, in the future, this method will automate the creation of requests for debugging and troubleshooting by integrating it with systems such as, for example, OTRS - an open request processing system. Another promising area is the use of this technology for the automated creation of problems (the cause or potential cause of one or more emerging incidents) in the Problem Management Console.

4. References

[1] Jelliti M, Sibilla M, Jamoussi Y and Ghezala H B 2010 A model based framework supporting ITIL Service IT Management *Enterprise, Business-Process and Information Systems Modeling - 11th International Workshop BPMDS 2010 and 15th International Conference EMMSAD 2010 Held at CAiSE 2010 Proceedings* **50 LNBIP** pp 208-219

[2] Shrestha A, Cater-Steel A, Toleman M, Behari S and Mehdi M R 2020 Development and evaluation of a software-mediated process assessment method for IT service management *Information & Management* **57** 4 p. 103213

[3] Edgington T M, Raghu T S and Vinze A S 2010 Using process mining to identify coordination patterns in IT service management *Decision Support Systems* **49(2)** pp 175-186

[4] Ahmad N, Tarek Amer N, Qutaifan F and Alhilali A 2013 Technology adoption model and a road map to successful implementation of ITIL *Journal of Enterprise Information Management* **26**(5) pp 553-576

[5] Ye J, Chow J H, Chen J and Zheng Z 2009 Stochastic gradient boosted distributed decision trees. *Proceedings of the 18th ACM conference on Information and knowledge management ACM* pp 2061-2064

[6] Zuev D, Kalistratov A and Zuev A 2018 Machine learning in IT service management *Procedia Computer Science* **145** pp 675–679

[7] Taherkhani A, Cosma G and McGinnity T 2020 Adaboost-CNN: an adaptive boosting algorithm for convolutional neural networks to classify multi-class imbalanced datasets using transfer learning *Neurocomputing* **404** pp 351-366
[8] Friedman J 2002 Stochastic gradient boosting *Computational Statistics & Data Analysis* **38** 4 pp 367-378
[9] Chen T and Guestrin C 2016 XGBoost: A Scalable Tree Boosting System *22nd ACM SIGKDD International Conference on Knowledge Discovery and Data Mining KDD* pp 785-794
[10] Indolia S, Goswami A K, Mishra S P and Asopa P 2018 Conceptual Understanding of Convolutional Neural Network *A Deep Learning Approach Procedia Computer Science* **132** pp 679-688
[11] Shibaikin S D, Egunova A I and Abbakumov A A 2020 Analysis of application of neural networks, gradient boosting and the method of nearest neighbors for classification of regulatory reference information *Cientific and technical volga region bulletin* **2** pp 54-58
[12] Nikulin V V, Abbakumov A A and Shibaikin S D 2020 Analysis of the application of machine learning methods of computer systems to improve security against fraudulent texts *Bulletin of the Astrakhan State Technical University. Series: Management, Computer Engineering and Informatics* **1** pp 29-40

Measuring the Continuity of a Gas-Liquid Flow in a Pipeline

G N Akhobadze

V. A. Trapeznikov Institute of Control Sciences of Russian Academy of Sciences
65 Profsoyuznaya street, Moscow 117997, Russia

E-mail: ahogur@yandex.ru

Abstract. The principles of building a meter for the continuity of gas-liquid flows in pipelines have been proposed, based on the polarized electromagnetic wave characteristics. A two-component flow in a pipeline has been analyzed considering the cross-sectional area of the flow in the pipeline. The judgment about the flow continuity by the phase difference between the electromagnetic fields polarized parallel and perpendicular to the wave propagation direction has been justified. The naturally and artificially anisotropic flow continuity measuring study results are given. A polarization technique of electromagnetic wave birefringence in the isotropic medium has been considered, based on placing an isotropic medium in an electric field formed by two capacitor electrodes. The possibility of eliminating the parameter measuring errors induced by temperature impact on the electromagnetic wave refractive index in controlled media is shown based on two electrodes of a capacitor creating an electric field in the dielectric pipeline measuring section as an option for implementing flat reflectors of an open resonator. The study results can be used in the instrument-making industry in the development and implementation of smart contactless measuring instruments in the gas-liquid flow analysis and control.

1. Introduction

Many processes associated with gas-liquid flows in pipelines often face the problem of measuring continuity. Knowledge of this physical parameter is especially important when transporting organic coolants of nuclear units, fueling propulsion systems, etc.

Continuity characterizing the uniformity of two-component media such as gas-liquid flows can be defined as follows:

$$S = V_L/(V_L + V_G), \quad (1)$$

where V_L and V_G are the liquid and gas volumes per unit length of the pipeline, respectively.
To date, different by the principle of operation devices are known for measuring the gas-liquid flow continuity, based on various physical phenomena. Among all the known ones, the most widespread are devices using the interrelation of the gas-liquid flow properties and the electromagnetic wave characteristics.

The paper proposes the principle for building a gas-liquid flow continuity meter based on using the polarization characteristics of electromagnetic waves passing through the controlled flow.

2. Measuring the Anisotropic Flow Continuity

Let an anisotropic medium with tensor magnetic and dielectric permeability flows through the pipeline. As a rule, the medium anisotropy (manifestation of different behavior depending on the electromagnetic field direction) cannot be simultaneously magnetic and dielectric [1].

According to electrodynamics, when electromagnetic oscillations interact, e.g., with an anisotropic medium, in the latter, an electromagnetic field self-polarization may occur, where the electric and magnetic field oscillations are not chaotic but strictly monodirectional. Therefore, an electromagnetic wave propagating through an anisotropic medium can be decomposed into two components of the same amplitude, polarized parallel and perpendicular to the wave propagation direction. Moreover, these polarized waves with different refractive indices propagate at different velocities. Consequently, due to the different propagation velocities, the oscillation phases of the electric vectors of these waves at the controlled medium exit do not coincide. As a result, at the anisotropic medium exit, a phase difference ψ arises between both waves, which can be calculated as follows:

$$\psi = \omega\ (n_1 - n_2)\ \iota/c, \tag{2}$$

where ω is the circular frequency of electromagnetic oscillations, c is the electromagnetic wave propagation velocity in free space, ι is the wave path length in an anisotropic medium, n_1 and n_2 are the refractive indices of polarized waves.

Equation (2) shows that at constant ω and $\Delta n = n_1 - n_2$ values (with a positive sign of the difference in refractive indices), the phase difference ψ will be determined by the wave path length in the anisotropic medium.

Analysis of equation (1) shows that in the pipeline free of a controlled liquid, $V_L = 0$ и $V_G = max$; and in the presence of a liquid flow without gas inclusions, $V_L = max$ и $V_G = 0$. Hence, it can be concluded that when monitoring the gas-liquid flow continuity in the pipeline, the measured parameter of the gas-liquid flow in the pipeline can be determined by the V_L и V_G values related to the pipeline cross-sectional area. In our case for, e.g., a V_L volume, the wave path length ι determines the liquid flow cross-sectional area. Therefore, determining the wave path length ι related to the V_L volume per the unit length of the pipeline by the flow cross-sectional area allows estimating the anisotropic gas-liquid flow continuity in the pipeline.

Since the ι parameter may vary from 0 (no flow) to the pipeline maximum inner diameter d when filling vertical to the flow direction, we can write for the V_L volume:

$$V_L = \pi d\iota h/4, \tag{3}$$

where h is the length of the pipeline horizontal measuring section through which the measured medium flows.

From equation (3), for ι, we obtain:

$$\iota = 4V_L/\ \pi dh.$$

Substituting the last relation into equation (2) and performing a transformation, for ψ, we have:

$$\psi = 4\omega\Delta n V_L/\pi dhc.$$

The obtained equation shows that the volume of liquid in the pipeline section can be determined by measuring the phase difference between the polarized electromagnetic waves at the anisotropic medium exit.

Practice shows that with a known pipeline measuring section volume V_o, it is expedient to express the continuity by the formula:

$$S = V_L/\ V_o,$$

where V_o is the pipeline measuring section volume, determined by the pipeline cross-sectional area and the pipeline measuring section length. As a result, for the continuity S, considering equation (3) and $V_o = \pi d^2 h/4$, we have:

$$S = \iota/d. \tag{4}$$

Having determined ι from (4) and substituting in (2), we obtain:

$$\psi = \omega\Delta n S d/c.$$

Hence, for the required S parameter, we find:

$$S = \psi c / \omega \Delta n d.$$

The resulting formula shows that at the minimum and maximum path length of electromagnetic polarized waves propagating through an anisotropic medium, corresponding to the absence and presence of a controlled flow in the pipeline, the flow continuity can be determined by changing the phase difference ψ from 0 to ψ_{max}. In this case, for $\iota = 0$ (S = 0), $\psi = 0$, and for $\iota = d$ (S = 1), $\psi = \psi_{max}$.

Figure 1 shows the block diagram of measuring the gas-liquid flow continuity in a pipeline.

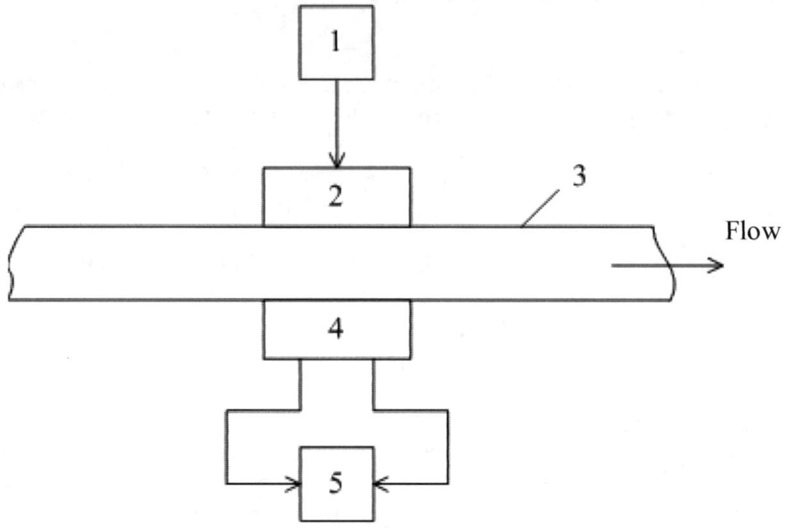

Figure 1. Block Diagram of Measuring the Anisotropic Flow Continuity:
1 – microwave oscillator; 2 – emitter; 3 – pipeline; 4 – receiver; 5 – phase difference meter.

3. Measuring the Isotropic Flow Continuity

Along with anisotropic media, there are those with the same properties in any direction. Such media are called isotropic [2].

To measure the isotropic medium continuity, the birefringence effect can be used, according to which, when an isotropic medium is placed in an electric field, the refractive index of an electromagnetic wave with a polarization plane parallel to the flux lines of this field will change according to the law [3]:

$$n_e = n_o + \lambda B E^2,$$

where n_o is the medium refractive index for a wave with a polarization plane orthogonal to the flux lines of the applied electric field, λ is the radiation wavelength, E is the electric field intensity, B is the coefficient depending on the medium properties, called the Kerr constant.

There are several ways to create the birefringence of electromagnetic waves in the isotropic medium. In practice, polarization is often used, where the polarization plane of a wave incident on the medium is oriented at $\lambda/4$ relative to the direction of the flux lines of the electric field applied to the controlled medium. Thereat, by analogy with anisotropic media, the feed wave can be decomposed into two components of the same amplitude, polarized parallel and perpendicular to the flux lines of the applied electric field. In the medium, due to the imposed electric field, these components propagate at different velocities since the refractive indices for them differ by $\Delta n = \lambda B E^2$. As a result, at the medium exit, a phase difference ψ will occur between the two polarized waves:

$$\psi = 2\pi B \iota E^2 \quad (5)$$

Equation (5) shows that for this case, at constant E and B, in a pipeline filled with the isotropic flow, the phase difference ψ will be determined by the polarized electromagnetic wave path length.
Solving jointly equations (4) and (5) allow calculating the continuity by the formula:

$$S = \psi/2\pi BdE^2 \qquad (6)$$

Therefore, when monitoring isotropic media, measuring the flow continuity is reduced to determining the phase difference of the electromagnetic waves at the medium exit.
Comparing the formulas expressing the continuity of anisotropic and isotropic flows in pipelines, we see that if for anisotropic media, the natural medium properties are decisive, then for isotropic media, it is the controlling electric field intensity causing artificial anisotropy in the isotropic medium. Comparison of these formulas also shows that at a wavelength determined by the Δn to BE^2 ratio, birefringence effects of the same nature can be obtained in both anisotropic and isotopic media.

The diagram for measuring the isotropic flow continuity, in contrast to that for measuring the anisotropic flow continuity, additionally requires creating a controlling electric field in the pipeline measuring section, causing the radioelectric Kerr effect in this medium.

4. Combined Continuity Measurement Principle

The above continuity measurement principles may provide errors caused by the changes in the dielectric properties (dielectric anisotropy) of anisotropic and isotropic media, associated with the ambient temperature instability.

The analysis of measuring the continuity of, e.g., isotropic flows in pipelines shows that under the temperature effect, the main error factor may be the B coefficient associated with the refractive index of the measured medium, through which the wave propagates.
In electrodynamics, the medium refractive index n is determined by the equation $n = \sqrt{\varepsilon\mu}$, where ε and μ are the dielectric and magnetic permeabilities of the controlled medium. For $\mu=1$, for the isotropic medium refractive index, we obtain $n = \sqrt{\varepsilon}$. So, in equation (6), $\sqrt{\varepsilon}K\lambda$ should be used instead of B, where K is a constant. Then, for the isotropic flow continuity, we have:

$$S = \frac{\psi}{2\pi\sqrt{\varepsilon}K\lambda dE^2}. \qquad (7)$$

The last equation shows that a change in ε may cause an error in the continuity measurement. In our case, this error can be eliminated by using the capacitor electrodes generating an electric field in the pipeline measuring section as the open resonator reflectors [4]. Let on the outer surface of the dielectric pipeline through which the measured isotropic medium flows, two plates are installed diametrically opposite one another (Figure 2), simultaneously performing the functions of capacitor electrodes and open resonator reflectors. Then, when exciting electromagnetic oscillations between these plates, the resonant frequency of the open resonator:

$$\omega = \frac{\pi q c}{2D\sqrt{M}},$$

where D is the dielectric pipeline outer diameter, M is the coefficient consi Flow multaneous impacts of the pipeline dielectric properties and the controlled isotropic flow on the wave propagation velocity. In this case, the M coefficient with a certain accuracy can be represented as the sum of the dielectric permeabilities of the pipeline ε_p and the measured flow ε. Since the ε_p value depends on the pipeline material, it can be considered constant in the measurement. Then, as follows from the last formula, at constant ε_p and D values, the change in the resonance frequency ω will be determined by only the change in the dielectric permeability ε. Therefore, the equation for the frequency ω takes the form:

$$\omega = \frac{\pi q c}{2D\sqrt{\varepsilon_p + \varepsilon}}$$

Solving the last equation, we obtain

$$\varepsilon = \frac{\pi^2 q^2 c^2 - 4D^2\omega^2\varepsilon_p}{4D^2\omega^2} \quad (8)$$

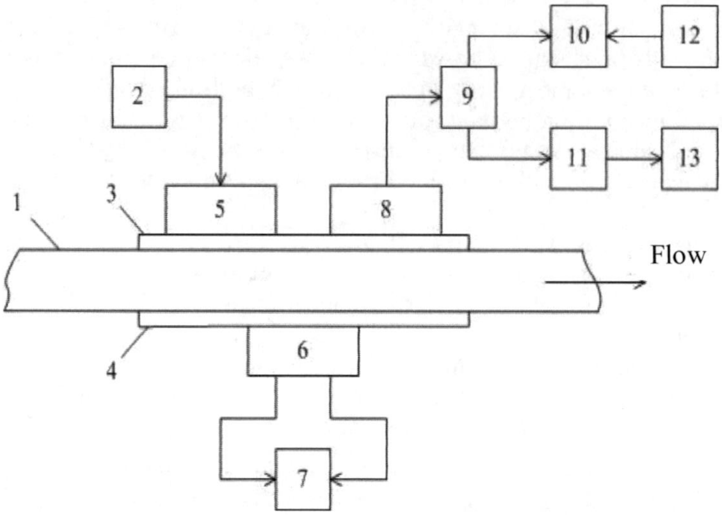

Figure 2. Block Diagram for Measuring the Isotropic Flow Continuity:
1 – dielectric pipeline; 2 – microwave oscillator; 3, 4 – plates;
5 – input coupler; 6 – receiver; 7 – phase meter; 8 – output coupler;
9 – tee; 10 – mixer; 11 – detector; 12 – reference oscillator; 13 – frequency response device.

Equation (8) shows that the resonance eigenfrequency of an open resonator formed by two plates can be used to judge the change in the dielectric permeability of isotropic flows in pipelines.
Substituting the ε value from (8) into (7), we have

$$S = \sqrt{\frac{D^2\omega^2\psi^2}{\pi^2K^2\lambda^2d^2\varepsilon^4(\pi^2q^2c^2 - 4D^2\omega^2\varepsilon_p)}}$$

Thus, with a change in ε, the problem of determining S is reduced to the simultaneous measurement of the phase difference of the electromagnetic waves at the isotropic medium exit and the resonant frequency of the open resonator.

The open resonator technique allows using the Q-factor [5] of the above resonant system to measure the dielectric permeability ε, along with the resonance eigenfrequency.
In our case, for the Q-factor of the open resonator, we have:

$$Q = \frac{D\omega\sqrt{M}}{\alpha c} \quad (9)$$

As is known, the resonator Q-factor can be determined by the ratio of the resonant frequency ω to the frequency bandwidth $\Delta\omega$ of the resonance curve at the half-power level, i.e., $Q=\omega/\Delta\omega$. Then, considering this and $M=\varepsilon_T+\varepsilon$, equation (9) will take the form:

$$\Delta\omega^2 = \frac{\alpha^2 c^2}{D^2(\varepsilon_p+\varepsilon)}.$$

Hence, for ε, we obtain:

$$\varepsilon = \frac{\alpha^2 c^2 - \Delta\omega^2 D^2 \varepsilon_p}{\Delta\omega^2 D^2}. \qquad (10)$$

This formula shows that the dielectric permeability of the controlled medium can be judged by the resonance curve bandwidth at the half-power level.
Substituting the ε value from (10) into (7), we have:

$$S = \sqrt{\frac{\psi^2 \Delta\omega^2 D^2}{4\pi^2 K^2 \lambda^2 d^2 E^4 (\alpha^2 c^2 - \Delta\omega^2 D^2 \varepsilon_p)}}.$$

Thus, in this case, to determine the flow continuity, the ψ and $\Delta\omega$ parameters should be known. Practice shows that in the measurement, along with the change in the ε parameter, the pipeline dielectric permeability ε_p may also change, causing an extra error in measuring the continuity. Calculations show that the use of the above two oscillatory characteristics (resonant frequency and Q-factor) of an open resonator may exclude an error associated with a change in ε_p. As can be seen, e.g., from the open resonator resonant frequency dependence on the dielectric permeabilities ε and ε_p, in the absence of flow, the ω value can be used to determine ε_p when it changes with the ambient temperature. Consequently, the drift of the resonant frequency depending on ε_p can always be considered when measuring the flow continuity in pipelines.

5. A Way to Reduce the Error in Measuring Continuity

When working with anisotropic flows, as can be seen from formula (2), the continuity measurement error may be caused by the difference in the polarized electromagnetic wave refraction indices of the medium.

As noted above, the anisotropic medium refractive indices n_1 and n_2 have different values due to the tensor nature of its permittivity, and can be commonly expressed as:

$$n_1 = \sqrt{\varepsilon_1};$$
$$n_2 = \sqrt{\varepsilon_2},$$

where ε_1 and ε_2 are the anisotropic medium dielectric permeabilities for one and another wave polarizations, respectively. Here it is assumed that the magnetic permeabilities $\mu_1=\mu_2=1$. Therefore, in this case, to determine the measurement error, data on the simultaneously changing ε_1 and ε_2 parameters are required.

Based on the open resonator technique, a new principle can be proposed to determine the simultaneously changing ε_1 and ε_2 parameters.

The proposed principle is based on forming two mutually perpendicular open resonant systems on the outer surface of a dielectric pipeline through which the anisotropic medium flows, and measuring their oscillation characteristics correspondingly related to the above dielectric permeabilities.

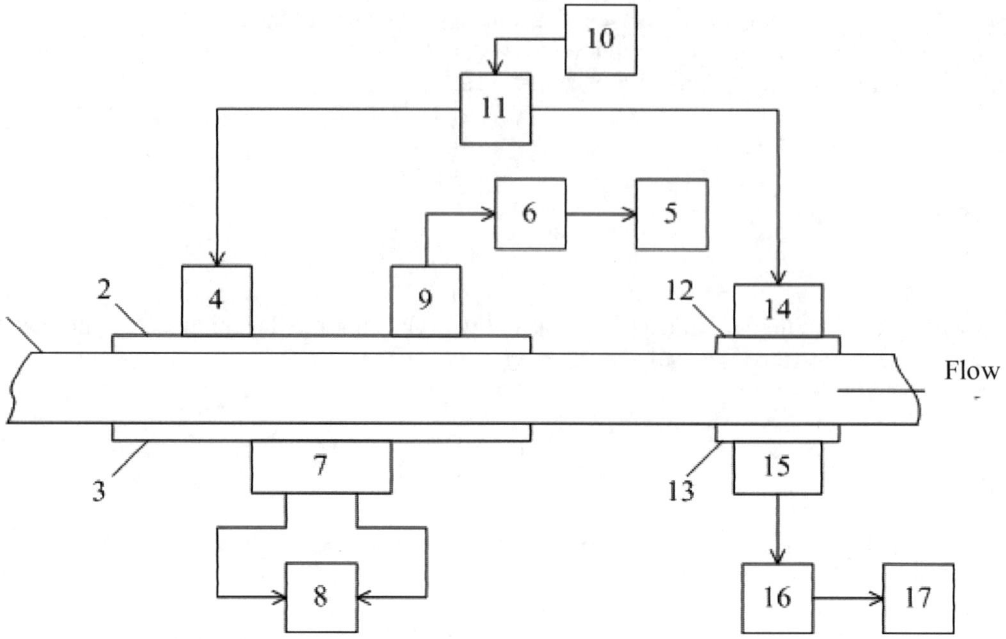

Figure. 3. Block Diagram for Measuring Continuity Based on Longitudinal and Transverse Open Resonators:
1 - dielectric pipeline; 2,3 - reflectors of the longitudinal open resonator;
4 – input coupler; 5 – frequency response device; 6 – detector; 7 – receiver; 8 – phase meter;
9 – output coupler; 10 – microwave oscillator; 11 – tee;
12, 13 – the transverse open resonator reflectors; 14 – input coupler;
15 – output coupler; 16 – detector; 17 – frequency response device.

When electromagnetic oscillations are excited in two open measuring resonators (Figure 3) their resonant frequencies can be expressed as:

$$\omega_1 = \frac{\pi q c}{2D\sqrt{M_1}}$$

$$\omega_2 = \frac{\pi q c}{2D\sqrt{M_2}} \tag{11}$$

where ω_1 and ω_2 are the resonance eigenfrequencies of the longitudinal (reflectors are installed along the pipeline) and transverse (reflectors are installed across the pipeline) open resonators, respectively, M_1 and M_2 are coefficients considering the impact of the dielectric permeabilities ε_1 and ε_2, as well as the pipeline dielectric permeability ε_p on the electromagnetic wave propagation velocity along and across the pipeline between reflectors of measuring open resonators. In this case, the generated waves have components, polarized parallel and perpendicular to the directions of the wave propagation into the anisotropic flow through the corresponding resonator reflectors. With the same configuration and dimensions of the reflectors, as well as the equality of the parameters of electromagnetic oscillations used to excite the longitudinal and transverse resonators, it can be assumed that:

$$\mathrm{E}_{\parallel}^{lr} = \mathrm{E}_{\perp}^{tr} \text{ and } E_{\perp}^{lr} = \mathrm{E}_{\parallel}^{tr},$$

where $\downarrow E_{||}^{lr}, E_{\perp}^{lr}$ is the intensities of the electric fields polarized in the space between the longitudinal wave resonator reflectors; $\downarrow E_{||}^{tr}, E_{\perp}^{tr}$ is the intensities of the electric fields polarized in the space between the transverse wave resonator reflectors. Note that the field (resonance phenomenon) in longitudinal and transverse resonators is the result of multiple reflections of waves with only the electric field intensities $E_{||}^{lr}$ and $E_{||}^{tr}$, respectively. As for the waves with the electric field intensities $E_{\perp}^{lr}$ and $E_{\perp}^{tr}$, they can be used to estimate the continuity of controlled flows in pipelines after the exit from the anisotropic medium.

In our case, due to the tensor nature of the controlled anisotropic flow dielectric permeability, its values should be different with respect to polarized waves [6]. Therefore, the notations ε_1 and ε_2 can be introduced for the medium dielectric permeability with respect to the waves with the electric field intensities $E_{||}^{lr}$ ($E_{\perp}^{tr}$) and $E_{\perp}^{lr}$ ($E_{||}^{tr}$), respectively.

Considering that in this case, waves with the electric field intensities $E_{||}^{lr}$ and $E_{||}^{tr}$ are propagating between the reflectors of the above resonators, the M_1 and M_2 coefficients:

$$M_1 = \sqrt{\varepsilon_T + \varepsilon_1}\,;\; M_2 = \sqrt{\varepsilon_T + \varepsilon_2}\,.$$

Substituting the M_1 and M_2 values into the system of equations (11) and performing the transformation for the frequencies ω_1 and ω_2, we obtain:

$$\omega_1^2 = \frac{\pi^2 q^2 c^2}{4D^2}\left(\frac{1}{\varepsilon_T + \varepsilon_1}\right)$$

$$\omega_2^2 = \frac{\pi^2 q^2 c^2}{4D^2}\left(\frac{1}{\varepsilon_T + \varepsilon_2}\right) \qquad (12)$$

The last formulas show that with the mutually perpendicular input of electromagnetic waves into an anisotropic flow, different values of the controlled medium dielectric permeabilities can be obtained by measuring the resonance frequencies of two open resonators.
The solution of the system of equations (12) gives for ε_1 and ε_2:

$$\varepsilon_1 = \frac{\pi^2 q^2 c^2 - \omega_1^2 4D^2 \varepsilon_p}{4D^2 \omega_1^2};$$

$$\varepsilon_2 = \frac{\pi^2 q^2 c^2 - \omega_2^2 4D^2 \varepsilon_p}{4D^2 \omega_2^2}.$$

Using these formulas, equation (2) can be rewritten as:

$$\psi = \frac{\omega}{c}\left[\sqrt{\frac{\pi^2 q^2 c^2}{4D^2 \omega_1^2} - \varepsilon_p} - \sqrt{\frac{\pi^2 q^2 c^2}{4D^2 \omega_2^2} - \varepsilon_p}\,\right] l. \qquad (13)$$

Here it is assumed that $\varepsilon_1 > \varepsilon_2$.

From the equation (13), it follows that when determining the phase difference between two polarized waves, related to the anisotropic flow continuity, the measurement error, caused by a change in the difference in the refractive indices (Δn) of these waves, can be considered and excluded.

According to the block diagram for measuring the anisotropic flow continuity with two open resonators, in this case, the longitudinal and transverse resonators are simultaneously excited by the microwave oscillator output signal. Then, by the frequency characteristics of modules 5 and 17 (Figure 3), the resonance eigenfrequencies are determined, related to the parameters ε_1 and ε_2, respectively. The circuit also allows determining the phase difference between two polarized waves generated, e.g., in the space between the longitudinal resonator reflectors.

Journal of Physics: Conference Series **2091** (2021) 012042 doi:10.1088/1742-6596/2091/1/012042

Calculations show that based on the above diagram, it is relatively easy to implement an algorithm (13) considering the impact of the anisotropic flow dielectric permeability on the measurement result with the possibility of reducing the continuity measurement error.

6. Conclusion

Thus, the interrelation of the polarized electromagnetic wave characteristics and the gas-liquid flow properties allows suggesting the principles for building continuity meters for naturally and artificially anisotropic flows in pipelines with the possibility of eliminating measurement errors.

7. References

[1] Zhukenov M K 2018 *Propagation of Electromagnetic Waves in Anisotropic Magnetoelectric Media* ed M K Zhukenov (Pavlodar: Kereku) p 105
[2] Ivlev D D, Ishlinsky A Yu and Maksimova L A 2000 On the isotropic medium flow properties *Reports of the Academy of Sciences* **375** pp 191–194
[3] Barykina E I and Brazhe R A 2008 Reflection and refraction of electromagnetic waves at the dielectric-electrically conductive medium interface *Allowing the Inversion of an Electronic Subsystem* UlSTU Bulletin **1** pp 37–39
[4] Kuzmichev I K, Ermaka V D, May A V and Troshlio A S 2017 Open resonator to add powers in the subterahertz and terahertz Ranges *Radio Physics and Radio Astronomy* **22** pp 67–77
[5] Kundenko N P 2012 Determining the Q-factor of open resonant system in weak coupling mode *Energy Saving Energy Energy Audit* **11** pp 66–71
[6] Kochereshko V P 2017 *Magneto-Optical Phenomena in Semiconductors and Semiconductor Nanostructures: Textbook* ed V P Kochereshko and A Yu Serov (St. Petersburg: Solo) p 86

Journal of Physics: Conference Series **2091** (2021) 012043 doi:10.1088/1742-6596/2091/1/012043

Application of the digital image correlation method for diagnosing the condition of railway parts

M O Sharikova[1], **A Yu Poroykov**[2] **and A Yu Marchenkov**[2]

[1] Scientific and Technological Center of Unique Instrumentation Russian Academy of Sciences, 15 Butlerova Str., Moscow 117342, Russia
[2] Moscow Power Engineering Institute, National Research University, 14 Krasnokazarmennya Str., Moscow 111250, Russia

E-mail: sharikova.mo@ntcup.ru

Abstract. One of the main factors contributing to accidents in railway transport are fatigue fractures of rails, wheel axles and other critical parts of the rolling stock. Despite the widespread use of various physical control methods (eddy current, ultrasonic, magnetic), the causes of accidents are, as a rule, missing defects in the structural elements of the car bogie. The paper considers the application of the digital image correlation method to measure the spatial distribution of deformations. The results of laboratory tests of railway parts using this method are presented. The possibility of visualizing fatigue cracks in the test specimens is demonstrated. The results can be in demand both by scientific research organizations of the railway industry when solving expert problems, and by production and repair enterprises when carrying out regular diagnostic procedures.

1. Introduction

During the operation of railway rails, elements of wheelsets and other highly loaded units of railway equipment under the influence of various factors, a gradual degradation of the physical and mechanical properties and structure of the metal occurs, leading to the occurrence and development of fatigue cracks. Fatigue fractures of critical parts of the rolling stock, such as railroad rails, are one of the main factors of train derailment. Despite the existence of various physical control methods (eddy current, ultrasonic, magnetic), the causes of accidents are, as a rule, missing defects in various railway facilities. In addition to the obvious reasons associated with the influence of the human factor, the omission of defects can be explained, firstly, by a wide variety of geometric shapes and sizes of the controlled elements, and secondly, by a rather low sensitivity to the detection of unfavorably located crack-like defects, which is inherent in traditional methods of control.

The idea of comparing digital image correlation (DIC) and finite element modeling has been successfully tested on samples of the rail sections [1]. In this paper, we develop this approach. The paper investigates the possibility of detecting crack-like defects in railway rails and wheel axles using the machine vision method [2-4]. To obtain information about the parameters of a defect, the DIC method [5-8] is used in the work, which makes it possible to determine with high accuracy the actual dimensions of rail defects.

2. Description of the method

Railway rail P43 was chosen as the first object under study. Samples for testing were rail fragments 10 ± 1 mm thick. After cutting, the surfaces of the cuts were subjected to milling and grinding to obtain the required quality of the end surfaces. Thin cuts were made on the cut fragments over the entire thickness of the samples at a given depth, simulating a defect. The cuts were made on an electric discharge machine using a wire of 0.25 mm diameter. The width of the finished cut is about 0.25 ... 0.3 mm. The notch depth is 2 ... 4 mm, depending on the sample. The location of the notch also depended on the sample - the notch was made either in the rail neck, or in the fillet part between the neck and the head, or in the fillet part between the neck and the rail foot (figure 1).

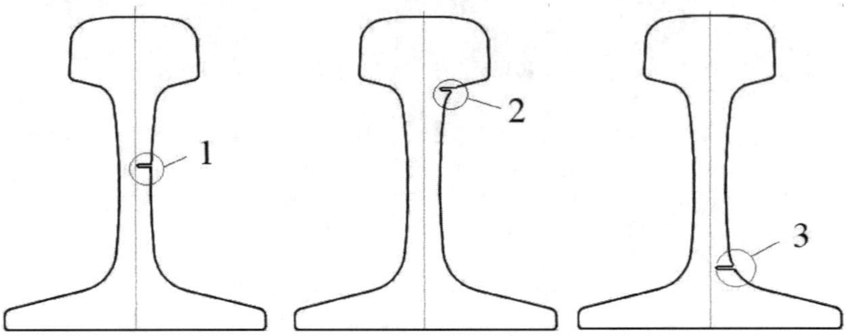

Figure 1. The location of the notch: the notch was made either in the rail neck (1), or in the fillet part between the neck and the head (2), or in the fillet part between the neck and the rail foot (3).

The second object under study is the axle of the RU-1Sh wheelset. Samples are made from the axle for three-point bending tests. To make samples from the central part of the RU-1Sh axis, having a nominal diameter of 172 mm, a blank 350 mm long is cut out on a band saw machine. From the obtained ingot, which has a cylindrical shape, a workpiece with a square cross-section with a side of 120 mm and a length of 350 mm is obtained by electroerosive cutting. The resulting workpiece is cut lengthwise into workpieces for samples with a section of 25 mm x 20 mm. A narrow notch with a depth of 3 or 5 mm (depending on the sample) is applied to the obtained blanks over the entire width of the sample, simulating a defect. The notch is made by electric discharge machining using a wire with a diameter of ∅0.25 mm. The notch width is 0.25 ... 0.3 mm.

To increase the contrast of the investigated surface and improve the measurement results using the DIC method, a random pattern was applied to one surface of the samples using white and black aerosol paints.

A fatigue crack of a given length was grown on the notched fragments of rails and axles with paint applied to the surface. Fragments of the rail were installed in the grips of the Instron 8801 testing machine by the end surfaces, and smooth cyclic loading was carried out by stretching in a sinusoidal cycle with a frequency of 10 Hz and a cycle asymmetry coefficient $R = 0$ (pulsation cycle). The maximum cycle load was selected for each specimen, depending on the notch length, so that at the moment of the cycle maximum, the average stress in the notched cross section was no more than 250 MPa, which guarantees the operation of the main part of the section in the elastic deformation region.

At the axis fragment, the crack growth was carried out according to the three-point bending test scheme. To do this, the fragment is installed on a static three-point bending test device, consisting of two lower supports separated by a distance of 250 mm and a cylindrical punch with a diameter of 10 mm. The punch is secured in the upper grip of the Instron 8801 testing machine, the supports in the lower grip. To grow a crack in a specimen, it is smoothly cycled loading with a sinusoidal cycle with a frequency of 5 Hz and a cycle asymmetry coefficient $R = 0$ (pulsation cycle). The maximum load of the cycle was selected in such a way that at the moment of the maximum of the cycle the average bending stress in the cross section with a notch did not exceed $\sigma \approx 250$ MPa, which is obviously lower than the stress at which the transition from elastic to elastoplastic deformation occurs. The lengths of the finished cracks were 1 ... 6 mm.

The dimensions of all cracks during growth were monitored visually. The actual crack size was determined after removing the sample from the grips using a Meiji RZ stereomicroscope. For a more accurate detection of the sizes of cuts and cracks, a Magnaflux Spotcheck magnetic particle inspection kit was used according to the 2nd sensitivity class. Longitudinal magnetization of the sample was carried out with the help of an electromagnet, after which a magnetic suspension was applied, and on a Meiji RZ stereomicroscope at a magnification of 15x, the crack length was measured along the traces of the magnetic suspension.

Specimens with a notch and fatigue crack were tested on an Instron 5982 universal testing machine in the compression loading mode (for rail fragments) and in the bending test mode (for axial specimens). Photos of the test setup during the tests are shown in figure 2. The loading of the rail samples was performed by cyclic compression according to a programmed cycle that simulates the passage of a car at a low speed (about 3 ... 4 km / h). Each rail was loaded with three such cycles. The maximum cycle load was 45 kN.

Figure 2. Photos of the test system

The axis was loaded with cyclic bending. During one test, loading was performed in ten triangular cycles with a maximum load of 1 kN. The deformation rate was 0.1 mm / min. The time of one loading cycle is about 30 seconds.

The tests were repeated several times (usually 7 ... 10 times) on each sample.

3. Experimental results

During the test, the samples were filmed on two video cameras Imager SX, which are an important part of the LaVision StrainMaster measuring system. This system makes it possible to apply the DIC (Digital Image Correlation) method, based on digital processing of images of the surface of a deformed sample.

For samples (a rail with a notch in the neck and axle), the results of processing the experimental images by the DIC method are presented in figures 3 and 4, respectively. Under the influence of the load, the surface areas of the samples are displaced. The nature and direction of this displacement can be understood from the displacement vectors. The offset is calculated between the first frame in the series, when no sample is loaded, and the current frame. The color in the figure shows the amplitude of the deformation.

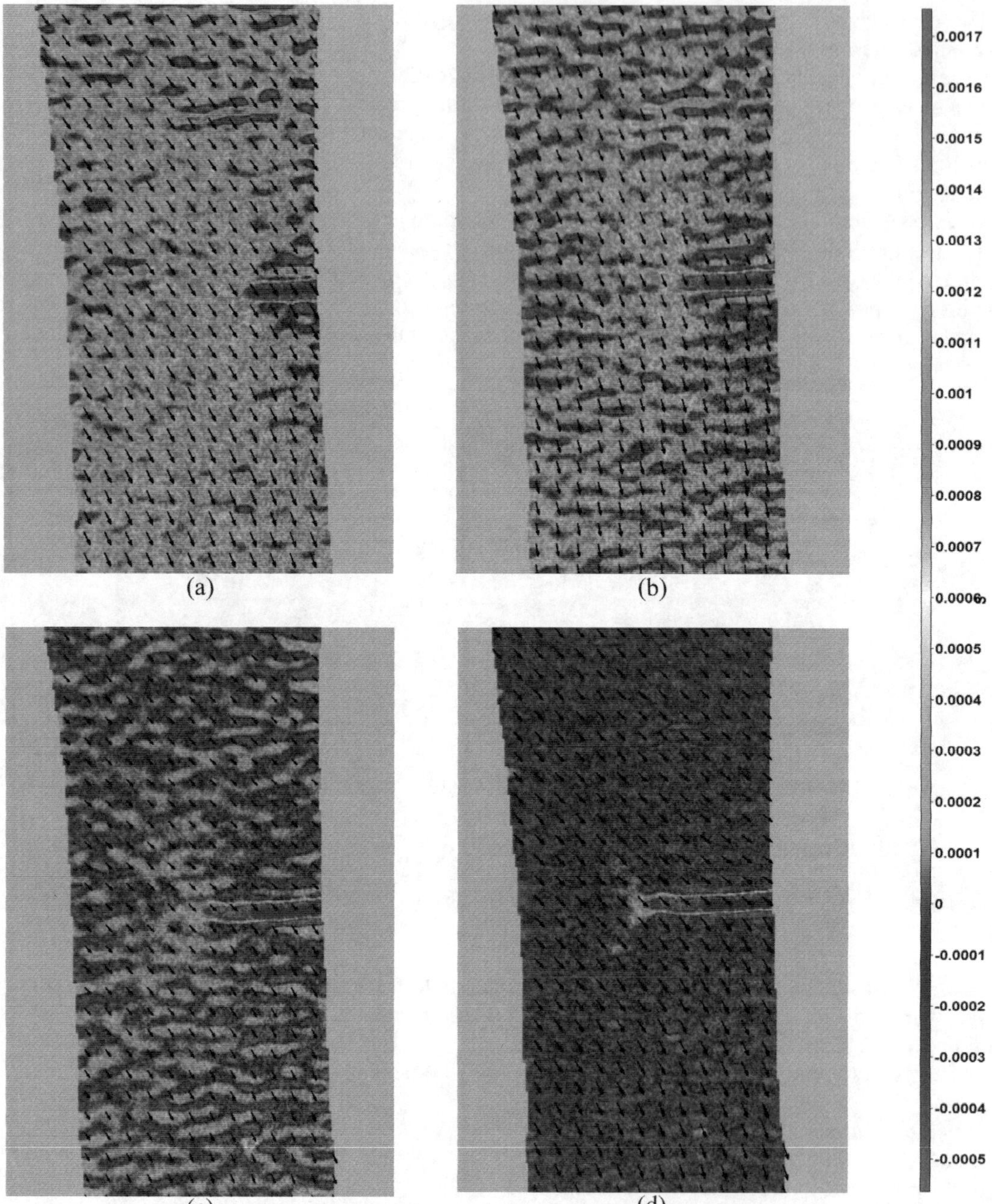

Figure 3. DIC inspection results for a rail sample with a notch in the neck at the beginning of loading (a), after 37 minutes (b), after 53 minutes (c) and after 59 minutes (d).

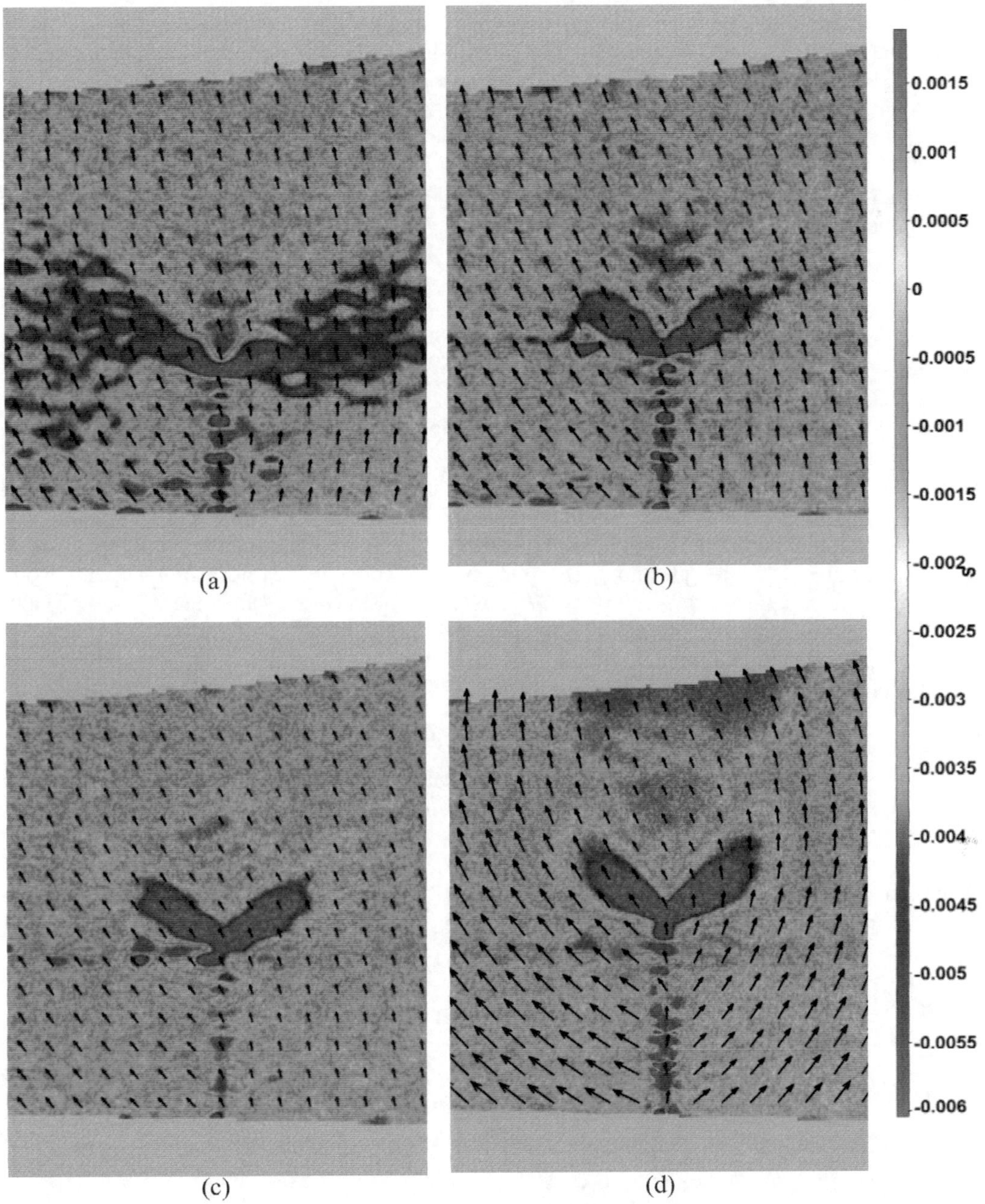

Figure 4. DIC inspection results for an axle sample at the beginning of loading (a), after 13 minutes (b), after 22 minutes (c) and after 38 minutes (d).

The pictures (figure 4) show the results of processing experimental images for testing axis, on which the size of the crack along the top of «the butterfly» in the deformation field is visualized. It can be seen how the crack grows with time.

4. Conclusion

The paper presents the results of laboratory tests for the strength of railroad rails and wheelset axles using the method of digital image correlation. The possibility of visualizing fatigue cracks in the test specimens is demonstrated. It is shown that the use of the DIC method makes it possible to quite accurately identify and evaluate the linear dimensions of defects arising in rails and axles, and to estimate the distribution of deformations in the section of the rail and the axis under its loading.

Acknowledgments
The Russian Foundation for Basic Research (project 20-38-51019).

References

[1] Bardakov V V, Marchenkov A Y, Poroykov A Y, Machikhin A S, Sharikova M O, Meleshko N V 2021 Feasibility of Digital Image Correlation for Fatigue Cracks Detection under Dynamic Loading *Sensors* **21** 19 pp 6457 DOI: 10.3390/s21196457
[2] Davies E R 2012 Computer and machine vision: theory, algorithms *London: Elsevier*
[3] Prince S J D 2012 Computer vision: models, learning, and inference *Cambridge: Cambridge University Press*
[4] Mohan A, Poobal S 2018 Crack detection using image processing: A critical review and analysis *AEJ* **57** 2 pp 787-98 DOI: 10.1016/j.aej.2017.01.020
[5] Hamrat M, Boulekbache B, Chemrouk M, Amziane S 2016 Flexural cracking behavior of normal strength, high strength and high strength fiber concrete beams, using Digital Image Correlation technique *Construction and Building Materials* **106** pp 678-692 DOI: 10.1016/j.conbuildmat.2015.12.166
[6] Alam S Y, Loukili A, Grondin F, Roziére E 2015 Use of the digital image correlation and acoustic emission technique to study the effect of structural size on cracking of reinforced concrete *Engineering Fracture Mechanics* **143** pp 17-31 DOI: 10.1016/j.engfracmech.2015.06.038
[7] Brooks W S, Lamb D A, Irvine S J 2015 IR reflectance imaging for crystalline Si solar cell crack detection *IEEE journal of photovoltaics* **5** 5 pp 1271-1275 DOI: 10.1109/JPHOTOV.2015.2438636
[8] Iliopoulos S, Aggelis D G, Pyl L, Vantomme J, Van Marcke P, Coppens E, Areias L 2015 Detection and evaluation of cracks in the concrete buffer of the Belgian Nuclear Waste container using combined NDT techniques *Construction and Building Materials* **78** pp 369-378 DOI: 10.1016/j.conbuildmat.2014.12.036

Diagnostics of solid-state microwave modules at the initial stage of technological tests

A Yu Kiryanov

Joint Stock Company 'Scientific and Production Enterprise 'Pulsar', 27, Okruzhnoy proezd, Moscow, 105187, Russia

E-mail: kiryanov29@yandex.ru

Abstract. Process testing is an integral part of the production of solid-state microwave modules used in radar stations. At the final stage of production, long-term technological tests are carried out. However, at the initial stage of testing, it is possible to carry out diagnostics of solid-state microwave modules, which makes it possible to identify malfunctions leading to failures of solid-state microwave modules.

1. Introduction

One of the main requirements for solid-state microwave modules used in radar stations is the requirements for reliability characteristics and long-term continuous operation. Technological tests are an integral part of the technological process and allow revealing hidden defects that lead to premature failure of solid-state microwave modules. The test duration is 72 hours. However, already at the initial stage of technological tests, it is possible to carry out diagnostics of solid-state microwave modules, which makes it possible to reveal hidden defects leading to the failure of solid-state microwave modules.

2. Main part

To carry out technological tests, a test bench is used, shown in 'figure 1'. The test bench is structurally made of the following blocks: a temperature setting block, a monitoring and control block, a load block, an instrument block, a generator block and a power supply block.

The duration of technological tests (detection of hidden manufacturing defects) for solid-state microwave modules is usually 72 hours. The operating time of the module in the switched-on state must certainly correspond to the time of reaching thermal equilibrium in the structure of the active elements of the microwave mode, and the time the module is in the switched-off state must correspond to the cooling time of the heat-generating element to the ambient temperature.

For technological tests, the following operating mode of the modules is provided: 45 minutes of operation of the module with the supply of supply voltage and input microwave signal and 15 minutes of standby mode without supply of supply voltage and input microwave signal. The modules are connected to the test bench by means of high-frequency cables equipped with connectors. The monitored values of the electrical quantities of the modules are displayed on the LCD display of the industrial computer, which also processes and records the measured data.

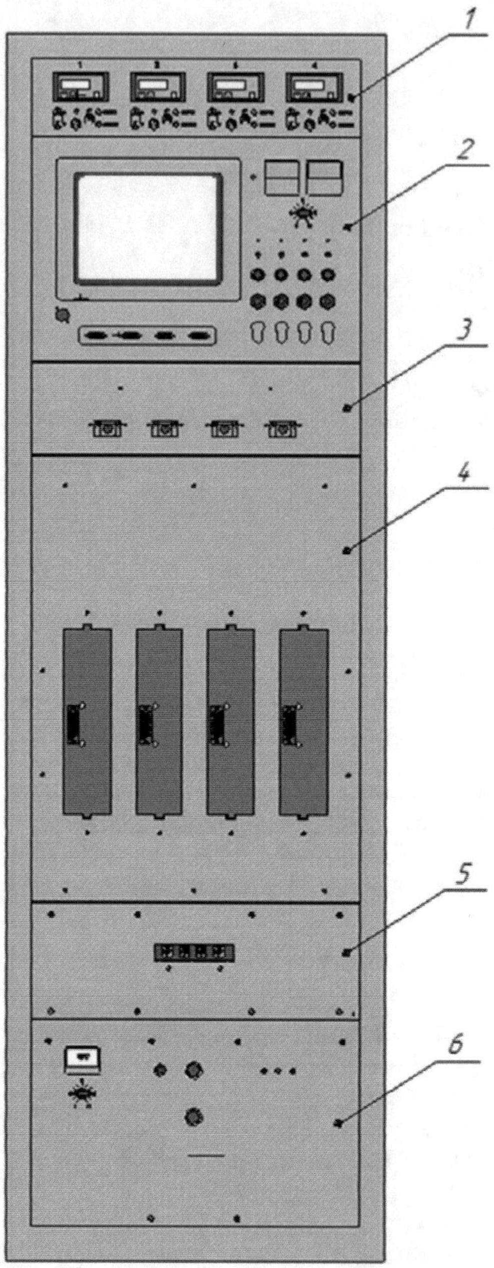

Figure 1. Appearance of the test bench, where: 1 - temperature setting block, 2 - monitoring and control block, 3 - load block, 4 - an instrument block, 5 - generator block, 6 - power supply block.

The temperature setting block is designed to control and regulate the temperature of the microwave module case.
The monitoring and control block is used to control the supply voltage and service voltage, measure the current consumed by the modules, the input and output power of the microwave signal of the modules,

control the operating modes of the modules, control the set parameters and display them in real time on the LCD.
The block of loads is used to ensure the dissipation of the output microwave power on the load.
The instrument block is designed to stabilize the temperature of the case of each microwave module.
The generator block contains a microwave generator and a power divider. The microwave signal from the power divider is fed to the microwave input of the module under test.
The power supply block is designed to operate the blocks of test bench and tested modules.
When choosing test modes for microwave modules, the following basic parameters are set:
- duration of microwave power pulses;
- duty cycle of microwave power pulses;
- is the power dissipated during the pulse;
- air temperature in the heating circuit;
- module supply voltage;
- time of periodical switching on and off of the module during the tests.
Technological test modes must be selected as the maximum permissible, in accordance with the technical conditions, in order to identify all modules with hidden defects [1].
The main technical characteristics of the test bench are shown in table 1.

Table 1. Main technical characteristics of the test bench

№п/п	Parameter name, unit of measurement	Parameter value
1.	Supply voltage	
1.1	Three-phase four-wire AC network, V	380 ± 10%
1.2	Current frequency, Hz	50 ± 1.5%
2.	Error of the control tools	
2.1	by supply voltage,%	< 2
2.2	by current consumption,%	< 3
2.3	by air temperature in the heating circuit,%	< 3
3	Number of simultaneously tested modules	4
4	Air temperature setting range in the heating circuit, ° C	22 ÷ 65
5	Overall dimensions (LxWxH), mm	700x700x2000

A feature of the test bench is the division of the instrument block into four independent sections shown in ‘figure 2’.

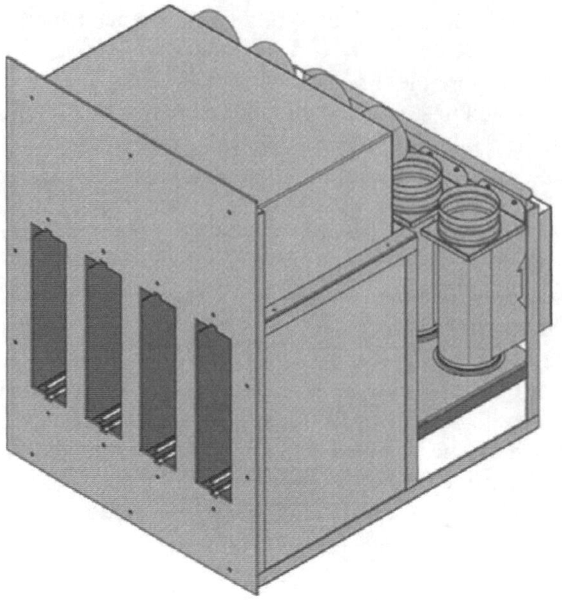

Figure 2. Instrument block, structurally divided into four independent sections.

Each section is a forced convective heat exchange circuit (further in the text - the thermal circuit), in which the tested module installed 'figure 3'. The thermal circuit contains: a microwave module, which, due to self-heating, heats the air in the circuit, a fan designed to create a forced air flow, a controlled heater that serves to maintain a given air temperature, and a thermocouple for measuring the air flow temperature [2]. The temperature of the module case is measured by a sensor built into the microwave module and transmitted via the serial interface to the monitoring and control block.

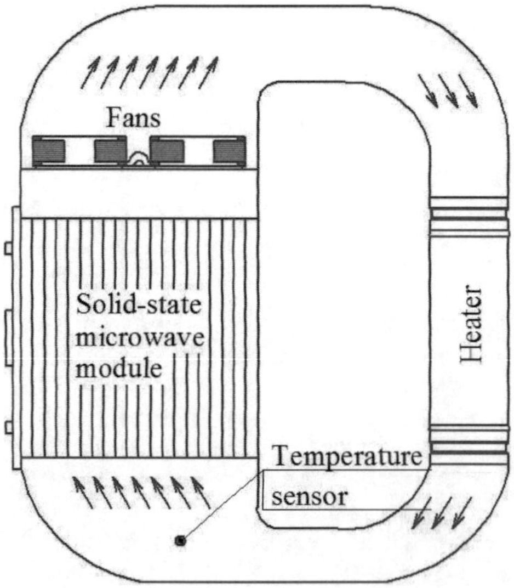

Figure 3. Thermal circuit

When studying the operating modes of the thermal circuit without a heater, it was found that the cooling efficiency of a solid-state microwave module at an air flow rate of more than 5 m / s decreases sharply, since thermal resistance of the heatsink - the environment takes on minimum values. In 'figure 4' a graph of the dependence of the temperature of the module heatsink on the air flow rate for modules with different power consumption is presented.

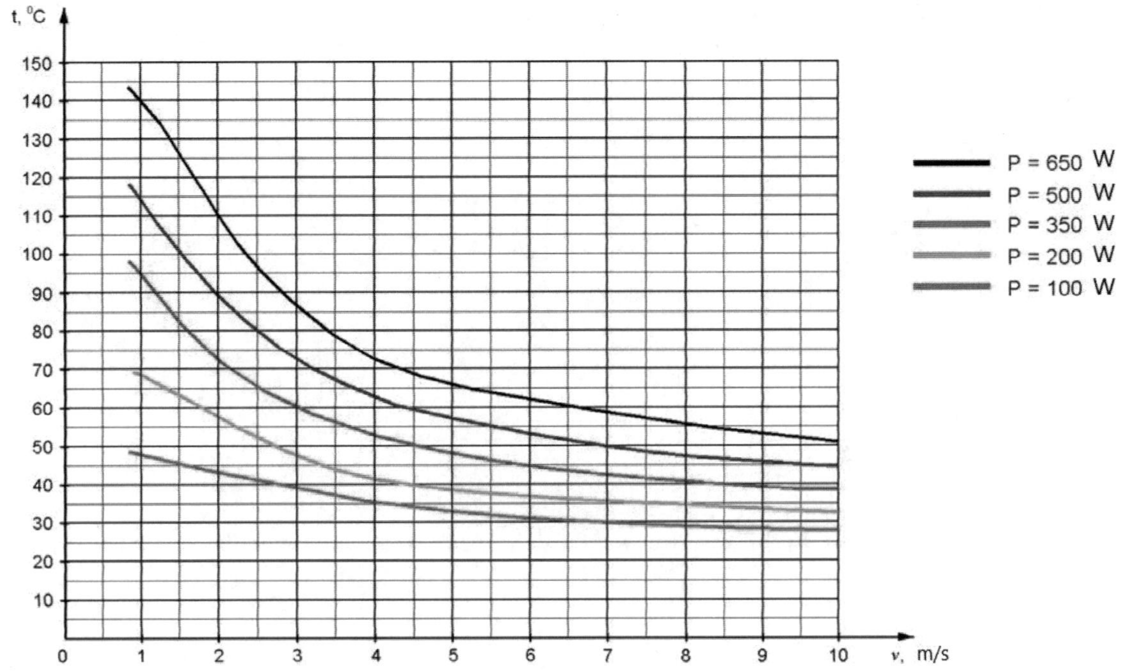

Figure 4. Dependence of the module heatsink temperature on the air flow rate.

It was also revealed that at an air flow rate of more than 5 m / s, the temperature of the air flow and the module case have a directly proportional relationship 'figure. 5'. Thus, by measuring the air temperature in the chamber, it is possible to control the heating of the microwave module housing. This circumstance allows us to conclude that it is possible to diagnose the module in the warm-up mode by measuring the temperature of the air flow of the thermal circuit.

The dependence of the temperature of the air flow in a closed circuit of convective heat transfer from time to time is shown in the graph 'figure 6'. In this case, the temperature of the air flow in the circuit increases due to the heating of the case of the module under test. Using the graph of the dependence of the air flow temperature with a known working (reference) module, it is possible to identify malfunctions of the tested solid-state microwave module associated with the failure of any functional units of the module or poor thermal contact of the amplifying elements with the module case. Thus, the heating graph of the faulty module (curve 2) will differ from the reference one (curve 1). The duration of the diagnostics of the solid-state microwave module will be 50 minutes.

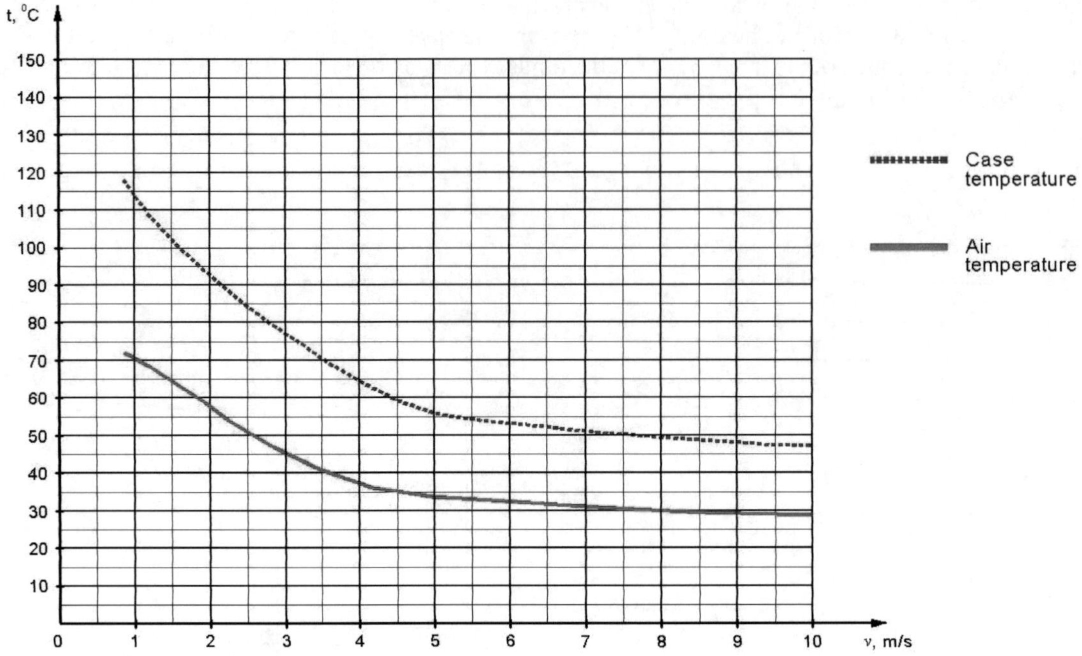

Figure 5. Dependence of the temperature of the air and the module case on the air flow rate.

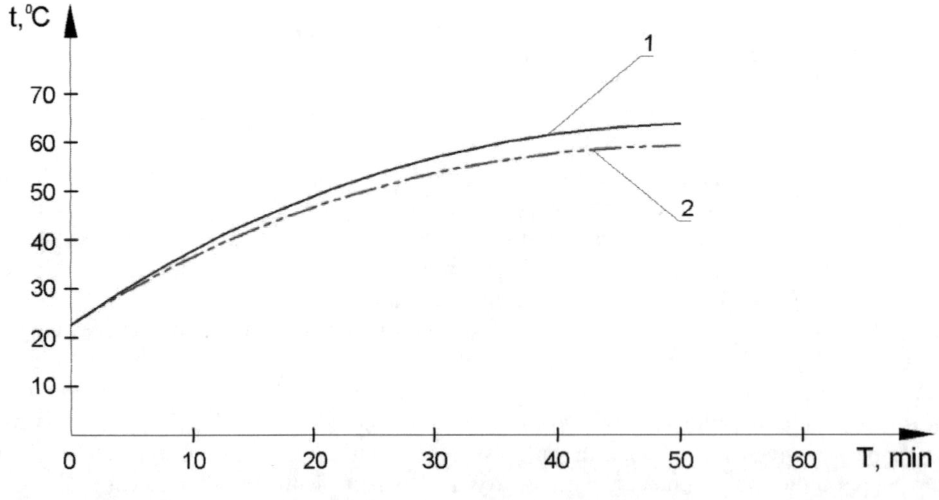

Figure 6. Dependence of the temperature of the air flow in a closed circuit of convective heat transfer from time.

3. Conclusions

Diagnostics of modules, at the initial stage of testing, allows to identify some of the types of failures in the mode of output of a solid-state microwave module to a stationary mode of operation. Such failures are associated with defects in installation of elements and assembly of boards, defects in soldering, or poor thermal contact of the amplifying elements with the case of the module under test. Thus, when these types of failures are detected, in the heating mode of the solid-state microwave module, there is no need to carry out technological tests for a long time. As a result of the application of this method, the total testing time is reduced, the labor productivity is increased, which ultimately provides a positive economic effect by reducing the number of equipment repairs under operating conditions. The cost of work to eliminate defects in the factory is lower than in the operating conditions, and the quality of repairs is significantly higher. Downtime of equipment in operation is reduced, penalties for equipment failures during the warranty period are reduced [3].

References

[1] Kiryanov A Yu 2020 Development of a methodology and design of a test bench for technological tests of solid-state microwave modules *Electronic equipment Series 2 Semiconductor devices No.2 (257)* pp 69-75

[2] Kiryanov A Yu 2020 Test bench for technological testing of solid-state microwave modules used in radar systems *Collection of reports Materials of the 13th Int. Scientific and Technical Conf. Acoustooptic and radar methods for information measurements and processing Moscow Russia* pp 81-84.

[3] Kiryanov A Yu 2020 The technical and economic effect of introducing test benches into production for technological testing of solid-state microwave modules *Defense complex - to the scientific and technological progress of Russia No. 1 (145)* pp 43-47

Application of acoustic microscopy for evaluation of numismatic material

Y S Petronyuk[1,2], S A Titov[1] and V M Levin[2]

[1] Scientific and Technological Centre of Unique Instrumentation of Russian Academy of Sciences, 15 Butlerov street, 117342 Moscow, Russia
[2] N.M. Emanuel Institute of Biochemical Physics of Russian Academy of Sciences, 4 Kosygin street, 119334 Moscow, Russia

E-mail: jps7@mail.ru

Abstract. A new non-destructive express method for assessing the authenticity of numismatic material is considered in the paper. Non-destructive methods of pulsed acoustic microscopy in the frequency range of 50-100 MHz were applied. Samples of silver coins dated 1924 and 1979 were examined. The characteristic values of the longitudinal and transverse sonic velocity of the samples were obtained. The method of hydrostatic weighing was used to measure the density of the metal. It was shown that assessment of the authenticity and the safety degree of coins can also be carried out by revealing the internal defects such as corrosion. Acoustic visualization in the reflection mode allows imaging the offset of the obverse relative to the reverse and to determine the centre of the coin deviation, which also confirms the quality of the minting.

1. Introduction

To determine the authenticity of coins, there are some criteria related to the technology of their manufacture, the composition of the metal and the degree of their preservation. Genuine coins are minted; this is an expensive and complex technological process that is rarely used by counterfeiters. As a rule the image on fakes is less clear than when minting. This is the main technological difference between fakes can be determined by visual inspection under a microscope. At the same time, the obverse, reverse and edge are examined with a certain step of notches, the accuracy of letters, etc. In addition to visual inspection, the traditional methods of coin examination include measuring the size and weight. Dimensions are measured with a special ultra-precise calliper with an accuracy of tenths or even hundredths of a millimetre, and weight with an accuracy of hundredths of a gram. For coins made of precious metal, the sample on the surface is determined, i.e., the percentage of metal and impurities is determined. To do this, determine the density of the metal by hydrostatic weighing and compare its value with the catalogue data on the same coins. Spectral analysis can also be used to study the metal composition [1]. Currently, there is no catalogue of spectral data for coins in Russia. During the examination, specialized catalogues and data from the National Numismatic Register (NPR) are used, which certifies numismatic material by entering information about coins into the information database [2]. Information about the coin in the NPR includes confirmation of authenticity, weight, safety assessment in accordance with accepted international standards (Sheldon scale), high-precision digital photography, and a variety of coins in catalogues. The difficulties caused by the fact that the older the coin, the less perfect the technology of its production and there may be large differences in the weight, size, density of the originals. Besides this a drawing and its offset relative to the centre of the coin and the edge, other defects and damages may be relevant for the coin cost estimation. Here, the expert's experience comes to the rescue. A number of problems can be solved by non-destructive diagnostic methods – X-ray fluorescence analysis and acoustic microscopy. Ultrasonic imaging is the safe rapid method that allows visualizing the pattern on the surface and its displacement, but also detecting elements of the internal structure – lamination of low-quality metal under the influence of the internal corrosion process, which is an absolute sign of the authenticity of coins, especially the 18th century, and which cannot be imitated. This paper shows the potentialities of the method for the application in expert diagnostics and certification of coins.

2. Experimental study

2.1. Principles of method

Method of pulsed acoustic microscopy is based on the use of high-frequency ultrasound. A short focused pulse (50 - 100 MHz) falls on a solid-state object from an immersion liquid and propagates in it in the form of longitudinal and transverse elastic waves. The signal recorded by the microscope contains echo pulses from the upper and lower surfaces of the object under observation; the echoes differ in the delay time and can be used to measure the sound velocity in the material as well as to display layer-by layer the bulk structure. Acoustic images are formed as a database with the coordinates x, y, z, where z is determined by the travel time Δt of the ultrasonic pulse in the object. As a result of image processing, the following display options are obtained – A, B, and C-scan. Velocities of longitudinal C_L and transverse C_T elastic waves in the object and depth of defects localisation are determined by the position of the echo pulses in A-scan echogram (figure 1a):

$$C_L = \frac{2d}{t_L}, \qquad C_T = \frac{d}{t_{LT} - 0{,}5t_L}, \qquad (1)$$

where d is the local thickness of the sample (inside the focal zone), t_L is delay time of longitudinal waves that are reflected from the specimen's bottom; and t_{LT} is delay time of the pulse caused by combination of longitudinal and transverse waves propagation.

Acoustic image B-scan allows analysing the cross-section of the object through the depth, to evaluate its structure uniformity, to identify the thickness variations, to reveal the presence of hidden defects and other interesting in the volume as well assess their localization and size. The C-scan mode allows observing the entire structure of the sample at a selected depth; estimating the defects size, to indicate the obverse to reverse as well the coin centre displacements. Principles of the method and its resolution are described more detailed in [3-5].

Figure 1. Acoustic microscopy of silver coin "fifty-kopecks": (a) – echogram of reflected pulses; (b) - surface-obverse; (c) – bottom-reverse, imaged from the obverse side. 1 – reflection from surface, 2 –

bottom echo-pulse of longitudinal waves ($t_L = 0{,}85\mu s$), 3 – bottom echo-pulse based on combination of longitudinal and transverse waves propagation ($t_{LT} = 1{,}31\mu s$).

2.2. Experimental setup
The study was carried out using a pulsed scanning acoustic microscope developed jointly by the STC UP RAS and IBCP RAS (figure 2). Acoustic lens with central operating frequency of 50 and 100 MHz, angular aperture of 22° and focal length of 13 and 5 mm in immersion was applied. Distilled water and a water-based gel were used as an immersion. The step of the scanning system was 30 microns that corresponded to the pixel size in the images. Thickness of the specimens was measured by desktop micrometre with a spherical tip of 3 mm diameter. Density of the coins was measured by the hydrostatic method using the precision balance (KERN, Swiss).

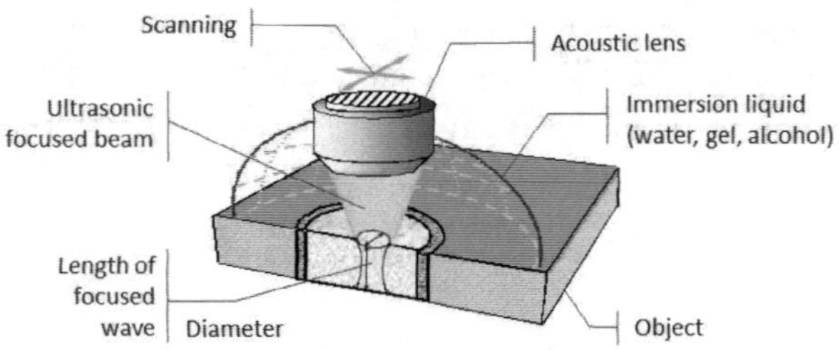

a

b

Figure 2. Experimental setup (a) and scanning impulse acoustic microscope SIAM-2 (b).

2.3. Materials under investigation
Two kind of silver coins were studied in the paper. The first one was Russian fifty-kopeck of 1924 issue and the second one was the one dollar USA of 1979 issue. Additionally, the bimetallic one euro coin from Spain was under investigation. The coins of fine condition were circulated and hadn't any visible defects.

2.4. Technique for measurement of the bulk wave velocities and the thickness
Very often the thickness of the coin under the test is difficult to measure because of its uneven surface. In this case a special technique for simultaneous measurement of the thickness and the velocities of the sound bulk waves can be used [6, 7]. The measurement scheme is presented in figure 3. In this technique the acoustic lens of the microscope is sequentially focused on the upper and lower surfaces of the plate and the received signal is recorded as a function of the vertical position z of the lens and time. Then, the unknown parameters can be calculated from the magnitude of the vertical displacement and the delay of reflected pulses.

As an example, figure 3 (b) shows the spatio-temporal signal $v(z,t)$, measured for an aluminum plate with a thickness of $d = 1{,}37 \pm 0{,}01$ mm. In the presented data format, the retarded time t is defined as follows:

$$t = t_0 - 2zC_W^{-1}, \qquad (2)$$

where the time t_0 was counted from the beginning of the probe pulse, and C_W is the wave velocity in the immersion medium (water). There are several distinct wavelets in the received signal (figure 3b). The $v_D(z,t)$ wave is formed by the D wave reflected from the sample surface (figure 3a). Its amplitude reaches a maximum F_D when the focus of the transducer is located on the sample surface $z \approx 0$. The

$v_L(z,t)$ signal is produced by a longitudinal wave L reflected from the bottom surface of the layer. It has a structure similar to the considered response v_D, but the maximum F_L is reached at a negative z, when the wave is focused on the lower surface of the layer. The $v_{LT}(z,t)$ wavelet is formed by longitudinal waves propagating in the layer in one direction, and shear waves passing through the layer in the other direction. If short enough pulses are used for measurements and the echoes reflected at the boundaries of the layer do not overlap, the superposition is valid:

$$v(z,t) = v_D(z,t) + v_L(z,t) + v_{LT}(z,t) + \cdots, \qquad (3)$$

The $v_{LT}(z,t)$ wavelet can be considered as a copy of the $v_D(z,t)$ response, shifted in the (z,t) plane by $\Delta z, \Delta t$:

$$v_L(z,t) = G_L R_D^{-1} v_D(z+\Delta z, t-\Delta t), \qquad (4)$$

where:

$$\Delta z = d\frac{C_L}{C_W}, \qquad \Delta t = \frac{2d}{C_L}, \qquad (5)$$

The obtained relation establishes a method for determining the layer thickness and the longitudinal wave velocity, based on measuring the relative displacement Δz, the delay Δt and solving the system of the equations (5). Considering that for the mixed mode wavelet $v_{LT}(z,t)$ is formed by both L and T waves, the shift and delay of the F_{LT} characteristic point relative to the F_D point can be estimated in a similar way (figure 3):

$$\Delta z = d\frac{(C_L + C_T)}{2C_W}, \qquad \Delta t = \frac{d}{C_L} + \frac{d}{C_T}, \qquad (6)$$

The results of the theoretical analysis were confirmed experimentally. The test sample study shows that the measurement error do not exceed 1% for the velocity of the longitudinal wave and the thickness of the layer, and do not exceed 2% for the velocity of the transverse wave.

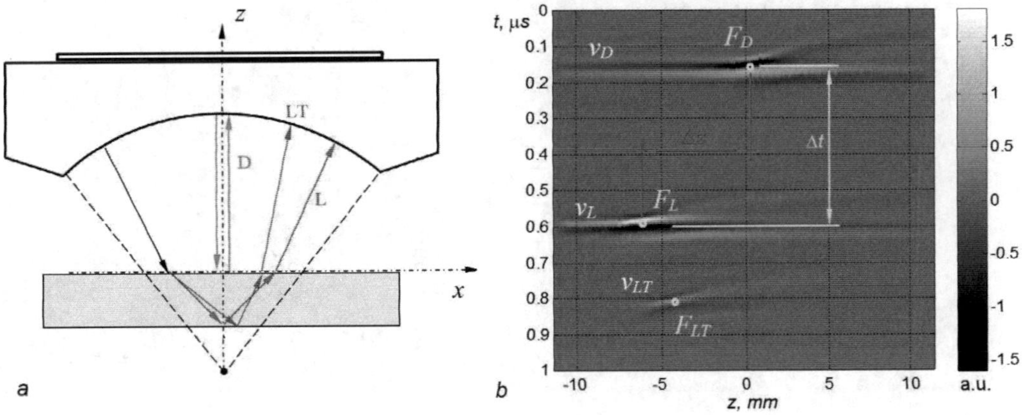

Figure 3. Scheme of the measurement (a) and recorded signal (b).

2.5. Results and discussion

Acoustic images for the Russian fifty-kopeck are presented in figure 1. The defects weren't revealed in the bulk structure of the sample. Velocity of longitudinal C_L and transverse C_T elastic waves were measured in the numismatic material. The measured values are shown in table 1. It was found that the material in available fifty-kopeck of 1924 issue is corresponds to the declared content in the coin edge – 9 g of silver. The measured density (10,23g·cm-3) and values of sonic velocities (3660 and 1760 m/s) were in correspondence with commonly known [8].

Figure 4 shows acoustic images of the one USA dollar coin internal structure. C-scan on the left part of the figure shows the coin structure at the depth of 1 mm. Acoustical contrast in the image is formed due to the coin pronounced relief and the damages in the coin volume. In the case an internal corrosion area (3) located in the center of coin disk was revealed in the specified depth. Presence of the damage is confirmed by the B-scan in figure 3 on the right. This cross-section of the specimen was obtained along the scanning line which is indicated in the C-scan on the left. The coin surface from the obverse side (1), the specimen bottom as a mirror of the reverse side (2) and an internal damage (3) are revealed in the B-scan section image. It should be noted that ultrasonic waves can pass through the damage (3) and reach the coin backside. So the damaging is similar to bulk corrosion in the form of clusters of microscopic defects which are visualized as bright areas in the acoustic image (C-scan).

Measured values of density and sonic velocities for the one-dollar coin are indicated in the table 1. The coin physico-chemical characteristics are a little bit different from the Russian fifty-kopeck silver coin ones. Velocities of the sonic waves are slightly higher (10%) than for the pure silver. Density of the coin is lower (12,5%) than for the pure silver. The results indicate a firming admixture in the coin material.

Table 1. Experimental data on physico-chemical characteristics of silver coins.

	Thickness, mm	*Density, g/cm³*	*Sonic velocity, m/s*	
			Longitudinal	*Transverse*
Fifty-kopecks	1,56 ± 0,01	10,23 ± 0,01	3660 ± 30	1760 ± 30
One-dollar	1,50 ± 0,01	8,95 ± 0,01	4010 ± 30	2000 ± 30

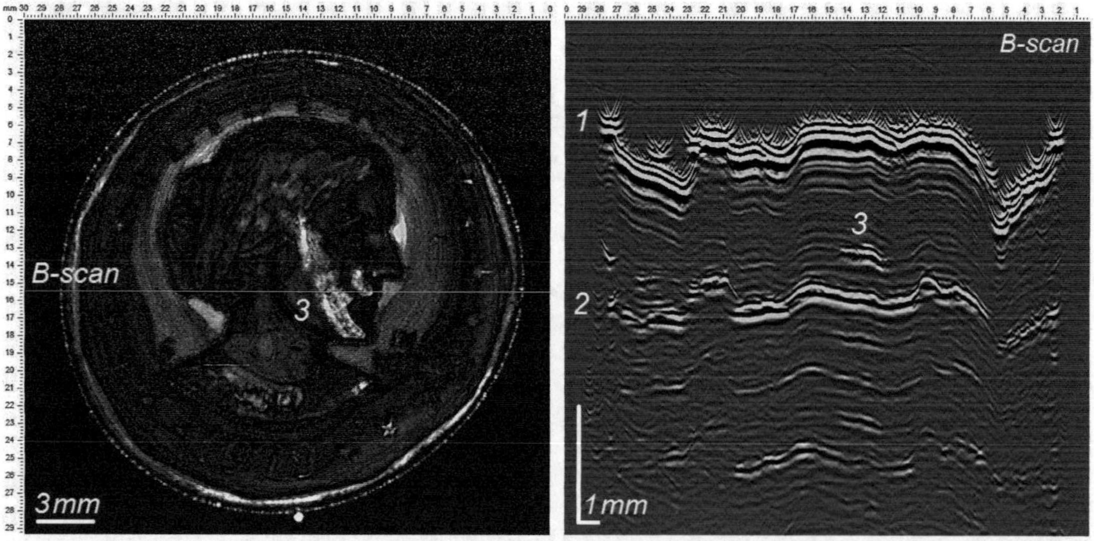

Figure 4. Acoustic microscopy of the silver dollar: C-scan image at the depth of 1 mm (left); coin cross-section B-scan (right). (1) – surface echo-line, (2) – bottom echo-line, (3) – defect inside the coin.

Finally, bimetallic one euro coin was studied (figure 5). It is known that the coin profile indicates the quality of coinage and the degree of its abrasion; it is an important parameter of the numismatic material estimation. Acoustic images in figure 5 illustrate the possibility for the task. In the C-scan the upper layer with thickness of 100 mkm is indicated. The highest points of the coin surface relief are seen as bright zones. Additionally, the B-scan shows the cross-section of the coin along the indicated line and the upper layer with 100 microns of depth is marked. On the left part of the B-scan cross-section the vertical border between two metals is seen. The difference between velocities of the elastic waves for them results in the drop of time delay for the echo pulse reflected from the coin reverse (2). Besides, a small defect at the bimetallic boundary is reveled at the depth of 0,7 mm from surface. It's hard to suppose how it is typical for such bimetallic coins.

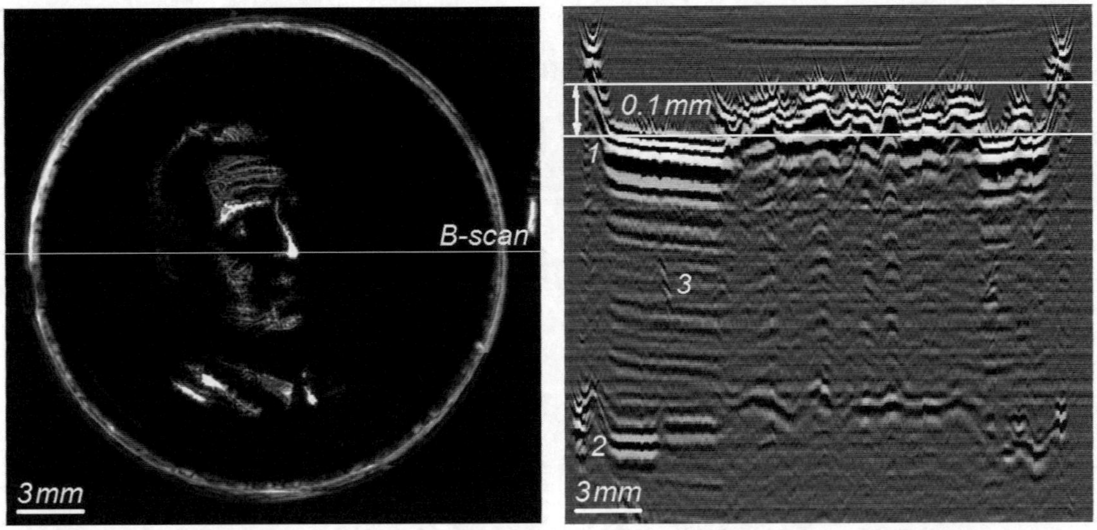

Figure 5. Ultrasonic imaging of the bimetal one euro coin: C-scan image of the top layer with thickness of 0.1 mm (left); cross-section B-scan (right). (1) – surface echo-line, (2) – bottom echo-line, (3) – defect on the bimetal vertical boundary inside the coin.

3. Conclusion

Experiments show that pulsed acoustic microscopy methods make it possible to perform non-destructive quantitative and qualitative express evaluation of numismatic materials in human safe conditions. Identification of the authenticity of coins can be carried out by the elastic wave velocities measurements. The degree the coin preservation can be estimated by the bulk structure visualization, defects revealing, surface profile measurement, analyzing obverse-to-reverse shifting. Hidden damages of the coin – internal corrosion and delaminating, can be identified by the proposed ultrasonic method.

Acknowledgments

The work was supported by the Ministry of Science and Higher Education of the Russian Federation. This work was performed using the equipment of the Centre for Collective Use of the Scientific and Technological Centre of Unique Instrumentation of the Russian Academy of Sciences (STC UI RAS) [http://ckp.ntcup.ru].

References

[1] Shaikhutdinova E F, Khramchenkova R H, Bakirov B A 2018 *Archaeology of the Eurasian steppes.* **5** p 242
[2] National Numismatic Register (NPR) https://nreestr.ru/

[3] Petronyuk Y S 2005 Application of ultrashort pulses of focused ultrasound for the characterization of modern carbon materials *Ph.D. thesis* (in Russian, Moscow: Moscow State University press)
[4] Zakutailov K V, Levin V M, Petronyuk Y S 2009 *Inorganic materials* **75** 8 p 28
[5] Petronyuk Y S, Morokov E S, Levin V M 2015 Methods of pulsed acoustic microscopy in industrial diagnostics *Bulletin of the Russian Academy of Sciences: Physics* **79** 10 p 268.
[6] Chen J, Bai X, Yang K, Ju B-F 2015 Simultaneously measuring thickness, density, velocity and attenuation of thin layers using V (z, t) data from time-resolved acoustic microscopy *Ultrasonics* **56** p 505
[7] Bai X, Sun Z, Sun A, Chen J, Ju B-F 2014 Determination of the multiple local properties of thin layer with high lateral resolution by scanning acoustic microscopy *Rev. Scientific Instrument* **85** p 094901
[8] Shaskolskaya M P 1982 Acoustic crystals (*Moscow: Nauka*) p 632

Journal of Physics: Conference Series **2091** (2021) 012046 doi:10.1088/1742-6596/2091/1/012046

Mathematical modeling of the eddy current testing of delamination cracks in multilayer CFRP objects

P N Shkatov

Federal State Budgetary Institution of Science «Scientific and Technological Centre of Unique Instrumentation» of the Russian Academy of Sciences, 15 Butlerova street, Moscow, 117342, Russia

E-mail: petr_shkatov@mail.ru

Abstract: We propose an approximate mathematical model making it possible to obtain estimates of the change in the eddy current signal under the influence of delamination in multilayer carbon fiber reinforced plastic (CFRP) objects. We give the dependences of the change in the eddy current signal on delamination in the CFRP objects for a new flaw detection method based on the comparison of eddy current signals from CFRP layers grouped by angular orientation.

1. Introduction

The increasingly widespread use of multilayer carbon-fiber materials (CFRP) in critical facilities determines the need for the development of non-destructive testing (NDT) means, taking into account the specifics of the CFRP. In the manufacture of products of this type, to achieve quality, labor-intensive precision control methods are used [1], which carry produce information about the main parameters of individual components. At the operation stage, for the timely detection of potential foci of destruction, express control based on non-contact NDT methods is required. This is due to a fairly large amount of testing to be done, combined with a strict time limit for examining most of the relevant objects – for example, an aircraft frame. The eddy current method is one of the most promising methods for meeting these requirements. It allows to detect operational defects containing ruptures of carbon fibers under shock impacts [2]. At the same time, for various reasons, defects in the form of delamination between the layers in the matrix without fiber breaks are formed in CFRP products during operation. Such defects are quite dangerous and lead to a significant decrease in the strength of the damaged area. The difficulty of detecting such defects by the eddy current method is due to the insignificant deformation of the eddy current current contours under the influence of delamination [3]. This is confirmed by the results of numerous studies carried out in recent years [4-5]. To increase the sensitivity of eddy current control, a method has been proposed based on comparing eddy current signals from groups of UPM layers with the same spatial orientation [6]. To evaluate the effectiveness of this method and the principled choice of control parameters, it is advisable to use mathematical modeling. The use of strict numerical methods based on the finite element method (FEM) for these purposes is associated with an unjustifiably high complexity of calculations [7]. An approximate mathematical model based on the smallness of changes in the eddy current distribution in individual layers under the influence of stratification can be used with adequacy acceptable in practice for the indicated purposes.

2. The technical essence of the eddy current method with increased sensitivity to defects such as delamination in the CFRP

In the known methods of eddy current flaw detection CFRP eddy current signal related to delamination is formed by the difference between the differentially connected identical eddy current probes (ECP).

In this case, the signal is produced only when one of the ECP interacts with the defective area, and the other with the defect-free one. This allow to only detect delamination with a clear boundary between the defective and defect-free areas, which in practice is not always the case. In the proposed method [6], the signal is also formed due to the simultaneous interaction of both ECP fibers oriented along the corresponding direction with a defect, growing stronger by the factor of about 2 when one of those becomes defect-free. Figure 1 shows an ECP with rectangular coils, and Figure 2 shows the dependence of its normalized output signal U^* on the angle γ between the plane of the frame and the direction of the fibers in the CFRP layers.

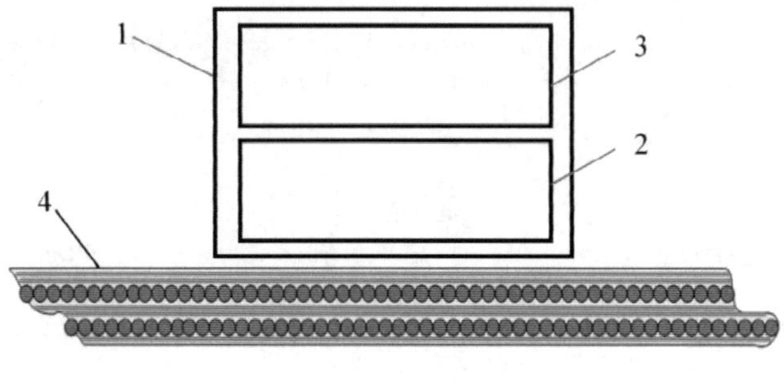

Figure 1. Eddy current probe:
1 - exciting coil; 2 - measuring coil; 3 - compensation exciting coil; 4 - sample

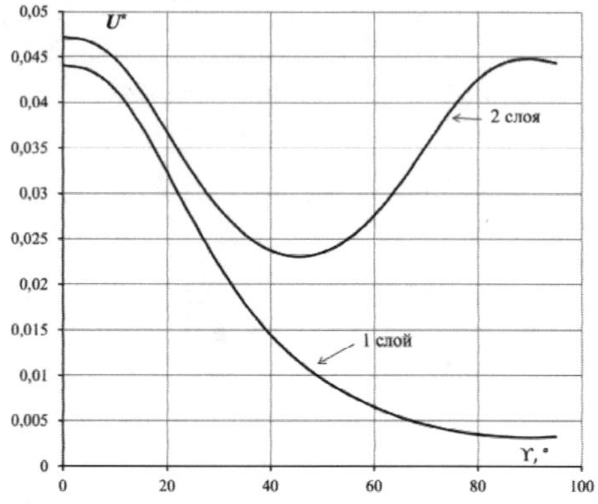

Figure 2. Dependence of U^* on the angle γ

Due to the compensation coil 3, the uninformative component of the initial voltage U_0, which arises due to the magnetic coupling between coils 1 and 2 in the absence of a controlled object, is suppressed. To weaken the influence of the working lift-off Z_0, the signal difference is normalized by the voltage of one of the ECPs, for example, by U_1, where U_1 and U_2 are eddy current signals ECP1 and ECP2 with different angular orientation.

3. Mathematical model of the effect of stratification on an eddy current signal

When constructing an approximate mathematical model, the following assumptions were made, due to the weak influence of stratification on the distribution of eddy currents in the CFRP layers:

- we assume eddy currents to be concentrated in the middle of each layer in the plane of the exciting coil;
- we assume that in the presence of stratification, due to its smallness, the magnitude of the currents in all layers does not change;
- we set the law of change of currents in the layers according to the exponential law, the exponent of which is determined using rigorous numerical calculations;
- we assume that when the plane of the turn is angularly displaced relative to the direction of the fibers by an angle γ, the changes in the eddy current signal are proportional to cosγ.

We are interested in the difference $\Delta U^* = (U_1 - U_2)/U_1$ normalized with respect to the voltage of one of the ECPs. Due to this, the value ΔU^* depends only on the law of layer-by-layer variation of currents I_k and does not depend on their absolute value. In this regard, the layer-by-layer law of current variation is set for the normalized values of $I_k^* = I_k / I_1$, where I_1 is the current in the upper 1st layer. Under the assumptions made, the value of $U_n\,(n=1,2)$ introduced into the corresponding ECP by eddy currents is determined by the expression

$$U_n = \frac{\mu_0}{4\pi}\sum_{k=1}^{K}\left(I_k^* \times \ln\left(R_{2,k} / R_{1,k}\right) \times abs\left[\cos\left(\varphi_n - \theta_k\right)\right]\right), \quad (1)$$

where K is total number of layers, are angular orientations of the plane of the ECP frame and carbon fibers in the *k*-th layer, and $R_{1,k}$ and $R_{2,k}$ are the distances from the middle of the *k*-th layer to the lower and upper conductors of the ECP measuring coil.

In accordance with [6], before the test, compensation is carried out on the defect-free section of the sensor, which consists of differentially connected ECP_1 and ECP_2, oriented along the angle in the corresponding directions. Therefore, after calculating the eddy current signals U_{10} and U_{20} in the defect-free area, the coefficient $q = U_{10} / U_{20}$ is determined. The q factor is used to correct the magnitude of the U_1 or U_2 signal before calculation.

4. Computer simulation results

Modeling was carried out for an eight-layer object with the direction of the layers shown in Figure 3. Figure 4 shows a version of the test when ECP_1 and ECP_2 were placed orthogonally to each other, with the plane of ECP_1 oriented in the direction of the fibers of the second layer.

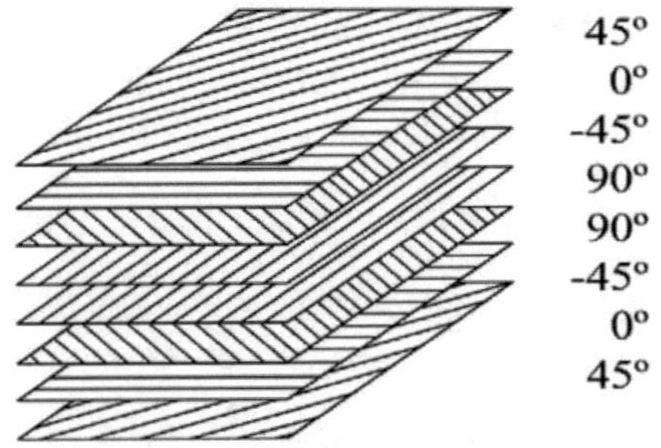

Figure3. Fiber direction in sample layers

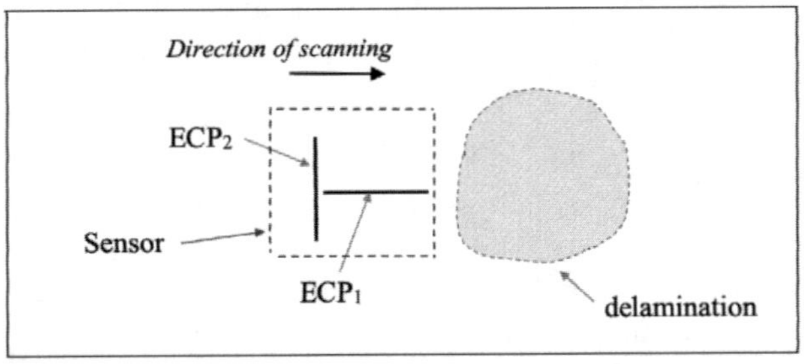

Figure 4. Scanning a sample with delamination by a sensor with two ECPs oriented orthogonally to each other (view from above)

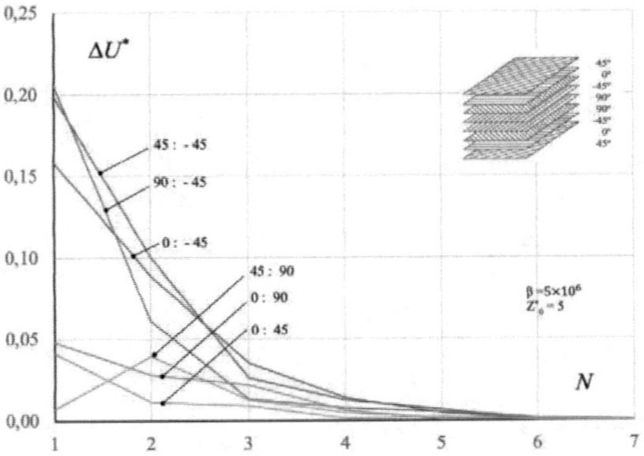

Figure 5. Dependences $\Delta U^* = \Delta U^*(N)$ on the number *N* of the layer before delamination

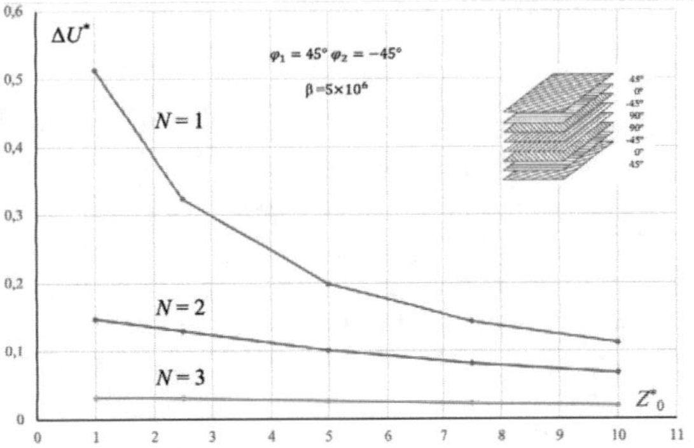

Figure 6. Dependences $\Delta U^* = \Delta U^*(Z_0)$ for several values of *N*

The thickness of the delamination T_p, located simultaneously under both ECPs, was taken to be equal to the thickness of the layers $T_c = 200$ μm. For generalization, the distance Z_0 from the center of the measuring coil was normalized to the value of T_c, i.e. $Z_0^* = Z_0 / T_c$.

For more detailed representation, Figures 5 - 6 show calculations for a measuring coil with a distance of $R_{12} = 1$ mm between the upper and lower conductors. Figure 5 shows dependence $\Delta U^* = \Delta U^*(N)$ of the amplitude of the eddy-current signal of differentially connected ECP pairs with different angular orientation on the number *N* of the layer before delamination. It can be seen from the graphs that the sensitivity of the sensor to delamination can also increase with increasing *N*. This is the case, for example, for sensor with the angular orientation of its two ECPs $\varphi_1 = 45^\circ$ and $\varphi_2 = 90^\circ$, where the angle *phi* is measured from the direction of the fibers of layer 2. For brevity, we introduced for the angular orientation of the ECP the notation $\varphi_1 : \varphi_2$, for example, 45:90. Figure 6 shows the dependences $\Delta U^* = \Delta U^*(Z_0)$ from which it follows that with an increase in the distance Z_0 from the delamination to the observation point, the effect of Z_0 decreases. In real structures, the distance R_{12} is more than 3 mm; as a result, the effect of variation in the working lift-off on the recorded signal is negligible. It should be noted that the assumptions made in the model are largely harmless due to the voltage normalization in the defect-free area. The reliability of the model was estimated by the FEM for a simplified 4-layer problem and experimentally.

5. Conclusions

1. The proposed mathematical model reflects, with adequacy acceptable in practice, the basic regularities in relation to the eddy current flaw detection method described in [6].
2. The model made it possible to estimate the change in the sensitivity to delamination depending on its depth, design parameters of the eddy current sensor, its distance from the surface, and the frequency of the exciting current.
3. The developed model can also be used to obtain the corresponding dependencies reflecting the effect of delamination in relation to other methods of eddy current testing of multilayer carbon fiber materials.

This work was performed using the equipment of the Centre for Collective Use of the Scientific and Technological Centre of Unique Instrumentation of the Russian Academy of Sciences (STC UI RAS) [http://ckp.ntcup.ru].

References

[1] Bulatova A N, Bulatov M F, Starov D V 2012 Investigation of the main parameters of multigraphene prepared by the cvd method *Russian Physics Journal* **54** p 1401

[2] Koyama K, Hoshikawa H, Hirano T 2011 Investigation of impact damage of carbon fiber rainforced plastic (CFRP) by eddy current non-destructive testing *Int. Conf. NDT (Montreal Quebec)* p 582

[3] Shkatov P N, Didin G A, Subbotin D E, Kokurov A M 2019 Study of interaction of eddy current probes and delamination in multidirectional CFRP *Materials Science Forum* **970** p 305

[4] Mizukami K, Syukri bin Ibrahim A, Ogi K et al. 2019 Enhancement of sensitivity to delamina tion in eddy current testing of carbon fiber composites by varying probe geometry *Compo site Struc tures* **226** 111227.

[5] Zeng Z, Tian Q, Wang H et al. 2019 Testing of delamination in multidirectional carbon fiber reinforced polymer laminates using the vertical eddy current method *Composite Structures* **208** p 314

[6] Shkatov P N, Didin G A *Russian patent N 2729457 with priority from 24.07.2019* (in Russian)

[7] Zeng Z., Jiao S., Du F. et al. 2018 Eddy current testing of delamination in carbon fiber reinforced polymer (CFRP): A finite element analysis *Journal Research in Nondestructive Evaluation* **29** 4 p 199

Measuring the depth of surface cracks using alternating current potential drop technique, with suppression of the influence of the electromagnetic properties of the metal

P N Shkatov

Federal State Budgetary Institution of Science «Scientific and Technological Centre of Unique Instrumentation» of the Russian Academy of Sciences, 15 Butlerova street, Moscow, 117342, Russia

E-mail: petr_shkatov@mail.ru

Abstract. Traditional measuring techniques often lead to errors due to the need to register signals in both defective and defect-free areas. In this paper, we introduce an alternating current potential drop technique with detuning from the influence of variations in the electromagnetic properties of the metal achieved by registering a signal only at the defective site. We show that, with an appropriate choice of measurement parameters, the use of the proposed technique leads to an increase in sensitivity to the crack depth as well as to an increase in the measurement range.

1. Introduction

In the process of taking direct or alternating current drop measurements, direct or alternating current is passed, using electrodes, through the object; at the same time, the voltage between the corresponding points on the surface of the object is measured with electrodes. The depth h of the crack is determined as the ratio U_r / U_0, where U_0 and U_r are voltages measured, respectively, at the defect-free and the defective sections of the object. The reliability of measurements is reduced due to the dependence of the recorded voltages on the electromagnetic properties of the metal. This is less of a problem for direct current measurements since, in this case, the ratio U_0 / U_r does not depend on the specific electrical conductivity and the values of U_0 and U_r are completely independent of the relative magnetic permeability μ. In practice, due to mechanical tension that caused the crack to develop, the values of both σ and μ in the defective area vary, which means that the problem of suppression of the variation of the properties of the metal when taking direct current measurements is not entirely eliminated. Moreover, measurements on direct current, in comparison with measurements on alternating current, are characterized by a significantly lower absolute sensitivity and a greater influence of the crack length, the thickness of the controlled area and the distance to the edges of the crack. With alternating current potential drop technique, the greater absolute sensitivity, as well as the lesser influence of the crack length and the metal thickness in the controlled area, are explained by the skin effect, which increases with increased current frequency; on the other hand, the value of the ratio U_r / U_0 is influenced to a greater extent by variations of both μ and σ [1].

When measuring on alternating current, a noise component arises due to the electromagnetic coupling between the current and the measuring circuits. To eliminate this effect, it is sufficient to measure the in-phase with the current passed through the current electrodes real component $\mathrm{Re}(U)$ of the recorded harmonic voltages [2]. The difference in the electromagnetic properties of the metal in the defective and defect-free areas when using alternating current is more noticeable due to the greater relative magnitude of changes in μ than in σ under the influence of mechanical tension. At the moment, we have no reliable quantitative data on the laws and magnitude of changes in the electromagnetic properties of the metal in the crack zone. Such data could be obtained with direct measurements -- for example, with high-resolution magneto-optical microscopy [3]. This would allow us to refine the available estimates of the measurement error under the influence of variations in the electromagnetic properties of the metal [4] while using traditional measurement methods [1].

2. A method for determining the depth of a surface crack with suppression of the influence of variations in the electromagnetic properties of the metal

The main idea behind our method [5] is that the crack depth is estimated using the ratio of the real components $\mathrm{Re}(\underline{U}_{P1})$ and $\mathrm{Re}(\underline{U}_{P2})$ of voltages $\underline{U}_{P1}$ and $\underline{U}_{P2}$, each measured between points located on the defective area. What is to be determined is the position of the points of contact of potential electrodes at which change in $\mathrm{Re}(\underline{U}_{P1})$ and $\mathrm{Re}(\underline{U}_{P2})$ differs from change in the crack depth h, while their ratio is uniquely related, and has the greatest sensitivity, to h.

To suppress the influence of electromagnetic properties of the metal, measurements of $\underline{U}_{P1}$ and $\underline{U}_{P2}$ should be carried out between points that are sufficiently close to each other (Figure 1 shows the probe setup for the required measurements). To solve this problem, we carried out an analysis of the distribution of the electric potential $\underline{V}$ when an alternating current $\underline{I}$ was passed through a section with a surface crack of depth h. The calculations were carried out using the finite element method and the computational model described in [4].

Figure 2 shows the distribution of the real component $\mathrm{Re}(\underline{V})$ of the potential along crack edges for various depth values h obtained at the indicated values of I, μ, σ, the current frequency f, the crack length L and distances between electrodes. In the figure, the graphs are shown for the symmetrical, relative to the crack, position of points T_1 and T_2. The distance R_{TP} along the crack from the line T_1T_2, as well as the depth h of the crack, for the purpose of generalization, are normalized by the distance $R_T = 5$ mm between the current electrodes, i.e. $h^* = h / R_T$. Since the potential distribution is symmetric relative to the crack center, the graphs are shown only for positive values of R_{TP} / R_T. The step between adjacent positions of contact installation points along the crack we consider is $0{,}5R_T = 2{,}5$ mm. At the same distance of $0{,}5R_T = 2{,}5$ mm, there are placed points of measurement of the potential relative to the crack, which satisfies the condition $R_P = R_T$.

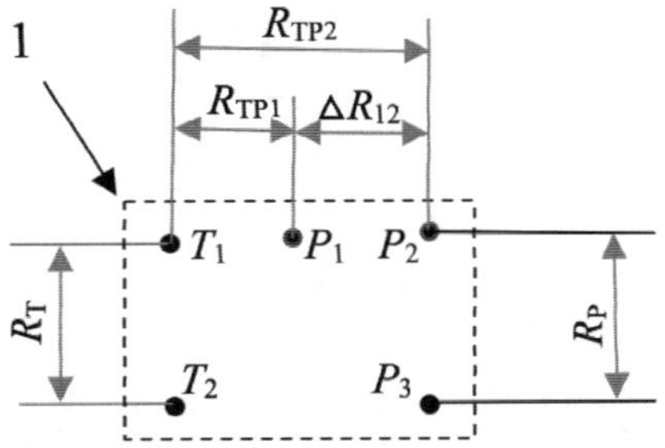

Figure 1. Notation for distances between the electrodes of the probe

The choice of the distances R_P and R_T is determined, on the one hand, by the sensitivity to the crack depth and, on the other, by the permissible error of needing to position the electrode symmetrically with respect to the crack trace. When the distance decreases, the sensitivity increases, but the error increases. Since this consideration holds true for both current and potential electrodes, it is advisable to choose $R_P = R_T$. The graphs show the $\mathrm{Re}(\underline{V})$ values for three potentials at points 1, 2, 3 and at points 1', 2', 3' with the same distance $\Delta R_{12} = 0.5R_T$ and two distinct values of, respectively, $R_{TP1} = 0,5R_T$ and $R_{TP1} = 1,5R_T$. The voltage $\underline{U}_{P1}$ is measured between points 1 and 3 (1' and 3'), while the voltage $\underline{U}_{P2}$ is measured between points 1 and 2 (1' and 2'). It can be seen from the shown dependences that the difference of the potentials at points 1 and 3 located on different sides of the crack significantly exceeds the difference of the potentials at points 1 and 2 located on one side of the crack. For this reason, the ratio $\mathrm{Re}(\underline{U}_{P1}) / \mathrm{Re}(\underline{U}_{P2})$ is quite big.

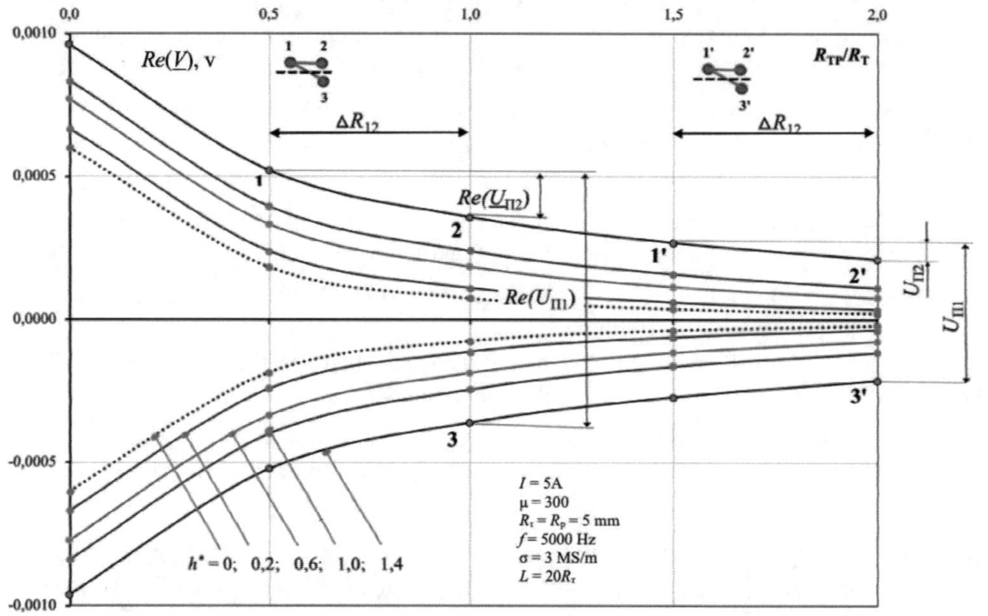

Figure 2. Distribution of the real component $\mathrm{Re}(\underline{V})$ of the potential along the crack edges for different values of the depth h

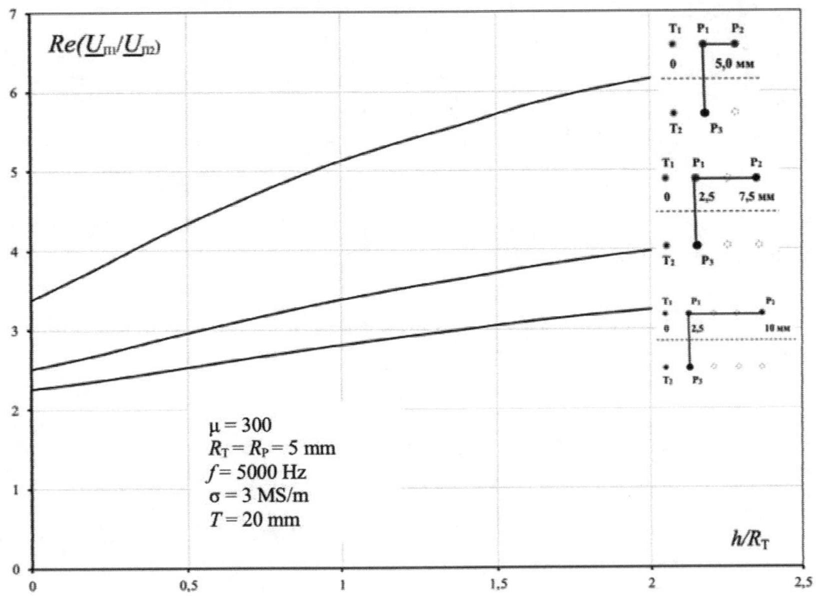

Figure 3. Calibration characteristics when positioning P_3 opposite P_1 (with $R_{TP3} = R_{TP1}$)

Figures 3 and 4 show the dependencies of the ratio $\mathrm{Re}(\underline{U}_{P1} / \underline{U}_{P2}) = \mathrm{Re}(\underline{U}_{P1}) / \mathrm{Re}(\underline{U}_{P2})$, for two distinct positions P3 of the potential electrode 6, on the relative depth $h^* = h / R_T$ for different values of the distance ΔR_{12} between the positions of the first potential electrode 4 and the second potential electrode 5. Calibration characteristics in Figure 3 are shown for the position of P3 opposite P1 ($R_{TP3} = R_{TP1}$), and in Figure 4 for the position of P3 opposite P2 ($R_{TP3} = R_{TP2}$). In both cases, the relative sensitivity to h^* increases as the distance ΔR_{12} decreases. This is due to the fact that, with decreased ΔR_{12}, the value of $\mathrm{Re}(\underline{U}_{\Pi 1})$ is going down more slowly than that of $\mathrm{Re}(\underline{U}_{\Pi 2})$. However, the decrease in $\mathrm{Re}(\underline{U}_{P1})$ and $\mathrm{Re}(\underline{U}_{P2})$, which accompanies the decrease in ΔR_{12}, leads to an increase in the margin of error in their measurement; this limits the possibility of decreasing ΔR_{12}. Therefore, we suggest choosing ΔR_{12} so that $0{,}9R_T < \Delta R_{12} < 1{,}1R_T$.

The comparison of the obtained calibration characteristics with those for traditional measurement techniques [1,2,4] shows that the technique proposed her, with an appropriate choice of parameters for the converter leads to increased sensitivity to the depth of the crack and expansion of the range of its measurement.

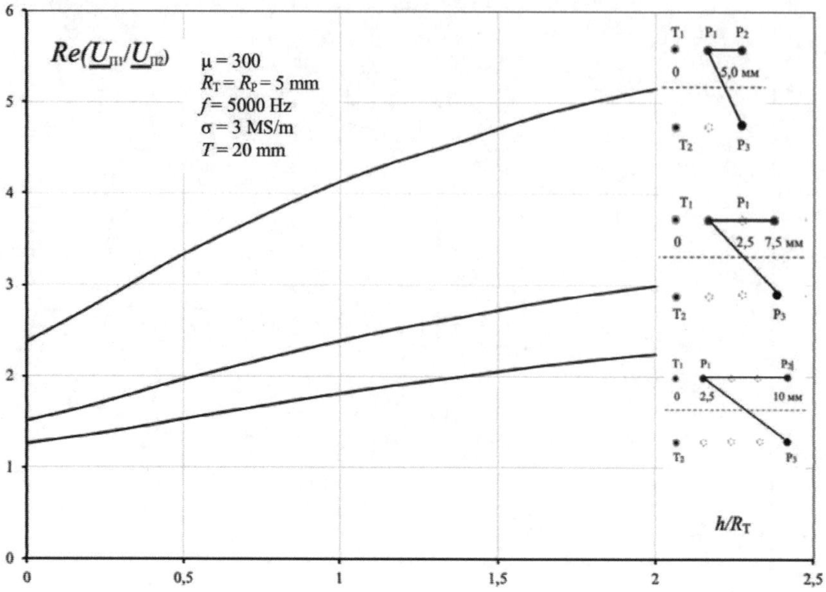

Figure 4. Calibration characteristics when positioning P_3 opposite P_2 (with $R_{TP3} = R_{TP2}$)

3. Conclusions

1. The proposed technique for determining the depth of a surface crack allows us to ignore the difference in the electromagnetic properties of the metal in the defective and the defect-free areas.
2. Additionally, it allows us to achieve increased sensitivity to the depth of the crack and expansion of the range of its measurement.

This work was performed using the equipment of the Centre for Collective Use of the Scientific and Technological Centre of Unique Instrumentation of the Russian Academy of Sciences (STC UI RAS) [http://ckp.ntcup.ru].

References

[1] Cost H, Deutsch V, Ettel P 1996 Crack depth measurement-Modern measuring technique for a well-known method *NDTnet* **1** 6
[2] Shkatov P N, Sandulyak A A, Kuzub I G, Ermolaev A A 2020 *Pribory* **6** 240 p 22 (In Russian)
[3] Vysokikh Y E, Krasnoborodko S Y, Kolomiytsev A S, Fedotov A A, Shaposhnikov A N, Bulatov M F, Churikov D V, Shevyakov V I 2019 Advanced aperture cantilevers made by focused ion beam for both magnetic domain structure and morphology investigation in high resolution magneto optical microscopy *Journal of Physics: Conference Series* **1410** 1 012187.
[4] Shkatov P N, Sandulyak A A, Kuzub I G, Ermolaev A A 2020 *Control. Diagnostika* **11** p 36 (*In Russian*)
[5] Shkatov P N *Russian patent N* 2746668 *with priority from* 10.08.2020 *(in Russian)*

Increase of the Informative Value of Magnetic Nondestructive Testing Using Vibration Induction Transducer on the Basis of Spectral Analysis

Yu L Nikolaev [1], P N Shkatov [1], A V Chernova [2], E F Akhmetshina [2] and A A Samorukov [3]

[1]Federal State Budgetary Institution of Science «Scientific and Technological Centre of Unique Instrumentation» of the Russian Academy of Sciences, 15 Butlerova street, Moscow, 117342, Russia
[2]Federal State Budgetary Educational Institution of Higher Education «MIREA–Russian Technological University», 20 Stromynka street, Moscow, 107996, Russia
[3]TMS RUS Limited Liability Company 20 Verkhnyaya Maslovka street, Bldg. 2, Moscow 127083, Russia

E-mail: petr_shkatov @mail.ru

Abstract. The ways of increasing the information content of magnetic flaw detection based on spectral analysis when registering magnetic fluxes of leakage from defects by a vibration induction probe (VIP) are considered. It is shown that the VIP signals, caused by the normal and tangential components of the strength of the magnetic scattering fluxes, are linearly independent, and the corresponding harmonic components of the spectra of these signals are weakly dependent on the noise component and carry information about the coordinates and parameters of the defect.

1. Introduction

The specific feature of magnetic nondestructive testing is containment of stray fluxes in small volumes and substantially nonlinear change of the respective magnetic field. To suppress the background that masks signals from defects or causes false signals, differential inclusion of identical, closely arranged magneto-sensitive elements [1] forming gradient transducers are used. Application of magneto-sensitive film elements [2] that allow to improve test resolution is promising. Induction transducers certainly have advantages [1]. While moving above a defective area, magnetic flux changes much more than while moving above a defect-free area. This makes it possible suppress the influence of background with the help of one sensitive element of a transducer. Another important advantage of an induction transducer is that its output signal can be analyzed spectrally [3].

It should be noted that the spectral analysis capabilities in induction transducer usage are rather limited because spectral analysis requires constant scanning rate that must not vary and is chosen based on testing performance considerations.

It is suggested to increase greatly the informative value and the range of application of the induction recording technique by recording magnetic stray fluxes using a vibration induction transducer (VIT). Magnetic fluxes focus $e(t)$ to VIT inductor thanks to almost harmonic vibration of the latter in the plane that is parallel to the surface being scanned. Vibration amplitude a matches the zone of significant

 2091 (2021) 012048 doi:10.1088/1742-6596/2091/1/012048

change in the respective spatial component of a defect's magnetic field strength H, while inductor's vibration frequency is selected based on the considerations of test sensitivity and informative value. The axis of the inductor can be aligned both in parallel (tangential component recording) and perpendicular to the surface being scanned (normal component recording).

2. Practical implementation of magnetic nondestructive testing using vibration induction transducer

The test object area with a crack on its surface is magnetized by a constant magnetic field with induction B in the plane that is orthogonal to the crack. This results in a locally distributed magnetic field of the defect and a background magnetic field. In the course of scanning of the test object's surface, VIT inductor 1 harmonically oscillates along x- coordinate that is parallel to the surface. The oscillatory motion of the inductor is geared from a piezoelectric or electrodynamic drive via a pendulum transmission element.

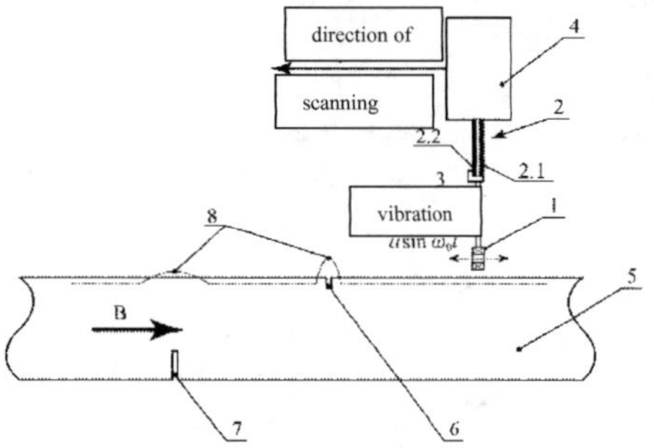

Figure 1. Schematic diagram of tangential component type of testing using a piezoelectric drive: 1 – inductor; 2 – bimorph; 3 – nonmagnetic rod; 4 – housing; 5 – test object; 6, 7 – defects, 8 – stray fluxes

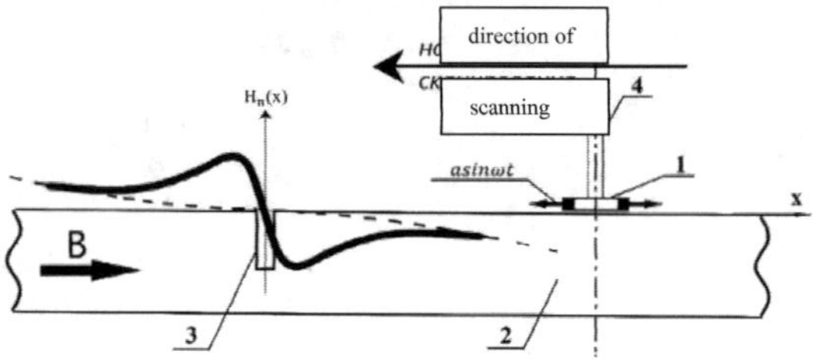

Figure 2. Schematic diagram of normal component type of testing using an electrodynamic drive: 1 – inductor; 2– test object; 3 - defect, 4 – pendulum element of an electrodynamic drive to transfer vibration to the inductor

The developed VIT with a piezoelectric drive [4] provides oscillation amplitude of up to 0.5mm at frequencies of the order of 500Hz and up to 2.0mm at frequencies of the order of 250Hz. The vibration motion amplitude of the VIT inductor with the selected electrodynamic drive design is adjustable within the range of up to 2.0 … 5.0mm at frequencies of the order of 200 and 150Hz, respectively. Figure 1 shows the schematic diagram of testing using a piezoelectric drive for vibration motion of the inductor targeted to recording the tangential component of magnetic field strength $H_\tau(x)$, figure 2 – that with an electrodynamic drive and inductor targeted to recording the normal component $H_n(x)$.

3. Mathematical model of vibration induction transducer output signal

Mathematical modelling was based on representation of the VIT output signal upon exposure to magnetic field of a defect and or the background represented as a sum of harmonic components. This was done using the a priori knowledge [5] of characteristic properties of functions $H_\tau(x)$ and $H_n(x)$ and background $H_\phi(x)$ for a respective component.

For example, the characteristic properties of function $H_\tau(x)$ include: parity, existence of a smooth maximum at point $x=0$ and two symmetric points of inflection in the segments of monotone increase and decrease. The margins of the range of a significant change $\pm\gamma$ are determined by points corresponding to the two symmetric minimums of function $H_\tau(x)$.

In general, the alternating-sign power polynomial as shown below meets these conditions

$$H_\tau(x)=\alpha_0+\sum_{n=1}^{m}(-1)^n\alpha_n x^{2n}, \tag{1}$$

where α_0 is the field strength at point $x=0$. According to calculations, an adequate reflection of the properties of $H_\tau(x)$ is achieved at $n\geq 2$.

For normal component $H_n(x)$, it makes sense to choose an approximation looking like an alternating-sign power polynomial containing only odd powers of variable x

$$H_n(x)=\sum_{n=1}^{m}(-1)^n\beta_n x^{(2n-1)} \tag{2}$$

At $n\geq 3$, the essential properties of function $H_n(x)$ are reflected: oddity, bimodality, existence of 3 points of inflection.

During VIT movement and oscillation along the tested surface in parallel to magnetic lines of the magnetizing field, the transmitted to inductor EMF $e(t)$ (VIT signal) can be represented as follows

$$e(t)=-k\cdot\frac{dH_Д(x)}{dx}\bigg|_{x=x(t)}\cdot\frac{dx(t)}{dt}, \tag{3}$$

where $H_Д(x)$ is the strength of the magnetic field of a defect in the respective component; t is the time; $x(t)$ is the equation of motion; k is the factor depending on the magnetic properties of the test object's material and testing environment, the number of windings and the sectional area of inductor, as well as irregularity of distribution of $H_Д$ in its volume.

The equation of vibration motion of the inductor looks like:

$$x(t)=x_0+\alpha\sin\omega_0 t, \tag{4}$$

where x_0 is the displacement of the inductor center from the plane of the defect; α,ω_0 are, respectively, the amplitude and the oscillation circular frequency of the inductor.

The transmitted to inductor EMF $e(t)$ (VIT signal) is reduced as follows:

$$e(t)=E_{1m}\cos\omega_0 t+E_{2m}\sin 2\omega_0 t+E_{3m}\cos 3\omega_0 t+E_{4m}\sin 4\omega_0 t+E_{5m}\cos 5\omega_0 t \tag{5}$$

where E_{im} are the amplitude values of harmonic components.

At inductor vibration motion with amplitude a and circular frequency ω_0, taking into account (1) and (2), the coefficients of $e(t)$ expansion into Fourier's series during recording in the normal and tangential components were obtained.

Figure 3 shows examples of experimental spectrograms of VIT output signal during recording in the normal and tangential components are shown, which agree well those calculated.

Analysis of calculated and experimental dependencies obtained has made it possible to identify a number of important regularities of VIT signal formation.

In case of tangential component recording of VIT signal:

- Presence of even harmonics in VIT signal informs about existence of a defect.
- Amplitudes E_2 and E_4 are maximal above the center of a defect and proportional to strength maximum, which amplitudes of odd ones are close to zero.

- Based on the position of the maximum of 2nd harmonic amplitude, the error of determining the crack coordinate amounts to 0,05–0,3mm.

In case of normal component recording of VIT signal:

- Amplitudes of odd harmonics E_1 and E_3 of VIT signal reach maximum when VIT is positioned within ±0,5mm relative to the plane of a defect, and amplitudes of even E_2 and E_4 are close to zero. Peak values of the output signal are a sum of amplitude values of odd harmonics.
- The amplitude of the 2nd harmonic of VIT signal becomes equal to zero when VIT is positioned directly above a defect whatever the amplitude of vibration might be.

The experimental spectrograms were made for several reference samples used in magnetic nondestructive testing. Width b of defects therein varied between 3 and 100 µm and depth h – between 0,1 and 4 mm. Preliminary analysis has shown that the spectral composition of the VIT signal changes with variation of both the depth and the width of defects. This suggests the possibility of determining the depth and the width of a crack separately based on totality of spectral components the VIT signal.

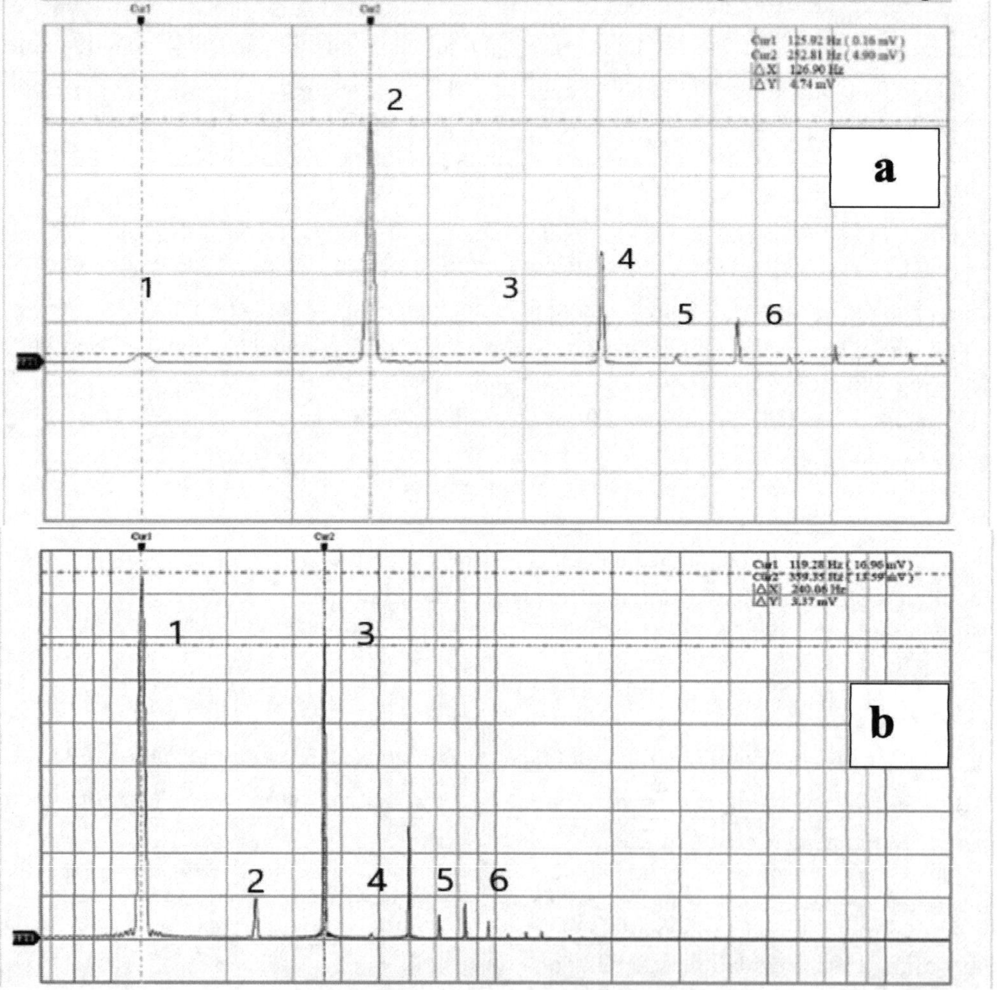

Figure 3. Spectrograms of the VIT output signal during recording of different strength components of the defect's magnetic field: a – tangential, b – normal. Numbers indicate numbers of the harmonic.

Conclusions

1. The spectral composition of the signal of a vibration induction transducer (VIT) provides additional information about existence of a defect, which allows to increase the signal/background ratio upon detection of a defect and improve the accuracy of determining defect coordinates.
2. There are premises for separate estimation of the depth and width of a defect based on the spectral composition of the VIT signal.

This work was performed using the equipment of the Centre for Collective Use of the Scientific and Technological Centre of Unique Instrumentation of the Russian Academy of Sciences (STC UI RAS) [http://ckp.ntcup.ru].

References

[1] Reutov Yu Ya, Shcherbinin V E, Volkov A V 2014 Possibilities for the selection of magnetic field transducers for nondestructive testing *Defectoscopia* **12** p 71 (In Russian)
[2] Smirnov V V, Alykova O M, Bulatov M F 2019 Applied Problems of Magnetism *Bulletin of the Russian Academy of Sciences* **83** 7 p 895 (In Russian)
[3] Pashagin A I, Benklevskaya N P, Shcherbinin V E 2005 Use of the harmonic analysis of an induction probe's signal for increasing the information content of test results *Defectoscopia* pp 15 – 19 (In Russian)
[4] Shkatov P N, Nikolaev Yu L, Chernova A V et al Studying a signal due to the tangential com ponent of stray fluxes from a surface flaw detected with a vibrating induction transducer *Russian Patent N 2600517 with priority from 13.02.2015* (In Russian)
[5] Klyuev V V, Muzhitsky V F, Gorkunov E S, Shcherbinin V E 2006 Nondestructive Testing: Reference Guide: in 8 volumes (*Moscow: Mashinostroenie*) (In Russian)

Crystallization Double-Layer Magneto-Active Films for Magnetophotonics

T V Mikhailova[1], Yu E Vysokikh[2], A N Shaposhnikov[1], V N Berzhansky[1], S Yu Krasnoborodko[2], M F Bulatov[2] and D V Churikov[2], A Karavainikov[2], V I Belotelov[2]

[1] V.I. Vernadsky Crimean Federal University, Simferopol, 295007, Russia
[2] Scientific and Technological Centre of Unique Instrumentation of the RAS, Moscow, 117342, Russia

visokihy@gmail.com

Abstract. Magneto-optics, magnetophotonics and magnetoplasmonics stay at the edge of scientific interests last years due to their unique features to manage the light and electromagnet field. Bi-substituted iron garnet (Bi:IG) is one of most promising magneto-optical material for these applications in order to its high efficiency in visible and infrared spectra. The possibility to integrate Bi:IG films to silicon semiconductor process leads to creation nanoscale hight performance magneto-optical devices. Bi:IG structures of different composition might be deposited by vacuum deposition on different substrates. The investigation of crystallization process of Bi:IG double-layer films at a different process parameter on gadolinium gallium garnet and fused quartz substrates allowing to determine dependences and suggestions for integration Bi:IG to semiconductor process or multicomponent optical nanostructures.

1. Introduction

Today scientific community actively develop the idea of design magneto-optic and optomagnonic structures that combine the optical effects of transmission / localization of optical waves and the dynamics of magnetization at femto- and picosecond times [1-6]. Rare-earth garnets and iron garnets have crystal structure which allows to achieve advanced features of magnetic system. Nowadays more and more efforts are paid to implement bismuth and cerium to the garnet lattice. It may significantly increase the magneto-optic (MO) performance. Possibility of garnet materials to be deposited on the crystals without garnet structure opens the door to semiconductor process. The MO films' integration to silicon semiconductor process is not easy to implement but important for fabricating new nanoscale and high-performance devices [7]. Real synthesis of magneto-active structures meets a lot of difficulties in the way of using different garnet composition as a part of nanoscale one-dimensional and two-dimensional lattices of other magnetic or nonmagnetic materials [8-14]. Crystallization process is key factor which significantly impact on films' quality. Authors fabricate double-layers films of different compositions and type of substrates. This work shows the investigation of deviations of films' microstructure compare to its composition, substrates and growth conditions.

2. Experimental

Bi:IG Reactive ion beam sputtering method (RIBS) was used to fabricate Bi:IG double-layers at a URM 3-279.014 setup in oxygen–argon mixture using an ion-beam Kholodok-1 source. The films were deposited from ceramic target on gadolinium gallium garnet (GGG) substrates with crystallographic orientation of (111) or fused quartz SiO2 substrates [15–18]. Crystallization annealing at air (atmosphere pressure) and the temperature range from 650°C to 710°C and time period from 20 min to 60 min formed garnet phase. Different compositions of Bi:IG were used: G1 – Bi1.0Lu0.5Gd1.5Fe4.2Al0.8O12, G8 – Bi2.5Gd0.5Fe3.8Al1.2O12, GN – Bi2.8Y0.2Fe5O12.

All investigated samples have two layers (G1/G1, G1/G2 or G1/GN) with thicknesses of first G1-layer, second G1-, G8- and GN-layers of 80 nm, 230 nm, 195 nm and 177 nm, respectively. Two key process modes (methods) of crystallization annealing were carryout. Each layer of the structure were crystallized separately in both modes. Temperature of annealing depended on the composition and was 700°C, 680°C and 650°C for G1, G8 and GN, respectively.

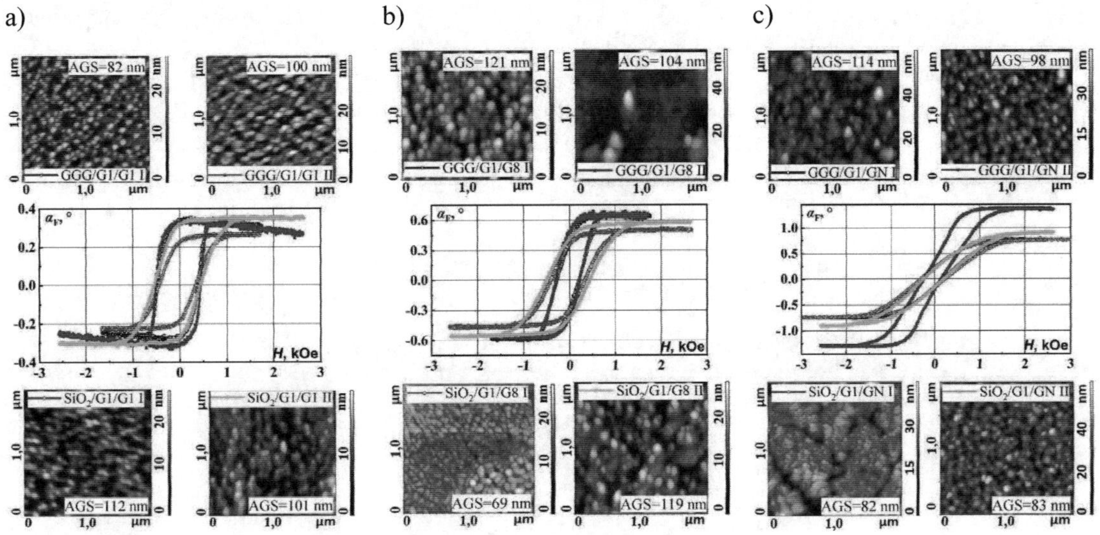

Figure 1. Topography and FHLs of double-layers garnet structures on fused SiO2 and GGG substrates with layers' compositions: G1/G1 (a), G1/G8 (b), G1/GN (c).

First method (dipping), hereinafter referred "I", description:

- The sample was almost immediately dipped in a thermal field at the annealing temperature.
- Crystallization annealing of the sample occurred within 20 min after reaching thermodynamic equilibrium.

Second method (slow), hereinafter referred "II", description:

The synthesis was carried out at very slow heating (heating rate of about 2-3°C / min) of the sample to the optimum temperature for a given composition and subsequent crystallization at a constant temperature for 60 min.

Crystallization features of double-layers films were revealed by surface structure which depends on process parameters (type of substrate, thickness of layers, temperature and time of annealing). Atomic force microscopy (AFM NTEGRA NT-MDT) was used to measure topography parameters of Bi:IG thin films. Active vibroisolation platform was used to protect the system from vibration noises. Average roughness of each film was measured to determine the process regime which allows to minimize this parameter. Tapping mode of AFM was used for investigations because it allows one to get the best value of resolution and minimize the negative influence of the probe on the sample. In this mode, the cantilever console oscillates with self-resonant frequency and high amplitude (approximately 50-100 nm). At these amplitudes, the needle contacts the sample surface at the moment of maximum deflection of the console down from its equilibrium position, which significantly changes the frequency, phase and amplitude of its oscillations. A tapping mode has a higher horizontal resolution compared to the contact mode. Cantilevers NSG10 (TipNano) were used with resonant frequency of 240 kHz and force constant of 11.8 Nm.

We used AFM images of 15x15 and 2x2 microns area for analysis. Parameters of height range P, root mean square roughness RMS and polycrystalline size AGS were determined. AGS was estimated by advanced watershed method.

Faraday rotation angle and Faraday hysteresis loops (FHLs) were measured by compensation method using automated magneto-polarimeter at the wavelength of 655 nm. Magneto-polarimeter includes the light source (semiconductor laser), polarizer, MO modulator, electromagnet, compensator, analyzer, photodetector, electronic system and a personal computer. Special holder of the sample and a Hall sensor to measure the intensity of magnetic field were placed in the gap of electromagnet. The range of available magnetic fields was from -5 to 5 kOe.

Faraday rotation angle αF, specific Faraday rotation ΘF, coercivity HC, squareness coefficient KS, saturation magnetic field HS were determined from FHLs. To calculate the specific angle ΘF, the total thickness of magnetic layers was taken into account. The film thickness h was calculated on the basis of sputtering velocity with control by optical (MII-4 microinterferometer) and AFM methods.Table 1 lists the determined structural and magnetic parameters of samples.

Table 1. Parameters of investigated double-layers films.

Composition & method	Θ_F, °/μm	*RMS*, nm	*P*, nm	H_C, *Oe*	K_S, *rel. units*
		$Gd_3Ga_5O_{12}$			
G1/G1 I	0,97	4,6	89	426	0,98
G1/G1 II	1,01	5,9	65	487	0,99
G1/G8 I	2,22	3,4	51	240	0,52
G1/G8 II	2,22	9,8	60	264	0,71
G1/GN I	2,92	7,8	52	230	0,24
G1/GN II	5,25	5,2	80	193	0,19
		SiO_2			
G1/G1 I	0,75	4,6	47	384	0,84
G1/G1 II	1,01	12	142	501	0,91
G1/G8 I	1,75	9,6	119	389	0,66
G1/G8 II	2,07	5,9	65	469	0,68
G1/GN I	2,92	8,1	65	289	0,27
G1/GN II	3,50	9,1	117	241	0,19

3. Results

Figures 1 demonstrates the correlation of topography and FHLs of three types of samples. All samples, regardless of their composition, have polycrystalline surface after the crystallization. As it was noted earlier in our works, the processes of crystallization proceed differently on substrates which has the garnet structure and without [19,20.]. Epitaxial growth of Bi:IG polycrystals occurs on garnet substrates or layers. But in case of SiO2 substrates garnet phase growth occurs by the mechanism of spontaneous crystallization. According to AFM data in both cases the film formation goes by an island mechanism. For the GGG substrate the growth of individual oriented polycrystals distributed evenly over the surface of the substrate is observed [21]. The typical size of polycrystals is the same as it shown in Figure1. In the case of a SiO2 substrate, the formation of polycrystals begins spontaneously in some places on the surface of the substrate, and the film formation goes by combining micron sizes aggregates of grains. It is also seen by increasing the saturation fields of samples from GGG substrate to SiO2 (Figure1., FHLs). Grains on SiO2 substrates have a larger anisotropy spread.

As a result, the film of G1 composition is formed on the surface of SiO2 only with a layer thickness bigger than 60-70 nm [22]. Garnet structures of the G8 and GN compositions are not formed on SiO2. So, the G1-layer with thickness of 80 nm was set for double-layer films. Its purpose is to be a seed for the crystallization of the second main layer G1, G8 or GN. These layers are considered by us to be functional of the two-layer film as a whole.

The structure of double-layer films depends on the initial structure of the sublayer G1. For instance, the surface of double-layer G1/G8 I films demonstrates different features on GGG and SiO2 substrates, respectively. The surface of the films on GGG is characterized by a uniform distribution of crystallites. The grains are approximately the same size, even on a large-area frames. Average size of the crystallites is about 90-100 nm. The surface of the films on SiO2 shows the presence of regions with different crystal structures – regions with large crystallites of about 200-500 nm of a clear rectangular or triangular shape, smooth regions with small "round" crystallites of the same size or regions clearly being the boundaries of intergrown conglomerates. A fragment of a 2x2 μm smooth region is shown in Figure 1.

The most uneven structure is observed on SiO2 substrates samples at first annealing mode. The grains have the greatest size variation and even nano-cracks (Figure 1, c). Films are formed at the second annealing mode have more homogeneous structure on SiO2. The longer annealing time (second mode) leads to re-formation of centers of crystallization into a large smooth polycrystalline film. In contrast, an increase of annealing time leads to the formation of large crystallites and the heterogeneity of their distribution on the GGG. Figure 2 shows the surface structure of such films on GGG and the histograms of their grain size distribution. Topography comparison on GGG substrate compare to annealing method shown on Figure 3.

Depending on the second layer composition we vary the magnetic characteristics – the type of anisotropy. The films G1/G1 has the maximum squareness ratio KS = 0,99 and low saturation field HS 0,59-1,10 kOe (Figure 1, FHLs). It confirms the implementation of «easy axis» anisotropy type. «Angle phase» and «easy plane» anisotropies are realized in G1/G8 (K_S = 0.66, H_S = 0,75-1,34 kOe) and G1/GN (K_S = 0.27, H_S = 0,74-1,60 kOe) films.

We should notice that studies other samples showed different results. The second annealing method led to the formation of a uniformly distributed polycrystalline structure for the SiO2 substrate and the formation of crystallites of various shapes for the GGG substrate.

So in order to synthesize a smoother film, we need to choose a faster synthesis mode (the first method) for growth on GGG substrates and a slower synthesis mode (the second method) for growth on SiO2 substrates. Table 2 shows the morphological parameters of all studied two-layer samples.

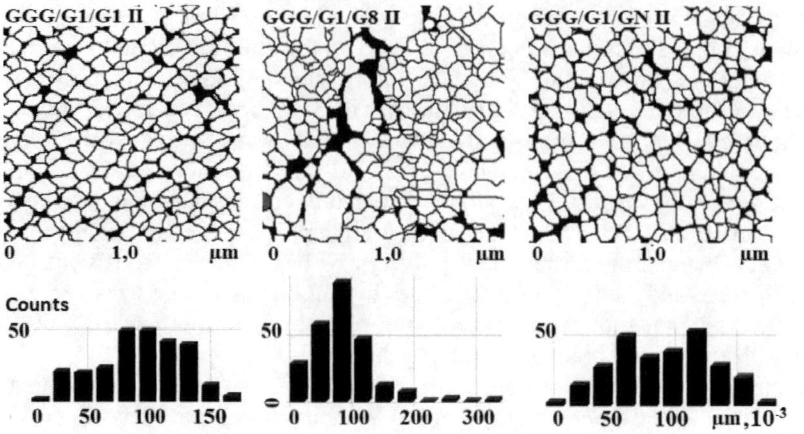

Figure 2. Grain analysis of garnet structure of different composition on GGG substrate. The films were annealed by second mode II.

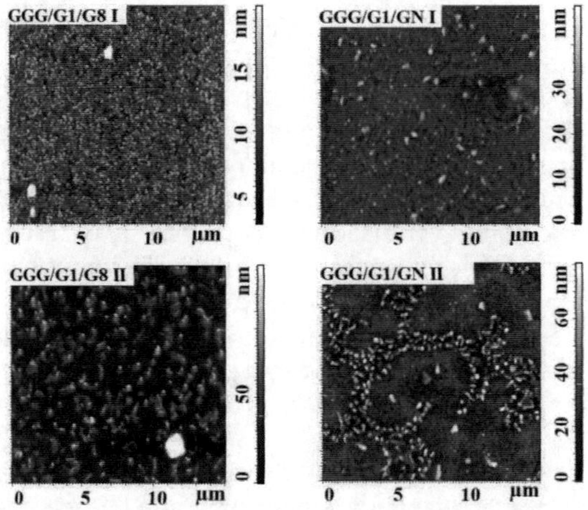

Figure 3. Topography of garnet structure of different composition on GGG substrate. The comparison of films annealed by first I and second II mode.

However, these parameters showed a dependence on the crystallization method. Thus, the films G1/G1 and G1/G8 demonstrate more pronounced uniaxial anisotropy in the second annealing method. This may also be due to the diffusion of garnet-forming elements from the substrate during long-term annealing in the case of a GGG substrate (Table 1).

The decreasing values of the Faraday rotation for SiO2 films occurs quite strongly when the first mode is applied (Table 1).

For samples G1/G1 and G1/G8 trends in the magneto-optical and magnetic parameters coincide. The films on GGG at the first and second annealing modes are identical.

The difference in annealing modes is most clearly visible for samples on SiO2. For all types of such films, the second annealing mode is most advantageous.

For G1/GN the first annealing mode leads to a decreasing films' quality for any substrate. This composition has the largest lattice parameter and its crystallization processes apparently require more specific conditions without a sharp change in the temperature gradient.

4. Conclusion

Thus, the influence of the crystallization annealing mode on the formation of the Bi:IG double-layers crystal structure is studied. Two methods, called briefly by the authors "dipping" and "slow", were compared. The both method is suitable for crystallization of Bi:IG films of composition G1/G1 and G1/G8 on GGG substrates. However, the second method is better suited for the formation of all investigated Bi:IG films on fused quartz SiO2 substrates which do not have the initial garnet structure and G1/GN regardless of the substrate type. It was found that smoother and more homogeneous samples on non-garnet substrates can be obtained using slow and prolonged annealing, while high-quality films on garnet substrates are formed with shorter heating and annealing.

Acknowledgments

The authors from V.I. Vernadsky Crimean Federal University TVM, ANSh, VNB, AVK and VIB acknowledge support by the Russian Ministry of Education and Science (Megagrant project N 075-15-2019-1934) for the synthesis of film samples with optimized magnetic parameters.

The authors from Scientific and Technological Centre of Unique Instrumentation of the RAS YuEV, SYuK, MFB, DVCh acknowledge support by the Ministry of Science and Higher Education of the Russian Federation under the State contract No. 0069-2019-0009 for the films measurements. This work was performed using the equipment of the Centre for Collective Use of the Scientific and Technological Centre of Unique Instrumentation of the Russian Academy of Sciences (STC UI RAS) [http://ckp.ntcup.ru].

References

[1] Pantazopoulos P A, Tsakmakidis K L, Almpanis E, Zouros G P, Stefanou N 2019 High-efficiency triple-resonant inelastic light scattering in planar optomagnonic cavities *New J. Phys* **21** 095001

[2] Ignatyeva D, Ozerov V, Hamidi M, Chernov A, Belotelov V 2020 Optical excitation of spin waves in all-dielectric nanostructured materials with iron garnet *AIP Conference Proceedings* **2300** 020045

[3] Ozerov V A, Sylgacheva D A, Kozhaev M A, Mikhailova T V, Berzhansky V N, Hamidi M, Zvezdin A K, Belotelov V I 2021 One-dimensional optomagnonic microcavities for selective excitation of perpendicular standing spin waves *Journal of Magnetism and Magnetic Materials* **168167**

[4] Mikhailova T V, Berzhansky V N, Shaposhnikov A N, Karavainikov A V, Prokopov A R, Kharchenko Yu M, Lukienko I M, Miloslavskaya O V, Kharchenko M F 2018 Optimization of one-dimensional photonic crystals with double layer magneto-active defect *Opt. Mater.* **78** 521

[5] Mikhailova T V, Shaposhnikov A N, Prokopov A R, Karavainikov A V, Tomilin S V, Lyashko S D, Berzhansky V N 2018 Hybrid states of Tamm plasmon polaritons in nanostructures with Bi-substituted iron garnets *EPJ Web of Conferences* **185** 02016

[6] Mikhailova T V, Lyashko S D, Tomilin S V, Karavainikov A V, Prokopov A R, Shaposhnikov A N, Berzhansky V N 2017 Magneto-optical microcavity with Au plasmonic layer *J. Phys. : Conf. Ser.* **917** 062053

[7] Bulatov M F, Parkhomenko Y N 2006 Conduction mechanisms of magnetic semiconductors with a garnet structure in relation to variable-valence impurity concentration *Semiconductors* **40** pp 169–171

[8] Inoue M, Baryshev A V, Goto T, Baek S M, Mito S, Takagi H, Lim P B 2013 Magnetophotonic Crystals: Experimental Realization and Applications *Magnetophotonics* **7** 163

[9] Armelles G, Cebollada A, García-Martín A, González M U 2013 Mimicking electromagnetically induced transparency in the magneto-optical activity of magnetoplasmonic nanoresonators *Adv. Optical Mater* **1** 10
[10] Lutsenko S V, Kozhaev M A, Borovkova O V, Kalish A N, Temiryazev A G, Dagesyan S A, Berzhansky V N, Shaposhnikov A N, Kuzmichev A N, Belotelov V I 2021 Multiperiodic magnetoplasmonic gratings fabricated by the pulse force nanolithography *Optics Letters* **46** 17 pp 4148-4151
[11] Bi L, Hu J, Jiang P, Kim H S, Kim D H, Onbasli M C, Dionne G F, Ross C A 2013 Magneto-optical thin films for on-chip monolithic integration of non-reciprocal photonic devices *Materials* **6** 5094–117
[12] Bulatov M F, Churikov D V 2019 On the formation of microheterogeneities in epitaxial films of nonstoichiometric ferrogarnets *Journal of Surface Investigation* **13** 2 pp 206-209
[13] Berzhansky V N, Shaposhnikov A N, Prokopov A R, Karavainikov A V, Mikhailova T V, Lukienko I N, Kharchenko Yu N, Golub V O, Salyuk O Yu, Belotelov V I 2016 One-dimensional magnetophotonic crystals with magnetooptical double layers *Journal of Experimental and Theoretical Physics* **123** pp 744-751
[14] Wehlus T, Korner T, Leitenmeier S, Heinrich A, Stritzker B 2011 Magneto-optical garnets for integrated optoelectronic devices *Phys. Status Solidi A* **208** 2 252
[15] Berzhansky V N, Karavainikov A V, Mikhailova T V, Prokopov A R, Shaposhnikov A N, Shumilov A G, Lugovskoy N V, Semuk E Y, Kharchenko M F, Lukienko I M, Kharchenko Yu M, Belotelov V I 2017 Nano-and micro-scale Bi-substituted iron garnet films for photonics and magneto-optic eddy current defectoscopy *JMMM* **440** 175
[16] Berzhansky V, Mikhailova T, Shaposhnikov A, Prokopov A, Karavainikov A, Kotov V, Balabanov D and Burkov V 2013 Magneto-optics of nanoscale Bi: YIG films *Appl. Opt.* **52** pp 6599-6606
[17] Berzhansky V N, Karavainikov A V, Milyukova E T, Mikhailova T V, Prokopov A R, Shaposhnikov A N 2010 Synthesis and properties of substituted ferrite-garnet films for one-dimensional magnetophotonic crystals *Functional materials* **17** 1 pp 120-126
[18] Shaposhnikov A N, Prokopov A R, Berzhansky V N, Karavainikov A V, Vysokikh Y E, Gerasimenko N N, Smirnov D I 2017 Magneto-optical, structural and surface properties of RIB sputtered (Bi, Ga)-substituted DyIG films *Materials Research Bulletin* **95** pp 115 – 122
[19] Shaposhnikov A N, Prokopov A R, Karavainikov A V, Berzhansky V N, Mikhailova T V, Kotov V A, Balabanov D E, Sharay I V, Salyuk O Yu, Vasiliev M, Golub V O 2004 Modification of Bi: YIG film properties by substrate surface ion pre-treatment *Materials Research Bulletin* **55** pp 19-25
[20] Berzhansky V N, Shaposhnikov A N, Prokopov A R, Karavainikov A V, Mikhailova T V, Semuk E Yu, Sharipova M I, Dolgova T V, Fedyanin A A, Kotov V A, Golub V O 2011 One-dimensional magnetophotonic crystals based on double-layer Bi-substituted iron garnet films *Mat.-wiss. u. Werkstofftech.* **42** pp 19-23
[21] Shaposhnikov A N, Fedorenko A A, Mikhailova T V, Osmanov S V, Karavainikov A V, Lyashko S D, Milyukova E T, Nedviga A S, Semuk E Yu, Berzhansky V N 2021 *Nanophysics and nanoelectronics. Proceedings of the XXV International Symposium* **1** pp 346-347
[22] Berzhansky V N, Shaposhnikov A N, Prokopov A R, Karavainikov A V, Mikhailova T V, Lukienko I N, Kharchenko Yu N, Miloslavskaya O V, Kharchenko N F J. *Magn. Soc. Jpn.* 2012 **36** pp 42-45

Dynamic strain measurement of propeller shaft vibrations

G A Kushner

G. A. Kushner Astrakhan state technical university 16 Tatishcheva street, Astrakhan 414056, Russia

E-mail: guriy.kushner@mail.ru

Abstract. The paper proposes an approach to the registration of vibrations parameters to increase the reliability and predict the durability of technical systems with a continuously rotating shaft. For systems with stochastic loads, such as a ship's shaft line, the actual measurement of shaft stresses and deformations is an actual way to prevent failures and non-destructive testing under operating conditions. The adaptation of the dynamic strain measurement method made it possible to develop a software and hardware complex for recording and analysing transverse, torsional and longitudinal vibrations of shafts. The design of a hardware complex consisting of a measuring mobile and stator modules connected by a wireless network that allows dynamic strain measurement is proposed. The connection diagram and the main metrological and technical characteristics of the modules are given. To test the operability of the hardware complex, an experimental installation was built that allows carrying out investigation of the shaft line vibrations in real operation conditions. Experimental data are presented, the analysis of which allows us to predict the durability of the system.

1. Introduction
The study of shaft vibrations is inextricably linked with the measurement of vibrations. In the established practice of measurements, there are several approaches: strain measurement [1-3], measurement of vibrations by accelerometers [4,5], measurement of the difference in rotation frequencies at nodal points [6,7], polarization-optical methods of photo elasticity, X-ray methods and methods of brittle and electroplating coatings.

The registration of vibrations parameters plays a special role in terms of increasing the reliability and predicting the durability of systems with stochastic loads. An example of such a system is a ship's shaft line, where variable and random loads act on the propeller shaft [8], which are difficult to predict. In this case, the factors that directly affect the reliability are not only the design with adequate reserve coefficients [9], but also the actual measurement of stresses and deformations of the shafts.

2. Construction of the strain measuring system
The proposed approach to improving the reliability of rotating shaft systems is based on the method of dynamic strain measurement. According to the measurement theory, the signals coming from strain gauges indicators are subject to separation into a constant and a variable component. The subtraction from the signal of the constant component will depend on the type of the strain gauge task being solved. In the developed hardware complex, the method of dynamic strain measurement of a rotating shaft is adopted, so only the variable component will be important for registration.

It should be noted that the registration of a variable component requires a high frequency of data collection and a broad band passing data from the component database of the device. In the proposed hardware complex, first of all, filtering of the low-frequency non-informational part of the spectrum is provided. This signal processing led to a decrease in interference and an increase in the signal enhancement of the measuring path. The wired connection of the resistors in a bridge scheme made it possible to supply voltage U, and receive a signal output voltage ΔU. The sensitivity coefficient of the scheme S was calculated depending on the passport value K_R according to Table 1.

Table 1. Characteristics of bridge scheme

Bridge scheme	Function	Sensitivity coefficient	Wire connection
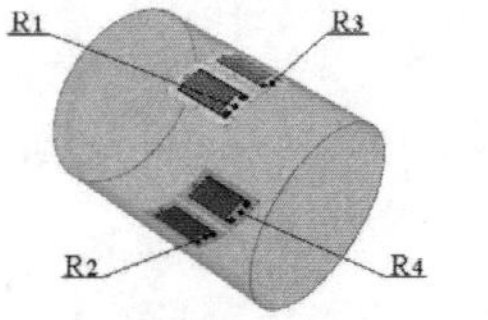	Measurement of transverse deformations, full-bridge scheme with thermal distortion compensation	$\frac{1}{K_R}$	
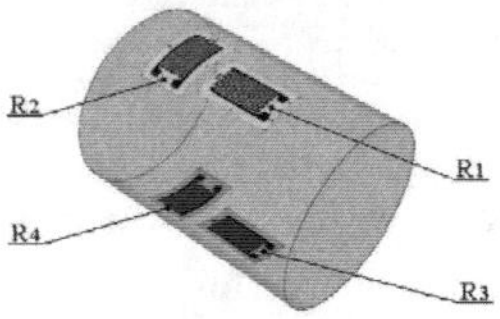	Measurement of single-axis longitudinal deformation, full-bridge scheme with thermal distortion compensation and camber	$\frac{2}{K_R(1+\vartheta)}$ where ϑ – Poison`s ratio	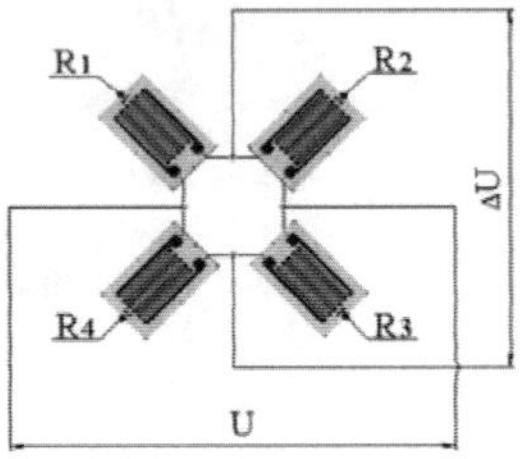
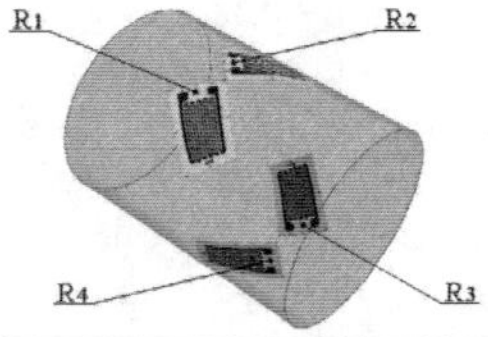	Measurement of torsional deformations, full-bridge scheme with thermal distortion	$\frac{1}{K_R}$	

The software and hardware complex consists of a measuring module (the rotor module) with a digital strain gauge indicator for measuring relative deformation using primary converters, which uses an information processing algorithm. The stator module, which is part of the software and hardware complex, is designed to record data received from the measuring module using Wi-Fi communication with wireless network technology together with devices of IEEE 802.11 standards. The device diagram shown in figure 1 allows for wireless dynamic strain measurement of shaft vibrations.

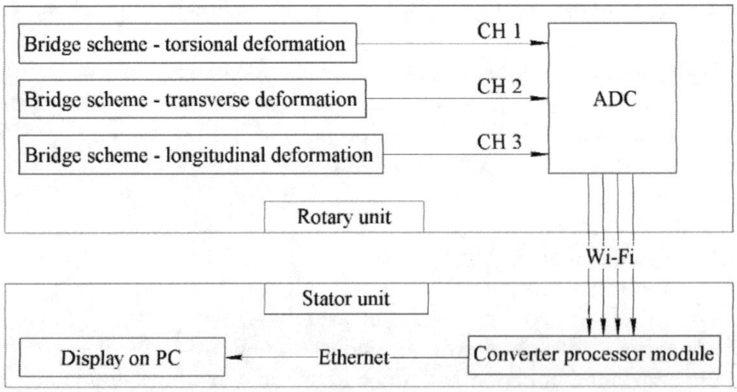

Figure 1. Schematic diagram of the hardware part.

A prototype of a system for registering torsional, transverse and longitudinal vibrations of propeller shafts, shown in figure 2 was built on the basis of an analog-to-digital converter with a built-in filter manufactured by «ZetLab».

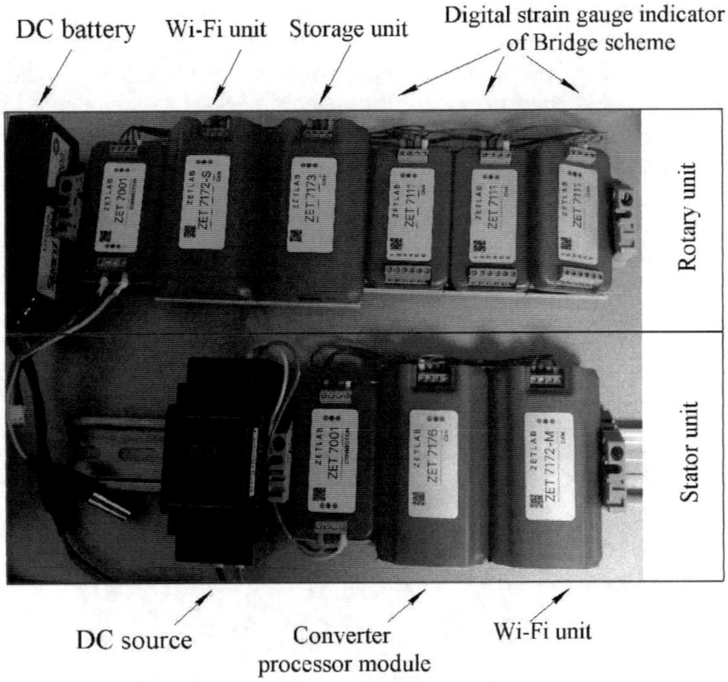

Figure 2. Prototype of a dynamic strain measurement device.

The proposed design of the dynamic strain measurement device also allows to connect and take data array from sensors and measuring transducers, such as vibrometers, infrared temperature sensors, rotational speed sensors and others, both on the rotating parts of the measurement object and on stationary ones. The use of Wi-Fi system for transmitting measurement data and power supply allows registering dynamic processes in the entire required frequency range. Data backup on a solid-state storage unit and transmission over a Wi-Fi radio channel makes it possible to use a telemetry complex for endurance tests, research on the performance of ship shafting systems. The main characteristics of the device's digital strain gauge indicator are shown in Table 2.

Table 2. Metrological and technical characteristics

Characteristics	Value
Measured physical quantity	Deformation, stress, torque
Measurement range of signals from resistive strain gages, mV/V	From -7 to 7
Threshold sensitivity, mV/V	0,0001
Absolute measurement error, %	0,05
Relative measurement error, %	0,05
Temperature measurement error, % / 10°C	0,05
Zero drift error, %	0,05
Output power to resistive strain gages, Unit amount	3 V; 1250 Hz; 10 mA
Updating frequency, Hz	Up to 1250

3. Experimental studies of shaft line vibrations

As a practical application of the developed strain gauge system, the measurement of the ship's shaft line vibrations is chosen as a responsible system, the failure of which can lead to significant economic losses. Strain measurement of the shaft line system of a transport vessel is associated with certain technical difficulties and increased complexity of testing. The use of experimental installations, usually manufactured on the basis of asynchronous electric motors, does not allow fully assessing and registering the stress-strain state of the propeller shaft operating in open water.

The paper proposes a floating experimental installation for testing a prototype of the developed strain gauge system, which allows studying the processes of shaft vibrations and resonance phenomena under controlled conditions. A special feature of the installation is the presence of a shaft line system assembled in the hull of a small vessel, shown in figure 3.

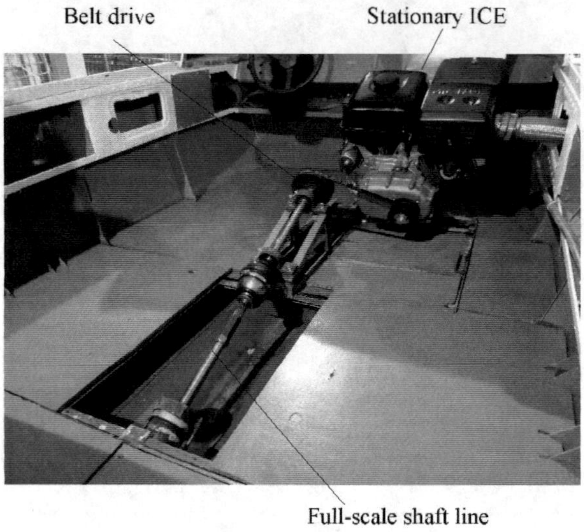

Figure 3. Power plant.

The proposed design of the installation makes it possible to study the longitudinal, transverse and torsional vibrations of a full-scale ship's shaft line, as well as to conduct multifactorial experiments to study the stress-strain state and vibrations of shafts using telemetric measuring systems. Conducting tests on the proposed device design allows us to obtain experimental data in real operation conditions without using a full-scale ship power plant. A fragment of experimental data is shown in figure 4.

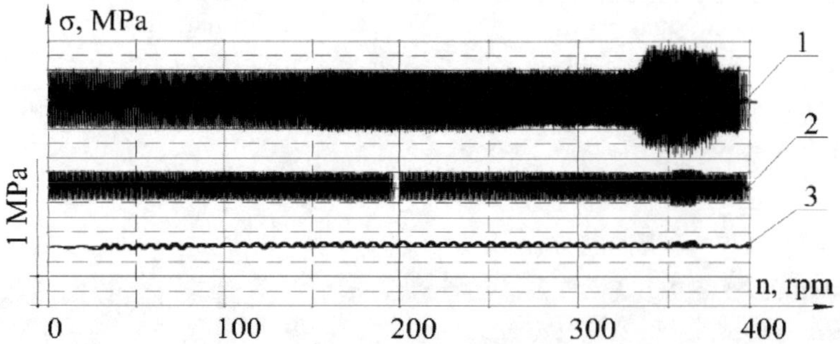

Figure 4. Experimental data.

Figure 4 shows an array slice of experimental data obtained using the device, where the numbers 1, 2 and 3 indicate the oscillograms of transverse, torsional and longitudinal vibrations obtained by strain measurement. In the given oscillograms, an increase in the oscillation amplitudes in all directions is observed in the range from 350 to 375 rpm, which indicates the presence of a connection between different types of vibrations in the full-scale shaft line.

Shafting failure prediction is possible using a wear damage model. Most failures are due to progressively increasing wear on parts such as stern tube bearings or torsional vibration dampers. In addition to the obvious wear of the units, which can be controlled, the increase in loads leads to the growth of microcracks that can break the shaft after a certain time of operation. The paper proposes to use a wear model based on stress readings obtained using the device. The wear of shafting elements leads to an increase in the actual stresses. In this case, the wear curve can be built from the points that characterize the maximum stress values. The dependence of maximum stresses σ_{max} on operating time t is shown in figure 5.

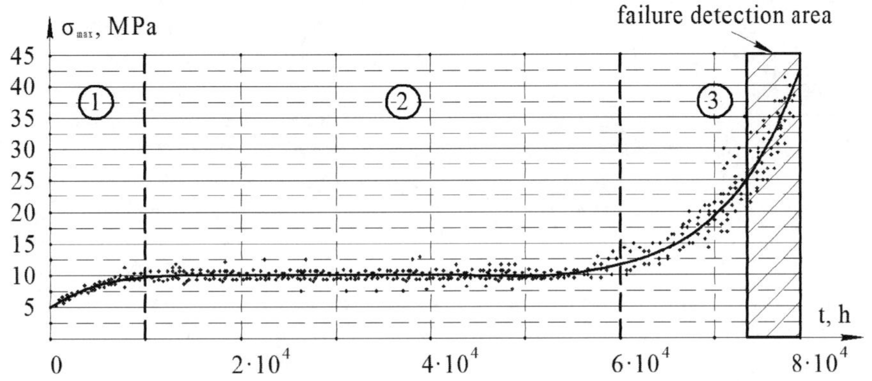

Figure 5. Dependence of mechanical stresses on the operating time of the shafting.

The wear curve can be divided into 3 stages. The first stage is running in. The second stage is steady-state wear, when there is a natural change in the shapes and sizes of parts. The third is intensive wear, characterized by a sharp increase in the wear rate due to unacceptable changes in the mating parts. During this period, the shaft fails. At stage 3, based on the operating experience of the ship shafting, the failure detection area should be determined. In this area, it is possible to apply an algorithm for calculating load cycles [10]. Based on the data of theoretical studies, it is necessary to develop a package of applied programs capable of predicting shaft line failure based on the analysis of vibrations

4. Conclusion

The hardware complex proposed in the work allows for diagnostics based on the actual condition of the shaft line and predicting shaft failures caused by excess of cyclic strength from transverse vibrations. Implemented in the ship's shaft line, the system will prevent the breakdown of the ship's shaft line, leading to serious economic losses.

The adapted method of dynamic strain measurement makes it possible to diagnose the actual condition of the shaft line and give a forecast assessment of its failure caused by excess of cyclic strength due to vibrations. The analysis of the research results has shown a sufficient number of works devoted specifically to experimental methods of studying vibration–based diagnostics, vibration loading and vibrations of shaft line systems and ship power plants. The question of interpretation of experimental data also remains open, which indicates the need for further study.

References

[1] Lomakin I V, Zagidullin L R, Ageeva K A 2016 Strain gauge channel for monitoring the shaft runout of agricultural machinery *Scientific notes of the Kazan State Academy of Veterinary Medicine named after N. E. Bauman* **225** pp 130-33 (In Russian)

[2] Krot P V, Prikhodko I Yu 2009 Active control of torsional vibration and vibration in rolling mills *Vibration of machines: measurement, reduction, protection: scientific, technical and production collection of articles* **3** pp 44-60 (In Russian)
[3] Pokusaev M N, Mamontov V A, Ruban A R, Gorbachev M M, Kovalev O P 2019 Study of the torsion vibrations of the river general service launch "RK-2091" project 376 propulsion system *Marine intelligent technologies* **1-4** pp 88-92 (In Russian)
[4] Kovalev V I, Romanenko D A, Kukushkin E V 2018 Device for measuring vibration of cardan gears *Reshetnev readings* **1** pp 493-5 (In Russian)
[5] Nikolaev N I, Gritsenko M V, Gerasidi V V 2019 Results of vibration monitoring of ship technical equipment in operation in accordance with the requirements of the RS rules and foreign standards *Marine intelligent technologies* **4-3** pp 96-103 (In Russian)
[6] Stolyarov A S 2019 Measurement of unsteady shaft speed of an internal combustion engine using optical sensors *Proc. Actual problems of marine energy* vol 1 (Saint-Petersburg) pp 187-190 (In Russian)
[7] Kudryavtsev M V 2006 Optical monitoring of torsional vibrations of shafts *Scientific and Technical Bulletin of Information Technologies, Mechanics and Optics* **28** pp 3-7 (In Russian)
[8] Alexandrov A V, Rybalko T R 2020 Numerical modeling of transient processes of propeller shaft deformation under the action of ice loads *Proceedings of the Krylov State Scientific Center* **4** pp 70-5 (In Russian)
[9] Platonov V V, Filatov A R 2020 Direct calculation of propeller shaft fatigue based on modern methods *Proceedings of the Krylov State Scientific Center* **2** pp 68-71 (In Russian)
[10] Richards F D, LaPointe N R, Wetzel R M 1974 A Cycle Counting Algorithm for Fatigue Damage Analysis *SAE Transactions* pp 1187-97

Journal of Physics: Conference Series **2091** (2021) 012051 doi:10.1088/1742-6596/2091/1/012051

Assessment automation of road traffic safety with account for road conditions of an individual itinerary

D A Skorobogatchenko[1], V V Borovik[2], A I Frolovichev[3]

[1]Volgograd State Technical University, Lenin avenue, 28, Volgograd 400005, Russia
[2]State Owned Public Enterprise of Volgograd region «Directorate of highways», Zhukova avenue, 90, Volgograd 400107, Russia
[3]Russian University of Transport, Obraztsova avenue, 9, Moscow 127055, Russia

E-mail: dmitryskor2004@gmail.com

Abstract. The paper substantiates the need to develop an automated system for traffic safety assessment in urban agglomerations, taking into account road conditions. The authors suggest a methodology for assessment of road traffic accidents, which makes it possible to take into account a wide range of factors affecting them. The methodology is based on complementing the traditional approach of final accident rate calculation with algorithms for collecting and analyzing data using Big Data tools, in particular, convolutional neural networks, fuzzy neural networks such as ANFIS, and cluster analysis using the k-means method. All accident rates are grouped according to the principle of homogeneity of acquisition of information for their calculation. Further, one of the data processing tools is applied to each group. As a result, labor intensity is reduced and the effectiveness of the application of the method of final accident rates increases. For practical calculations, the authors have developed a client-server application that uses data on geometric characteristics, current traffic situation, weather and climatic effects at the time of the trip along a specific itinerary. By means of application use, the analysis of traffic safety on a number of routes in Volgograd was carried out and the results are presented in comparison with the calculations made via the traditional method. It is shown that the use of information about the current situation on a specific section of the road network in terms of the current time significantly increases the accuracy of calculations.

1. Introduction

Road safety in Russia is a complex social and economic problem. Although the accident rate on Russian roads has been decreasing in recent years, it remains quite high (figure 1). Thus, over the past ten years, more than 200.000 people died in road accidents in our country, which is equivalent to the population of a small town. More than a quarter of the dead are people of the most active working age, which is a serious demographic problem. A significant amount (up to 3% of the country's GDP) is made up of economic losses from injury to people in road accidents [1].

In addition to direct losses associated with the death and injury of people, when assessing the social and economic damage from road accidents, it is necessary to take into account the damage to vehicle owners, costs of the road maintenance service, loss of cargo owners, as well as the cost of the work of specialized law enforcement and medical institutions, costs of social security [2], as well as a number of additional indirect losses (figure 2-3).

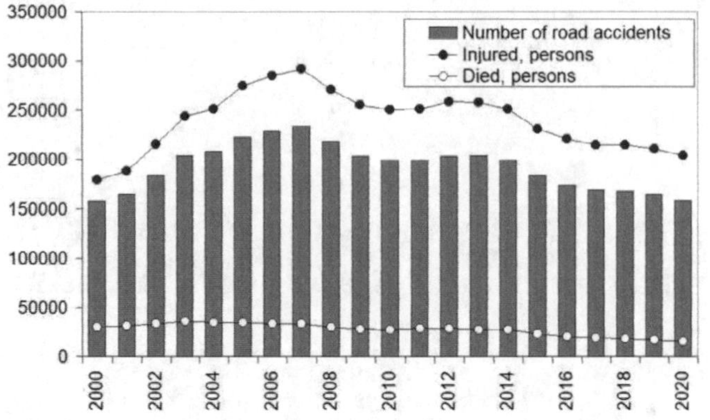

Figure 1. Analysis of road accidents' level and injuries in Russia in the dynamics for the period 2000-2020.

One of enough effective areas of activity which reduces the number of accidents and the severity of their consequences is the development of organizational measures in the field of traffic safety, as well as automated systems for assessment and accident prevention [3].

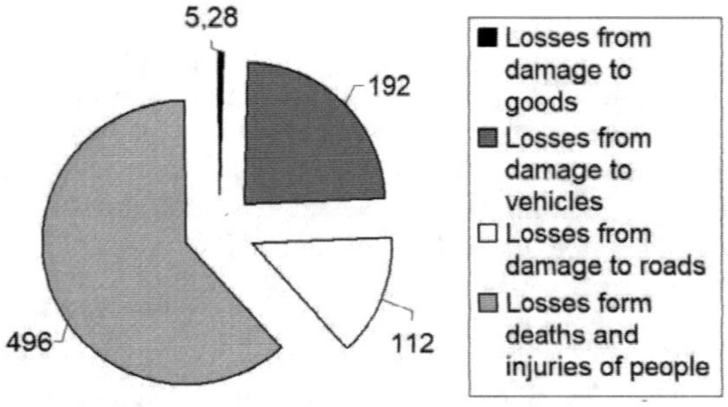

Figure 2. Assessment of damage to a person from an accident, million rubles.

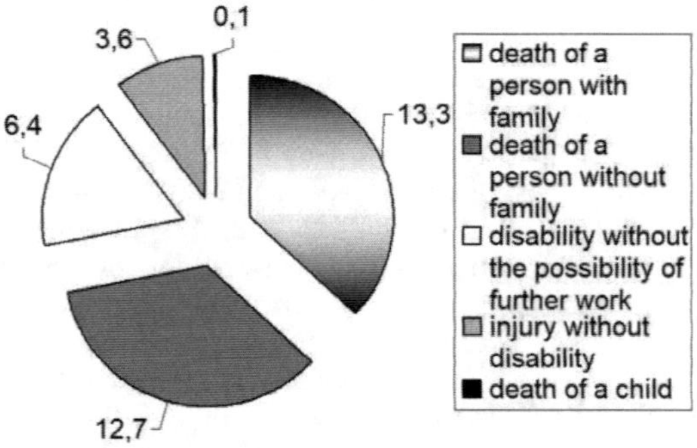

Figure 3. Analysis of economic losses from road accidents, billion rubles.

Due to the fact that the accident rate is largely determined by the technical parameters of the road, its actual state, speed and intensity of road and pedestrian traffic, as well as the availability of technical means of traffic organization, the most common approach to traffic safety determination is the assessment of the accident rate taking into account road traffic conditions [4]. However, in addition to road conditions, the accident rate is caused by such factors as driver's behavior, technical condition of the car, transport and operational state of the road, as well as peculiarities of weather and climatic impact on both road conditions and car operation, as well as behavior of drivers and pedestrians [5].

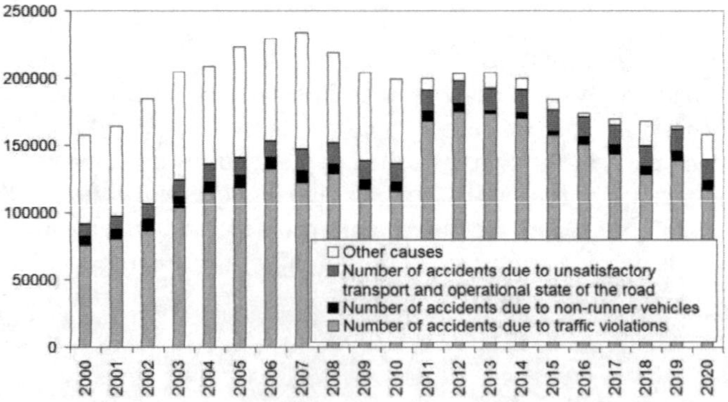

Figure 4. Analysis of the causes of road accidents in Russia in the dynamics for the period of 2000-2020.

Statistics show a steady increase in the number of accidents due to violation of traffic rules by the drivers (figure 4).

Thus, today, in the design of a smart and comfortable urban environment, issues connected with the development and design of systems for assessment and improvement of traffic safety, taking into account traffic conditions in urban agglomerations, which allow, on the basis of the accident rate forecast for a specific route, to develop recommendations for a safe high-speed movement mode of the vehicle.

The aim of the study is to develop an integrated automated system for assessment of individual traffic safety along a given route of the road network within the framework of a set of factors which determine road conditions.

2. Incisive analysis

When analyzing the main methods for road safety determination, five main groups can be mentioned (figure 5).

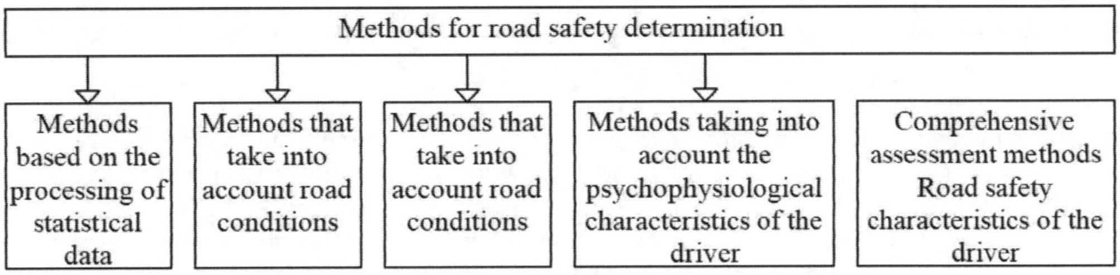

Figure 5. Analysis of the main methods for road safety determination.

One of research areas of road safety assessment, based on the processing of statistical data, presents the method of area identification of concentration of road accidents and methods for assessment of road accident by severity level [6], which was been developed recently, using tools for processing large amounts of data [7]. This scientific field provides acceptable accuracy in determination of locations of road accidents concentration. However, taking into account road conditions, the methods of this direction are based mainly on data on traffic intensity, which does not allow establishing the causes of road accidents and, as a result, to develop measures for their elimination.

The research area of road safety assessment, taking into account road conditions, is presented by domestic and foreign methods of final accident rates calculation [8, 9]. This scientific field was further developed on the basis of simulation of road conditions and traffic flows [10]. The main disadvantage of this area is the high complexity of calculations and the need to use a large number of probabilistic indicators.

In the scientific area connected with the analysis of conflict situations, the conflict points' method [11] and its development in the form of an assessment of the probable number of road accidents [12] should be mentioned. Currently, this research area is being developed taking into account the relationship between transport conflicts and accidents [13].

The research area of traffic safety assessment which is based on the psychophysiological and behavioral characteristics of drivers is presented by a set of methods for the assessment of the behavior of road users and its simulation [14]. The main disadvantage of this direction is the complexity of extrapolation of the models to collective traffic and the need to take into account the individual characteristics of a particular road user.

The set of methods for a comprehensive assessment of road safety includes a qualimetric method and a method based on minimization of total social losses from road accidents. The modern development of this research area can be found in the paper [15]. This approach is appropriate for a comprehensive systematic analysis of road traffic in the region. However, it should be noted that the assessment obtained on the basis of the methods of this group does not fully reflect the peculiarities of traffic flows in the city and on roads outside settlements.

Next, let's move to the analysis of tools used to automate road traffic safety assessment (figure 6).

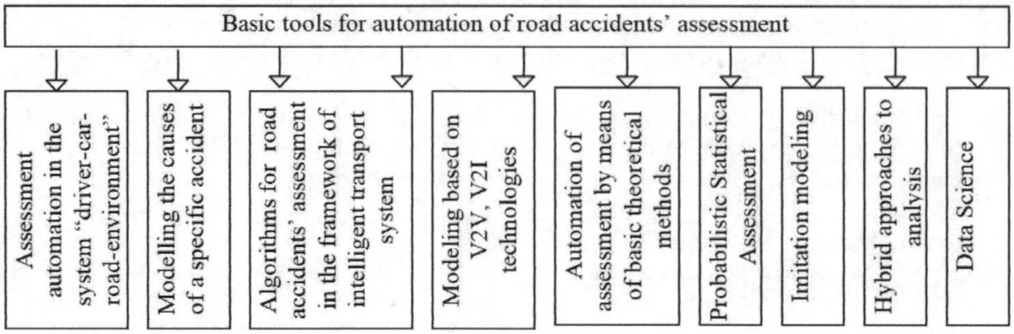

Figure 6. Analysis of the main instruments used to automate the assessment of road accidents and traffic safety.

Automation of road traffic safety assessment, taking into account the influence of "driver-car-road-environment" factors, is founded on the use of statistical analysis methods together with prediction of the occurrence of an accident, localization of hazardous areas and identification factors that have the greatest impact on an accident. Data for analysis is supplied either by automated road data banks, or directly by users, i.e. participants in road accidents, while filling in is available on mobile applications [16]. The approaches in this area are aimed at prediction of potential number of accidents in each region and identification of potentially dangerous factors, the analysis of which is necessary for the design of road infrastructure and improvement of traffic management.

The crash assessment mechanism in the design of intelligent transport systems is used to ensure road safety, reduce environmental problems and optimize urban traffic. This approach is a global trend for assessment of road accidents in modern urban agglomerations and allows us improving the quality of road traffic dramatically [17].

Within the framework of research area for automation of the main theoretical methods, a set of mathematical models was developed, which makes it possible to analyze and predict the dynamics of the main indicators of road traffic safety. The most common approach is to automate the assessment of the causes of road accidents for insurance companies.

The scientific field of modeling of a specific incident is founded on the assessment of various factors that led to the accident, such as braking distance, speed, state of the braking system, etc. As an example, we can mention such software systems as CARAT, AUTO-GRAF, PC-Crash, ANALYSER PRO [18, 19].

Simulation of accidents based on V2V, V2I technologies is founded on the fact that the state of the parameters of a car (speed, steering, braking system, etc.) under control is communicated wirelessly to neighboring cars or piece of road infrastructure [20]. Automation of accident assessment, based on the processing of data from vehicles equipped with communication means connected with infrastructural facility or self-driving vehicles, allows reducing the influence of the human factor, as the main source of accidents, and to homogenize the characteristics of road traffic.

Probabilistic-statistical tools for the assessment of road accidents are presented by regression models for the evaluation of the number of road accidents, as well as the number of deaths and injuries in them, based on a set of social and economic indicators. Currently, statistics on road traffic accidents with injured and dead persons are mainly used to predict road accidents. Analysis methods of big data can be considered as a modern continuation of the approach presented in epy paper [21]. This research area should also include machine learning tools and soft computing. A review of approaches using machine learning methods in the analysis of road accidents is presented in the paper [22]. The use of a multilayer perceptron is considered as one of the most suitable structures for the solution of issues of accident forecasting and transport control strategies [23]. An overview of models for the implementation of soft computing is presented here [24].

Such software products as VISSIM, SUMO and MATSim are successfully used for the simulation. A feature of the scientific field is the visualization of data from the model of traffic flows in an accident cluster and the desire to create a three-dimensional model of a recognizable environment of an accident cluster of the road. As an example of the implementation of automated safety assessment systems based on “driver-car-road-environment” factors, we can cite the study [25].

Hybrid approaches include a set of methods combining several of the listed tools. In particular, it should be noted the synthesis of expert and simulation modeling, mechanisms of interaction between expert systems and decision support systems, systems for automation of the assessment of road accidents, combining GIS technologies, analytical models and databases [26, 27].

Summing up the analysis of approaches and tools for the assessment of road traffic safety, the prevalence of an integrated approach and the desire to take into account the maximum number of factors reflecting the impact of road conditions should be noted. In this regard, we believe that the most promising is the construction of an automated system for the assessment of individual traffic safety based on taking into account the actual road conditions using Big Data tools. This approach will allow, on the one hand, to take into account in the calculations of the maximum number of factors affecting road accidents, and on the other hand, to increase the accuracy of the calculations.

3. Models and results

A distinctive feature of the author's research is the synthesis of the traditional methodology for road accident assessment [28] with a mechanism for the evaluation of the current state of road conditions along a specific route. The authors believe that the implementation of the assessment of the impact of road conditions along a specific route in relation to time will eliminate the drawbacks connected with averaging which is inherent in the assessment method of accident rate currently used.

In accordance with ODM 218.4.005-2010, the assessment of road traffic safety today is based on the calculation of the index of final accident rate:

$$K_{ав} = \prod_{i=1}^{n} K_i \,, \tag{1}$$

where $K_{ав}$ - final accident rate; K_i - partial accident rates. Let's define them: K_1 - intensity, cars/day; K_2 - percentage of passenger cars in the traffic flow, %; K_3 - carriageway width, m; K_4 - safe flow speed, km / h; K_5 – number of lanes; K_6 - lighting, lx; K_7 - intersection type; K_8 - total intensity at the intersection of cars / day; K_9 - total intensity of pedestrians, persons / day; K_{10} - visibility of cross roads from the adjoining roads, m; K_{11} – location of public transport stop; K_{12} – location of pedestrian crossings outside cross roads; K_{13} - pedestrian traffic outside cross roads, persons / day; K_{14} - sidewalk location; K_{15} - longitudinal slope, %; K_{16} - horizontal curve radius in the ground plan, m; K_{17} - tramway location; K_{18} - adhesion coefficient.

Further, the entire list of partial accident rates was conventionally divided into four groups in relation to the procedure for data acquisition (figure 7).

The first group includes coefficients, the value of which is assigned according to geometric parameters of the road network. They include the width of the carriageway, number of lanes, geometric characteristics of intersections, presence of public transport stops, pedestrian crossings, etc. The value of the coefficients of this group is supposed to be obtained on the basis of the definition of graphical objects on the route map by the neural network. The technique of using artificial neural networks used by the authors is described in detail in the papers [29, 30, 31].

The second group includes the coefficients characterizing the intensity of traffic both along the route and at cross roads. Their values are determined mainly by the current road traffic situation and depend on the time of day, seasons, road conditions. The connection here is nonlinear, besides there are many factors, the influence of which is rather difficult to take into account quantitatively. In this regard, it was suggested to use a fuzzy neural network [32].

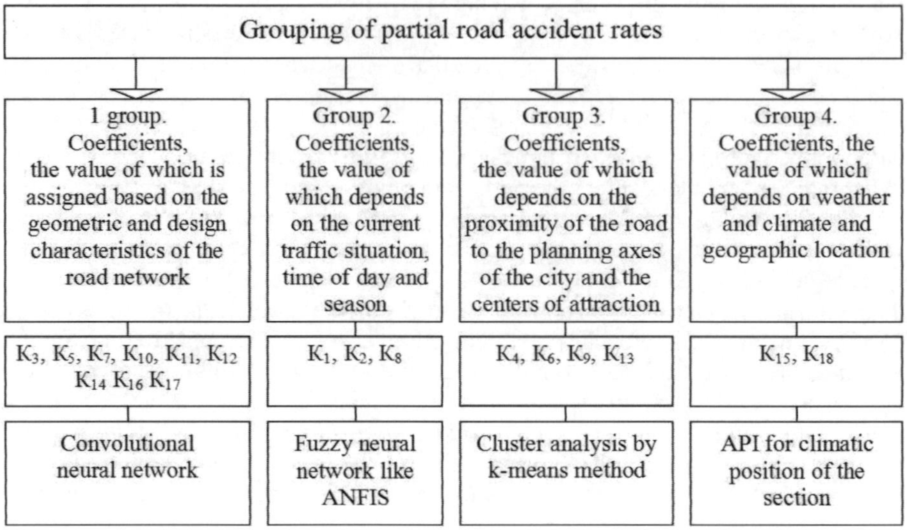

Figure 7. Grouping of partial accident rates in relation to data collection.

The third group includes coefficients, the values of which are largely determined by the position of the road section in terms of the administrative center of the city, the main planning axes and the centers of attraction of the urban agglomeration. They include the percentage of cars in the traffic, the level of road illumination, and the intensity of pedestrian traffic. To obtain values for this group of coefficients, it is suggested to use cluster analysis using the *k*-means method.

The fourth group includes coefficients, the values of which depend on the current weather and climatic situation. This is mainly the coefficient of adhesion. The data collection mechanism is based on API drawn from the website with a calculation mechanism according to the algorithm set out in the regulatory literature on road climatology [33]. On the basis of these data in order to determine the coefficient for the state of road pavement, the authors suggest using a decision tree [34]. This fact will allow taking into account the conditions for the appearance of ice by means of use of the statistics of weather phenomena observation.

A mobile application for practical calculations using the Android Studio development environment was designed. The main menu was supplemented with components for choosing between routes, a map, a button for the calculation of the final accident rate, and buttons for display in the text format of the final value of the accident rate and the level of traffic safety, as well as the recommended speed (figure 8).

As an illustration of the work of the automated assessment system of road traffic safety assessment, we will analyze the change in the accident rate along several routes in Volgograd at different times of the day and seasons of the year, and also compare the results obtained with the calculations using the ODM 218.4.005-2010 method, traditionally used in the industry.

Figure 9 shows the route from the bus stop "Shopping Center" in Voroshilovsky district of Volgograd to the bus stop "SuperMAN store" in Tsentralny district. In the system, this route was divided into seven sections with identical road conditions. Further, for this route, the calculation of accident rates was made taking into account the influence of current road conditions and according to the ODM methodology 218.4.005-2010. The calculations were made for summers and winters during the decline of the morning peak of intensity (11 a.m.) and in the evening peak of the growth of road traffic intensity (17 p.m.). The calculation results are shown in figures 10-13.

Figure 8. General view of a client-server application for road traffic safety assessment taking into account road conditions.

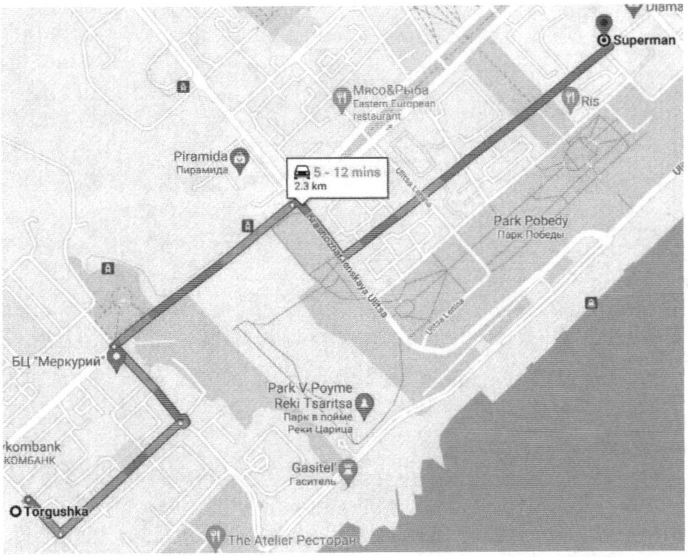

Figure 9. Route "Shopping center - SuperMAN store" along which comparative calculations were made assessment taking into account road conditions.

Analyzing the results obtained, it can be concluded that the approach suggested by the authors gives significant differences in determination of the final coefficient in comparison with the traditional method of ODM 218.4.005-2010. Taking into account the influence of current road conditions, in particular the intensity of automobile and pedestrian traffic, composition of road traffic flow, as well as weather, the calculated value of the final accident rate characterizes more accurately the safety of road traffic along the route in comparison with the traditional approach.

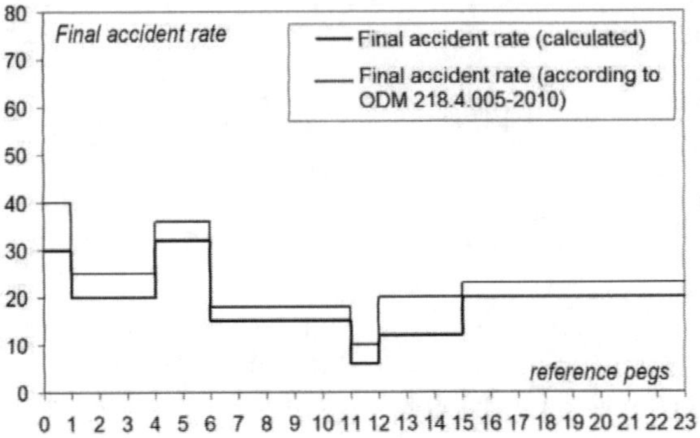

Figure 10. Calculation of the final accident rate along the route "Shopping center - SuperMAN store" made in the automated system and in accordance with ODM 218.4.005-2010 at 11:00 in June 2020.

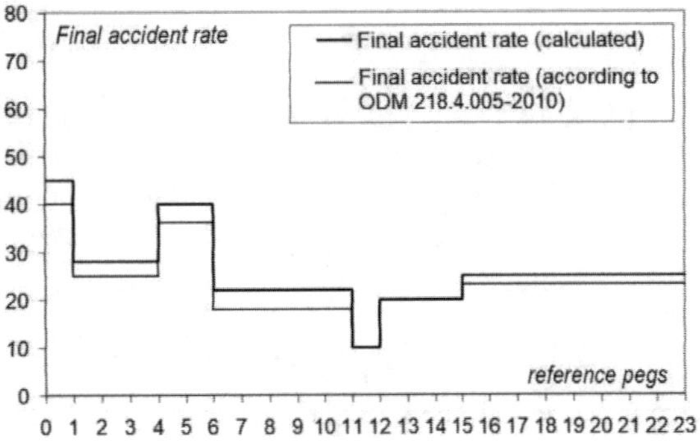

Figure 11. The same route at 17.00 in June 2020.

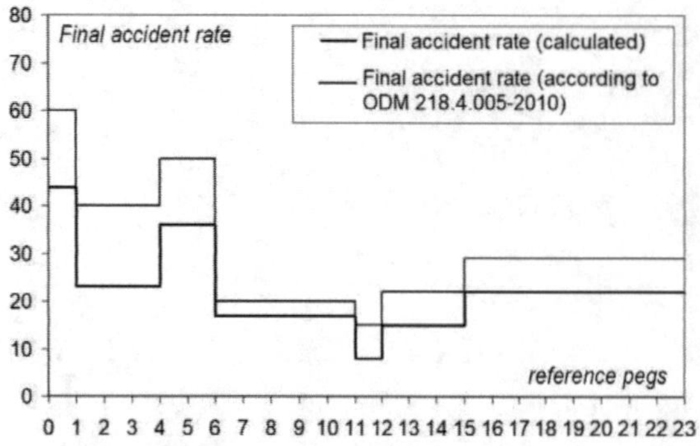

Figure 12. The same route at 11:00 in December 2020.

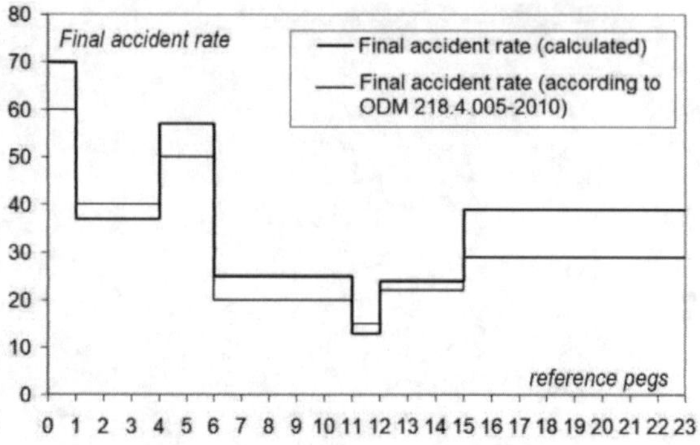

Figure 13. The same route at 17:00 in December 2020.

The values of the final accident rate obtained according to ODM 218.4.005-2010 technique do not differ within the season and are sufficiently averaged within the year, what is acceptable when developing general recommendations for traffic management, but not enough for individual road safety assessment.

Since the traffic intensity, as one of the factors determining the accident rate, it fluctuates to a large extent not only by the hours of the day, but also by the days of the week, and the authors present an example of calculation of the final accident rate according to route shown in Figure 11, made for the same time and season on Monday and Sunday (figures 14-15).

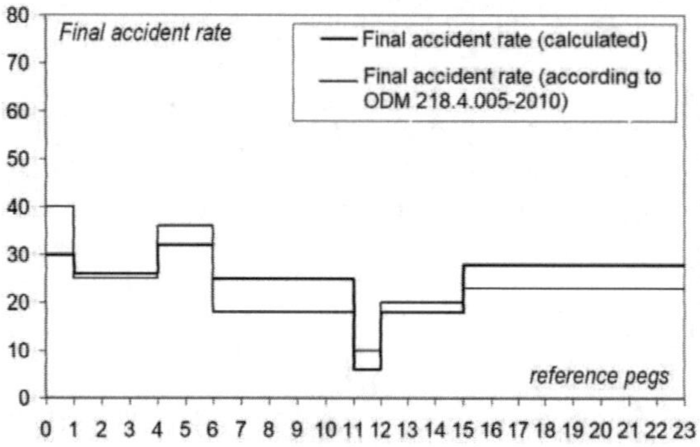

Figure 14. Calculation of the final accident rate along the route "Shopping center - SuperMAN store", made in the automated system and in accordance with ODM 218.4.005-2010 technique on Monday at 16:00 in April 2021.

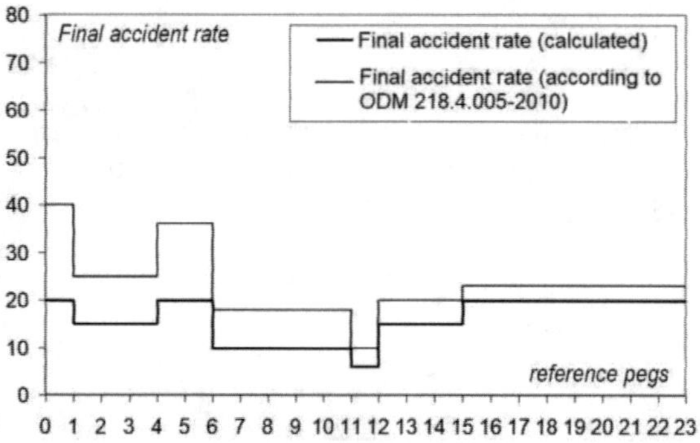

Figure 15. The same route on Sunday at 16:00 in April 2021.

A feature of the developed system is the current road safety assessment when the itinerary is changed. Let us illustrate the calculation of the final accident rate when choosing the alternative routes from the point "Shopping center" to the point "SuperMAN store" (figures 16-18).

The route laid through the central highway is the most dangerous. The weighted average accident rate is about 18.7 (figure 16). An alternative route built along the shortest route through a less occupied, second street, gives a lower value of the calculated final accident rate, i.e. 16.7 (figure 17).

The third alternative route, shown in Figure 18 is the safest. The weighted average of the final accident rate is about 12.7 with almost the same travel time, but a significant increase in mileage compared to the second route.

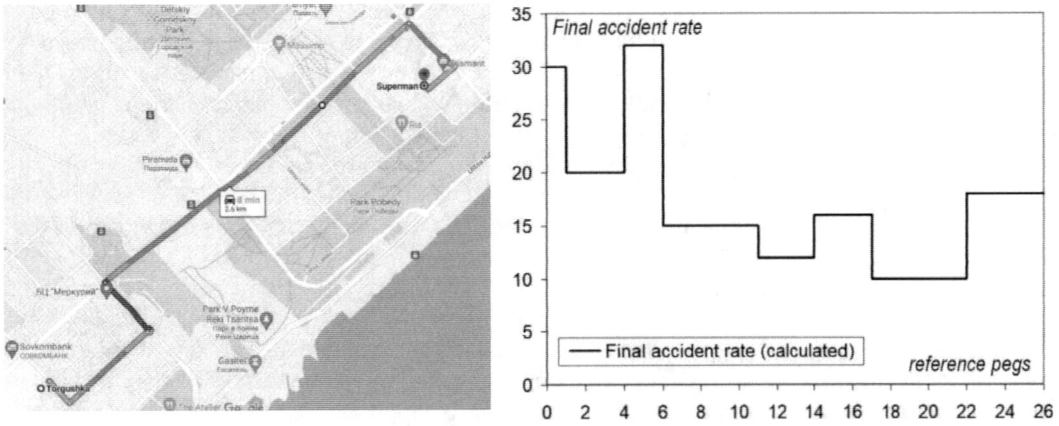

Figure 16. Assessment of road traffic safety on the first route from the point "Shopping Center" to the point "SuperMAN store", made in the automated system at 11:00 in April 2021.

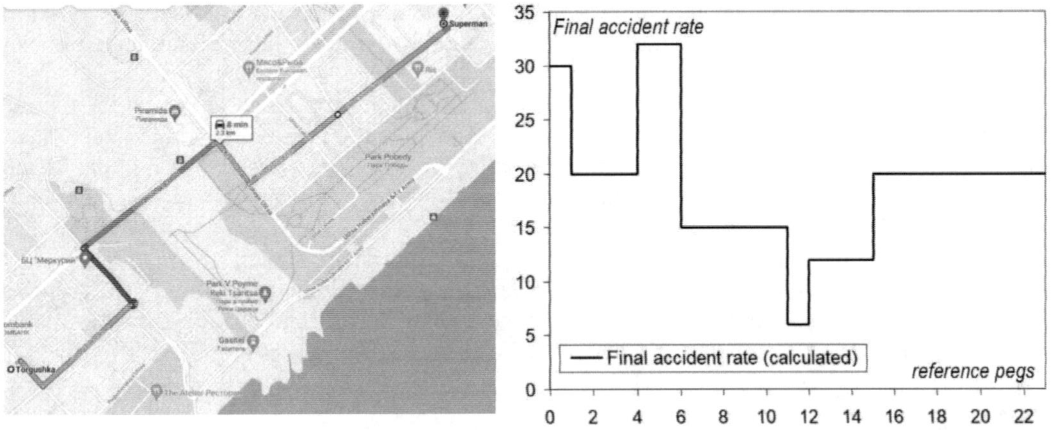

Figure 17. Traffic safety assessment on the second route from the point "Shopping Center" to the point "SuperMAN store", made in the automated system at 11:00 in April 2021.

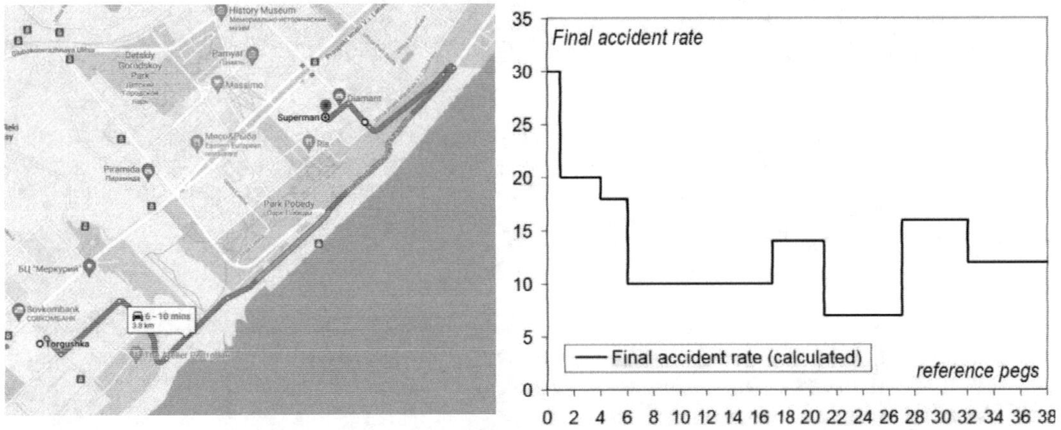

Figure 18. Road traffic safety assessment on the third route from the point "Shopping Center" to the point "SuperMAN store", made in the automated system at 11:00 in April 2021.

4. Conclusions

Summing up the results, it can be stated that the authors suggested an approach to the assessment of road traffic safety, complementing the traditional method of accident rate calculation, using data on the current traffic situation, as well as information on the actual weather and climatic situation at the time of the trip, within which the following results are obtained and measures are suggested.

1. The mechanism was developed for the calculation of data on the road traffic intensity of vehicles in relation to the route. The use of a fuzzy neural network of the ANFIS type is suggested as a calculation tool. The use in the calculations of not averaged data, but actual data at the time of the trip contributes to an increase in the accuracy of the assessment of road traffic safety.

2. An algorithm for the determination of values of geometric and design parameters of the road network along the selected route was developed. The use of this algorithm will significantly simplify the application of the accident rate method and reduce the complexity of the calculations. The assessment of geometric parameters does not require access to automated banks of road data, but will allow determining the parameters of structural elements of a highway directly from the map along the route by means of use of neural networks.

3. In order to determine the values of accident rate ratio depending on the proximity of the road to the planning axes of the city and the centers of attraction, it is suggested to cluster urban areas by population density and administrative significance. The use of cluster analysis makes it possible to group the road network according to the intensity of pedestrian traffic, illumination and composition of the traffic flow.

4. In order to take into account the state of the road pavement, it is suggested using the normative assessment method with program access to weather data on the corresponding website. Thus, the classical approach to the assessment of road traffic safety in unfavorable weather conditions or taking into account the seasonality is significantly specified by using data on the actual state of road pavement along the route at a specific time.

5. As a tool for the implementation of the suggested approach, a client-server application was designed that makes it possible to assess the safety of road traffic along a specific route. It should be noted that the application uses the mechanism for the assessment of road traffic safety of alternative routes to the point of destination.

References

[1] Federal State Statistics Service: URL: https://rosstat.gov.ru/ (reference date: 20.12.2020)
[2] Belogreben A A, Matantseva O Yu, Titov A E 2019 *Transport business of Russia* **6** 24–8
[3] Skorobogatchenko D A, Erokhin A V 2014 *Bulletin of Volgograd State University of Architecture and Civil Engineering. Series: Building and architecture* **36 (55)** 174-181
[4] Rolison J J, Regev S, Moutari S, Feeney A 2018 *Accident Analysis & Prevention* **115** 11-24
[5] Haque M Ohidul, Haque Tariq H 2018 *Transportation Research Part A: Policy and Practice* **118** 594–607
[6] Batmanov E Z 2008 *Bulletin of Dagestan State Technical University. Technical sciences* **14** 73-8
[7] Pradhan B, Ibrahim Sameen M 2020 Predicting Injury Severity of Road Traffic Accidents Using a Hybrid Extreme Gradient Boosting and Deep Neural Network Appro*ach // Laser Scanning Systems in Highway and Safety Assessment. Advances in Science, Technology & Innovation (IEREK Interdisciplinary Series for Sustainable Development)* (Cham: Springer)
[8] Babkov V F 1982 *Road conditions and traffic safety* (Moscow: Transport)
[9] Rankin V U, Cloughy P , Halbert S et al. 1981 *Automobile transportation and organization of road traffic. Translated from English* (Moscow: Transport)
[10] Drosu A, Cofaru C. & Popescu M V 2020 *International Journal of Automotive Technology.* **21**, 309–317
[11] Rappoport H. A. 1955 *Strassen und Tiefbau* **8** 499–510
[12] Lobanov E M 1980 *Road design and traffic organization taking into account the psychophysiology of the driver* (Moscow: Transport)
[13] Zheng L, Sayed T, Mannering F 2020 *Analytic Methods in Accident Research* **100142**
[14] Fan X, Pan G, Mao Y et al. 2020 *The Visual Computer* **36** 1203–1218
[15] Chen F, Lyu J, Wang T 2020 *Accident Analysis & Prevention* **147** 105752

[16] Tyulkin E V 2015 *Bulletin of civil engineers* **5** 248-253.
[17] Torrisi V, Ignaccolo M, Inturri G 2018 *Innovative Transport Systems to Promote Sustainable Mobility: Developing the Model Architecture of a Traffic Control and Supervisor System*. In: Gervasi O. et al. (eds) Computational Science and Its Applications – ICCSA 2018. ICCSA 2018. Lecture Notes in Computer Science, **10962**. (Cham: Springer)
[18] Azemsha S A, Galushko V N, Skirkovsky S V 2015 *Science and technology* **4** 18-24
[19] Evtyukov S A, Vasiliev Ya V 2012 *Examination of road accidents: methods and technologies* (Saint Petersburg: SPbGASU)
[20] Barik S, Mohanty S, Agarwal R, Pramanik J, Samal A K 2020 *A Proposed Wireless Technique in Vehicle-to-Vehicle Communication to Reduce a Chain of Accidents Over Road*. In: Swain D., Pattnaik P., Gupta P. (eds) Machine Learning and Information Processing. Advances in Intelligent Systems and Computing, **1101** (Singapore: Springer)
[21] Hu L, Zhao Y 2018 *Research on the Architecture of Road Traffic Accident Analysis Platform Based on Big Data*. In: Abawajy J., Choo KK., Islam R. (eds) International Conference on Applications and Techniques in Cyber Security and Intelligence. ATCI 2017. Advances in Intelligent Systems and Computing, **580** (Cham: Edizioni della Normale)
[22] Singh J, Singh G, Singh P, Kaur M 2019 *Evaluation and Classification of Road Accidents Using Machine Learning Techniques*. In: Shetty N., Patnaik L., Nagaraj H., Hamsavath P., Nalini N. (eds) Emerging Research in Computing, Information, Communication and Applications. Advances in Intelligent Systems and Computing, **882**. (Singapore: Springer)
[23] Kouziokas G N 2019 *Neural Network-Based Road Accident Forecasting in Transportation and Public Management*. In: Nathanail E., Karakikes I. (eds) Data Analytics: Paving the Way to Sustainable Urban Mobility. CSUM 2018. Advances in Intelligent Systems and Computing, **879** (Singapore: Springer)
[24] Kannojiya A K, Maurya R, Rajitha B 2020 *Survey on Soft Computing Methods for Accident Condition and Severity Predictions*. In: Pandian A., Ntalianis K., Palanisamy R. (eds) Intelligent Computing, Information and Control Systems. ICICCS 2019. Advances in Intelligent Systems and Computing, **1039** (Singapore: Springer)
[25] Eremin V M 2014 *CAD and GIS of highways* **1 (2)** 90–93
[26] Eliseev M E, Mazunova L N, Eliseeva I N 2020 *Proceedings of Nizhegorodsky state technical university named after R.E. Alekseev* **2 (129)** 28-41
[27] Lazarev S N, Oreshin N A, Mamleev D A 2019 *Management of activities to ensure road safety: state, problems, ways of improvement* **1(2)** 278-84
[28] ODM 218.4.005-2010 2011 *Recommendations for traffic safety on highways* (Moscow: FGUP Informavtodor)
[29] Skorobogatchenko D A, Stepanova P Yu 2019 *Bulletin of Volgograd State University of Architecture and Civil Engineeering. Series: Building and architecture* **1(74)** 62-71
[30] Skorobogatchenko D A, Zabaznov A S 2018 *Bulletin of Perm National Research Polytechnic University. Construction and architecture* **9(2)** 98-109
[31] Kravets A G, Skorobogatchenko D A , Agazadyan A R , Nedostupov D O 2018 *Bulletin of Astrakhan State Technical University. Series: Management, computer technology and informatics* **1** 49-61
[32] Dmitry S, Sergey V 2019 *Use of Fuzzy Neural Networks for a Short Term Forecasting of Traffic Flow Performance*. In: Kravets A., Groumpos P., Shcherbakov M., Kultsova M. (eds) Creativity in Intelligent Technologies and Data Science. CIT&DS 2019. Communications in Computer and Information Science, **1083** (Singapore: Springer)
[33] ODM 218.2.003-2009 2009 *Methodical recommendations for a specialized forecast of the condition of the road surface* (Moscow: FGUP Informavtodor)
[34] Skorobogatchenko D A, Agureev I A 2018 *Bulletin of Volgograd State Technical University* **13 (223)** 59-65

Decision Support System for Evaluating Elements of Data Transmission Systems

A. A. Oleinikov[1], A. A. Sorokin[1]and I. A. Beresnev[2]

[1]Astrakhan State Technical University, Tatischevastr, 16, Astrakhan 414056, Russia
[2]Joint Stock Company "South Center for Shipbuilding and Ship Repair", Astrakhan Region, p. Red barricades 416356, Working str, 1

E-mail: a.oleynikov.astu@mail.ru

Abstract. The expansion of the coverage areas of telecommunications networks poses the task of improving the quality of service for telecom operators. Currently, in the conditions of transferring employees to a remote mode of operation, the access equipment of individuals is beginning to require much more attention to failures, as well as the communication equipment of legal entities. In both cases, the unavailability of the data transmission network associated with the failure can lead to various kinds of financial and administrative liability. In this paper, we propose algorithms that underlie the developed decision support system for evaluating the operational states of data transmission system elements. The system is designed for the analysis and formation of recommendations on conducting scheduled preventive maintenance or responding to pre-emergency and emergency incidents.

1. Introduction

The development of data transmission networks entails an increase in the number of competent employees in the staff of telecom operators who are able to make timely informed decisions on modernization and repair. The scale of the networks of some operators, especially at the federal level, has a significant coverage area. The equipment used on such networks for the most part has similar operational parameters, in this case there is a need to automate the adoption of the same type of decisions [1, 4-6, 8]. The research proposed for consideration is based on algorithms for making operational state of data transmission system elements.

2. Acceptance of access network development decisions

According to the results of the analysis of the data describing the state of the EDTS, when assessing the operational load, it is possible to make erroneous decisions about the urgency of modernization, based on the classes of operational states (table 1). The algorithm for assessing the state of the EDTS (figure 1) provides for the use of feedbacks to reduce the level of making erroneous decisions:

- decisions on the development of the telecom operator's access network and monitoring the state of data transmission system elements (EDTS) after the implementation of the modernization project [2, 3, 7, 9, 10].The purpose of the work is to develop a decision support system for assessing the
- the first feedback (figure 1), in case of incorrect classification, based on data received in real time from a specific EDTS, provides for repeated clustering to determine new cluster labels, the number of which corresponds to the number of classes of operational states (table 1);

- the second feedback connects the monitoring results processing unit and the fuzzy classification unit. Here, the operational parameters of the upgraded equipment are monitored. If there is no excess of the indicators in the specified period, the project is put into final operation and removed from monitoring. The algorithm provides for re-classification, and in case of an increase in the level of influence of other operational parameters on re-clustering.

3. Algorithm of enhanced monitoring of the status of data transmission elements after modernization

When operating data transmission systems, it is necessary to pay attention to changes in the level of demand for services on upgraded equipment, depending on changes in social factors. The need to monitor the operational state of the EDTS arises when the demand for the provision of services increases when connecting subscribers through an upgraded network element. According to experts in the field of DTS operation, the most up-to-date data can be obtained during peak load hours. This data is used when conducting enhanced monitoring based on the algorithm shown in figure 2. With enhanced monitoring, the EDTS is allocated to the group of equipment that requires priority attention to assess the operational qualities and success of the modernization carried out.Upon completion of the modernization work, the DTS is put on enhanced monitoring. Real-time data from the analyzed equipment is compared with the forecast values. If the obtained indicators meet the required standards within a specified period of time, the EDTS is removed from enhanced monitoring. Otherwise, the EDTS state is analyzed by classification. If it is not possible to assign the EDTS to one of the operational classes, a new grouping of EDTS is performed by fuzzy clustering, accompanied by filling the database with information used to determine the timing of modernization.

4. The composition of the decision support system

EDTS diagnostics is the establishment of the current state of an element, by correlating it with a certain operational class. The class determines the urgency of carrying out the modernization of the EDTS on the basis of the specified criteria in the previously generalized form of a set of states. At the initial stage, the data is analyzed and recognized. At this moment, a state vector is formed. The peculiarity of the classification algorithm, which makes it difficult to take into account changes in the assessment of EDTS in transition states, when changing the class, is low sensitivity. This feature requires an additional analysis of the operational characteristics of the EDTS.

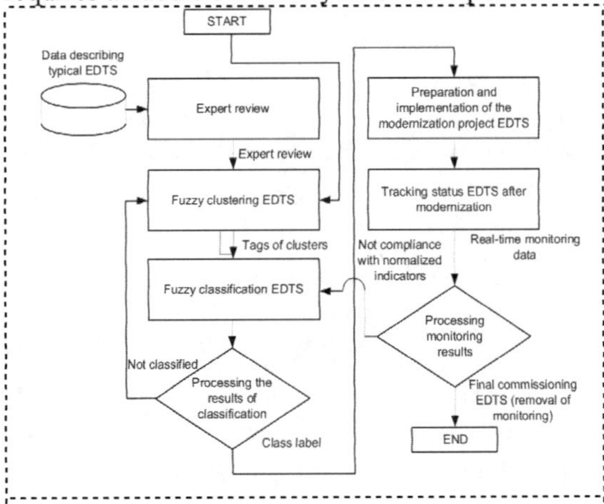

Figure1. Algorithm for estimating the status of data transmission elements.

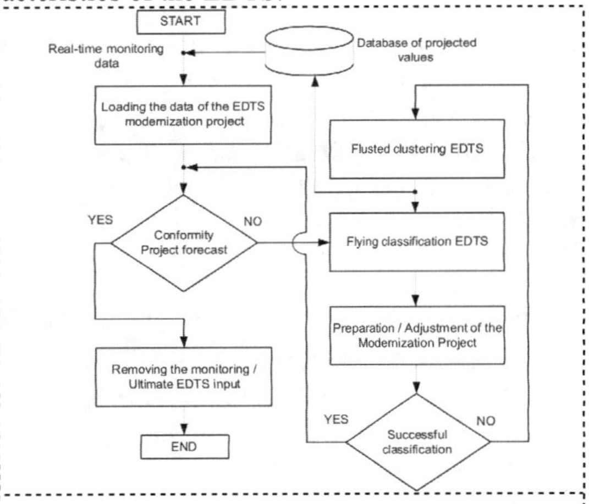

Figure 2. Algorithm of enhanced monitoring of the status of data transmission elements after modernization.

At the next stage, the analyzed and verification data are compared and a decision is made to correlate one of the classes that characterize the state with the analyzed EDTS. Classification of elements with boundary states occurs due to the use of fuzzy clustering, which allows determining whether the classified DTS belong to several classes with different degrees.

Based on this, a decision support system (DSS) is proposed, the structure of which is shown in figure 3. The DSS consists of the following blocks:

The "Pre-Project Evaluation Unit" collects data on the state of the EDTS from technical performance monitoring systems and from commercial systems (billing) and fills the database.

The "Expert Assessment Unit" uses data of a retrospective nature, processed with the involvement of expert groups. It forms classes for subsequent grouping by the fuzzy clustering block according to the operational states of the EDTS, which require measures to maintain the operability of the equipment. The list of classes is presented in table 1.

Table 1. The Classes of operational states of data transmission elements.

	Class name	Description	Priority of providing control actions
1	Class 1	no work required not required	low
2	Class 2	modernization is necessary for 3 years	average
3	Class 3	modernization within 1 year	high
4	Class 4	immediate upgrades required	very high

The duration of the time periods was chosen by the expert group on the basis of parameters that are of high importance during the operation of the network. The block also generates a database of production rules used for fuzzy classification. Next, the factors describing the operational state of a particular type of EDTS are selected and formalized as an array of input variables:

$$A = \{a_1, a_1, a_1, ..., a_n\} \tag{1}$$

where n - is the number of input variables.

The "fuzzy clustering block" processes data of a retrospective nature and data received in real time. These data are grouped based on the recommendations received in the expert evaluation block. When working with the fuzzy clustering algorithm, fuzzy c-means, cluster labels are formed, in the form of an array:

$$R_{cluster}^{a \times b} = R_{cluster}^{1 \times 4} \tag{2}$$

where $a = 1$ – is the number of the data vector; $b = \overline{1,4}$ – is the cluster number.

The fuzzy clustering block, at the output, provides an array of data describing the cluster centers. They serve as one of the input parameters for the operation of the fuzzy classification block.

The previously obtained array of cluster centers is passed to the "fuzzy classification block", in which, based on the data contained in it, membership functions are constructed for each of the input variables:

$$\forall t^{a_{iDSS}} \subset T_{a_{iDSS}} \exists m_{t,a_{iDSS}}(a_{iDSS}) \tag{3}$$

where $a_{i_{DSS}}$ - is the input variable with the number i_{DSS}.

The state of the EDTS is correlated with one of the four classes of the operational condition of the equipment with the dates of modernization set by experts in advance, as shown in table 1.

In the" block of preparation and implementation of the modernization project", the label of the operational status class of the EDTS is transferred and the preparation of work begins taking into account a certain class of urgency.

The approved and implemented modernization project is put on enhanced monitoring by the "EDTS state tracking unit after modernization". It monitors the status of the EDTS by re-classifying it.

In the monitoring process, the means of the "visualization block" are used, based on Kohonen maps using weighted adders:

$$f_j = h_j + \sum_{i=1}^{m} s_{ij} t_{\hat{e}\tilde{o}\,i} \tag{4}$$

where f_j is the output of the adaptive adder, h_j is the threshold, s_{ij} weight per j neuron on i input, $t_{\hat{e}\tilde{o}\,i}$ is a multidimensional input vector.

If the state of the EDTS corresponds to the class of equipment that does not require modernization, the project can be classified as successful and remove the EDTS from enhanced monitoring. Otherwise, it will require additional evaluation based on repeated clustering and classification.

5. Structuring of the decision support system for assessing the state of data transmission network elements.

The developed DSS consists of blocks containing the following functional groups: the main program blocks that use algorithms and techniques based on fuzzy logic in their work and knowledge bases formed with the involvement of an expert group and a database containing information about the state of the ESPD and replenished in real time (figure 3).

The work of the proposed DSS for assessing the state of the EDTS involves the participation of several groups of specialists of the telecom operator and takes into account the following features:

1) The expert evaluation unit is used at the initial stage, as well as at the stages of checking and adjusting the system operation in connection with the study of the effectiveness of its operation or the introduction of new operational parameters that allow for a more complete assessment of the EDTS.

2) The decision-maker (DM) performs periodic monitoring of the preparation and implementation of the EDTS modernization project. This functionality becomes available through the use of feedbacks between the DSS blocks, which make it possible to control the classification process and monitor the network after the modernization stage, automatically sending the implemented modernization project for re-analysis.

The main working time of the DSS is required by employees with different skill levels who will exchange information with the system through a human-machine interface. We will refer this group of specialists to the DSS operators.

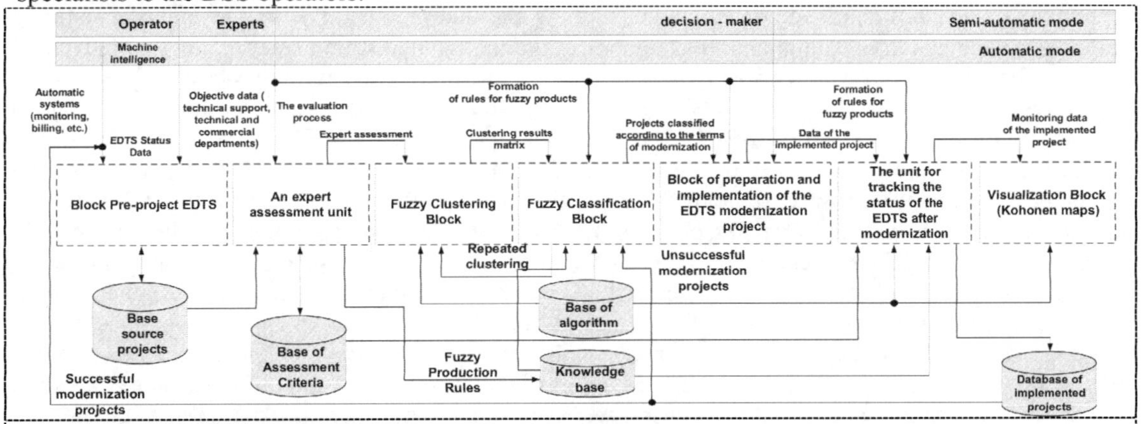

Figure 3. Block diagram of the DSS for assessing the state of EDTS.

Highly qualified specialists of the main activities of the communications enterprise, acting as experts, evaluate existing modernization projects and fill in the knowledge bases of the fuzzy models used. The DM monitors the work of the DSS, including at the stage of implementation and adaptation to a specific data transmission system, providing recommendations for improvement and configuration. The final list of personnel involved includes: DSS operators; Experts; DM.

6. Databases of the decision support system for assessing the state of data transmission system elements and their composition.

The DSS for assessing the state of the EDTS in the process of work is associated with five databases. The first database of the initial projects includes the EDTS type and retrospective operational data or real-time data. The second database with the EDTS evaluation criteria stores the threshold values of the factors that have the greatest impact, in the form of input variables for each of the EDTS. It should be taken into account that the type of equipment can affect the number and type of criteria. Depending on the type of equipment, the number and type of criteria may vary.

The third database contains software based on the algorithms used in the work of the DSS to evaluate the EDTS. Including an algorithm for calculating expert assessments to determine, by ranking, the most significant factors for the subsequent filling of the criteria database. Additionally, the database contains software based on decision-making algorithms for the development of the telecom operator's access network, monitoring the state of the EDTS after modernization, an algorithm for fuzzy clustering and classification. In the fourth database of implemented projects, data on EDTS that have undergone modernization is stored. There is information about their success, the current state and the state of the node before the upgrade. This information is further used to automate the adoption of the same type of decisions on similar EDTS. The knowledge base stores the results obtained during clustering in the form of an array containing the values of cluster centers used as an objective function in the process of classification of EDTS. The database also contains the production rules used in the evaluation of theEDTS.

7. Software components of the decision support system for assessing the state of data transmission system elements

Figure 3 shows the DSS blocks for assessing the state of the EDTS, which can be conditionally divided into subsystems for storing and processing data. The block "pre-project evaluation of the EDTS" uses an interface for interaction between the operator and the DSS for evaluating the EDTS. It enters the initial data on the state of the EDTSduring the initial registration or adjusts existing ones. In the" expert assessment block", using expert opinions, the main indicators of the operational condition are selected and evaluated according to the type of the analyzed EDTS. In the" fuzzy clustering block", the retrospective data is clustered by the fuzzyc-means algorithm according to qualitative and quantitative parameters. The array of cluster centers obtained in the clustering process is passed as an objective function to the "fuzzy classification block", which makes decisions, based on the Sugeno algorithm, on the assignment of the EDTSto a specific cluster describing the state of the EDTS. The "Block of preparation and implementation of the EDTSmodernization project" works with the data obtained during clustering, correlating them with the classes of EDTSstates formed with the participation of experts and prepares projects in accordance with the specified priorities by classes (table 1). In the "block of tracking the state of EDTSafter modernization", the modernization project is put on enhanced monitoring to determine the level of success and entering the implemented projects into the database for subsequent use. The "visualization block" uses Kohonen maps and is auxiliary, when analyzing the states of the EDTS, by specialists with different skill levels. At the initial stage, interaction with the DSS for the evaluation of the EDTStakes place with the involvement of an expert group and operators who fill the database with up-to-date information. The DM monitors and coordinates this process. When filling the database with data on implemented projects and other retrospective information received from the monitoring systems of the state of the EDTS, billing and other systems of the telecom operator, the work of the DSS goes into semi-automatic mode. In this mode, specialists monitor the implementation of modernization projects, and the analysis of incoming data and decision-making on standard projects is carried out by the DSS in automatic mode.

Providing personnel with different levels of qualification in the field of EDTSoperation with a tool that makes it possible to assess and monitor the current state of network infrastructure elements allows reducing the number of stages in the process of coordinating the restoration of node equipment operability by automating the decision-making process for typical situations. The proposed DSS for assessing the state of the EDTSon the basis of fuzzy logic allows us to take into account the borderline and pre-emergency conditions of equipment when predicting the timing of work on the modernization of the EDTS. In addition, it becomes possible to plan the level of filling of the operator's resource base based on the recommended terms of modernization.

8. Conclusion

Reducingthe level of involvement, in the process of assessing the operational conditions of typical equipment, by highly qualified personnel is achieved through the use of a decision support system based on fuzzy clustering and fuzzy classification. In this paper, a DSS was proposed for assessing the operational state of data transmission system elements using formalized expert knowledge, in the form of production rules, for equipment with similar operational characteristics. The use of the proposed DSS makes it possible to involve personnel with different skill levels, reduce the time for assessing the state of the ESPD and the costs of purchasing and installing equipment.

References

[1] Vishnevskij V M 2003 *Theoretical Foundations of Computer Network Design* (Moscow:Tekhnosfera) (In Russian)
[2] Gitman M B, Stolbov V YU, Gilyazov R L2011 *Management of socio-technical systems based on fuzzy preferences* (Moscow: Lenand) (In Russian)
[3] Goldshtejn B S, Kucheryavyj A E 2014 *Communication Network Post_NGN* (Saint Petersburg: BHV_Peterburg) (In Russian)
[4] Jyrki T J 2015 The Telecommunications Handbook: Engineering Guidelines for Fixed, Mobile and Satellite Systems. (Wiley Publ.)
[5] Rajli D Kriner M 2007 *NGOSS.Building effective support systems and network operation for the operator* (Moscow, Alpina Business books Publ.) (In Russian)
[6] Shtovba S D 2007 *Design of fuzzy systems by means of MATLAB*(Moscow, Goryachayaliniya-Telekom Publ.) (In Russian)
[7] Sorokin A A, Oleinikov A A 2018 *Support system for decision making in the process of modernization of data transmission network elements* (Samara: Infocommunication technologies vol. 16) pp 74 – 81 (In Russian)
[8] Kiseleva O I, Shubat O M2017 *Forecasting the company's revenue: methodological approaches and principles* (Materials of the International Conference Russian Regions in the Focus of Changes, Ekaterinburg) pp 597-604 (In Russian)
[9] Rogozin O V 2009 *Method of fuzzy inference solutions in the task of selecting software based on the qualitative characteristics of this security as an investment* (Quality, innovation, education, no 3) pp 43-49 (In Russian)
[10] Shtovba S D, Pankevich O D 2004 *Designing fuzzy classifiers in the MATLAB system," Designing scientific and engineering applications in the MATLAB environment* (Moscow, Proceedings of the II All-Russian Scientific Conference "Design of Scientific andEngineering applications in Matlab environment) pp 1318-1335 (In Russian)

Journal of Physics: Conference Series **2091** (2021) 012053 doi:10.1088/1742-6596/2091/1/012053

Visual odometry in local underwater navigation problems

I.M. Azhmukhamedov, P.I. Tamkov , N.D. Svishchev , A.V. Rybakov
Astrakhan State University, Tatisheva 20a, Astrakhan, Russia

E-mail: svishchev.nikolai@gmail.com

Abstract. The work processes of the ORB-SLAM algorithm are presented. The results of experimental studies on temporal comparisons of the operation of the algorithm with different parameters and cameras are presented. The necessity of forming a visual odometry (VO) system as a local navigation of remote-controlled and autonomous underwater robots has been substantiated. The two most suitable odometry methods in the underwater environment are described, such as their advantages and disadvantages. The work processes of the ORB-SLAM algorithm are presented. The results of experimental studies on temporal comparisons of the operation of the algorithm with different parameters and cameras are presented. The procedure for preparing video data is described: processing a video stream, adjusting camera parameters for calibration. The experiments represent the testing of the ORB-SLAM3 algorithm on a sample of video filmed as part of the ecological monitoring of the Caspian shelf in 2020.

1. Introduction

In the field of navigation with the use of computer vision, many methods have been developed for constructing dense three-dimensional maps based on depth cameras. The SLAM (simultaneous localization and mapping) visual system algorithms have gained great popularity in the last decade, the key factor being the system's ability to obtain metric, brightness, color, texture information about the environment and create accurate and realistic 3D maps. However, environmental mapping methods can be inaccurate due to inadequate camera projection models.

Descriptors form the core of many machine vision algorithms with real-time requirements such as object detection, visual odometry, or SLAM tasks. It is usually assumed that the resulting images, and therefore the areas highlighted around the key points, come from a perspective projection that ignores image distortion, or projections inherent in wide-angle lenses. Deviations from the ideal perspective projection are often corrected using standard compensation equations. The latter, however, cause artifacts if the camera's field of view is enlarged. In addition, many applications (eg. monocular odometry) require keypoints only on an undistorted projection.

2. Literature review

In June 2020, one of the extreme modifications of the visual odometry algorithm was published, called: ORB-SLAM 3 [1]. This method has three main parallel streams: a tracking stream, a local mapping stream, and a map compensation and merging stream. Most modern works do not take into account the features of optical distortions of lenses of used wide-angle cameras, this is evidenced by a number of works on the study of some of the most popular methods: ORB-SLAM, ORB-SLAM 2, ORB-SLAM 3 [1-3].

2.1. Camera models in SLAM algorithms

Since the first publication of the ORB-SLAM algorithm in 2015 [1] and at its extreme configuration, with the addition of data from the inertial navigation module, in 2021 [3], the authors of the works have made a significant breakthrough in the investigated problems of SLAM. This model projects hemispherical lens distortions onto a flat projection of the image, thereby generating artifacts in the frames caused by this transformation [4]. Distortion compensation methods in modern research suggest the presence of robust methods that eliminate interference when using computer vision algorithms [5]. This approach shows good results in the speed and reliability of the entire system by reducing computations and artefacts in the output image.

2.2. Current state of Visual Odometry research

Significant research interest is aimed at ensuring the autonomy of mobile systems using computer vision, namely, at self-determination of location and assessment of movement, as evidenced by a number of published works [6 - 12]. According to the tendency of publications, automatic retention of the position relative to the visually tracked object is of less interest to scientists in comparison with navigation and solving SLAM problems in conditions of an optically inhomogeneous medium [7, 9, 12]. The works related to the classification and relative orientation of autonomous mobile complexes were mainly aimed at visual tracking of objects of interest by means of digital and analog cameras [6, 11]. Basically, the authors of the works operated both with a three-dimensional cloud of points obtained using a stereo pair of cameras and using monocular vision using additional navigation aids, such as inertial navigation modules, echo sounders, laser pointers and other navigation systems [13 - 20]. The analysis of the presented works revealed the lack of results on the application of the algorithms under study in conditions of dynamic structural noise. In the overwhelming number of works, numerous methods and algorithms for recovering from interference of underwater images (color correction, contrast, separation of object boundaries, restoration of transparency and image illumination) were considered [21 - 28]. In works devoted to the restoration of the quality of underwater images, both artificial neural networks [21, 28] and classical methods of image preprocessing, as well as their combination, were used.

2.3. The future of underwater visual odometry algorithms

Due to the limited underwater visibility range, capturing images at close range is often the only solution for obtaining large visible areas of a textured surface. This limitation has motivated research to improve visual odometry techniques over the past several years. In the works mentioned above, an installation was proposed for creating two-dimensional and three-dimensional maps by obtaining images and fixing landmarks in them, along which the map is glued or the distance travelled is calculated. A common problem when mosaicking SLAM problems arises from the presence of distortion and dynamic structural noise in the environment. Strategies for dealing with such occlusions are being discussed. Another difficulty arises from the proliferation of alignment errors.

3. Investigation of methods for constructing a camera model and distortion

All SLAM algorithms require accurate camera calibration. First of all, the internal parameters of the camera / lens system are extracted, which consists of the focal length matrix and the optical centre of the camera (f_x, f_y, c_x, c_y), the rotation matrix R and displacement t, and 5 parameters of the distortion function $(k_1, k_2, k_3, p_1, p_2)$. Where k_1, k_2, k_3 are the radial distortion coefficients. Radial distortion - image distortion as a result of the non-ideal parabolic shape of the lens and the optical properties of the medium. Distortions caused by the radial distortion of the lens are equal to 0 at the optical center of the sensor and increase towards the edges, while the consideration of the radial distortion of the medium will be described below. As a rule, radial distortion contributes the most to image distortion. p_1, p_2 - coefficients of tangential distortion. Tangential distortion - image distortion caused by errors in the installation of the lens parallel to the image plane [29].

3.1. Projection of a 3D point onto a 2D image surface

In order to find the projection of a point in space onto the image plane, you first need to translate the point (X_w, Y_w, Z_w) from the world coordinate system to the camera coordinate system, taking into account external parameters (rotation matrix R and displacement t) (Figure 1).

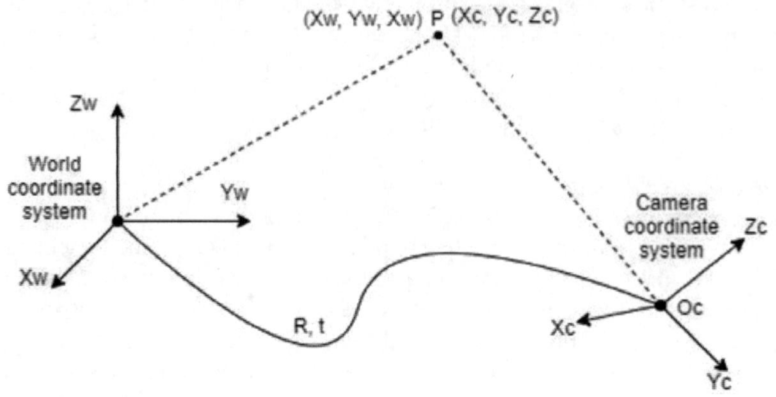

Figure 1. The world coordinate system and the camera coordinate system are linked by rotation and movement. These six parameters (3 for rotation and 3 for movement) are called external camera parameters.

Next, using the internal parameters of the camera, we project the point onto the image plane. The equations that relate a 3D point (X_w, Y_w, Z_w) in world coordinates with its projection in image coordinates (u, v) are written below:

$$\begin{bmatrix} u' \\ v' \\ w' \end{bmatrix} = P \begin{bmatrix} X_w \\ Y_w \\ Z_w \\ 1 \end{bmatrix}, \tag{1}$$

where the coordinates of the image are:

$$u = \frac{u'}{w'}, v = \frac{v'}{w'} \tag{2}$$

where P is a 3×4 projection matrix consisting of two parts: the inner matrix K, which contains the internal parameters, and the outer matrix $(R|t)$, which is a combination of the 3×3 rotation matrix R and the 3×1 transformation vector t.

$$P = K[R|t], \tag{3}$$

where the internal matrix K, is equal:

$$K = \begin{bmatrix} f_x & \gamma & c_x \\ 0 & f_x & c_y \\ 0 & 0 & 1 \end{bmatrix}, \tag{4}$$

f_x, f_y - focal lengths x and y in pixels, as a rule, have the same value. c_x, c_y - coordinates x and y of the optical center in the image plane (usually located in the center of the image). Using the center of the image is usually a good enough approximation [30], γ - the misalignment between the axes. Usually equal to 1, in most calibration packages such as *OpenCV* this factor is equal 0.

Thus, if the camera image is zoomed in or out due to certain factors, all of these parameters (f_x, f_y, c_x, c_y) will be scaled (multiplied or divided) to the same scale. The matrix of internal parameters does not depend on the type of scene, and after calculation it can be reused (provided that the focal length is fixed).

3.2. Lens distortion model

To find the parameters of the distortion function, researchers generally resort to the classical Brown-Conrady model, which describes the transition from an undistorted image to an image with distortion by the following transformations of coordinates (x_u, y_u) of a point on an undistorted image:

$$\begin{cases} x_d = x_u(1+k_1r^2+k_2r^4+\cdots)+(p_2(r^2+2x_u^2)+2p_1x_uy_u) \\ \qquad (1+p_3r^2+p_4r^4+\cdots); \\ y_d = y_u(1+k_1r^2+k_2r^4+\cdots)+(p_1(r^2+2y_u^2)+2p_2x_uy_u)' \\ \qquad (1+p_3r^2+p_4r^4+\cdots), \end{cases} \quad (5)$$

where (x_d, y_d) - coordinates of the point (x_u, y_u) on the distorted image; $r = ((x_u - c_x)^2 + (y_u - c_y)^2)^{\frac{1}{2}}$ - distance from the optical center (c_x, c_y) to point (x_u, y_u); k_i - radial distortion coefficients; p_i - tangential distortion coefficients.

In practical problems, the coefficients of tangential distortion are considered up to the coefficient p_2 inclusive, and the radial - up to coefficient k_3 or k_6 inclusive. Thus, the radial component in the tangential distortion is almost always omitted.

The *OpenCV* library (which is the world standard for computer processing of images and video in the field of free software) the following distortion model is used (6).

It should be especially noted that in *OpenCV* the coefficients of the radial distortions $k_{4...6}$ are not used as coefficients for higher degrees of the polynomial (as in the classical description (5)), and as the coefficients of a similar polynomial in the denominator. This type of radial distortion function reduces the probability of exceeding the maximum the possible value of the variable when calculating the twelfth power numbers r:

$$\begin{cases} x' = \Delta_r x + \Delta x_t; \\ y' = \Delta_r y + \Delta y_t; \end{cases} \quad (6)$$

where $\Delta_r = \frac{1+k_1r^2+k_2r^4+k_3r^6}{1+k_4r^2+k_5r^4+k_6r^6}$ - coordinate displacement caused by radial distortion; $\Delta x_t = p_2(r^2+2x^2)+2p_1xy$ - offset of the x coordinate caused by tangential distortion; $\Delta y_t = p_1(r^2+2y^2)+2p_2xy$ - displacement of the y coordinate caused by tangential distortion; $r = (x^2+y^2)^{\frac{1}{2}}$.

Thus, for modeling [31] (for the purpose of subsequent compensation) the distortion requires the following input data (7). It can get using the *calibrateCamera* function of the *OpenCV* library by images of the calibration template obtained from the calibrated camera:

$$A = \begin{pmatrix} f_x & 0 & c_x \\ 0 & f_y & c_y \\ 0 & 0 & 1 \end{pmatrix}; D = \begin{pmatrix} k_1 \\ k_2 \\ p_1 \\ p_2 \\ k_3 \\ \cdots \end{pmatrix}, \quad (7)$$

where A - camera matrix (used to normalize pixel coordinates); D - vector of distortion coefficients; (f_x, f_y) - reduced lens focal lengths along two axes.

The Brown – Conrady model allows obtaining distorted coordinates of a point from undistorted ones, and this is enough to obtain an image with compensated distortion from a distorted image.

To eliminate distortion in the image, you can simulate the distortion on a uniform rectangular point map, associate it with the distorted image, and then transform the image so that the point map becomes uniformly rectangular again.

However, the problem often arises to obtain the coordinates of a point in an image with a compensated distortion by the coordinates of a point in a distorted image without transforming the entire image (since this operation has a high computational complexity).

Ideally, it is necessary to find a solution to the polynomial, but its degree is too high, which makes an analytical solution impossible. However, a numerical solution can be found using the iterative algorithm described above. With each iteration, the values (x, y) approach to (x_u, y_u).

4. Experiments

In the visual SLAM system, one of the main problems is the selection of key points in the image that can be used as reliable reference points. These points must be invariant to rotation, movement, scaling, projection distortion, and uneven lighting. In practice, these requirements are not always met, and the problem arises of finding a match for the same key points on different images [7 - 10].

4.1. Investigation of camera calibration in underwater visual odometry

To investigate the effect of the camera model on the performance of visual odometry underwater, a SLAM algorithm called ORB-SLAM3 was used. This algorithm was chosen based on the following considerations:

- Easy to configure internal camera parameters;
- Visual influence of camera parameters on the work of visual odometry;
- Easy to customize your database;
- Options for running the algorithm in both online and offline modes;
- Various variations of the cameras used, monocular camera, stereo pair, joint work with inertial navigation system to compensate for camera turns in three corners.

The video from the GoPro 4 camera was taken as a basis, the movement of the camera was carried out by means of a remote-controlled unmanned underwater vehicle moving in the water column with a direct line of sight to the relief of the bottom of the Caspian Sea (Figure 2). The robot moved along a relatively straight trajectory, with a periodic change in the distance to the ground ± 0.5 m. The ground is a mixture of a bat and a whole shell with small bas-reliefs 10-20 cm. The transparency of the water in the thickness at the time of shooting was 2 meters, the transparency was assessed using a disk Secchi.

Figure 2. A frame of underwater video taken for testing the ORB-SLAM3 algorithm.

Given the fact that the video was shot in a high density environment, the distortion coefficients should be different from those obtained during the calibration of the GoPro camera using the *OpenCV* library.

This paper analyzes two methods for eliminating distortion in underwater images. The first, numerical, is to calibrate the camera in a standard model directly under water. The second, analytical one, consists in correcting underwater radial distortion using parameters obtained to eliminate

distortions above water, and applying a correction that takes into account the refractive index of the medium different from air.
The numerical method assumes calibration directly under water, while the second assumes the presence of only the value of the density of the medium, which is the optimal method within the framework of our problem, but within the framework of constructing a trajectory based on visual odometry, this method will lead to a rapid accumulation of errors.

After obtaining the calibration values of the GoPro 4 camera using the *OpenCV* library and entering these data into the camera configuration file of the ORB-SLAM3 algorithm (Table 1).

Table 1. GoPro Camera Model Coefficients 4.

Parameter	Parameter value
Lens model	Pinhole
f_x	458.654
f_y	457.296
c_x	367.215
c_y	248.375
k_1	-0.283408
k_2	0.073959
p_1	0.000193
p_2	1.761871

The data presented above does not take into account the radial distortion of the medium, so you can clearly see the effect of this distortion in the field (Figure 3).

Figure 3. Black dots are the key points of the frame selected for constructing the trajectory, red dots are the key points of the current frame, along which the trajectory is calculated, and the dashed line is the calculated trajectory of movement along the frames of the video stream.

Given the fact that the underwater robot is moving straight in the video, nevertheless, when building a trajectory using visual odometry, the movement curve has a strong curvature with a bulge downward, which tells us that we are moving inside the pit, which is not true. This effect can be explained by the fact that although we have the true calibration data of the camera lens, we nevertheless did not take

into account the radial distortion of the environment, which gives such an error in the analysed SLAM algorithm.
With a known refractive index of water, the correction of the distortion that occurs when the camera is immersed is described by transforming the coordinates of the image in the plane of the pinhole camera screen:

$$\begin{bmatrix} x' \\ y' \end{bmatrix} = \frac{1}{(1-(n^2-1)(x^2-y^2))^{\frac{1}{2}}} \begin{bmatrix} x \\ y \end{bmatrix}, \quad (8)$$

where $(x\ y)^T$ and $(x'\ y')^T$ are the coordinates of the object on the screen of the pinhole camera when shooting over water and under water, respectively, and n is the refractive index of water.

The calculated coefficients of the radial distortion of the medium, together with the lens distortion, give new values $k_1 = 0.262383$ and $k_2 = -0.9531$, respectively. When the algorithm is restarted, taking into account the coefficients, we will obtain the following trajectory (figure 4).

Figure 4. Black dots are the key points of the frame selected for constructing the trajectory, red dots are the key points of the current frame, along which the trajectory is calculated, and the dashed line is the calculated trajectory of movement along the frames of the video stream.

However, the accuracy of the correction calculated in this way, as well as its resistance to the error in determining the refractive index, require more detailed study. Including in the context of the most significant case from a practical point of view - underwater photography.

5. Conclusions and further research prospects

In this work, a study was carried out of the influence of the optical properties of the aquatic environment on the visual odometry algorithm of an underwater robot. This study proves the lack of robustness of the classical methods of camera calibration and accounting for environmental distortion in the operation of underwater visual odometry. Further developments involve the introduction of a generative adversarial neural network (GAN) into existing algorithms, the rejection of the camera distortion transformation model in favor of using a distortion compensation model, the rejection of the current least squares methods that are not resistant to spurious outliers when constructing graph adjacency matrices in SLAM problems in favor of improving the positioning algorithm based on graphs SLAM, this approach will allow developing robustness to outliers in SLAM problems. To achieve this goal, it is proposed to conduct the following series of complex studies on the selected methods.

The first study is aimed at general improvement of 3D mapping algorithms in an optically heterogeneous and noisy environment, with special attention to improvement methods based on graph based SLAM graphs. The second approach is to develop methods for compensating for distortion that occurs when using wide-angle lenses of a stereo pair of digital cameras and the features of an external optically inhomogeneous medium. The main method that can be relied on in this approach is the existing high-speed detector FAST (Features from Accelerated Segment Test), which, within the framework of our study, needs to be modified for wide-angle lenses. In addition, it is necessary to develop a descriptor based on the existing ORB (Oriented FAST and rotated BRIEF), taking into account distortion. Typically a fisheye camera is used to capture wide-angle and full-screen images, but such wide images are not invariant to camera rotation, that is, the distortion of the scene captured by the camera depends on the image area. On the other hand, such distortions are static, and they can be taken into account in computer vision algorithms without attempting to project onto a flat projection, preserving the integrity of the data obtained and increasing the speed of the algorithm.

The third stage is the improvement of algorithms for filtering dynamic structural noise arising due to the presence of suspended particles in the medium that reduce the computer vision field. The solution to the problem of removing image artifacts from a stereo camera is based on the use of a GAN. The traditional way of evaluating algorithms for image reconstruction and enhancement tasks is to measure distortion, the difference between true images and reconstructed images, using metrics such as mean square error (MSE), peak signal-to-noise ratio (PSNR) and structural similarity index (SSIM). However, these methods are not able to solve the problem of filtering dynamic structural noise arising due to the presence of particles of different sizes in the medium. Structural interference of this type can be divided into two classes, the first class of interference consists of very small particles that are not detected by the SLAM algorithm as key points, but strongly narrow the field of view of the camera. The second type of structural interference is large dynamically moving particles that introduce errors in SLAM navigation by creating new feature points. To eliminate the first type of interference, it is proposed to use image recovery algorithms based on GAN, while one of the difficult tasks is to minimize the computational resources required to use this algorithm. The second type of interference leads to emissions during SLAM operation and can be compensated for by an adaptive controller that takes into account the expected direction of travel and the readings of the inertial navigation system.

References

[1] Mur-Artal R, Montiel J and Tardos D 2015 ORB-SLAM: a Versatile and Accurate Monocular SLAM System *IEEE Transactions on Robotics* **31** pp 1147 – 1163
[2] Mur-Artal R and Tardos D 2017 ORB-SLAM2: an Open-Source SLAM System for Monocular, Stereo and RGB-D Cameras *IEEE Transactions on Robotics* **33** pp 1255 – 1262
[3] Campos C, Elvira R, Rodrıguez G, Montiel J and Tardos D 2021 ORB-SLAM3: An Accurate Open-Source Library for Visual, Visual-Inertial and Multi-Map SLAM *IEEE Transactions on Robotics* pp 1 – 17
[4] Kannala J and Brandt S 2006 A Generic Camera Model and Calibration Method for Conventional, Wide-Angle, and Fish-Eye Lenses *IEEE Transactions on Pattern Analysis and Machine Intelligence* **28** pp 1335 – 1340
[5] Urban S, Weinmann M and Hinz S 2017 mdBRIEF - a fast online-adaptable, distorted binary descriptor for real-time applications using calibrated wide-angle or fisheye cameras *Computer Vision and Image Understanding* **162** pp 71–86
[6] Xu Z, Liu C and Dou L 2019 3D visual sensing technique based on focal stack for snake robotic applications *Results in Physics* **12** pp 1520-1528

[7] Seonghun H, Ridao P and Sanz P 2019 Three-Dimensional Visual Mapping of Underwater Ship Hull Surface using View-based Piecewise-Planar Measurements *IFAC PapersOnLine* **52-21** pp 384-389
[8] Marco L and Stahl A 2018 Convolutional Autoencoder aided loop closure detection for monocular SLAM *IFAC PapersOnLine* **51-29** pp 159-164
[9] Lai K, Wu L, Wei Y and Lao S 2017 Two-view underwater 3D reconstruction for cameras with unknown poses under flat refractive interfaces *Pattern Recognition* **69** pp 251-269
[10] Yinghao W, Taa X, Xiaoa R and Wei Y 2019 Survey of underwater robot positioning navigation *Applied Ocean Research* **90**
[11] Jianbin H and Baoguo L 2020 Swimming target detection and tracking technology in video image processing *Microprocessors and Microsystems* **80**
[12] Ozog P, Johnson-Roberson M and Eustice R 2017 Mapping underwater ship hulls using a model-assisted bundle adjustment framework *Robotics and Autonomous Systems* **87** pp 329-347
[13] Klemen I and Gracias N 2020 Automatic scale estimation of structure from motion based 3D models using laser scalers in underwater scenarios *ISPRS Journal of Photogrammetry and Remote Sensing* **159** pp 13-25
[14] Oleari F, Fabjan K and Rizzini L 2014 Performance Evaluation of a Low-Cost Stereo Vision System for Underwater Object Detection *Proceedings of the 19th World Congress*
[15] Himri K, Ridao P and Gracias N 2018 Semantic SLAM for an AUV using object recognition from point clouds *IFAC PapersOnLine* **51-29** pp 360-365
[16] Hansen N, Christensen D and Blanke M 2015 Short-Range Sensor for Underwater Robot Navigation using Line-lasers and Vision *IFAC-PapersOnLine* **48-16** pp 113-120
[17] Bonin-Font F, Campos M and Codina G 2016 Towards Visual Detection Mapping and Quantification of Posidonia Oceanica using a Lightweight AUV *IFAC PapersOnLine* **49-23** pp 500-505
[18] Deepak K, Subudhi B, Veerakumar T and Chaudhury S 2018 Spatio-contextual Gaussian mixture model for local change detection in underwater video *Expert Systems With Applications* **97** pp 117-136
[19] Gracias N and Santos-Victor J 2014 Underwater Video Mosaics as Visual Navigations Maps. *Computer Vision and Image Understanding* **79** pp 66-91
[20] Xin S, Shi J, Liu L, Dong J, Plant C and Wang X 2018 Transferring deep knowledge for object recognition in Low-quality underwater videos *Neurocomputing* **275** pp 897 – 908
[21] Qin J, Chen Y and Wang G 2020 A novel deep neural network for noise removal from underwater image *Signal Processing: Image Communication* **87**
[22] Deepti H, Desaia C, Tabiba R, Ujwala B, and Mudenagudia U 2020 Adaptive Cubic Spline Interpolation in CIELAB Color Space for Underwater Image Enhancement *ScienceDirect Procedia Computer Science* **171** pp 52 – 61
[23] Hou G, Pan Z, Wang G, Yang H, Duan J 2019 An efficient nonlocal variational method with application to underwater image restoration *Neurocomputing* **369** pp 106-121
[24] Wang Y, Na L, Zongying L, Zhaorui G, Zheng H, Zheng B and Mengnan S 2018 An imaging-inspired no-reference underwater color image quality *Computers and Electrical Engineering* **70** pp 904 – 913
[25] Shahrizan A, Abdul G and Ashidi G 2017 Automatic system for improving underwater image contrast and color *Computers and Electronics in Agriculture* **141** pp 181-195
[26] Zeba P, Desai C, Ramesh T, Medha B, Ujwala P and Mudengudi Uma 2020 Framework for Underwater Image Enhancement *Procedia Computer Science* **171** pp 491 – 497
[27] Xiaodong L, Zhi G and Chen M 2020 Integrating physical model and generative adversarial network for underwater image enhancement *Neurocomputing* **72** pp 256 – 268

[28] Huimin L, Yujie L, Uemura T, Kim H and Serikawa S 2018 Low illumination underwater light field images reconstruction using deep convolutional neural networks *Future Generation Computer Systems* **82** pp 142 -148

[29] Kannala J 2006 A Generic Camera Model and Calibration Method for Conventional, Wide-Angle, and Fish-Eye Lenses *IEEE Transactions on pattern analysis and machine intelligence* **28** 8

[30] Zhang Z 2000 A flexible new technique for camera calibration *IEEE Transactions on pattern analysis and machine intelligence*

[31] Konovalenko I 2018 Analysis and compensation of geometric distortions, appearing when observing objects under water *Pattern Recognition and Image Analysis* **28** pp 379–392

Algorithm for dynamic formation of a single raster photo map of the seabed during the movement of an autonomous underwater vehicle

A A Timoshenko[1,2], A V Zuev[1] and E S Mursalimov [1]

[1]Institute of Marine Technology Problems of Far Eastern Branch of the Russian Academy of Sciences, 5a Sukhnova str., Vladivostok 690091, Russia
[2]Far Eastern Federal University, 8 Sukhnova str., Vladivostok 690091, Russia

E-mail: {timoshal, zuev, mursalimov} @ marine.febras.ru

Abstract. An algorithm has been developed for creating a single raster map of the seabed from photos obtained from vertically downward cameras of autonomous underwater vehicles (AUV) using tile graphics. The images obtained during the movement of AUV are combined into a single scalable photo map, divided into square segments (tiles). This representation of graphical information allows to quickly access the images with specialized tools after lifting the AUV to the surface and reduce the time spent by the operator to analyze the results of the mission. The images were combined using simple geometric transformations based on the data received from the navigation systems of the underwater vehicle and the parameters of its camera. The efficiency of the algorithm was tested on real data taken from a marine expedition.

1. Introduction

Autonomous underwater vehicles (AUVs) are already being used to perform a number of tasks under water, for example, inspection of pipelines and underwater oil production facilities, search for objects on the bottom surface, archaeological research of coastal waters, etc. To solve these problems, the AUVs is equipped with technical vision systems (TVS) (optical, hydroacoustic, etc.). The TVS helps to carry out the survey of the specified by the operator sea area. The images obtained during survey are saved on the onboard storage device. After lifting the AUV to the surface the volume of accumulated data during the execution of many hours of shooting missions is such that copying them to the operator's computer for further analysis even using wired data transfer interfaces takes several hours. The loss of this time can be critical when quickly searching for objects sunk in accidents.

For communication with a carrier vessel or a coastal control post, almost all modern AUVs are equipped with wireless data transmission interfaces [1-3], which can be divided into three main groups. The first group includes radio frequency interfaces (Wi-Fi, radio modems), which are used only when the AUV rises to the surface of the water to transmit control commands and data previously prepared on board the AUV. Due to the low penetrating ability of radio waves in the aquatic environment, it is not possible to transmit data for their visual analysis from the TVS directly to the control post through the water. In addition, even with direct visibility between the remote antennas of the control post and the AUV, the transmission of the entire volume of information collected during the mission will require much more time than using wired interfaces.

The second group includes hydroacoustic data transmission interfaces that have a sufficiently long range of communication with the AUV under water, but significantly low bandwidth. They are mainly used to monitor the state of the AUV during the execution of missions, as well as to control the vehicle using special commands. Attempts are also being made to transmit data from the on-board TVS to the control station through these communication interfaces. For example, in [4], an approach is described in which only the part of the images that contains the objects of interest to the researcher is transmitted through the hydroacoustic communication channel. It compensates the low bandwidth of the channel, but the maximum data transmission distance in the experiment was 590 m, and the control station was able to receive only 22% of the selected images, which is not enough to get a general picture of the seabed.

The third group of wireless data transmission includes underwater optical communication interfaces developed in recent years. They have a high data transfer rate, but they strongly depend on the distance between the source and the receiver, the turbidity of the water and the direct visibility [5]. Some examples allow to reach data transfer speeds of up to 12.4 Gbit / s, while the distance is only 1.7 m [6]. The paper [7] describes the exchange between the AUV and the underwater sensor node using an LED emitter at a distance of about 100 m, while the data transfer rate is about 10 Mbit/s in clear water with a straight line of sight. This technology allows real-time analysis of data from on-board TWS to the control station, but only for shallow dives.

The data collected by AUVs data is often stored in the form of a catalog with many files of photo images received from the TWS, which are viewed manually during the analysis, which requires a lot of time. Therefore, for the convenience of viewing, it is advisable to stitch all the images into a single raster map [8-10], which is used for aerial, underwater and satellite surveys. The main effective methods for combining intersecting images are the Scale Invariant Feature Transform (SIFT) [11], in conjunction with the random sample consensus algorithm (RANSAC) [12], and the Fourier-Mellin Invariant descriptor (FMI) [8, 9]. These methods make it possible to combine images using the search and comparison of common features in these images or analyzing their spectral characteristics, but their application on board the AUV is difficult due to the limited performance of the onboard computer, and the error accumulated in the process of combining. Therefore, they are used after lifting the vehicle to the surface, for example, there is a method proposed in [13] which allows to construct raster map from video frames taken at different altitudes to increase the accuracy of mosaicing. But at the same time, the AUV has to shoot the same area several times, which takes even more time.

Another example of combining images is the construction of a transect for studying the biocenosis of water spaces using specialized equipment. In [14], a system of six cameras is presented. Stretching at some depth behind a floating vehicle, it allows automatically construct a transect in the form of a photo image and a 3D model. With the help of this system, it is possible to obtain high-quality images and a model of the surveyed area, but all processing of the image sequence occurs after mission on separate computers. In addition, for the effective use of the system, it is necessary to avoid disturbances and undercurrents, which can cause errors in determining the position of the cameras in space.

Thus, the task of quick access to the photographic images accumulated during AUV's missions remains relevant. At the same time, for the convenience of data analysis, it is necessary to form a single raster map of the seabed directly on board the vehicle during the mission, which should be available without copying immediately after the AUV rises to the sea surface, when wireless radio frequency communication interfaces are usually activated. A new algorithm has been developed that uses the navigation data of the AUVs and the parameters of their cameras to ensure greater accuracy of comparing images in long-term missions, and which is able to work on standard onboard computers. At the same time, the map is planned to be stored as a set of small frames for each scale. This method of separation is used in tile graphics, where tiles are small images of the same size, from which a large one is made up. Tile graphics are widely used in geographic information systems (GIS). The map presented in this way, at first, is stored as separate images, and, at second, can be easily scaled and conveniently viewed using Web interfaces. A schematic image of the map is shown in figure 1.

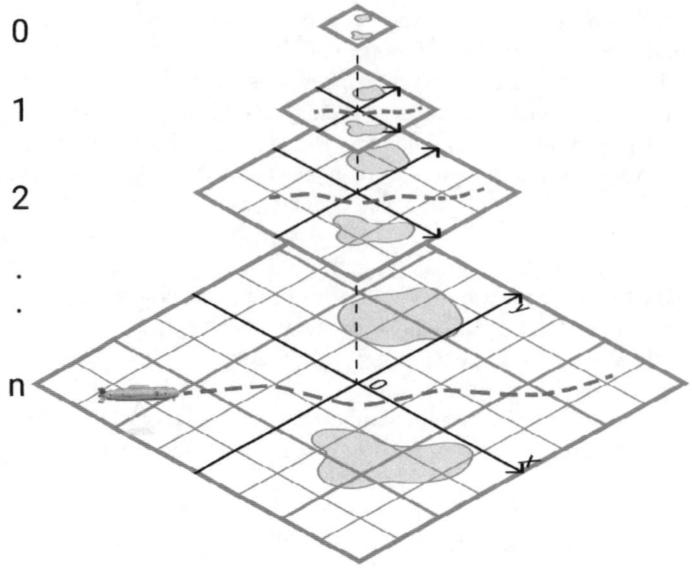

Figure 1. General view of the formed photomap.

2. Description of the algorithm

The main purpose of the algorithm is to merge individual images of a certain area of seabed into a single scalable photo map. The images that come from the TWS are combined using the navigation data of the AUV, namely the position in the horizontal plane, the height above the bottom, the angle of the course, as well as the camera parameters that determine the characteristics of the images. This is achieved by scaling, rotating, and shifting each new image, then dividing it into tiles of the same size and overlaying them on the generated map. To be able to quickly zoom the map, it is represented in several layers, each is a four times reduced copy of the previous layer. Figure 2 shows the general scheme of the algorithm. The dotted line on the diagram indicates the processes that are performed cyclically. Next, we will consider all the stages of the algorithm in more detail.

At the input, the algorithm receives an image that needs to be transformed (rotated and scaled) before being divided into tiles, so that all objects on two separate intersecting images are equally oriented and have the same size.

In general, an image is a matrix of pixels $C = R^{[k \times n]}$, the coordinates of any pixel in the image can be expressed as a vector $\mathbf{r} = \begin{bmatrix} x & y \end{bmatrix}^T$. The image is rotated and scaled with the next rotation matrix:

$$A^{[2\times3]} = \begin{bmatrix} \alpha & \beta & x_c(1-\alpha) - y_c\beta \\ -\beta & \alpha & x_c\beta - y_c(1-\alpha) \end{bmatrix}, \tag{1}$$

where $0 \le x_c \le n-1$, $0 \le y_c \le k-1$ are coordinates of the point around which the image is rotated; $\alpha = s\cos\theta$; $\beta = s\sin\theta$; θ is the angle of rotation of the image, in this case, the reverse angle of the AUV's course; s is the scale factor.

The scale factor of the image in this case depends on the height of the vehicle above bottom and can be calculated using the formula:

$$s = \frac{H_{nom}}{H} = \frac{fm}{H} \tag{2}$$

where H is the height at which the each picture was taken, $H_{nom} = fm$ is the nominal shooting height, f is the camera focus, m is the selected map scale [15].

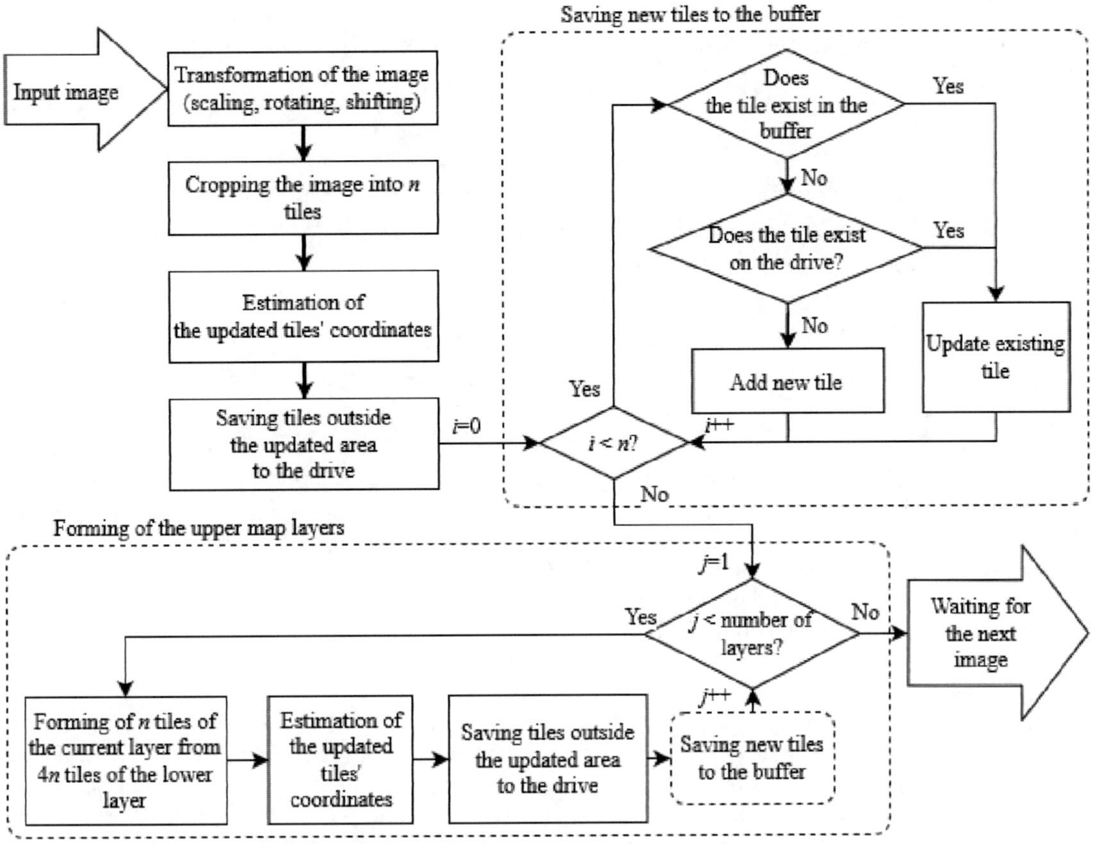

Figure 2. General scheme of an algorithm.

In computer graphics, in particular when using OpenCV tools for image processing, the corners of the image can be cropped during rotation, since the rotation matrix (1) is applied to the coordinates of the pixels of the image *C*, but the size and orientation of the image itself does not change. Therefore, in order to avoid cutting corners when rotating the image, the center of the original image is mentally aligned with the center of the square image $C = R^{\lfloor sd \times sd \rfloor}$, where *d* is the diagonal of the original image, *s* is the scale factor. Then the coordinates of the pixels of the image *C* in the coordinate system of the image C will take the form $\tilde{\mathbf{r}} = [\tilde{x} \;\; \tilde{y}]^T = [x - x_c + \tilde{x}_c \;\; y - y_c + \tilde{y}_c]^T$, where $\tilde{x}_c$, $\tilde{y}_c$ are the coordinates of the center of the image C. To rotate and scale the image, the vector $\tilde{\mathbf{r}}$ is supplemented with a 1 and the rotation matrix is applied (1):

$$r' = \begin{bmatrix} x' \\ y' \end{bmatrix} = A^{[2\times3]} \begin{bmatrix} \tilde{\mathbf{r}} \\ 1 \end{bmatrix} = \begin{bmatrix} \alpha & \beta & \tilde{x}_c(1-\alpha) - \tilde{y}_c\beta \\ -\beta & \alpha & \tilde{x}_c\beta - \tilde{y}_c(1-\alpha) \end{bmatrix} \begin{bmatrix} \tilde{x} \\ \tilde{y} \\ 1 \end{bmatrix}, \quad (3)$$

After converting the image, its location has to be found out on the photo map. To do this, the coordinates of the AUV, expressed in meters, must be converted to pixels. The number of image pixels per meter of the map at the selected scale is calculated using the formula

$$P = 1000\frac{p}{m}, \quad (4)$$

where p is the number of pixels per millimeter of the camera matrix, P is the number of pixels per meter of the photo map at the nominal scale m. coordinates of the center $\mathbf{r}_C$ of the image converted by the formula:

$$\mathbf{r}_C = [x_C \quad y_C] = \lfloor Px_m \quad -Py_m \rfloor^T, \tag{5}$$

where x_m, y_m are the coordinates of the center of the image, expressed in meters.

The tile coordinates are calculated and recorded using the transformed AUV's coordinates $\mathbf{r}_{ul}$, which contains the upper-left corner of the image:

$$\mathbf{r}_{ul} = [x_{ul} \quad y_{\mathrm{ul}}]^T = \left\lfloor \frac{x_C - \frac{d}{2}}{s} \quad \frac{y_C - \frac{d}{2}}{s} \right\rfloor^T, \tag{6}$$

where s is the length of one side of the tile in pixels. Then the image is expanded so that it fits into the tile grid and divides it into an integer number of tiles. Knowing the coordinates of the upper-left tile, later it will be easy to calculate the coordinates of the other tiles into which the image is divided.

The transformed image can now be divided into tiles and saved to the vehicles' physical storage. After splitting, each tile is assigned coordinates within its own scale. Then the resulting tiles should be written to a physical drive, but during the experiments it was shown that the process of saving tiles directly to hard drive is the slowest process in the algorithm, so instead of direct recording, a different approach was used.

Each new input image may intersect with the previous one, and when recording directly, the tiles describing the intersection area will be updated and overwritten with each new image. To reduce the number of overwrites and speed up the processing of images, only those tiles that will not be updated with a new incoming image should be saved to the drive. For this purpose, a temporary storage (buffer) is implemented in the RAM of the AUV's computer, where tiles are stored after processing the incoming image. If the tiles stored in the buffer do not fall within the area that the new image describes, they are written to the drive.

Next, tiles for smaller map scales should be formed. Separate buffers are created for each scale level. The tiles of the previous layer are used to compose each new one. First, the tiles of the previous layer are combined by four, and then their size is reduced by four times. Coordinates are assigned to the resulting tiles and, taking into account the principle described above, they are saved to the drive.

After forming a new set of basic-level tiles, the corresponding buffer is cleared: all tiles outside the area of the incoming image are saved to the drive, since these tiles are considered finished. Then a new set of tiles is added to the corresponding buffer according to the following principle. First, it checks whether there are tiles with the same coordinates in the buffer. If yes, the tiles are combined. If not, the presence of such files on the hard drive is checked. If the result is positive, new tiles should be combined with tiles on the drive, and they are added to the buffer. Otherwise, new tiles are simply added to the buffer. After that, similar actions are performed for tiles of a smaller scale. The described algorithm can be implemented effectively applied even on medium-performance AUV's onboard computers.

3. Experimental verification of the developed algorithm

For the experiments, a series of images taken by the underwater vehicle MMT-3000 at a depth of approximately 35 m was used. The shooting was carried out using the onboard Videoscan-285 camera with a frequency of 5 frames per second, an ICX285 matrix with a resolution of 1392x1040 pixels and a focal length of 8 mm. The physical size of the matrix is 8.8 mm by 6.6 mm. Thus, the number of pixels per millimeter of the matrix is approximately 158. The images are black and white, the resolution is reduced by half, i.e. to 696x520 pixels. The width and height of the tiles are set to 256 pixels. The mapping starts from the 17th scale level and goes up to 0. For each image, information about the position, orientation and speed of the underwater vehicle was saved. The operation of the algorithm has been verified on a personal computer with a 2.3 GHz processor. Images were read and written on a solid-state drive (SSD).

The table 1 shows a sample of the results of the algorithm operation. The average image processing time is 0.176 seconds, which is enough for dynamic formation of photo map with shooting frequency of the described camera. Figure 3 shows a fragment of the map obtained from experimental images.

Table 1. Results of algorithm operation

№	Image name	Basic layer (sec)	Higher layers (sec)	General time (sec)	Tiles written
1	09203831.jpg	0.141	0.027	0.168	0
2	09203835.jpg	0.155	0.034	0.189	6
3	09203839.jpg	0.225	0.024	0.249	4
4	09203843.jpg	0.19	0.033	0.223	0
5	09203847.jpg	0.223	0.042	0.265	9
6	09203851.jpg	0.209	0.029	0.238	9
7	09203855.jpg	0.204	0.044	0.248	7
8	09203859.jpg	0.207	0.034	0.241	4
9	09203903.jpg	0.147	0.051	0.198	9
10	09203907.jpg	0.139	0.03	0.169	4

4. Conclusion

There in this paper was proposed a new algorithm for combining images into a single photo map from a sequence of individual images or video frames using tile graphics and simple transformations of input images based on the navigation data of the AUV. The use of tiles allows to present the generated map in a convenient form both for a person and for the on-board information and control system of the AUV, and the use of navigation data for comparing images instead of existing methods, in addition to reducing the processing time of images, also it allows to link objects on the map to the global coordinate system with high accuracy.

Figure 3. A fragment of the generated map.

As a result of the experiments, it is clear that the developed algorithm with its implementation in the C++ language makes it possible to use it on the onboard standard computers of the AUV. At the same time, it should be noted that the quality of comparing images on a single photo card is not high, but when searching quickly, it is important to reduce the access time to the data collected during the mission with the possibility of viewing them by analysts for further decision-making. In future, it is planned to use tiles of different scales for automatic recognition of objects and determining their location directly during the movement of the AUV to inform the operator about the current progress of missions in conditions of limited bandwidth of hydroacoustic data transmission channels.

Acknowledgments

This work was supported by the Russian Foundation for Basic Research, project 20-38-70161 (development of the algorithm and its mathematical description) and the Ministry of Science and Higher Education of the Russian Federation, grant number 13.1902.21.0012 “Fundamental Problems of Study and Conservation of Deep-Sea Ecosystems in Potentially Ore-Bearing Areas of the Northwestern Pacific” (agreement no. 075-15-2020-796)” (implementation of the algorithm and conducting experimental researches).

References

[1] Johansson B, Siesjö J and Furuholmen M 2010 Seaeye sabertooth a hybrid AUV/ROV offshore system *OCEANS 2010 MTS/IEEE* (Seattle: IEEE) pp 1–3
[2] Albiez J, Joyeux S, Gaudig C, Hilljegerdes J, Kroffke S, Schoo C et al 2015 Flatfish – a compact subsea-resident inspection AUV *OCEANS 2015-MTS/IEEE* (Washington: IEEE) pp 1-8
[3] Gornak, V E, Inzartsev A V, Lviv O Yu, Matvienko Yu V, Shcherbatyuk A F and Inzartsev A V 2007 MMT-3000 is a new small-sized autonomous uninhabited underwater vehicle of the Institute of Marine Technologies Problems of the FEB RAS. *Underwater Research and Robotics* **1** pp 12-20
[4] Ahn J, Yasukawa S, Sonoda T, Nishida Y, Ishii K and Ura T 2018 An optical image transmission system for deep sea creature sampling missions using autonomous underwater vehicle *IEEE J. of Oceanic Eng.* **45** 2 pp 350-61
[5] Miroshnikova N E, Petruchin G S and Sherbakov A V 2019 Problems of Underwater Optical Links Modeling *2019 Systems of Signal Synchronization, Generating and Processing in Telecommunications (SYNCHROINFO)* pp 1-9
[6] Wu T C, Chi Y C, Wang H Y, Tsai C T and Lin G R 2017 Blue laser diode enables underwater communication at 12.4 Gbps *Sci. reports* **7** 1 pp 1-10
[7] Pontbriand C, Farr N, Hansen J, Kinsey J C, Pelletier L P, Ware J and Fourie D 2015 Wireless data harvesting using the AUV Sentry and WHOI optical modem *OCEANS 2015-MTS/IEEE* (Washington: IEEE) pp 1-6
[8] Bülow H and Birk A 2009 Fast and robust photomapping with an unmanned aerial vehicle (UAV) *2009 IEEE/RSJ Int. Conf. on Intelligent Robots and Systems* pp 3368-73
[9] Bulow H, Birk A and Unnithan V 2009 Online generation of an underwater photo map with improved fourier mellin based registration *OCEANS 2009-EUROPE* pp 1-6
[10] Ait-Aoudia S, Mahiou R, Djebli H, and Guerrout E H 2012 Satellite and aerial image mosaicing-a comparative insight *2012 16th Int. Conf. on Information Visualization* pp 652–7
[11] Fischler M A and Bolles R C 1981 Random sample consensus: a paradigm for model fitting with applications to image analysis and automated cartography *Communications of the ACM* **24** 6 pp 381–95
[12] Lowe D G 2004 Distinctive image features from scale-invariant keypoints *Int. j. of computer vision* **60** 2 pp 91–110
[13] Gracias X and Negahdaripour S 2005 Underwater mosaic creation using video sequences from different altitudes *Proc. of OCEANS 2005 MTS/IEEE* pp 1295–300
[14] Mizuno K, Tabeta S, Matsumoto Y, Sakamoto S, Sugimoto Y, Ogawa T et al 2018 Development of a towed optical camera array system (SSS: Speedy Sea Scanner) for sea environmental monitoring *2018 OCEANS-MTS/IEEE Kobe Techno-Oceans (OTO)* pp 1-5
[15] Krasnopevtsev B V 2008 *Photogrammetry* (Moscow: SCP "Reprography" MSUGaC) pp 160

Methodology for improving the decision support system in order to reduce the probability of environmental accidents at gas production industry

N M Amshinov, A M Likhter and I M Azhmukhamedov

Astrakhan State University, 20a Tatischev St. Astrakhan, 414056, Russia

E-mail: amshinov.nika@yandex.ru

Abstract. The aim of this work is to establish the key correlations between the technological, economic and environmental characteristics of gas production industry. To solve this problem, methods of mathematical modeling, fuzzy logic, and management theory were used. A generalized criterion was constructed that serves to assess the cost characteristics of measures for the implementation of management decisions developed by the decision support system and aimed at reducing the probability of environmental accidents.

1. Introduction

An important problem in the process of operation of existing and development of new gas production complexes (GPC) is to ensure their environmental safety. There are many normative documents of the federal and industry level, aimed at supporting decision-making to ensure environmental safety [5-10]. The problems of ensuring environmental safety of GPC are devoted to the works of both domestic: Glebova E. B. [12], Abrosimov A. A. [11], Dadonov Yu. A. [13], Kulechev V. M. [14], and foreign authors: Kahraman S. [15], Jiaxin Fu [16], Andie Graham [17], Dinca C. [18], Gao X. [19].

Since the 1990s, the number of technogenic accidents in the world and in Russia has been increasing because of the growing share of fire-, explosion-, and chemically hazardous technologies, which are potential sources of major industrial accidents. For example, the catastrophe at Kumzhinskoye gas-condensate field in 1980, where a gas blowout occurred during drilling and resulted in a fire. This accident resulted in subsidence and flooding of the land surface with the area of about 50 thousand m^2 by the river water, formation of three large craters with griffins of gas and condensate. The accident could not be completely eliminated, the field was mothballed [16]. Another example is the accident at the Lisichansk-Tikhoretsk oil pipeline. In 1993 a pipe burst and 408 tons of crude oil spilled out. As a result of the fire some of the oil burned, the other part accumulated in the underground and surface waters, soils and rocks, bottom sediments [15]. One of the most large-scale accidents in recent years at a hydrocarbon production facility is a fuel leak in Norilsk. The damage is estimated in billions of rubles, and it will take at best 10 years to restore the natural environment.

The above examples show that the existing in practice system of measures to monitor the environmental state of the environment, preventing and eliminating environmental incidents is not an absolute guarantee of preventing them, since eliminating their consequences is very expensive, long and not always final process.

For this reason, it is necessary to improve Decision Support System, aimed at preventing environmental accidents and ensuring environmental safety of GPC.

The importance of solving this problem is confirmed by the "Green Deal" of many countries, including the Russian Federation. Active work is currently carried out to find and implement information management systems (IMS) in order to reduce the risk of emergency situations at hazardous production facilities (HPF), which include GPC. It has been said about environmental issues in the message of the President of the Russian Federation at the April 22, 2021 climate summit: "...methane accounts for 20%

of anthropogenic emissions, and each ton of it creates a greenhouse effect 25-28 times greater than 1 ton of CO2... In this regard, it would be very important to establish broad and effective international cooperation to calculate and monitor the volume of emissions of all types of harmful emissions into the atmosphere."

2. Problem setting

The objective of this work is the creation of the Decision Support System (DSS), which provides for the establishment of relationships between technical, environmental and economic parameters of GPC facilities.

3. Relevance

Practice shows that a number of environmental accidents occur as a result of insufficient training of operating personnel, as well as the lack of sufficient modern tools of operational control of technological and environmental parameters. Consequently, in order to reduce the probability of environmental accidents in the process of construction and operation of GPC, both personnel training and the creation of a modern information-measuring system characterized by a higher, comparing with the existing one, efficiency of providing measurement information for the DSS are required. In spite of a significant number of works which consider different aspects of preventing environmental accidents, this task has not been solved to the full extent as applied to the analysis of processes which lead to extraordinary situations.

Thus, the task of developing a decision support system to reduce the risks of environmental accidents at gas production enterprises is relevant and the results obtained in the research process are new.

4. DSS Development Methodology to reduce environmental risks of GPC

Let us consider the main typical elements of GPC, namely:

1. Well;
2. Combined feed (CF);
3. Integrated Gas Treatment Unit (IGTU).

At the stage of construction and operation, each element of GPC has negative impact on the environment in a varying degree (figures 1,2,3).

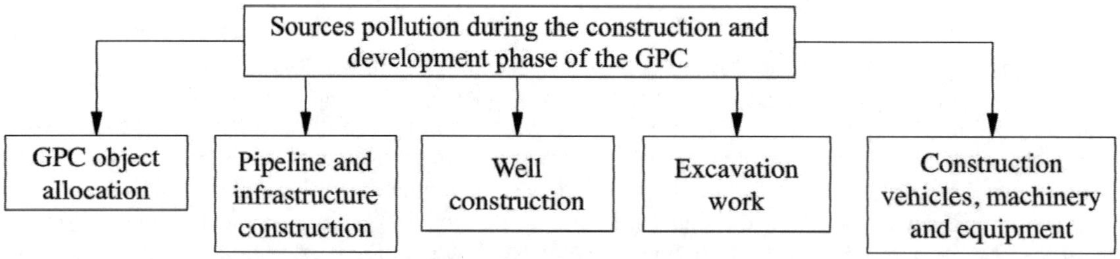

Figure 1. Sources of environmental pollution in the GPC at the stage of construction and development.

During the operation of the GPC, the main air pollutants of gas-producing enterprises are hydrocarbons. Sources of air pollution are Integrated Gas Treatment Unit, wells during their blowing, wellhead equipment, and defective internal pipelines. Also, in addition to the release of polluting substances into the atmosphere, there are a waste-water discharge, which affects living marine and river bioresources, and significant solid waste generation during emergencies and repairs, which have a negative impact on the soil, vegetation, rivers, and streams in the region.

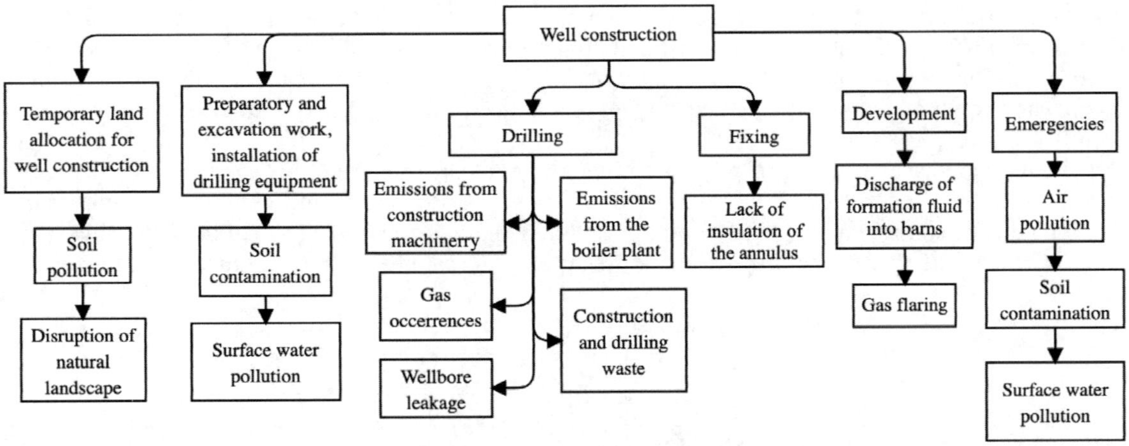

Figure 2. The list of causes and consequences of the negative impact of the well on the environment at the stage of construction.

Figure 2 lists the causes of the negative environmental impact of the well during its construction phase.

Figure 3 shows the possible sources of environmental pollution at the stage of GPC operation, created by its main elements: well, CF, IGTU.

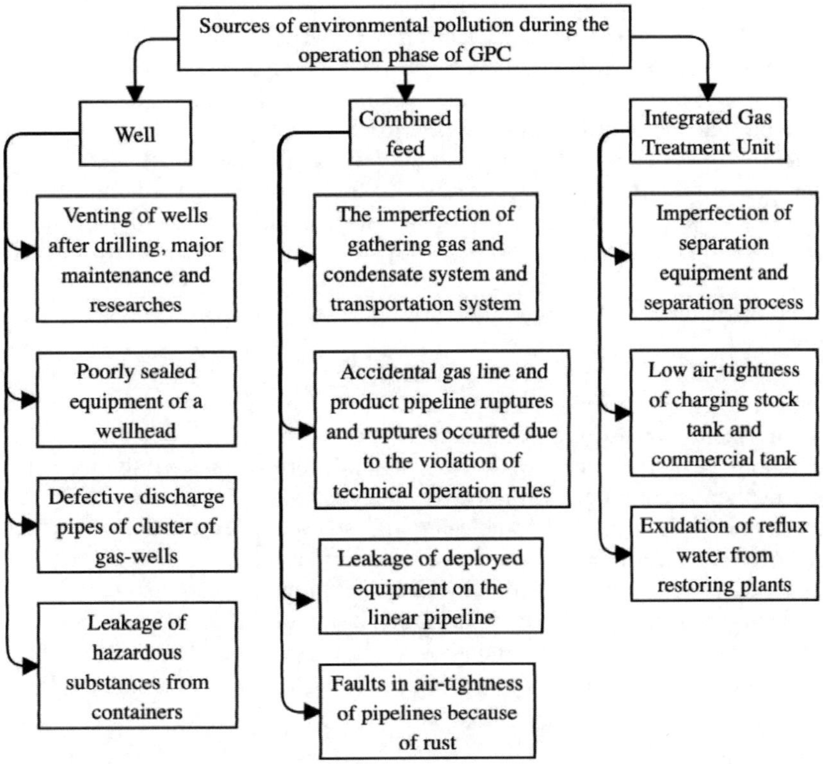

Figure 3. Sources of environmental pollution at the stage of operation of GPC facilities.

The existing information-measuring systems for monitoring the technological indicators of the gas production enterprise at the stage of its construction and operation, as well as the processes of storing, transferring and processing information are characterized by insufficient automation, digitalization and efficiency, as a result of which it can be made a managerial decision, which may lead to an environmental accident.

Despite the fact that the ecological impact on the environment from the GPC is almost inevitable, in order to prevent the transmission of environmental impact into accidents, the technological parameters of production must be controlled in a better way. Currently, the monitoring of the state of the GPC objects is carried out by sensors that transmit the results of technological parameters measurements to the control panel (Table 1).

Table 1. Main sensors for each object of the GPC.

Sensors	GPC targets		
	Well	CF	IGTU
Temperature sensor	+	+	+
Pressure sensor	+	+	+
Flow meter	+	+	+
Air pollution sensor	+		+
Sight glass			+

The currently used means of monitoring technological parameters at the main facilities of the GPC have a number of significant disadvantages, for example, the speed of response, as well as a significant measurement error. For example, pressure and temperature sensors have measurement error between 0.25 and 0.5%, ultrasonic flow meter has measurement error between 1 and 5%, air pollution sensors have measurement error in the range of 0.25-5% of the LFL (lower flammability level), sight glass - in the range of 3 mm. Obviously, this is one of the factors affecting the time, which is necessary for a decision maker (DM) to form the control action in order to prevent environmental accidents. Hence the need to modernize measuring and controlling systems as a whole. In addition to this reason, the quality of management is affected by the fact that some information for the DM comes on paper or verbally, which can lead to the ambiguity of transmitted production information and, consequently, to making insufficiently effective decisions. To eliminate this drawback, it is necessary to ensure the full digitalization of production information flows.

The proposed DSS Development Methodology to reduce environmental risks of GPC consists of the following sections:

1) Structuring the GPC;
2) Forming the causes of environmental accidents of GPC;
3) Improving the information-measuring complex of GPC;
4) Researching and improving the system of transferring and storing managerial information;
5) Developing the mathematical model of the DSS, which establishes the connection between system of gathering measurement information about technological parameters at the main facilities of GPC, environmental specifications of the enterprise and economic efficiency;
6) Creating the algorithm underlying the DSS, as well as computer program, implementing this algorithm;
7) Estimating the economic efficiency from the implementation of the developed DSS.

To fulfill point 3 of the proposed methodology, let us consider the existing model scheme of management of gas production enterprises (figure 4).

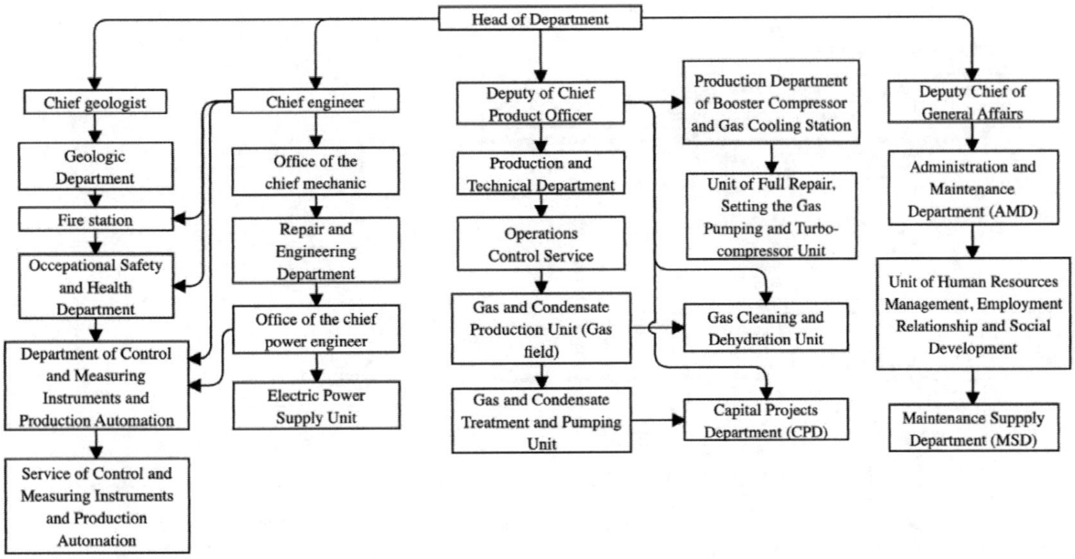

Figure 4. Model structure of management of gas production enterprises management.

The given model scheme of GPC management structure is characterized by a number of significant drawbacks:
1. A large volume of information coming to the head of the department - DM;
2. Insufficient digitalization of information flows coming from all services of GPC.

Possible ways to improve GPC management models will be described later.

Let us consider a model scheme of the information-measuring complex (figure 5). Information-measuring information at GPC is transferred with the help of the Integrated Automation Control System of the gas production enterprise (IACS EP).

IACS EP of a gas production enterprise has a multilevel territorially distributed structure. The first level is the central control office of the gas production enterprise, the second level is management information systems of gas fields, the third level is control station of units (ACS EP of gas treatment units (GTU), of the main technological component of fields; ACS EP of GTU, ACS EP of booster compressor stations (BCS), etc.) and the last level is local automation control systems (ACS of gas pumping unit (GPU), etc.). There is a volume of performed functions for each level of control. The structure of functional subsystems for each level of control is determined, through the analysis of the volume of performed functions. Possible ways of improving the information-measuring complex of GPC will be considered in the following works.

Mathematical model should include criterion, linking such parameters as pressure (P, MPa), temperature (T, K), flow rate (Q, cub. m/day), degree of automation and digitalization (ε), environmental emissions (V, m^3), assessment of economic characteristics for exact solutions (S, ths. rub.). As a result, the necessary solution to prevent the environmental accidents (EA) will be formed. In general way, it can be written:

$$\varphi' = \varphi\,(P, T, \varepsilon, V, S) \tag{1}$$

φ' – control action, φ – the function, linking technological, environmental and economic parameters.

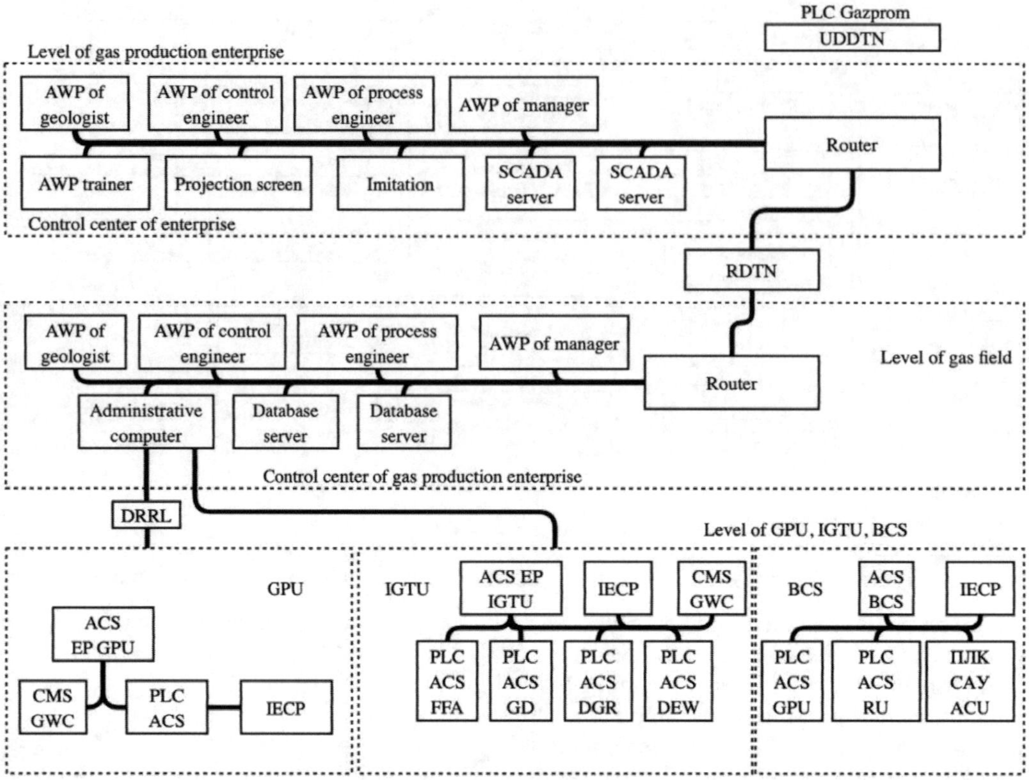

Figure 5. The structural scheme of the IACS EP of GPC

The mathematical model, being developed, will form the basis for creation of the algorithm for the DSS and the program for its implementation.

The criterion for estimating the effectiveness of implementation of the developed DSS is the difference between the DSS implementation costs for reducing the probability of environmental accidents and the actual costs for the prevention and elimination of environmental accidents (EA) in case they occur:

$$\Delta = \text{DSS implementation costs} - \text{Costs for the prevention and elimination of EA.} \quad (2)$$

To prevent environmental accidents, the DM (Head of Department) must process all incoming information, but due to its large volume there is a risk of making an incorrect decision (figure 6).

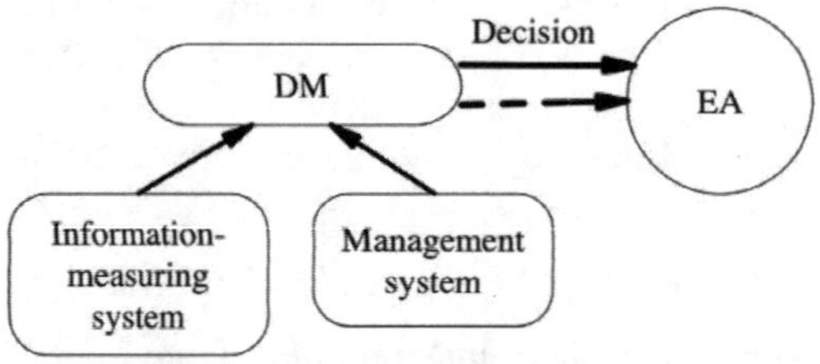

Figure 6. Illustration of the decision-making process under the existing GPC management organization

For this reason, it is necessary to introduce the DSS, which will reduce the probability of making an incorrect decision by the DM and, as a consequence, will not allow an environmental accident (figure 7).

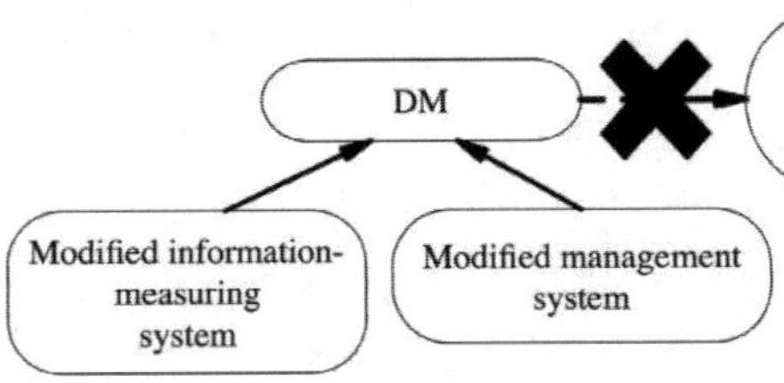

Figure 7. Illustration of the decision-making process after implementing suggested DSS into GPC.

5. Results

The above methodology can be used in the construction of a DSS aimed at preventing and eliminating the consequences of environmental accidents, as well as in the development of a mathematical model that establishes the relationship between the system for collecting measurement information about technological parameters at the main facilities of the GPC, the environmental indicators of the enterprise and its economic efficiency.

6. Conclusion

The study of existing regulatory documents and practical measures shows that for the effective prevention of environmental accidents and the elimination of their consequences, it is necessary to 1. study the possibility of transferring some management decisions to lower levels that are close to the sources of environmental risks; 2. develop methodological recommendations for actions aimed at preventing and eliminating environmental accidents, based on the analysis of expert assessments and materials on previous environmental incidents in the gas production industry at the stages of construction and operation of the GPC.

References

[1] Kudryavtseva O V, Ledashcheva T N and Pinaev V E 2014 Development of environmental audit in Russia *St. Supervision* No **3** 201-203
[2] Petrova E N, Potravny I M 2015 Environmental audit in the integration of management systems: development trends *Bulletin of Mininsky University* No **2** p 1-9
[3] Strakhova N A, Bobarykina Yu E and Smorgunova M V 2012 Methodological approaches to environmental and economic decision-making during environmental audit *New tech.* No **3** 113--118
[4] Ali Mansur Noman Marhub 2011 Expert decision support system in the intelligent system of environmental monitoring of atmospheric air in an industrial region *abstract of the dissertation of the Candidate of Technical Sciences Moscow* p 147
[5] Federal Law No. 7-FL of January 10, 2002 «On Environmental Protection»
[6] Decree of the President of the Russian Federation No 176 dated 19.04.2017 «On the Strategy of Environmental Safety of the Russian Federation for the period up to 2025»
[7] «Composition and structure of documented information on the state of the environment and its pollution» By order of the Ministry of Natural Resources of Russia dated 30.07.2020 N 510
[8] «Requirements for monitoring the state of the environment, its pollution». Approved. By order of the Ministry of Natural Resources of Russia dated 30.07.2020 N 524
[9] «Methodology for the development (calculation) and establishment of standards for permissible emissions of pollutants into the atmospheric air» By order of the Ministry of Natural Resources of the Russian Federation No 581 of 11.08.2020
[10] «Methodological guidelines for the development of draft standards for waste generation and limits on their placement» By order of the Ministry of Natural Resources of the Russian Federation No. 1021 dated 07.12.2020
[11] Abrosimov A A 2002 Ecology of processing of hydrocarbon systems M.: Chemistry p 608

[12] Glebova E V, Glebov L S 2000 Course of ecology M.: Gubkin Russian State University of Oil and Gas p 184
[13] Accidents and accidents in the Russian oil and gas industry 2001 Edited by Dadonov Yu A, Karshenbaum V Ya - ANO «tEchnoneftegaz» p 213
[14] State report «On the state of industrial safety of hazardous production facilities, rational use and protection of the subsoil of the Russian Federation in 2000» 2001 Edited by Kulecheva V M M.: State Unitary Enterprise «Scientific and Technical Center for Industrial Safety of Gosgortehnadzor of Russia» p 196
[15] Podovalov Yu A 2010 Ecology of oil and gas production. - Moscow: Infra-Engineering, - p 416
[16] Bogoyavlensky V I, Perekalin S O, Boychuk V M etc. 2017 The disaster at the Kumzhinsky gas condensate field *Causes, results, ways to eliminate the consequences Arctic: ecology and economics* No 1(25) pp 32-46
[17] Kahraman C, Uçal Sari I 2017 Intelligence Systems in Environmental Management: Theory and Applications *Intelligent Systems Reference Library 113 Springer* p 488
[18] Jiaxin Fu, Yingqi Liu, FeiHong Sun 2021 Identifying and Regulating the Environmental Risks in the Development and Utilization of Natural Gas as a Low-Carbon Energy Source. *Frontiers in Energy Research*
[19] Andie Graham, Douglas A Wilcox 2021 The impacts of Marcellus Shale gas drilling accidents on amphibians in a Pennsylvania fen. Wetlands Ecology and Management, 29 (1) pp 155-167
[20] Dinca C, Rousseaux, P, Badea A, 2007 A Life Cycle Impact of the Natural Gas Used in the Energy Sector in Romania *Journal of Cleaner Production* 15(15) pp 1451–62
[21] Gao X H, Dong G W etc. 2018 Groundwater Environmental Impact and Supervise Recommendation in Process of Shale Gas Development in China *IOP Conference Series: Earth and Environmental Science* p 191
[22] Dinca C Rousseaux P Badea A 2007 A Life Cycle Impact of the Natural Gas Used in the Energy Sector in Romania *Journal of Cleaner Production* 15(15): pp 1451–62
[23] Zhang, N Y, Zhang, Z, Rui, Z H, et al., 2018 Comprehensive Risk Assessment of High Sulfur Containing Gas Well *Journal of Petroleum Science and Engineering* 888–897
[24] Sun Y Q, Wang D, Tsang D etc. 2019 A Critical Review of Risks, Characteristics, and Treatment Strategies for Potentially Toxic Elements in Wastewater from Shale Gas Extraction *Environment International* pp 452-469
[25] Monika Wójcik, Wojciech Kostowski 2020 Environmental Risk Assessment for Exploration and Extraction Processes of Unconventional Hydrocarbon Deposits of Shale Gas and Tight Gas: Pomeranian and Carpathian Region Case Study as Largest Onshore Oilfields *Journal of Earth Science* 31(1) pp 215-222
[26] MacKinnon M A, Brouwer J and Samuelsen S 2018 The role of natural gas and its infrastructure in mitigating greenhouse gas emissions, improving regional air quality, and renewable resource integration. *Prog. Energy Combust* pp 62–92
[27] Mackay D J C and Stone T J 2013 Potential greenhouse gas emissions associated with shale gas extraction and us*e* London, United Kingdom Department of Energy and Climate Change 50
[28] McKenzie L M, Witter R Z, Newman L S and Adgate J L 2012 Human health risk assessment of air emissions from development of unconventional natural gas resources. *Sci. Total Environ* 424 pp 79–87
[29] Shonkoff S B C, Hays J and Finkel M L 2014 Environmental public health dimensions of shale and tight gas development *Environ. Health Perspect* 122 (8) pp 787–795
[30] Annika W Walters, Carlin E Girard, Richard H Walker, Aïda M Farag, David A Alvarez 2019 Multiple approaches to surface water quality assessment provide insight for small streams experiencing oil and natural gas development. Integrated Environmental Assessment and Management *15* (3) pp 385-397

Journal of Physics: Conference Series **2091** (2021) 012056 doi:10.1088/1742-6596/2091/1/012056

Building a decision support system based on fuzzy logic to reduce the total harmful emissions from a fossil fuel boiler or power plant

L I Filinkov, A M Likhter, A V Rybakov, K S Varlamov

Astrakhan State University, 20a Tatishcheva street, Astrakhan 414056, Russia

E-mail: rybakov_alex@mail.ru

Abstract. This article discusses the software implementation using the theory of fuzzy sets. To build the model, we use data on wind direction, wind strength, the distance of residential buildings and the population of the expected area of pollution distribution. To do this, the most critical parameters must be ranked such as size of population. wind speed and wind direction. We will consider them as input data. We use Scilab as the processing environment.

1. Introduction

The energy produced at thermal power plants and boiler houses (hereinafter referred to as stationary sources) is very much in demand by consumers. However, these enterprises become sources of pollution. Finding such parameters of the operation of stationary sources, in which the impact on human health and the state of the environment will be minimal, is an urgent task [1].

The analysis of environmental policy issues and decision-making is traditionally carried out using the methods of multi-criteria analysis [2]. Zadeh [3] was the first to propose fuzzy logic as a decision-making methodology. Since then, several studies have used this methodology on environmental policy issues, such as sustainability [4], environmental impact assessment [5] or environmental vulnerability [6]. In addition, Silvert [7] proposed fuzzy indices of environmental conditions, and Dimitrov [8] used fuzzy logic to manage social complexity and find consensus on environmental policy issues. In the field of environmental economics, Kunch and Springael [9] studied a system of fuzzy logic for tax policy that will reduce CO_2 emissions, as well as a fuzzy methodology for evaluating sold permits for CO_2 emissions.

Therefore, to solve such problems, it is advisable to use the theoretical apparatus of fuzzy sets, which was developed by L. Zadeh in 1965 [3]. The traditional way to represent an element of the set A is to use the characteristic function $\mu A(x)$, which is equal to 1 if this element belongs to the set A, or 0 otherwise. In fuzzy systems, an element can partially belong to any set A. The degree of belonging of an element to the set A is determined by the function $\mu A(x)$, the interval of which is equal to [0,1]. The specific value of the membership function is called the degree or coefficient of membership, which can be defined as a functional dependence or discretely as a numerical sequence. In the theory of fuzzy sets, there are linguistic variables with values attributed to them. For example, the linguistic variable temperature can take different values: "positive", "negative", "high", "low", zero". Mathematical operations have been developed for fuzzy sets, which are a generalization of similar operations performed on "clear" sets. The fuzzy inference system, on the basis of which various fuzzy control, monitoring and diagnostics systems are built for various fields of technology, uses fuzzy inference rules. The basic rule of inference of the "if – then" type is called a fuzzy implication that takes the form

If x is A, then y is B,

where A and B are linguistic values.

Fuzzy reasoning is a procedure that allows you to get a conclusion that follows from a set of "if then" rules.

Mamdani-zadeh fuzzy inference systems. The system also consists of sequentially connected blocks: a fuzzifier, a solution output block with a base of output rules, and a defuzzifier. The fuzzifier converts the set of input data into a fuzzy set at the input, and the defuzzifier converts the fuzzy sets into a specific value of the output variable. Since the system can use many fuzzy rules, it contains an aggregation block, which is implemented in the form of a logical adder (the Max operator). The Mamdani-zadeh system contains the following operators [10]:

1. The Logical or Arithmetic Product Operator for determining the resulting activation level.
2. The Logical or Arithmetic Product Operator for determining the membership function of the entire implication $A \rightarrow B$.
3. The Logical Sum Operator as an aggregator of equivalent results of the implication of many rules.
4. The Defuzzification Operator that converts a fuzzy result into a clear value of the output variable.

To solve the problem of developing a fuzzy diagnostic system, all numerical ranges of measured values (partial discharge energy densities, temperatures, concentrations of characteristic gases) must be translated into input linguistic variables, for example, "very low", "low", "average", "high", "very high". For this task of diagnosing power transformers, the output linguistic variables should be as follows:" normal mode"," maintenance (maintenance) is necessary"," repair is necessary", "emergency mode and shutdown is necessary". The rules for processing information of a fuzzy system are formulated in the following form:

If x_1 is A_1 and ...and xn is A_p, then y is B,

where A_i, B are the values of input and output linguistic variables, respectively.

Rules and algorithms for information processing are developed on the basis of existing functional analytical dependencies, statistical data and operational experience. It should be noted that it is practically quite difficult to completely obtain the processing algorithm analytically. The defuzzifier transforms a fuzzy set into a deterministic point solution, which is possible in several ways [11]:

1. Defuzzification relative to the center of the area.
2. Defuzzification relative to the middle center.
3. Defuzzification relative to the average maximum.
4. Defuzzification in the form of selecting the minimum from the maximum values.
5. Defuzzification in the form of selecting the maximum of the maximum values.

2. Materials and methods

Meanwhile, the estimate of the magnitude of the emission from a stationary source itself may not really reflect how strong and weak the impact of the emission is. For example, if the source is located far from the city, then its impact on the environment will be significantly less than from a source located in a densely populated area of the city. Similarly, the situation of the source's impact on the environment will depend on the properties of the environment at the moment to spread this pollution in space, which, first of all, will be determined by the wind speed at a given time. It follows that one of the most critical parameters is the direction of the wind, since it will depend on how much more people fall under the influence of the flue gas plume from the source of emissions.

Similarly, a number of other parameters of various kinds can be proposed, on which the environmental impact depends. Due to the fact that these parameters are qualitatively different, we suggest using the theory of fuzzy sets (or fuzzy logic) to build a decision support system. To do this, the most critical parameters, which were mentioned above, must be ranked. We will consider them as input data. We use Scilab as the processing environment.

As a stationary object, consider the Astrakhan CHP-2. According to the literature, the zone of influence of the plume is usually divided into three zones: up to 1 km, 1-5 km, and more than 5 km.

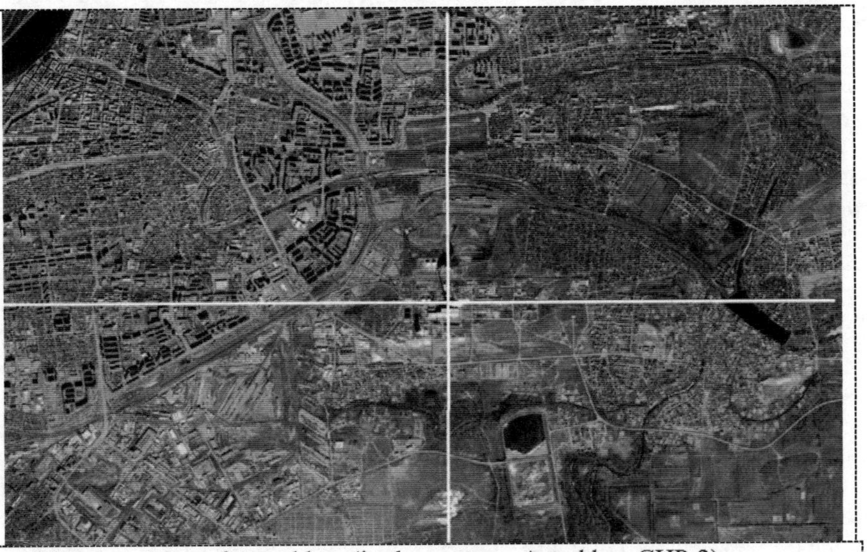

Figure 1. Map of Astrakhan (in the center – Astrakhan CHP-2).

The following zones can be identified on the map:

0-90-medium population,

91-250-small population,

251-359-densely populated area.

Next, we rank the wind speed:

0-1 m / s is the most unfavorable state, since the concentration of harmful substances is concentrated in a small area of space, at such times, the reduction in the total emission value should be especially critical;

1-10 – average speed,

more than 10 – the most favorable state (Low).

Then you need to rank the magnitude of the outlier. In order to get information about the current wind speed and direction, a program was written that implements an API request for an electronic resource (https://openweathermap.org/).

Next, we formulate a set of rules. It is obvious that if the wind speed, direction, distance and magnitude of the emission are in the "Many" zone, then these will be the most unfavorable parameters and the output signal will be "Dangerous!". Between the input data, you can choose the rule "And", or "Or", which allows you to determine the most critical parameters. If all the input parameters are at the " Low " level, the output signal "Not dangerous"is generated.

After entering all the necessary data and setting the rules connecting the inputs and outputs, it is possible to perform aggregation and defuzzification in order to obtain an accurate signal ("Yes/No") about the need to reduce the emission value of a stationary source.

Aggregation is a procedure for determining the degree of truth of conditions for each of the rules of the fuzzy inference system. In this case, the values of the membership functions of the terms of linguistic variables that make up the above-mentioned conditions (antecedents) of the cores of fuzzy production rules obtained at the fuzzification stage are used.

If the condition of a fuzzy production rule is a simple fuzzy statement, then the degree of its truth corresponds to the value of the membership function of the corresponding term of the linguistic variable.

If the condition is a composite statement, then the degree of truth of a complex statement is determined on the basis of the known truth values of the elementary statements that make up it using the previously introduced fuzzy logical operations in one of the bases specified in advance.

Defuzzification in fuzzy inference systems is the process of transition from the function of belonging to the output linguistic variable to its clear (numeric) value. The purpose of defuzzification is to use the results of accumulation of all output linguistic variables to obtain quantitative values for each output variable, which is used by devices external to the fuzzy output system (actuators of intelligent ACS).

The considered stages of fuzzy inference can be implemented in an ambiguous way: aggregation can be carried out not only in the basis of the fuzzy logic of the Task, activation can be carried out by various methods of fuzzy composition, at the accumulation stage, unification can be carried out in a different way from max-unification, defuzzification can also be carried out by various methods. Thus, the choice of specific ways to implement individual stages of fuzzy inference determines one or another fuzzy inference algorithm. At the moment, the following algorithms are most often used in fuzzy inference systems.

The Mamdani algorithm found application in the first fuzzy automatic control systems. It was proposed in 1975 by the English mathematician E. Mamdani for controlling a steam engine [12].

The formation of the rule base of the fuzzy inference system is carried out in the form of an ordered consistent list of fuzzy production rules in the form of "IF A THEN B ", where the antecedents of the cores of the fuzzy production rules are constructed using logical bundles "And", and the sequents of the cores of the fuzzy production rules are simple.

Fuzzification of input variables is carried out in the way described above, as well as in the general case of constructing a fuzzy inference system.

Aggregation of the sub-conditions of the fuzzy production rules is carried out using the classical fuzzy logical operation "And" of two elementary statements A, B : $T(A \cap B) = \min\{ T(A);T(B) \}$.

Activation of the sub-conclusions of the rules of fuzzy production is carried out by the method of min-activation $\mu (y) = \min\{c; \mu (x) \}$, where $\mu (x)$ and c are, respectively, the membership functions of the terms of linguistic variables and the degree of truth of fuzzy statements forming the corresponding consequences (consequents) of the cores of fuzzy production rules. The accumulation of sub-conclusions of the rules of fuzzy production is carried out using the classic for fuzzy logic max-combining membership functions. Defuzzification is carried out by the method of the center of gravity or the center of the area. In our case, this particular method was used.

Let's estimate the value of the concentration of nitrogen oxides in the atmosphere and compare it with the legally established norms.

According to [1], the maximum surface concentration of nitrogen dioxide, $mg \cdot m^{-3}$, from the emissions of a single point source can be calculated by the formula:

$$C_M = \frac{A \cdot M \cdot F \cdot m \cdot n \cdot \eta}{H^2 \cdot (V \Delta T)^{\frac{1}{3}}}, \qquad (1)$$

where A is the coefficient depending on the temperature stratification of the atmosphere;

M is the mass of nitrogen oxide emissions per unit of time (emission power), $g \cdot sec^{-1}$;

F is a coefficient that takes into account the sedimentation rate of harmful substances;

m, n-coefficients that take into account the conditions for the exit of combustion products from the chimney;

η - a dimensionless coefficient that takes into account the influence of the terrain;

H – height of the chimney, m;

ΔT is the difference between the temperature of the emitted gas-air sweep Tg and the temperature of the atmospheric air Tv, °C;

V is the volume of the removed combustion products, $m^3 \cdot sec^{-1}$.

According to the operating maps of the boilers of the Astrakhan CHPP-2, the concentration of nitrogen oxides in the flue gases is on average 150 $mg \cdot m^{-3}$ at various operating modes. At the maximum capacity of the power plant, the flue gas consumption is 1 million $m^3 \cdot h^{-1}$ ($278\ m^3 \cdot sec^{-1}$).

Then the power of the emission of nitrogen oxides will be:

$$M = 150\ mg \cdot m^{-3} \cdot 10^6 = 150\ kg \cdot hour^{-1} = 42\ g \cdot sec^{-1},$$

and the maximum surface concentration of nitrogen dioxide is:

$$C_M = \frac{200 \cdot 42 \cdot 1 \cdot 1{,}49 \cdot 14{,}3 \cdot 1}{220^2 \cdot (277 \cdot 120)^{\frac{1}{3}}} = 0{,}115\ mg \cdot m^{-3}$$

According to the Sanitary Rules and Norms of SanPiN 1.2.3685-21 approved on 01.03.2021 "Hygienic standards and requirements for ensuring the safety and (or) harmlessness of environmental factors for humans", the maximum permissible concentration (MPC) of nitrogen dioxide NO_2 (concentration that provides acceptable (acceptable) risk levels when exposed to at least 24 hours – the average daily) is 0.1 mg/m^3. Thus, the calculation showed that there are risks of exceeding the average daily maximum permissible concentration of nitrogen oxides in the urban area of the city of Astrakhan.

3. Results

A user of the DSS, for example, a boiler room operator, can use the regulators to set the current state of the input parameters and see what the value of the output signal will be.

If you select only two input characteristics (for example, wind speed and direction), it becomes possible to visualize the output surface of the fuzzifier depending on the input values using the built-in Scilab graphics tools. The command window allows you to generate code that corresponds to the mathematical model that is introduced into the Scilab environment

4. Discussion

As it was shown above, the rules received from the domain expert are presented in the form " If {condition}, then {action}". The rules include recommended actions to be taken when indicators exceed certain values. Below is one of these produced rules:

IF
"wind speed" = "a lot"
And "wind direction" = "a lot"
And "emission value" = "a lot",
THEN
Recommendation: "Dangerous! It is necessary to reduce the emission power of a stationary source"

After entering all the necessary data and setting the rules connecting the inputs and outputs, it is possible to perform aggregation and defuzzification in order to obtain an accurate signal ("Yes/No") about the need to reduce the emission value by a stationary source.

5. Conclusion

The possibility of reducing the current magnitude of the impact of stationary sources on the environment depends on a number of parameters, among which we can distinguish both weather and technological. Depending on the contribution of a particular parameter, often unformalized mathematically, at the level of expert opinion, you can build the logic of decision-making that the owner of the process (for example, the operator of the boiler house). The article describes an approach that allows you to create a DSS based on fuzzy logic. If necessary, the proposed mathematical model can be easily supplemented if the expert reasonably suggests taking into account and ranking many of the still unaccounted for parameters: for example, the height of the chimney, the noise impact from a stationary source, whether the surface of the earth is flat or with height differences, etc.

6. References

[1] Filinkov L I and Likhter A M 2019 *Analiz podkhodov k sozdaniyu sistemy podderzhki prinyatiya reshenii po razmenii v vzdukh v otkotel'noi* [Analysis of approaches to creating a decision support system for reducing air emissions from boiler houses] **3** pp 10-17

[2] Greening L A and Bernow S 2004 *Design of coordinated energy and environmental policies: use of multi-criteria decision-making* Energy Policy**32** pp 721–735

[3] Zadeh L A 1976 *The concept of a linguistic variable and its application to making approximate decisions* Moscow.: Mir 165

[4] Andriantiatsaholiniaina L A, Kouikoglou V S and Phillis Y A 2004 *Evaluating strategies for sustainable development: fuzzy logic reasoning and sensitivity analysis* Ecol. Econ.**48** pp 149–172

[5] Li Y P, Huang G H, Yang Z F and Nie S L 2009 *IFTCIP: an integrated optimization model for environmental management under uncertainty* Environ. Model. Assess.**14** pp 315–332

[6] Tran L T, Knight C G, O'Neill R V, Smith E R, Riitters K H and Wickham J 2002 *Fuzzy decision analysis for integrated environmental vulnerability assessment of the mid-Atlantic region* Environ. Manage **29** pp 845–859

[7] Silvert W 2000 *Fuzzy indices of environmental conditions* Ecol. Modell.130 pp 111–119

[8] Dimitrov V 1997 *Use of fuzzy logic in dealing with social complexity* Complex. Int.4 http://www.csu.edu.au/ci/vol4/dimitrov1/dimitrov.htm.

[9] Kunsch P L and Springael J 2005 *A fuzzy methodology for evaluating a market of tradable CO2-permits in Energy and Environment* SpringerUS pp 149–173

[11] Novak V and Perfileva I 1999 *Mathematical principles of fuzzy logic* Mochkorzh, J.Dordrecht: Kluwer Academic ISBN 978-0-7923-8595-0

[12] Mamdani E H and Assilian S 1975 *An experiment in linguistic synthesis with a fuzzy logic controller* Int. J. Man-Mach. Stud.7 pp 1–13

Control System of Assembly Production Organizational and Technological System of Automotive Cluster Factories

A A Shabanov [1], E E Averchenkova[2]

[1]JSC "Moscow Production Association "Metallist", 2-nd Donskoy proezd, 4, Moscow, 119071, Russia

E-mail: aashabanov86@mail.ru

[2]Bryansk State Technical University Bryansk, boulevard 50-let Oktyabrya, 7, Bryansk, 241035, Russia

E-mail: lena_ki@inbox.ru

Abstract. New term "Assembly Production Organizational and Technological System" is proposed. The model of Assembly Production Organizational and Technological System developed by the authors is described. In the model, this system is considered as a set of subsystems included in it: Assembly Tools Subsystem, Calibration Equipment Subsystem, Spare Tools and Parts Subsystem. Control system of the Assembly Production Organizational and Technological System is described. Developed by the authors model of Assembly Tools Set Element, which is a part of the Assembly Tools Subsystem is described. Developed by the authors classification of elements of assembly tools set is provided. The authors divide the assembly tools set into the following groups of equipment: torque wrenches, non-programmable screwdrivers and nutrunners, multispeed programmable screwdrivers and nutrunners, tooling and adaptors. Assembly tools for tightening threaded joints are divided into 3 groups according to the level of automation of the tightening process. These groups are divided into subgroups according to the principles developed by the authors. Developed classification of elements of assembly tools set is expedient to be used in automotive cluster assembly factories and other factories which have a similar assembly tools set structure.

1. Introduction

Assembly operations in mechanical engineering and instrumentation make up a significant, and in some cases, such as in car assembly, the main part of manufacturing process. If it is necessary to ensure the maintainability of products, detachable joints are used, of which threaded joints are most common. As noted by Gulia N V, Klokov V G and Yurkov S A in work [1], the widespread use of threaded joints is due to such advantages as manufacturability, interchangeability, versatility, reliability and mass character. The main technological equipment used in large assembly factories, including automotive clusters factories, is the assembly tools.

As noted by Lamin I I in work [2], the possibility of implementing the basic requirements for ensuring product quality, a given productivity, and obtaining the maximum economic effect is largely determined by the structure of the fleet and the parameters of technological equipment. According to the research of Abbyasov V M, Bukhteeva I V and Elkhov P E, when developing assembly technological processes, it is necessary to choose assembly equipment with an optimal level of reliability, which will provide the required productivity and economic results [3].

The issues of equipment and automation of mechanical assembly shops were considered in the works of Voronenko V P, Sokolova Y V, Skhirtladze A G and Bryukhanova V N [4, 5]. The issues of development of assembly technological processes, calculation of parameters of assembly equipment and tightening of threaded joints were considered in works [6-9]. However, in modern studies, there is no model and control system of Assembly Production Organizational and Technological System and a formalized classification of assembly equipment used in the automotive cluster factories. A review of the achievements of world research on assembly equipment for threaded joints and automated assembly factories confirms the above [10-15]. Therefore development of model and control system of Assembly Production Organizational and Technological System and formalized classification of assembly equipment used in the automotive cluster factories is important and topical task that proves the relevance of the paper. Scientific novelty lies in the development of model and control system of Assembly Production Organizational and Technological System of automotive cluster factories, model of Assembly Tools Set Element and formalized classification of Assembly Tools Set of automotive cluster factories.

2. Model of Assembly Production Organizational and Technological System of Automotive Cluster Factories

On the basis of previous works [16, 17], new term and model of Assembly Production Organizational and Technological System are proposed by authors. Assembly Production Organizational and Technological System is considered as a combination of three subsystems:

1) Assembly Tools Subsystem that includes Assembly Tools Set used in production (elements of Assembly Tools Set are screwdrivers and nutrunners, torque wrenches, adapters, bits, sockets and tooling) and IT System for Assembly Tools Set management. The Assembly Tools Subsystem directly provides the assembly process. IT System must ensure the effective functioning of the Assembly Tools Subsystem. IT System includes hardware and software that allows saving the information described in the model in subsection 5 about each element of the Assembly Tools Set.

2) Calibration Equipment Subsystem that includes Calibration Equipment and IT System for Calibration Equipment management. Calibration Equipment includes measuring benches and sensors for calibration of the assembly tools and for selective checking of tightened joints. IT System includes hardware and software intended for interaction with Assembly Tools Subsystem for recording the results of Assembly Tools Set Elements calibration, as well as controlling the measuring benches and sensors for measuring the tightening torque.

3) Spare Tools and Parts Subsystem that includes Spare Tools and Parts and IT System for Spare Tools and Parts management. Spare Tools and Parts are necessary to provide a production without stops and used in case of failure elements of Assembly Tools Set. IT System includes hardware and software that manage the stock of spare tools and parts.

IT systems of each subsystem differ in tasks and algorithmic. In large Assembly Production Organizational and Technological System, in order to ensure the efficiency of the functioning of the System as a whole, IT systems of all subsystems should be interconnected. Model of Assembly Production Organizational and Technological System is described below:

$$APOTS = \{ATS, CES, STPS\} \tag{1}$$

where

$$ATS = \{AT, IT\ AT\} \tag{2}$$

$$CES = \{CE, IT\ CE\} \tag{3}$$

$$STPS = \{STP, IT\ STP\} \tag{4}$$

where

$APOTS$ – Assembly Production Organizational and Technological System

ATS – Assembly Tools Subsystem;

AT – Assembly Tools Set;
IT AT – IT System of Assembly Tools Subsystem;
CES – Calibration Equipment Subsystem;
CE – Calibration Equipment;
IT CE – IT System of Calibration Equipment Subsystem;
STPS – Spare Tools and Parts Subsystem;
STP – Spare Tools and Parts;
IT STP – IT System of Spare Tools and Parts Subsystem.

3. Control System of Assembly Production Organizational and Technological System

Control system of Assembly Production Organizational and Technological System of automotive cluster assembly factories is shown in figure 1. The subject of management is the head of the relevant department (at assembly factories of the automotive cluster, usually it is a Head of Assembly Shop) or another decision-maker on the management of Assembly Production Organizational and Technological System. The object of management is Assembly Production Organizational and Technological System. The state of the object depends on the decisions made by the decision maker.

The decision-making strategy is as follows: decisions are made on the basis of analysis of external influences (production plan, quality level, and costs), the current state of Assembly Production Organizational and Technological System, analysis of the accumulated database about the state of Assembly Production Organizational and Technological Systems of other factories of the automotive corporation and guided by decision-making criteria. The methods and algorithms developed by the authors for the optimal management of the subsystems included in the Assembly Production Organizational and Technological System provide support for decision-making. In the process of functioning of the System, a database of information is accumulated due to feedback on the state of the subsystems. The task of the subject of management is to carry out such control actions (bold line in figure 1), taking into account all the information received, to ensure the required state of the object of management based on the developed criteria. Management is carried out on the basis of the developed criteria for each of the subsystems of Assembly Production Organizational and Technological System.

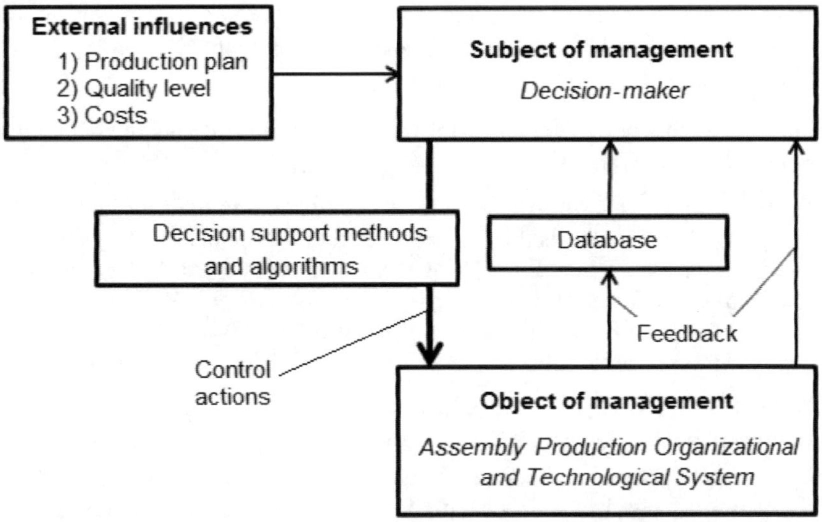

Figure 1. Control system of Assembly Production Organizational and Technological System

As shown by the author in work [17], the use of the developed control system of Assembly Production Organizational and Technological System used at automotive cluster factories can reduce production costs by more than 40% due to the optimal choice of spare tools and spare parts.

4. Model of Assembly Tools Set Element of Automotive Cluster Factories

Assembly tools directly provide assembly process. The model of Element of Assembly Tools Set is described below:

$$AT_k = \{a_k, b_k, c_k^{rp}, d_k^s, e^k, f_k, g_k^w, h_k^x, i_k, j_k^v\} \quad (5)$$

where

AT_k – k-th Element of Assembly Tools Set;

a_k – model of k-th assembly tool, $a = 1 \ldots l$, l – total number of models of assembly tools used in this Assembly Tools Set;

b_k – serial number of k-th assembly tool, serial number is the number assigned by the tool manufacturer;

c_k^{rm} – location of the k-th assembly tool in the workshop, which is characterized by the r-th sector and the m-th work station where the assembly tool is used;

d_k^s – operations of the technological process performed by the k-th assembly tool, which are characterized by the s-th code (or the name of the technological processes);

$e^k = \{e^{1k}, e^{2k}\}$ – numbers assessing the suitability of the k-th assembly tool for use, where $e^{1k} = C_p$, $e^{2k} = C_{pk}$, where C_p and C_{pk} are the process capability ratios.

f_k – k-th assembly tool torque measured during calibration;

g_k^w – history of the w-th use of the k-th assembly tool, which contains information about the date and time of entries and withdrawals from the m-th work station in the shop, $w = \{t_{\text{entry}}, t_{\text{withdrawal}}, m\}$, the parameters t_{entry} and $t_{\text{withdrawal}}$ are specified in format. DD.MM.YY, $m \in N$;

h_k^x – history of the x-th breakdown of the k-th tool, which contains information about the date and time of tool breakdown, z-th type of breakdown, $x = \{t_{breakdown}, z\}$;

i_k – degree of wear of the k-th assembly tool is described by the linguistic variable $\mathrm{LV}_k(t) = \langle \mathbb{A}_k, \mathbb{E}_k, \mathbb{W}_k, \mathbb{O}_k, \mathbb{V}_k, y \rangle$, where $\mathbb{A}_k$ – is the internal number of the k-th assembly tool; $\mathbb{E}_k$ – is a set of values $\mathrm{LV}_k(t)$ representing the terms: $\mathbb{E}_k$ = {"low wear", "moderate wear", "high wear"}; $\mathbb{W}_k$ – domain of the wear of the k-th assembly tool, set by intervals: "low wear" = [0 ... 10000] tightenings, "medium wear" = [10001 ... 100000] tightenings, "high wear" = [100001 ... +∞] tightenings ; $\mathbb{O}_k$ – is a syntactic rule; $\mathbb{V}_k$ – is a semantic rule for specifying fuzzy subsets of the set $\mathbb{W}_k$, y – is the number of the tightening, $y \in N$.

j_k^v – v-th technological devices used with the k-th tool, $\in \{G_o, U_p, K_j\}$, где G_o - o-th model of adapter, U_p - p-th model of bit or socket of k-th assembly tool, K_j - j-th model of tooling, without which the use of assembly tool is impossible. Technological devices are also an element of Assembly Tools Set, but since they are usually used with assembly tools, they are included under the index j in the model of the assembly tool with which they are currently used. This model allows accumulating and storing all the

Journal of Physics: Conference Series **2091** (2021) 012057 doi:10.1088/1742-6596/2091/1/012057

necessary information about the elements of the Assembly Tools Set for the effective assembly production management.

The model of Assembly Tools Set is shown below:

$$AT = \{AT_k | k = 1 \dots n\} \tag{6}$$

where

AT_k – k-th Element of Assembly Tools Set,

n – total number of Assembly Tools Set Elements.

5. Classification of Assembly Tools Set of Automotive Cluster Factories

In large Assembly Production Organizational and Technological System in automotive cluster factories, a large number of different assembly equipment is used. The number of Assembly Tools Set Elements is usually several thousand elements. In classification of Assembly Tools Set Elements developed by the authors, identified the following groups of assembly equipment:

1 – Torque Wrenches,

2 – Non-Programmable Screwdrivers and Nutrunners,

3 – Multi-speed Programmable Screwdrivers and Nutrunners,

4 – Adapters, Bits, Sockets and Tooling.

Assembly tools for tightening threaded joints used in automotive cluster factories are divided by the authors into groups 1st, 2nd and 3rd, depending on the level of tightening process automation and design of the tool. Different types of failures are typical for these groups. Types of failures are differently distributed in time and differently affecting the torque – main parameter of the assembly joint. Correspondence of real torque to its calculated nominal value ensures the quality of assembly joint. The equipment of each group is designed to perform certain operations of the technological process with different parameters. Multi-speed programmable screwdrivers and nutrunners are usually used to make the most critical threaded joints that have a great impact on such important product properties as safety and reliability. Non-programmable screwdrivers and wrenches are used to make less critical joints that do not have a big impact on the safety and reliability of the product. Torque wrenches are the main joint checking means in Assembly Production Organizational and Technological System, which do not use multi-speed programmable tools. Torque wrenches are used to tighten and check critical assembly joints. Thus, the proposed classification of equipment allows dividing the used assembly equipment into groups according to the types and importance of the threaded joints performed. The 4th group can be singled out separately, since adapters, bits, sockets and tooling are used with all tools and are part of assembly tool in the process of product assembling.

Non-programmable screwdrivers and nutrunners were divided into subgroups according to the following criteria:

1 – Design,

2 – Engine type.

Classification of non-programmable screwdrivers and nutrunners is shown in table 1.

Table 1. Classification of non-programmable screwdrivers and nutrunners (second group) used in Assembly Production Organizational and Technological System of automotive cluster assembly factories

	Index		Index	
1 - Design	A	Screwdrivers	A_1	With pistol handle
1 - Design	A	Screwdrivers	A_2	With straight handle
1 - Design	B	Nutrunners	B_1	With pistol handle
1 - Design	B	Nutrunners	B_2	Straight tool
1 - Design	B	Nutrunners	B_3	Angle tool
2 - Engine type	A	Battery		
2 - Engine type	B	Pneumatic precise		
2 - Engine type	C	Pneumatic impact		

For example, battery non-programmable angle nutrunner in the developed classification will have the designation $2_{1B\_1}{}^{2A}$, in this designation 2 – group of non-programmable screwdrivers and wrenches, 1B_1 – angle nutrunner, 2A – battery tool.

Multi-speed programmable screwdrivers and nutrunners were divided into subgroups according to the following criteria:

1 – Design,

2 – Type of power supply and connection with controller.

Torque wrenches are wrenches with a built-in torque gauge. Torque wrenches were divided by authors into the following 4 subgroups:

A – torque wrenches used with adapter;

B – torque wrenches with built-in adapter;

C – torque wrenches with scale;

D – measuring torque wrenches.

For example, a torque wrench with a built-in adapter in the developed classification will have the designation 1B. In this designation, 1 is a group of torque wrenches (first one), B is the type with built-in adapter. Torque wrenches are divided into subgroups in terms of their functional use. In conveyor production, torque wrenches types A and B are mainly used. Torque wrenches with adapter (subgroup A) are used for most joints. Wrenches with built-in adapter (subgroup B) are used to tighten hard-to-reach connections. In many factories, torque wrenches are the primary means of final tightening of threaded joints, and these tools are used to check all joints.

Classification of multi-speed programmable screwdrivers and nutrunners is shown in table 2.

Table 2. Classification of multi-speed programmable screwdrivers and nutrunners (third group) used in Assembly Production Organizational and Technological System of automotive cluster assembly factories

	Index		Index	
1 - Design	A	Screwdrivers	A_1	With pistol handle
1 - Design	A	Screwdrivers	A_2	With straight handle
1 - Design	B	Nutrunners	B_1	With pistol handle
1 - Design	B	Nutrunners	B_2	Straight tool
1 - Design	B	Nutrunners	B_3	Angle tool
2 – Type of power supply and connection with controller	A	Battery wireless		
2 – Type of power supply and connection with controller	B	Cable power supply and connection		

For example, a programmable angle nutrunner with connection between the tool and the controller via cable in the developed classification will have the designation $3_{1B\_3}^{2B}$. In this designation, 3 - multi-speed programmable screwdrivers and wrenches, 1B_3 - angle nutrunner, 2B - and cable power supply and connection. Multi-speed programmable screwdrivers allow programming the speed of joint tightening. These tools are electrically driven and consist of controller and assembly tool. The main difference between a programmable assembly tool and a non-programmable one is that this type of assembly tool allows tracing the history of all tightenings performed, because it allows storing data of parameters of several thousand last tightening. Traceability of the parameters of the screwing operation performed, which is present in programmable tools, makes it possible to identify a possible defect at the production stage and therefore make it possible to correct it at the lowest cost, according to the rule of tenfold costs. This rule says that if an error is made at one stage of the product's production which is revealed at the next stage, then it will take 10 times more costs to fix it than if it was found in time. If an error was discovered after one stage, then 100 times more costs will be required, etc. [18]. Therefore, programmable tools, when used correctly, are able to provide a high level of product quality and reduce the cost of identifying and correcting defects.

The above classification of assembly tools confirms that Assembly Tools Set should be considered as a complex system, with a large number of different elements, characterized by varying automation level.

6. Conclusions

The tasks of managing Assembly Production Organizational and Technological System are the efficient use of financial resources for the purchase of equipment and spare parts to provide production without stops ensuring a successful result for the headquarter of the automotive corporation that owns the automotive cluster factory. The developed control system and models of Assembly Production Organizational and Technological System, Assembly Tools Set Element and the classification of Assembly Tools Set make it possible to solve these tasks by simplifying the management work at the automotive cluster factories and reducing the negative influence of factors associated with the insufficient level of knowledge of decision-makers. The developed Classification of Assembly Tools Set Elements is recommended for use primarily at automotive assembly factories. Also, this Classification is advisable to use at other enterprises with a similar composition of Assembly Tools Set. As new design options for assembly tools are developed, this classification can be easily expanded, which also confirms its ease of use.

Scientific novelty lies in the development of model of Assembly Production Organizational and Technological System of automotive cluster factories, control system of Assembly Production Organizational and Technological System, model of Assembly Tools Set Element of automotive cluster factories, formalized classification of Assembly Tools Set of automotive cluster factories.

7. Reference

[1] Gulia N V, Klokov V G and Yurkov S A 2004 *Detali mashin (Machine Parts)*, Moscow: Izd. Tsentr Akademiya

[2] Lamin I I 2008 Optimization of park structure and parameters of assembly equipment in automotive manufacture *Izvestiya MGTU MAMI* No. **2 (6)** pp 279-287

[3] Abbyasov V M, Bukhteeva I V and Elkhov P E 2009 Preselection and optimization of reliability of automatic assembly equipment *Assembly in mechanical engineering, instrument making* No. **5** pp 28-33

[4] Voronenko V P and Sokolova Ya V 2011 Heuristic algorithm for facility layout problem in object-locked areas in machining workshops with serial production *Mining information and analytical bulletin (scientific and technical journal)* No. **S4-6** pp 43-47

[5] Voronenko V P, Skhirtladze A G and Bryukhanov V N 2001 *Machine-building Production* (Moscow: Higher school, Publishing Center "Academy")

[6] Bukhteeva I V, Abbyasov V M and Elkhov P E 2013 Directed selection of a configuration of flexible automated assembly line for assembling of gearbox shafts *Izvestiya MGTU MAMI* vol. **2** No. 1 (15). pp 134-140.

[7] Lamin I I and Abramchenkov N V 2008 Simulation model of preproject calculation of assembling equipment parameters *Izvestiya MGTU MAMI* No. **2** (6) pp 274-279

[8] Grigoryeva N S 2009 Flexible module assembling technologies *Assembly in mechanical engineering, instrument making* No. **4** pp 11-19

[9] Chechuga A O 2019 Features of development of technological processes of automated and robotized assembly *Izvestiya of the Tula State University. Technical science* No. **9** pp 555-559

[10] Zhang Q, Xie Z, Liu Y and Liu H (2019) Development of Bolt Screwing Tool Based on Pneumatic Slip Ring In: Yu H, Liu J, Liu L, Ju Z, Liu Y, Zhou D (eds) *Intelligent Robotics and Applications. ICIRA 2019. Lecture Notes in Computer Science* **11740**. *Springer, Cham.* https://doi.org/10.1007/978-3-030-27526-6_40

[11] Chumakov R (2008) Optimal control of screwing speed in assembly with thread-forming screws. *Int J Adv Manuf Technol* **36** pp 395–400 https://doi.org/10.1007/s00170-006-0839-1

[12] Perween S, Zaheer A and Khalid R (2013) Classification and Balancing of an Automotive Assembly Line. In: Qi E, Shen J, Dou R (eds) *International Asia Conference on Industrial Engineering and Management Innovation (IEMI2012) Proceedings. Springer, Berlin, Heidelberg* https://doi.org/10.1007/978-3-642-38445-5_44

[13] Poparad H (2017) Advanced Automotive Assembly Line Trends as Tools in Optimizing Production Line Performance. In: Chiru A, Ispas N (eds) *CONAT 2016 International Congress of Automotive and Transport Engineering. CONAT 2016. Springer, Cham.* https://doi.org/10.1007/978-3-319-45447-4_65

[14] Bedeoui A, Benhadj R, Trigui M and Aifaoui N (2021) Evaluating Assembly Sequences with the Assembly Tools Operation Space Criteria. In: Kharrat M, Baccar M, Dammak F (eds) *Advances in Mechanical Engineering, Materials and Mechanics. ICAMEM 2019. Lecture Notes in Mechanical Engineering. Springer, Cham.* https://doi.org/10.1007/978-3-030-52071-7_16

[15] Mishra A, Deb S (2016) An Intelligent Methodology for Assembly Tools Selection and Assembly Sequence Optimisation. In: Mandal D K, Syan C S (eds) *CAD/CAM, Robotics and Factories of the Future. Lecture Notes in Mechanical Engineering. Springer, New Delhi.* https://doi.org/10.1007/978-81-322-2740-3_31

[16] Shabanov A A 2015 Solution of the optimization problem of equipment parameters inspection frequency *Questions of radio electronics / General technical series* Issue **10**, *Moscow: AO TsNII Elektronika* pp 105-114

[17] Shabanov A A 2015 Development of rational choice method of structure and composition of spare assembly tools and accessories for mechanical assembly system *Questions of radio electronics / General technical series* Issue **10**, *Moscow: AO TsNII Elektronika* pp 115-124

[18] Gissin V I 2003 Quality Management (2nd edition) *(Moscow: ICC "Mart", Rostov-n / D: Publishing Center "Mart")*

3D-modeling of Augmented Reality objects using Shi-Tomasi corner detection algorithms

O M Demidenko, N A Aksionova, A V Varuyeu and A I Kucharav

F. Skorina Gomel State University 104 Sovetskaya street, Gomel 246024, Belarus

Abstract. This article covers development of the Python-based software module for Blender 3D, as well as it covers research of Shi-Tomasi corner detection algorithm using the developer's construction documents. The corners detected may be used for further three-dimensional modelling, replanting not requiring adjustment of the construction documents, or may be used for retrieval of the accurate data.

1. Introduction

Blender 3D uses "GoodFeaturesToTrack" modelling support module derived from the elaborated utilities library of BluePrint SDK to facilitate the process of 3D-modelling. The basis of the module lies in the function of corner detection of developer's layouts for further modelling of architectural designs, buildings, houses, apartments, and etc. This process provides for facilitation of development of a three-dimensional model, ensures introduction of initial adjustments in the further three-dimensional model not requiring changes in the construction documents themselves. One further peculiarity of this product is that the final layouts serve as the AR marks, which are used for implementation of the mobile application for viewing the prepared three-dimensional model.

2. Shi-Tomasi algorithm

This algorithm is used for search for and detection of view angles, which are to be in the grayscale. View angles are mainly identified as regions with large changes in the gradient intensity in all possible dimensions and directions. When analyzing a digital image, variation in pixel light intensity (x,y) is suggested to be measured by shifting the square window with the center in (x,y) by one pixel in each of eight directions (up, down, rightwards, leftwards and in four diagonal directions). The window size is often selected equal to 3x3, 5x5 or 9x9 pixels. Thus, for each of the directions (u,v), where

$$(u,v)\in\{(1,0),(1,1),(0,1),(-1,0),(-1,-1),(0,-1),(1,-1)\}.$$

The light intensity variation is calculated as follows:

$$V_{u,v}(x,y)=\sum_{\forall a,b}\left(I(x+u+a,y+v+b)-I(x+a,y+b)\right)^2,$$

where I(x,y) is the pixel light intensity (x,y) of the original image.

Window W for Image I (usually its size is selected equal to 5x5 pixels, but it may be selected different depending on the image size) with the center in point (x,y), as well as the shift of the window by (u,v) (figure 1) are considered below.

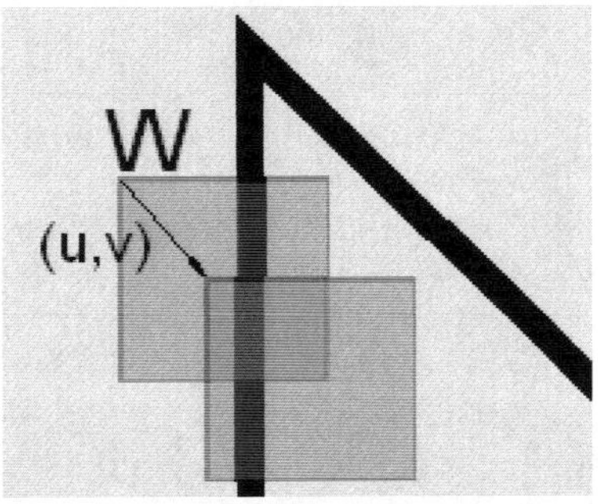

Figure 1. Example of the viewed window W and its shift (u,v).

Then the weighted total of the squared difference between window W and window W shifted by (u,v) (i.e. change in the vicinity of point (x,y) after shift by (u,v)) is:

$$E(u,v)=\sum_{(u,v)\in W} w(x,y)(I(x+u,y+v)-I(x,y))^2 \approx \sum_{(u,v)\in W} w(x,y)(I_x(x,y)u-I_y(x,y)v)^2,$$

where W(x,y) is the weight function (Gaussian function or binary window is usually used).

Function E (u,v) needs to be maximized for corners detection. To determine the derivatives Taylor's theorem is applied to the above equation and, by using certain mathematical steps, the bellow is received:

$$E(u,\upsilon)\approx\begin{bmatrix} u & \upsilon \end{bmatrix} M \begin{bmatrix} u \\ \upsilon \end{bmatrix},$$

where M is the autocorrelation matrix:

$$M=\sum_{(u,v)\in W} w(u,v)\begin{bmatrix} I_x^2 & I_x I_y \\ I_x I_y & I_y^2 \end{bmatrix},$$

where Ix and Iy are the derivatives in the directions.

The corner is characterized with large changes in function E(x,y) with respect to numerous directions (x,y), which is equivalent high in module proper values of matrix M. The arrangement of proper values is demonstrated in figure 2.

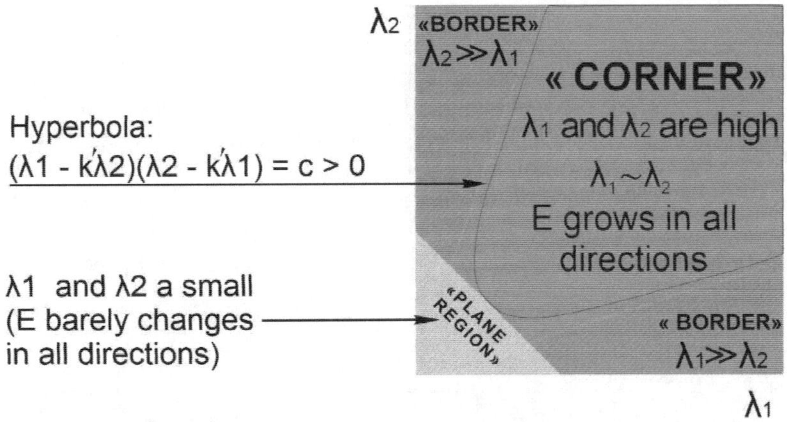

Figure 2. Changes in the values of function E (x,y) at corner detection.

Since direct calculation of proper values is demanding regarding the resources, the following response measure is suggested:

$R=\min(\lambda_1,\lambda_2)$.

Thus, the values of such proper values determine, if the region is a corner, a border or a plane by the following criteria:

1. When | R | is small, which occurs, when λ1 and λ2 are small, the region is a plane.
2. When R < 0 at λ1 >> λ2 or vice versa, the region is a border.
3. When R is high at high λ1 and λ2 and λ1~λ2, the region is a corner.

If the value is above the predetermined threshold, then the point is deemed to be a corner and thus to be a point of interest. Figure 3 demonstrates that only when λ1 and λ2 are above the minimum value λmin, the point is deemed to be a corner (green region).

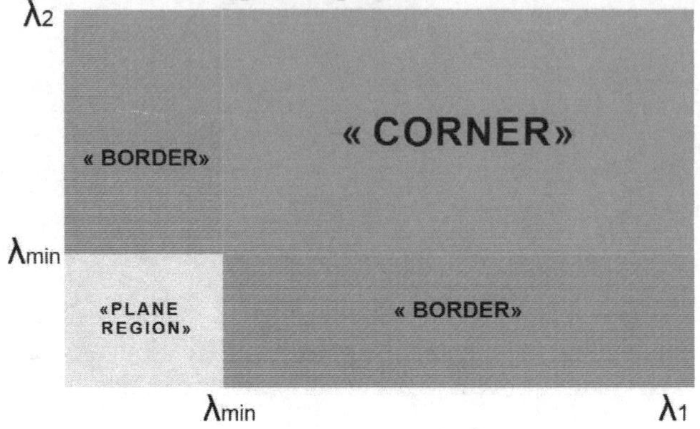

Figure 3. Determination of the values at corner detection.

3. Using OpenCV library

In order to make computer vision easier to use, OpenCV library was developed. This library contains functions that allows to process an image with a set of operations, without need to keep in mind of how images are stored in memory. Examples of these processes: get gray scale version of the image, resize an reposition part of images, filter images by colors, detect contours, and many other. Using this library enables a programmer to build a computer vision algorithm with ease. Because of library itself is written in C++, it is extremely efficient, but it doesn't require a programmer to use C++ language, as there are bindings to other languages. Bindings are libraries that delegate function calls from one programming language to another.

One of possibilities that this library provides is corner detection. It is introduced as two functions: "Harris corner detention" and "Shi-Tomasi feature tracker". By the simplest possible definition, a corner is a point in space (two-dimensional in our case), where two edges are joined.

Harris corner detection algorithm is implemented as mathematical functions, that can be explained on high level. First, algorithm calculates a function which is able to determine in which direction the uniqueness of the window changes the most. After that, algorithm finds direction in which it should move the window to maximize uniqueness difference. It also find direction in which uniqueness changes the least. Having that, algorithm calculates, how much uniqueness change in both of these two directions.

At last, algorithm has to determine how different these values are, and how big they are, at the same time. When both values are small, this means that image almost doesn't change in any direction, which indicates a flat region. When one of values is much greater than the other, it means that by moving window in one direction changes uniqueness a lot, while moving in second direction doesn't change uniqueness of the window at all, which, in turn, indicates an edge: moving along the edge doesn't change uniqueness, moving across the edge changes uniqueness a lot. When both values are big, it indicates that moving in both of these directions effectively changes uniqueness of the image, which indicates a corner.

In order to calculate difference between these two values while also considering their size, algorithm must provide a scoring function, which returns a number denoting how much a certain part of an image looks like a corner. This is where Harris and Shi-Tomasi algorithms differ.

Shi-Tomasi algorithm suggests a different scoring function that is not that precise, but is a good approximation and is calculated much faster on computer processors. Speed is important because this function needs to be calculated for each pixel on the image. Function itself is pretty simple: it just takes the lowest of two values (figure 4):

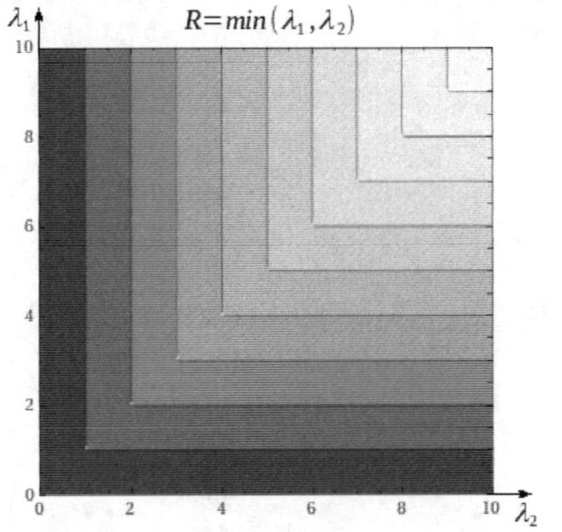

Figure 4. Plot for Shi-Tomasi algorithm corner scoring function.

Obviously, some information is lost due to such approximation. For example, values that correspond to flat regions are not that well defined now, and mixed along with values that correspond to edge regions. But because this algorithm is aimed on corner detection, such inaccuracy can be omitted.

In OpenCV library, Shi-Tomasi corner detection algorithm is implemented and wrapped into "GoodFeaturesToTrack" function. First function parameter is an object of a gray scale image. This function adds some logic to the algorithm. First of all, it picks N best corners detected on the image. Amount of desired corners to be detected is supplied as a second function parameter. Another words, it converts image matrix to a list of coordinates. Second, it applies score threshold value to remove worst corners from the list. Threshold value is supplied as a third function parameter. Third, this function is capable of removing corners that are too close to each other, and this value is supplied as a fourth function parameter.

Before image is processed by corner detection algorithm, it must be converted to gray scale color space. This is also done by one of the OpenCV function, which is capable of converting images between color spaces, considering gamma conversion and color intensities.

4. "GoodFeaturesToTrack" module operation

Modelling assistance is implemented in form of addon for Blender 3D. Belnder 3D provides a python library "Blender Python" ("bpy" for short). This library is imported into python code and enables a developer to interact with all Blender functions programmatically.

In order to create an addon, developer must implement a specific addon interface that is provied by bpy. This implementation should provide Blender with information about addon's name, description, type, author, and a functions to execute. In return, blender provides us a context object as a function parameter, which represesnts current state of everything in this Blender session. If these requirements are not satisfied, Blender refuses to import a script as an addon. After creating 3D corner points, the program tries where the walls of the building are.

```
class ImportImageWithCorners(bpy.types.Operator):
    """Script to track corners on a reference image""" # Tooltip
    bl_idname = "object.track_image_corners"
    bl_label = "Track corners"
    bl_options = {'REGISTER', 'UNDO'}
```

First step is to check if scene is set up properly for this addon to be used. Addon assumes that scene has an imported reference image, and this image is selected. Addon executes a couple of checks:

whether there's a selected object, whether this object is a reference object or not, does this object has a reference to image, and whether this image is imported correctly. Reference objects in Blender 3D are implemented as extended "Empty" objects. Empty object in blender is something that takes place in 3D space, but doesn't represent a 3D model. Image reference is one of possible types of empty objects. Even though that selected object might me an image, it might not have an actual image attached to it, so this must also be checked.

```
def execute(self, context):
        empty = context.active_object
        if not empty:
            self.report({"WARNING"}, "Nothing is selected!")
            return {"CANCELLED"}
        if empty.type != 'EMPTY':
            self.report({"WARNING"}, "This object is not an empty!")
            return {"CANCELLED"}
        if empty.empty_display_type != 'IMAGE':
            self.report({"WARNING"}, "This empty object is not an image!")
            return {"CANCELLED"}
        try:
            filepath = empty.data.filepath
            dimensions = empty.data.size
        except AttributeError:
            self.report({"WARNING"}, "Object has no img ref!")
            return {"CANCELLED"}
```

Second step, addon gets filepath to the image and loads it into the OpenCV library. It converts this image to greyscale, and executes corner detection function. This function recieves three parameters: number of corners, quality of corners and minimal distance between corners.

```
img = cv2.imread(filepath)
gray = cv2.cvtColor(img,cv2.COLOR_BGR2GRAY)
corners = cv2.goodFeaturesToTrack(gray,
self.corners, self.quality, self.distance)
if corners is None or not np.size(corners): return {'FINISHED'}
```

These parameters are also declared as addon's properties, so when a user executes the function, he's able to alter these parameters by entering values into the appeared dialog. Each value has either decimal or floating point type, maximum, minimum and step values:

```
corners: bpy.props.IntProperty(
        name="Number of corners",
        min=1, default=25
    )
    quality: bpy.props.FloatProperty(
        name="Quality of corners",
        min=0.00001, max=1, default=0.04
    )
    distance: bpy.props.IntProperty(
        name="Distance between corners (px)",
        min=1,  default=10
    )
```

When this addon is used, the following dialog appears to enable user to adjust values used for corner detection (figure 5):

Journal of Physics: Conference Series **2091** (2021) 012058 doi:10.1088/1742-6596/2091/1/012058

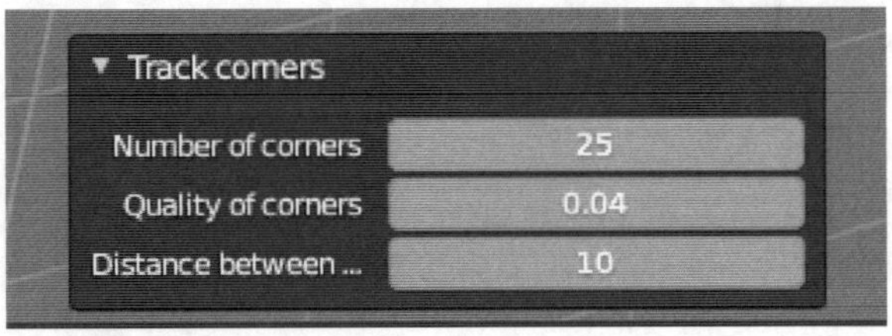

Figure 5. User interface for corner detection.

Corner detection function returns an array of coordinates, that represent corners in space of image's pixel coordinates, having point of origin at the top right corner. In Blender 3D space, for X and Y coordinates, in metric, and point of origin is historically at the bottom right corner. This means that corner coordinates must be mapped into the Blender space.

In order to do that, size of the reference image must be known. However, Blender 3D API only provides one dimensional size of the reference image. Through means of documentation and experiments, image was discovered to have its longest dimension align with the size of the dimension image. For example, when a square image is loaded, it's size is set to 5 by default, meaning at the image will be 5 units long and 5 units high, regardless of its resolution. If the image's width is two times larger than it's height, then in 3D space image will be 5 units wide and 2.5 units high. This means we will have to determine the bigger dimension of the image.

It should be also kept in mind that by default, image has offset to its origin point, 50% by height and 50% by width, making image's point of origin in it's center. This offset can also be read from Blender API.

The formula has the following form:

$$\boldsymbol{f}(\boldsymbol{x}) = \frac{(\boldsymbol{x}/\boldsymbol{d_X} + \boldsymbol{O_X}) \cdot \boldsymbol{S} \cdot \boldsymbol{d_X}}{\boldsymbol{max}(\boldsymbol{d_X}, \boldsymbol{d_Y})}$$

$$\boldsymbol{f}(\boldsymbol{y}) = \frac{(\boldsymbol{y}/\boldsymbol{d_Y} + \boldsymbol{O_Y}) \cdot \boldsymbol{S} \cdot \boldsymbol{d_Y}}{\boldsymbol{max}(\boldsymbol{d_X}, \boldsymbol{d_Y})}$$

In this formula, x and y are input coordinates, dX and dY are image dimensions in pixels, OX and OY is image offset in percent, S is size of the reference image. This formula maps into the following code, considering that additional dimensions provided by corner detection function should be stripped, and than each point should also have a zeroed Z coordinates as we map points from 2D space to 3D space:

```
e_size = empty.empty_display_size
offset = empty.empty_image_offset
dim = dimensions
f = lambda x: (x / dim + offset) * e_size * dim / np.max(dim)
corners = np.apply_along_axis(, 1, corners)
corners = np.append(corners * [1,-1], np.zeros((len(corners), 1)), axis=1)
```

Corners are now mapped into 3D space. However, these coordinates are local, meaning that transformations like moving, rotating or scaling, are not applied yet. However, this can be achieved by assigning a parent object to the created object. Blender 3D will automatically apply all transformations of the parent image to the created object.

During object creation, a couple of utility operations must be executed: mesh should be created and assigned to the new object, the new object mast be put into the main and default collection of objects, object must be selected by plugin convention:

```
mesh = bpy.data.meshes.new(mesh_name)
obj = bpy.data.objects.new(mesh.name, mesh)
col = empty.users_collection[0]
col.objects.link(obj)
obj.select_set(True)
bpy.context.view_layer.objects.active = obj
```

```
mesh.from_pydata(corners, [], [])
obj.parent = empty
return {'FINISHED'}
```

When object is created, parent object is assigned, and corners are imported into the mesh. Because we do not create edges or faces based on the vertices, empty arrays are supplied as arguments. At last, addon must return an exit code indicating that addon execution has gone successfully.

Due to current addon implementation, each time user changes one of the parameters, addon's main function is executed. It means that while user changes the slide input, this function is executed continually which can cause performance issues as the addon uses some performance heavy operations like reading file from disk into video memory. In order to improve addon performance, image should only be read once by introducing lazy static space for storing reference to this image, but this lies ot of scope of this project.

First, the user specifies the number of corners to be detected. Then the user specifies the quality level from 0 to 1, which means the minimum quality of the corner, all the corners bellow which are rejected. The minimum distance between the corners is ensured by Euclidean space. Upon specification of such information, the function detects corners on the image. All corners below the quality level are rejected, the function then sorts the remaining corners by quality from highest to lowest.

The function then takes the first strongest corner, discards all the corners in the vicinity within the range of the minimal distance and returns N strongest corners. Thereafter, the software places for each detected corner a primitive point, the vertex, in the three-dimensional space, which is within the specified building layout, and visually straightens with the layout corners. The user shall use it as a basement for grounding the individual building model.

Once three-dimensional corner points are created, the software tries to determine, where the building walls are. If the space on the line between two points is of dark color, the space is deemed to have the wall. The software creates borders between the points, where necessary (figure 6). Now, the user may interact with the created model: specify the thickness of all or of certain walls using modifiers, squeeze the building in vertical direction to reach the required height, place windows and door frames using logical operators or repeat processes for each building level. Thus, the exact model of any complex building may be created in less than a day, if the user has access to the layout diagram.

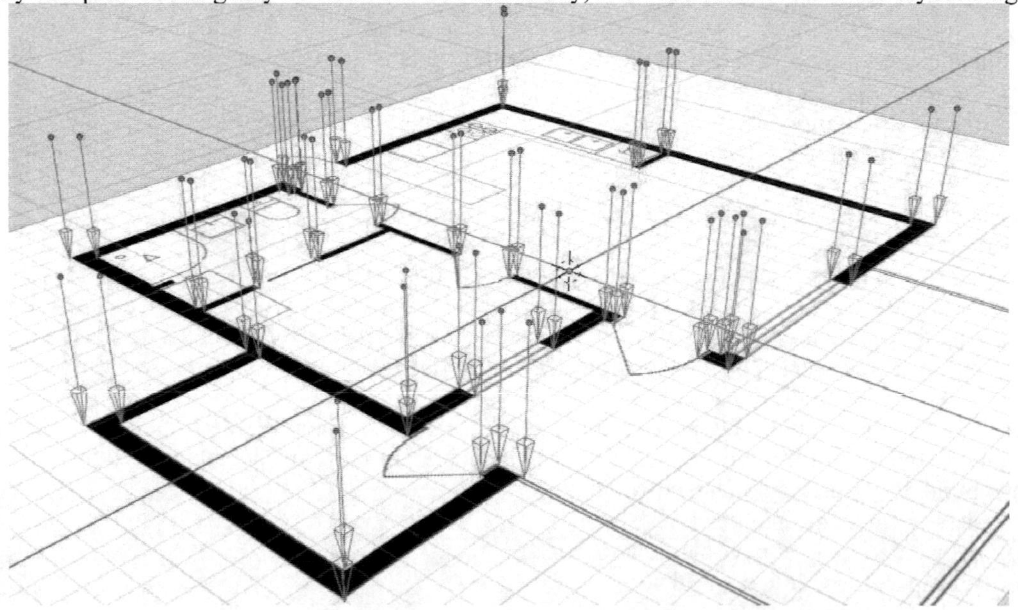

Figure 6. Results of operation of "GoodFeaturesToTrack" module in Blender 3D.

5. Conclusions

The conducted research allowed for design and development of the software module for performance of three-dimensional modelling in Blender 3D based on Shi-Tomasi corner detection algorithm. "GoodFeaturesToTrack" module is included in the developed BluePrint SDK and is one of the main modules of the mobile application of augmented reality.

Acknowledgments

The reported study was funded by State Research Programs Belarus, project number 20212185.

References

[1] Stefan Leutenegger, Margarita Chli and Roland Siegwart 2011 BRISK: Binary Robust Invariant Scalable Keypoints *ICCV* pp 2548-55

[2] Harris C and Stephens M A combined corner and edge detector 1988 In Fourth Alvey Vision Conference, Manchester, UK pp 147-151

[3] Aksionova N A and Kucharav A I 2021 Development of SDK for a mobile application using augmented reality technology *Bulletin of F. Skorina Gomel State University* **3(126)** pp 81-4.

[4] Varuyeu A V 2021 Information capacity of objects in conjunction with augmented reality *Bulletin of F. Skorina Gomel State University* **3(126)** pp 92-95

[5] Aksionova N A and Kucharav A I 2021 Presentation and implementation of educational material using augmented reality technology *Electronic collection of materials of the International scientific conference «Trends and prospects for the development of science and education in the context of globalization»* **71** pp 235-39

Control system of intelligent buildings based on convolutional neural networks

Yu. A. Lezhnina[1], N. S. Maltseva[2]

[1] *Moscow Polytechnic University, Bolshaya Semyonovskaya str., 38, Moscow, Russia*
[2]*Astrakhan State Technical University, Tatishcheva 16, Astrakhan, Russia*

E-mail: lejninau@mail.ru

Abstract. In recent years, with the development of "Internet of things" appeared another concept – smart home environment, which is defined as the physical infrastructure to operate of ambient intelligence. For the effective operation of such facilities requires to develop of new methods and approaches for adaptive control. In this paper, we analyzed the parameters characterizing the state of the control object, identify the necessary set of measured parameters obtained from the sensors, and an informative set of input parameters. We design method of training of neural network.

1. Introduction

At the moment there is progress in the application of deep learning networks. This progress has been just in the last 3-5 years. Many of the experts characterize this development as a breakthrough. On a neural network replaced more of components and systems of recognition of speech and images of giant commercial companies as Google, Apple, Facebook. In Google's neural networks make some responsible handling of the search and identify the relevance of information.

Based on the results of Russian and foreign researchers [1, 2, 3, 4, 5, 6, 7, 8, 9, 10, 11, 12, 13, 14], given them from 2005 to 2020, the use of AI technology in building management allows to save energy on average from 21.81 to 44.36%, and increase comfort on average from 21.67 to 85.77%. A complete overview of the methods most widely used in the management of smart buildings is given in [15]. Neural networks are most often used researchers among scientific papers. Fuzzy logic is also widely used to save energy and increase thermal comfort, as it allows to linguistically describing the sensations of thermal comfort sa fuzzy rules. Recently, to increase comfort while reducing energy costs, researchers have been developing hybrid methods (for example, by combining FL and ANN; ANN and GA/or PSO; FL and GA/or PSO)

Another important component of intelligent building management systems are sensors. Feedback from a large number of sensors is provided to collect a sufficient amount of data and coordinate the entire operation of the building management system. Recently, sensors have also become smarter, cheaper and more compact. They need to be selected for certain functions and develop new ones, which is the topic of separate research [13, 14, 16, 17].

The application of artificial intelligence methods and personalized comfort models in smart buildings is a promising area of research. And improving the productivity of AI-based management is still an urgent task associated with collecting a large amount of high-quality real data, which is not enough in buildings, especially in the energy sector. [18, 19, 20, 21].

2. Control system

Providing more convenience and comfort of the people lead to the rapid developing of intelligent buildings, in which control system is one of the main systems. In recent years, with the development of "Internet of things" appeared another concept – smart home environment, which is defined as the physical infrastructure to operate of ambient intelligence. For the effective operation of such facilities requires to develop of new methods and approaches for adaptive control. With the advent of methods for training deep network, it became possible to use them for control of complex objects. Thus the task of developing of control systems on based on deep neural networks is an actual task.

Understanding the concept of thermal comfort and finding ways to quantify a fuzzy situation as comfortable or uncomfortable led to a standardized analytical approach. With this approach, thermal comfort is assessed by the Predicted Mean Vote (PMV), which predicts the average value of thermal sensation on a standardized scale for a large group of people. Now the PMV index is adopted by the ISO 7730 standard. As the best level of thermal comfort, PMV at level 0 with a limit of 0.5 is recommended. The range of values is assumed to be from -3 to +3, which corresponds to the following values of sensations: Cold, Cool, Slightly Cool, Neutral, Slightly Warm, Warm, Hot. A key feature of the control system that takes into account the subjective feelings of residents is feedback from sensors. A network of sensors is being developed to collect various types of variables that are stored in a database for later use by the system for decision-making. These sets differ from the sets of intelligent sensors that are used for intelligent control (figure 1).

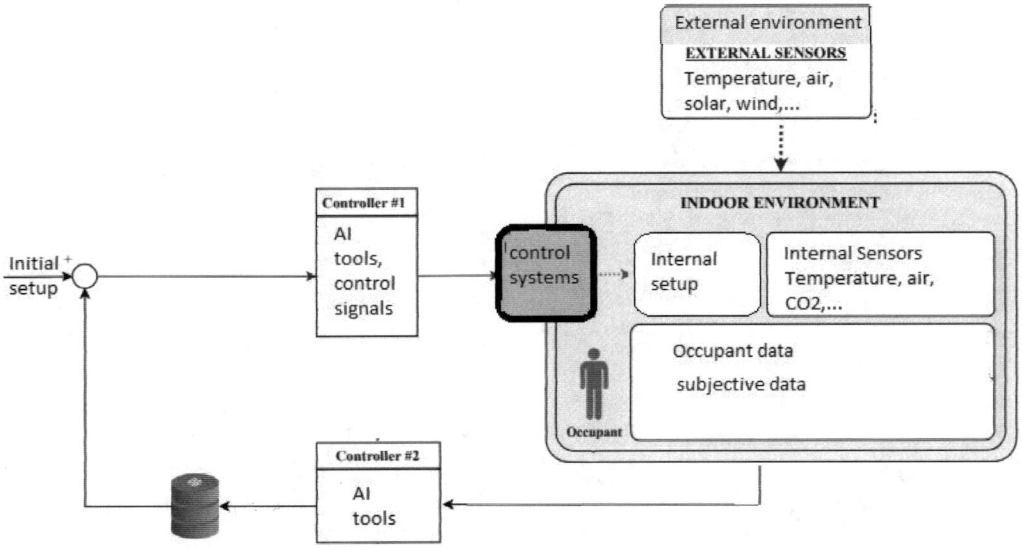

Figure 1. The diagram of the intelligent control system.

Considering the proposed scheme, it is possible to distinguish inputs and outputs in the context of thermal comfort and energy efficiency management. Inputs are integrated into seven groups: heating, ventilation and air conditioning (HVAC) systems, indoor and outdoor climatic conditions, occupant-related variables, building geometry and components, among others. The outputs can be categorized into four general groups: Comfort indices (PMV, PPD, and others), Microclimate indicators (temperature, CO_2 concentration), Energy/Load (HVAC, cooling/heating, cool/hot water), and Others (including cost and time efficiency). The

Journal of Physics: Conference Series **2091** (2021) 012059 doi:10.1088/1742-6596/2091/1/012059

relationship between the selected inputs and outputs shown as numbers the times when a specific input is used to approximate certain outputs (figure 2).

Among the approaches that use automatic feature selection, we choose convolutional neural networks. Their main advantages are the ability to work with the original signal and a great generalization ability, which can contribute to the stability of the classifier to outliers. Automation of feature selection is achieved through a combination of convolutional and fully connected layers responsible for feature selection and classification.

In a convolutional neural network, in addition to fully connected layers, there are convolutional layers, as well as pooling layers. Combinations of such layers make it possible to efficiently extract features from the input data, significantly reducing the amount of information processed, while preserving information specific to the classification task. When working with thermal comfort data, thermal diagrams compiled according to the frequency of use of inputs and outputs of smart home sensors can be used as input of the convolutional network (figure 2). In this case, the task is reduced to classifying images corresponding to the levels of perceived comfort.

Inputs		Outputs										
		Comfort Index			Microclimate		Energy				Value	
		PMV	PPD	Others	Tempe -rature	CO2	HVAC	Air	Water	Gas	Cost	Ti
space	Component Efficiency	18	2	1	10		12	9	2	1	2	4
	Set-point temperature	16	2	11	11		30	6	2	1	3	4
Climat	Air temperature (indoor)	40	5	30	40	3	57	32	4	15	9	6
	Relative humidity (indoor)	34	7		22	1	35	12			6	2
	Air Velocity (indoor)	21	5		4	2	11	9			2	1
	CO2 (indoor)	19	4		20	4	39	14			2	
	Air temperature (outdoor)	32	1		34		36	15	8	2	2	4
	Relative humidity (outdoor)	6	6		10		14	4	2	1		
	Air Velocity (outdoor)	4	3		5		2					
	CO2 (outdoor)	1			8		3	2	1		2	
Occupant variable	Clothing	15	1		1		9	8		5		1
	Activity	13	1		4	4	9	6		8	2	
	Preference	10	4		20	1	15	10	3	8	5	
Building Component	Window	10	1		4		8	6		1	1	
	Door	9			4		1	3				
	window blind	5			1		1	3		1		
Building Property	Location	1					2	2		1		1
	Geometry	4	3		1		4	7				
	Configuration	2					1	3				
Others	Energy Information	3			2		6	2		2	1	
	Power consumption	2			5		6	2		4		3
	Time	6	3		9		11	9		1		2

Figure 2. Thermal diagrams of inputs and outputs.

The design of a deep model is carried out as follows:

- formation of a system of requirements for the model: determination of inputs and outputs, performance characteristics and accuracy;
- the basic architecture of the model is based on a block approach;
- decomposition of blocks into sequential or branching structures;
- managing the problem of reducing the accuracy of classification and retraining by heuristically configuring hyperparameters of a deep model and adding layers of normalization, subdescritization and regularization;
- optimization of the model by forming convolutional layers;

Journal of Physics: Conference Series **2091** (2021) 012059 doi:10.1088/1742-6596/2091/1/012059

- model training with the selection of parameters and the number of training epochs;
- evaluation of the quality of the resulting solution based on the construction of error matrices and calculation of accuracy and error metrics based on the results of the model.

The proposed chain of actions leads to obtaining an instance of the model with certain properties. If, as a result, the model does not meet the requirements imposed on it, then it is necessary to roll back a few steps along the trajectory of creating a deep model (up to the first stage) and repeat the search in the heuristically adjusted direction. As a result, the process of searching for an effective classification model can be formalized in the form of a tree, the root node of which precedes the first stage of the search algorithm corresponds to the solution of the problem of formulation of the research problem.

It should be noted that when solving classification test problems, split testing of models with classical convolutional layers and with deep separable convolution was carried out, which confirmed the effectiveness of using the flow approach. The next layer of the feature extraction block was the batch normalization layer, which allows to achieve regularization and stability of the model. The ReLU function was selected to perform the activation operation. The subdescritization layer completes the feature extraction block. The subdescritization operation applied to the elements *x_(i.j)* of the original matrix X results in a smaller matrix Y, for which the value of each element *y_(i,j)* at the size of the subdescritization window d is calculated according to the expression

$$y_{i,j} = \max_{0 \le a \le d}(x_{i+a,j+b}). \quad (1)$$

Note that the number of output filters in the convolution and the size of the convolution kernel are proposed to be selected according to the principle of their minimization within a given classification accuracy. With each new step consisting in extracting features of the next level, it is recommended to increase the number of output filters for deep separable convolution.

The module for the allocaction of hierarchical features is implemented in two levels, the values of the number of output filters in convolutions for these levels are experimentally set to 64 and 64 (a decrease in this value led to a decrease in accuracy, and an increase led to instability of the learning process in early epochs and an increase in its resource intensity). For each 3-by-3 window, the values in the window are averaged over time.

The feature fusion module of the first level has a layer performing a deep separable convolution operation with the number of output filters equal to 128, of the second level - 128. The convolution filter has a square shape with side 3.

The multilayer perceptron for the classification of heat maps has layers with a capacity of 128, 64 and 10 elements. The deep model has been trained for 150 epochs.

The model shows low accuracy, achieving almost zero classification accuracy for the first epochs of learning. A light convolutional neural network in two layers and a deep model from the first epoch achieve accuracy above 30%. After 10 epochs of training, the resulting model will reach the expected accuracy of 86%, which is 5% higher than the result of a deep model and 3% higher than the result of a light convolutional network. A low standard deviation from the mathematical expectation was obtained when training on a small data set. Thus, the high ability of the model to correctly generalize information was obtained with a high stability of the learning process.

3. Conclusion

In this paper, we analyzed the parameters characterizing the state of the control object, identify the necessary set of measured parameters obtained from the sensors, and an informative set of input parameters. We modified the method of training of neural networks without a reinforcement teacher. A feature of the proposed approach is the unification and simplification of the neural network architecture. The application of the modified method allowed creating a highly adaptable control system that recognizes the current state of a wide range of objects and forms the appropriate actions.

4. Reference

[1] Danassis P, Siozios K, Korkas C, Soudris D and Kosmatopoulos E 2017 A low-complexity control mechanism targeting smart thermostats. *Energy and Buildings* pp 340–50.

[2] Javed A, Larijani H, Ahmadinia A and Gibson D. 2017 Smart Random Neural Network Controller for HVAC Using Cloud Computing Technology. *IEEE Transactions on Industrial Informatics* pp 351–60.

[3] Chaudhuri T, Soh Y C, Li H and Xie L 2019 A feedforward neural network based indoor-climate control framework for thermal comfort and energy saving in buildings *Applied Energy* pp 44–53.

[4] Chen Y, Chandna V, Huang Y, Alam M J E, Ahmed O and Smith L 2019 Coordination of Behind-the-Meter Energy Storage and Building Loads: Optimization with Deep Learning Model *Proceedings of the Tenth ACM International Conference on Future Energy Systems - e-Energy '19,* pp. 492–9.

[5] Gao G, Li J and Wen Y 2019 Energy-Efficient Thermal Comfort Control in Smart Buildings via Deep Reinforcement Learning. arXiv: 1901.04693 [cs.SY].

[6] Ghofrani A, Nazemi S D and Jafari M A. 2020 Prediction of building indoor temperature response in variable air volume systems *Journal of Building Performance Simulation* 13:34–47.

[7] Zhao Y, Genovese P V and Li Z. 2020 Intelligent Thermal Comfort Controlling System for Buildings Based on IoT and AI. *Future Internet* 12:30.

[8] Marantos C, Lamprakos C P, Tsoutsouras V, Siozios K and Soudris D 2018 Towards plug&play smart thermostats inspired by reinforcement learning *Proceedings of the Workshop on INTelligent Embedded Systems Architectures and Applications - INTESA* pp 39–44.

[9] Zhang Z, Chong A, Pan Y, Zhang C, Lu S and Lam K P 2018 A deep reinforcement learning approach to using whole building energy model for hvac optimal control. *Building Performance Analysis Conference and SimBuild* pp 9.

[10] Hosseinloo A H, Ryzhov A, Bischi A, Ouerdane H, Turitsyn K and Dahleh M A 2020 Data-driven control of micro-climate in buildings; an event-triggered reinforcement learning approach. ArXiv:200110505 [Cs, Eess].

[11] Zou Z, Yu X and Ergan S 2020 Towards optimal control of air handling units using deep reinforcement learning and recurrent neural network. *Building and Environment* 168:106535.

[12] Dibley M, Li H, Rezgui Y and Miles J 2012 Cost effective and scalable sensor network for intelligent building monitoring *International Journal of Innovative Computing, Information and Control ICIC International* **8** 12 pp 8415–33

[13] Petrova I, Zaripova V, Zolina T, Lezhnina Y and Karpenko 2020 A Intelligent Indoor Microclimate Control *International Multi-Conference on Industrial Engineering and Modern Technologies, FarEastCon* 9271315

[14] Su K, Li J and Fu H 2011 Smart city and the applications *International Conference on Electronics, Communications and Control (ICECC)* 6066743 pp 1028-31.

[15] Farzaneh H, Malehmirchegini L, Bejan A, Afolabi T, Mulumba A and Daka P P 2021 Artificial Intelligence Evolution in Smart Buildings for Energy Efficiency *Appl Sci* 11 763.

[16] Zaripova V, Petrova I and Lezhnina Y 2019 Biosensors application for the life systems quality in a smart city *E3S Web of Conferencesthis* **208** 02005.

[17] Jaradat M, Jarrah M, Bousselham A, Jararweh Y and Al-Ayyoub M 2015 The internet of energy: Smart sensor networks and big data management for smart grid *Procedia Computer Science* 56 (1) pp 592-7.

[18] Nait-Abdesselam F and Titouna C 2020 Data Quality Improvements for Internet of Things Using Artificial Neural Networks *International Symposium on Advanced Networks and Telecommunication Systems (ANTS)* 9342762

[19] Titouna C, Nait-Abdesselam F and Darwaish A 2020 A data cleansing approach in smart home environments using artificial neural networks *International Conference on Wireless and Mobile Computing, Networking and Communications* 9253418

[20] Hashem I A T, Chang V, Anuar, N B , (...), Ahmed E and Chiroma H 2016 The role of big data in smart city *International Journal of Information Management* 36(5) pp 748-58

[21] Tang B, Chen Z, Hefferman G, (...), HeH and Yang Q 2015 A hierarchical distributed fog computing architecture for big data analysis in smart cities *ACM International Conference Proceeding Series* a28

Analysis of transport supply and demand based on convolutional neural networks

Yu. A. Lezhnina[1], N. S. Maltseva[2]

[1] *Moscow Technological University (MIREA), 78 Vernadsky Avenue, Moscow, Russia*
[2]*Astrakhan State Technical University, Tatishcheva 16, Astrakhan, Russia*

E-mail: lejninau@mail.ru

Abstract. The paper describe a solution of the problem of increasing the accuracy of calculating potential transport and pedestrian activity in the context of a shortage of marked data by using a geosocial approach. The advantages of the proposed model are the presence of a high number of degrees of freedom, which allows to flexibly configure the model based on the problem being solved. The authors note that the application of the approach according to the methodology proposed in the article to solve the problem of improving the efficiency of classifying pedestrian transport network data requires an individual project approach to the formation of initial data for analysis. We formulate a hypothesis that the accuracy of the forecast of transport demand and supply can be increased if the classifying model takes into account and analyses not only the properties of a particular territory, but also the characteristic features of the systems with which it interacts. Hexagons and heat maps were chosen for the classification of territories. A feature of the proposed approach is the unification and simplification of the neural network architecture..

1. Introduction

According to the World Bank for 2018, the annual growth of the urban population is 70 million people and by 2050 could reach 6.5 billion people. This kind of urbanization brings many problems, and the problem of mobility is one of the most significant. The approach to solving infrastructure problems involves extensive use of the possibilities of digitalization, which leads to a rapid change in mobility. Automated spatial data analysis can provide solutions and process large amounts of data on both traditional hard and soft computing. The second case involves extensive involvement of fuzzy logic, artificial neural networks, evolutionary modeling [1, 2, 3].

When studying the urban transport systems of various countries and cities, it is possible to identify a tendency to reduce the convenience of living with an unlimited increase in the use of personal cars. This is due to the following transport problems of the city:

1. Increasing the complexity of transport services: there are problems with network capacity, as a result of which network congestion increases, travel time increases and transportation conditions worsen.
2. The growth of the negative impact of transport on the environment: a high degree of accidents (the main criterion is the mortality of pedestrians and drivers in road accidents), ecology, including noise impact on adjacent territories.

The main causes of transport problems include:

1. Urbanization and urban population density growth
2. Settlement system and urban planning policy, including the balance of population settlement and placement of territories of concentration of jobs

3. Motorization, which implies the growth of a fleet of personal vehicles and a shortage of urban areas for its placement and storage
4. The level of development of transport infrastructure at which the access and carrying capacity of the transport network is insufficient.

The urban population still loves personal transport and now the fleet of personal vehicles in Russia is growing, while causing a decline in the development of the public transport system, the number of routes is decreasing, and the corresponding infrastructure is falling into disrepair. However, world experience shows that the attitude of the urban population to public transport increases with the continued growth of motorization. After reaching the level of personal vehicle ownership of 300-400 cars per 1000 people, the capacity of the city's road network is completely exhausted. The usual practice is the application of restrictions on the use of cars in the city and the introduction of a paid travel system. There is a revival of public transport. With the further growth of the level of personal transport, there is a tendency for the development of all types of public passenger transport.

The priority directions of urban transport development are:

- Development of high-speed comfortable transport
- Introduction of intelligent transport systems
- Automation of vehicle and system management

Data monitoring for territorial and transport planning is one of the most important factors. Now the city generates a huge amount of big data that can be used in analyzing traffic flows and planning pedestrian transport infrastructure so that it is stable, smart, slow, affordable, clean and generally improves the quality of life [4, 5, 6]. These include:

- "Big Data" of transport enterprises - organizers of passenger transportation in cities (collection and storage are better organized in large cities);
- "Big data" in monitoring traffic and pedestrian flows;
- Data from GLONASS and GPS tracks of vehicles;
- Data from transport laboratories;
- Data from video recording complexes of traffic violations;
- Data from "Safe City" video surveillance cameras;
- Data from transport detectors;
- Data from special studies;
- Data from cellular operators.

All this amount of data can be used for decision-making and management of smart cities [7, 8, 9].

2. Control system

Modeling makes it possible to visually depict complex processes of transport activity, predict the redistribution of traffic flows as a result of external influences on sections of the street-road network [10, 11] such as:

- construction of new sections of the network;
- reconstruction (expansion) of network sections;
- closure of individual sections of the network– - changes in traffic conditions in the network–
- changes in the route network and timetables of the public transit port.

From the point of view of the geosystem approach, the state and properties of each territorial unit are determined by its interaction with neighboring objects, the characteristics of the enclosing geospatial system and the behavior of its constituent systems [12, 13, 14].

Based on this, we formulate a hypothesis that the accuracy of the forecast of transport demand and supply can be increased if the classifying model takes into account and analyzes not only the properties of a particular territory, but also the characteristic features of the systems with which it interacts and, in particular, into which it enters.

Hexagons were chosen for the classification of territories (figure 1).

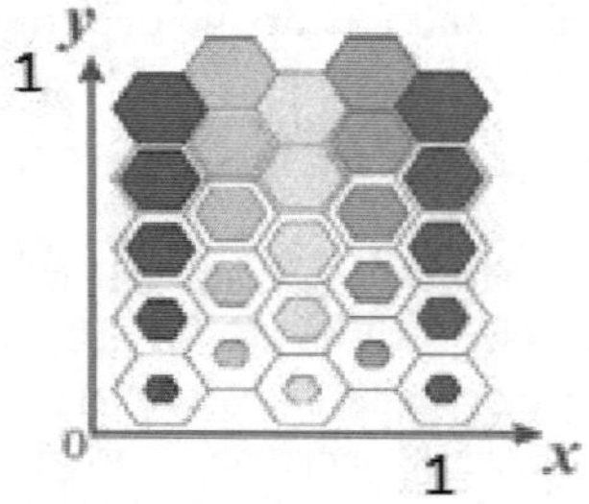

Figure 1. Legend to the cartographic scheme of spatial-statistical classification of territories, where y is transport demand, x is transport supply

The process of obtaining neural network input data is potentially subject to full automation. This is the big data that will be the input data of the model. The deep classification model is a black box that accepts data from cellular operators of the territory (L0) and its host geosystems (Li), as well as synthetic maps with required layers (LS). The congestion of the area with potential passengers is presented based on the congestion of the cellular network in the form of a heat map (figure 2).

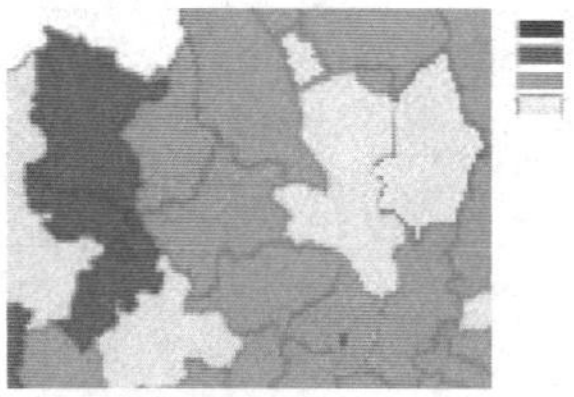

Figure 2. Heat map of the territory

The number of inputs can vary based on the number of levels of the geosystem model of the territory (figure 3). At the same time, with an increase in their number, the capacity of the model increases [15]. The model has one vector output, each *i*-th element of which determines the maximum probability that the territory belongs to the *i*-th class.

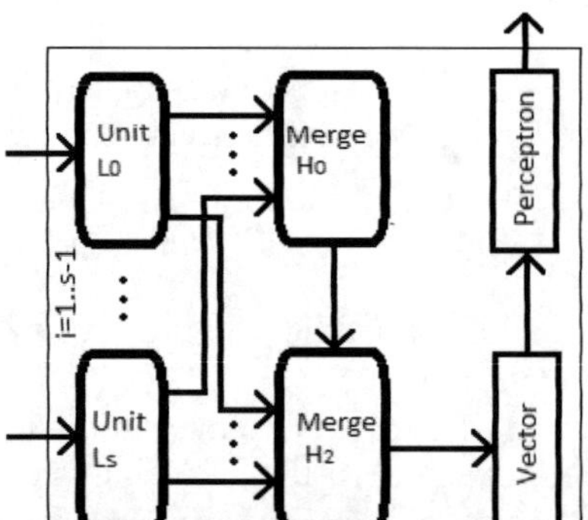

Figure 3. General architecture of a neural network model with s inputs

For the primary allocation of the feature vector based on the data of each input, we introduce the Unit Lx block, which allocates hierarchical features of the input data. Then the architecture is decomposed into N feature extraction blocks, each of which has an external output. The structure of each block represents a chain of layers. The operation of the layer is based on a two-dimensional convolution operation with a kernel W of size K represents a linear transformation in which each value y_(i,j) of the output matrix Y is calculated based on the values of x, the original matrix X by the formula:

$$y_{i,j} = W * X = \sum_{a=0}^{K-1} \sum_{b=0}^{K-1} W_{a,b} x_{i+a,j+b} . \quad (1)$$

The batch normalization layer is used for regularization and stability of the model [16]. To perform the operation of activating the output of the input and hidden layers, the ReLU function is used, which performs a transformation of the form x=max(0,x). The block of extracting characteristic features is completed by a layer of subdescritization, which applies the maximum operation to reduce the size of the resulting representations and has external outputs. The output of the last feature fusion module is converted into a vector and fed to the input of a multilayer perceptron. The number of tightly coupled layers of a multilayer perceptron and their power are selected experimentally.

The module for the allocation of hierarchical features is implemented in two levels, the values of the number of output filters in the convolutions for these levels are experimentally set to 128 and 64. For each 3-by-3 window, the values in the window are averaged over time.

The feature fusion module of the first level has a layer performing the operation of a deep separable convolution with the number of output filters equal to 64, of the second level - 64. The convolution filter has a square shape with side 3.

The multilayer perceptron for the classification of heat maps has layers with a capacity of 128, 64 and 10 elements. The deep model has been trained for 150 epochs.

The model shows low accuracy, reaching almost zero classification accuracy for the first epochs of learning. A light convolutional neural network in two layers and a deep model from the first epoch achieve accuracy above 45%. After 20 epochs of training, the resulting model will reach the expected accuracy of 86%, which is 7% higher than the result of a deep model and 5% higher than the result of a light convolutional network.

The resulting visualization (figure 4) allows you to identify areas that correspond to dense buildings, but sparse settlement. They allow you to analyze public transport routes and predict the density of the traffic schedule, according to the dynamics of the heat map.

Figure 4. classification of the allocated territory

3. Conclusion

The creation of such a model is focused on the organization of public transport and increasing the comfort of organized flows.

The use of data from mobile operators will allow the matrix of transport demand and supply to take into account underage residents and groups of citizens who are not included in official statistical systems (population census, central election commissions).

The proposed visualization in the form of hexagons makes it possible to identify problem areas of the transport model and conduct a detailed analysis of the selected territories in order to find a solution.

4. Reference

[1] Su K, Li J, Fu H 2011 Smart city and the applications *International Conference on Electronics, Communications and Control, ICECC 2011 - Proceedings* 6066743 pp 1028-31.

[2] Hashem I A T, Chang V, Anuar, N B , (...), Ahmed E and Chiroma H 2016 The role of big data in smart city *International Journal of Information Management* 36(5) pp 748-58

[3] Tang B, Chen Z, Hefferman G, (...), HeH and Yang Q 2015 A hierarchical distributed fog computing architecture for big data analysis in smart cities *ACM International Conference Proceeding Series* a28

[4] Jaradat M, Jarrah M, Bousselham A, Jararweh Y and Al-Ayyoub M 2015 The internet of energy: Smart sensor networks and big data management for smart grid *Procedia Computer Science* 56 (1) pp 592-7.

[5] Zaripova V, Petrova I and Lezhnina Y, 2019 Biosensors application for the life systems quality in a smart city *E3S Web of Conferencesthis* **208** 02005.

[6] Sivanathan A, Sherrat D, Habibi Gharakheili H, Radford A, Wijenayake C, Vishwanath A and V. Sivaraman 2017 Characterizing and Classifying IoT Traffic in Smart Cities and Campuses *IEEE INFOCOM Workshop on SmartCity: Smart Cit-ies and Urban Computing*

[7] Fouladgar M, Parchami M, Elmasri R and Ghaderi A 2017 Scalable deep traffic flow neural networks for urban traffic congestion prediction. *In Proceedings of the International Joint Conference on Neural Networks (IJCNN)* pp 2251–8.

[8] Yamansavascilar B, Guvensan M A, Yavuz A G and Karsligil M E 2017 Application identification via network traffic classification *International Conference on Computing, Networking and Communications (ICNC)* pp 843-8

[9] Gupta A, Gurrala G and Sastry P S 2019 An Online Power System Stability Monitor-ing System Using Convolutional Neural Networks *IEEE Transactions on Power Systems* **34** 2 pp 864-72

[10] Wang W.-X, Guo R.-J and Yu J 2018Research on road traffic congestion index based on comprehensive parameters: taking Dalian city as an example *Advances in Mechanical Engineering* **10** 6 168781401878148

[11] Le Cun Y, Bengio Y and Hinton G 2015 Deep learning *Nature* **521** 7553 pp 436

[12] Helber P, Bischke B, Dengel A and Borth D 2019 Eurosat: A novel dataset and deep learning benchmark for land use and land cover classification *IEEE Journal of Selected Topics in Applied Earth Observations and Remote Sensing* **12** 7 pp 2217–26

[13] Adetiloye T and Awasthi A 2019 Multimodal big data fusion for traffic congestion prediction *Multimodal Analytics for Next-Generation Big Data Technologies and Applications*, pp 319-35

[14] Kussul N, Lavreniuk M, Skakun S and Shelestov A 2017 Deep learning classifi-cation of land cover and crop types using remote sensing data *IEEE Geoscience and Remote Sensing Letters* **14** 5 pp 778–82

[15] Hu X, Weng Q 2009 Estimating impervious surfaces from medium spatial resolution imagery using the self-organizing map and multi-layer perceptron neural networks *Remote Sensing of Environment* **113** 10 pp 2089–02

[16] Ioffe S., Szegedy C. (2015) Batch normalization: Accelerating deep network training by reducing internal covariate shift // arXiv: 1502.03167 [cs.LG].

The decision support system for conducting fire-prevention measures in woodlands

A A Sorokin, N S Maltseva, D S Puzankov, V F Shurshev and L V Galimova

Astrakhan State Technical University, Tatishcheva 16, Astrakhan, Russia

E-mail: maltsevans@mail.ru

Abstract. Purpose of the work is to propose provisions for the aggregation of parameters of a climatic and anthropogenic nature. Regulations are necessary to determine the rating of forest areas in terms of fire hazard level. As a result of the study, theoretical provisions for the aggregation of variables have been proposed. Variables are described on different metric scales and affect the final result in different ways. The provisions are based on the methods of the theory of fuzzy sets and fuzzy inference. The computational complex for determining the rating of the forest areas is implemented in the form of a hierarchical system of fuzzy inference. It is based on the Takagi-Sugeno algorithm. Simulation of the complex operation was carried out. As a result, it was concluded that the generated output pattern has a stepped nature. This makes it possible to classify the analyzed elements into state groups. This allows you to solve the problems of planning work on the implementation of preventive and / or preparatory measures to eliminate or prevent forest fires. The prospect of further research is the formation of decision support systems in the control processes organization of forest areas.

1. Introduction

Forest fires annually damage the ecosystems and economies of the regions [1,2]. In the process of woodland burning, many harmful substances are released into the environment, a carbon dioxide is also released [3,4]. The «greenhouse effect» is an important problem in global economic relations. The "greenhouse effect" is caused by carbon dioxide emissions [5]. Decarbonization of the atmosphere is an important component of reducing the negative impact of human activity on the planet's ecosystem [6]. Solving the problems of forest fire prevention and the introduction of ESG (Environmental, Social, and Corporate Governance) principles [7] into the activities of corporations is important for the development of the "Green" economy.

Monitoring systems allow you to know in advance about the occurrence of a fire hazard or fire. An analysis was carried out [8,9]. It is concluded that the most widespread monitoring systems are based on the use of video surveillance cameras and software systems for analyzing the video stream. Their disadvantage is the reactive principle of operation. The fire is registered after its occurrence. While the control effect on the fire is being formed, the spread of the fire continues. One of the methods to eliminate this limitation is to take preventive measures to prevent fire as well as strengthening of preparation methods for carrying out measures to extinguish the fire. Work in the field of obtaining special coefficients is the basis for the development of methods for determining the level of fire hazard. An example of coefficients are: a complex fire hazard indicator (CFHI) by Nesterov V.G., Russia [10]; forest fire hazard index FFDI (English Forest Fire Danger Index), Australia [11,12]; National fire hazard rating NFDRS (National Fire-Danger Rating), USA [11,13]; natural fire hazard index FWI (Fire Weather Index), Canada [11,14]; in addition, there are a number of additional methods for determining such indices [11,15].

Mathematical operations help in defining such indices. But with the help of these mathematical operations it is not possible to generalize the parameters measured on different metric scales. It is not possible to generalize parameters using different units of measurement or having an uneven level of influence on the final result. It is necessary to formulate provisions that will eliminate the identified limitations.

2. Review of literary sources and formulation of the research problem

As the analysis shows [10], the main method for calculating of the fire hazard is the complex fire hazard indicator (CFHI) by Nesterov V.G.. The definition of Nesterov's indicator is based on the use of the formula:

$$K_{CFHI} = \sum_{t=1}^{n} t_i \cdot (t_i - r_i), \tag{1}$$

where is the air temperature, r is the dew point temperature, n is the number of days since the last rain (precipitation d_{rain}<2,5mm). In [11] CFHI is determined using a formula of the form:

$$K_{CFHI_i} = K_{CFHI_{l-1}} \cdot K_{rain} + t_i \cdot (t_i - r_i), \tag{2}$$

where $K_{CFHI_{i-1}}$ – the previous day's fire hazard ratio, K_{rain} – precipitation correction factor, where in $K_{rain} = 1$ if the amount of precipitation is less than 3 mm and $K_{rain} = 0$, if the amount of precipitation is more than 3 mm. The development of the method was carried out in the works [16,17]. In this case, additional indicators are used. They allow you to assess more fully the meteorological situation at the observation site.

Consideration of such parameters is carried out when determining the fire hazard in Australia [11,12].

In [13] the calculation of the fire hazard index K_{FHI} is described. This index is used in the United States. To obtain it, use the formula:

$$K_{FHI} = 0{,}71 \cdot (K_{BC}^2 + (K_{HF} + K_{TA})^2)^{1/2}, \tag{3}$$

where K_{BC} – burning coefficient of materials, K_{HF} – human factor, K_{TA} – coefficient of influence of thunderstorm activity.

The review [11,14,18,19] showed that the method of calculating of the fire hazard index used in Canada has become widespread. It is called FWI (Fire Weather Index). Several parameters are considered in the definition. These parameters characterize the state of the climate and forest litter in the monitored area. Taking into account [11,14], the FWI is determined using the formula:

$$K_{FWI} = K_{ER} \cdot K_{FFM_{1,2}} \cdot k_{FFM_3}, \tag{4}$$

where K_{ER} – energy release, kW/m^2, $K_{FFM_{1,2}}$ – coefficient uniting the moisture content indices of combustible materials to a depth of 18 cm and a weight per unit area up to $2kg/m^2$ – k_{FFM_1} and at a depth of up to 7 cm and a weight per unit area up to *5 kg/m²*–k_{FFM_2}, k_{FFM_3} determines the moisture content of combustible materials at a depth of 1.2 cm and a weight per unit area up to *0,025kg/m²*(FFM is forest fire material).

In addition to relations (1) - (4), when determining the fire hazard, the influence of the anthropogenic factor can be taken into account.

Analysis [10-19] established a pattern which shows an increase in the number of parameters during the development of models. These parameters are taken into account when determining the fire hazard index. The parameters can be divided into groups: (g_1) climatic parameters (temperature, air humidity, wind speed, thunderstorm activity, etc.); (g_2) parameters of the state of combustible materials (type of vegetation, forest litter, depth of peat occurrence, soil moisture, etc.); (g_3) parameters characterizing the anthropogenic impact on the environment (the proximity of industrial facilities, product pipelines, settlements and other places of temporary or permanent residence of people). Aggregation of parameters is carried out using information processing operators (IPO). In relation to relations (1) - (4), various "classical" mathematical operations are used as the IPO. The analysis of the features of the parameters (g_1) - (g_3) showed that, depending on belonging to the group, they can be assessed on different metric scales, have different levels of influence on the final result, while the level of influence itself may depend on the area where the analyzed parameter is used.

The analysis of methods for determining the fire hazard index [10-19] has revealed a contradiction between: (a) the need to increase the number of aggregated parameters in order to assess the fire hazard in the observed forest area; (b.) the complication of the IPO of the parameters used for aggregation due to the fact that these parameters are estimated on different metric scales and have a different level of influence on the value of the final variable. To eliminate the contradiction, it is advisable to develop generalized provisions for ranking forest areas according to the level of fire hazard. In this case, it is necessary to take into account various factors of natural and anthropogenic nature.

The aim of the work is to propose theoretical methods for aggregating various parameters,and also to increase the number of aggregated parameters to determine the rating of forest areas according to the level of fire hazard.

3. Proposal of theoretical provisions for determining the index of fire hazard

3.1. Analysis of information processing methods based on the provisions of the theory of fuzzy sets and fuzzy inference

Taking into account [20-22], the formation of theoretical methods that allow the aggregation of various variables can be presented as a problem of constructing a universal approximator. Practice and analysis [20-24] show that at present, methods based on the use of neural networks [22,23] have become widespread for the aggregation of heterogeneous parameters with different metric scales, various levels of influence on the final result. Methods based on fuzzy inference have also become widespread [20-22, 24].A feature of the methods based on the use of neural networks is the need to form a representative training sample [23]. Taking into account [23, 25, 26], the task of forming such a training sample is not trivial enough.An alternative to systems based on neural networks are systems based on the provisions of fuzzy inference [21,24]. Historically, such computational methods have become a theoretical basis for formalizing the functioning of expert systems. As it is known from [27,28,29], expert systems allow aggregating information under conditions of uncertainty in the values of parameters, when information about part (or all) of the analyzed parameters is presented in a qualitative (verbal) form.

The essence of using the provisions of fuzzy inference is to separate the range of values of input and output variables $x \in [x_{\min}; x_{\max}]$ into the set of overlapping intervals, $\{i_j\}$, where j is the number of the interval . The number of intervals i, into which the range of values of the variable is divided, is determined by the expert group. This group is engaged in the formation of a computing complex. The ability of the intervals to overlap each other results in "fuzziness". Each $i \subset \{i_j\}$ corresponds to a certain semantic value t, called a term. Each term verbally describes the state of a variable at a given interval. The total number of terms corresponds to the number of intervals i into which the range of values of the variable x has been divided. The total number of terms t describing the state of the variable x is called the term set $\{t_j\}$. Each term $t \subset \{t_j\}$ is associated with a membership function (MF) $\mu_{t_j}(x)$. The domain of definition of the MF corresponds to the range of values of the variable x on the interval i. The area of the MF value varies in the range from 0 to 1. The implementation of the aggregation operator for the values of the input variables and the subsequent conversion to the output variable are performed using production rules. The production rule is formed by an expert in the relevant field of knowledge. The production rule shows how the semantic values of the output variable change when the semantic values of the input variables change. Provided that the fuzzy inference system (FIS) has several input variables and one output, the production rule can be represented as:

$$p : if (x_1 = t_{x_1}) \wedge ... \wedge (x_k = t_{x_k}) \Rightarrow (y_p = t_y)_\xi \qquad (5)$$

where $x_1, ..., x_k$– a set of variables that describe the parameters that characterize the modeling object, t_{x_1} – one of the terms of a term set that describes the state of a variable x_1, t_{x_k} – one of the terms of a term set that describes the state of a variable x_k, y_p – an output variable that describes the general state of the modeling object within a given rule p, t_y – one of the terms of the term set describing the state of the variable y, ξ – coefficient of confidence (CC) to a rule. The set of rules that describe all the states of input variables necessary for reproducing the object of modeling form the rule base (RB). A similar thing, taking into account the assumption that rules of the form (5) are used to form theRB, can be represented as follows:

$$\mathrm{Y}_{p=1}^{p_{\max}} [\mathrm{I}_{k=1}^{k_{\max}} (x_k = t_{x_k}) \Rightarrow (y_p = t_\psi)_\xi] = y, \qquad (6)$$

where y is the value of the output variable characterizing the state of the simulation object. If a rule has a structure similar to rule (5), and the FIS itself operates on the basis of the model described in (6), then it has the structure MISO (Multiple Input Single Output).

Taking into account [24,29], the most widespread are the algorithms of Mamdani, Takagi-Sugeno, etc. As a rule, such algorithms include the following operations: (o_1) fuzzification - presentation of clear information in a fuzzy form, performed using MF; (o_2) - aggregation of variables - is performed using logical operators used in rules of the form (5); (o_3) activation of rules - determines the value of the output variable obtained using the rule, taking into account its CC; (o_4) - accumulation of the results of the rules, the union of the output values is carried out; (o_5) - defuzzification, converting a fuzzy output value into a numerical estimate. Depending on the type of algorithm, individual operations can be combined. For example, the Takagi-Sugeno algorithm combines the operations (o_4) and (o_5). In the framework of further research, strategic offensive weapons operating on the basis of the Takagi-Sugeno algorithm are considered. As described in [21,24], one of the problems of constructing FIS is the "curse of dimension" associated with an increase in the number of rules of the form (5) in the rule base (6) with an increase in the number of aggregated variables and the number of terms used to describe the state of the variable. To eliminate the "curse of dimension", [24] the use of hierarchical FIS (HFIS) is described. The essence of the HFIS consists in: (i) dividing the set of input variables X into several disjoint subsets of variables; (ii) construction for each subset of a separate FIS, as a result of the functioning of which a set of intermediate variables is formed; (iii) the use of subsequent levels of the hierarchy for aggregation of intermediate FIS output variables. The formation of FIS of subsequent levels of the hierarchy continues until a set of target output variables is obtained.

3.2. Formation of provisions for the analysis of the state of forest areas

A feature of the research object (forest) is the presence of various parameters on which the probability of fire depends. Some of the parameters have a numerical value of the groups (g_1) and (g_2), and some of the parameters from the group (g_3) have a verbal, qualitative value. Depending on the location of the forest area, the number of parameters and the degree of their importance may vary. The main task in determining the rating of forest areas according to the level of fire hazard is to prioritize the implementation of preventive measures to prevent a fire or liquidation measures to eliminate a fire.As the initial data for the developed model, it is proposed to aggregate parameters from groups (g_1) - (g_3). Taking into account the analysis [1-5, 10-19], it is proposed to use various parameters to form a rating of forest areas by the level of fire hazard of a controlled area. The list of parameters is given in table 1. The ranges of parameter values are semantically related to the values of the corresponding terms.

Table 1.List of parameters for rating formation woodlands by fire hazard level.

Parameter name	Unit of measure-ment	Range of values				
		Bad	Rather bad	The average	Rather good	Good
Air humidity, x_1	%	0 ÷ 10	10 ÷ 30	30 ÷ 60	60 ÷ 90	90 ÷ 100
Air temperature, x_2	^{0}C	+35÷ +50	+25 ÷ +35	+20 ÷ +25	+10 ÷ +20	-50 ÷ +10
Wind speed, x_3	m/s	20 ÷ 15	15 ÷ 10	10 ÷ 5	5 ÷ 2.5	2.5 ÷ 0
Precipitation amount for the previous day, x_4	mm	0 ÷ 1	1 ÷ 2	2 ÷ 6	6 ÷20	20 ÷ 30
The amount of precipitation during the current day, x_5	mm	0 ÷ 1	1 ÷ 2	2 ÷ 6	6 ÷20	20 ÷ 30
Proportion of dry trees, x_6	%	50 ÷ 100	20 ÷50	10 ÷20	5 ÷ 10	0 ÷ 5
Proportion of partially dried trees, x_7	%	70 ÷ 100	50 ÷70	40 ÷ 50	10 ÷ 40	0 ÷ 10
Humidity of FFM$_1$, x_8	%	0 ÷ 6	6 ÷ 12	12 ÷ 18	18 ÷ 22	22 ÷ 100
Temperature of FFM$_1$, x_9	^{0}C	+35 ÷ +50	+25 ÷ +35	+20 ÷ +25	+10 ÷ +20	-50 ÷ +10
Humidity of FFM$_2$, x_{10}	%	0 ÷ 9	9 ÷ 15	15 ÷ 20	20 ÷ 25	25 ÷ 100
Temperature of FFM$_2$, x_{11}	^{0}C	+35 ÷ +50	+25 ÷ +35	+20 ÷ +25	+10 ÷ +20	-50 ÷ +10
Humidity of FFM$_3$, x_{12}	%	0 ÷ 11	11 ÷ 18	18 ÷ 25	25 ÷ 33	33 ÷ 100
Temperature of FFM$_3$, x_{13}	^{0}C	+35 ÷ +50	+25 ÷ +35	+20 ÷ +25	+10 ÷ +20	-50 ÷ +10
Peat moisture, x_{14}	%	0 ÷ 11	11 ÷ 18	18 ÷ 25	25 ÷ 33	33 ÷ 100
Peat temperature, x_{15}	^{0}C	+65 ÷ +100	+40 ÷ +65	+25 ÷ +40	+10 ÷ +25	0 ÷ +10
Peat layer thickness, x_{16}	m	0,5 ÷ 1	1 ÷2	2 ÷ 3	3 ÷4	4 ÷ 5
Assessment of fire hazard of production facilities, x_{17}	points	80 ÷ 100	60 ÷ 80	40 ÷60	20 ÷ 40	0 ÷ 20
Assessment of the fire hazard of product pipelines, x_{18}	points	80 ÷ 100	60 ÷ 80	40 ÷60	20 ÷ 40	0 ÷ 20
Assessment of the fire hazard of transport routes, x_{19}	points	80 ÷ 100	60 ÷ 80	40 ÷60	20 ÷ 40	0 ÷ 20

Assessment of fire hazard of social facilities (settlements, households) , x_{20}	points	80 ÷ 100	60 ÷ 80	40 ÷60	20 ÷ 40	0 ÷ 20
Assessment of fire hazard of tourist sites, x_{21}	points	80 ÷ 100	60 ÷ 80	40 ÷60	20 ÷ 40	0 ÷ 20

If necessary, the ranges of values of variables can be adjusted depending on the type of forest area. Computing complexes based on HFIS allow adding or excluding variables that describe the required set of parameters. This is due to a kind of "modular" design. Adding or removing taken into account parameters will require adjusting or adding elements of the computing complex with the help of which the aggregation of new variables is carried out. The previously used elements for aggregating variables can be used in a new computing complex. The structure of a computing complex that aggregates variables from the set shown in table. 1 is shown in figure 1.

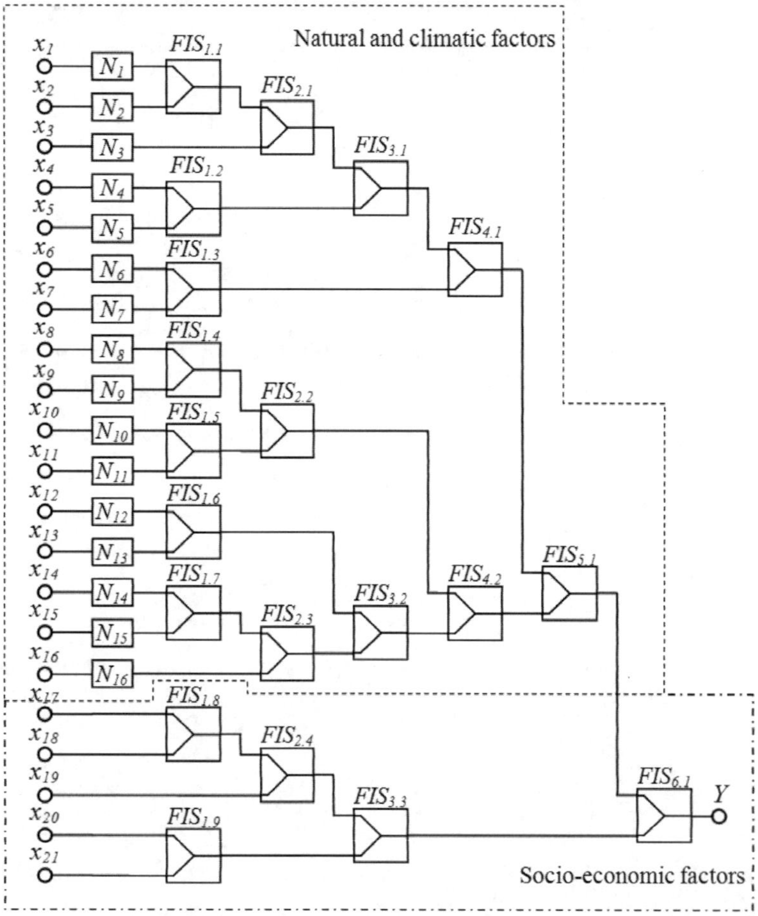

Figure 1. The structure of the hierarchical fuzzy inference system for determining the rating of forest areas by the level of fire hazard.

A feature of the structure of the HFIS under consideration is the presence of normalization blocks, which are used for preprocessing parameters, information about which is presented on various metric scales. The use of such a component in the formation of FIS is recommended in [21, 30]. The use of normalization is justified by the fact that the parameters before processing FIS are reduced to a single metric scale with the same range of values (for example, from 0 to 1, or from -1 to 1, from 0 to 100, etc.). As noted in [30], after adding normalization blocks, FIS becomes invariant to changes in the range of values in natural units.

In FIS, there is no need to change the description of input variables using FP. In [31], the use of piecewise functions was proposed to improve the normalization process. This made it possible to bring the parameters not only to the general range of values, but also to ensure the same correlation between the variables. These are the variables that are directly fed to the FIS input and the variable that is formed as a result of the FIS operation. Practical application of this normalization method makes it possible to simplify the process of forming the FIS knowledge base. This may happen because terms describing the same ranges of values of input and output variables have the same semantic meaning. Applied to the table. 1 parameters $x_1 \div x_{16}$ are described using various metric scales, parameters $x_{17} \div x_{21}$ are described using a scale of a rank scale in the range from 0 to 100 points. Taking into account [31], the parameters $x_1 \div x_{16}$ were normalized to the range from 0 to 100 points.

The form of membership functions of FIS variables, to which the values of the input parameters are supplied, after normalization are shown in figure 2. The terms used to describe the values of a variable have the following semantic meaning: NB - "bad"; NM - "rather bad"; Z - "average"; PM - "rather good"; PB - "good" influence on the value of the output variable.

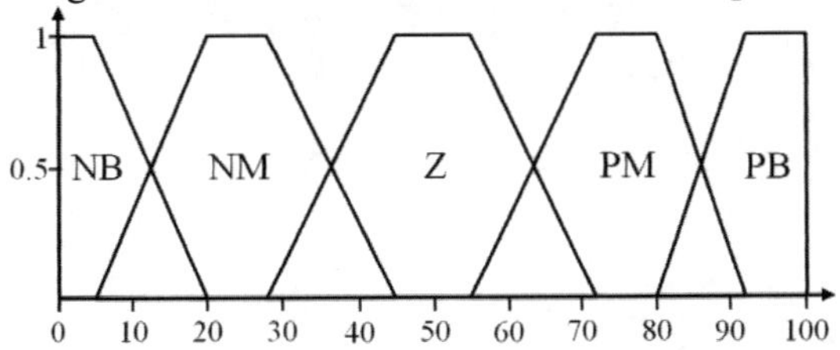

Figure 2. Appearance of MF input variables.

A similar set of terms is used to describe the output variables. The semantic meanings of the terms are as follows: NB - “bad”; NM - "rather bad"; Z - "average"; PM - "rather good"; PB is the "good" value of the variable. Singletons are used as MF to describe the terms of output variables. The term NB corresponds to the value "0", the term NM - "25", the term Z - "50", the term "PM" - 75, and the topic "PB" - the value of 100 points. The FIS output regularity, implemented using production rules of the form (5) and mathematical operations of the Takagi-Sugeno algorithm, are described using a three-dimensional surface that displays the values of the output variable for given pairs of values of the input variables. In relation to the FIS included in the computational complex under consideration, the surfaces are shown in figure 3, namely: in figure 3 a. shows the surfaces FIS11, FIS14, FIS15, FIS16, FIS17; in figure 3 b. - surfaces FIS18, FIS19, FIS22, FIS24, FIS32, FIS33, FIS42, FIS51, FIS61; in figure 3 c. - FIS12 surface; in figure 3 years - FIS13 surface; rice. 3 days - FIS21 surface; in figure 3 f. - FIS23 surface; rice.3 g. - FIS31 surface; in figure 3 h. - FIS41 surface.

The performance check of the computing complex was carried out using numerical simulation.

4. Checking the functionality of the proposed provisions

To check the proposed provisions, a test set was formed, consisting of 173 vectors of values. It is assumed that each vector describes the area of the forest being monitored. The task of modeling is to show how the computing system ranks the analyzed objects. The values of the variables were selected at random from the ranges shown in Table. 1. After the simulation, the results were sorted in ascending order. The resulting pattern is shown in figure 4.

It can be seen from the graph that most of the output values were distributed into 5 groups - classes of states of forest areas. Within the framework of the proposed model, the possibility of ignition in a controlled area increases with a decrease in the output value of the variable. Thus, it is proposed to: (a) group 1 should include the areas with the highest fire hazard, the output value is 0 points; (b) group 2 should include areas with increased fire hazard, the output value is 25 points; (c) group 3 should include areas with an average level of fire hazard, the output value is 50 points; (d) group 4 should include areas with low fire hazard, the output value is 75 points; (e) group 5should assign the areas with the lowest fire hazard, the output value is 100 points.

The study of the reasons for the formation of the pattern allows you to determine the parameters that served as the reason for the formation of the assessment. If the value of the parameter influenced the formation of a low final grade, then this is a signal for the formation of control actions. These actions will be aimed at eliminating the cause of the formation of such a value or taking preventive measures.

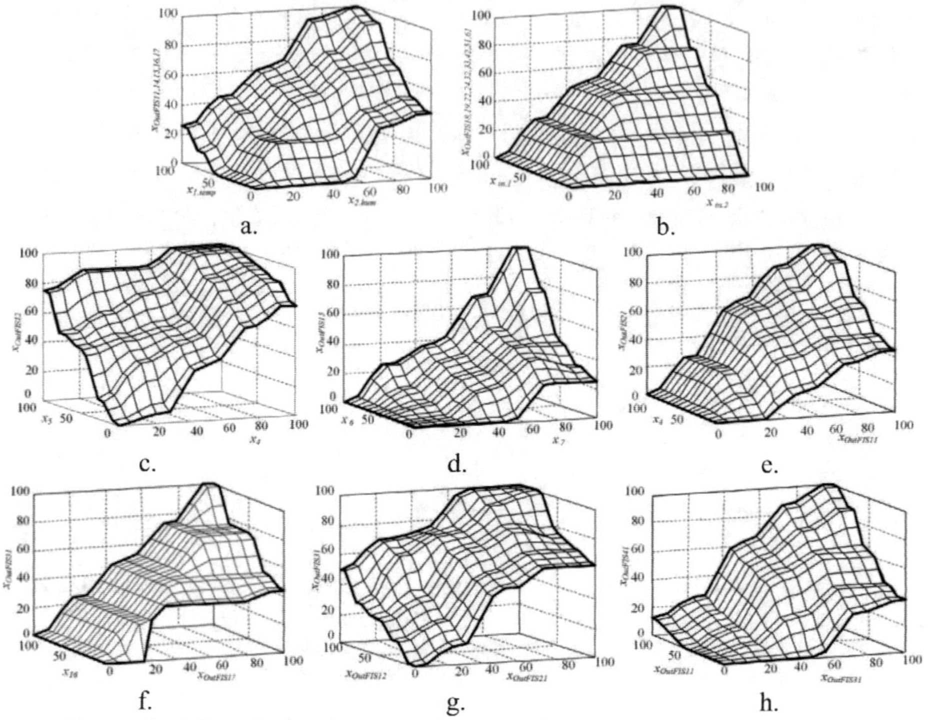

Figure 3. FIS surfaces of a computing complex for determining the rating of forest areas according to the level of fire hazard.

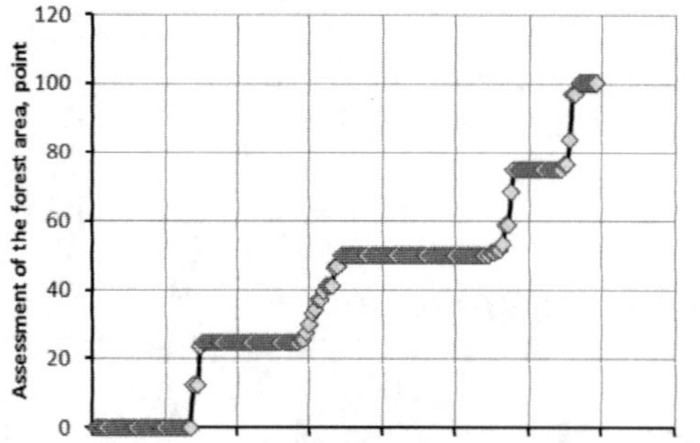

Figure 4. Output regularity of the investigated model.

The implementation of the measures will reduce the damage from a possible fire.

5. Conclusion

In the course of the research, theoretical provisions were proposed, these provisions allow the aggregation of variables presented on different metric scales and having different levels of influence on the final result. The provisions are based on the methods of the theory of fuzzy set and fuzzy logical inference. They represent one of the directions for the development of the artificial intelligence field. The implementation of the system for determining the rating of the forest is carried out by means of the formation of a hierarchical system of fuzzy inference. As the study of the formed computing system has shown, the output pattern is predominantly stepwise. This makes it possible to solve the problem of distributing the elements of the analyzed sample into classes. This classification allows you to use the results of the analysis for planning preventive or preparatory measures to eliminate forest fires.

References

[1] Nektegyaev G G and Borisov A I 2019 Evaluation of economic damage from forest fires *Moscow economic journal* **11**pp 601-05. (In Russian)

[2] Zharinov S N and Golubeva E I 2017 Economic evaluation of forest fires effects (case study of the tver oblast) *Moscow University Bulletin* Series 5. Geography **3** pp 61-69. (In Russian)

[3] Korotkov V N, Romanovskaya A A, Karaban R T and Smirnov N S 2012 Estimation of the carbon budget of Russian forests in the framework of reporting under the Kyoto Protocol *Bulletin of the Moscow State Forest University - Lesnoy Bulletin* **7** pp 58-64. (In Russian)

[4] Zakharchenko A V, Pasko O A and Tigeev A A 2020 Impact of natural factors on mercury in lakes of the north of western Siberia *Bulletin of the Tomsk Polytechnic University. Geo Assets Engineering* **331 5** pp 64–76 (In Russian)

[5] Nikberg I I 2017 Some health and environmental problems in Australia *Hygiene and sanitation* **96** 3 pp 243-47.

[6] Gazman V 2019 Socio-economic efficiency of the leasing in renewable energy *Higher school of economics economic journal* **23** 2 pp 238-63. (In Russian)

[7] Miroshnichenko O S and Brand N A 2021 Banks financing the green economy: a review of current research *Finance: Theory and Practice* **25** 2 pp 76-95. (In Russian)

[8] Lesnoy Dozor, Description of the Complex. http://lesdozor.ru/opisanie-kompleksa/

[9] Monitoring and control system Forest Guard (Sistema monitoringa i upravlenija Lesohranitel') http://service.lesohranitel.ru/finist/docs/present.pdf

[10] GOST R 22.1.09-99 1999 Safety in emergency situations. Monitoring and forecasting forest fires. General requirements // Gosstandart of Russia (Moscow)

[11] Gubenko I M and Rubinstein K G 2012 Comparative analysis of methods for calculating fire hazard indices *Proceedings of the Hydrometeorological Center of Russia* **347** pp 207-22 (In Russian)

[12] Dowdy A J, Mills G A, Finkele K and Groot W 2009Australian fire weather as represented by the McArthur Forest Fire Danger Index and the Canadian Forest Fire Weather Index *CAWCR Technical Report* **10** 91 p.

[13] Cohen J D and Deeming J E 1985 The National Fire Danger Rating System: Basic equations *General Technical Report PSW-82. Pacific Southwest Forest and Range Experiment Station, Berkeley, California* p 23

[14] Wotton M B 2009 Interpreting and using outputs from the Canadian Forest Fire Danger Rating System in research applications *Springer Science + Business Media, LLC Environmental and Ecological Statistics*, **16(2)** pp 107–31.

[15] Sharples J J, McRae Rh D, Weber R O and Gill A M 2009 A simple index for assessing fire danger rating *Environmental Modelling& Software* **24** 6 pp 764-74

[16] Dolgov A and Sumina E 2007 Methodology of forest fire risks estimation - basis for decision making support in crisis situations, caused by the forest fire *Civil security technologies*. **4** 3 (15) pp 79-83. (In Russian)

[17] Sofronov M A, Sofronova T M and Volokitina A V 2004 Assessment of fire hazard by weather conditions using meteorological forecasts *Forestry* **6** pp 31-32. (In Russian)

[18] Van Wagner C E 1987 Development and structure of Canadian forest fire weather index system *Minister of Supply and Services Canada* p. 48

[19] Pinto M M, Da Camara C C, Hurduc A, Trigo R M and Trigo I F Enhancing the fire weather index with atmospheric instability information *Environmental Research Letters* **15 (9)** pp 2-11

[20] Kosko B. 1994 Fuzzy systems as universal approximators *IEEE Transactions on Computers* **43** 11 pp 1329–33.

[21] Pegat A 2017 Fuzzy modeling and control (*Binom, Laboratory of Knowledge*, *Moscow*) p 800. (In Russian)

[22] Shvedov A S 2018 Functions approximating by neural networks and fuzzy systems *Control sciences* **1** pp 21–29. (In Russian)

[23] Komashinsky V I and Smirnov D A 2003 Neural networks and their application in control and communication systems *(Hotline-Telecom, Moscow)* p 93 (In Russian)

[24] Shtovba S D 2007 Design of fuzzy systems by means of MATLAB (*Goryachayali-niya-Telekom, Moscow)* p 288 (In Russian)

[25] Paklin N and Oreshkov V 2009 Business analytics. (*From data to knowledge, Peter)* p 622 (In Russian)

[26] Vityaev E E and Martynovich V V 2017 Probabilistic formal concepts in a natural language processing task *The bulletin of irkutsk state university. series: mathematics* **22** pp 31-49. (In Russian)

[27] Glushenko S A 2017 Analysis of software for the implementation of fuzzy expert systems *Software systems and computational methods* **4** pp 77-88. (In Russian)

[28] Kopyrin A S and Kopyrina A O 2021 Development of the generic system of inference rules by knowledgebase *Software systems and computational methods* **1** pp 1-9. (In Russian)

[29] Leonenkov A V 2005 Fuzzy modeling in MATLAB and fuzzyTech *(BHV, Petersburg)* p 736 (In Russian).

[30] Lee K H 2004 First Course on Fuzzy Theory and Applications. *Springer Science & Business Media* p 335.

[31] Sorokin A 2020 Improvement of information-analytical complexes based on hierarchical systems of fuzzy output *Large-scale systems control* **88** pp 99-123. (In Russian)

[32] Nafchi H Z and Cheriet M 2018 Efficient No-Reference Quality Assessment and Classification Model for Contrast Distorted Images *IEEE Transactions on Broadcasting* **64** 2 pp 518-23.

[33] Leont'ev V K 2009 On measures of similarity and distances between objects *Computational Mathematics and Mathematical Physics* 11 **49** pp 2041-58. (In Russian).

Predicting of cyber attacks on critical information infrastructure

I M Kosmacheva[1], N V Davidyuk[1], SV Belov[1] , I Yu Kuchin[1], I Yu Kvyatkovskaya[1], M F Rudenko[1] and V I Lobeyko[2]

[1]Astrakhan State Technical University, Tatischeva str, 16, Astrakhan, 414056, Russia
[2]Astrakhan State University, Tatischeva str, 20A, Astrakhan, 414056, Russia

E-mail: ikosmacheva@mail.ru

Abstract. According to modern statistics and analytical reviews, targeted computer attacks (cyber attacks) are becoming more and more numerous. Attackers began to use non-standard schemes for implementing attacks, using employees of organizations as intermediaries, which reduces the efficiency of detecting violations. At the same time, the targets of attackers are increasingly critical information infrastructure (CII) objects. The number of cyberattacks on the critical infrastructure of the Russian Federation increased by 150%. Successful attacks on CII are associated with a lack of software updates for industrial equipment, personnel errors, incorrect configuration of protection tools and can potentially lead to disasters. Prediction of computer attacks on CII based on a comprehensive analysis of the characteristics of incidents and system users can significantly increase the efficiency of incident detection, since it is obvious that technical and anthropogenic characteristics in this case should be taken into account together.It is difficult to classify computer incidents due to the volume and heterogeneity of the data about them. The paper proposes approaches that provide for the initial systematization of system log data and user characteristics, an assessment of their informativeness. This will reduce the complexity of further data processing and increase the performance of the computer attack forecasting system by excluding some uninformative data from a single secure storage. The second important task is to create test systems based on available platforms for analyzing and detecting computer incidents in order to train future information security specialists in big data analysis technologies.

1. Introduction

The information security incident management process becomes the foundation of an organization's information security system. For the operational detection of computer incidents in the IT- systems of critical information infrastructure (CII) objects, the CII subject needs, within the framework of creating a security system for significant CII objects, to set up security audit of information security events sources, centralized automated collection, processing and analysis of information security (IS) events, and also the response to them [1-3].

The relevance of the research topic is due to the fact that Russia is introducing systems for predicting computer attacks on CII objects, which require prompt restoration of full-fledged work in case of failures [1], operational response to incidents. Incidents are most often associated with the human factor, therefore, the concept of behavioral analysis (User and Entity Behavior Analytics - UEBA, "behavioral analysis of users and entities") is important - technologies for identifying cyber threats based on the analysis of user behavior, the formation of patterns of typical user behavior, and devices, applications and other objects in the information system for the timely detection of deviations in the behavior of objects from the typical, normal. An effective method for detecting insiders is analyzing user behavior, building user profiles and identifying anomalies in their behavior [4-7], which is not fully implemented in popular systems of the SIEM (Security Information and Event Management) class, which are quite expensive and difficult to configure.

In general, SIEM systems provide:

- consolidation and filtering of data from various sources (network devices, applications, OS, firewalls, IDS, anti-virus tools, DLP);
- centralized storage of event logs in a unified format and enrichment of events with missing information;
- identification of the correlation of events (search for relationships and patterns) and their processing according to the rules, which makes it possible with a high probability to determine anomalies, potential threats, failures in the operation of IT infrastructure, attempts of unauthorized access, external attacks;
- automatic notification of information security incidents, maintaining a database of incidents and generating reports; - provision of tools for analyzing events and analyzing incidents [8].

SIEM systems use correlation, methods for identifying patterns of incidents, but algorithms and rules for the operation of incident management systems are not available for study, and the used signature methods are characterized by a high percentage of false positives. During the process of adding rules for managing incidents in the system, It's possible to harm the logic of other critical rules.

SIEM systems for implementation in small and medium-sized businesses of a certain industry must have their own specifics and functionality, comparing certain events from certain sources for high-quality attack detection. Due to the complexity, requirements, price of projects for managing information and events in the security system using SIEM often do not meet the expectations of customers and users.

Modern attacks, which are becoming more and more camouflaged, are difficult to distinguish from regular, legitimate user behavior. The paper proposes an algorithm for constructing a test subsystem based on a freely distributed version of the Deductor information and analytical system for organizing an event data store, systematizing the characteristics of computer incidents and / or users, processing them in order to establish links between the attack and the cause of the threat. Big Data technology, implemented in the Deductor system, allows generating rules (security policies), configuring them in accordance with user profiles for more effective detection of attempts to implement attacks [9].

2. Main part

2.1. Data collection and systematization

The composition of system log data is of great importance for analyzing the danger of information security incidents: operating systems (OS), antiviruses, firewalls, various services and applications containing heterogeneous information, including which accounts were granted access or what actions were taken executed at a certain point in time. Records of various logs, called "security events", are the main source data used by SIEM, UEBA systems.

So, for example, from the Access.log file (figure 1) you can get information in the form of technical characteristics T_i ($i = 1,2, ...$) about the browser used and the user's IP address, data on the client's location (country and city), etc., registered at a certain point in time or during a session.

IP-адрес	Data	Data transfer method	Server code	count of bytes returned by the server	url	used browser and operating system
199.15.234.66	07/Mar/2017:18:24:31	get	200	3033	http://www.google.com	Mozilla/5.0 (Windows; U; Windows NT 5.1; en-US; rv
175.44.24.82	07/Mar/2017:18:44:36	get	200	2334	http://www.google.com	Mozilla/5.0 (compatible; MSIE 9.0; Windows NT 6.1; \
175.44.24.82	07/Mar/2017:18:44:37	get	200	1184	http://www.buttercupgames.com/oldlink?itemId=EST	Mozilla/5.0 (compatible; MSIE 9.0; Windows NT 6.1; \
175.44.24.82	07/Mar/2017:18:44:38	get	200	1814	http://www.buttercupgames.com/category.screen?ca	Mozilla/5.0 (compatible; MSIE 9.0; Windows NT 6.1; \
175.44.24.82	07/Mar/2017:18:44:39	get	200	1073	http://www.buttercupgames.com/oldlink?itemId=EST	Mozilla/5.0 (compatible; MSIE 9.0; Windows NT 6.1; \
175.44.24.82	07/Mar/2017:18:44:40	POST	200	3067	http://www.buttercupgames.com/product.screen?pro	Mozilla/5.0 (compatible; MSIE 9.0; Windows NT 6.1; \

Figure 1. Access.log file fragment

Certain user characteristics C_i (i = 1,2, ...) may indicate an insider-attacker: skill level, admission level, professional field of activity, age, gender, personal qualities, activity on social networks, etc. It is proposed to use encoded files as input data about users (figure 2), where **A, B**, ... is a certain fact revealed when observing a given user within a transaction / session with a certain number (ID). This fact may be:

- launching a specific list of applications;
- excess traffic volume;
- access to the system at unusual times;
- making changes to the system configuration;
- incorrect password when entering, etc.

If the fact is fixed, then it is assigned 1, otherwise - 0, as shown in figure 2.

ID	A	B	C	D	E	F	G	H	I	K	...
1001	1	0	0	1	1	0	0	0	0	0	...
1002	1	0	0	0	0	1	0	0	0	0	...
1003	1	1	1	0	0	0	0	0	1	0	...
...	...	...	...	...	...	...	...	...	...	...	...

Figure 2. Data about system users.

After the aggregation of data on technical and user characteristics in the download file, it is proposed to upload them to the Deductor system. These data are heterogeneous and not always systematized, which affects the further result of the analysis of incidents, especially in the face of a lack of time for decision-making when restoring services.

It is necessary to convert data about events in the system to a convenient data type, format, fill in the existing gaps (which are caused by a lack of information, technical errors in data collection), identify and then exclude or edit anomalies in the data, as well as set uninformative parameters (the values of which are stable and are unchanged within the group).

2.2. Preparing and filtering data

Monitoring and detection of attacks is based on anomaly analysis, semantic, signature, statistical analysis. When automating the information security management process for CII, it is important to use algorithms that speed up data analysis to make a decision about attributing an event to a possible attack (incident), because processing is carried out in real time in the presence of a large amount of data and in conditions of limited system resources. To do this, it is necessary to regroup the data, to identify uninformative, contradictory, anomalous data in order to narrow the search area for events with characteristics that have signs of a computer attack. The example of this approach is shown schematically in figure. 3.

It's possible to use pre-preparation algorithms (normalization, filtering, filling in the gaps) in the Deductor system. The detection of anomalies and outliers for quantitative data is available in the system, and for qualitative, categorical (of which there are quite a few in the database of attack indicators) - you can use the probability distribution of the values of the features.

By registering all events over a period of time, objects accumulate huge arrays of events. Each such event (incident) is a set of features that characterize it at a certain point in time.

Association rules search methods can be used to study the probability of a combination of various parameters, signs of an event, an incident.

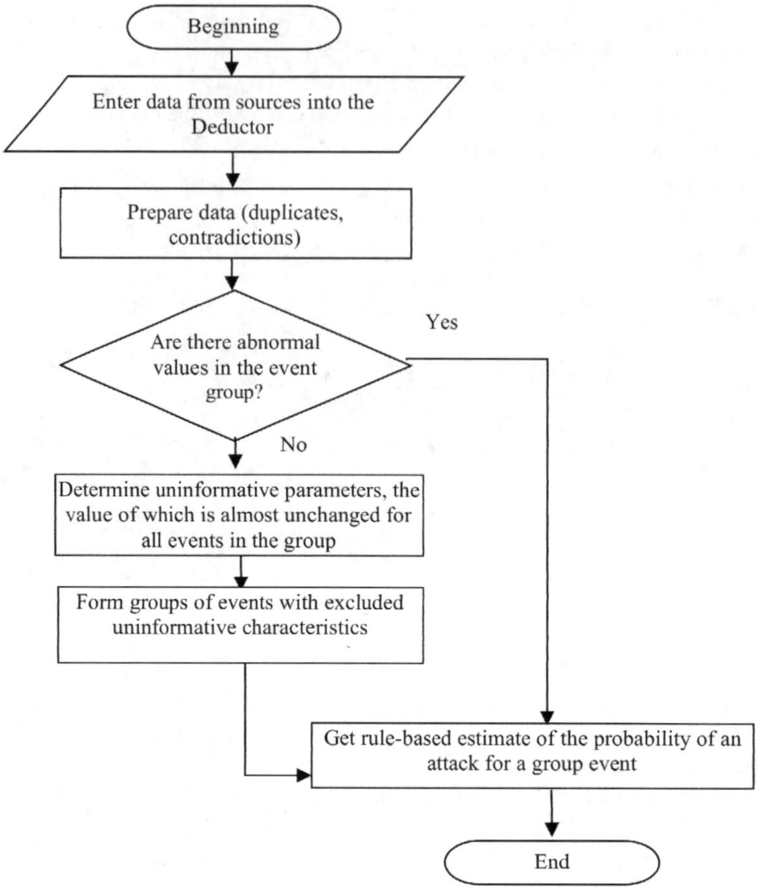

Figure 3. Algorithm of sequential selection of events in the database when examining them for signs of an attack.

2.3. The example of processing input data

As a result of studying the data of system logs with about 13,000 records using the Deductor information and analytical system, initial statistics were obtained (figures 4-5).

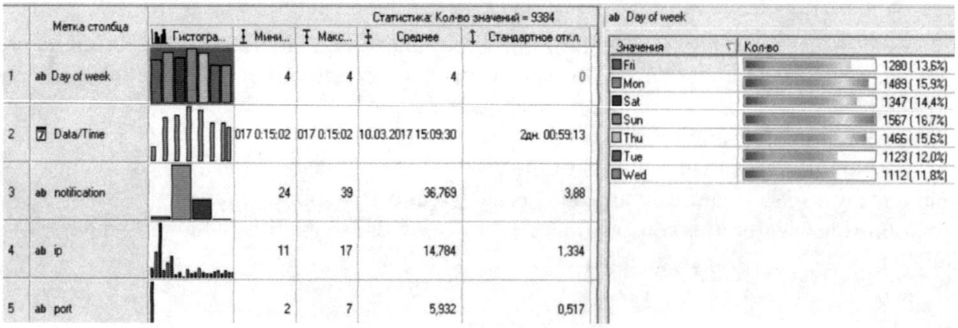

Figure 4. Distribution system files parameters values.

This statistics makes it possible to determine the time at which the most activity was observed, which notifications, IP addresses or PORTs are more frequent were used when registering system events, and other attributes critical for the process of predicting attacks on CII. The use of statistical analysis methods is the most common type of implementation of detection technology, including abnormal behavior, which is important for the formation of patterns, profiles of system and user behavior.

Checking the abnormal data indicated their presence for parameters such as URL request and Number of bytes transmitted by the server (figures 5-6).

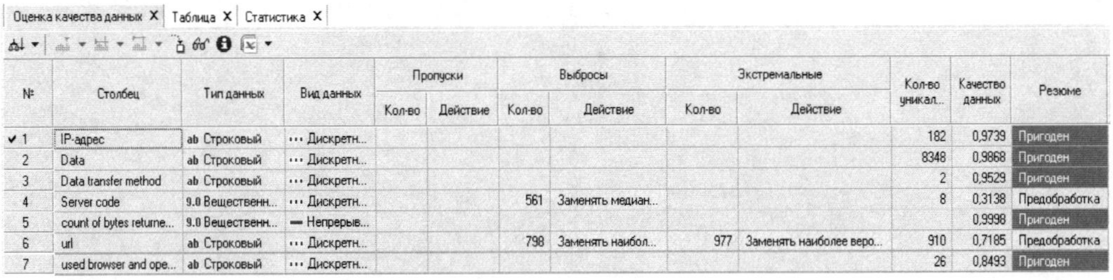

Оценка качества данных X | Таблица X | Статистика X

№	Столбец	Тип данных	Вид данных	Пропуски		Выбросы		Экстремальные		Кол-во уникал...	Качество данных	Резюме
				Кол-во	Действие	Кол-во	Действие	Кол-во	Действие			
✔ 1	IP-адрес	ab Строковый	··· Дискретн...							182	0,9739	Пригоден
2	Data	ab Строковый	··· Дискретн...							8348	0,9868	Пригоден
3	Data transfer method	ab Строковый	··· Дискретн...							2	0,9529	Пригоден
4	Server code	9.0 Вещественн...	··· Дискретн...			561	Заменять медиан...			8	0,3138	Предобработка
5	count of bytes returne...	9.0 Вещественн...	— Непрерыв...								0,9998	Пригоден
6	url	ab Строковый	··· Дискретн...			798	Заменять наибол...	977	Заменять наиболее веро...	910	0,7185	Предобработка
7	used browser and ope...	ab Строковый	··· Дискретн...							26	0,8493	Пригоден

Figure 5. Check for abnormal data.

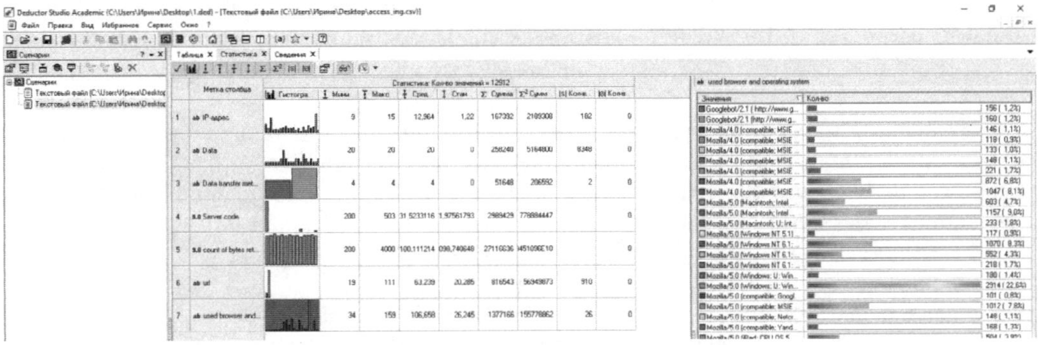

Figure 6. Abnormal URL request.

The location of suspicious IP addresses for which the amount of transmitted data differs can be seen from the dependence graph in figure 7.

So, at the initial stage of building a test subsystem for identifying events with signs of attacks, it is possible to determine informative attributes that will serve as a criterion for classifying events. This will make it possible to use them when constructing the first hypotheses regarding the possible connections between the input and output parameters, expressed using the rules.

2.4. Formation a system of rules for attributing an event to an attack on CII

It's possible to form rules for predicting computer attacks, classifying events in the Deductor system using Decision Trees, association rules, neural networks and other approaches.

1. When choosing further algorithms for data processing, it should be borne in mind that not all of them are designed to work with high-quality data or with data with a large variability of values.

2. When constructing rules, there is also the problem of obtaining a large number of them, some of which have no practical significance. There are criteria for assessing the value of a rule, the level of its usefulness (support, reliability).

3. If the value of the rule support is too large, then as a result of the operation of the algorithm, obvious and well-known rules will be found. A too low support value will lead to finding a large number of rules that are not obvious to the analyst. If the level of confidence is too low, then the value of the rule is questionable and also leads to an increase in the number of rules [10].

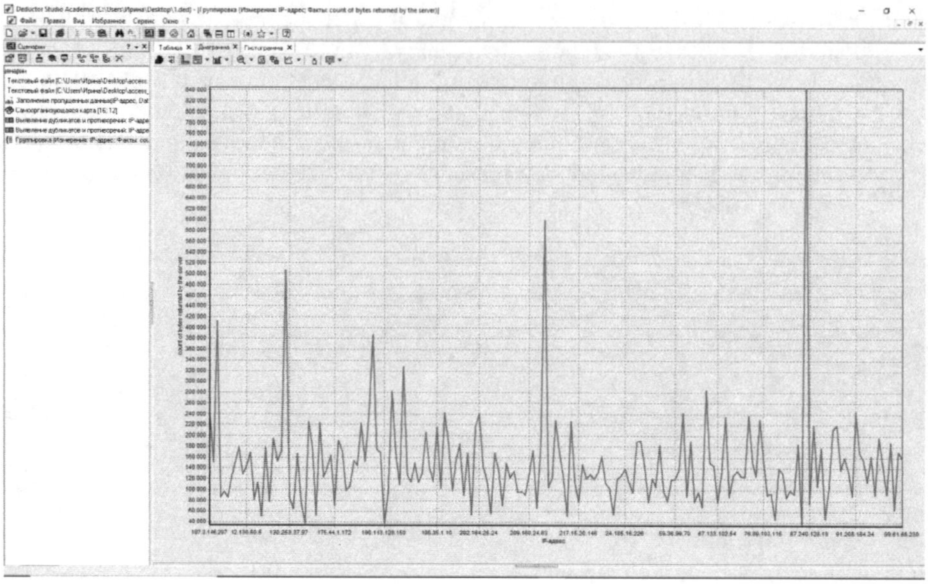

Figure 7. Location of suspicious IP-addresses for which the volume of transmitted data differs in comparison with others.

It is possible to investigate the rules for compatibility (absence of contradictions), determine their value, level of usefulness, thresholds in order to reduce the risks of errors of the 1st and 2nd kind, if there is a training base with the characteristics of the event and information about the attribution (or non-attribution) of this event to an attack.

3. Conclusion

Personal data and characteristicsof users along with collected data about system events in logs, DLP, SIEM systems, etc. should be aggregated and used in the development of predictive and classification models to identify a warning, rank incidents and establish links between an incident and its cause. This will make it possible to adapt the analysis algorithms based on artificial intelligence methods for the specific specifics of the environment in which the attack is being implemented, including the characteristics of the user and his behavior in the system.

In the paper the authors analyzed the capabilities of SIEM systems for solving problems of detecting computer attacks on CII objects. Widespread SIEM systems do not always provide effective detection of incidents; they require highly qualified personnel both for deployment and further support in the face of time pressure. This requires considering alternative solutions for the use of information security analytics.

The paper shows the main difficulties in preparing input data for processing and analysis, presents the main stages in the implementation of tools for building a system for detecting attack signs in the event database based on the Deductor system.

Thus, in order to identify incidents at CII at an early stage, it is necessary to improve approaches to predicting attacks, implementing algorithms for detecting malicious activity and predicting potential unauthorized impact based on the analysis of both technical and anthropogenic characteristics, training personnel in advance for this kind of analytical work. For this purpose, the paper proposes the test subsystem for the analysis and detection of computer incidents at the CII.

References

[1] Panin D N, Bobkov E O and Balashova E A 2020 Analysis of berrating attack on critical information and infrastructure in technologies *Personal autonomy* **2(22)** pp 55-64

[2] Sibikina I, Davidyuk N, Kosmacheva I, Kuchin I and Belov S 2020 The algorithm for controlling the use of big personal data *Proceeds of Int. Conf. En and T 2020* doi:10.1109/EnT50437.2020.9431268
[3] Vasilyeva I N 2019 *Investigation of information security incidents: a tutorial* (SPb.: Publishing house of St. Petersburg State University of Economics) p 113
[4] Abdulorizov A N 2020 Critical information infrastructure of a security facility *Young Scientist* **20(310)** pp 16-19
[5] Doynikova E V, Fedorchenko A V and Kryukov R O 2020 Determination of features of cyber-attack goals based on analysis of data in open security data sources *IOP Conference Series: Materials Science and Engineering* (Krasnoyarsk Science and Technology City Hall of the Russian Union of Scientific and Engineering Associations) p 12160
[6] Ahmed M, Mahmood A N and Hu J 2016 A survey of network anomaly detection techniques *J. Netw. Comput. Appl.* **60** pp 19–31
[7] Alpeev E V, Stadnik A N, Korolev I D and Konyshev E A 2020 Verification of false operations in the system for detecting computer attacks by using a complex method based on intellectual analysis of data and decision tree *Information systems and technologies IST-2020* (Nizhny Novgorod State Technical University) pp 462-467
[8] Fedorchenko A V, Levshun D S , Chechulin A A and Kotenko I V 2016 An analysis of security event correlation techniques in siem-systems. Part 1 *SPIIRAS Proceedings* **4 (47)** pp 5-27
[9] Feshina E V, Ovcharov A P and Labintseva V R 2018 Possibilities of the Deductor system as a event tool for intellectual data analysis *Colloquium-journal* **12-6 (23)** pp 49-51
[10] *Common patterns or samples*: https://intuit.ru/studies/courses/6/6/lecture/186?page=2

Application of a computer vision system for recognizing tomato fruits and determining their position relative to the gripper device of the harvesting robot

A.V. Rybakov[1], A.N. Marenkov[1], V.A. Kuznetsova[1], A.V. Stanishevskaya[1]

Astrakhan State University, 20a Tatishcheva street, Astrakhan 414056, Russia

E-mail: rybakov_alex@mail.ru

Abstract. The article presents a method for recognizing tomato fruits covered with foliage, determining their centers and boundaries using the OpenCV computer vision library and a hardware complex based on Raspberry Pi 4. The methods for solving the inverse kinematics problem for the five-link robotic manipulator designed by the authors, installed on a mobile platform, in order to create a robot for collecting fruits are considered. The simulation of the manipulator movement in the Scilab environment is performed.

1. Introduction

The introduction of agricultural robots with technical vision systems can significantly increase the productivity and profitability of the industry and reduce the cost of production. In addition, the use of robots makes it possible to exclude a person from a number of monotonous operations, reduces the loss of working time associated with the human factor and contamination of the greenhouse with fungi and microorganisms [1]. The task of developing manipulation systems with precise positioning for agricultural robots is therefore of great importance. One of the ways to control manipulators is to position them using computer vision. Computer vision solves several tasks at once, and object recognition is considered one of the highest priorities. For example, in [2], a Mitsubishi Movemaster RV-M1 robotic manipulator with a built-in webcam for processing images of round objects is described. This system was successful in the presence of ideal lighting conditions. In [3], a mechatronic system with image processing for sorting objects is discussed, and in [4], the use of two CCD cameras for solving lemon sorting problems was discussed. Visual Basic 6.0 software was used to evaluate the RGB color space of lemons. For three different lemon varieties, the system showed success in the range from 86.6% to 100%. The RGB analysis method was also successful in the work [5]. Here, mirrors were used to evaluate three images of wheat, which allowed to increase the number of analyzed features. The results obtained show that an increase in the number of functions processed by a computer increases the success of the system operation practically without spending computing power. The system showed 10% greater accuracy than conventional sorters by one attribute.

The paper [6] describes a sorting system that uses automated visual control to remove dirt from wool. In the article [7], various image recognition methods for sorting raisins were studied. The system capable of classifying fruits at high speeds [8] included a camera that uses a system for determining shades, saturation and color intensity. The results of the work show that colors can be accurately classified with only one parameter-hue, which provides 3 orders of magnitude faster classifications than the usual RGB system.

The paper [9] presents an autonomous object sorting system based on vision using a robotic manipulator. The Scorbot-ER 3 robotic arm was modified with an additional camera and markers, which allowed to expand the system's capabilities. In the course of this study, an industrial robot Scorbot - ER 9Pro was used, and a USB camera was installed on a camera holder made on a 3D printer. The camera holder was mounted on the gripper. The tasks of the project were as follows: to determine the specification of the object using image processing methods; set the interface between Matlab and ACL Scrobot; select and place objects based on size, color and shape.The photos of the objects were taken using the image processing tool in Matlab. After the detection of the main bodies, the coordinates of their centroids were determined. Controlling the robot's position was a key task in this study, as it was integrated with several software systems. Image processing is carried out using RGB and HSV methods, which are processed into a single image showing the location of pollutants. It is reported that this approach provides an accuracy of 96%.

2. Experiments

There are three main ways to get information about a three-dimensional scene:
1. Based on a two-dimensional image, an estimate of the third coordinate is given based on the relative apparent location of individual objects. Such systems are called 2.5-dimensional (2.5 D).
2. The monocular CVS (Computer Vision System) is supplemented by a rangefinder that determines the third coordinate-the range (distance) to the object.
3. The stereo CVS containing two video cameras separated by a certain distance (base) in a plane perpendicular to the third coordinate — the distance to the object.

The first method of estimating the depth of the scene is the simplest. It is based on the use of the following techniques:

- estimation of the distance to the object based on knowledge of its dimensions;
- the use of the obvious fact that the object covering the other object is closer when overlapping images of opaque objects on each other.
- the use of the perspective effect, which consists in the visual convergence of two parallel lines going into the distance;
- analysis of the light effects of shadows from objects, glare, etc.

The second method is aggregation, i.e. when the third coordinate is determined not by the SCR, but by another system.

The third method -stereo systems. Two video cameras separated by a certain distance-the base, allow you to form a general three-dimensional image of the visible scene. A quantitative estimate of the distance to individual objects can be defined as the height of the triangle formed by the base and the directions from each camera to this object. To do this, you only need to know the angle between these directions.

Based on the results of the analysis of these three methods, we selected a monocular SCR.

To analyze an image in order to determine the coordinates and dimensions of objects on it, for example, the m function can be used, which calculates the image moments with an accuracy of up to the 3rd order of a vector or raster shape. Image moments in computer vision, image processing and related fields are some particular weighted average values of the pixel intensities of an image, or functions of such values. [4]

The concept of robotic harvesting using mobile robots equipped with manipulators can be explained using Figure 1. Earlier in [1], the features of manipulator control of such a robot were presented.

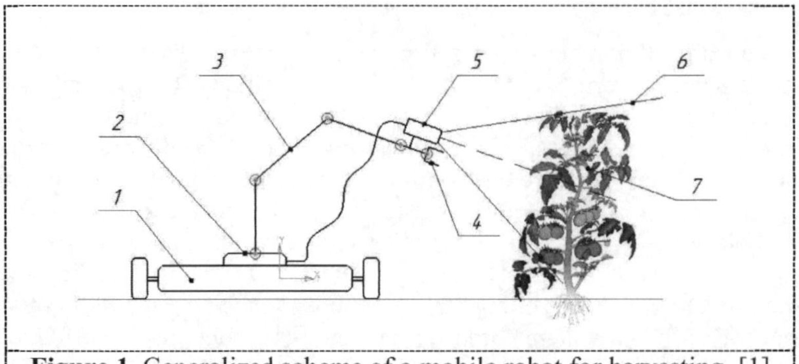

Figure 1. Generalized scheme of a mobile robot for harvesting. [1]

Figure 1 shows a mobile robotic platform 1 with a manipulator 3 installed on it, equipped with a specialized grip - 4 and a computer vision module - 5. The latter is a video camera and an ultrasonic rangefinder connected to the control unit - 2. The video camera and rangefinder have a certain viewing angle -6 and are designed to ensure an accurate approximation of the tong with the fruit located on the stem of the plant - 7.

The main tasks in this case are the following: accurate determination of the positions of the fruits; trajectory control of the movement of the manipulator's grip; implementation of the operation of separating the fruit from the stem.

The edge detection of the tomato was implemented using a high-pass filter using the Canny operator [2]. Figure 2 shows the result of applying the Canny operator to the captured image and detecting edges.

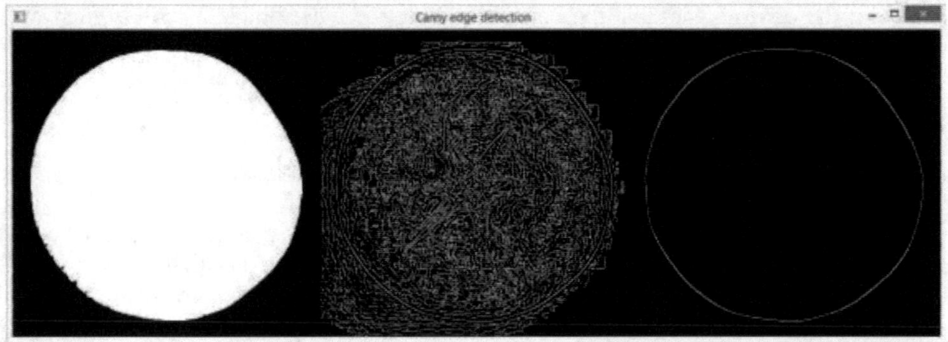

Figure 2. Detecting the edges of a tomato using the Canny operator; a) binary image; b) fetal edges obtained without hysteresis thresholds; c) applying hysteresis thresholds and obtaining a contour.

To determine the contour of the fruit, the expansion and erosion operators were used, applied to the resulting edges. This procedure fills in the gaps between the pixels of the tomato edges that were incorrectly sealed. Figure 3 shows the dilation and erosion operators.

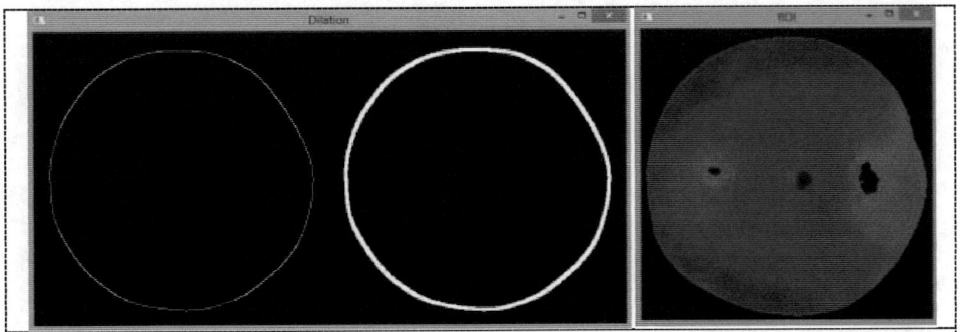

Figure 3. The result of the program for determining the contours of tomatoes.

For understanding the conception of distance measurement it is necessary to start with a simple one-dimensional case presented in Figure 4. Nevertheless, this one-dimensional model is applicable to a number of real problems, for example, to describe obtaining images of a flat earth's surface from an airplane using a camera whose optical axis is directed vertically downward. The observed point B is projected into the image point E. The origin of the camera coordinate system is located at point O.

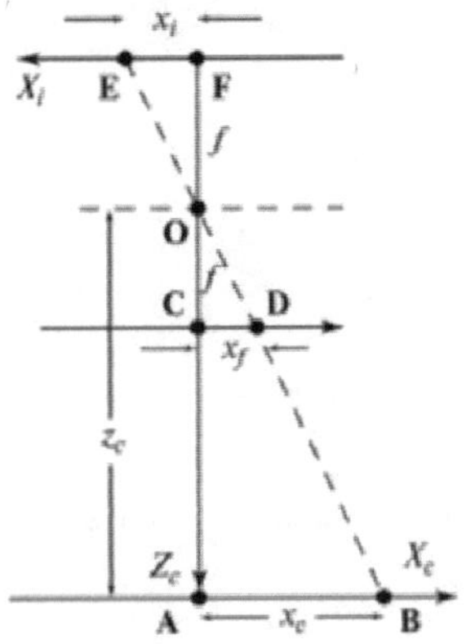

Figure 4. The simple model of a perspective projection with an indication of the actual and front planes of the image.

The distance between points O and B measured along the optical axis is equal to the distance zc between points O and A. The image of point B is shifted from the center of the image by a distance xi along the Xi axis. The focal length of the lens is f. Taking into account the similarity of triangles, we write equation 1. It follows from it that the actual coordinate of a point on a two-dimensional image (the size of the object) is equal to the spatial coordinate of the point (the size of a one-dimensional object), taken at a scale equal to the ratio of the focal length to the distance to the object

$$x_i f^{-1} = \frac{x_c}{z_c}$$

$$x_i = (f z_c^{-1}) x_c \quad (1)$$

The projection mapping of a three-dimensional space onto a plane is not unambiguous. All three-dimensional points lying along one projecting ray will be projected into the same point of the two-dimensional image. Therefore, a significant amount of spatial information is lost on a two-dimensional image.

$$\frac{x_i}{f} = \frac{x_c}{z_c} \; or \; x_i = (f z_c^{-1}) x_c$$
$$\frac{y_i}{f} = \frac{y_c}{z_c} \; or \; y_i = (f z_c^{-1}) y_c \qquad (2)$$

The mathematical model in the form of equations (7) can be used in computer algorithms to determine the set of all three-dimensional points on the projecting ray passing through the image point (x_i, y_i).

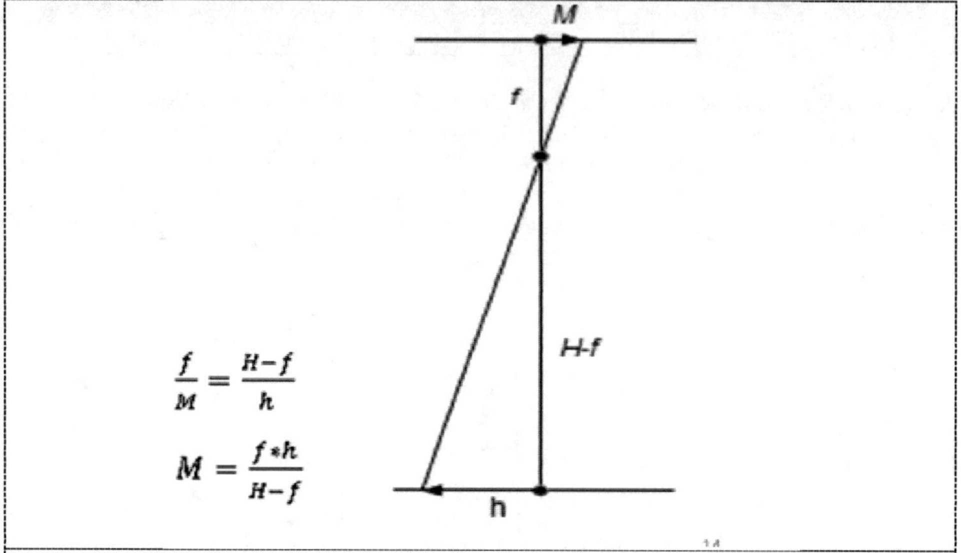

Figure 5. Geometric model of image projection from the camera.

We have developed a 3D model of a mobile robot for tomatoes harvesting with a Puma-type manipulator placed on it (Figure 6).

Figure 6. 3D-model of robot for tomatoes harvesting.

For further calculations and modeling, a model was built in Scilab using the Robotic Toolbox library (Figure 7). The library is a set of tools that allow you to create a model of a robotic manipulator, to carry out its behavior and control.

The library's set of tools allows you to:

- Create a robotic model in accordance with the Denavit-Hartenberg notation.
- Solve direct and inverse kinematics
- To control the forward and reverse dynamics
- Display of the graphical schematic model of the manipulator and its animation
- Use Xcos blocks to simulate the control of a robotic model
- Perform geometric transformations.

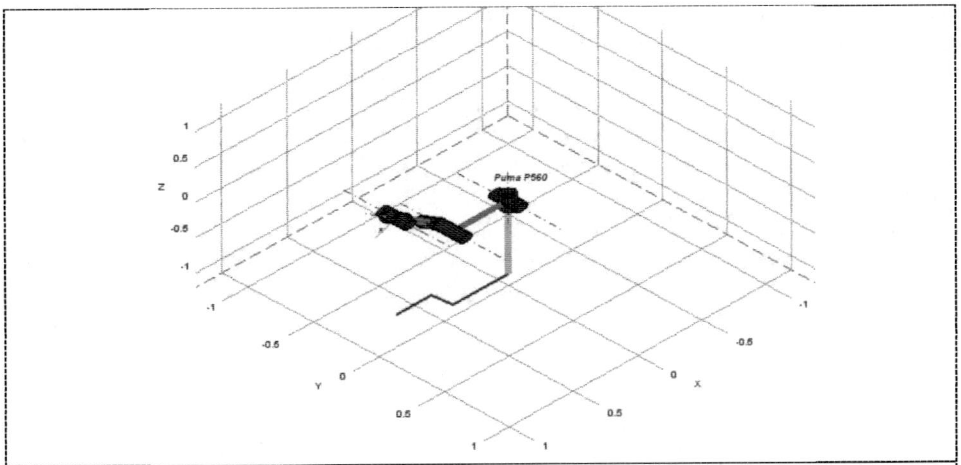

Figure 7. Scilab model of manipulator for harvesting.

3. Conclusions

To harvest tomatoes, it can be use robotic complexes equipped with multi-link manipulators. In this case, the accuracy of determining the coordinates of the center of the object detected by the STZ depends on many factors, such as illumination, camera parameters, and so on.

In this article, we proposed an approach to determining the coordinates of the center of the tomato fruit; we estimated the dependence of the error in determining the coordinates of the center of the object on the distance between it and the camera lens.

Based on the results obtained, a program for positioning the manipulator was developed. In it, the values of the coordinates of the center of the fetus are fed to the input, taking into account the error and the coordinates of the robot's grip; the direct and inverse problem of the manipulator's kinematics is solved – in order to bring the grip to the fetus.

4. References

[1] Rybakov A V, Lihter A M, Pogozheva A B, Mihailova A V and Dusaliev A B. 2020 *Proektirovanie robototekhnicheskih manipulyatorov s sistemoi komp'yuternogo zreniya dlya sbora tomatov* [Design of robotic manipulators with a computer vision system for picking tomatoes] Prikaspiiskii zhurnal: upravlenie i vysokie tekhnologii. [Caspian Journal: Management and High technologies] № 3 (51) 135-147

[2] Djajadi A, Laoda F, Rusyadi R, Prajogo T and Sinaga M 2017 *A model vision of sorting system application using robotic manipulator* Journal.uad.ac.id, [Online] Available: http://journal.uad.ac.id/index.php/TELKOMNIKA/article/view/615/424

[3] 2013 *Object Sorting System Using Robotic Arm* Journaldatabase.info, [Online] Available: http://journaldatabase.info/articles/object_sorting_system_using_robotic.html

[4] Khojastehnazhand M *Development of a lemon sorting system based on color and size* [Online] Available: http://www.academicjournals.org/article/article1380110185_Khojastehnazhand%20et%20al.pdf
[5] Pearson T, Brabec D and Haley S 2008 *Color image based sorter for separating red and white wheat,* Sensing and Instrumentation for Food Quality and Safety vol. 2 no. 4 280-288
[6] Zhang L, Dehghani A, Su Z, King T, Greenwood B and Levesley M 2005 *Development of a Mechatronic Sorting System for Removing Contaminants From Wool* IEEE/ASME Transactions on Mechatronics vol. 10 no. 3 297-304
[7] Abbasgholipour M, Omid M, Keyhani A and Mohtasebi S 2011*Color image segmentation with genetic algorithm in a raisin sorting system based on machine vision in variable conditions* Expert Systems with Applications vol. 38 no. 4 3671-3678
[8] Kay G and G de Jager 1992 *A versatile color system capable of fruit sorting and accurate object classification* Proceedings of the 1992 South African Symposium on Communications and Signal Processing
[9] Ali Md K, Aizat K, Yerkhan T and Zhandos 2018 O *Vision-based Robot Manipulator for Industrial Applications* Procedia Computer Science 133 205-212 10.1016/j.procs.2018.07.025.
[10] Kennet D A 2014 *Practical Introduction to Computer Vision with OpenCV* 237-241

Journal of Physics: Conference Series **2091** (2021) 012064 doi:10.1088/1742-6596/2091/1/012064

Development of a fuzzy expert system for power transformer diagnostics

A P Khlebtsov[1], A N Shilin[2], A V Rybakov[1] and A Yu Klyucharev[1]

[1]Astrakhan State University, Tatishcheva 20a, Astrakhan 414056, Russia,
[2]Volgograd State Technical University, 28 Lenin Ave., Volgograd 400006, Russia

E-mail: gergines@mail.ru

Annotation. *In this paper, an expert information system for assessing the technical condition of a power transformer is developed. The system will work on the basis of the fuzzy logic device, and provide operational information about the state of the power transformer. The paper uses fuzzy inference algorithms. The R programming language is used to write a program that uses fuzzy logic. We analyzed the data of chromatographic analysis of gases dissolved in oil, as well as the data of thermal imaging images, identifying the most heated points in power transformers. A database of fuzzy logic rules has been formed. Several examples of defuzzification of the results obtained by the center of gravity method are given. As a result of the program, a three-dimensional graph was obtained that characterizes the surface of the fuzzy output. The developed software package allows you to detect defects in working electrical equipment at an early stage of their development, which not only prevents a sudden shutdown of production as a result of an accident, but also significantly reduces the cost of repairing equipment and increases its service life*

1. Introduction

Currently, substations in Russia operate physically outdated equipment, which can lead to losses of electricity for consumers and failures in the electrical networks. This creates a tendency for visible and hidden failures of electrical equipment in the power supply system of consumers. Emergency situations are manifested primarily due to the long-term operation of electrical equipment, including the use of power transformers and switches with worn-out inputs. At the moment, the assessment of the technical condition of electrical equipment is carried out by electrical personnel on the basis of the system of planned preventive repairs (PPR) [1]. The main drawback of the PPR system is the lack of an integrated approach to the maintenance of electrical equipment, which can lead to unjustified labor intensity of equipment repairs [2, 3]. Also, practice shows that the maintenance of obsolete substations must be performed much more often than it is regulated in the rules of the PPR. All this leads to economic costs for the repair and maintenance of electrical equipment.

The solution of the problem of assessing the technical condition of power electrical equipment of electric networks is largely associated with the introduction of effective methods of control and technical monitoring of equipment. Diagnostics is necessary and mandatory for reliable and trouble-free operation of power electrical devices [4, 5].

In this paper, an expert information system for assessing the technical condition of a power transformer is developed. The system will work on the basis of the fuzzy logic device, and provide operational information about the state of the power transformer. Based on the data obtained, the operational staff will be able to take measures in advance to eliminate emergency situations at the substation.

2. Methods for constructing a fuzzy expert diagnostic system

This article uses fuzzy inference algorithms. Fuzzy logic is a form of classical logic and set theory in which the truth value of variables can be any real number from 0 to 1 inclusive. It is used to handle the concept of partial truth, where the true value can range from completely true to completely false [6, 7].

In this paper, we used the R programming language to write a program that uses fuzzy logic. The program reads data from a csv file generated using the sensor readings included in the hardware complex based on the Raspberry Pi microcomputer. The program analyzes data on the state of the transformer oil and the temperature of the inputs. Its purpose is to support decision-making about the state of the transformer. The presented sets package provides a flexible and configurable basic infrastructure for basic fuzzy logic operations in the R database [8]. Plotly R's graph library creates interactive graphs. Creating line graphs, point graphs, area diagrams, three-dimensional diagrams, etc. [9].

Oil diagnostics

Long-term exposure to electromagnetic fields due to prolonged operation of the equipment leads to a significant change in the chemical and electrophysical properties of the oil, which leads to a deterioration in its electrical insulation qualities, and can also cause changes in the occupied volume. All this poses a significant danger not only for the substation equipment itself, but also for the maintenance personnel on its territory.

Thus, early diagnostics of the oil condition is a necessary condition for trouble-free operation of oil-filled electrical equipment [10].

There are four types of transformer oil analysis:

- Analysis of the electrical strength of the oil,
- Chromatographic analysis,
- Analysis of the dielectric characteristics of the oil,
- Analysis of oil moisture content [11].

Chromatographic analysis of oil is the most common method for assessing the insulation condition of oil-filled electrical equipment. The reason for the popularity of chromatographic analysis of oil in the electric power industry is the ability to detect slowly developing defects at an early stage of their development [12].

Methods of sampling, preparation and chromatographic analysis of gases dissolved in oil are given in [13].

In our study, we analyzed the following oil indicators:

- Moisture content;
- Acid number (KOH);
- Flash point in a closed crucible;
- Cleanliness class;
- Dielectric loss angle tangent (tgS);
- Water-soluble acids and alkalis;
- Ionol.

The membership functions are built on the basis of expert evaluation. Experts of one of the energy companies, who are employees of the laboratory of the "service of isolation and protection against overvoltage", were asked to give an appropriate numerical assessment of the elements of the term set of the linguistic variable on a ten-point scale. The membership functions based on this survey are shown in Figure 1. Each indicator has a degree of quality. Good condition, risk area, marriage. Therefore, if at least one of the indicators is in the area of marriage, then the oil is considered not suitable [14]. Similarly, it is considered when the indicator is in the risk area. Based on this, the characteristics of the transformer oil evaluation parameters were determined as:

1. Good condition (good). Described by the triangular membership function. good = fuzzy_triangular (corners=c(0,2)
2. The initial degree of wear (control). Described by a trapezoidal function. control=fuzzy_trapezoi (corners=c(1,2,6,7)
3. Emergency state (crush). Described by the triangular membership function. good = fuzzy_triangular (corners=c(6,9,10)

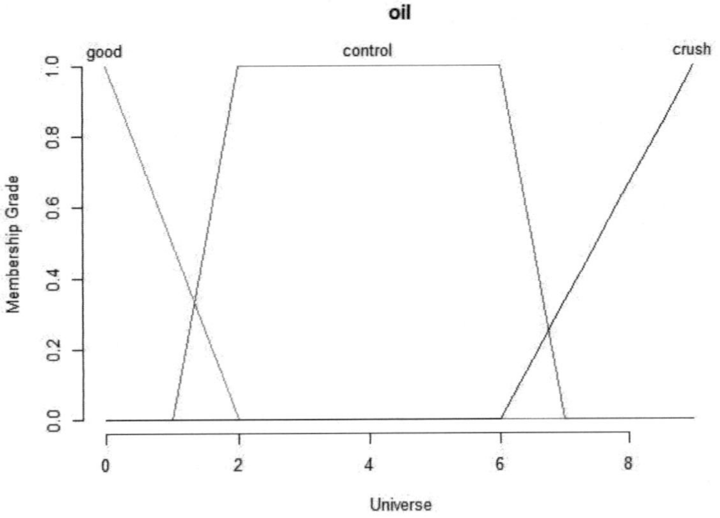

Figure 1. Appearance of the transformer oil condition accessory functions.

Thermal control method

Thermal control methods (TCM) are based on the measurement, analysis and evaluation of the temperature of the equipment under study. The main criterion for the use of TCM in diagnostics is the presence of radiation of heat flows in the controlled object [3]. There are two main types of temperature measurement applicable to the diagnosis of power electrical equipment:

Contact contacts: they are installed directly on the equipment body. Contact devices include thermometers based on the temperature dependence of the thermo-EMF, thermometers based on the temperature dependence of the electrical resistance of a substance.

Non-contact: the temperature of a heated body can be estimated based on the measurement of the characteristics of its thermal radiation, which is a different length of electromagnetic waves [15].

In our study, we analyzed data from thermal imaging images, identifying the most heated points in power transformers. This method made it possible to quickly and informatively accumulate data on the heating of transformers.

According to [16], the classification of the detected defect by the value of excess temperature is carried out according to 3 degrees of malfunction (criteria).

1. The initial degree of development of the defect, which should be kept under control (temperature difference of 10-20 °C)
2. Developed defect, frequent monitoring 1 time per month. Elimination of the defect as soon as necessary (temperature difference of 20-40 °C)
3. Emergency defect. Requires immediate elimination (temperature difference □ 40°C)

In our study, we added another criterion. It implies that the equipment is fully functional and there is no temperature difference between the contact connections. The rules are set based on expert assessments of the cause of the fault. Thus, the specified parameters for the fuzzy logic machine are defined as:

1.Good condition. Described by the triangular membership function good = fuzzy_triangular(corners=c (0,1));
2.The initial degree of development of the defect (control). Described by the triangular membership function control=fuzzy_triangular (corners=c(0,1,3));
3.Developed defect (critical). Described by the trapezoidal membership function critical=fuzzy_trapezoid(corners=c(1,2,3,4));
4.Emergency defect (crush). Described by the trapezoidal membership function crush =fuzzy_trapezoid(corners=c(3,4,9,10).

Figure 2 shows the appearance of the temperature parameter membership functions.

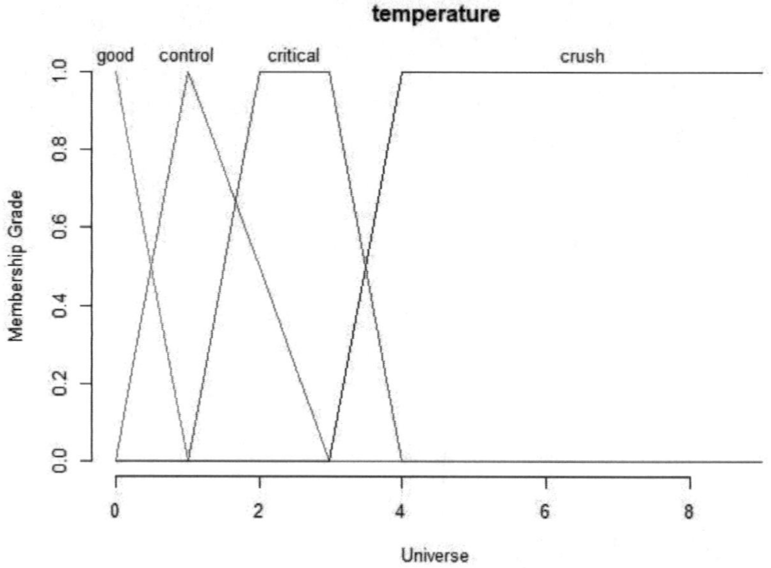

Figure 2. Appearance of the accessory temperature functions.

Based on the described conditions of the membership functions of the temperature parameters and the state of the transformer oil, it is possible to describe the general condition of the fuzzy logic apparatus:

- Excellent condition (perfect). The state in which the transformer is operating normally.
- The initial degree of the defect (to check). A developing defect that should be kept under control.
- Developed defect (to check month). A developing defect that should be kept under control once a month.
- Emergency defect (switch off). An emergency condition of the transformer that requires immediate shutdown of the transformer.

The base of fuzzy rules in the R language looks like this:

fuzzy_rule(temperature %is% good && oil %is% good, condition %is% perfect),
fuzzy_rule(temperature %is% good && oil %is% control, condition %is% tocheck),
fuzzy_rule(temperature %is% good && oil %is% crash, condition %is% switchoff),
fuzzy_rule(temperature %is% control && oil %is% good, condition %is% tocheck),
fuzzy_rule(temperature %is% control && oil %is% control, condition %is% tocheckmonth),
fuzzy_rule(temperature %is% control && oil %is% crush, condition %is% switchoff),
fuzzy_rule(temperature %is% critical && oil %is% good, condition %is% tocheckmonth),
fuzzy_rule(temperature %is% critical && oil %is% control, condition %is% tocheckmonth),
fuzzy_rule(temperature %is% critical && oil %is% crush, condition %is% switchoff),

fuzzy_rule(temperature %is% crush && oil %is% good, condition %is% switchoff),
fuzzy_rule(temperature %is% crush && oil %is% control, condition %is% switchoff),
fuzzy_rule(temperature %is% crush && oil %is% crush, condition %is% switchoff)

The states corresponding to these rules can be summarized in Table 1.

Table 1. Fuzzy Logic rule base.

№	Fault description	Temperature	Oil
1	Excellent condition (perfect)	Working condition	Good condition
2	Initial degree of defect (tocheck)	Working condition	Initial wear rate
		Initial degree of defect development	Good condition
3	Developed defect (tocheck month)	Initial degree of defect development	Initial wear rate
		Developed defect	Good condition
		Developed defect	Initial wear rate
4	Emergency defect (switchoff)	Working condition	Emergency condition
		Initial degree of defect development	Emergency condition
		Developed defect	Emergency condition
		Emergency defect	Good condition
		Emergency defect	Initial wear rate
		Emergency defect	Emergency condition

The appearance of the accessory function is shown in Figure 3.

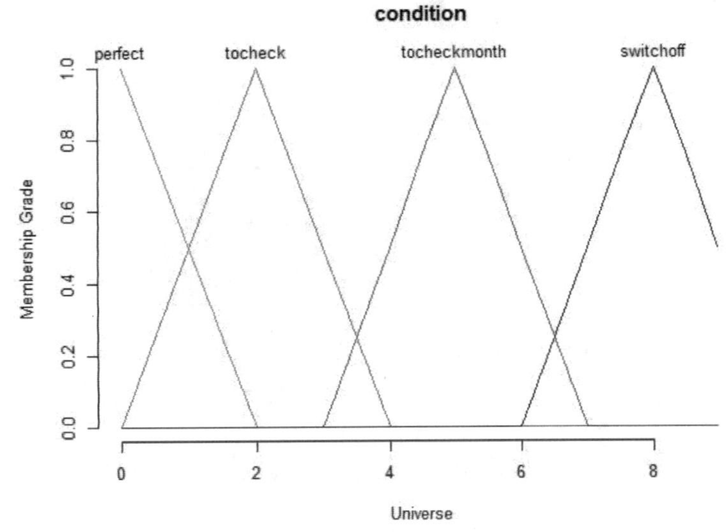

Figure 3. Membership functions of the output variable.

3. Method of conducting the experiment

The program reads data from a csv file formed on the basis of chromatographic analysis of transformer oil and thermal imaging inspection of power transformer inputs.

An example of data from the chromatographic analysis of power transformer oil is presented in Table 2.

Table 2. Chromatographic parameters of transformer oil.

Breakdown voltage	Moisture content	Acid number	Flash point	Cleanliness class	The tangent of the angle diel. losses	alkalinity	Ionol	Conclusion
good ≥ 35-40 > marriage	good ≤ 20 - 25 < marriage	good. ≤ 0,05 - 0,15 < marriage	good ≥ 125 > marriage	good ≤10-11< marriage	good. ≤ 12 - 15 < marriage	good ≥5,5	good ≥0,1	
38	18	0,12	144	6	0,1	5,98	0,057	suitable, risk area

An example of the temperature parameters is shown in Table 3

Table 3. Substation temperature analysis.

Year	Region	Station	StClass	Device	Tdif A	Tdif B	Tdif C	Tdif 0
2012	ARES	PS Pirogovka	110/10	T-2 Input of 10 kV, heating of the bolt contact of the hardware clamp	20		11	

This data is processed by the program and allows you to form the initial conditions in terms of fuzzy logic. As a result of the program, a three-dimensional graph was obtained that characterizes the surface of the fuzzy output shown in Figure 4.

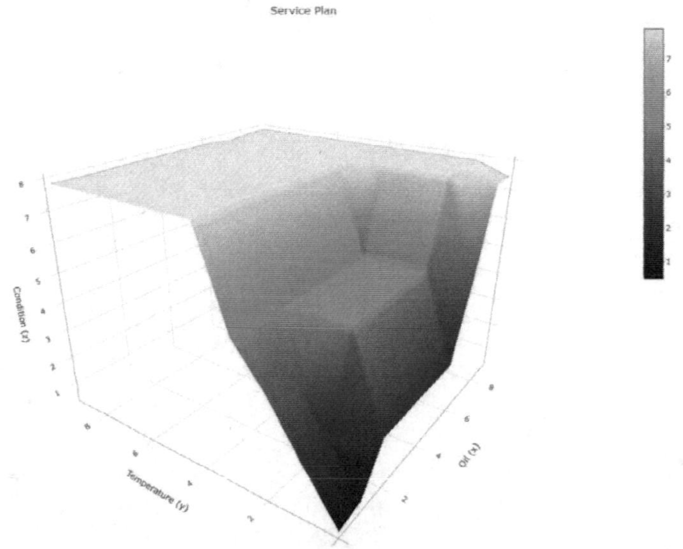

Figure 4. Visualization of the surface of the fuzzy inference system.

4. Results and discussion

Several examples of defuzzification of the results obtained by the center of gravity method are given. In Figure 5, the oil condition is the initial degree of oil wear (2), the temperature of the inputs is the initial degree of defect development (2). The defuzzification output is the developed defect (tocheck month)

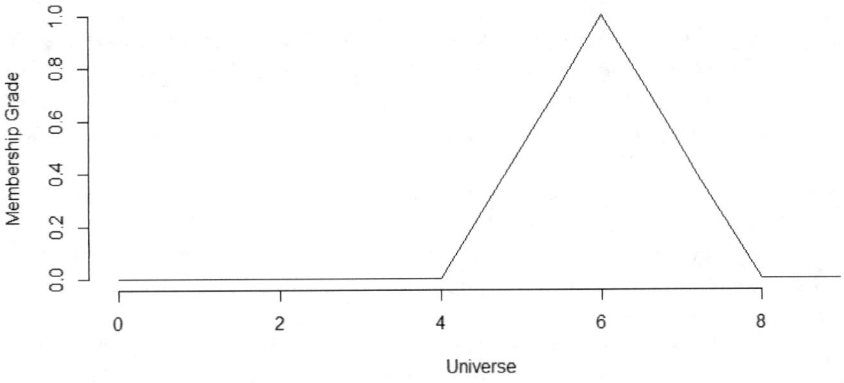

Figure 5. Example of the expert system operation.

In Figure 6, the oil condition is the initial degree of oil wear (2), the temperature of the inputs is the initial degree of defect development (0). The defuzzification output is the developed defect (tocheck month)

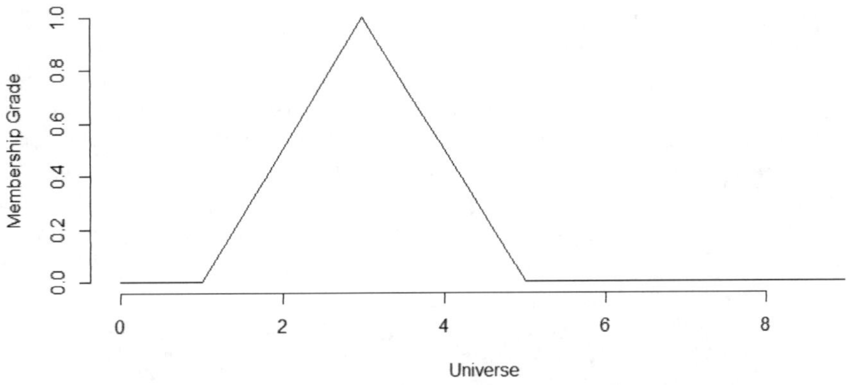

Figure 6. Example of the expert system operation.

These examples are based on statistical data collected from the acts of preventive diagnostics of an oil-filled power transformer based on chromatographic analysis of dissolved gases and thermal imaging. The object of diagnostics is power transformers of the 110 kV voltage class. The quality assessment was carried out by comparing the results of the transformer condition assessment obtained on the basis of the developed model with the actual expert assessments specified in the acts of preventive diagnostics. The comparison showed a fairly large number of their matches, which indicates the adequacy of the developed fuzzy system.

5. Conclusion

The fuzzy expert system is developed using the "R" program. The developed software package allows you to detect defects in working electrical equipment at an early stage of their development, which not only prevents a sudden shutdown of production as a result of an accident, but also significantly reduces the cost of repairing equipment and increases its service life. In the future, the system can be used by the energy services of enterprises to diagnose power transformers with a voltage of 110 kV.

References

[1] Grudinsky P G, Mandrykin S A, Ulitsky M S 1974 *Technical Operation of the Main Electrical Equipment of Stations and Substations* (Moscow, "Energy") p 376
[2] Khalyasmaa A I 2015 *Development of a System for Assessing the Technical Condition of Electric Grid Equipment Based on Neuro-Fuzzy Logical Inference*: dis. cand. of chemical sciences (named after the first President of Russia B. N. Yeltsin. Yekaterinburg) p 147
[3] Khalyasmaa A I 2015 *Diagnostics of electrical equipment of electric stations and substations : a textbook* (Yekaterinburg: Ural Publishing House. un-ta) p 64
[4] Khlebtsov A P, Zaynutdinova L Kh, and Shilin A N 2020 Development of methods and devices for diagnostics of electric power equipment of transformer substations *Electrotechnical and information complexes and systems* **16** 1631427 pp. 14-27
[5] Khlebtsov A P, Rybakov I A, Svishchev N D, and Shilin A N 2020 Expert system for predicting the state of transformers based on fuzzy logic *IOP Conf. Series: Materials Science and Engineering* **976** 012001
[6] Zadeh L A 1976 *The concept of a linguistic variable and its application to making approximate decisions* (Moscow: Mir) p 165
[7] Novak V, Perfileva I and Mochkorzh J 1999 *Mathematical principles of fuzzy logic* (Dordrecht: Kluwer Academic) p 320
[8] Meyer D, Hornik K 2009 Generalized and Customizable Sets in R *Journal of Statistical Software*, **31**, 1018637 pp 1-27
[9] Sievert C 2020 Interactive Web-Based Data Visualization with R, plotly, and shiny. *Chapman and Hall* (CRC Florida)
[10] Shilin A N 2017 *Control and diagnostic devices in the electric power industry: textbook. manual* (VolgSTU. – Volgograd) p 132
[11] Khlebtsov A P, Zaynutdinova L Kh, and Shilin A N 2019 Sensors for monitoring the technical condition of power transformers *Problems of obtaining, processing and transmitting measurement information: materials of the II International Scientific and Technical Conference* 2112982 pp 38-44
[12] Ovsyannikov Yu M, Popov G V 2015 Predicting the state of a power transformer based on the results of chromatographic analysis of dissolved gases *News of higher educational institutions. Electro-mechanics* pp 53-60
[13] Guidelines for the diagnosis of developing defects in transformer equipment based on the results of chromatographic analysis of gases dissolved in oil: RD 153-34. 0-46. 302-00. (VNIIE, Moscow) 2001 28 p
[14] Guidelines for the preparation and conduct of chromatographic analysis of gases dissolved in the oil of power transformers: RD 34.46.303-98. (RAO "UES of Russia") of 13.03.1998
[15] Nazarov V I 2017 *Teplotekhnicheskie izmereniya i pribory* (Minsk: Higher School) 280 p
[16] Methods of infrared control of electrical equipment and overhead lines: RD 153-34. 0-20. 363-99 (Publishing house: ORGRES) 1999 136 p

The Single Page Application architecture when developing secure Web services

D V Kornienko, S V Mishina and M O Melnikov

Bunin Yelets State University, 28 Kommunarov street, Yelets 399770, Russia

Abstract. The article is devoted to the development of a prototype of a secure single page-application (SPA) web service for automating user information accounting. The relevance of this study is very high due to the use of web services. The article provides a classification of web applications and shows the features of the architecture of a single page application used in the development of the service. Based on a comparative analysis of the architectural styles of the API, the most appropriate style was selected. Considered and taken into account the key points of the development of a secure application programming interface (API), the requirements that must be met by RESTful API services. The work used popular authentication schemes (methods). A comparative characteristic of web frameworks of the Python programming language is given, on the basis of which a tool for implementing a web service is selected. Shows the main advantages of using Python when developing paged web services and the security tools included in the standard package of the Flask web services development framework. Shows how to securely prototype a Python RESTful SPA Web Service API using Flask. An example of using the Swagger tool to describe the specifications of the developed API is given. The process of setting up the application is considered in detail. The main recommendations for securing a web application, setting up a database and a web server are listed. The key points of ensuring the protection of the developed web application are considered. Conclusions are made regarding the choice of the architectural style of the application API, the most suitable tools and technologies for the software implementation of the service.

1. Introduction

In the context of the global digitalization of modern society, the trend of a massive transition of software solutions to online is becoming more and more noticeable [1]. Earlier developed software is being rebuilt for the format of interaction on the Web. Web solutions are preferred over mobile or desktop platforms when creating new services. In particular, the field of educational services is moving to the concept of a single information online space. This transition is motivated by a more convenient and accessible mechanism of interaction between the software and the target audience of the product. Any online resources do not impose strict hardware requirements on the user (storage memory, Random Access Memory, central processor unit power), do not tie the interaction to a specific type of platform (personal computer, smartphone, specialized device). They organize the work exclusively through the Internet browser. In addition, the share of online education and e-commerce platforms is growing, which also leads to increased attention and demand for websites and web applications [2].

However, the complexity of the logic of websites and information platforms is constantly increasing. Therefore, ordinary static websites or dynamic solutions based on content management systems (Wordpress, Joomla, 1C-Bitrix) can no longer satisfy the needs of customers when it comes to non-linear education platforms or e-commerce segment, and not trivial landing pages. Many questions arise both in functional terms and in relation to how efficiently the process of development, expansion, protection and maintenance of a software product proceeds. As a result, static websites have been almost completely replaced by full-featured web applications.

A web application is no longer a pre-built collection of HTML (HyperText Markup Language) pages, style files, scripts, and media files. Web applications are based on a client-server architecture based on the separation of a functional block and an external presentation. The business logic of the application is executed on a dedicated server (back-end), while the client interacts through the user interface in their Internet browser (front-end). This approach allows you to generate HTML markup content in real-time, with a minimum number of reloads on the viewed web page. Almost all e-commerce sites, social networks, instant messengers, online programs, forums and search engines are web applications [3].

E.V. Voevodina and Mikishanina E.A. in their article note the relevance of using web applications in the educational environment [4]. They describe a web application, which is a graphical editor that organizes the work with functions and their graphs. This service can simplify the solution of mathematical problems of high complexity when preparing students for exams.

Gladun A. is considering another way to use web applications in education. The work describes a web service that allows you to automatically control the knowledge gained by students in the framework of e-learning [5].

An equally urgent issue is the development of a secure service to automate the maintenance of information about users [6]. This article is devoted to the development of such a solution.

2. Materials and methods

The purpose of the research is to develop a web service-an indicator of user information. This article discusses the use of the service as an indicator of students ' professional competencies. The service application programming interface (further API) must conform to the principles of REST (Representational State Transfer) and meet the requirements of a RESTful architecture.

The following tasks precede the implementation of this application:

- analyze current technologies and approaches used in the development of web applications;
- consider the architectural styles of the API;
- study the limitations and requirements arising from the use of a RESTful architecture;
- analyze the key principles of ensuring the security of REST API web services;
- select the appropriate tools for software implementation of the software product.

At the first stage, it is necessary to decide which category the developed web application will belong to.

Web applications can be classified into several types (by the type of main components): backend, frontend and SPA (single page application).

Backend applications (server-side applications) collect the main functionality and logic of the application, interact with the database (hereinafter referred to as the database). They are developed using a number of programming languages, the most popular of which are: Python, Ruby, PHP, C # (.NET), Node.js. However, production-ready solutions are almost impossible to realize without the use of special tools called web frameworks. Each of the programming languages listed above has its own set of frameworks for creating web applications:

- Python: Django, Flask, FastAPI, Tornado;
- PHP: Laravel, Simfony, Yii2;
- Ruby: Ruby on Rails, Sinatra;
- Node.js: Express.js, Meteor.js;
- C#: ASP .NET Core.

Frameworks take on some of the developer's responsibilities and help in ensuring application security, working with typical redirection, authentication, registration and other functionality. The application is deployed on a specially configured web server and independently generates the HTML representation of the pages (however, the browser will reload the page in most cases).

Frontend applications (client-side applications) run directly in the user's web browser. Developed using JavaScript, TypeScript and Vue.js, ReactJS, AngularJS frameworks. They are applicable in cases when the application does not need to work with the database and store client information for longer than one session. Fully responsible for the presentation of the HTML code.

SPA is a combination of solutions, dividing a web application into two (frontend and backend) and organizing interaction between them. The backend part deals with logic, processes requests, works with the database, and the frontend part forms an external view based on the data received from the backend and displays the result in the browser. The exchange of information between the client and server parts of the application, as a rule, occurs through a specially developed application programming interface. Almost all modern web applications are based on the SPA concept. Therefore, it is this concept that forms the basis of the competency accounting service.

An API is a set of methods and rules by which applications exchange messages and transfer data. As a rule, these are classes, functions, structures of one program, interacting with other programs. The API not only helps to organize interactions, but it also plays an important role in securing that interaction. Such an interface hides the specifics of the service implementation and works within the framework of what the developers provide. Using the application API, the operations that are available to the user are defined, restrictions are imposed, inputs and outputs are specified [7]. In addition, the API has a number of architectural styles that reflect certain aspects of how the interface works.

Applications of this type have complex logic and many components, which significantly affects the security of the system [8]. There are many factors to consider when organizing security. In addition to securing the client and server side, you should focus on developing a secure API. Intrusion or accidental client error can result service unavailability, data loss, or complete application failure. This can critically affect the user experience, and as a result, bring massive losses to the business. Therefore, ensuring security is a topical issue when developing any web service.

The implementation of the API is based on the use of protocols and software specifications that build syntactic and semantic rules for transmitted messages. Such specifications shape the architecture of the API.

There are many styles containing standardization schemas and data exchange rules. The most common architectural styles are RPC (gRPC), REST, SOAP, and GraphQL.
The architectural style, represented by the RPC (Remote Procedure Call) specification, allows a function to be called remotely in the context of an HTTP API. The client gets most of the capabilities when calling a particular procedure remotely, by means of personal formation of a special message for the server. The server deserializes the incoming message [9], runs the requested function and returns the result.

RPC is a new version of RPC designed to address the accumulated problems of the previous version of the specification. gRPC also adds a wide range of features like load balancing, message tracing, and more. In addition, the new standard has started using the concept of protobuffs and runs on top of HTTP / 2. Protobuf is a. proto file, which is a dictionary on the basis of which data is encoded and decoded in a special binary format. That is, it is a kind of new serialization format (alternative to JSON / XML) for exchanging information between client and server [10]. The innovation allows you to maintain high performance in an environment of heavy data exchange in real time. It is also possible to use SSL / TLS, OAuth 2.0, tokens for authentication, or write your own implementation of secure authentication.

RPC (gRPC) is most applicable in the development of high-performance microservices [11] for internal communication, as well as in the creation of APIs for remote control systems, for example, in the IoT (Internet Of Things) field.

GraphQL is another spec introduced by the developers at Facebook. GraphQL is focused on representing data in a graph format. Unlike the creators of a tabular view from relational databases, the developers of GraphQL started from the document structure borrowed from document-oriented databases like MongoDB. The implementation is based on a graph of entities (nodes) that refer to each other (edges).

First of all, a schema is built that describes (using the special syntax of the SDL language) the exact queries and data types in the expected responses. A client with a drawn up scheme, even before sending a request, can make sure that the server has an answer to this request. On the backend, the operation is interpreted according to the scheme and sends a response in JSON format, with the requested data. As a result, we have a technology that, based on the request scheme, can receive data from multiple endpoints and issue a single response in a specific form determined by the requestor.

In addition, GraphQL supports a subscription mechanism that allows you to receive information in real time from a web server.

GraphQL is a great choice when you need to encapsulate the complex logic of the systems on which the API is developed, which is important when designing bulky projects and microservices. In addition, GraphQL provides the ability to work more efficiently with data on mobile devices (mobile API) by optimizing the payload of a single message.

SOAP or Simple Object Access Protocol is a protocol specification for building web communication. It is a W3C industry standard and uses the XML (eXtensible Markup Language) format as the underlying interoperability format. SOAP implements a stateful messaging mechanism that is used in transactions involving multiple parties or in complex secure transactions. Works with HTTP (HyperText Transfer Protocol), SMTP (Simple Mail Transfer Protocol) and FTP (File Transfer Protocol) protocols. SOAP integrates with WS-Security protocols to guarantee the integrity and confidentiality of processed transactions with additional encryption [12]. By using XML, SOAP is the most verbose API style. Despite this, the XML interchange format has many implications that prevent SOAP from being a universal architectural style.

Thus, SOAP is most often used to provide reliable access within a company or directly with a client. Strong security capabilities make SOAP the preferred choice in the corporate and financial sectors.

REST (View State Transfer) is a style of API architecture for creating web services and services. It was presented in 2000 in a doctoral dissertation [13] by one of the creators of the HTTP protocol, Roy Fielding. REST is a simple interface for transferring information that does not use third-party software layers. That is when sending data, there is no conversion stage, the information is delivered in its original form, which has a beneficial effect on the client load, but adds a load on the network part.

Working with data is organized in JSON or XML formats. A web service that meets all the requirements and conditions of the REST architecture is called a RESTful web service.

RESTful architectural requirements (Fielding Constraints):

- Do not contain state (stateless): the state for processing a request can only be contained in the request itself, so the server does not store any session information.
- Caching.
- Common Interface: Allows for a consistent, application-independent interaction with the web server. It imposes a number of sub-restrictions: manipulating resources through their representations, identifying resources, "self-sufficiency" of messages, working with hypermedia.
- Focus on client-server architecture.
- Ability to deliver executable code to the client side.
- Independence from the number of levels / layers of the application.

In REST, all communication is based on the use of HTTP methods:

- GET - request to receive data;
- POST - sending / adding data;
- PUT - data editing;
- PATCH - partial data update;
- DELETE - deleting data.

The advantages and disadvantages of each of the schemes are presented in table 1.

Table 1. Advantages and disadvantages of the most popular schemes.

Schemas	Benefits	Disadvantages
RPC (gRPC)	simple interaction mechanism the ability to add functions with subsequent conversion to the end point high performance the ability to interact with security tools	low level of abstraction, which leads to difficulty in reuse difficulties in scaling problems with introspection of queries reduced security due to the inability to completely hide the details of the API implementation
GraphQL	to retrieve data of any complexity, one request is enough availability of a typed request data schema works effectively with related data detailed error log flexible customization of requests (format definition, portioning, single version)	Possible low performance if there are too many nested fields in one query difficulties with caching high threshold of entry, it is necessary to deal with SDL
SOAP	the ability to work with various transport protocols implementation of built-in error handling (Retry messages) security enhancements. detailed message format (lots of metadata)	support for a single XML format the need for a large service bandwidth due to the heavy weight of XML requires deep knowledge of protocols and data transfer rules
REST	openness of interaction simple implementation data caching at the HTTP level work with several formats of data presentation; stability due to the high level of abstraction	lack of a unified standardized structure high load on the network excessive or insufficient information, which may lead to the need to send an additional request

REST is most often used as a management API for CRUD (Create, Read, Update, Delete) to implement interaction with resources in lightweight scalable services. A resource is usually a data model object or a database table.

Of all the specifications, REST implements the highest level of abstraction and is best suited for developing simpler CRUD APIs. Maintains a balance of reliability and ease of use. The client-server bias and caching propensity (important for listing competencies, sorting and filtering) are also advantages for services like the one discussed in this article. The large, in comparison with other specifications, the load on the network is equalized by the specifics of the problem being solved (one student with a low degree of probability will have thousands of entries and add them to the platform in batches of several thousand, and the development of a separate mobile application is not initially planned). Therefore, it is REST that forms the basis of the architectural style of the project prototype being developed.

Building secure RESTful APIs also imposes certain standard requirements:

- Using the HTTPS protocol: encryption is required to ensure the integrity of the transmitted data.
- Rate-limits: It is necessary to monitor the load on the API. Dropping requests in case of overload or connecting additional systems for balancing.
- Authentication: User / application / device identification.
- Audit log: Recording actions by creating an entry in the logs log.
- Control of access rights (authorization): Determination of access rights for working with resources.
- Access to the business logic of the application.

When working with the REST architecture, it is customary to distinguish two levels of security, since the REST API is an interface for network interaction with a web application:

- the first level - getting access to the API;
- second level - getting access directly to the application.

Protection of the API level implies proper organization of authentication, authorization, registration and separation of access rights. The developer must ensure that only authorized clients can use the API and only have access to authorized operations.

Application layer security includes vulnerability checking of service endpoints - the URLs responsible for interacting with the interface.

The OWASP (Open Web Application Security Project) community has identified (at various levels) about ten API development vulnerabilities. The most dangerous are:

- Broken Object Level Authorization: No separation and enforcement of resource access control.
- Broken User Authentication: Vulnerabilities related to user authentication.
- Lack of Resources & Rate Limiting: Mismatch of checks and limits.
- Broken Function Level Authorization: Lack of access control.
- Mass Assignment: Vulnerabilities related to deserialization of received objects.
- Security Misconfiguration: Errors in working with application security settings.
- Insufficient Logging & Monitoring: Absence or incorrect vision of the logs and system monitoring logs.

First of all, it is necessary to pay attention to the problems associated with access control, authentication and authorization [14]. By project, the REST API is stateless, so access is controlled through local endpoints.

There are several common methods (authentication schemes): Basic Authentication, API Key, JSON Web Tokens, OAuth 2.0, Token-Base Authentication, Cookie-Based Authentication, and OpenID.

Basic Authentication (BA) is the simplest HTTP-based authentication scheme. The client or application forms an HTTP request, the header of which contains the "Authorization" field. A string of the form: Basic <username: password> (encoded in Base64) is passed as the value of this field. To ensure security and preserve the confidentiality of data, in combination with Basic Authentication, it is always necessary to use the HTTPS / TLS protocols [15], since the user ID and password enter the Network as plain text encoded in Base64 (which is easy to decode).

Journal of Physics: Conference Series **2091** (2021) 012065 doi:10.1088/1742-6596/2091/1/012065

Cookie-Based Authentication is a method based on checking the content of cookies, inside which all the necessary information about the session is stored. The user initiates a login request. In case of a successful login, the server sends a response, in the header of which the Set-Cookies field is included, containing the name of the cookie-field, value, cookie expiration, and so on. The next time the user needs to access the API, he will pass in the request header the value of the saved Cookie-field JSESSIONID with the key "Cookie": Cookie: JSESSIONID = 0112. Cookie-based authentication is possible only if the server has a mechanism for storing and checking user sessions. As with Basic Authentication, use encrypted Internet protocols for security. In addition, Cookie-Based Authentication has a number of vulnerabilities, which can be secured by passing a special CSRF security token in the request header: X-Csrf-Token: token_value.

API Key – a key in the form of a character string that the user passes along with the request to the server. The server will confirm the identity of the client if its key is contained in the application database. The key itself is issued by the application when registering a user. This scheme is used to protect against unauthorized access and allows you to impose a limit on API calls. API key can be passed: as a request parameter, as a request header with a field, as a cookie value.

The API Key should only be accessible to the owner user and the server. Therefore, APIs using API Key as their primary authentication scheme should only use secure HTTPS / SSL protocols.

The Token-Base Authentication or Bearer Authentication scheme is based on the use of a "token signed by the server". This token is then sent to the server inside the Authorization request header. However, unlike Basic Authentication, the "Basic" keyword is replaced by "Bearer": Authorization: Bearer <token>. After receiving the token, the server validates it (does a user exist with such a token, the validity period has not expired, etc.). Token-Base Authentication can be used independently, when the server itself generates tokens for new users, or be part of a more complex authentication mechanism like OAuth 2.0 or OpenID Connect. To prevent the interception of the token by attackers, you must use the secure HTTPS protocol. The difference between tokens and API Key is that the key can only provide access to API methods without the ability to get the user's personal data.

JSON Web Tokens (JWT) is an authentication mechanism based on the use of a special type of token - JWT token. It is a JSON data structure. Such a token contains a header with general information, a body with a payload (user-id, group, data) and a cryptographic signature. This scheme is one of the most secure mechanisms for transferring data between two parties, therefore it is considered the preferred method for controlling REST API access. The user for working with the API, when sending a request, adds to it the personal JWT previously issued by the server.

OAuth 2.0 is a more complex protocol responsible for user authorization. Allows a web application to acquire rights to use client resources on another service. This makes it possible to provide a third-party application with limited access to the controlled resources of the user without directly transferring the login and password to this application. OAuth can be used on any platform, as the protocol relies on the underlying web technology stack, namely HTTP requests, responses, URL redirects [16], etc.

The scheme of the web service using OAuth 2.0 looks like this:

- The application is being authorized. As a result of successful passage, an access token is created (a key consisting of a set of characters). Presentation of this token indicates that the application has access to protected resources. As in classic authentication schemes, the access token can be presented via an HTTPS request with the appropriate header or as one of the GET request parameters.
- The application requests and receives protected resources.

This approach is the most complex authorization option, but only when it is used is the service able to uniquely identify the application requesting authorization. Otherwise, authorization is done entirely on the client side. The schema is often used when developing REST APIs that interact with third-party services.

OpenID is a standardized method for a decentralized authentication scheme. A distinctive feature is the ability to create a single account for authentication on several services at once (creating a unique digital identifier) through the services of an intermediary service. In terms of the mechanism of operation, this method is similar to OAuth 2.0, but OpenID is intended solely for user authentication. The new version of OpenID Connection was transformed into an authentication add-on over OAuth 2.0, received a reliable encryption mechanism and digital signatures.

It was decided to use JWT as the main access control scheme for the developed application, since the service does not use third-party resources in its work, and such tokens are easy to use, have a convenient data description format and do not create unnecessary network load. The use of the HTTPS protocol in conjunction with a cryptographic signature provides an optimal level of protection for the service.

When securing a web service, input control deserves special attention. You need to make sure that all the data that the application will subsequently operate on is valid and conforms to the API specification.

The developer community has formed a number of recommendations when validating input data:

- conduct data validation both on the client side and on the server side;
- you should not use your own validation mechanisms, you should use the built-in functions for checking a programming language or framework;
- it is always necessary to check the content type, size and length of the request;
- do not create manual queries to the database on the backend, use only parameterized queries;
- use allow lists on the server;
- keep logs of errors, attempts to enter data, etc.

An equally important task when developing a REST API is choosing the right tools for the backend. The Python programming language has long established itself in the web development market. A lot of high-quality modules, packages, frameworks and utilities have been created for "python", which greatly simplifies the development process. Over the years, the industry has developed a rich code base and a huge international community of developers. An important factor is the relatively low threshold of entry, the simple and concise syntax of the language. This and much more has affected the fact that the development of application prototypes, proof-of-concept and minimum viable products (MVP) is faster and most efficiently done in Python.

There are many powerful and versatile frameworks for developing services and web applications in the Python ecosystem. The most popular framework for API development is Flask.

Flask is a WSGI (Web Server Gateway Interface) micro-framework that provides minimal basic functionality with the ability to add modules as needed and at the discretion of the developer. Due to the modular approach, the programmer can assemble only those tools that will definitely be used in the development of the project [17]. The flexibility of the architecture and the presence of a large number of ready-made extensions make it possible to use Flask both in simple projects and in the development of complex platforms. In the developer community, it is often used to create RESTful web service APIs. All of this became a determining factor in choosing Flask as the basis for the development of an application prototype.

For developing secure APIs, Flask provides a number of loadable modules [18]:

- Flask-Login and Flask_simplelogin for managing user sessions. Due to these extensions, registration, authentication and authorization mechanisms are integrated.
- Flask-Admin - an extension that adds an analogue of the administrator's personal account to manage the service and monitor systems.
- Flask-WTF– module for creating secure forms (validation, csrf tokens, etc.).
- Flask_jwt - extension for working with JWT authentication scheme.
- Flask_restful / Flask_restplus - Sets of helper objects and functions that simplify REST API development.
- Flask-Security - a package for managing user roles and groups.

The disadvantage of many developed services is that the description of the API itself is rigidly tied to the implementation in a specific language or its framework [19]. The open source tool Swagger is used to solve this problem and create a flexible unified API.

Swagger is a language and technology stack agnostic framework for describing the specification and documentation of RESTful APIs. It allows you to configure REST API without access to the source code of the application, through a special file with the YML (Yandex Market XML) or JSON extension [20]. A documentation package is also automatically generated based on this file.

3. Results

The server on which the developed web application will be launched is also a potentially vulnerable point for attackers. To reduce the risk of hacking, you need to disable root logins of the linux server by editing the / etc / ssh / sshd_config. In the PermitRootLogin option, change the value from yes to no.
The next step is installing and configuring the firewall. It is necessary to close access to all ports except 80 (HTTP), 443 (HTTPS) and 22 if we work with the server via SSH: *sudo ufw allow ssh hhtp 443 / tcp –force enable.*

In addition, you should replace the standard Flask application server with a more advanced WSGI server, gunicorn. It is a robust production server that many developers take as a standard when developing Python web applications. Configure it so that the service listens for requests from private port 8000: *gunicorn -b localhost: 8000 -w 4 restapiservice: app*. However, application web traffic can only enter ports 80 and 443 open on the firewall. At the same time, to ensure the security of the application, all traffic arriving on HTTP port 80 must be redirected to encrypted port 443 (HTTPS). To work with the HTTPS protocol, you need to obtain an SSL security certificate (for example, Let's Encrypt): *sudo Apt install python-certbot-nginx / sudo certbot –nginx -d <application_domain> -d www. <application_domain>*

An Nginx proxy will be used to redirect traffic. Listening port settings, HTTPS connection and traffic redirection occur through the setting of the configuration file / etc / nginx / sites-enabled / default [21].

The default SQLite database is also not an acceptable solution. The application will use the reliable and efficient PostgreSQL DBMS (Database Management System), which is the de facto standard on the Python + Django / Flask technological stack.

However, no DBMS can efficiently service the large number of independent processes that access it. Too many requests (real users or those created by an intruder) will lead to the failure of the database. To solve this problem, we will use the PgBouncer connection puller. Puller proxies all incoming requests in small chunks, creating specialized queues.
Initially, when developing a web application, you need to pay attention to the form of storing confidential application data: keys, addresses, database passwords, etc. The simplest and most reliable option is to use environment variables. The programmer creates a bat / bash script that, when launched on the server, will write the necessary data to the variables of the current environment. You can then use the os python module to read the required information and configure your Flask application. Due to this, secret data does not appear openly in the code.

When creating a user model, it is important to pay attention to the work with the "Password" field. Passwords should not be explicitly stored in the database. For security reasons, it is not the passwords themselves that are used to record and compare passwords, but their hashes. To do this, inside the set_password and check_password methods of the User model, we will use the functions of the security package - generate_password_hash and check_password_hash.

Next, let's look at the method responsible for adding a new competency. In this case, the user must be authorized. The function will be called every time the API receives a POST request like this:
curl -X POST -H "Content-Type: application / json" \ -d '{"title": "Achievement Ttitle", "type": "Achievement Type", "date": "Achievement Date", "proof" : "Achievement Proof URL"} '\ https: // site / api / achievement /

Getting the contents of the fields of the model should be done through a call to the get method, and in the absence of the corresponding field, the value of the variable is set to None. In addition, in order to ensure security, the work with the database is carried out through the objects of the built-in ORM (Object-Relational Mapping), and not by direct SQL (structured query language) queries.

The description of the API request specification itself, the setting of the handler and security restrictions are moved to the swagger.yml file. In this file, you need to specify the authentication scheme that the application will use, as well as a special function that contains the implementation of this scheme in Python:

components:

securitySchemes:
bearerAuth:
type: https
scheme: bearer
bearerFormat: JWT
securityDefinitions:
bearerAuth:
type:bearerAuth
x-basicInfoFunc: "auth.jwt_auth"

In the case of a developed web service, the schema handler (jwt_auth function) is located in a separate auth.py file.

In the swagger.yml file, each request to the API is represented by a separate block, which contains all the service options in a declarative style. In the specification block, you must specify: path, HTTP request type, response and security.

A description of the specification for adding competencies:

paths:
/achievement:
post:
operationId: " achievement.create"
tags:
responses:
201:
description: "Successfully created achievement in list"
406:
description: "achievement already exists"
security:
- bearerAuth: []

Now, when trying to access the POST /achievement API method, the system will ask the user for authorization.

The rest of the API methods are created in the same way.

After completing all the procedures for securing the server side, you must also take into account a number of considerations related to securing the client side of the application.

A web application developed and configured in this way closes most of the vulnerabilities of the frontend / backend side and allows the client to safely use all the functionality of the service.

4. Discussion

As it turned out during testing, the developed web service, with a rapid increase in the number of incoming requests (on the order of several hundred), may experience a number of performance problems. This is due to the technology stack used, the choice of which was justified by the priority in the simple and fast development of an application prototype. Using the synchronous Flask framework like a bottleneck for system speed and responsiveness.

Flask, as a web service development tool, is often criticized by the Python developer community.
In 2018, Sebastian Ramirez developed a Python web framework called FastAPI. According to Uber developers Pierre Molino, Yaroslav Dudin and Sai Sumant Miryal, FastAPI should be the best framework for developing REST API services and a more performant analogue of Flask [22]. Ramirez's framework is completely asynchronous and easily integrates with OpenAPI-schema, making it easier to work with Swagger and ReDoc. Despite the relatively short term on the market, FastAPI has established itself as a quality tool. Many large companies have started using it to develop their APIs. So in the project discussion thread on GitHub, Microsoft Lead Engineer Khan Kabir posted the following comment: "I have been using FastAPI very often in recent days. I definitely plan to use it for all ML services of my team at Microsoft." [23].

Also, as an alternative to Flask, many developers, in particular Adam Hopkins, recommend considering another relatively "young" framework - Sanic [24]. It was also conceived as a multithreaded, asynchronous counterpart to Flask. It is a micro-framework with a modular structure and a number of features such as basic routing of static files, special extensions for creating CRUD, mechanisms of non-blocking operations, and others.

In addition, Rhea Mutafis criticizes the use of Python because of its relatively poor performance, dictated by the interpretability of the language [25]. As an alternative, Google engineers offer their own development - the Golang programming language. Its main advantages are strong typing, the presence of a compiler, and goroutines are an efficient analogue of threads. In addition, Go provides the ability to work with static typing when developing web applications, which makes it much easier to test and maintain the product in the future.

The transition to an alternative technology stack, for example, the FastAPI framework, will help to further improve the performance of the service, as well as expand the potential for scaling the system.

5. Conclusions

Thus, SPA APIs have become a staple in the development of modern web and mobile applications. They allow you to organize interactions between applications, services, and software platforms. Due to their versatility and ease of use, REST architectural style interfaces account for about 80% of all public and proprietary APIs.

In the course of the research, a prototype of a protected SPA RESTful API web service was developed, with the help of which users will be able to keep track of personal information related to various areas of their professional, educational or personal activities. The Python programming language and its Flask web microframework significantly increased the efficiency, quality and speed of MVP development, and the Swagger framework made it possible to achieve independence of the developed REST API from the technology stack.

The project has a high potential for expanding and scaling by adding new functionality, replacing the used database with a more flexible No-SQL document-oriented version (for example, MongoDB), implementing advanced caching based on the in-memory Redis database, authorization through social networks. In addition, it is possible to switch to a more efficient asynchronous backend framework, refine and optimize the UX / UI of the client side of the application.

The prospect of further research on this issue is to find ways to restructure the application by moving to a modular architecture. This will allow you to flexibly customize the project for the needs of customers, including organizations that are not tied to working with clients from any one field of activity.

References

[1] Kornienko D V 2020 Organization of a system of digital education practices in the municipal sphere of general education *J. Phys.: Conf. Ser.* **1691(1)** 012108

[2] Mishina S V 2020 Strategies for students' lean competencies formation in the educational process of the university *J. Phys.: Conf. Ser.* **1691(1)** 012213

[3] Shcherbatykh S V and Mishina S V 2019 Development of professionally significant qualities of future economists by means of the hidden curriculum | Desarrollo de cualidades profesionalmente significativas de futuros economistas mediante el currículum oculto *Utopia y Praxis Latinoamericanat* **24** pp 387–95

[4] Volodina E V and Mikishanina E A 2020 Using the developed web-application for solving problems of increased complexity in mathematics *J. Phys.: Conf. Ser.* **1691** 012198

[5] Gladun A, Rogushina J, Garcı´a-Sanchez F, Martínez-Béjar R and TomásFernández-Breis J 2009 An application of intelligent techniques and semantic web technologies in e-learning environments *Exp. Sys. App.* **36**

[6] Kornienko D V, Shcherbatykh S V, Mishina S V and Popov S E 2020 The effectiveness of the pedagogical conditions for organizing the educational process using distance educational technologies at the university *J. Phys.: Conf. Ser.* **1691(1)** 012090

[7] Shevat A, Jin B and Sahni S 2018 *Designing Web APIs: Building APIs That Developers Love* (O'Reilly Media, City of Newton)

[8] Jiao J, Yang Y, Feng B, Wu S, Li Y and Zhang Q 2017 Distributed rateless codes with unequal error protection property for space information networks *Entropy* **19(1)** 38

[9] Abbaspour E, Fani B and Heydarian-Forushani E 2019 A bi-level multi agent-based protection scheme for distribution networks with distributed generation *Int. J. Electrical Power and Energy Sys.* **112** pp 209–20
[10] Cong W, Zheng Y, Zhang Z, Kang Q and Wang X 2017 Distributed Storage and Management Method for Topology Information of Smart Distribution Network *Dianli Xitong Zidonghua/Automation of Electric Power Sys.* **41(13)** pp 111–18
[11] Bonér J 2016 *Reactive Microservices Architecture: Design Principles for Distributed Systems* (O'Reilly Media, City of Newton)
[12] Song X, Zhang Y, Zhang S, Song S, Ma J and Zhang W 2018 Active distribution network protection mode based on coordination of distributed and centralized protection *Proc. 2017 China Int. Electrical and Energy Conf.* pp 180–83
[13] Fielding R 2000 *Architectural Styles and the Design of Network-based Software Architectures* (University of California, California)
[14] Tong B B, Zou G B and Shi M L 2013 A distributed protection and control scheme for distribution network with DG *Advanced Materials Research* pp 732–33 628–33
[15] Zhong S, Liu C, Yang Z and Yan D 2009 Privacy protection model for distributed service system in converged network *Int. Conf. E-Business and Inf. Sys. Security* (Institute of Electrical and Electronics Engineers, Piscataway)
[16] Maximov R V, Ivanov I I and Sharifullin S R 2017 Network topology masking in distributed information systems *CEUR Work. Proc.* **2081** pp 83–7
[17] Greenberg M 2016 *Developing web applications using Flask in Python* (DMK Press, Moscow)
[18] Dwyer G 2016 *Flask By Example* (Pack Publishing Ltd, UK)
[19] Percival G 2018 *Test Driven Development* (DMK Press, Moscow)
[20] Swagger https://community.sanicframework.org/t/how-do-i-find-out-which-companies-use-sanic/724
[21] Melnikov M O 2021 Setting up a Debian server for developing Python and Django web applications *Modern Science* **2(2)** https://www.elibrary.ru/item.asp?id=44789469
[22] Uber Engineering https://eng.uber.com/ludwig-v0-2/
[23] GitHub https://github.com/tiangolo/fastapi/pull/26
[24] Sanicframework https://swagger.io/
[25] Moutafis R 2020 Why Python is not the programming language of the future *Towards Datascience* https://towardsdatascience.com/why-python-is-not-the-programming-language-of-the-future-30ddc5339b66

Advantages of graphene technologies in the design of antenna systems for the Internet of Things in the decimeter range of cellular operators (LTE, GSM, 5G)

A G Cherevko, Y V Morgachev

Siberian State University of Telecommunications and Information Science, 86 Kirov Street, Novosibirsk 630102, Russia

E-mail: cherevko@mail.ru, morgachev.yury@gmail.com

Abstract. The characteristics and resistance to external influences of flexible graphene antennas are considered. In particular, the analysis of the influence of single and cyclic bends at different cycle times is carried out. Resistance to water of such antennas is analysed. The results of characteristics comparison of the literature data with the results obtained by the authors are presented. The design of a flexible graphene eco-friendly two-layer antenna on a paper substrate operating in the LTE frequency range has been proposed. The technology discussed in this report will help fill the gap in the literature in eco-friendly double-sided graphene printing technologies. The literature review shows that the development of graphene inks for printing on flexible substrates has made a breakthrough in the development of flexible electronics due to their unique advantages. The influence of edge effects on the change in the graphene surface resistance at the edges of the manufactured sample is estimated.

1. Introduction

Flexible antennas in the design of elements of the Internet of Things are used in the following applications: military industry, security, logistics and a number of other areas. Typically, in the design of flexible antennas, silver, aluminum and copper are used as the conductive material in the applied ink. However, this ink is expensive and non-biodegradable. Environmental safety is currently one of the important parameters of the developed electronics due to the imposed restrictions on the disposal of electronic devices. For example, the Waste Electrical and Electronic Equipment (WEEE) directive states that 55% of electronics must be recycled. A similar policy is being pursued in the territories of many countries of the world.

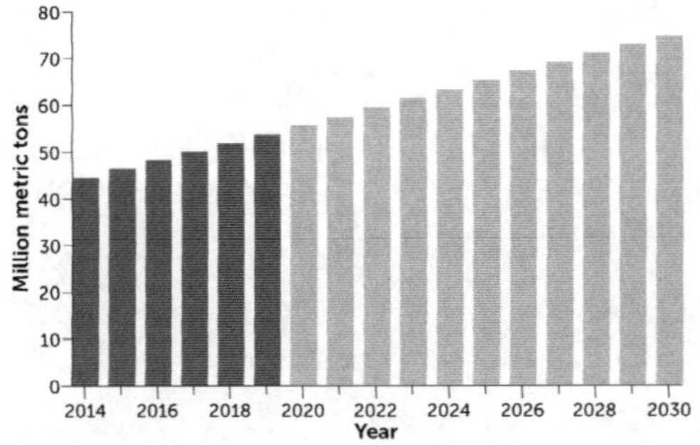

Figure 1. The amount of e-waste in the world by year.

The issues of environmental friendliness and cost reduction of flexible antennas are solved by replacing metals with graphene and using graphene printing technologies.

Currently, the surface resistance of graphene inks is higher than that of similar inks that use metals. However, these inks do not rust [2], have a lower price [1], have less fatigue [4], are more resistant to mechanical stress [3] and do not need processing [5]. Due to the low drying temperature [2], this ink allows the use of temperature-sensitive substrates [6]. Flexible devices that use graphene antennas are being developed for the following applications: security [8], logistics [9], medicine [7] and many others [10-12].

2. Main part

For printing with graphene ink, it is currently possible to use a wide range of methods: aerosol, inkjet, screen, soft lithography, flexography and rotogravure [13]. Also available are 3D printing [2] and the graphene transfer method [3] for the manufacture of graphene antennas. Screen printing is one of the most used methods for applying graphene inks due to the high quality of the design achieved and the maturity of the technology. It is possible to apply this method, including for conformal surfaces [14]. Examples of successful application of this type of printing on paper and fabric substrates are presented in [15] and [16], respectively.

One of the most important factors that should be considered when choosing a method for manufacturing wearable devices is the resistance of a conductive material to external influences. After all, elements of wearable electronics are exposed to more external influences than conventional gadgets such as a smartphone or laptop. The impact of edge effects on the change in the graphene surface resistance at the edges of the manufactured sample is estimated.

The influence of a large number of bending cycles on the change in electrical resistance is considered in [4]. The authors cut a strip 11.3 cm long and 0.9 cm wide for this experiment and secured with electrical contacts on a bending machine. The resistance of the strip was measured during each of 32,500 bending cycles (with a bending radius of 42.5 mm). No significant effect on the measured graphene strip resistance was obtained, as seen in figure 2.

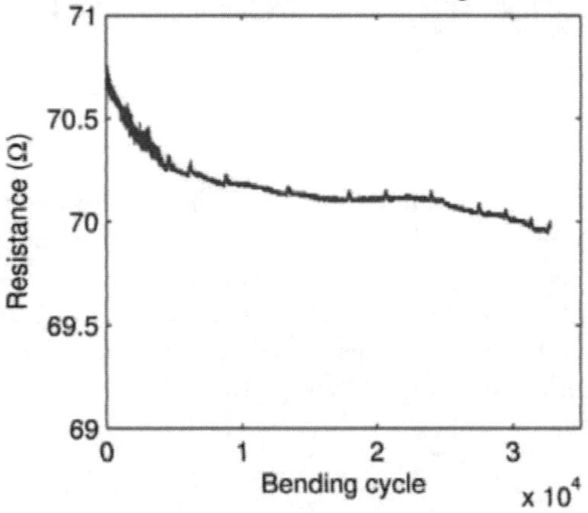

Figure 2. Change in the graphene strip resistance depending on bending for 32500 cycles.

A similar study on the effect of a large number of bending cycles on the graphene resistance was carried out in [17]. For this, the authors fabricated patches by inkjet printing with graphene and silver inks onto a polyimide substrate using a DMP-2800 printer. Both patches have the same thickness - 1 micron. First, a bend was made in the valley form (figure 3 (a)) at 90 ° for 2000 bending cycles, then 2000 cycles in the valley form >135 °, 1000 cycles of a bend in the mountain form (figure 3 (b)) at 90 ° and 500 mountain form bend cycles >135 °. The resistance of the silver patch increased by about 1993% after 5500 bending cycles, while the resistance of the printed graphene patch increased by about 15%.

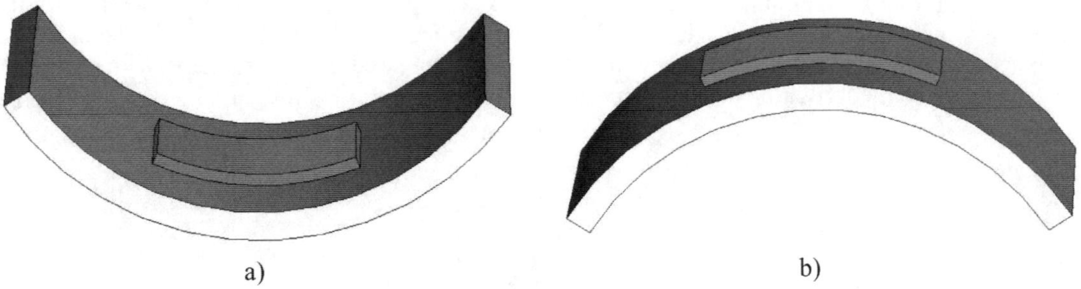

Figure 3. The shape of the bend of the samples: a) valley bend, b) mountain bend.

Figure 4 shows the normalized change in resistance versus the number of bending cycles of inkjet graphene and silver patches.

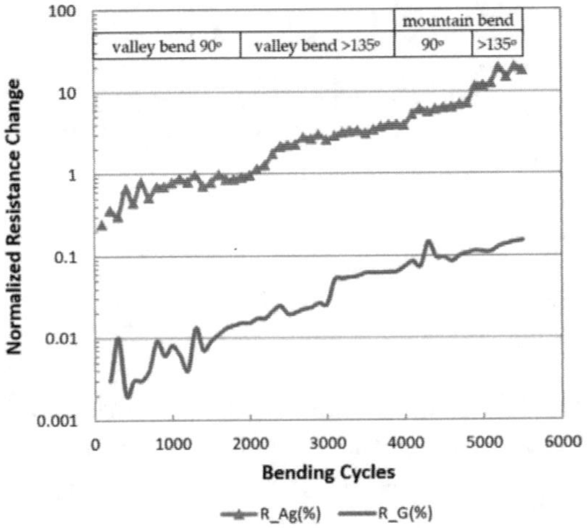

Figure 4. Dependence of the normalized change in the resistance of the patches depending on the bending cycles.

The surface of the patches was photographed through a microscope (figure 5) after the experiment on the effect of a large number of bends on the normalized change in the resistance of the patches. The light falls in such a way that a glare is formed in these photographs. A significant change in the structure of silver patch (figure 5 (a)) and relative constancy of the graphene structure (figure 5 (b)) can be seen in this glare. These results show that this type of printed graphene paves the way for flexible and wearable electronic devices.

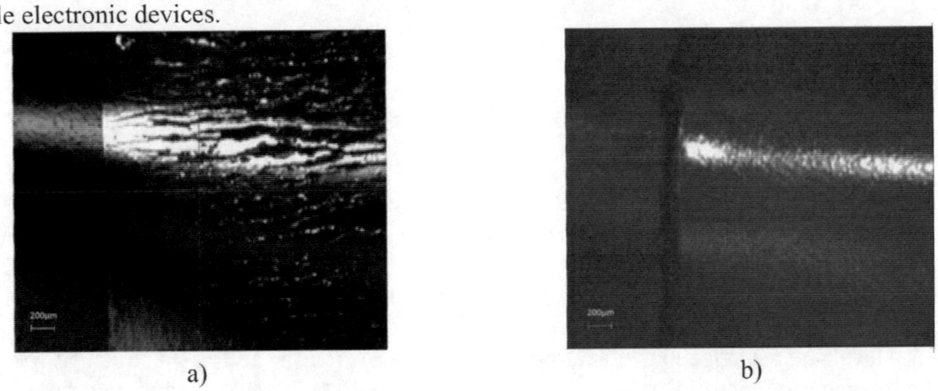

Figure 5. Photographs of silver (a) and graphene (b) patches after 5500 bending cycles.

Another important property of flexible wearable devices is resistance to high temperatures. The authors of the article [17] conducted tests on the resistance of conductive materials to rather harsh conditions. Each patch was heated to 120 ° C for one hour. Upon reaching this temperature, the resistance was measured and then the temperature was increased in 10 °C steps with sequential resistance measurements up to 290 °C. The results of this test are shown in figure 6.

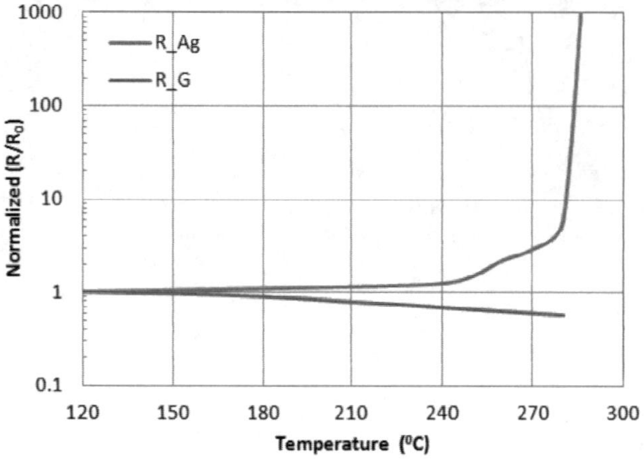

Figure 6. Dependence of the normalized change in resistance as a function of temperature.

Evidently, the resistance of the silver patches begins to increase sharply after 250 °C and changes more than 1000 times at a temperature of 280 °C. This is due to the significant oxidation of silver patch. The normalized resistance of graphene patch hardly changes with increasing temperature. decrease in patch resistance is observed, when in heated to 280 °C

Another impact that flexible devices constantly face is the impact of water. The article [8] discusses the effect of wetting and its consequences on an RFID tag. The antenna of this RFID tag is made by 3d printing with graphene ink on 100% cotton. The reading range was measured before getting wet, one minute after immersion in water, then after 1 hour, 1 day and 1 week drying. The results of this test are shown in figure 7.

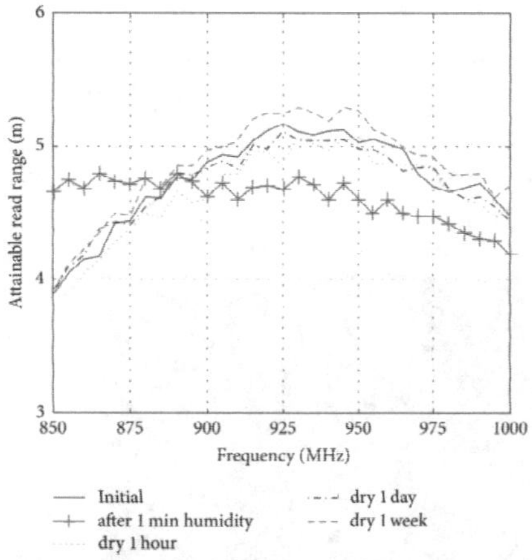

Figure 7. Influence of water on the reading range of a graphene RFID tag.

Evidently, the reading range deteriorates by 0.5 m after 1 minute drying in the 880-1000 MHz range. The authors attribute this to two factors: an increase in the conductivity of graphene with an increase in the humidity level and a change in the electrodynamic properties of the cotton substrate. However, after 1 hour, the reading range returns to its original position, and after 1 week, the reading range is higher than before getting wet.

The presented analysis confirms the feasibility of studying graphene deposition technologies and emphasizes its advantages in comparison with traditional metallic inks in the application of flexible wearable devices.

The dipole antenna operating at a frequency of 1800 GHz (GSM technology) was made on a photographic paper substrate to test the screen-printing technology used by us. Its 3D model is shown in figure 8.

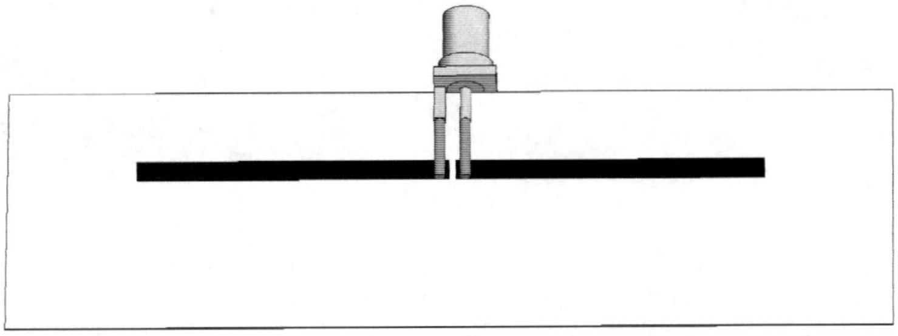

Figure 8. 3D model of a dipole graphene antenna on a paper substrate. White - photo paper backing, orange - copper wire and gold - connector.

We have obtained the standing wave ratio for the voltage of this antenna. We compared the results obtained with the results presented in the article [6] in order to determine the effectiveness of the technology used. The characteristics of the antennas developed by us compare favourably with similar analogues. An example is a comparison of the fundamental characteristic - voltage standing wave ratio (VSWR) of dipole antennas [6] and ours [18] (figure 9). The normalized frequency (f/fmin) is plotted along the abscissa axis to take into account the difference in the frequency ranges of the two antennas. Here (fmin) is the frequency of the VSWR minimum.

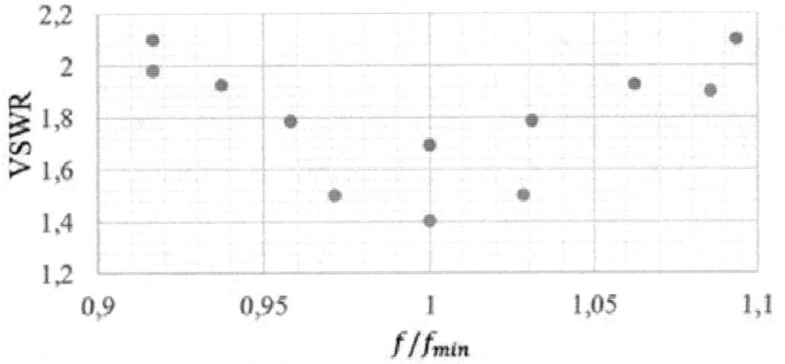

Figure 9. Comparison of the VSWR of the dipole antenna developed by us (• , [18]) with the VSWR (•) of the dipole antenna presented in [6].

It can be noted that the minimum VSWR value obtained by our antenna is 0.28 less than in [6].

To eliminate the influence of the cable on the characteristics of the connected dipole antennas through unbalanced lines, we have developed a dipole antenna with a J-shaped balun. The design of this antenna is described in more detail in [19]. This antenna operates at 2.6 GHz, which is part of the LTE frequency spectrum. Photos of the manufactured dipole two-layer antenna from both sides are shown in figure 10.

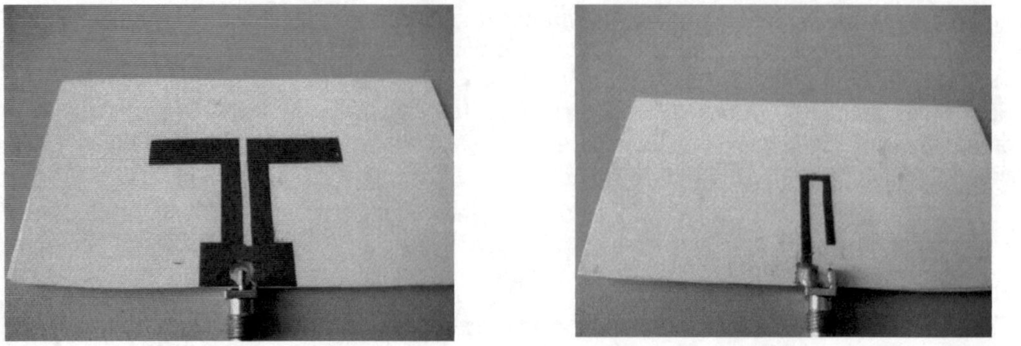

a) b)

Figure 10. Design of a dipole graphene antenna with a J-shaped balun operating at a frequency of 2.6 GHz: a) top view, b) bottom view.

The S11 parameters and the shape of the radiation pattern (RP) in the E-plane of this antenna obtained by modelling are shown in figure 11.

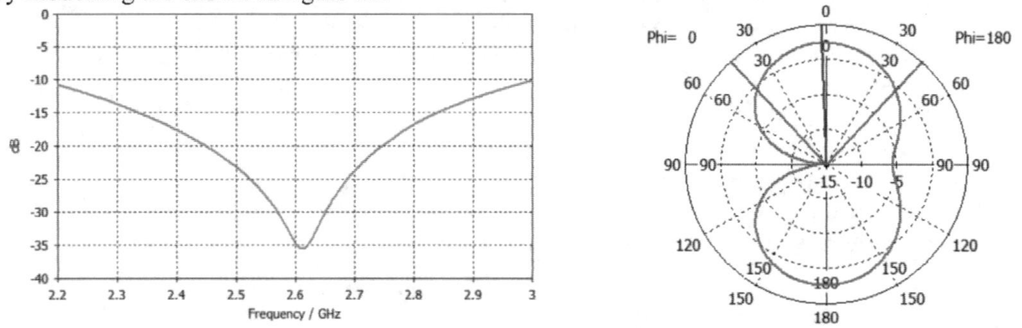

a) b)

Figure 11. S11 (a) and the RP in the E-plane of a dipole graphene antenna with a J-shaped balun.

The surface resistance of the manufactured sample was measured, it was 3.88 Ω/sq. However, at the edges, it was not constant and increased to 4.2 Ω/sq. Therefore, based on the measurement data, the influence of edge effects was assessed by changing the surface resistance of graphene at the edges of the manufactured sample. For this, a different surface resistance in the range from 3.88 to 8.33 Ω/sq was set at the edges of the graphene layer over a length of 1 mm. The result of the shift of the minimum of the S11 parameter in both frequency and amplitude is shown in figure 12.

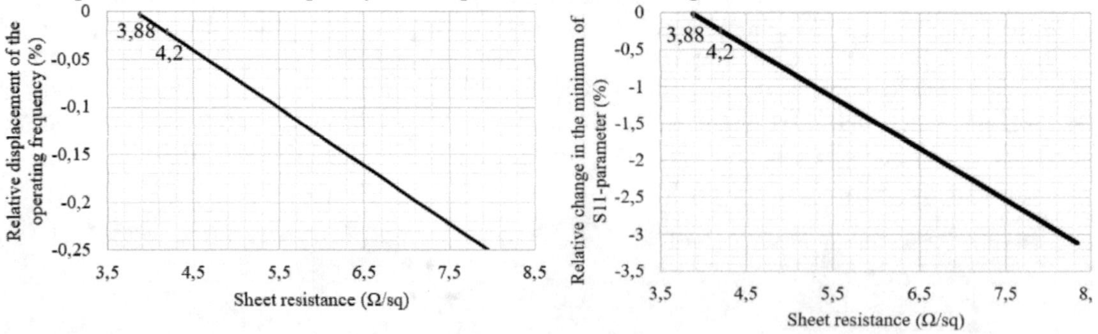

a) б)

Figure 12. Relative displacement of the operating frequency (a) and the change in the minimum of the S11-parameter (b) depending on the surface resistance of the edges.

Evidently, the change in the surface resistance of the edges of the sample has little effect on the relative shift of the operating frequency. A 2.15-fold increase in surface resistance shifts the operating frequency by 0.25%. The strongest change in the surface resistance of the edges affects the value of the minimum S11 parameter. With the same change in surface resistance, the minimum increases by 3.5%.

For our case, when the resistance at the edge's changes from 3.88 Ω/sq to 4.2 Ω/sq, the relative shift of the operating frequency and the relative change in the minimum of the S11 parameter is 0.019% and 0.22%, respectively. Because for our sample, the considered case is the worst, i.e. the maximum surface resistance observed at the edges is taken and when approaching the edge by 1 mm it changes stepwise, the influence of this effect can be considered insignificant.

3. Conclusion

Based on the literature analysis, it has been shown that graphene ink is more resistant than silver ink to various external influences characteristic of flexible wearable devices. The results of testing the used stencil technology for applying graphene ink are presented. It is shown that the VSWR obtained by us is 0.28 less than that of similar samples by other authors using the example of a dipole graphene antenna on a paper substrate operating in the GSM-1800 frequency range. The design of a two-sided dipole graphene antenna with a J-shaped balun operating in the LTE band is considered. Its S11 parameters and RP are presented. An inhomogeneity of the surface resistance at the edges of the antenna was detected with the help of the fabricated sample. Therefore, we evaluated the influence of edge effects on the change in the surface resistance of graphene at the edges of the manufactured sample. It was found that when the surface resistance at the edge's changes from 3.88 to 8.33 Ω/sq, the operating frequency shifts by 0.25% and the minimum of the S11 parameter increases by 3.5%.

4. References

[1] Scidà A, Haque S, Treossi E, Robinson A, Smerzi S, Ravesi S, Borini S and Palermo V 2018 Application of graphene-based flexible antennas in consumer electronic devices *Materials Today* **21** 223–30

[2] Akbari M, Khan M W A, Hasani M, Bjorninen T, Sydanheimo L and Ukkonen L 2016 Fabrication and Characterization of Graphene Antenna for Low-Cost and Environmentally Friendly RFID Tags *IEEE Antennas and Wireless Propagation Letters* **15** 1569–72

[3] Lamminen A, Arapov K, de With G, Haque S, Sandberg H G O, Friedrich H and Ermolov V 2017 Graphene-Flakes Printed Wideband Elliptical Dipole Antenna for Low-Cost Wireless Communications Applications *IEEE Antennas and Wireless Propagation Letters* **16** 1883–6

[4] Wang W, Ma C, Zhang X, Shen J, Hanagata N, Huangfu J and Xu M 2019 High-performance printable 2.4 GHz graphene-based antenna using water-transferring technology *Science and Technology of Advanced Materials* **20** 870–5

[5] Kumar J, Basu B, Talukdar F A and Nandi A 2017 Graphene: A possible low-cost eco-friendly solution for antenna applications *2017 IEEE Applied Electromagnetics Conference (AEMC)*

[6] Huang X, Leng T, Zhang X, Chen J C, Chang K H, Geim A K, Novoselov K S and Hu Z 2015 Binder-free highly conductive graphene laminate for low cost printed radio frequency applications *Applied Physics Letters* **106** 203105

[7] He H, Akbari M, Sydänheimo L, Ukkonen L and Virkki J 2017 3D-Printed Graphene Antennas and Interconnections for Textile RFID Tags: Fabrication and Reliability towards Humidity *International Journal of Antennas and Propagation* **2017** 1–5

[8] Huang X, Leng T, Zhu M, Zhang X, Chen J, Chang K, Aqeeli M, Geim A K, Novoselov K S and Hu Z 2015 Highly Flexible and Conductive Printed Graphene for Wireless Wearable Communications Applications *Scientific Reports* **5**

[9] Soontornpipit P 2018 Flexible Graphene Waffle-Typed Antenna for Skin-Typed Medical Applications *2018 International Electrical Engineering Congress (iEECON)*

[10] Sajal S Z, Braaten B D and Marinov V R 2015 A microstrip patch antenna manufactured with flexible graphene-based conducting material *2015 IEEE International Symposium on*

Antennas and Propagation & USNC/URSI National Radio Science Meeting

[11] Kumar J, Basu B, Talukdar F A and Nandi A 2018 Multimode-Inspired Low Cross-Polarization Multiband Antenna Fabricated Using Graphene-Based Conductive Ink *IEEE Antennas and Wireless Propagation Letters* **17** 1861–5

[12] Kumar J, Basu B, Talukdar F A and Nandi A 2017 Graphene-based wideband antenna for aeronautical radio-navigation applications *Journal of Electromagnetic Waves and Applications* **31** 2046–54

[13] Cruz S M F, Rocha L A and Viana J C 2018 Printing Technologies on Flexible Substrates for Printed Electronics *Flexible Electronics*

[14] Ng L W T, Zhu X, Hu G, Macadam N, Um D, Wu T, Le Moal F, Jones C and Hasan T 2019 Conformal Printing of Graphene for Single- and Multilayered Devices onto Arbitrarily Shaped 3D Surfaces *Advanced Functional Materials* **29** 1807933

[15] Leng T, Huang X, Chang K, Chen J, Abdalla M A and Hu Z 2016 Graphene Nanoflakes Printed Flexible Meandered-Line Dipole Antenna on Paper Substrate for Low-Cost RFID and Sensing Applications *IEEE Antennas and Wireless Propagation Letters* **15** 1565–8

[16] Karim N, Afroj S, Malandraki A, Butterworth S, Beach C, Rigout M, Novoselov K S, Casson A J and Yeates S G 2017 All inkjet-printed graphene-based conductive patterns for wearable e-textile applications *Journal of Materials Chemistry C* **5** 11640–8

[17] Monne M A, Grubb P M, Stern H, Subbaraman H, Chen R T and Chen M Y 2020 Inkjet-Printed Graphene-Based 1 × 2 Phased Array Antenna *Micromachines* **11** 863

[18] Cherevko A G, Morgachev Y V, Kotin I A, Yakimchuk E A, Soots R A and Antonova I V 2018 Graphene Antenna on a Biodegradable Substrate for Frequency Range of Cellular Operators *2018 XIV International Scientific-Technical Conference on Actual Problems of Electronics Instrument Engineering (APEIE)*

[19] Cherevko A and Morgachev Y 2019 Ecological graphene antennas modeling for multi-channel systems for transferring atmospheric data and oceanological information in the range of cellular communications *25th International Symposium on Atmospheric and Ocean Optics: Atmospheric Physics*

Algorithm for optical characterization of dielectric gradient index nanofilm by surface plasmon resonance spectroscopy

A V Anisimov, I Sh Khasanov

Scientific and Technological Centre of Unique Instrumentation of the Russian Academy of Sciences, Butlerova, 15, Moscow, 117342, Russia

E-mail: khasanov@ntcup.ru

Abstract. A numerical algorithm for determining the refractive index profiles of gradient nanofilms is proposed. A physical justification for the necessity of using spectroscopic measurements in surface plasmon resonance sensing in addition to angular measurements for the unambiguous reconstruction of the shape of the gradient refractive index profile is given. The proposed approach can be effective for nanofilms made of dielectric materials transparent in the IR and THz range.

1. Introduction

With the development of nanotechnology, research on composite materials with unusual properties known as metamaterials, is actively developing. A special class of metamaterials - gradient-index materials - promises interesting physical effects such as invisibility cloak, non-local dispersion, resonant tunneling of EM waves for effective energy transfer by evanescent waves, etc. [1]. In practice, the challenge is to precisely control the multilayer films manufacturing process required to achieve the desired optical properties. Reliable determination of layer thicknesses d_i of multilayer films during and after the sputtering process is one of the key factors for improving the quality of optical coating production. Existing methods of determining thicknesses by spectrophotometric measurements are no longer valid for ultra-thin films where $d \ll \lambda$, since the measured reflection coefficients R and transmittance T do not differ significantly from the uncoated substrate [2]. This generally applies to nanofilms made of materials used in the infrared (IR) and terahertz (THz) ranges of electromagnetic spectrum, and which are opaque in the visible range. In this case, the use of surface electromagnetic waves (SEWs) is preferred. Because the amplitude of the local electric field strength of the SEWs at the interface can be several orders of magnitude greater than that of the incident wave. And since SEWs are very sensitive to the surface properties of thin films [3], we hypothesize that surface plasmon resonance (SPR) sensing can be an effective method for non-destructive testing in the manufacturing of gradient refractive index nanofilms.

2. Algorithm for determination of the refractive index profile of gradient nano-film

To excite surface plasmon resonance, p-polarized light is required, incident at an angle exceeding the angle of total internal reflection at the interface between semi-infinite (compared to a thin film) media. The simplest and most well-known methods for exciting surface plasmon resonance are prism coupling schemes in the Kretschmann-Raether or Otto geometries [4]. There are several scanning modes for observing SPR: by angle, by wavelength (spectroscopic mode), and also in the case of the Otto scheme, scanning by the size of the air gap. The most frequently used mode, angle scanning, allows to determine the resonance angle and, in the case of a single layer, it is quite easily to determine its optical constants. However, as is known from photometry of thin films [5], there are multiple solutions corresponding to a one resonance curve R. In order to distinguish the physical solution from multiple solutions, spectral photometry of thin films is carried out, which sometimes also supplemented with ellipsometric measurements of the p- and s- components of the radiation [6]. In order to do so, the dispersions of the optical constants found are subjected to analyticity requirements and the satisfaction of the Kramers-Kronig relations.

To substantiate the sufficiency of spectral measurements for reliable reconstruction of optical constants, the correspondence between the results obtained by the inverse solution and the experimental data is usually indicated. In the following we will try to provide a physical justification for the motivation of using spectral values for the unambiguous reconstruction of the film profile using surface plasmon resonance.

The reflection coefficient from the metal-dielectric interface in the case of surface resonance, as well as for thin dielectric films, can be described within the theoretical framework of Fresnels equations. Although for gradient-index films the equations can be solved analytically. The reverse solution for arbitrary thin film is generally problematic to obtain. A more general approach and fundamental for numerical reverse engineering is the transfer matrix method.

2.1. Matrix Formalism

To determine the profile of a gradient-index layer, we represent it as a multilayer planar stack of finite thickness. To calculate the shape of the SPR curve of that multilayer structure, we use the derivation based on the matrix formalism [7,8]. At the interface of each layer, the luminous fluxes are reflected with Fresnel coefficients:

$$r_i = \frac{k_z^i \varepsilon_{i+1} - k_z^{i+1} \varepsilon_i}{k_z^i \varepsilon_{i+1} + k_z^{i+1} \varepsilon_i}, \tag{1}$$

where $k_z^i = \sqrt{\varepsilon_i \frac{\omega}{c} - k_0 \sin(\theta)}$ z-component (perpendicular to the interface) of the wavevector in *i*-th layer, ε_i is the dielectric constant of the *i*-th layer, ω is the frequency of the incident light, c is the speed of light, θ is the angle of incidence. The zero layer (i=0) corresponds to the prism. In the Kretschmann geometry the 1st layer corresponds to metal, the N-th layer corresponds to the external medium (e.g. air), in the Otto geometry the N-th layer corresponds to the metal, N-1 — to the external medium.

The optical fluxes between the *i*-th layer and the $(i + 1)$ layer are related by means of a characteristic matrix S_i (see Figure 1):

$$\begin{pmatrix} v_i \\ w_i \end{pmatrix} = S_i \begin{pmatrix} v_{i+1} \\ w_{i+1} \end{pmatrix}, \tag{2}$$

where

$$S_i = \begin{pmatrix} \frac{e^{-Ik_z^i d_i}}{1-r_i} & \frac{r_i e^{-Ik_z^i d_i}}{1-r_i} \\ \frac{r_i e^{Ik_z^i d_i}}{1-r_i} & \frac{e^{Ik_z^i d_i}}{1-r_i} \end{pmatrix}, \tag{3}$$

where d_i – thickness of the *i*-th layer.

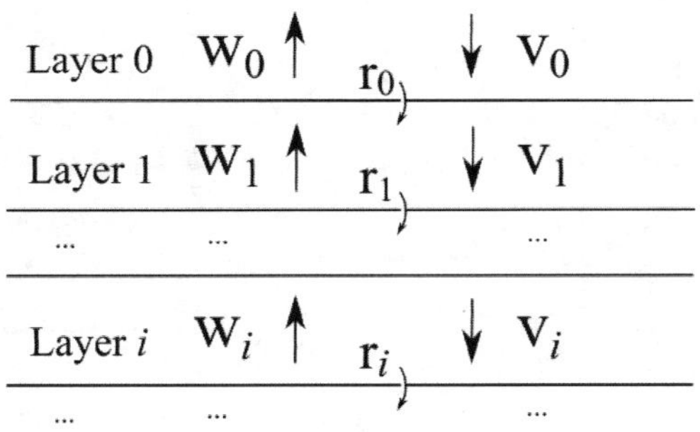

Figure 1. A multilayer stack. The labels next to the arrows indicate wave amplitudes.

Moreover

$$S_0=\begin{pmatrix} \frac{1}{1-r_0} & \frac{r_0}{1-r_0} \\ \frac{r_0}{1-r_0} & \frac{1}{1-r_0} \end{pmatrix}. \quad (4)$$

Thus, for an N-layer system, we get:

$$\begin{pmatrix} v_0 \\ w_0 \end{pmatrix} = \prod_{i=0}^{N} S_i \begin{pmatrix} v_{N+1} \\ w_{N+1} \end{pmatrix}. \quad (5)$$

For convenience, setting $v_0 = 1$, we obtain $w_0 = r$. Then reflectance can be found as

$$R = |\frac{S_{21}}{S_{11}}|^2 = |r|^2. \quad (6)$$

2.2. Reverse engineering algorithm

The analytical form of the obtained solutions for R is extremely cumbersome for direct mathematical analysis due to the non-linear form of interaction between layers. To find the refractive indices of the gradient-index profile sublayers, we use the curve-fitting procedures. For this purpose, discrepancy function DF are considered:

$$DF = \sum_{j=1}^{M}(R(X,\lambda,\theta_j) - \hat{R}(\lambda,\theta_j))^2. \quad (7)$$

Where $\hat{R}$ is modeled reflectance data, X – is vector of form $\{n_1, \ldots, n_N\}$, characterizing stack of dielectric sublayers with the thickness equal to $\frac{d}{N}$. To minimize discrepancy function DF, the program was written using built-in minimization procedures of SciPy package (e.g. Powell's method).

A common problem in minimizing high-dimensional functions is a larger number of components. Therefore, some adequate approximations have to be made. First, we consider all sublayers to be of equal thickness. Secondly, we use the method of successive approximations, starting the simulation with a small number of dielectric sublayers (N=2), which allows us to more accurately choose the initial approximation for the minimization method. Then we can sequentially increase the number of layers (N=4,…,N=100), decreasing their thickness and interpolating refractive index values, the more closely describing the form of the gradient-index profile (Figure 2).

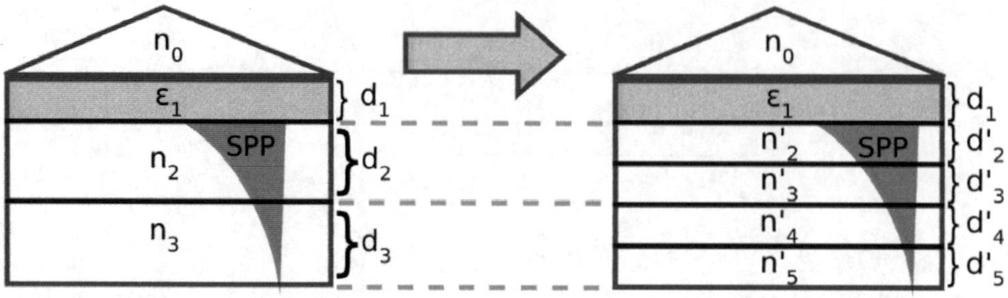

Figure 2. Illustration of the step of the algorithm for successive approximation to an index gradient profile by splitting dielectric layers into smaller thicknesses (from N=2 to N=4)

2.3. Example

As an example, we chose a real gradient-index profile from [9] with a thickness of 140 nm, where its dependence of the refractive index on the layer thickness is described by the formula (see Figure 3a):

$$n(x) = n_0(1 + (\frac{n_0}{n_{min}} - 1)(x - x^2)) \,. \quad (8)$$

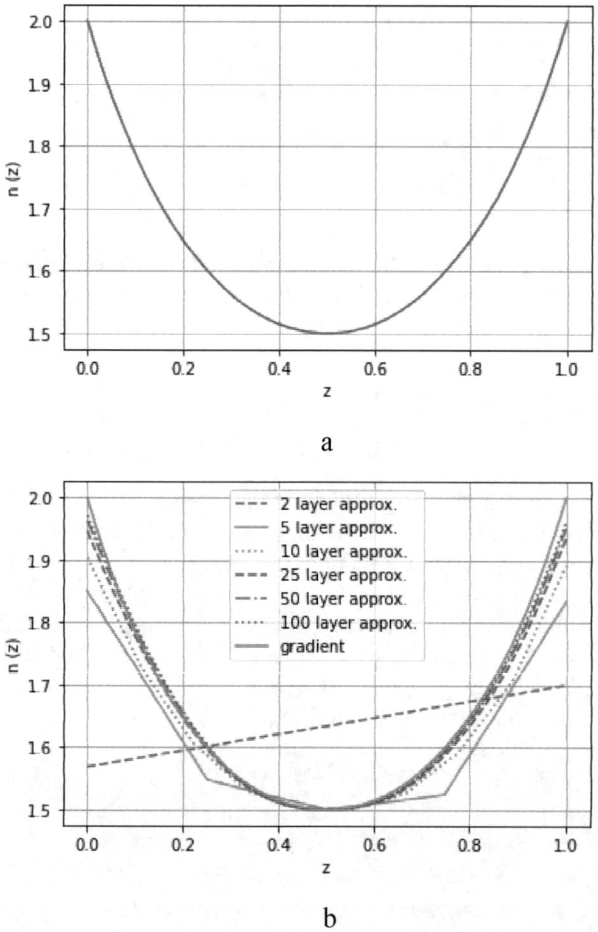

Figure 3. An example of a gradient layer described by equation (1) (a) and its successive approximation by the multilayer approximation algorithm (b)

In the case, when the profile is represented by a convex function, successive approximation from a two-layer to a multilayer model gives the correct determination of the given profile (see Figure 3b).

2.4. Algorithm reliability

However, it can be shown it is not enough for us to use only one scanning mode to unambiguously reconstruct the distribution of the refractive index of a thin film over its thickness (profile). Different gradient-index profiles (Figure 4a) can have matching SPR curves for the same parameters (wavelength and air gap size) (Figure 4b), but at other wavelengths or air gaps values their SPR curves will differ (Figure 4c-d).

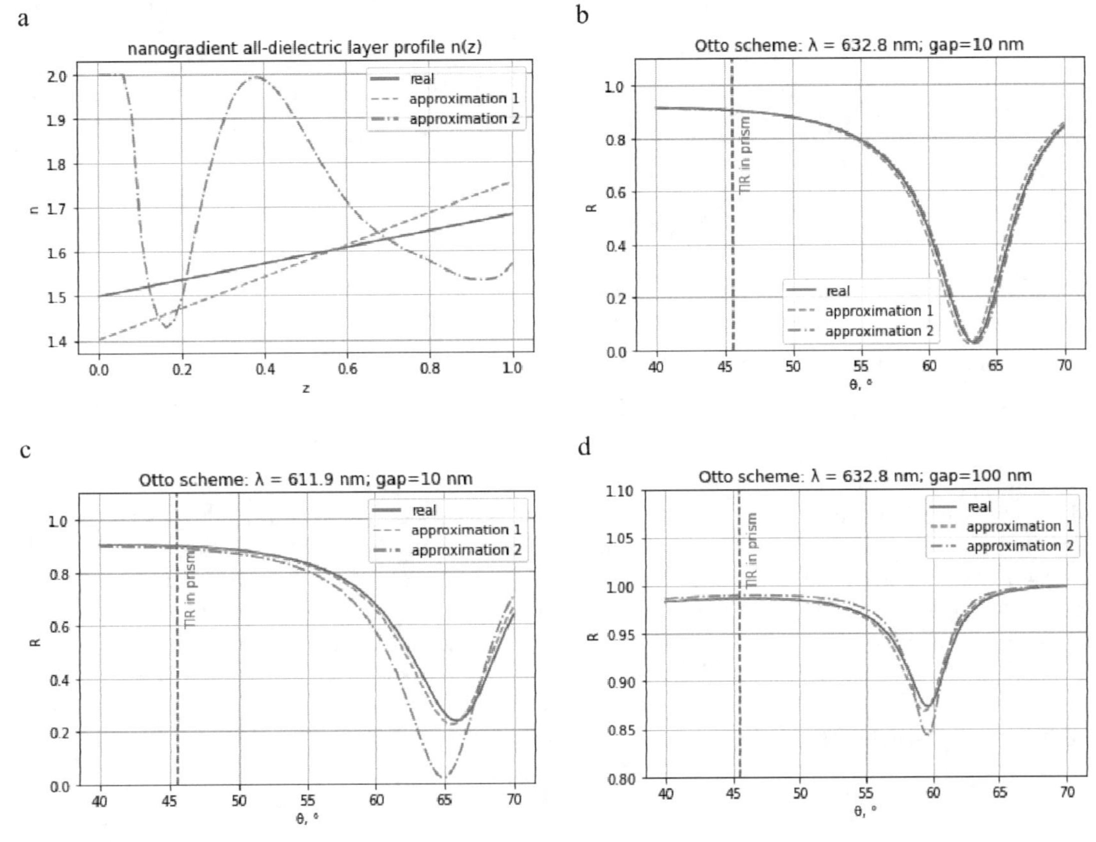

Figure 4. An example of the profiles of a gradient layer and found by its approximation by a multilayer model (2a). Divergence of approximation models from the real gradient profile (2b-d) when changing the initial parameters (wavelength and size of the air gap)

The weakness of all minimization methods is the inability to find the global minimum because the DF function is ill-conditioned quadratic function. In order to unambiguously determine the gradient-index profile, we need provide more data. For that purpose, we propose using the angle scanning mode to find the minimum of the SPR curve R (θ) and scanning by the wavelength, to study the dependence of the resonance angle and reflection coefficient R on the wavelength. Then, considering the dispersion of the refractive indices in the gradient layer to be small (or known in advance), we can unambiguously determine the shape of the refractive index profile of the gradient-index nanofilm by supplementing to the algorithm spectroscopic data. It can be shown that the dependence of the resonance angle on the wavelength is unique for each gradient profile (Figure 5). The presented algorithm works both in the Otto and Kretschmann configurations.

The proposed spectroscopic measurements can be implemented using acousto-optics, which is much easier to implement compared to X-ray reflectometry done in [9]. This method can be used for non-destructive testing in the manufacture of gradient-index thin films.

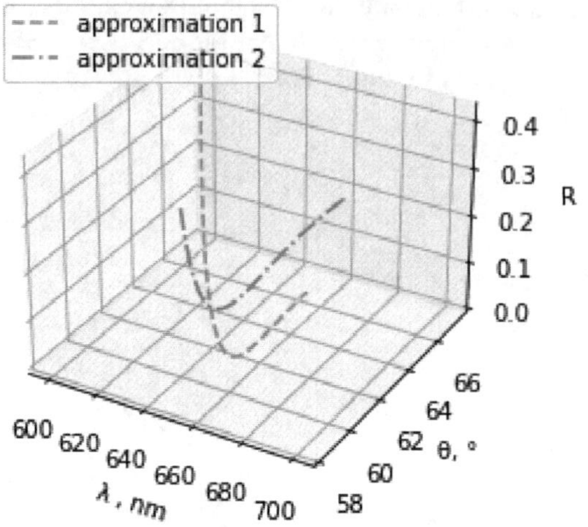

Figure 5. The dependence of the resonance angle on the wavelength for approximations 1 and 2 (they have one coincidence point)

3. The physical reasoning behind the algorithm

The physical reason why surface resonance spectrophotometry can uniquely identify the form of the gradient-index profile of thin film is due to the fact that under conditions of surface plasmon resonance the sub-layers contribute non-linearly and unequally to the reflectivity R. This is due to the fact that the evanescent wave field non-uniformly penetrates into a thin layer, exponentially decaying in the dielectric as $e^{-|k_z||z|}$[10]. Therefore, the contribution of each sub-layer of the multilayer stack to total reflectance is nonlinear. And given that the depth of penetration into the layer also depends on the frequency of the incident light, the resonance curve changes in a complex nonlinear manner with a wavelength.

This can be shown analytically by considering Kretschmann geometry with 2 dielectric layers. Then we can write the reflection coefficient R in the following form:

$$R = \left| \frac{e^{2id_3k_3}r_2 + r_3 + e^{2id_2k_2}r_1\left(e^{2id_3k_3} + r_2r_3\right) + e^{2id_1k_1}r_0\left(r_1\left(e^{2id_3k_3}r_2 + r_3\right) + e^{2id_2k_2}\left(e^{2id_3k_3} + r_2r_3\right)\right)}{e^{2id_1k_1}\left(r_1\left(e^{2id_3k_3}r_2 + r_3\right) + e^{2id_2k_2}\left(e^{2id_3k_3} + r_2r_3\right)\right) + r_0\left(e^{2id_3k_3}r_2 + r_3 + e^{2id_2k_2}r_1\left(e^{2id_3k_3} + r_2r_3\right)\right)} \right|^2. \quad (9)$$

Both the sub-layer thickness d_i and the refractive index n_i of the sub-layer contribute to the optical path and optical density. The latter also determines the magnitude of the interaction (energy exchange) between the sub-layers.

To show the effect of layer thickness d_i on reflectivity, we need to find derivate:

$$\frac{\partial R}{\partial d_i} = \frac{\partial r}{\partial d_i}r^* + \frac{\partial r^*}{\partial d_i}r. \quad (10)$$

To show that thickness of dielectric sub-layers makes a disproportionate nonlinear contribution to the total reflectance R of the thin film, we can find the difference between their partial derivatives:

$$\frac{\partial R}{\partial d_2} - \frac{\partial R}{\partial d_3} = \frac{A}{B}. \tag{11}$$

where $A = 2ie^{2i(d_1k_1+d_2k_2)}(-1+r_0^2)(-1+r_1^2)\,(e^{2id_3k_3}k_3(-1+r_2^2)r_3 + k_2(e^{2id_3k_3}r_3 + e^{2id_3k_3}r_2^2r_3 + r_2(e^{4id_3k_3} + r_3^2)))$, $B = (e^{2id_1k_1}(r_1(e^{2id_3k_3}r_2 + r_3) + e^{2id_2k_2}(e^{2id_3k_3} + r_2r_3)) + r_0(e^{2id_3k_3}r_2 + r_3 + e^{2id_2k_2}r_1(e^{2id_3k_3} + r_2r_3)))^2$.

It can be seen that the numerator A of this expression vanishes only at some isolated points, in other cases, it is a complex nonlinear function. The same conclusion can be reached for the refractive indices by analogy with thickness d_i. This unevenness in the contributions of the layers to the total reflectivity can be demonstrated numerically. Figure 6-8 shows that the dependence of the reflectance R on the refractive indices of the dielectric layers n_i is asymmetrical, indicating a disproportionate role of the layers. This is valid for all wavelengths and geometries. But this effect may not always be pronounced enough to reveal the difference between the layers.

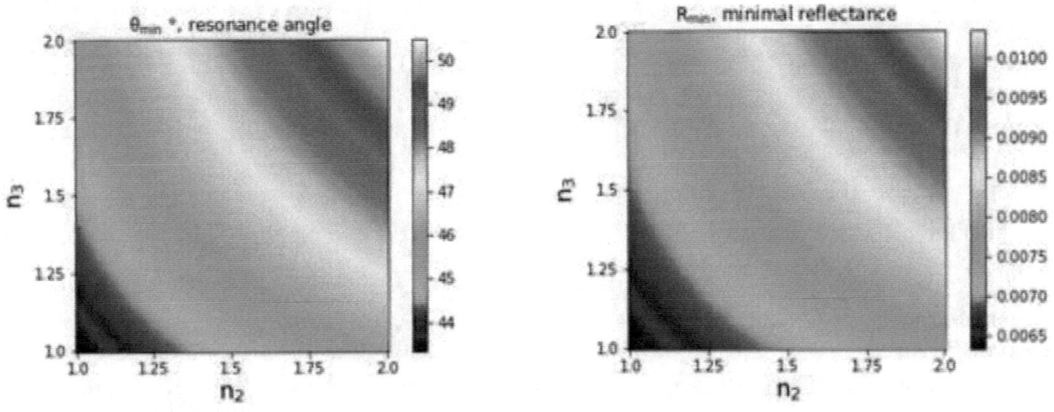

Figure 6. The dependence of the resonance peak R_{min} and resonance angle θ_{min} on the refractive indices n_2 and n_3 of the dielectric layers of equal thickness (d = 26 nm) for Kretschmann geometry: n_0 = 1.5 (glass), ε_1 = -18.30 + 0.48 I, d_1 = 52 nm (Ag), n_4 = 1 (air), λ = 632.8 nm (He-Ne).

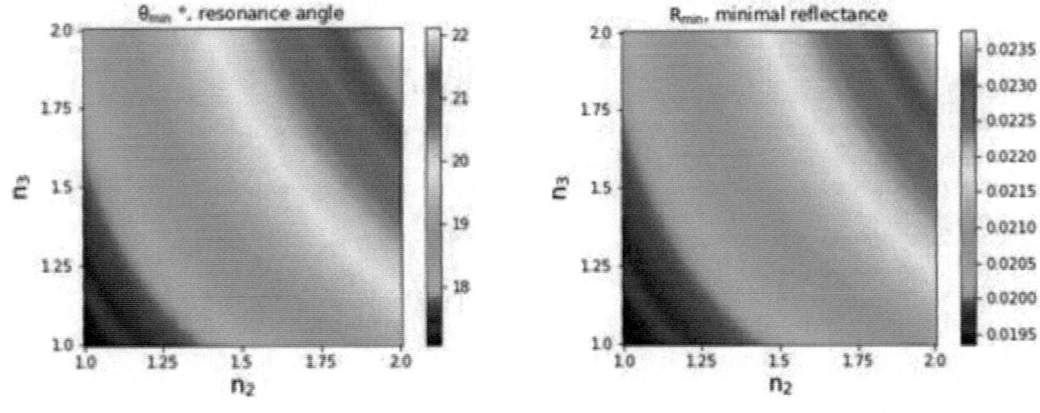

Figure 7. The dependence of the resonance peak R_{min} and resonance angle θ_{min} on the refractive indices n_2 and n_3 of the dielectric layers of equal thickness (d = 1 μm) for Kretschmann geometry: n_0 = 3.47 (Si), ε_1 = -179.16+87.538 I, d_1 = 50 nm (Pb), n_4 = 1 (air), λ = 10.6 μm (CO_2).

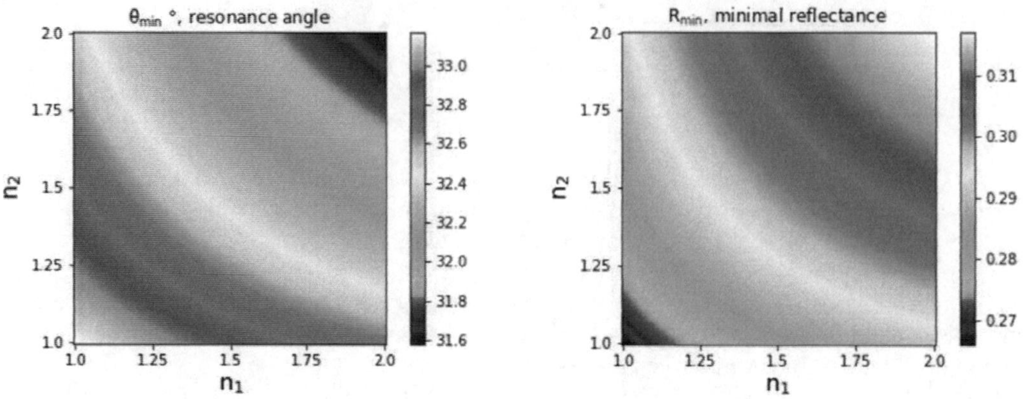

Figure 8. The dependence of the resonance peak R_{min} and resonance angle θ_{min} on the refractive indices n_1 and n_2 of the dielectric layers of equal thickness (d = 500 nm) for Otto geometry: n_0 = 3.47 (Si), ε_4 = -1.614+2.936 I (InSb), n_3 = 1, d_3 = 600 nm (air), wavelength λ = 140 μm (THz).

Taking all of the above into account, it can be assumed that surface resonance spectrophotometry for thin films can provide more informative measurements of film structure compared to conventional spectrophotometry.

4. Conclusions

We have shown that surface plasmon resonance spectrophotometry is fundamentally capable of reconstructing the distribution of the dielectric refractive index over the layer thickness (profile) in thin gradient-index dielectric films. We have developed an algorithm for determination of refractive index profile of the thin dielectric film by sequential approximation from a two-layer to a multilayer model, using SPR curves both in angle and in wavelength scanning modes. We have shown that it is possible to reconstruct a thin film profile without resorting to commercial applications (like OptiLayer) using open source library (SciPy). The physical justification for supplementing angular measurements with spectral measurements to unambiguously reconstruct the gradient-index film was given. Unlike spectrophotometric methods, SPR spectrophotometry allows to study ultra-thin films with thickness $d \ll \lambda$, what is especially important for nanofilms made of materials which are opaque in UV/visible range, but transparent in IR and THz range.

Acknowledgments

The theoretical basis of this study was carried out with the support of the Ministry of Science and Higher Education of the Russian Federation under the State contract No. 0057-2019-0006. A program for numerical calculations of surface plasmon resonance curves for multilayer optical nanofilms was written with the support by RFBR according to the research project 20-52-54004.

References

[1] A. B. Shvartsburg; A. A. Maradudin 2013 *Waves in gradient metamaterials* (World Scientific)
[2] Stenzel O 2016 *The Physics of Thin Film Optical Spectra: An Introduction* (Springer International Publishing)
[3] Nikitin A K, Khitrov O V, Gerasimov V V, Khasanov I S and Ryzhova T A 2019 In-plane interferometry of terahertz surface plasmon polaritons *J. Phys.: Conf. Ser.* **1421** 012013

[4] Klimov V V 2010 *Nanoplazmonika* (Moscow: Fizmatlit) (In Russian)
[5] Amotchkina T V, Trubetskov M K, Pervak V, Romanov B and Tikhonravov A V 2012 On the reliability of reverse engineering results *Appl. Opt., AO* **51** 5543–51
[6] Tikhonravov A V, Amotchkina T V, Trubetskov M K, Francis R J, Janicki V, Sancho-Parramon J, Zorc H and Pervak V 2012 Optical characterization and reverse engineering based on multiangle spectroscopy *Appl. Opt., AO* **51** 245–54
[7] Byrnes S J 2020 Multilayer optical calculations *arXiv:1603.02720 [physics]*
[8] Barchiesi D 2012 A classroom theory of the surface plasmon polariton *Eur. J. Phys.* **33** 1345–57
[9] Shvartsburg A B, Obod Y A, Kuzmichev A I, Volpian O D and Parkhomenko Y N 2014 Nanogradient all-dielectric films: technology of fabrication and the first experiments *Opt. Mater. Express, OME* **4** 2250–61
[10] Maier S A 2007 *Plasmonics: Fundamentals and Applications* (Springer US)

Numerical simulation of non-Newtonian Carreau fluid in a lid-driven cavity

Li Shuguang

School of Science, Dalian Maritime University, Dalian, 116026, China

Abstract. In this work, the 2D lid-driven cavity flow of non-Newtonian Carreau fluids has been studied by finite difference method on a staggered grid. A finite-difference algorithm on staggered grid based on projection method is adopted to solve the lid-driven cavity flow, which includes a second-order central difference scheme for the non-Newtonian viscous stress term. This study has been conducted for the certain pertinent parameters of Reynolds number (Re=100-1000), power-law index (n=0.6-1.4). The results show that as the Reynolds number increases, the influence of the power-law index on the flow increases. As the power-law index decreases, the flow field becomes more complicated.

1. Introduction

The lid-driven cavity problem has always been a very classic problem in fluid mechanics and has been extensively studied. Since some pioneering works gave good results for stable solutions a few decades ago, the classic lid-driven cavity problem has attracted the interest of many scholars [1]. The rise of non-Newtonian fluids has benefited from the development of nonlinear science, especially the large number of nonlinear flow characteristics of polymer solutions in the chemical industry and composite materials [2].

Numerical simulation is one of the most common methods to study the lid-driven cavity problem. Because the finite difference method is simple and efficient in numerical simulation, it has attracted much attention [3-5]. Recently, we have used the finite difference method on staggered grids to solve the natural convection problem of non-Newtonian Carreau fluids and obtained good results [7]. In solving the 2D problem, the calculation cost of the projection method and the vorticity-flow function method are roughly the same [8]. In addition, although some recent studies have used high-order compact difference methods on staggered grids to study the flow of Newtonian fluids. However, for non-Newtonian fluids, the more general second-order central difference scheme is more efficient and stable than the higher-order compact scheme [4].

In this work, we use a second-order center scheme on a staggered grid to simulate the lid-driven cavity problem for non-Newtonian Carreau fluids.

2. Mathematical model

The physical model shown in figure 1 consists of a square cavity with the height of L. The domain is filled with the non-Newtonian fluids. The horizontal walls are impermeable as the top wall is driven from the left to the right by a constant speed of v_0.

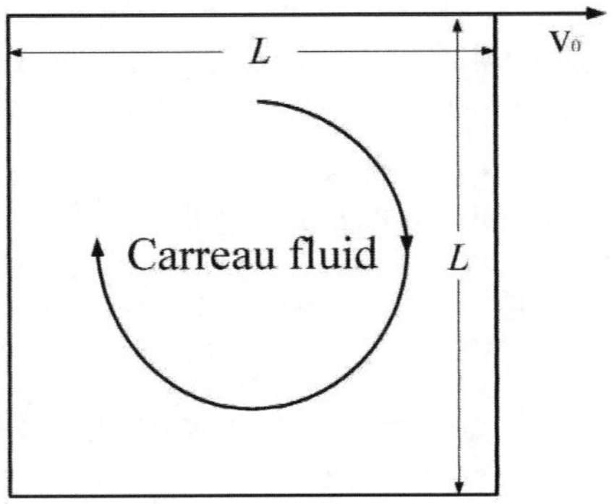

Figure 1. Geometry of the present study.

To derive the governing equations, some important assumptions are clarified as follows:

1. The flow in the square cavity is laminar, incompressible and unsteady.
2. The Carreau model is taken for the non-Newtonian fluid.
3. Mass force is not considered.

Based on the above assumptions, the governing equations of mass and momentum in dimensional form are expressed as follows, respectively:

$$\frac{\partial v_i}{\partial x_i} = 0 \tag{1}$$

$$\rho_0 \left(\frac{\partial v_i}{\partial t} + v_j \frac{\partial v_i}{\partial x_j} \right) = -\frac{\partial p}{\partial x_i} + \frac{\partial \tau_{ij}}{\partial x_j} \tag{2}$$

where v_i are the velocity components, ρ_0 is the density, p is the isotropic pressure, t is the time, and τ_{ij} are the components of the extra stress tensor.

For an incompressible fluid obeying the Carreau phenomenological constitutive model as its constitutive equation, the extra (or deviatoric) stress tensor can be written as [2, 7]:

$$\tau_{ij} = 2\mu D_{ij} = \mu \left(\frac{\partial v_i}{\partial x_j} + \frac{\partial v_j}{\partial x_i} \right), \tag{3}$$

where in the Carreau model, g (i.e. the shear viscosity) is related to the second invariant of the deformation-rate tensor as [2]:

$$\frac{\mu - \mu_\infty}{\mu_0 - \mu_\infty} = \left[1 + (\lambda I_2)^2\right]^{(n-1)/2}. \tag{4}$$

Here D_{ij} indicates the rate-of-deformation tensor and μ is apparent viscosity of the Carreau model, $I_2(D_{ij})$ is related to the second invariant of the deformation-rate tensor as [2]:

$$I_2(D_{ij}) = \sqrt{2 D_{ij} D_{ij}} \tag{5}$$

Here μ_0 is the zero-shear-rate viscosity, μ_∞ is the infinite-shear-rate viscosity, λ is a time constant, n is the power-law exponent, and a is a dimensionless parameter that describes the transition region between the zero-shear-rate region and the power-law region. Therefore, the deviation of n from unity indicates the degree of deviation from Newtonian behavior. With $n \neq 1$, the constitute (6) represents pseudoplastic (shear-thinning) fluid ($n<1$) and for ($n>1$) it represents a dilatant (shear-thickening) fluid, respectively. Pseudoplastic fluids are characterized by an apparent viscosity which decreases with increasing shear rate, however in dilatants fluids the apparent viscosity increases with increasing shear rate.

The following non-dimensional variables are introduced [7].

$$(\bar{x},\bar{y})=\frac{(x,y)}{L}\ ,\ \bar{t}=\frac{t}{t_0}\ ,\ (\bar{v}_1,\bar{v}_2)=\frac{(v_1,v_2)}{v_0}\ ,\ \bar{p}=\frac{p}{\rho v_0^2}\ ,\ v_0=\frac{L}{t_0} \tag{6}$$

The following non-dimensional governing equations are obtained (the small mark is removed):

$$\frac{\partial v_i}{\partial x_i}=0 \tag{7}$$

$$\frac{\partial v_i}{\partial t}+v_j\frac{\partial v_i}{\partial x_j}=-\frac{\partial p}{\partial x_i}+\frac{1}{\mathrm{Re}}\frac{\partial \tau_{ij}}{\partial x_j} \tag{8}$$

$$\tau_{ij}=\mu_a\left(\frac{\partial v_j}{\partial x_i}+\frac{\partial v_i}{\partial x_j}\right) \tag{9}$$

$$\mu_a=s+(1-s)\left[1+(CuI_2)^2\right]^{(n-1)/2} \tag{10}$$

where $\mathrm{Re}=\frac{\rho v_0 x_0}{\mu_0}$ is Reynolds number, n is the "power-law exponent", the ratio s has been fixed at 0,001, $Cu=\frac{\lambda v_0}{L}$ has been fixed at 3.

3. Numerical Methods

The problem (7)-(10) is solved numerically using the finite difference method, which reduces the continuum problem to a discrete problem prescribed by a system of algebraic equations.

In [6], Chorin proposed an explicit projection method to solve Navier-Stokes equations. We directly apply the explicit projection method to solve the problem (7)-(10). The finite difference scheme based on the projection method consists of the following three steps.

First, the dimensionless middle velocities are calculated by

$$\frac{v_i^*-v_i^n}{\Delta t}=\frac{1}{\mathrm{Re}}\frac{\partial \tau_{ij}^n}{\partial x_j}-\left[v_j\frac{\partial v_i}{\partial x_j}\right]^n \tag{11}$$

Second, the dimensionless pressure is obtained by solving the following Poisson equation based on the middle velocities:

$$\frac{v_i^{n+1}-v_i^*}{\Delta t}=-\frac{\partial p^{n+1}}{\partial x_i} \tag{12}$$

$$\frac{\partial v_i^{n+1}}{\partial x_i}=0 \tag{13}$$

$$-\left(\frac{\partial^2 p^{n+1}}{\partial x_1^2}+\frac{\partial^2 p^{n+1}}{\partial x_2^2}\right)=-\frac{1}{\Delta t}\left(\frac{\partial v_1^*}{\partial x_1}+\frac{\partial v_2^*}{\partial x_2}\right) \tag{14}$$

Finally, the dimensionless velocities and temperature are obtained from the following equations:

$$\frac{v_i^{n+1} - v_i^*}{\Delta t} = -\frac{\partial p^{n+1}}{\partial x_i} \tag{15}$$

Next, we solve the equation (7)-(10) by finite difference method on a staggered grid. The first momentum equation is approximated at point $\left(i+\frac{1}{2}, j\right)$, the second one at the point $\left(i, j+\frac{1}{2}\right)$. The continuity equation and pressure are approximated at the point (i, j).

For the discretization of non-Newtonian viscous stress, we use the numerical method proposed in [7]. Then, the viscous diffusion term of the first momentum equation can be discretized as:

$$\frac{\partial}{\partial x}\left(2\mu\frac{\partial u}{\partial x}\right) + \frac{\partial}{\partial y}\left[\mu\left(\frac{\partial u}{\partial y} + \frac{\partial v}{\partial x}\right)\right]$$

$$= \mu\left(\frac{\partial^2 u}{\partial x^2} + \frac{\partial^2 u}{\partial y^2}\right) + 2\frac{\partial \mu}{\partial x}\frac{\partial u}{\partial x} + \frac{\partial \mu}{\partial y}\left(\frac{\partial u}{\partial y} + \frac{\partial v}{\partial x}\right)$$

$$= \frac{1}{2}\left(\mu_{i+\frac{1}{2},j+\frac{1}{2}} + \mu_{i+\frac{1}{2},j-\frac{1}{2}}\right)\left[\frac{1}{\Delta x^2}\left(u_{i+\frac{3}{2},j} - 2u_{i+\frac{1}{2},j} + u_{i-\frac{1}{2},j}\right) + \frac{1}{\Delta y^2}\left(u_{i+\frac{1}{2},j+1} - 2u_{i+\frac{1}{2},j} + u_{i+\frac{1}{2},j-1}\right)\right]$$

$$+\left[\frac{1}{2\Delta x}\left(\mu_{i+\frac{3}{2},j+\frac{1}{2}} - \mu_{i+\frac{1}{2},j+\frac{1}{2}}\right) + \frac{1}{2\Delta x}\left(\mu_{i+\frac{3}{2},j-\frac{1}{2}} - \mu_{i+\frac{1}{2},j-\frac{1}{2}}\right)\right]\frac{1}{2\Delta x}\left(u_{i+\frac{3}{2},j} - u_{i+\frac{1}{2},j}\right)$$

$$+\frac{1}{2\Delta y}\left(\mu_{i+\frac{1}{2},j+\frac{1}{2}} - \mu_{i+\frac{1}{2},j-\frac{1}{2}}\right)\left[\frac{1}{\Delta y}\left(u_{i+\frac{1}{2},j+1} - u_{i+\frac{1}{2},j-1}\right) + \frac{1}{\Delta x}\left(v_{i+1,j+\frac{1}{2}} - v_{i,j+\frac{1}{2}}\right) + \frac{1}{\Delta x}\left(v_{i+1,j-\frac{1}{2}} - v_{i,j-\frac{1}{2}}\right)\right]$$

where the non-Newtonian viscosity can be written as:

$$\mu_{i+\frac{1}{2},j+\frac{1}{2}} = s + (1-s)\left\{1 + Ca^2\left[\frac{1}{16}\left(\frac{1}{\Delta x}\left(u_{i+\frac{3}{2},j} - u_{i-\frac{1}{2},j}\right) + \frac{1}{\Delta x}\left(u_{i+\frac{1}{2},j+1} - u_{i-\frac{1}{2},j+1}\right)\right)^2\right.\right.$$

$$+\frac{1}{16}\left(\frac{1}{\Delta y}\left(v_{i,j+\frac{3}{2}} - v_{i,j-\frac{1}{2}}\right) + \frac{1}{\Delta y}\left(v_{i+1,j+\frac{1}{2}} - v_{i+1,j-\frac{1}{2}}\right)\right)^2$$

$$\left.\left.+\left(\frac{1}{\Delta x}\left(v_{i+1,j+\frac{1}{2}} - v_{i,j+\frac{1}{2}}\right) + \frac{1}{\Delta y}\left(u_{i+\frac{1}{2},j+1} - u_{i+\frac{1}{2},j}\right)\right)^2\right]\right\}$$

For the discretization of convection terms, the numerical technique introduced in [9] is used.

4. Numerical simulation results

Some results of numerical simulation for different power law exponents are given. Figure 2 shows the flow field distribution of non-Newtonian Carreau fluids with power law exponents $n = 0{,}6$ and $n = 1{,}4$ when $\mathrm{Re} = 400$. Figure 3 shows the viscosity field distribution of non-Newtonian Carreau fluids with power law exponents $n = 0{,}6$ and $n = 1{,}4$ when $\mathrm{Re} = 400$.

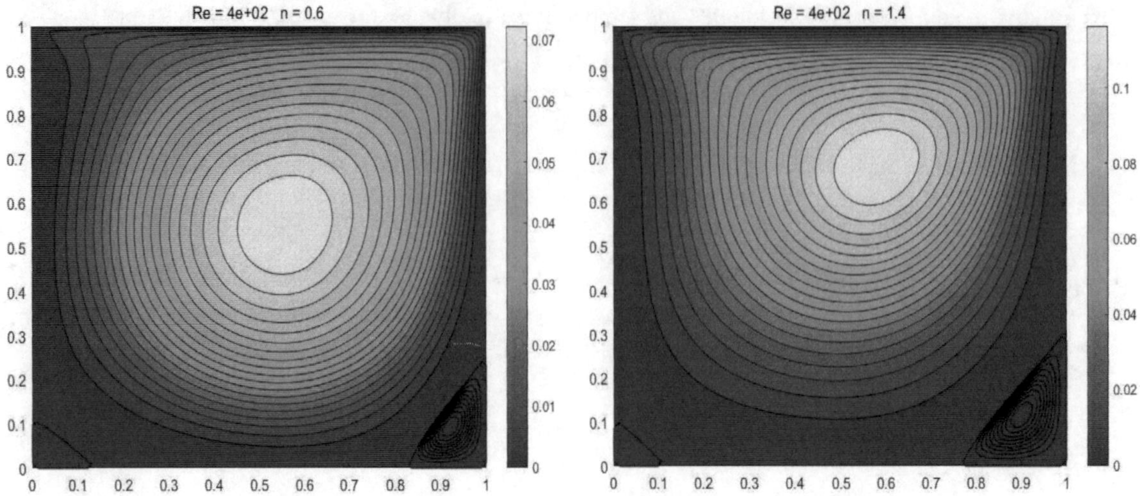

n=0,6 n=1,4

Figure 2. Flow field distribution of Carreau fluids with power law exponents $n = 0,6$ and $n = 1,4$ when $\mathrm{Re} = 400$.

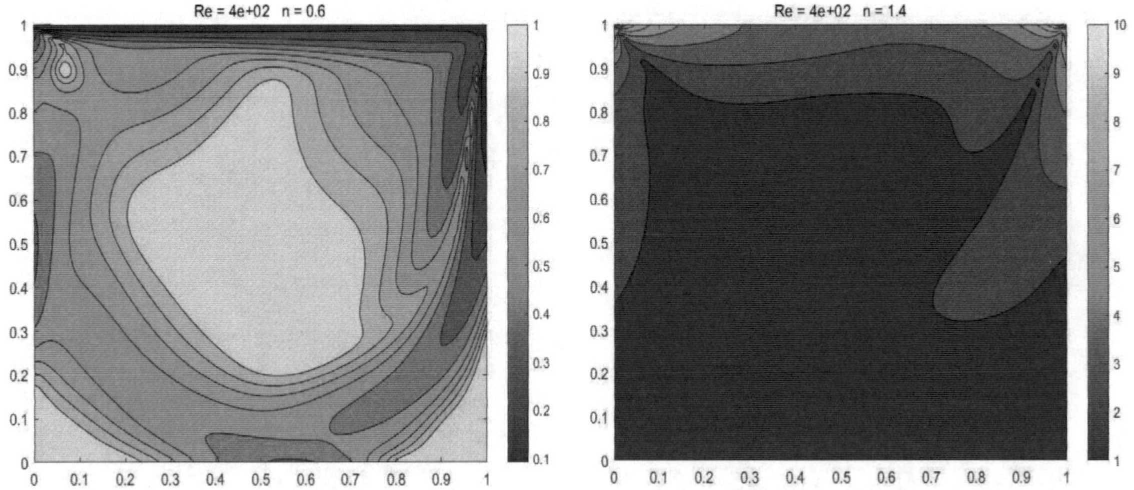

n=0,6 n=1,4

Figure 3. Viscosity field distribution of Carreau fluids with power law exponents $n = 0,6$ and $n = 1,4$ when $\mathrm{Re} = 400$.

Figure 4 and figure 5 show the flow and viscosity field distribution of non-Newtonian Carreau fluids with power law exponents $n = 0,6$ and $n = 1,4$ when $\mathrm{Re} = 1000$, respectively.

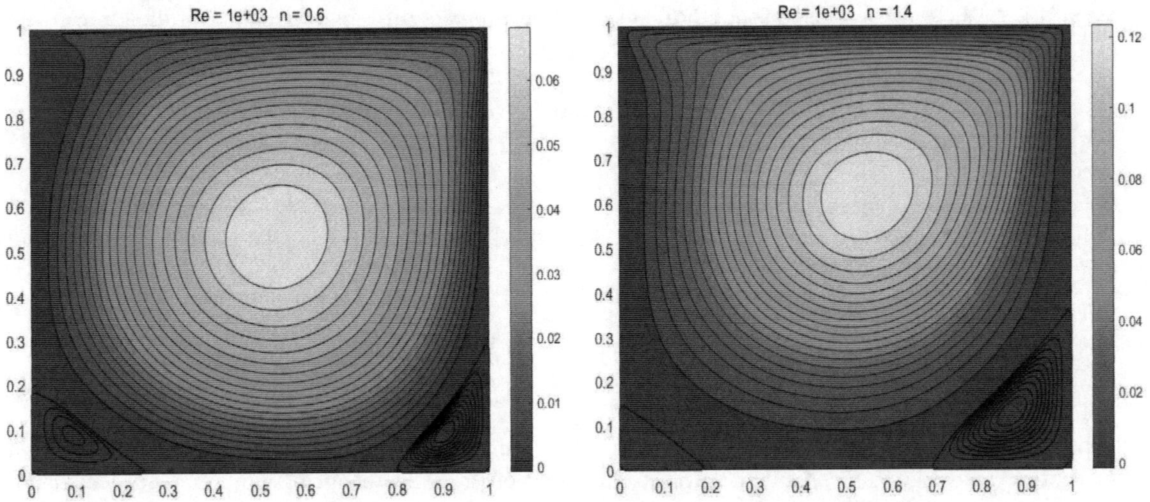

n=0,6 n=1,4

Figure 4. Flow field distribution of Carreau fluids with power law exponents $n = 0,6$ and $n = 1,4$ when $\mathrm{Re} = 1000$.

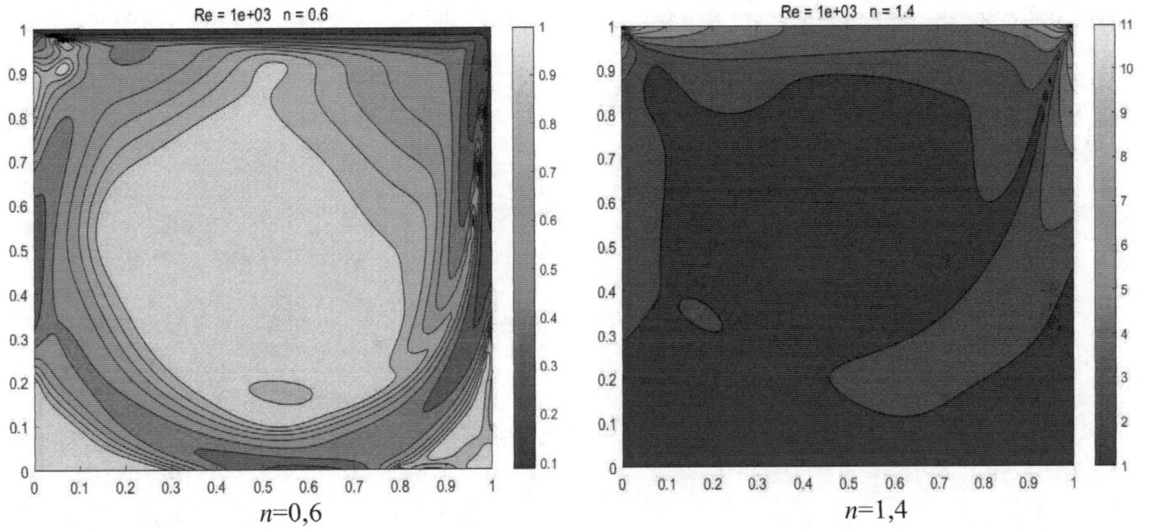

n=0,6 n=1,4

Figure 5. Viscosity field distribution of Carreau fluids with power law exponents $n = 0,6$ and $n = 1,4$ when $\mathrm{Re} = 1000$.

5. Conclusion

Numerical results show that the new finite difference algorithm can effectively simulate the lid-driven cavity problem of non-Newtonian Carreau fluids. The lid-driven cavity problem for non-Newtonian fluids is much more complicated than that for Newtonian fluids. .

References

[1] Bruneau C H, Saad M 2006 The 2D lid-driven cavity problem revisited *Computers & fluids* **35** 3 pp 326-348
[2] Bird R B, Armstrong R C, Hassager O 1987 Dynamics of polymeric liquids. Vol. 1: Fluid mechanics. (*John Wiley*) 649

[3] Li S, Wu X 2018 L^∞ error bound of conservative compact difference scheme for the generalized symmetric regularized long-wave equations *Computational and Applied Mathematics* **37** 3 pp 2816-36

[4] Li S 2019 Numerical study of a conservative weighted compact difference scheme for the symmetric regularized long wave equations *Numerical Methods for Partial Differential Equations* **35** 1 pp 60-83

[5] Dimitrienko Y I, Li S, Niu Y 2021 Study on the dynamics of a nonlinear dispersion model in both 1D and 2D based on the fourth-order compact conservative difference scheme *Mathematics and Computers in Simulation* **182** pp 661-689

[6] Chorin A J 1968 Numerical solution of the Navier-Stokes equations *Mathematics of computation* **22** 104 pp 745-762

[7] Dimitrienko Y I, Li S 2020 Numerical simulation of MHD natural convection heat transfer in a square cavity filled with Carreau fluids under magnetic fields in different directions *Computational and Applied Mathematics* **39** 4 pp 1-26

[8] Reis G A, Tasso I V M, Souza L F, Cuminato J A 2015 A compact finite differences exact projection method for the Navier - Stokes equations on a staggered grid with fourth-order spatial precision *Computers & Fluids* **118** pp 19-31

[9] Seibold B 2008 A compact and fast Matlab code solving the incompressible Navier-Stokes equations on rectangular domains mit18086 navierstokes. m. *Massachusetts Institute of Technology*

Dynamics of media with stationary parameters and telegraph equations

A G Kushner[1,2,3], E N Kushner[4]

[1] Institute of Control Sciences of RAS, 65 Profsoyuznaya street, Moscow 117997, Russia;
[2]Lomonosov Moscow State University, Faculty of Physics, GSP-2, 1-2 Leninskie Gory, Moscow 119991, Russia;
[3] Moscow Pedagogical State University, M. Pirogovskaya Str., 1/1, 119991 Moscow, Russia
[4] Moscow State Technical University of Civil Aviation, 20 Kronshtadtsky blvd, Moscow 125993, Russia.

E-mail: kushner@physics.msu.ru, ekushner@ro.ru

Abstract. The paper proposes an approach for constructing exact solutions of differential equations of mathematical physics, in particular, the telegraph equation. The method is based on the theory of finite-dimensional dynamics of systems of evolutionary differential equations. This theory is a natural extension of the theory of dynamical systems to partial differential equations. It allows one to construct exact solutions of partial differential equations even in the case when equations do not have symmetry algebras sufficient for integration.

1. Introduction

Methods of finite-dimensional dynamics make it possible to distinguish finite-dimensional submanifolds in an infinite-dimensional solutions' space of of evolutionary partial differential equations. These submanifolds can be used to construct exact solutions of evolutionary equations even in the case when the equation does not have a sufficient supply of symmetries, as well as to study the stability of solutions and construct new numerical methods.

The main idea of this method is as follows (see [1-3]). Consider ordinary differential equations

$$\boldsymbol{y}^{(k+1)} = \boldsymbol{f}(\boldsymbol{x}, \boldsymbol{y}, \boldsymbol{y}', \dots, \boldsymbol{y}^{(k)}). \tag{1}$$

This equation generates a distribution $\mathcal{F}$ of straight lines on the space $J^k(\mathbb{R})$ of k-jets [4]. Therefore, symmetries of equation (1) are diffeomorphisms of $J^k(\mathbb{R})$ that preserve this distribution. Infinitesimal symmetries of equation (1) are represented by vector fields that form Lie algebra Symm $\mathcal{F}$. This Lie algebra has an ideal Char $\mathcal{F}$ consisting of vector fields, the shifts along which preserve each integral curve of the distribution $\mathcal{F}$. These are the so-called *characteristic symmetries*. The corresponding quotient Lie algebra Shuff $\mathcal{F}$ is called Lie algebra of *shuffling symmetries*. This term is justified by the fact that shifts along vector fields from this factor algebra reshuffle the integral distribution curves $\mathcal{F}$.

Each vector field from Shuff $\boldsymbol{\mathcal{F}}$ is uniquely determined by a function on the space of k-jets. This function is called *generating* for a given vector field. By $\boldsymbol{S_\phi}$ denote a vector field of shuffling symmetry with generating function $\boldsymbol{\phi}$.

Let $\boldsymbol{\Phi}_t$ be the shifts along a vector field $\boldsymbol{S} \in \mathbf{Symm}\,\boldsymbol{\mathcal{F}}$ and let $\boldsymbol{L}$ be an integral curve of the distribution $\boldsymbol{\mathcal{F}}$. The transformation $\boldsymbol{\Phi}_t$ generates a surface $\cup_t \boldsymbol{\Phi}_t(\boldsymbol{L})$ consists of other integral curves. This surface

can be interpreted as a graph of some evolutionary partial differential equation with two independent variables $\boldsymbol{t}, \boldsymbol{x}$.
The described reasoning is reversible. Consider the following evolutionary differential equation

$$u_t = \phi(x, u, u_x, u_{xx}). \tag{2}$$

Since the right-hand side does not explicitly depend on time t, this equation describes processes in media with stationary parameters.
Let $\boldsymbol{\phi}(\boldsymbol{x}, \boldsymbol{y_0}, \boldsymbol{y_1}, \boldsymbol{y_2})$ be a generating function of the shuffling symmetries for equation (1). Here $\boldsymbol{x}, \boldsymbol{y_0}, \boldsymbol{y_1}, \boldsymbol{y_2}$ are canonical coordinates on 2-jet space $\boldsymbol{J}^2(\mathbb{R})$. Solving equation (1) and shifting a $\boldsymbol{k}$-graph of its partial solution along the vector field $\boldsymbol{S_\phi}$ we obtain a surface. This surfase is a $\boldsymbol{k}$-graph of a solution to equation (2).
The following statements are equivalent:

- equation (2) describes evolutions on solutions of equation (1);
- the function $\boldsymbol{\varphi}$ is a generating function for the symmetries of equation (1)

The finite-dimensional dynamics method has been successfully used to study solutions of evolutionary differential equations. For example, dynamics of the Rapoport – Leas equation were constructed in [5]. These dynamics have been used to build attractors. Dynamics of the Burgers – Huxley equation were constructed in [6]. The dynamics of the generalized Korteweg – de Vries equations were constructed in [7]. Third order dynamics was used for a numerical method [8]. The theory of finite-dimensional dynamics was extended to systems of evolutionary equations [9], as well as to non-evolutionary equations [10].

2. Symmetries of ordinary differential equations

First, consider scalar equations. Let x, y_0, $y_1, \dots, y_k$ be coordinates on $\mathbb{R}^{k+2}$. Let $h(x)$ be a function of class C^k. A curve $j^1(h) = \{y_0 = h(x),\ y_1 = h'(x), \dots, y_k = h^{(k)}(x)\} \subset \mathbb{R}^{k+2}$ is called a 1-*jet* of a function $h(x)$. The space $J^k = \mathbb{R}^{k+2}$ is called the *space of k-jets*.
Equation (1) defines a 1-dimensional distribution $\mathcal{F}$ on the space J^k: each integral curve of this distribution is the k-jet of some solution to equation (1). This distribution is generated by the vector field

$$D = \partial_x + y_1\partial_0 + \dots + y_k\partial_{k-1} + f\partial_k,$$

where $\partial_i = \partial_{y_i}$ $(i = 0, \dots, k)$.
Now let's consider systems of differential equations. For simplicity, we restrict our consideration to systems of two equations:

$$\begin{cases} p^{(k+1)} = P(x, p, q, p', q', \dots, p^{(k)}, q^{(k)}), \\ q^{(k+1)} = Q\big(x, p, q, p', q', \dots, p^{(k)}, q^{(k)}\big). \end{cases} \tag{3}$$

Here $\boldsymbol{x}$ is an independent variable and $\boldsymbol{p}, \boldsymbol{q}$ are unknown functions.
Introduce a $(\boldsymbol{2k+3})$-dimentional space with coordinates $\boldsymbol{x}, \boldsymbol{p_0}, \boldsymbol{q_0}, \dots, \boldsymbol{p_k}, \boldsymbol{q_k}$. This space we denote by $\boldsymbol{J^k}$, too. System (3) generates a 1-dimensional distribution $\boldsymbol{\mathcal{F}}$ on this space. This distribution is generated by the vector field (here we use the same notation as above)

$$\boldsymbol{D = \partial_x + p_1\partial_{p_0} + q_1\partial_{q_0} + \dots + p_k\partial_{p_{k-1}} + q_k\partial_{q_{k-1}} + P\partial_{p_0} + Q\partial_{q_k}}.$$

A diffeomorphism of the space J^k that preserves the distribution $\mathcal{F}$ is called the *symmetry* of equation (1) or system (3).
Two cases are possible:

1) a symmetry preserves each integral curve of the distribution $\mathcal{F}$;
2) a symmetry shuffles integral curves of this distribution.

A symmetry is called *characteristic* in the first case and *shuffling* in the second one. The concepts of characteristic and shuffling symmetries carry over to vector fields. For example, the vector field D is a characteristic infinitesimal symmetry.
All infinitesimal symmetries form Lie algebra with respect to Lie bracket. Let us denote this algerbra by Symm $\mathcal{F}$. Characteristic symmetries form an ideal in this Lie algebra, which we denote by Char $\mathcal{F}$. The quotient Lie algebra

$$\text{Shuff}\,\mathcal{F} = \text{Symm}\,\mathcal{F} / \text{Char}\,\mathcal{F}$$

is called Lie algebra of *shuffling symmetries*.
For equation (1) infinitesimal shuffling symmetries are uniquely determined by functions (see [4]):

$$S_\psi = \psi\partial_0 + D(\psi)\partial_1 + D^2(\psi)\partial_2 + \cdots + D^k(\psi)\partial_k$$

For system (3) we have to use pairs of functions:

$$S_{(\varphi,\psi)} = \varphi\partial_{p_0} + \psi\partial_{q_0} + D(\varphi)\partial_{p_0} + D(\psi)\partial_{q_0} + \cdots + D^k(\varphi)\partial_{p_k} + D^k(\psi)\partial_{q_k}.$$

Here $D^i(\psi) = D(D^{i-1}(\psi))$, $(i = 2, \dots, k)$. The function ψ (or the pair (φ, ψ)) is called *generating function* of the vector field S_ψ (or $S_{(\varphi,\psi)}$).

3. Dynamics of evolutionary equations

Equation (1) is called a *finite-dimensional dynamics* of equation (2) if the function $\boldsymbol{\phi}(\boldsymbol{x}, \boldsymbol{y_0}, \dots, \boldsymbol{y_k})$ is a generaing function for some shuffling symmetry of equation (1). The number k is called an order of the dynamics.

Theorem 1. Equation (1) is a finite-dimensional dynamics of equation (2) if and only if

$$\boldsymbol{D^{k+1}(\phi) - \sum_{i=0}^{k} D^i(\phi)\frac{\partial f}{\partial y_i} = 0}.$$

This theorem allows one to construct finite-dimensional dynamics of any order for equation (2). The following theorem allows one to construct finite-dimensional families of solutions to equations (2) from the general solution to equation (1).

Theorem 2. Let $\boldsymbol{y(x, C_1, \dots, C_k)}$ be a general solution of equation (1) and $\boldsymbol{\Phi_t}$ is a shift transformation along the vector field $\boldsymbol{S_\varphi}$. Then the function $\boldsymbol{u(t, x)} = \left(\boldsymbol{\Phi_t^{-1}}\right)^* (\boldsymbol{y(x, C_1, \dots, C_k)})$ is a k-parameter solution to equation (2).

Now consider equations containing the second time derivative:

$$\boldsymbol{u_{tt} = f(x, u, u_t, u_x, u_{tx}, u_{xx})}. \quad (4)$$

This equation can be written as a system of two evolutionary equations:

$$\boldsymbol{u_t = v, \qquad v_t = f(x, u, v, u_x, v_x, u_{xx})}.$$

Introduce two functions $\boldsymbol{\phi = p_0}$, $\boldsymbol{\psi = f(x, p_0, q_0, p_1, q_1, \dots, p_k, q_k)}$.
System (3) is called *finite-dimensional dynamics* of equation (4) if the pair $(\boldsymbol{\varphi}, \boldsymbol{\psi})$ is a generating vector function for shuffling symmetries.

Theorem 3. System (3) is the finite-dimensional dynamics of equation (6) if and only if the pair (φ, ψ) the following conditions hold:

$$\begin{cases} D^{k+1}(\varphi) &= S_{(\phi,\psi)}(P), \\ D^{k+1}(\psi) &= S_{(\phi,\psi)}(Q). \end{cases}$$

Let a curve L be a k-jet of a solution to system (3). Shifting it along the trajectories of the vector field $S_{(\phi,\psi)}$ we get a surface representing a k-graph of a solution to equation (4).

4. Telegraph equation

In this section, the method will be illustrated for the telegraph equation

$$u_{tt} - u_{xx} = au + bu_t + c, \tag{5}$$

where a, b, c are constants. The systems

$$\begin{cases} p'' &= \frac{p'}{x+\alpha}, \\ q'' &= \frac{q'}{x+\alpha} \end{cases} \tag{6}$$

and

$$\begin{cases} p'' = \frac{2b\alpha-\alpha^2(x+\beta)}{4b^2+16a-\alpha^2(x+\beta)^2} p' - \frac{4\alpha}{4b^2+16a-\alpha^2(x+\beta)^2} q', \\ q'' = -\frac{4a\alpha}{4b^2+16a-\alpha^2(x+\beta)^2} p' - \frac{2b\alpha+\alpha^2(x+\beta)}{4b^2+16a-\alpha^2(x+\beta)^2} q' \end{cases} \tag{7}$$

are second order dynamics for equation (5). Here α, β are arbitrary constants. The general solution to system (6) is

$$\begin{cases} p(x) = C_3 + C_4(x+\alpha)^2, \\ q(x) = C_1 + C_2(x+\alpha)^2, \end{cases}$$

and the general solution to system (7) is

$$\begin{cases} p(x) = \frac{1}{2}C_2x^2 + C_3x + C_4, \\ q(x) = \frac{1}{8\alpha}\left(x(C_2\beta - C_3)(2\beta + x)\alpha^2 + \alpha(8C_1 + 2bC_2x^2 + 4bC_3x) - 32C_2\left(a + \frac{b^2}{4}\right)x\right). \end{cases}$$

Applying the shift transformations Φ_t^{-1} to the constructed general solutions, we find 4-dimensional family of solutions of equation (5). For example, the function

$$u(t,x) = -1 + \frac{e^{\frac{1-\sqrt{5}t}{2}}}{10}\left(\frac{5}{2}x^2 + 5 + \sqrt{5}(10x - t + 1)\right) + \frac{e^{-\frac{1-\sqrt{5}t}{2}}}{10}\left(\frac{5}{2}x^2 + 5 - \sqrt{5}(10x - t + 1)\right)$$

is a solution to equation (5). It correspond the case when $a = b = c = 1$, $\alpha = 1$, $\beta = 0$, and $C_2 = 1$, $C_1 = C_3 = C_4 = 0$.

References

[1] Kruglikov B S, Lychagina O V 2005 Finite dimensional dynamics for Kolmogorov – Petrovsky – Piskunov equation *Lobachevskii Journal of Mathematics*. **19**, pp 13-28.

[2] Lychagin V V, Lychagina O V. 2007 Finite Dimensional Dynamics for Evolutionary Eguations *Nonlinear Dyn.* **48**, pp 29-48.
[3] Kushner A, Matviichuk R. 2021 Finite Dimensional Dynamics of Evolutionary Equations with Maple *Differential Geometry, Differential Equations, and Mathematical Physics. The Wisła 19 Summer School.* Eds: M. Ulan, E. Schneider. Springer Birkhauser pp 123-149
[4] Kushner A G, Lychagin V V, Rubtsov V N. 2007 *Contact geometry and non-linear differential equations* (Cambridge: Cambridge University Press) p 496.
[5] Akhmetzyanov A V, Kushner A G, Lychagin V V 2017 Attractors in models of porous media flows *Doklady Mathematics* **95**, no. 1, pp 72–75.
[6] Kushner A, Matviichuk R. 2020 Exact solutions of the Burgers – Huxley equation via dynamics *Journal of Geometry and Physics* **151**, pp 1-12.
[7] Kushner E, Kushner A, Samohin A. 2019 Differential invariants of third order evolutionary non-linear PDEs *2019 Twelfth International Conference Management of large-scale system development (MLSD).* – IEEE Piscataway, NJ, United States.
[8] Salnikov A, Akhmetzyanov A, Kushner A, Lychagin V. 2019 A numerical method for constructing attractors of evolutionary filtration equations, *1st International Conference on Control Systems, Mathematical Modelling, Automation and Energy Efficiency*, IEEE, pp 22-24.
[9] Gorinov A, Kushner A. 2020 Dynamics of Evolutionary PDE Systems *Lobachevskii Journal of Mathematics* **41** no. 12, pp 2448-57.
[10] Kushner A, Matviichuk R. 2021 Dynamics and exact solutions of non-evolutionary partial differential equations *Differential Geometry and its Application* **76**, 101761

Application of Brillouin scattering in optic-fiber sensors

I V Bogachkov[1], N I Gorlov[2], E T Kitova[3] and T I Monastyrskaya[2]

[1] Department of Communications and Information, Omsk State Technical University, 11 Pr. Mira, Omsk 644050, Russia
[2] Siberian State University of Telecommunications and Computer Science, 86 Kirova Street, Novosibirsk 630102, Russia
[3] Novosibirsk State Technical University, 20 Pr. K. Marksa, Novosibirsk 630073, Russia

E-mail: gorlovnik@yandex.ru

Abstract. The article analyses possible applications for Brillouin scattering in optic-fiber sensors. They make it possible to do temperature and physical strain measurements with high accuracy for long line sections. The functional dependences of the output signals characteristics on parameters measured by the sensors are studied. Of particular interest are the results of the analysis of the spectral component shifts in the Brillouin light scattering depending on the fiber elongation and temperature. After a brief review of the basic theoretical principles the results of some researches aimed to expand the dynamic range and to increase spatial resolution. The results of simulation in professional design software environment OptiSystem 17.1. are described. The results are obtained by experimental research based on Brillouin reflectometer AQ8603 manufactured by "ANDO" (Japan). They show that it is possible to implement optic-fiber sensors based on Brillouin scattering in telecommunication systems, oil and gas industry, in electric-power industry, construction, aviation and space industry. The objectives for further research are to perform metrological analysis at all stages of the method implementation, to complete the base of Brillouin spectrograms for optical fiber of various types and to improve algorithms for automated processing spectra in order to expand functionality of the systems.

1. Introduction

Optic fibers are widely used as communication channels in which light waves can be transmitted over long distances. In this situation fiber lines are isolated from external disturbances by means of cable technologies. However, by increasing environmental influences on the properties of light penetrating into the waveguide, fiber can be used for detection, monitoring and even measurement of external disturbances (measured values) in the integrated or distributed format. When optical power exceeds the prescribed power threshold, nonlinear phenomena, such as Brillouin scattering due to its strong dependence on external environmental variables (deformation and temperature), then the waveguide can be used successfully in optic sensor systems. In these cases, the optic fiber is a medium in which interaction occurs, acting simultaneously as both a distributed converter and an optical channel. These sensors can measure changes in particular parameter along the whole fiber converter. Therefore, dynamic range, correlated to the maximum fiber length and spatial resolution of the converter (minimal fiber length required for measurement during serial disturbances or events) are key factors which are significant and they should be investigated [1].

Fiber-optic sensors have the following advantages compared to their electrical analogs [2]:

- fiber-optic line is explosion- and fireproof;
- optic fiber has high resistance to the influence of corrosive media, pressures and temperatures;
- fiber sensors are distributed data collection system with the remote information processing devices;
- they easily operate in combination with optical information processing systems;
- small size and weight of optic fiber;
- high corrosion resistance, especially to chemical solvents, oil and water;
- free of induction;
- low cost.

The above advantages make it possible to design, manufacture and perform technical operation of distributed systems for monitoring of long lines along the whole length in real-time.

2. Theoretical background

A distributed fiber-optic sensor is a kind of measuring instrument. It transforms a physical quantity which is measured into the optical signal. This signal is transmitted through the optical fiber to a processing device to process the optical signals. Algorithms used for processing the intermediate data received are provided by the required metrological sensor characteristics. Furthermore, they provide the information presentation in tabular or graphical format.

Rayleigh, Brillouin and Raman scattering spectra are shown in Figure 1 [3].

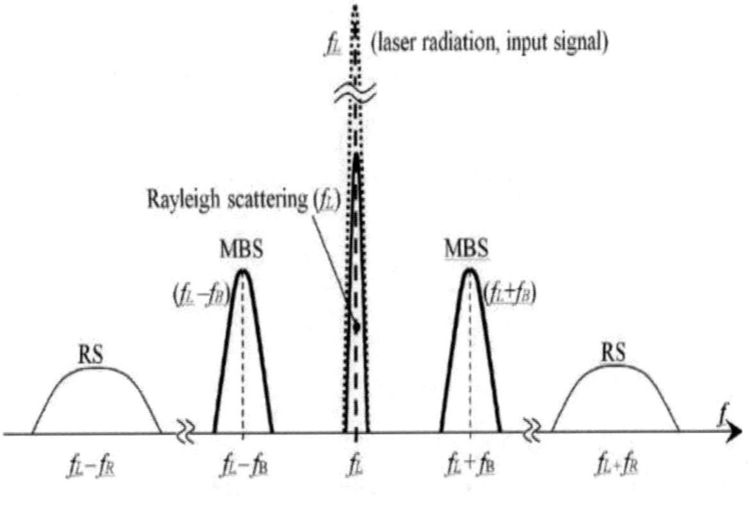

Figure 1. Spectrum of light scatterings in the fiber (fB ~ 10 ... 11 GHz, fR ~ 13 THz).

The system of the distributed sensing, which is based on the Brillouin scattering can simultaneously measure temperature and voltage along the fiber. Typically, the intensity of light and the frequency shift in the Brillouin scattering is affected by the temperature and deformation in the fiber. These are defined by the following formula [4]:

$$\begin{bmatrix}\Delta v_B(T,\varepsilon)\\ \Delta P_B(T,\varepsilon)\end{bmatrix} = \begin{bmatrix}C_{v,T} & C_{v,\varepsilon}\\ C_{P,T} & C_{P,\varepsilon}\end{bmatrix}\begin{bmatrix}\Delta T\\ \Delta\varepsilon\end{bmatrix}, \tag{1}$$

where:
ΔP_B (T, ε) - variations of power in the Brillouin scattering spectrum;
ΔV_B (T, ε) - shift of the central frequency in the Brillouin scattering;
ΔT – fluctuations in fiber temperature;
Δ ε - fluctuations in fiber strain;
$C_{P,T}$ and $C_{P,\varepsilon}$ are temperature coefficients and the strain coefficient for the Brillouin scattering power respectively;
$C_{V,T}$ and $C_{V,\varepsilon}$ are also the coefficients of temperature and strain for the shift of the Brillouin frequency depending on temperature and strain.
These preset values of the coefficients are given in table1.

Table 1. Relationship of Brillouin light scattering intensity and frequency shift and temperature coefficient.

	Strain coefficient	Temperature coefficient
Brillouin scattering optical power variation ΔP_B	$C_{P,\varepsilon}$ =-8 ± 1×10-4%/με	$C_{P,T}$ = +0.33 ± 0.3 % /K
Brillouin scattering optical frequency shift ΔV_B	$C_{V,\varepsilon}$ =- +0.052 ±0.004 MHz/με	$C_{V,T}$ = + 1.09 ± 0.083 MHz / K

On rearrangement, the formula (1) can be written as:

$$\begin{bmatrix}\Delta\varepsilon\\ \Delta T\end{bmatrix} = \frac{1}{\left|C^{\varepsilon}_{B_v}C^{T}_{BI} - C^{\varepsilon}_{BI}C^{T}_{B_v}\right|} \times \begin{bmatrix}C^{\varepsilon}_{BI} & -C^{T}_{B_v}\\ -C^{T}_{BI} & C^{\varepsilon}_{B_v}\end{bmatrix}\begin{bmatrix}\Delta v_B\\ \Delta I_{B_{A.S}}\end{bmatrix} \tag{2}$$

Linear coefficient for temperature and strain can be accurately determined from the inverse matrix in (1) and can be written as [5]:

$$\Delta T = \frac{|C_{P,\varepsilon}|\cdot\Delta v_B + |C_{v,\varepsilon}|\cdot\Delta P_B}{|C_{P,T}C_{v,\varepsilon} - C_{P,\varepsilon}C_{v,T}|} \tag{3}$$

$$\Delta\varepsilon = \frac{|C_{P,T}|\cdot\Delta v_B + |C_{v,T}|\cdot\Delta P_B}{|C_{P,T}C_{v,\varepsilon} - C_{P,\varepsilon}C_{v,T}|} \tag{4}$$

Measurement unit for strain in the fiber is typically determined as relative fiber stretching expressed as relation 1μℇ = 1 μm of stretching per 1m of fiber.
The dependence of frequency shift on temperature and strain is presented in Figure 2 and Figure 3.

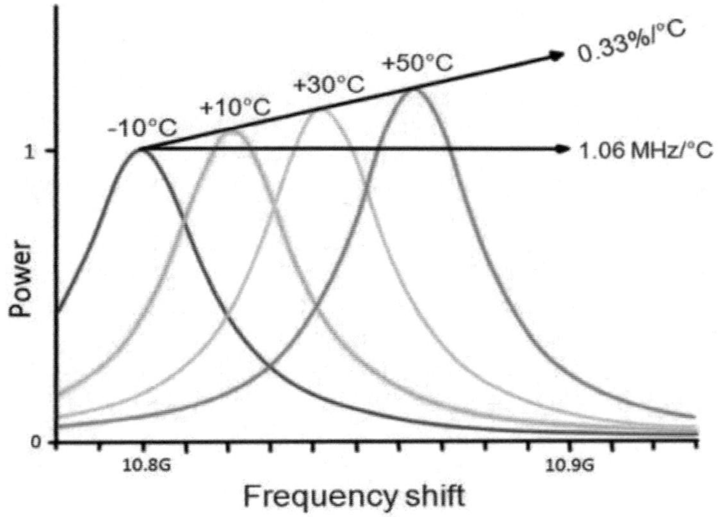

Figure 2. Dependence of frequency shift on temperature.

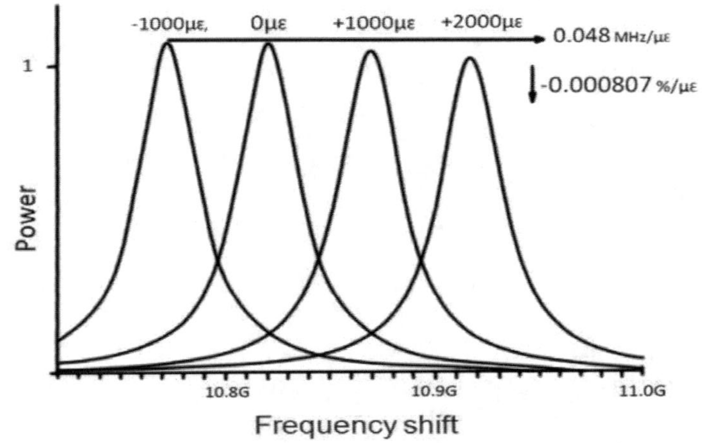

Figure 3. Dependence of frequency shift on strain.

By measuring the power of the scattered signal and the shift of the Brillouin frequency, it is possible to obtain the temperature and strain distribution along the fiber.

Distributed Brillouin fiber-optic sensors provide innovative solutions to control the temperature and strain in distributed constructions. The effective range for these sensors is about 20-30 km, which is limitation for their use in some applications where the distance is much longer. In order to increase the operating range, the methods are proposed in paper [6], based on distributed Raman amplification.

The paper [7] proposes a combined amplification of the second order and optical pulse coding to expand the real dynamic range for distributed fiber-optic sensor. The analysis presented and the experimental results show that the appropriate optimization of these two methods makes it possible to enhance the signal/noise ratio measurements when using a superlong and sensitive fiber. This solution increases the sensing distance to 120 km with a spatial resolution of 5 m.

In order to increase the spatial resolution, the paper [8] proposes and shows a new differential method of optical reflectometry. By analysing spatiotemporal property of the pulse excited by Brillouin spontaneous light scattering, it made it possible to obtain the distribution of the Brillouin weighting coefficient along the fiber. Based on this distribution a method of two-step subtraction is offered. A pair of pulses with a small difference in width is used as a probe pulse. When performing the two-stage subtraction on these two pairs of Brillouin spectrum it is proved theoretically and experimentally that the differential Brillouin spectrum is spatially associated with the difference in width of pulse pair. The spatial resolution of 0.4 m is obtained experimentally when using pulse pairs 60/56 ns for the probing length of 7.8 km with accuracy of the Brillouin frequency of 4.1 MHz.

The analysis results of the noise by using Monte-Carlo method to correct the dependences between the frequency resolution, quality, signal/noise ratio and the frequency step in the distributed Brillouin fiber-optic sensors are presented in [9]. Quantitative estimation of the Brillouin amplification spectrum is of great importance for the distributed sensors to increase the Brillouin frequency resolution and corresponding Brillouin strain and temperature resolutions. To estimate the error in determination of the Brillouin central frequency spectrum two analytical expressions were obtained with polynomial second order fitting and without it.

In paper [10], an equivalent Rayleigh criterion is offered to measure stressed cross section and smaller spatial resolution for the Brillouin distributed sensor. According to this criterion, at any preset probing length the minimum allowable strain length is 1/2 of the pulse length at Brillouin frequency uncertainty of 5%.

3. Results and discussions

The simulation was carried out in the professional design environment OptiSystem 17.1. It is a comprehensive program software Design Suite, which allows users to plan, test and simulate processes in optical fibers in current telecommunication systems. The computer simulation scheme is shown in Figure 4.

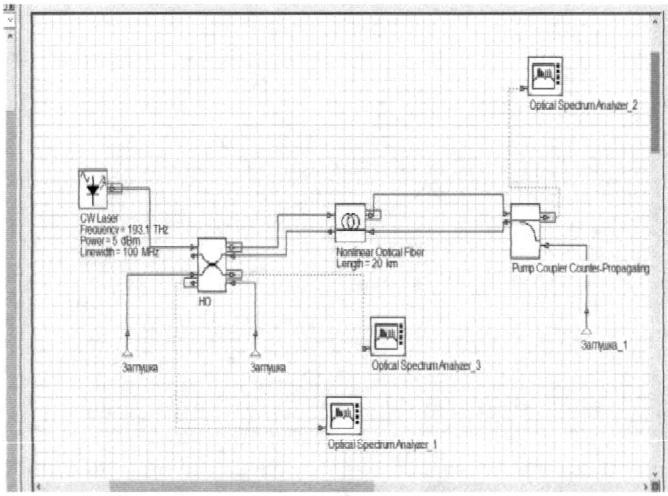

Figure 4. Simulation scheme.

The following component parameters were set.

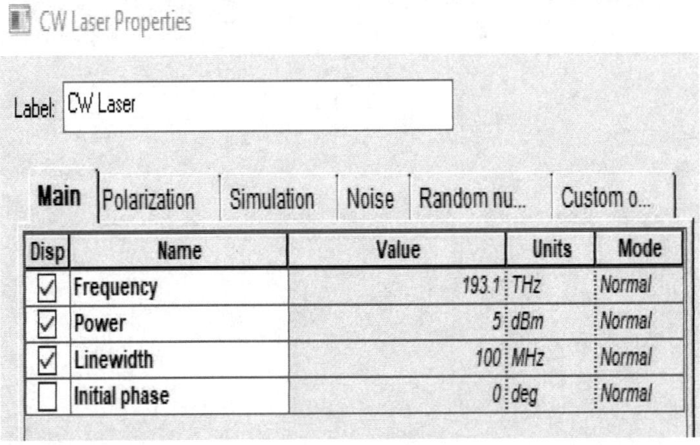

CW Laser Properties

Label: CW Laser

Main | Polarization | Simulation | Noise | Random nu... | Custom o...

Disp	Name	Value	Units	Mode
☑	Frequency	193.1	THz	Normal
☑	Power	5	dBm	Normal
☑	Linewidth	100	MHz	Normal
☐	Initial phase	0	deg	Normal

Figure 5. Laser parameters.

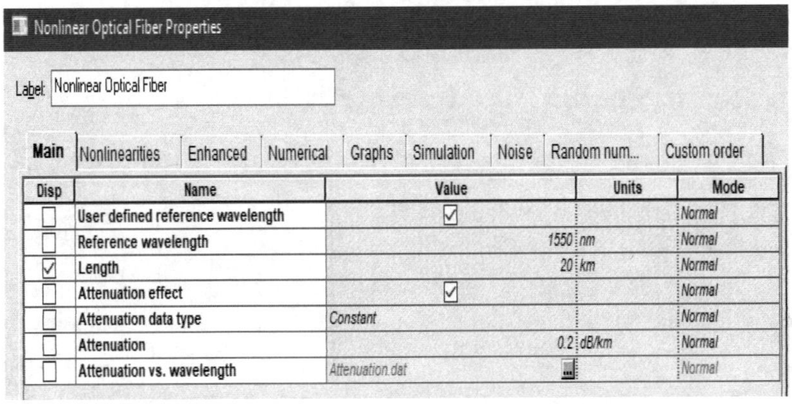

Nonlinear Optical Fiber Properties

Label: Nonlinear Optical Fiber

Main | Nonlinearities | Enhanced | Numerical | Graphs | Simulation | Noise | Random num... | Custom order

Disp	Name	Value	Units	Mode
☐	User defined reference wavelength	☑		Normal
☐	Reference wavelength	1550	nm	Normal
☑	Length	20	km	Normal
☐	Attenuation effect	☑		Normal
☐	Attenuation data type	Constant		Normal
☐	Attenuation	0.2	dB/km	Normal
☐	Attenuation vs. wavelength	Attenuation.dat		Normal

Figure 6. Optical fiber parameters.

The simulation results are shown in Figures 7-9.

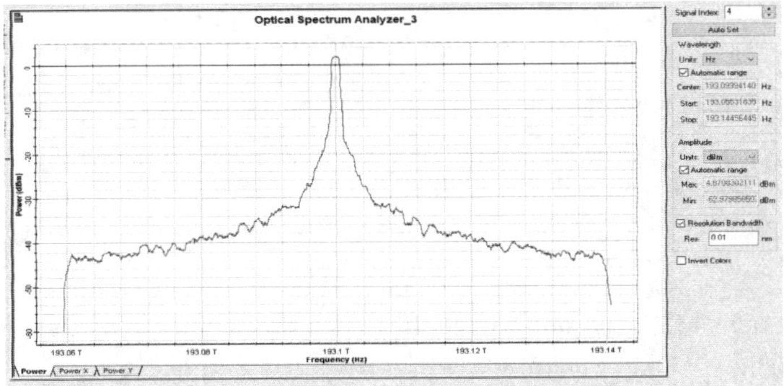

Figure 7. Rayleigh scattering spectrum.

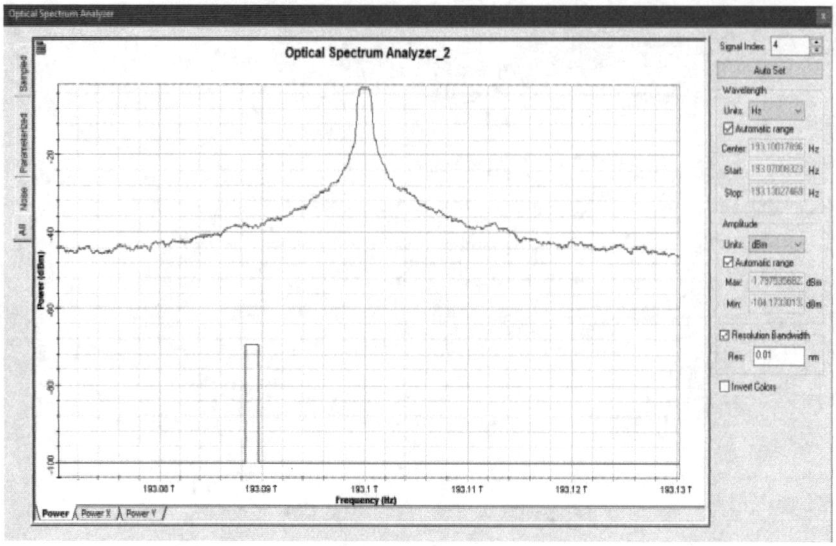

Figure 8. Brillouin and Rayleigh scattering spectra.

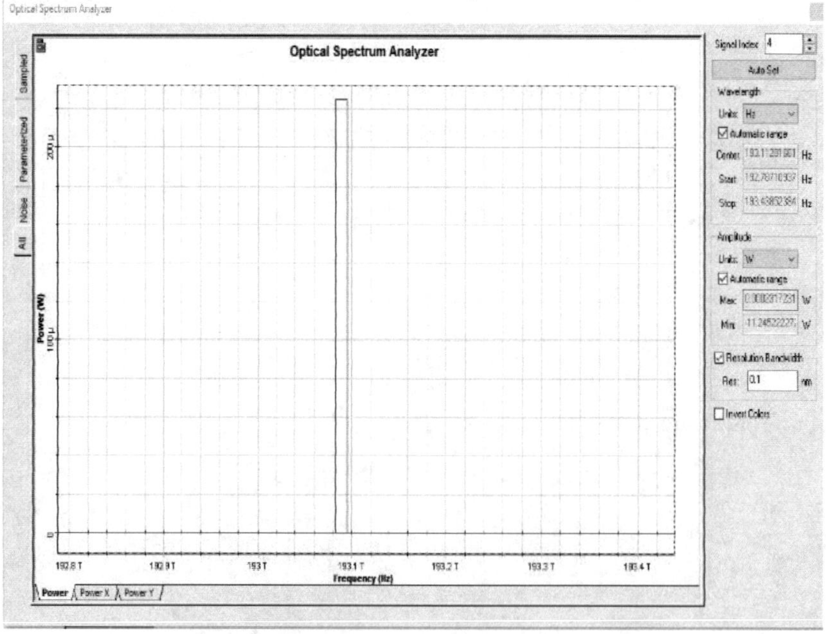

Figure 9. Brillouin scattering spectrum shift.

Experimental research was conducted by using Brillouin optical reflectometer manufactured by Japanese company Ando AQ 8603 for one mode optical fiber of G.652 standard, the sections of which were heated and stretched longitudinally. The results are shown in Figure 10.

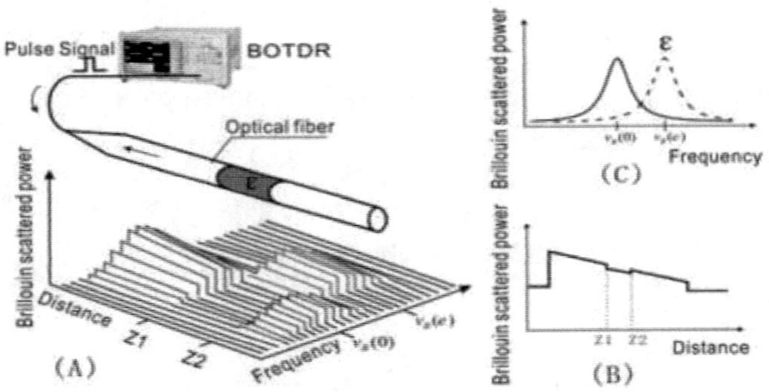

Figure 10. Experiment scheme.

The experiment results are presented in Figure 11 and Figure 12.

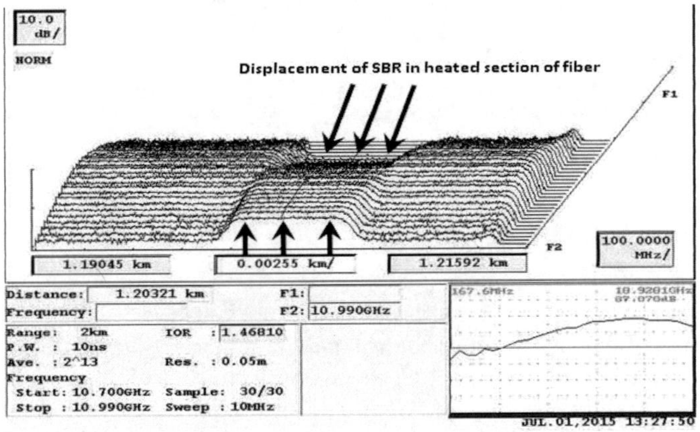

Figure 11. Spectrum shift when optical fiber heated to 100°C.

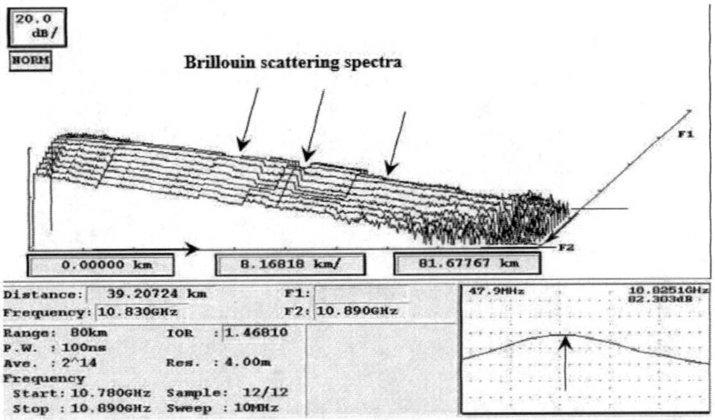

Figure 12. Spectra for OF in fiber-optic cable.

Algorithm processing of the spectrogram shows the strain propagation profile along longitudinal coordinate which is presented in Figure 13.

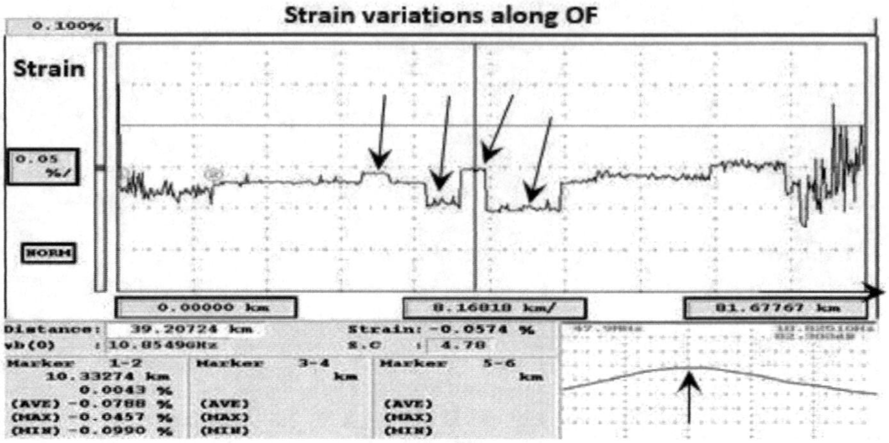

Figure 13. Strain profile in optical fiber.

4. Conclusions

Fiber-optic sensors based on the principle of Brillouin scattering can be widely used in telecommunication systems, mechanical engineering, power industry, construction, aviation and space industries. Important role they could play in control and automation systems in technological processes and plants. The objectives for further research are to perform metrological analysis at all stages of measuring information transformation and increasing the base of Brillouin spectrograms for optical fiber of various types. Of great importance is the improvement of algorithms for automated spectra processing in order to expand the functionality of the sensors under investigation.

5. References

[1] Bernini R, Minardo A and Zeni L 2007 Self-demodulated heterodyne frequency domain distributed Brillouin fiber sensor *IEEE Photonics Technol. Letters* **19** pp 447-449

[2] Li Y, Bao X, Dong Y and Chen L 2010 A novel distributed Brillouin sensor based on optical differential parametric amplification *J. of Lightwave Technol.* **28** pp 2621-26

[3] Urricelqui J, Fernandino F, Sagues M and Loayssa A 2015 Polarization diversity scheme for BOTDA sensors based on a double orthogonal pump interaction *J. of Lightwave Technol.* **33** pp 2633-37

[4] Belal M and Newson T P 2013 Experimental examination of the variation of the spontaneous Brillouin power and frequency coefficients under the combined influence of temperature and strain *J. of Lightwave Technol.* **30** pp 1250-55

[5] Luo J, Hao Y, Ye O and Li L 2013 Development of optical fiber sensors based on Brillouin scattering and FBG for on-line monitoring in overhead transmission lines *J. of Lightwave Technol.* **31** pp 1559-65

[6] Barrios F, López S, Sanz A, Corredera P and Castañón J 2010 Distributed Brillouin fiber sensor assisted by first-order Raman amplification *J. of Lightwave Technol.* **28** pp 2162-72

[7] Lee H, Hayashi N I, Mizuno Y and Nakamura K 2017 Slope-assisted Brillouin optical correlation-domain reflectometry using polymer optical fibers with high propagation loss *J. of Lightwave Technol.* **35** pp 2306-10
[8] Li Q, Gan J, Wu Y, Zhang Z, Li J and Yang Z 2016 High spatial resolution BOTDR based on differential Brillouin spectrum technique *IEEE Photonics Technol. Letters* **28** pp 1493-96
[9] Yu Y, Luo L, Li B, Soga K and Yan J 2016 Frequency resolution quantification of Brillouin-distributed optical fiber sensors *IEEE Photonics Technol. Letters* **28** pp 2367-69
[10] Ravet F, Bao X, Yu O and Chen L 2005 Criterion for subpulse-length resolution and minimum frequency shift in distributed Brillouin sensors *IEEE Photonics Technol. Letters* **17** pp 1504-06

Isolation of the singular part of the electromagnetic field scatteredby objects with sharp edges and conic points

A N Bogolyubov, I E Mogilevsky, V V Rovenko
Lomonosov Moscow State University, Faculty of Physics

E-mail: bogan7@yandex.ru, imogilevsky@mail.ru, rovenko.vladimir@physics.msu.ru

Abstract. In this article we propose a method for overcoming a number of problems that may arise when solving scattering problems on objects of complex structure. This approach can be used for mathematical and numerical modeling of waveguide, emitting and scattering systems of various complexity.

1. Introduction

In this paper we consider several problems of electromagnetic field diffraction on complex dielectric structures in the presence of sharp edges at their boundaries. Such problems are both of purely mathematical and of practical interest. For example, complex-section waveguides are widely used in the design of various microwave devices, such as antennas, filters, polarizers, multiplexers, etc. Analysis of problems related to complex-section waveguides and various features of scattering bodies makes it possible to create devices with parameters superior to their counterparts based on rectangular and round waveguides [1].

However, presence of sharp edges on the boundaries and surfaces of permittivity discontinuities leads to the appearance of features in the electromagnetic field either in the vicinity of a singularboundary point or within inhomogeneity of the filling [1-4]. The generalized solution can have singularities even when the right-hand side of the equation system is smooth [5]. This also complicates the numerical solution of scattering problems for the complex-section waveguides and objects with sharp edges and conic points [1].

One of the ways to overcome these problems is to isolate the singular electromagnetic field in the vicinity of the singular pointusing results on the asymptotics of the solution to elliptic boundary value problems presented in the works of V. A. Kondratiev [5], as well as S. A. Nazarov and B. A. Plamenevsky [8].

2. Method description

Below we consider isolation of the singular part of the electromagnetic field in the following range of problems:

1. Diffraction on a perfectly conducting body in the vicinity of a conical point [1].
2. For a waveguide in the vicinity of the sharp edges of the boundary, the discontinuity surfaces of the permittivity, in the vicinity of the sharp edge of the metal-dielectric wedge in the waveguide [1].
3. Electromagnetic field diffraction on bodies with dielectric edges in the vicinity of the edge.

Despite the fact that physical and mathematical formulations of these problems differ significantly,

Journal of Physics: Conference Series **2091** (2021) 012071 doi:10.1088/1742-6596/2091/1/012071

construction of the singular part of the electromagnetic field is based on the general method of isolating the singular part of the elliptic boundary value problem solution, first proposed by V. A. Kondratiev [5]:

1) For good solvability, the problem is considered in an infinite domain, and after constructing the asymptotics, it is reduced to a problem in a finite domain using a slicing function.

2) Transformation of the original statement of the problem to a more convenient form for the study of asymptotics. This can be done either by switching to a different coordinate system, or by using special functions.

3) The imposition of a number of conditions on the functions included in the equations, in particular, on their right-hand sides. In particular, it is important that they belong to the Kondratiev space with a special type of norm.

4) Performing a special logarithmic transform, and then the Fourier transform, to use its properties in constructing the asymptotics.

5) The transition from the solution of the problem in the space of Fourier images on a straight line to the solution in a band, which is necessary for constructing the asymptotics using the deduction theorem. This condition is achieved by imposing a constraint on the constructed solution on a straight line. Namely, it must belong to the intersection of Kondratiev spaces with a certain relation between these spaces.

6) Construction of the asymptotics and transition using the inverse Fourier transform to construct the asymptotics of the original problem in an infinite domain.

7) Using the cutting function to move from a problem in a finite domain or a diffraction problem on a finite body to the considered problem in an infinite domain.

We present main results for the considered problems.

3. Body featuring a conical point

Consider the three-dimensional problem of diffraction of an electromagnetic wave on a bounded ideally conducting body featuring a conical point (figure 1). When considering this problem, we will give the method in the most detail. Subsequent tasks will be solved similarly.

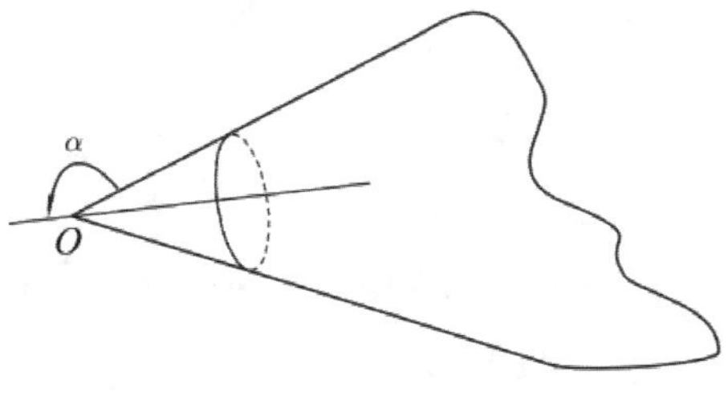

Figure 1. Conical surface.

$$rot\bar{H} = -ik\bar{E} + \frac{4\pi}{c}\bar{j}, \operatorname{supp} j \subset D_0 \subset \mathbb{R}$$

$$rot\bar{E} = ik\bar{H}, \tag{1}$$

$$div\bar{H} = 0,$$

$$div\bar{E} = 0.$$

The boundary conditions have the form:

$$\left(\bar{H},\bar{n}\right)|_{\partial\Omega} = 0, \left[\bar{E},\bar{n}\right]|_{\partial\Omega} = 0, \tag{2}$$

where $\bar{n}$ is the normal vector to the conic surface. Meixner condition at the top of the cone:

$$\bar{E}, \bar{H} \in \left(L_2^{loc}\right)^3 \tag{3}$$

For a detailed study of the field in the vicinity of a conical point, this problem is first considered in an infinite cone. In the future, the use of the slicing function will allow the original diffraction problem to be reduced to the problem under consideration. A spherical coordinate system with a center at a conical point is introduced. This approach makes it possible to significantly simplify the boundary conditions, which makes it possible to study the properties of the desired solution in more detail. The main equation of the problem (1) is transformed using the representation:

$$\bar{E} = rotrot\left(\bar{e}_r U_r\right) + ikrot\left(\bar{e}_r V_r\right), \tag{4}$$

$$\bar{H} = -ikrot\left(\bar{e}_r U_r\right) + rotrot\left(\bar{e}_r V_r\right), \tag{5}$$

where U and V are the electric and magnetic Debye potentials, for which we obtain the following problems [11], leaving the Laplace operator in the left part, and denoting the right part as functions $f_{1,2}(M)$:

$$\Delta U = f_1(M), \tag{6}$$

$$U|_{\theta=\alpha} = 0$$

$$\Delta V = f_2(M), \tag{7}$$

$$\frac{\partial V}{\partial \theta}\Big|_{\theta=\alpha} = 0$$

First, we find the asymptotics for problem (6) with Dirichlet conditions.

Following the results of V.A. Kondratiev [5], we introduce the space of functions V_γ^l with the norm:

$$\|u\|_{V_\gamma^l}^2 = \sum_{s=0}^{l} \int_K r^{2\gamma-2(l-s)\left|\frac{\partial^2 u}{\partial x^s}\right|^2} d\Gamma \tag{8}$$

where K is the outer part of an infinite cone, $l \geq 0$ is an integer, γ is a real number.

Next, it is assumed that the right part $f_1(M) \in V_\gamma^l$.

Let's replace the variable $\tau = \ln\frac{1}{r} \Rightarrow \tau \in (-\infty; +\infty)$ and do the Fourier transform on the variable τ. Therefore, task (6) will take the form:

$$-\lambda^2 \hat{U} + i\lambda \hat{U} + \Delta_{\theta,\varphi} \hat{U} = \hat{F}(\lambda, \theta, \varphi), \tag{9}$$

$$\hat{U}|_{\theta=\alpha} = 0$$

$$\Delta_{\theta,\varphi} \hat{U} = \frac{1}{\sin\theta} \frac{\partial}{\partial\theta}\left(\sin\theta \frac{\partial \hat{U}}{\partial\theta}\right) + \frac{1}{\sin^2\theta} \frac{\partial^2 \hat{U}}{\partial\varphi^2}$$

Here $\hat{U}(\lambda, \theta, \varphi)$ and $\hat{F}(\lambda, \theta, \varphi)$ are the images of the function U and the right side of the equation, respectively. From the properties of the Fourier transform for the functions $\hat{F}(\lambda, \theta, \varphi)$ an estimate follows:

$$\sum_{k=0}^{l}\int_{-\infty+ih}^{+\infty+ih}|\lambda|^{2k}\left\|\hat{F}(\lambda,\theta,\varphi)\right\|_{H^{l-k}(0,\alpha)\times(0,2\pi)}^{2}d\lambda\le C\|f_1\|_{V_\gamma^l(K)}^2, h=l+\frac{1}{2}-\gamma \quad (10)$$

The solution of problem (5) is sought in the form of a series of spherical functions $P_{nm}^{(m)}(\cos\theta)e^{im\varphi}$. Here $m=0,\pm1,\pm2...$, and n_m is determined from the relations $P_{nm}^{(m)}(\cos\alpha)0$ [10]. Let us consider in more detail the question of the existence of a solution to this equation after finding the asymptotics. Taking into account the above, the solution of problem (9) will take the form:

$$U(\lambda,\theta,\varphi)=\sum_{n,m}\frac{\hat{F}_{nm}(\lambda)}{(i-\lambda)\lambda+n_m(n_m+1)}P_{nm}^{(m)}(\cos\theta)e^{im\varphi}. \quad (11)$$

Here $\hat{F}_{nm}$ are the coefficients of expansion over spherical functions of the corresponding functions.

If the function $U(\lambda,\theta,\varphi)$ has no poles on the line $Im\lambda=h=l+\frac{1}{2}-\gamma$, we get the following estimate:

$$\|U(\lambda,\theta,\varphi)\|_{V_\gamma^{l+2}(K)}\le C\|f(r,\theta,\varphi)\|_{V_\gamma^l(K)} \quad (12)$$

So far, the solution (11) constructed in the space of Fourier images is defined only on the line $h=-\gamma+1+1/2$. To construct its asymptotics, it is necessary that the function (11) be defined in some band $h_1<h<h_2$.

Let $h_1<h<h_2 f(r,\theta,\varphi)\in V_{\gamma_1}^{l_1}(K)\cap V_{\gamma_2}^{l_2}(K),\gamma_1>\gamma_2$, l_1 and l_2 be such that $h_1=l_1+\frac{1}{2}-\gamma_1$ and $h_2=l_2+\frac{1}{2}-\gamma_2$ satisfy the condition $h_2>h_1$, the function $\hat{U}(\lambda,\theta,\varphi)$ has no poles on the lines $Im\lambda=h_1$ and $Im\lambda=h_2$.

Using the deduction theorem, we move from integration along the straight line $Im\lambda=h_1$ to integration along the straight line $Im\lambda=h_2$, and the poles of solution (11) located between them allow us to obtain an asymptotic representation of the solution using slicing function.

Finally, the asymptotic expansion of the solution of the three-dimensional problem of diffraction of an electromagnetic wave on a bounded body containing a conic point is constructed using the Kondratiev method (an expression for the Debye electric potential):

$$U(r,\theta,\varphi)=\chi\left\{\begin{array}{l}\sum_{-\frac{1}{2}<n_m<\frac{3}{2}}C_{n.m}r^{n_m}P_{nm}^{(m)}(\cos\theta)e^{im\varphi}+\\ \sum_{-\frac{1}{2}<n_m+1<\frac{3}{2}}D_{n.m}r^{n_m+1}P_{nm}^{(m)}(\cos\theta)e^{im\varphi}\end{array}\right\}+\Re(r,\theta,\varphi) \quad (13)$$

where n_m – solutions of equation $P_{nm}^{(m)}(\cos\alpha)=0$, $C_{n,m}$ and $D_{n,m}$ are constants, $\Re(r,\theta,\varphi)$ – is the smooth part of the function $U(r,\theta,\varphi)$, $\chi(r)$ – slicing function:

$$\chi(r)=\begin{cases}1, r\le d/2\\ 0, r>d\end{cases}, \chi(r)\in C^\infty.$$

For the problem of the magnetic Debye potential, a completely similar representation will take place. Only n_m will be solutions of the equations $\frac{\partial P_{nm}^{(m)}(\cos\theta)}{\partial\theta}|_{\theta=\alpha}=0$.

The solvability of problems for finding n_m is shown based on the approach described in [10] for $P_{nm}^{(m)}(\cos\alpha)=0$. The main idea is the decomposition of $P_{nm}^{(m)}(\cos\theta)$ and $\frac{\partial P_{nm}^{(m)}(\cos\theta)}{\partial\theta}$ in the ranks with respect to $(2\sin\theta)^{-\frac{k}{2}}, k=1,3,5\ldots$ Further, using the method of mathematical induction, we can prove the solvability of $P_{nm}^{(m)}(\cos\alpha)=0$ and $\frac{\partial P_{nm}^{(m)}(\cos\theta)}{\partial\theta}\Big|_{\theta=\alpha}=0$ and find the corresponding n_m.

4. Waveguide with sharp edges

Consider the allocation of the singular part of the electromagnetic field of the waveguide in the vicinity of the edge of the dielectric wedge. It is assumed that permittivity is a piecewise-continuous scalar real function.

To study the problem, we use the mathematical formulation from [7] for eigenvectors, supplemented by the conjugation conditions on the surfaces of the dielectric permittivity discontinuity:

$$-\text{grad div } H_\perp - k^2\varepsilon\text{rot } E_z = -\gamma^2 H_\perp,$$

$$-ik \text{ rot}\varepsilon H_\perp - \text{div}\varepsilon\text{grad } E_z = -\gamma^2\varepsilon E_z,$$

where $k=\frac{\omega}{c}$ is a wavenumber, γ is a spectral parameter. It is assumed that all functions have a dependence on z of the form $e^{i\gamma z}$. The boundary conditions and the conjugation conditions have the form:

$$(\mathbf{H}\cdot\mathbf{n})\big|_{\partial\Omega}=0,\ E_z\big|_{\partial\Omega}=0,$$

$$\left[(\mathbf{H}\cdot\mathbf{n})\right]\big|_C=0,\ [E_z]\big|_C=0,\ (\mathbf{H}\times\mathbf{n})\big|_C=0,$$

$$[\text{div}H_\perp]\big|_C=0,\ \left[\varepsilon\left(\text{grad}E_z+ik(\mathbf{H}\times\mathbf{i}_z)\right)\cdot\mathbf{n}\right]\big|_C=0.$$

Here C is the permittivity discontinuity line, $\mathbf{n}$ is the normal to the boundary of the region or to the discontinuity line.

The following representation of the electric field in the vicinity of the angular point of the discontinuity line is obtained using the method of distinguishing the features of the electromagnetic field in the vicinity of the edge:

$$E_z(r,\varphi)=\chi\sum_{-\delta<\nu_k^{(1)}<1} r^{\nu_k^{(1)}}\left\{C_k^{(1)}\cos\left[(\pi-\varphi)\nu_k^{(1)}\right]+D_k^{(1)}\cos\left[(\pi-|\omega_0-\varphi|)\nu_k^{(1)}\right]\right\}+$$

$$+\chi\sum_{-\delta<\nu_k^{(2)}<1} r^{\nu_k^{(2)}}\left\{C_k^{(2)}\cos\left[(\pi-\varphi)\nu_k^{(2)}\right]+D_k^{(2)}\cos\left[(\pi-|\omega_0-\varphi|)\nu_k^{(2)}\right]\right\}+\Re(r,\varphi),$$

Where $C_k^{(i)}, D_k^{(i)}$ are numbers, $\nu_k^{(1)}$ and $\nu_k^{(2)}$ — solutions of equations:

$$\sin\pi\nu_k^{(1)}-\alpha\sin\left(\pi\nu_k^{(1)}-\nu_k^{(1)}\omega_0\right)=0,$$

$$\sin\pi\nu_k^{(2)}+\alpha\sin\left(\pi\nu_k^{(2)}-\nu_k^{(2)}\omega_0\right)=0,$$

(except $\nu_k=0$), $\chi(r)=\begin{cases}1, & r\le d/2,\\ 0, & r>d,\end{cases}$ $\chi(r)\in C^\infty$ is a slicing function, $\Re(r,\varphi)$ is the smooth part of the solution for which an estimate in the corresponding norm is obtained.

5. Body with dielectric edges

Study of the singular part of the electromagnetic field in the vicinity of the sharp edge of a metal-dielectric wedge in a waveguide, provided that the permeability inside the waveguide $\mu \equiv 1$, and that the dielectric constant ε is real, depends only on the transverse coordinates and has limited first derivatives, leads to a mathematical formulation of the problem for the eigenvectors of the field components $\vec{A} = \{H_\perp, E_z\}$ and eigen values β^2 (β — the propagation constant) is given in [9]:

$$-\text{grad div } H_\perp - k^2 \varepsilon \text{rot } E_z = -\gamma^2 H_\perp,$$

$$-ik \text{ rot}\varepsilon H_\perp - \text{div}\varepsilon\text{grad } E_z = -\gamma^2 \varepsilon E_z,$$

where $k = \dfrac{\omega}{c}$ — wavenumber, γ is a spectral parameter. It is assumed that all functions have a dependence on z of the form $e^{i\gamma z}$. The boundary conditions and the conjugation conditions have the form:

$$(\mathbf{H}\cdot\mathbf{n})\big|_{\partial\Omega} = 0,\ E_z\big|_{\partial\Omega} = 0,$$

$$\left[(\mathbf{H}\cdot\mathbf{n})\right]\big|_C = 0,\ \left[E_z\right]\big|_C = 0,\ (\mathbf{H}\times\mathbf{n})\big|_C = 0,$$

$$\left[\text{div}H_\perp\right]\big|_C = 0,\ \left[\varepsilon\left(\text{grad}E_z + ik\left(\mathbf{H}\times\mathbf{i}_z\right)\right)\cdot\mathbf{n}\right]\big|_C = 0.$$

Here C is the permittivity discontinuity line, $\mathbf{n}$ is the normal to the boundary of the region or the discontinuity line.

Using the Kondratiev method, we finally obtain the following representation of the longitudinal component of the electric field:

$$E_z(r,\varphi \quad \chi \sum \quad [\quad] \quad [\quad]$$

where ν_k — solutions of equation: $\alpha \quad [\quad]$ (except $\nu =$), $\Re$ is the smooth part of the solution for which an estimate in the corresponding norm is obtained, $\chi =$ $\in$ is a slicing function.

6. Conclusion

Based on the obtained asymptotic representation of the electromagnetic field in the vicinity of the scattering body singularities, it is possible to significantly increase the convergence rate of the finite element method by introducing singular functions into the space of trial functions that have singularity of a given form and that accurately approximate the singular part of the solution [9].

7. References

[1] Bogolyubov A N, Mogilevsky I E, Rovenko V V 2019 *Phys. bases of instr.* **8** 1 pp 6-12
[2] Bogolyubov A N, Mogilevsky I E, Sveshnikov A G 2015 *Comp. math. and math. phys.* **55** 3 p 446
[3] Bogolyubov A N, Mogilevsky I E, Rovenko V V 2021 An Asymptotic Expansion of the Solution to the Problem of the Electromagnetic Theory of Diffraction on Objects with Conical

Points *Bulletin of the Russian Academy of Sciences: Physics* **85** 1 pp 45-49 DOI: 10.3103/S106287382101007X

[4] Bogolyubov A N, Delitsyn A L, Mogilevsky I E, Sveshnikov A G 2003 Singularities of normal modes in a waveguide with an inhomogeneous filling and reentering edges *Journal of communications technology & electronics* **48** 7 pp 715 – 722.

[5] Kondratiev V A 1967 Boundary value problems for elliptic problems in domains with conical or corner points *Trudy Moskov. Matem. Obshch.* **16** pp 209 – 292

[6] Bogolyubov A N, Mogilevsky I E 2016 *Phys. bases of instr.* **5** 2 pp 72-79

[7] Delitsyn A L 1999 Application of the finite element method to the calculation of modes of dielectric waveguides *Computational mathematics and mathematical physics* **39** 2 pp 298 – 304

[8] Nazarov S, Plamenevsky B A 2011 Elliptic problems in domains with piecewise smooth boundaries *de Gruyter* DOI: 10.1515/9783110848915

[9] Bogolyubov A N, Erokhin A I, Mogilevskii I E, Svetkin M I 2017 A hybrid method for numerical solution of Poisson's equation in a domain with a dielectric corner *Computational Mathematics and Mathematical Physics* **57** 8 pp 1310 – 1319 DOI: 10.1134/S096554251708005X

[10] MacDonald H M 1899 *Proc. Lond. Math. Soc.* pp 264-281

[11] Budak B M, Samarskiy A A, Tikhonov A N 2004 Collection of problems in mathematical physics (*M.: "FIZMATLIT"*) 688

Diffraction propagation of vortex diffractals

P Korolenko[1,2], R Kubanov[1], N Pavlov[1], A Zotov[1]
[1] Lomonosov Moscow State University, Faculty of Physics, Department of Optics, Spectroscopy and Nanosystems Physics, Moscow 119991, Russia
[2] P.N. Lebedev Physics Institute, Russian Academy of Sciences Moscow 119991, Russia

E-mail: pvkorolenko@rambler.ru

Abstract. A brief retrospective analysis of studies of fractal light radiation is carried out. To assess the prospects of this scientific direction, new original results of studying the diffraction propagation of vortex wave beams with a fractal structure (vortex diffractals) are also presented. For this purpose, computational algorithms and related software have been developed. In calculating the amplitude-phase and scaling characteristics of diffractals two-dimensional Weierstrass functions and multistructures of Gaussian beams were used. The results indicate a high information capacity of vortex diffractals and explain their resistance to the influence of turbulence in the propagation medium.

1. Introduction

The results of studying the characteristics of light beams with an initially complex amplitude-phase profile significantly improved the characteristics of laser information systems. Thus, beams with a vortex wavefront structure have a number of unique properties that provide an increased degree of stability of characteristics when propagating in a turbulent atmosphere [1, 2]. Waves with a fractal wavefront structure (diffractals) have also attracted the attention of researchers [3]. Their valuable quality is the manifestation of structural self-similarity during the propagation process. However, in the literature, the case when radiation has both vortex and fractal properties is insufficiently considered [4]. The aim of this work is to develop and use specially developed algorithms and computational programs to determine the specific characteristics of vortex fractal beams. Particular attention is paid to assessing the practical significance of the properties of such beams.

2. Retrospective analysis of fractal beam studies

There are several ways to create fractal beams. The most promising and widespread of them were covered in the review [3]. Fractal wave beams can be obtained by passing a plane wave through various kinds of masks, plates and screens with a fractal configuration. The fields of these beams have a number of remarkable properties. Thus, annular axisymmetric masks, obtained by rotating the one-dimensional Cantor system, form self-similar distributions in both the transverse and longitudinal directions. Moreover, a whole system of radiation focusing regions is formed along the optical axis, the distance between which obeys the scaling dependence [6]. Fractal wave structures can also be obtained as a result of the passage of a plane wave through apertures, the boundaries of which are described by fractal curves. In particular, the diffraction pattern at the aperture described by the Koch curve has a self-similar form, and the fractal dimensions of the diffracting and initial fields are interrelated [7]. This makes it possible to determine the fractality of an object based on the properties of a particular diffraction pattern. When analyzing the transmission of radiation through the plates, the transmission of which is determined by the two-dimensional Weierstrass function, a high correlation was established between the fractal properties of the Fourier transforms of the plate structures and images of light beams. This result is also important for improving methods for diagnosing objects with fractal features, as well as for interpreting the aesthetic properties of fractal images [5]. The disadvantage of the method for producing fractal beams using specially made plate masks is their low radiation strength. In this regard, the possibility of generating waves with a fractal structure directly in lasers with unstable cavities has acquired a topical character [8].

In a number of works on the use of fractal radiation in communication systems, fractal vortex beams with a helicoid structure of the wave front were studied. These tube-shaped beams producing a circular energy flow are characterized by the quantum orbital momentum and its related order (topological charge) of the screw axis dislocation [3, 9].

Waves with a vortex structure have become the subject of much investigation due to the variety of their possible practical applications, including optical communication lines. Their valuable property is rather high stability against the effect of atmospheric inhomogeneities [1,10]. The degree of stability increases with increasing angular momentum and dislocation order (topological charge). In this relation, of particular interest is the work [4] where it is proposed to form a fractal vortex field using superposition of Gaussian beams. Varying their phase relations, one can widely vary the effective orbital momentum of the total beam. Fractality of the latter manifests itself in transverse distributions of both the intensity and the phase.

There is another method for obtaining a fractal vortex beam, which uses a special type of fractal phase plate based on properties of the Cantor set [11]. This plate, unlike an ordinary spiral zone plate producing one vortex, gives rise to a whole system of vortex axis formations after the passage of a plane wave.

To characterize fractal properties of radiation more completely, it should be mentioned that evolution of illumination along the propagation axis reproduces the fractality of the lens, and diameters of partial vortices are proportional to the topological charge. The authors believe that their developed system can be used for manipulation and three-dimensional monitoring of microobjects and for improvement of lithographic processes and quality of multifocal contact lenses.

The results obtained in [11] confirm the data of the extensive theoretical and experimental investigations with various types of aperiodic fractal plates [12]. It is stressed that plates with an aperiodic structure admit of fractional topological charges which break symmetry of screw phase dislocations and form chains of anisotropic vortices.

3. Results of application of the two-dimensional Weierstrass function

Two approaches were used to analyze the characteristics of vortex fractal beams. The first approach was based on the properties of the two-dimensional Weierstrass function. In that case, the following expression was used:

$$w_{k,m} = \sqrt{2}\sigma \frac{(1-b^{2D-4})^{1/2} \sum_{v=0}^{V} \sum_{n=0}^{N} \left[\left(b^{(D-2)n} \sin(2\pi s b^{n} (k \sin(\alpha v) + m \cos(\alpha v))) \right) e^{i\alpha v} \right]}{(1-b^{(2D-4)(N+1)})^{1/2}}. \quad (1)$$

Formula (1) is the result of the rotation of the coordinate system in which the one-dimensional Weierstrass function was previously specified. Here α is the angle of single rotation, V is the number of repetitions, σ is the standard deviation, $N+1$ is the number of harmonics, D is the fractal dimensionality, b and s are the scaling parameters, k and m are the transverse indices that correspond to mutually perpendicular coordinates. If $v = 0$, the wavefront does not have a helical structure and the beam is not vortex. Let us compare this case with the case $v = 1$, where there is a smooth azimuthal phase distribution, which is characteristic of an optical vortex. The initial amplitude-phase structure of the beam if parameter v in exponent indicator is equal to zero is shown on figure 1*a*, *b*. In the numerical simulations the parameters were set as follows: $D = 1.25$, $\sigma = 3.3$, $V=23$, $N=5$, $b = 2$ and

s = 0.03. The range was $k * m$ = 128 * 128. Amplitude W and phase Φ were determined from the relations: $W_{k,m}=|w_{k,m}|$, $\Phi_{k,m} = \arg(w_{k,m})$.

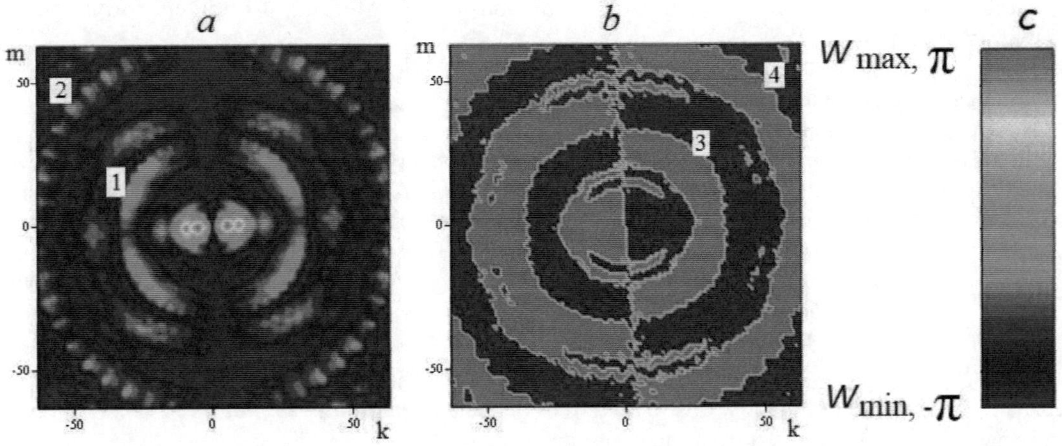

Figure 1. Distribution of the amplitude $W_{k,m}$ (a) and phase $\Phi_{k,m}$ (b) in the cross section of the fractal beam – v = 0. (c) is palette for phase and intensity range.

The amplitude and phase distributions obey the same scaling law. The scaling factor η is equal to the parameter b = 2. It is clearly seen that the amplitude maxima form segments of circles, the radii of which differ by a factor of 2. In particular, the radius of the circle of segment 1 is 2 times less than the radius of circle of segment 2. The phase distribution has a different form (figure 1b). The phase values are distributed according to a binary law, since the Weierstrass function is an alternating function. Circles with the same phase values obey the same scaling law. In the figure, the radius of circle 3 is 2 times less than radius of circle 4. Circular zones with identical phases are separated from each other by edge dislocations, where the phase abruptly changes by π.

Figure 2a,b corresponds to v = 1. In this case, the amplitude distribution is axisymmetric (Figure 2a), and the phase distribution is characterized by smooth azimuthal changes by 2π (Figure 2b). In the center there is a helical dislocation of the wave front. This distribution of amplitude and phase gives the beam a vortex tubular appearance.

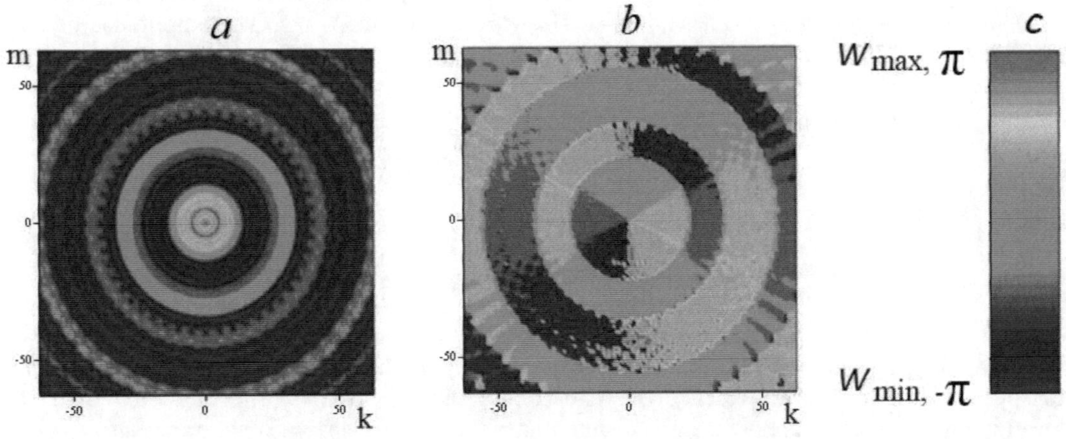

Figure 2. Distribution of the amplitude $W_{k,m}$ (a) and phase $\Phi_{k,m}$ (b) in the cross section of the fractal beam – v = 1. (c) is palette for phase and intensity range.

Calculations show that the tubular shape of the beam is retained in the far propagation zone. This form of the light field provides a high resistance of radiation to the influence of turbulent formations on atmospheric optical paths [2].

4. Properties of beams with a screw system dislocation

Another approach was used to obtain additional information on the properties of vortex fractal beams. The calculations were carried out based on the approach based on the addition of the light fields of a fractal system of Gaussian beams with a stepwise changing configuration [2,4]. A step-by-step calculation of the change in the shape of the distribution of the amplitude and phase of the light field was carried out by the iteration method using the following expression for the initial Gaussian beam:

$$g^{<1>}(x,y)=\exp\left(-\frac{(x-c_x)^2+(y-c_y)^2}{w^2}\right), \tag{2}$$

where x, y - transverse coordinates, w is the width of the beam centered at the point (c_x, c_y). Amplitude W and phase Φ were determined from the relations:
Each of these beams decayed into N daughter beams, thus

$$g^{<n+1>}(x,y)=\sum_{k=0}^{N-1}g^{<n>}\left(x+\frac{R}{\tau^n}\cos\left(\frac{2\pi k}{N}\right),y+\frac{R}{\tau^n}\sin\left(\frac{2\pi k}{N}\right)\right). \tag{3}$$

Here $\frac{R}{\tau^n}$ - characterizes the distance of the displacement of the daughter beams from the parent at the next iteration (τ is the scaling coefficient), cos ($2\pi k/N$) and sin ($2\pi k/N$) are the components of the displacement vector, the exponent gives the azimuthal phase incursion in the case fractal helical beams. In the calculations, the values $\tau = 2$ and $N = 3,4,5,6$ were used. The R value was considered a constant that could vary.
The propagation of the beams was simulated by the method of decomposing the field into plane waves. In the paraxial approximation, the phase incursion of the wave with indices p, q at a distance z was

$$\varphi_{p,q}=\frac{2\pi z}{T}\left(f(p)^2+f(q)^2\right). \tag{4}$$

Here $T = 2a^2/\lambda$ is the Talbot distance, a is the full size of the grid on which the field is specified, λ is the wavelength. $f(p) = \mathrm{mod}\,(p + K/2, K) - K/2$ is an auxiliary function that allows one to take into account the positive and negative harmonics defined from 0 up to $K - 1$. The amplitudes of plane waves at distance z, thus, had the form $S_{p,q}\exp(i\varphi_{p,q})$. To obtain the final field distribution, an inverse FFT was used.

This algorithm made it possible to simulate fractal structures with several screw dislocations of the wavefront. As examples, two such structures with a scaling coefficient $\eta = 3$ are shown in Figures 3*a*, *b* and 4*a*,*b*.

Figures 3*a* and 4*a* show the images of the beams in the initial plane, and figures 3*b* and 4*b* - in the far propagation zone. Cross dots mark the position of dislocations with a topological charge of +1, and triangular ones - with a topological charge of -1. In the distant zones, the images are morphologically similar to the initial ones. They are characterized by one scaling factor $\eta = 3$. For clarity, in Figure 3*b*, numbers 1 - 4 and 5 - 8 mark the location of the amplitude maxima, which corresponds to the geometry of similar elements. The sizes of the marked fragments differ in accordance with the scaling factor by a factor of 3. The presence of dislocations leads to local "vortices" of the field. The dislocation present in the center of the image determines the general vortex type of the beam.

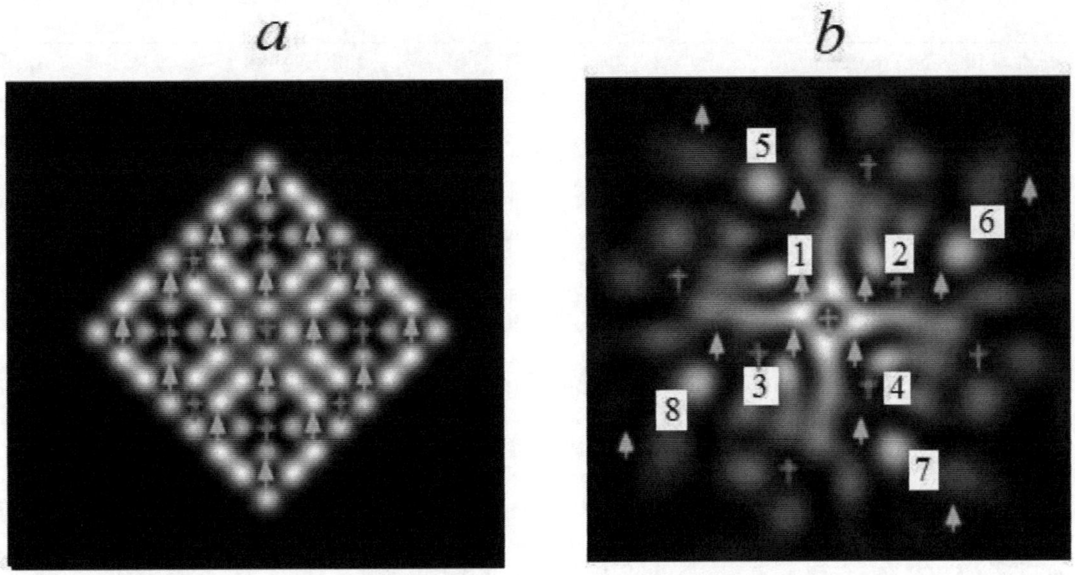

Figure 3. Example 1. Beam structure in the initial plane (*a*) and in the far zone (*b*).

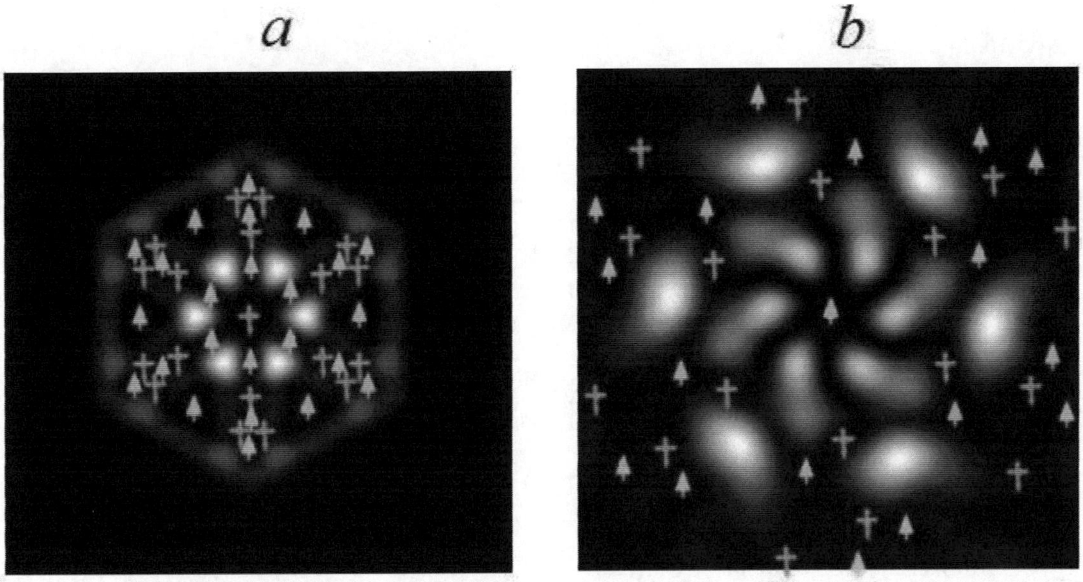

Figure 4. Example 2. Beam structure in the initial plane (*a*) and in the far zone (*b*).

Despite the structural difference, the beam in Figure 4*a* is transformed in a similar way. However, in the process of propagation, the central dislocation, in contrast to the previous case, changes its sign (figure 4*b*). This feature is explained by the diffraction transformation of the amplitude-phase profile of the fractal beam during its propagation. Thus, the method of forming fractal waves in the form of a superposition of Gaussian beams can significantly expand the understanding of the properties of fractal vortex beams.

5. Conclusions

The performed analysis of studies of the propagation of radiation with a fractal and vortex structure shows that the problem of determining the properties of light beams, which simultaneously have a fractal and vortex form, is urgent. Such beams have a high information resistance to the influence of inhomogeneities of the propagation medium. The used method of transforming the two-dimensional Weierstrass fractal structure made it possible to determine the effect of the vortex shape of the radiation on its amplitude and phase characteristics. Characteristics indicate that the vortex shape of the beam provides a practically valuable tubular type of radiation while maintaining the scaling parameters. The use of an alternative method for calculating the structure of fractal vortex beams with a system of screw dislocations in the cross section showed that in this case the morphological features of the amplitude-phase distributions are preserved. The developed and applied design scheme, which has proven its effectiveness, can be used to calculate the characteristics of other types of waves. The significance of the results obtained is largely determined by the fact that they were obtained for the practically important case of beam propagation in a turbulent atmosphere. The discovered structural stability of fractal vortex beams can play a positive role when used in atmospheric communication lines. The number of dislocations at the wave front at different distances from the initial plane remains constant. In this case, the balance between the number of positive and negative dislocations doesn't change. Their location determines the general morphology and symmetry in the distribution of the field amplitude.

The obtained results will be useful in the development of optical information systems based on fractal vortex beams.

Acknowledgments

This work was financially supported by the Russian Science Foundation, project no. 19-12-00310.

References

[1] Averchenko A V, Zotov A M, Korolenko P V, Pavlov N N 2020 Helical beams in information systems with open channels of radiation spread *Bulletin of the Russian Academy of Sciences: Physics* **84** 1 15-18 DOI: 10.3103/S1062873820010050

[2] Aksenov V, Dudorov V, Kolosov V 2019 *Optika atmosfery i okeana* **32** 792

[3] Korolenko P V 2020 Wave Beams with a Fractal Structure, Their Properties and Applications: A Literature Review *Physics of Wave Phenomena* **28** 4 pp 313-325. DOI: 10.3103/S1541308X2004007X

[4] Cho Y K, Kim K 2010 Increasing the orbital angular momentum of a fractal beam *In 10th IEEE International Conference on Nanotechnology* pp 312–315. DOI: 10.1109/NANO.2010.5697898

[5] Zotov A M, Korolenko P V, Mishin A Y, Ryzhikova Y V 2019 Physical Basics of Neuroaesthetics *Moscow University Physics Bulletin* **74** 6 pp 625–630 DOI: 10.3103/S0027134919060262

[6] Muzichenko Ya, Zinchik A, Stafeev S 2010 *Sci.Tech. J. Inf. Technol. Mech. Opt.* **6** 22

[7] Horváth P, Šmíd P, Vašková I, Hrabovský M 2010 Koch fractals in physical optics and their Fraunhofer diffraction patterns *Optik* **121** 2 pp 206–213 DOI: 10.1016/j.ijleo.2008.06.010

[8] Sroor H, Naidoo D, Miller S W, Nelson J, Courtial J, Forbes A 2019 Fractal light from lasers *Physical Review A* **99** 1 013848 DOI: 10.1103/PhysRevA.99.013848

[9] Aksenov V P, Dudorov V V, Kolosov V V, Venediktov V Y 2019 Probability distribution of intensity fluctuations of arbitrary-type laser beams in the turbulent atmosphere *Optics express* **27** 17 pp 24705-24716 DOI: 10.1364/OE.27.024705

[10] Aksenov V P, Pogutsa C E 2013 The effect of optical vortex on random Laguerre-Gauss shifts of a laser beam propagating in a turbulent atmosphere *Atmospheric and Oceanic Optics* **26** 1 pp 13 – 17 DOI: 10.1134/S1024856013010028

[11] Zang H, Zheng C, Ji Z, Liu X, Tian Y, Wei L, Fan Q, Wang X, Cao L and Liang E 2018 *Optik - International Journal for Light and Electron Optics* **175** 344
12. Machado F, Monsoriu J and Furlan W 2017 *INTECH Open Souse Chapter* **10** 257

Journal of Physics: Conference Series **2091** (2021) 012073 doi:10.1088/1742-6596/2091/1/012073

Application of methods of multidimensional statistical analysis for the classification of signal fragments during cutting processing

A V Belov[1], N N Trufanov[2], D V Churikov[2,3] and O V Kravchenko[1,2,4]

[1]Bauman Moscow State Technical University, 2 Baumanskaya Street, Moscow, 105005, Russian Federation
[2]Scientific and Technological Center of Unique Instrumentation of the Russian Academy of Sciences, 15 Butlerova Street, Moscow, 117342, Russian Federation
[3]Kotel'nikov Institute of Radio-Engineering and Electronics of RAS, 11-7 Mokhovaya Street, Moscow, 125009, Russian Federation
[4]Federal Research Center "Computer Science and Control" of RAS, 40 Vavilova Street, Moscow, 119333, Russian Federation

E-mail: trufan0vn@yandex.ru

Abstract. The preliminary processing of the signal coming from the vibration sensor is performed. Spectral analysis methods allow identifying the distinctive zones corresponding to three parameters: the average amplitude of vibrations, the average value of the spectrum, and the complex parameter of the vibration density obtained using interpolation methods. Hence, the vector contains the listed parameters as components produce a vector space, and the equations of separating surfaces in three-dimensional space and classification errors are calculated.

1. Introduction

The addition of industrial equipment with innovative computing resources allows for real-time equipment diagnostics for early detection and troubleshooting, which allows us to improve the quality of the production process by reducing unforeseen situations [1]. The timely performance of equipment maintenance, depending on its actual condition, minimizes downtime and equipment setup costs, but at the same time, it requires a large amount of data to build a model for analyzing the system as a whole [2]. A literature survey shows that an accurate system for monitoring the condition of industrial equipment can increase productivity by 10–50% [3–6].

During the experiment, the vibration sensor generates real-time data, which is then processed. The spectral analysis methods allow us to identify distinctive zones, each corresponding to three parameters: the average oscillation amplitude, the average value of the spectrum, and the complex parameter of the oscillation density obtained using interpolation methods. Static methods allow us to calculate classification errors. Namely, the equations of separating surfaces in three-dimensional space are found, considering the listed parameters.

Currently, the critical point in metalworking is controlling the flow of the technological process and its analysis. One of the essential links in this chain is non-destructive testing in general and vibration diagnostics in particular. In this paper, we consider the problem of classifying a fragment of a signal coming from an accelerometer placed on the front panel of the Dobot Mooz 2 machine is shown in figure 1, by methods of multidimensional statistical analysis.

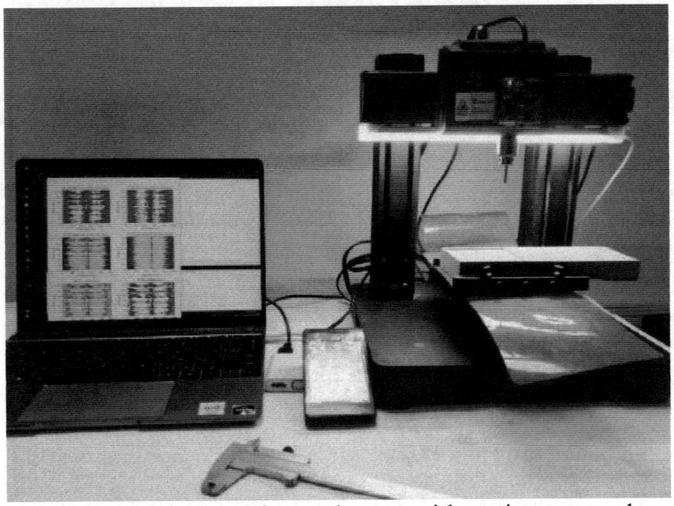

Figure 1. General view of the experimental setup with stationary accelerometer board.

The received signal undergoes preliminary processing, as a result of which it becomes dimensionless and also normalizes on a segment [0,1]. Hence, one may get the signal image shown in figure 2.

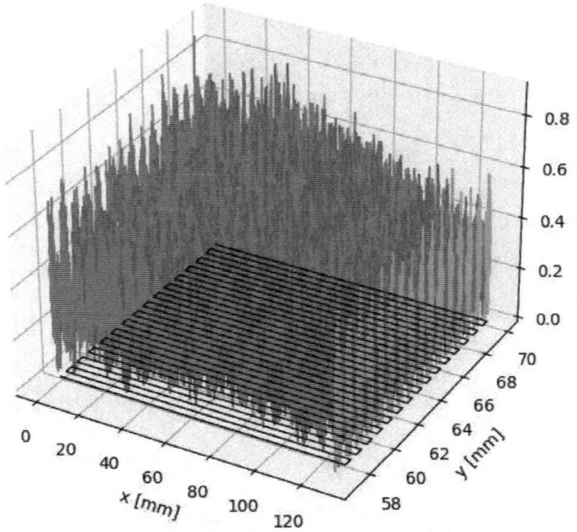

Figure 2. CNC control program trajectory in the X-Y plane.

Classification is the assignment of a sample to one of several pairwise non-intersecting sets. In our case, we divide the processing signal into distinctive zones (figure 3) and construct a separating plane using discriminant analysis methods. The zoning mentioned above allows one to highlight the distinctive parts in the signal [7].

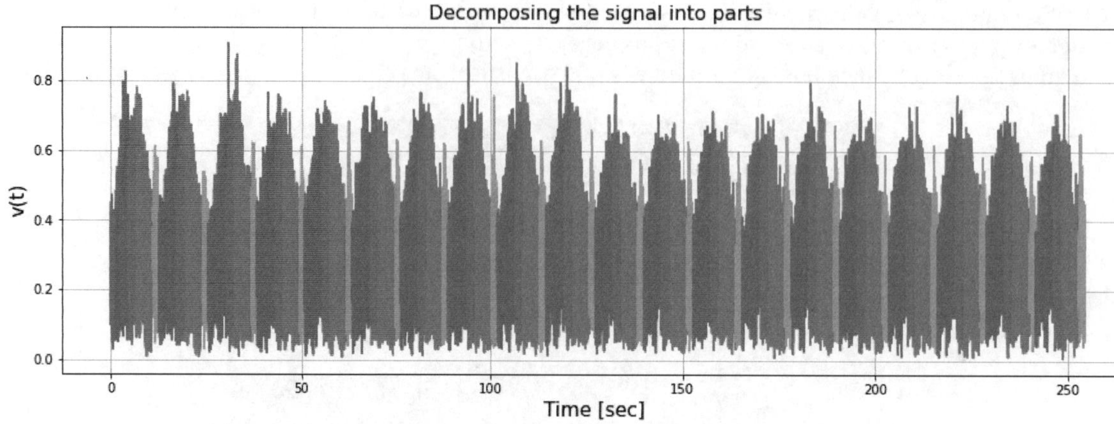

Figure 3. Distinctive parts in the signal.

2. Methodology

Let’s assign three parameters to each zone:

1. Average oscillation amplitude

$$S_i = \frac{1}{n_i}\sum_{k=1}^{n_i} v_{ik},$$

where n_i is a number of points in the i-th zone and v_i is a fragment of the signal corresponding to the i-th zone [8].

2. The average amplitude of the Fourier transform

$$F_i = \frac{1}{2N_j}\sum_{k=0}^{N_j-1}\sum_{n=0}^{N_j-1} v_{jn} \exp\left(-2\frac{\pi i}{N_j-1}kn\right).$$

3. The average area between the envelopes.

For the initial signal, we calculate the upper and lower envelopes are introduced in figure 4.

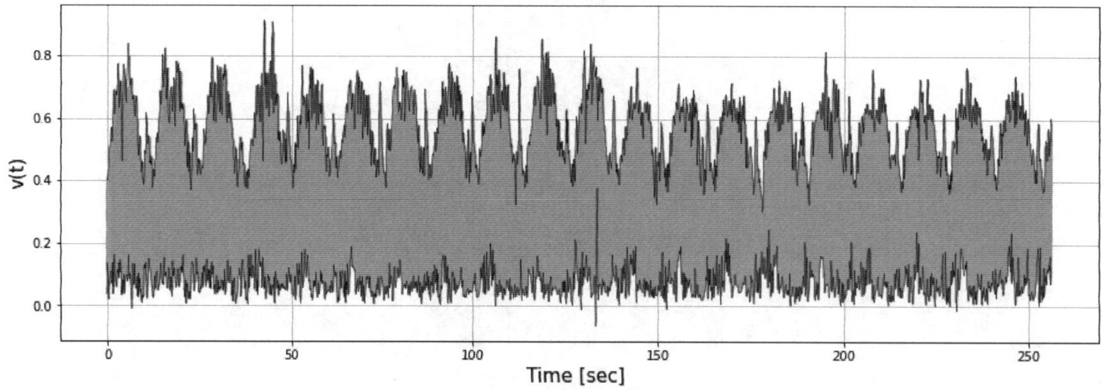

Figure 4. Envelopes for the signal.

Further, for each zone we define the parameter

$$T_i = \frac{1}{n_i}\int_{t_i}^{t_{i+1}} (f_1 - f_2)dt,$$

where n_i is a number of points in the i-th zone, f_1 and f_2 are upper and lower envelopes obtained by cubic interpolation, t_i and t_{i+1} are borders for the i-th zone.

Now, consider random variables

$$\xi_i = \begin{pmatrix} S_i \\ F_i \\ T_i \end{pmatrix}, \eta_j = \begin{pmatrix} S_j \\ F_j \\ T_j \end{pmatrix},$$

provided that the index i corresponds to points from the first set, and the index j corresponds to points from the second set. Then, from the assumptions of normality and homoscedasticity of the obtained vectors, one can obtain

$$\vec{\xi} \sim N(\mu_1, D), \vec{\eta} \sim N(\mu_2, D).$$

Let's introduce the notation

$$n_p(\vec{\gamma}, \vec{\mu}, D) = \frac{1}{(\sqrt{2\pi})^p \cdot \sqrt{|D|}} e^{-\frac{1}{2}(\vec{\gamma}-\vec{\mu})^T D^{-1}(\vec{\gamma}-\vec{\mu})}.$$

is the normal multidimensional density, where p is the dimension of the vector. Then the classification is be based on the likelihood ratio. If the inequality is met

$$\frac{n_p(\vec{\gamma}, \vec{\mu}_1, D)}{n_p(\vec{\gamma}, \vec{\mu}_2, D)} > 1,$$

then one can refer the vector $\vec{\gamma}$ to the first set, otherwise to the second. By transforming the likelihood ratio, we obtain the equation of the separating plane as follows [9,10]. Let the inequality is satisfied

$$\frac{n_p(\vec{\gamma}, \vec{\mu}_1, D)}{n_p(\vec{\gamma}, \vec{\mu}_2, D)} = \frac{(\sqrt{2\pi})^p \cdot \sqrt{|D|}}{(\sqrt{2\pi})^p \cdot \sqrt{|D|}} e^{-\frac{1}{2}(\vec{\gamma}-\vec{\mu}_1)^T D^{-1}(\vec{\gamma}-\vec{\mu}_1)+\frac{1}{2}(\vec{\gamma}-\vec{\mu}_2)^T D^{-1}(\vec{\gamma}-\vec{\mu}_2)}.$$

Then the following inequality is valid

$$-\frac{1}{2}(\vec{\gamma}-\vec{\mu}_1)^T D^{-1}(\vec{\gamma}-\vec{\mu}_1)+\frac{1}{2}(\vec{\gamma}-\vec{\mu}_2)^T D^{-1}(\vec{\gamma}-\vec{\mu}_2) > 0.$$

Opening the brackets one may get

$$-\frac{1}{2}(\vec{\gamma}-\vec{\mu}_1)^T D^{-1}(\vec{\gamma}-\vec{\mu}_1)+\frac{1}{2}(\vec{\gamma}-\vec{\mu}_2)^T D^{-1}(\vec{\gamma}-\vec{\mu}_2) = -\frac{1}{2}\vec{\gamma}^T D^{-1}\vec{\gamma}+\frac{1}{2}\vec{\mu}_1^T D^{-1}\vec{\gamma}+$$
$$\frac{1}{2}\vec{\gamma}^T D^{-1}\vec{\mu}_1-\frac{1}{2}\vec{\mu}_1^T D^{-1}\vec{\mu}_1+\frac{1}{2}\vec{\gamma}^T D^{-1}\vec{\gamma}-\frac{1}{2}\vec{\mu}_2^T D^{-1}\vec{\gamma}-\frac{1}{2}\vec{\gamma}^T D^{-1}\vec{\mu}_2+\frac{1}{2}\vec{\mu}_2^T D^{-1}\vec{\mu}_2 =$$
$$\vec{\gamma}^T D^{-1}\vec{\mu}_1-\vec{\gamma}^T D^{-1}\vec{\mu}_2-\frac{1}{2}\vec{\mu}_1^T D^{-1}\vec{\mu}_1+\frac{1}{2}\vec{\mu}_2^T D^{-1}\vec{\mu}_2.$$

After some algebra, one may achieve the following inequality

$$\vec{\gamma}^T D^{-1}(\vec{\mu}_1-\vec{\mu}_2)-\frac{1}{2}\vec{\mu}_1^T D^{-1}\vec{\mu}_1+\frac{1}{2}\vec{\mu}_2^T D^{-1}\vec{\mu}_2+\frac{1}{2}\vec{\mu}_1^T D^{-1}\vec{\mu}_2-\frac{1}{2}\vec{\mu}_1^T D^{-1}\vec{\mu}_2 =$$
$$\vec{\gamma}^T D^{-1}(\vec{\mu}_1-\vec{\mu}_2)-\frac{1}{2}\vec{\mu}_1^T D^{-1}(\vec{\mu}_1-\vec{\mu}_2)+\frac{1}{2}(\vec{\mu}_1+\vec{\mu}_2)^T D^{-1}\vec{\mu}_2 = \vec{\gamma}^T D^{-1}(\vec{\mu}_1-\vec{\mu}_2)$$
$$-\frac{1}{2}(\vec{\mu}_1-\vec{\mu}_2)^T D^{-1}(\vec{\mu}_1+\vec{\mu}_2) > 0.$$

Then the equation of the separating surface is as follows

$$\vec{\gamma}^T D^{-1}(\vec{\mu}_1-\vec{\mu}_2)-\frac{1}{2}(\vec{\mu}_1-\vec{\mu}_2)^T D^{-1}(\vec{\mu}_1+\vec{\mu}_2) = 0.$$

3. Simulation results

Let's construct separating planes for different clusters are shown in figure 5.

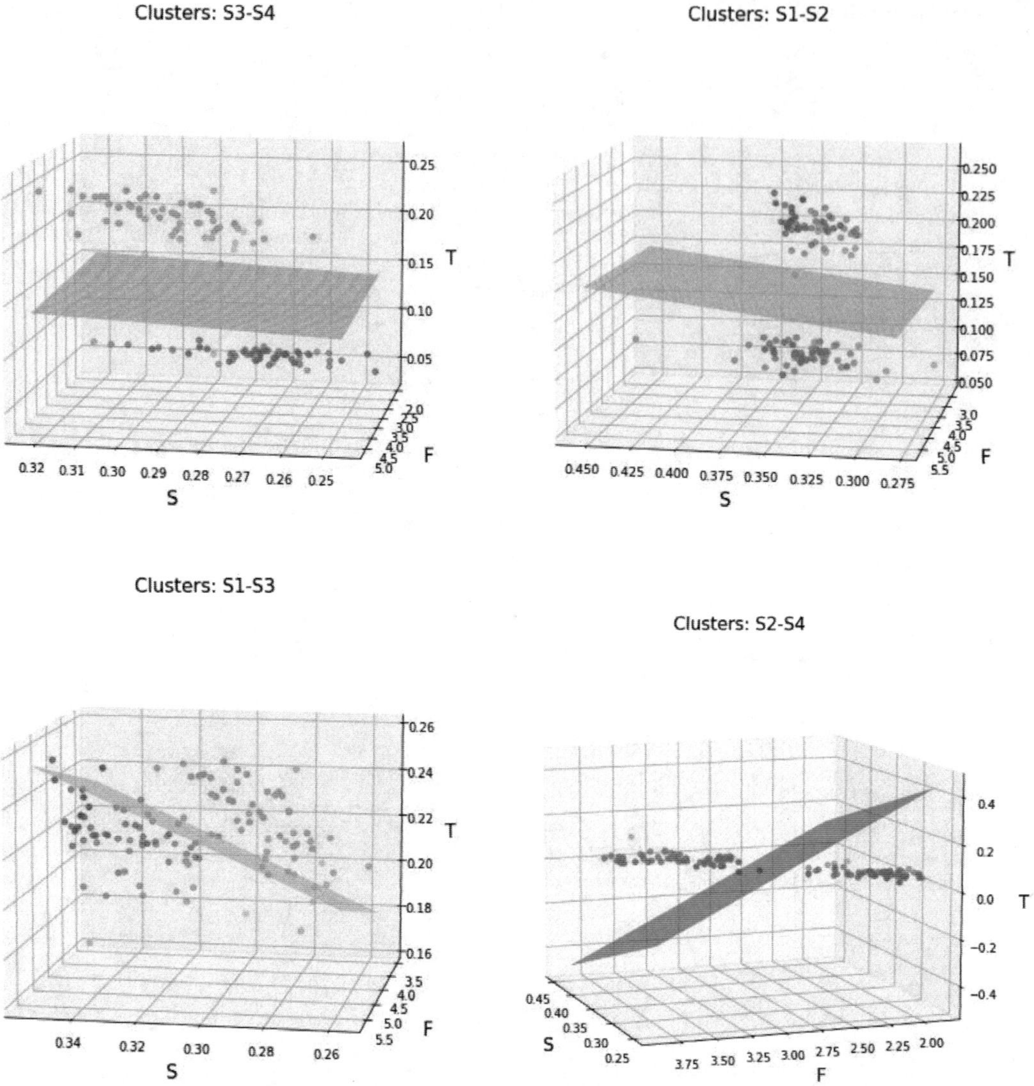

Figure 5. Separating planes. The green color shows the points S1, red is S2, blue is S3, gray is S4.

The classifying function is considered in the form of

$$y = a_1x_1 + a_2x_2 + a_3x_3 + c .$$

where in our case $x_1 = S_i$, $x_2 = F_i$, $x_3 = T_i$. The main goal of the discriminate analysis is to find such a linear combination of discriminant variables that would optimally divide the groups under consideration, and therefore maximize the ratio of inter-group variation to intra-group variation. With the help of the STATISTICA package, the corresponding classification functions f_1, f_2, f_3, f_4 were calculated, which are shown in figure 6.

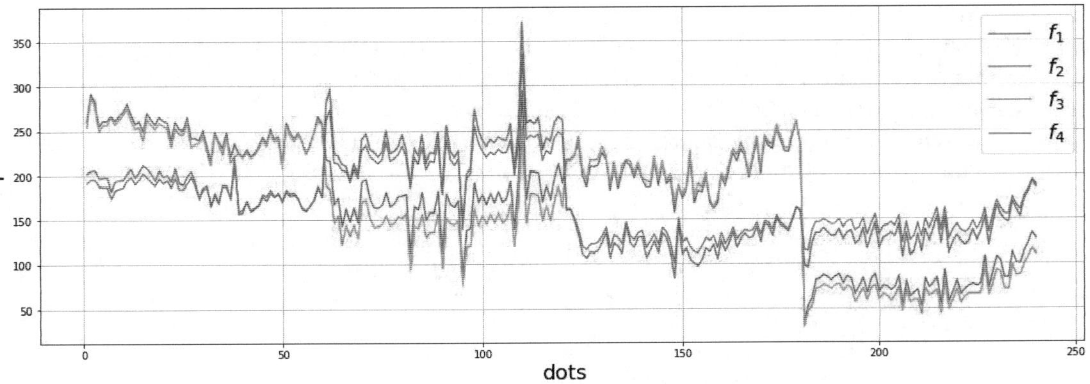

Figure 6. LDA classification functions for 4 clusters.

Thus, having calculated the value of the classification function for each vector, we refer it to the set whose function has the most significant value. With the help of this model, one may get the opportunity to determine exactly what process is taking place on the considered signal fragment, which in turn allows us to analyze the technological process.

4. Conclusion

The conducted statistical analysis allows us to make a classification model based on the selected features. Due to averaged parameters *S*, *F* and *T*, it becomes possible to apply this model to signal fragments of different lengths. The resulting data clusters correspond to different physical processes: there is a separation by the type of milling and by the input and output of the tool into the workpiece in different positions. Despite the simple shape of the trajectory of the control program, the analysis makes it possible to generalize the results obtained to the selection of specified features when processing on more complex trajectories. The expansion of the feature space and the verification of the model on actual industrial signals is the subject of further research.

This work was performed using the equipment of the Centre for Collective Use of the Scientific and Technological Centre of Unique Instrumentation of the Russian Academy of Sciences (STC UI RAS) [http://ckp.ntcup.ru].

References

[1] Hsu C-Y and Liu W-C 2021 Multiple time-series convolutional neural network for fault detection and diagnosis and empirical study in semiconductor manufacturing J. Intell. Manuf. 32 823–836

[2] Li Y, Mou W, Li J, Liu C, and Gao J 2021 An automatic and accurate method for tool wear inspection using grayscale image probability algorithm based on bayesian inference RCIM 68 102079

[3] Qin A, Guo L, You Z, Gao H, Wu X, and Xiang S 2020 Research on automatic monitoring method of face milling cutter wear based on dynamic image sequence Int. J. Adv. Manuf. Technol. 110 3365–3376

[4] Masalimov K A, Munasypov R A and Kudoyarov R G 2021 Russian Engineering Research 41 3 252–256

[5] Masalimov K A and Munasypov R A 2020 Diagnostic of the state of the cutting tool of metal-cutting machines using bidirectional recurrent neural networks with a long short-term memory Russian Engineering Research 1093–1097

[6] Rozhkov S V, Trushin N N and Shadsky G V 2016 Monitoring of the technical condition of spindle assemblies of metal-cutting machines Technical sciences 8 (2) 188–193

[7] Jiang J-S R and Sun C-T 1995 Neuro-fuzzy modeling and control Proc. of the IEEE 83 3 378–406
[8] Kravchenko V F, Fedorov I V, and Churikov D V 2005 Rvachev's V.L. R-Functions and the atomic functions in problems of complex-shaped contour objects description and digital image processing EW&ES 10 7 70–80. In Russian.
[9] Anderson T W 1963 *An Introduction to Multivariate Statistical Analysis* (New Jersey: Wiley)
[10] Ivchenko T I 2019 Mathematical Statistics (Moscow: KD Libkom)

Selection of window functions for predicting the frequency pattern of vibrations of the technological process using an artificial neural network

N N Trufanov[1], D V Churikov[1,2] and O V Kravchenko[1,3,4]

[1]Scientific and Technological Center of Unique Instrumentation of the Russian Academy of Sciences, 15 Butlerova Street, Moscow, 117342, Russian Federation
[2]Kotel'nikov Institute of Radio-Engineering and Electronics of RAS, 11-7 Mokhovaya Street, Moscow, 125009, Russian Federation
[3]Bauman Moscow State Technical University, 2 Baumanskaya Street, Moscow, 105005, Russian Federation
[4]Federal Research Center "Computer Science and Control" of RAS, 40 Vavilova Street, Moscow, 119333, Russian Federation

E-mail: trufan0vn@yandex.ru

Abstract. The frequency pattern of the process is investigated by analyzing spectrograms constructed using the window Fourier transform. A set of window functions consists of a rectangular, membership, and windows based on atomic functions. The fulfillment of the condition for improving the time localization and energy concentration in the central part of the window allows one to select a window function. The resulting spectrograms are fed to the input of an artificial neural network to obtain a forecast. Varying the shape of the window functions allows us to analyze the proposed spectrogram prediction model.

1. Introduction

The application of neural network forecasting methods for industrial data is a trend of our time and allows us to create models that underlie decision-making systems [1-3]. In the present paper, we consider a signal that represents the level of vibration acceleration at the current time. Vibration analysis is a promising method for diagnosing equipment and the technological process as a whole, allowing for non-destructive monitoring of the process in real-time [4-7]. The signal received from the accelerometer fixed on the front panel of the spindle assembly of the modular machine Dobot Mooz 2 is considered. The software of the board on which the sensor is placed allows you to save data in a file with the resolution ". dat". The resulting file consists of three columns with the values of the projections of the vibration acceleration vector on the coordinate axes. The obtained values are pre-processed and normalized in the range [0,1]. Hence, the dependence is obtained, the graph of which is shown in figure 1.

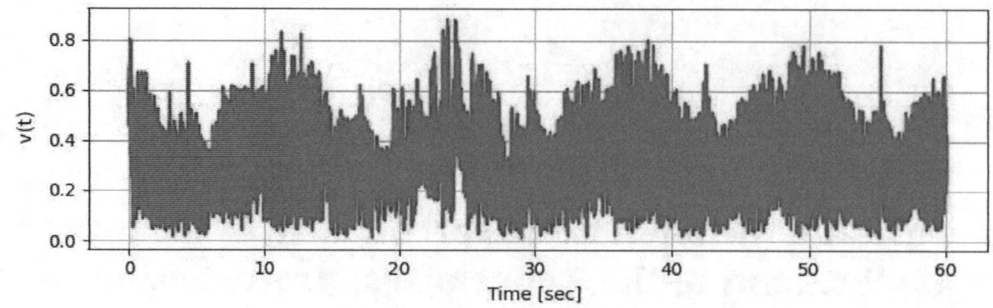

Figure 1. The raw signal from the experiment.

Further, the windowed Fourier transform is performed according to the eq. (1)

$$F(m,w)=\sum_{n=-\infty}^{\infty} v(n)\cdot w(n-m)\cdot e^{-iwn}, \tag{1}$$

where $w(n-m)$ is a window function. This paper considers five types of window functions: rectangular windows, windows constructed using hyperbolic tangents, an atomic function up(x), and a Gaussian function.

2. Methodology

Taking the window width of $N = 50$ points, we will construct the listed window functions (2)-(7) according to the formulas below.

1. Rectangular window

$$\varphi_1(x)=\begin{cases}0, & x<0.2N,\\ 1, & 0.2N \le x \le 0.8N,\\ 0, & x>0.8N.\end{cases} \tag{2}$$

2. Windows constructed using hyperbolic tangents [8]

$$\varphi(a,b,x)=\frac{1}{4}\left(1-a\cdot\tanh(\tilde{x}-b)\right)\left(1+a\cdot\tanh(\tilde{x}+b)\right) \tag{3}$$

where $\tilde{x}=x-N/2$. Figure 2 illustrates the construction of the desired function through the production of two smooth functions.

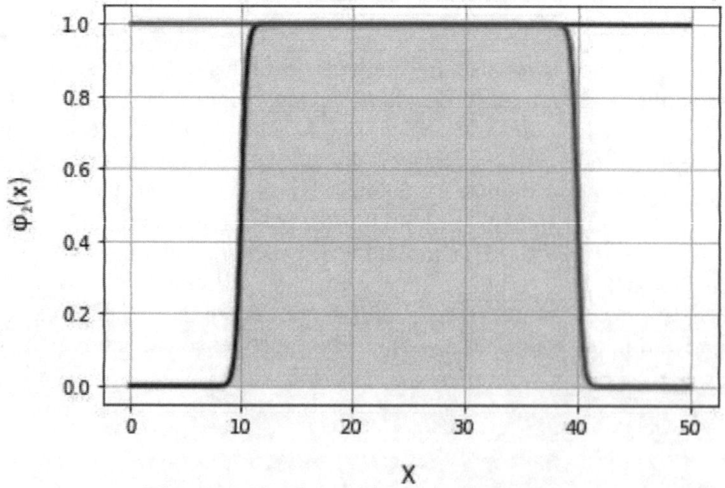

Figure 2. Product of hyperbolic tangents.

Applying this approach, one may build two windows with different parameters, according to formulas (4) and (5)

$$\varphi_2(x) = \mu(0.2,15)\,. \tag{4}$$

$$\varphi_3(x) = \mu(2,15)\,. \tag{5}$$

3. Atomic window [9,10]. Set by the expression (6)

$$\varphi_4(x) = \mathrm{up}(x/30)\,. \tag{6}$$

where

$$\mathrm{up}(x) = \frac{1}{2\pi}\int_{-\infty}^{\infty} \exp(jux)\prod_{k=1}^{\infty}\mathrm{sinc}(u \cdot 2^{-k})\,du\,.$$

4. The Gaussian window has the form (7)

$$\varphi_5(x) = \exp(-(x/18)^2)\,. \tag{7}$$

Normalizing window functions by width at the level of 0.5 allows selecting the parameters of window functions, shown in figure 3 (left). The listed weight functions were selected under two criteria: the highest concentration of energy in the central part and the best temporal localization. To analyze the energy criterion, consider the parameter γ

$$\gamma = \frac{\left\| f(t) \right\|_{t\in[-a/2;a/2]}}{\left\| f(t) \right\|},. \tag{8}$$

The obtained dependencies obtained from eq. (8) are shown graphically in figure 3 (right), highlighting the area of interest separately in figure 4.

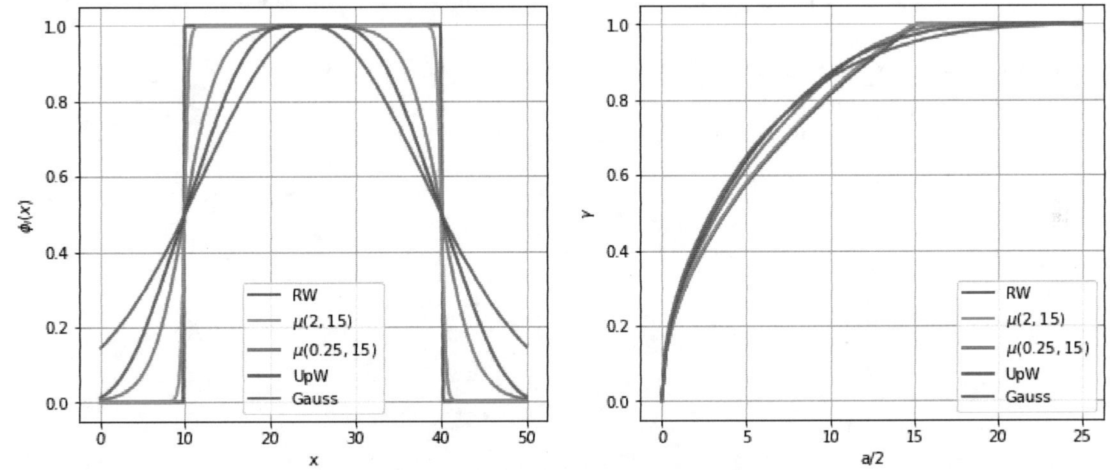

Figure 3. Different types of weight functions and their energy concentration in the central part.

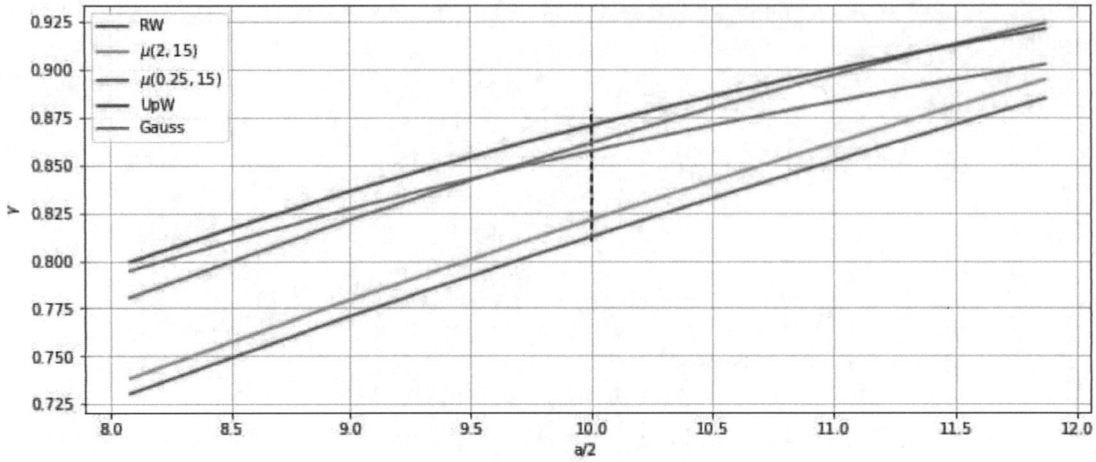

Figure 4. The concentration of energy on the interval of interest.

The temporal localization is defined in terms of the uncertainty constant Δ_f

$$\Delta_f = \frac{1}{\|f(t)\|}\left(\int_{-\infty}^{\infty}(t-t_f^*)^2|f(t)|^2dt\right)^{1/2},$$

where $t_f^* = \frac{1}{\|f(t)\|^2}\int_{-\infty}^{\infty} t|f(t)|^2 dt$. The diagram in coordinates $\gamma - \Delta_f$ allows one to perform a comparative analysis of the characteristics of the considered window functions are shown in figure 5.

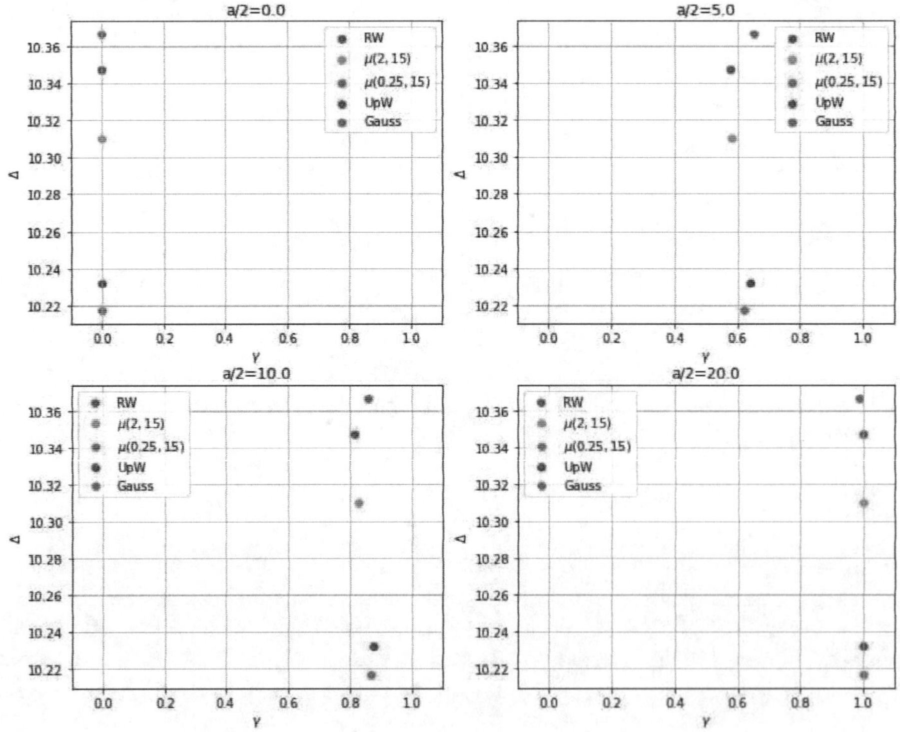

Figure 5. Comparison diagram of weight functions.

Journal of Physics: Conference Series **2091** (2021) 012074 doi:10.1088/1742-6596/2091/1/012074

The zoomed area of interest corresponds to the parameter's value $a/2 = 10$ is introduced in figure 6.

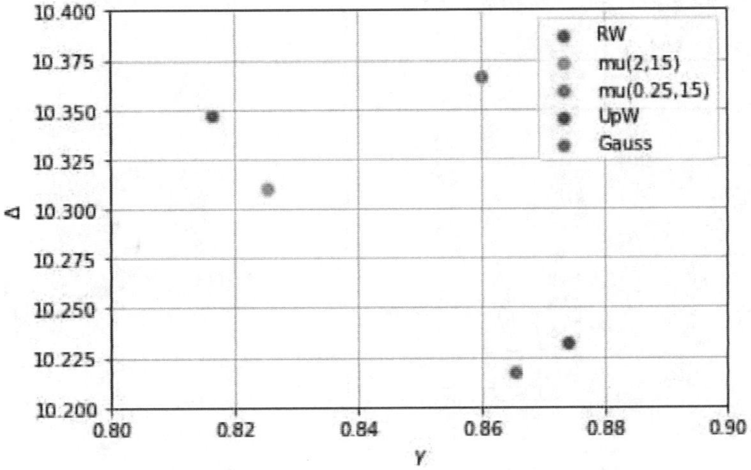

Figure 6. Comparison diagram of weight functions with a/2=10.

The analysis of the obtained dependences indicates the optimality of the atomic function up(x) in terms of the energy concentration in the central part, and the time-frequency localization, which determines the choice of this function for further consideration. Using the selected window, according to the eq. (1), we will construct the spectrogram shown in figure 7.

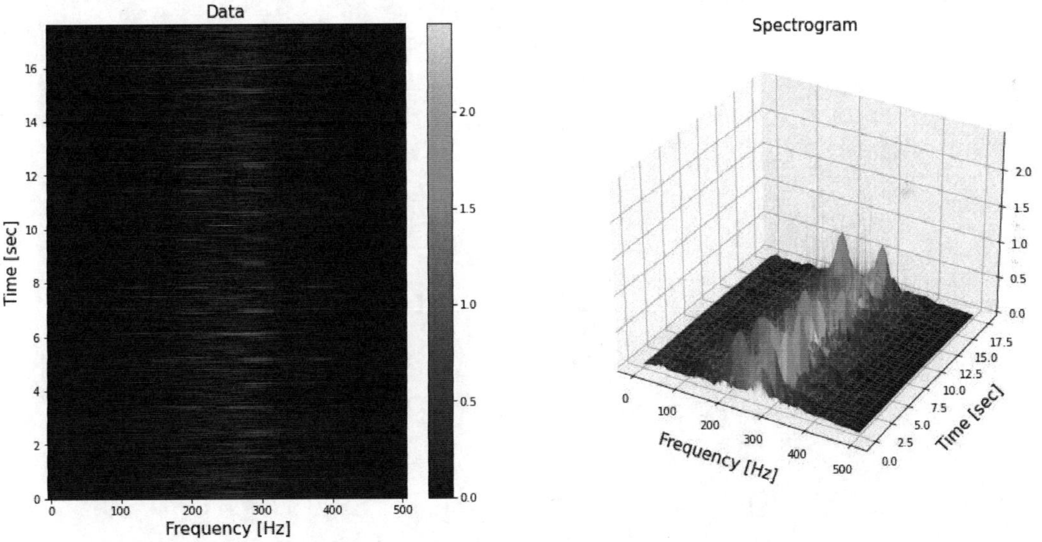

Figure 7. A spectrogram constructed using an atomic window.

3. Simulation results

The spectrogram analysis shows many frequency peaks associated with the technological process of processing the workpiece, so we will choose the zone corresponding to the technological process's ordinary course for training the neural network. Note that the prominent peaks are concentrated in the central frequency domain so that we will select only the central part of the fragment for forecasting (figure 8).

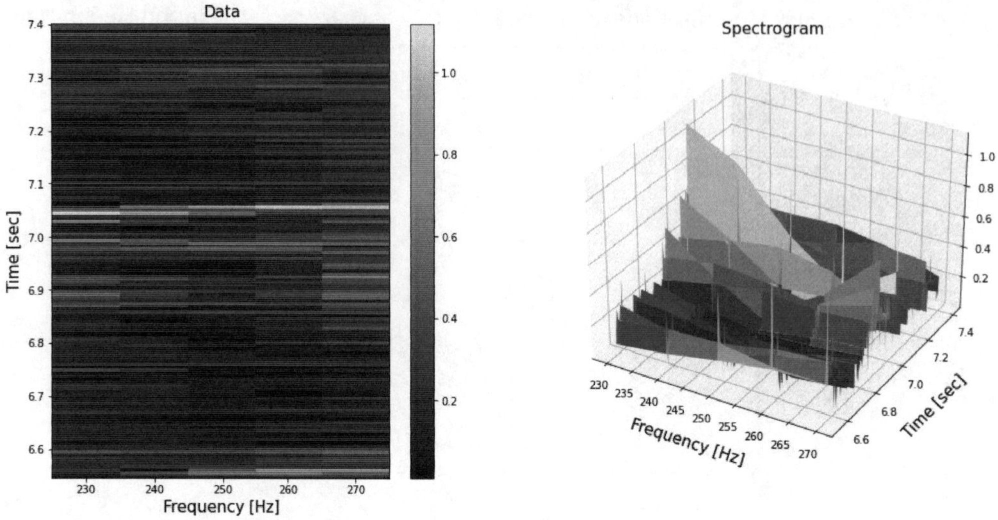

Figure 8. A fragment of training data for neural network.

Figure 9a shows the dependence of the spectrum amplitudes on time at fixed frequencies in the central part: $f_1 = 230$ Hz, $f_2 = 240$ Hz, $f_3 = 250$ Hz, $f_4 = 260$ Hz, $f_5 = 270$ Hz. Next, a merged array is utilized to train a neural network (figure 9b).

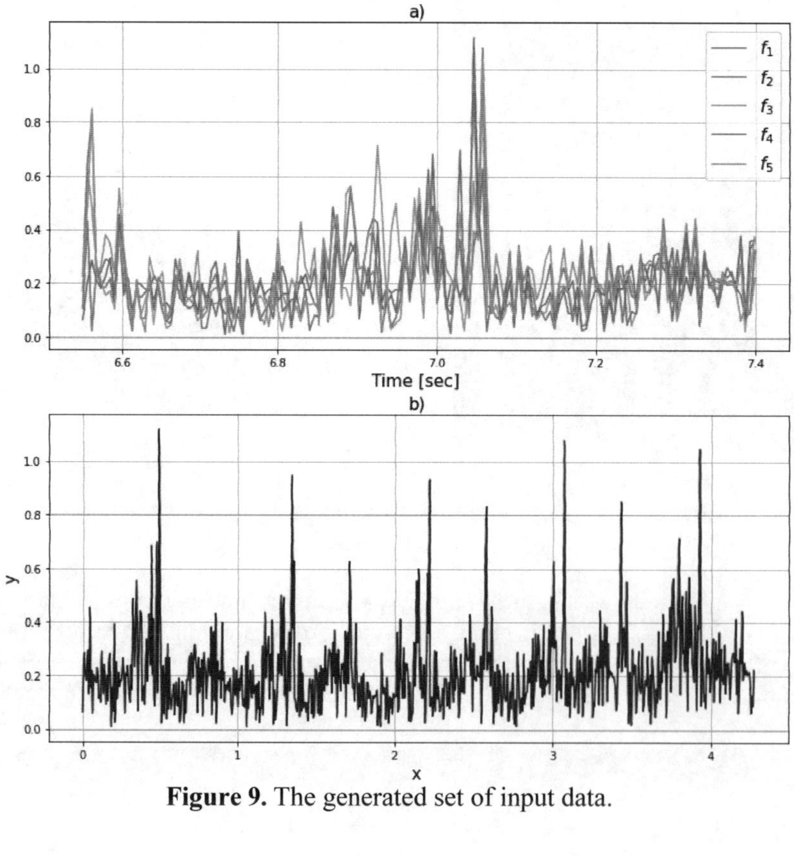

Figure 9. The generated set of input data.

Since we are interested in the maxima of amplitudes when analyzing the technological process and reduce the input data's dimension during preliminary preparation, we will construct envelopes. Data augmentation is performed by adding normally distributed noise. The resulting dataset and the division into training and test samples are shown in figure 10.

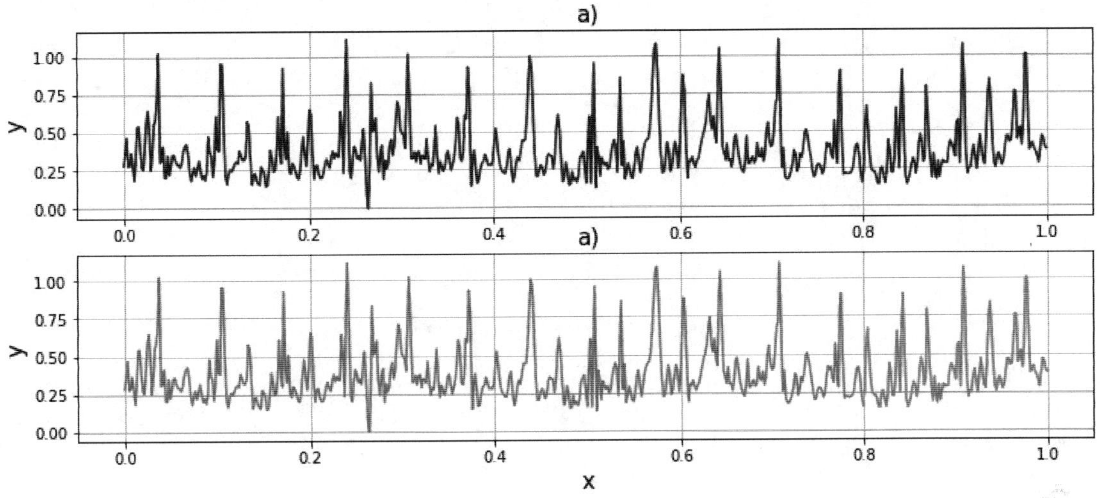

Figure 10. Training and test samples.

The neural network is a multilayer perceptron with two hidden layers, and we use ReLU as an activation function. Optimization is carried out using the Adam algorithm with the parameter $\lambda = 0.008$ for 1000 epochs. The loss function is shown in figure 11.

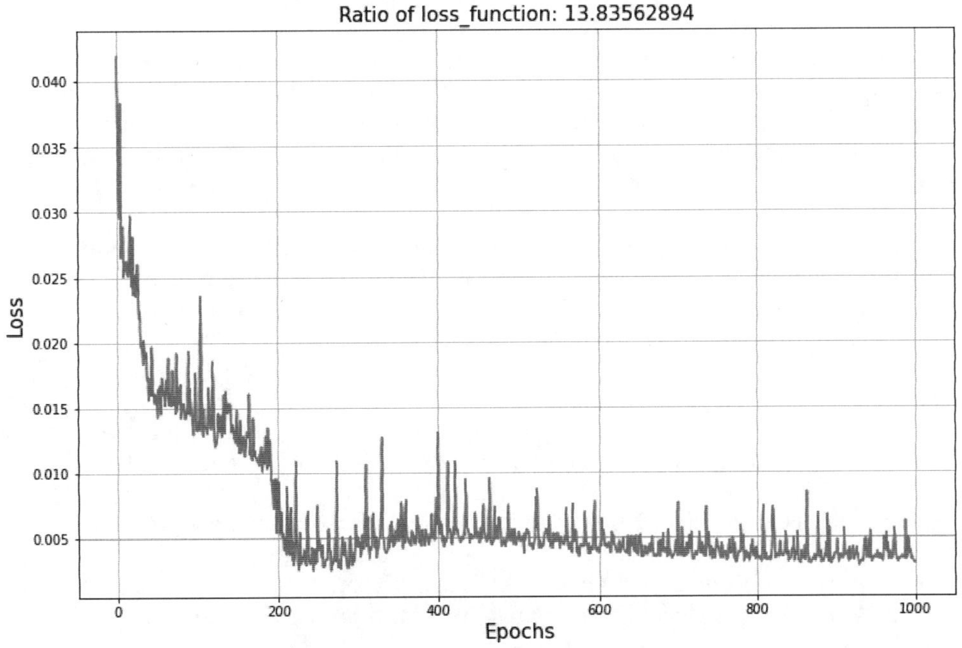

Figure 11. Training and test samples.

The forecast results are shown in figure 12 with a red dotted line, and the test data is shown in green.

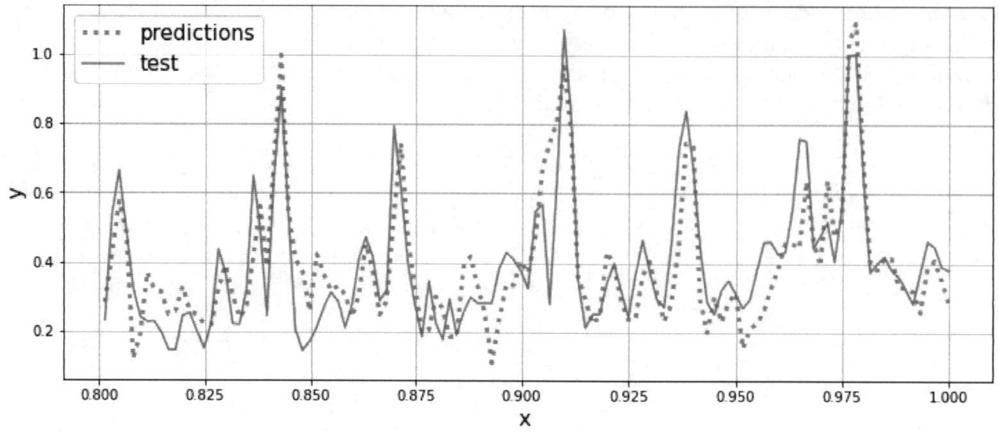

Figure 12. Forecasting results.

The received forecast allows us to conclude the adequacy of the received model. In addition, characteristic peaks were identified, which allows using the obtained model to make a spectrum forecast when monitoring the progress of the technological process. The received forecast allows us to conclude the adequacy of the received model. In addition, characteristic peaks were identified, which allows using the obtained model to make a spectrum forecast when monitoring the progress of the technological process.

4. Conclusion

The paper shows the practical possibility of analyzing this type of technical system by spectral analysis methods using the mathematical apparatus of machine learning. The constructed spectrograms make it possible to analyze the processes occurring in the cutting zone during the processing of the workpiece. The used methods of data preprocessing and augmentation allow us to apply this model to actual industrial data. Scaling the described model implies a deeper analysis of the type and architecture of the neural network, which in turn requires a significant expenditure of computing resources, but the high relevance of the problem leaves no doubt about the payback of the funds spent. Furthermore, the adequacy of the obtained forecast indicates the possibility of building intelligent decision-making systems, which is a prerequisite for the transition to digital production.

This work was performed using the equipment of the Centre for Collective Use of the Scientific and Technological Centre of Unique Instrumentation of the Russian Academy of Sciences (STC UI RAS) [http://ckp.ntcup.ru].

References

[1] Hsu C-Y and Liu W-C 2021 Multiple time-series convolutional neural network for fault detection and diagnosis and empirical study in semiconductor manufacturing J. Intell. Manuf. 32 823–836
[2] Marei M, El Zaatari S, Li W 2021 Transfer learning enabled convolutional neural networks for estimating health state of cutting tools Robot. Comput. Integr. Manuf. 71 102145
[3] Li Y, Mou W, Li J, Liu C, and Gao J 2021 An automatic and accurate method for tool wear inspection using grayscale image probability algorithm based on bayesian inference RCIM 68 102079
[4] Qin A, Guo L, You Z, Gao H, Wu X, and Xiang S 2020 Research on automatic monitoring method of face milling cutter wear based on dynamic image sequence Int. J. Adv. Manuf. Technol. 110 3365–3376
[5] Masalimov K A, Munasypov R A and Kudoyarov R G 2021 Russian Engineering Research 41 3

252–256
[6] Masalimov K A and Munasypov R A 2020 Diagnostic of the state of the cutting tool of metal-cutting machines using bidirectional recurrent neural networks with a long short-term memory Russian Engineering Research 1093–1097
[7] Rozhkov S V, Trushin N N and Shadsky G V 2016 Monitoring of the technical condition of spindle assemblies of metal-cutting machines Technical sciences 8 (2) 188–193
[8] Jiang J-S R and Sun C-T 1995 Neuro-fuzzy modeling and control Proc. of the IEEE 83 3 378–406
[9] Kravchenko V F, Fedorov I V, and Churikov D V 2005 Rvachev's V.L. R-Functions and the atomic functions in problems of complex-shaped contour objects description and digital image processing EW&ES 10 7 70–80. In Russian.
[10] Kravchenko O V, Samoilov V B, Trufanov N N, and Churikov D V 2021 Generation of the data training selection for vibration diagnostics of the cutting processing technological process Phys. Bas. of Instr. 10 1(39) 82–93. In Russian. doi: 10.25210/jfop-2101-082093

AUTHOR INDEX